D1228454

INTRODUCTION TO COMPUTATIONAL SCIENCE

Second Edition

INTRODUCTION TO COMPUTATIONAL SCIENCE

MODELING AND SIMULATION FOR THE SCIENCES

Second Edition

Angela B. Shiflet and George W. Shiflet

PRINCETON UNIVERSITY PRESS
PRINCETON AND OXFORD

Published by Princeton University Press, 41 William Street,
Princeton, New Jersey 08540
In the United Kingdom: Princeton University Press, 6 Oxford Street,
Woodstock, Oxfordshire OX20 1TW

press.princeton.edu
Cover art by Linda Stoudt, *Untitled* 2005 (detail), 4 3/16 × 3, oil on paper

Library of Congress Cataloging-in-Publication Data

Shiflet, Angela B.
Introduction to computational science : modeling and simulation for the sciences / Angela B. Shiflet and George W. Shiflet. — Second edition.
pages cm
Includes bibliographical references and index.
ISBN 978-0-691-16071-9 (hardcover : alk. paper) 1. Computer science. 2. Computer simulation. 3. Mathematical models. 4. Computational complexity. I. Shiflet, George W. II. Title.
QA76.6.S54143 2014
004—dc23
2013024877

British Library Cataloging-in-Publication Data is available
This book has been composed in Times Roman
Printed on acid-free paper. ∞
Printed in the United States of America
10 9 8 7 6 5 4 3 2 1

Dedicated to

Robert K. Cralle, Theodore H. Einwohner, and George A. Michael
Lawrence Livermore Laboratory
and
Robert M. Panoff
The Shodor Education Foundation,
whose friendship and guidance we have treasured

CONTENTS

PREFACE

Although only recently thought of as an emerging field of study, computational science has now become a fundamental and integrated aspect of almost every type of scientific research. Without question, the power of our computers and computational tools has advanced rapidly, but the complexity of the problems we face and the questions we ask may seem to be more than a match. The astounding volume of data generated in our pursuit to understand ourselves and our surroundings demands all the computational power we can muster. Without scientists who can appreciate and manage this power, we will not be able to apply our computational capabilities to their best advantage. We need scientists who are able to see things from a variety of perspectives and who can talk with those from other disciplines. In fact, boundaries between disciplines are less defined than traditionally thought, and many of the most important scientific questions are at crossroads of various disciplines. For the foreseeable future, it is essential to have scientists who can work at those intersections.

At our institution, we have used this text for our course, Modeling and Simulation for the Sciences, which is one of the required courses in an Emphasis in Computational Science program. Established at Wofford in 1998, this program requires four courses besides Modeling and Simulation—Calculus I, Programming and Problem-Solving, Data Structures, and one of the following: Data and Visualization, High-Performance Computing, or Bioinformatics. Additionally, to qualify for the Emphasis, students must major in a science, mathematics, or computer science with a bachelor of science and complete an internship that employs computational methods. We designed the program and the text to be a launching pad for computationally literate scientists who will fearlessly work in interdisciplinary efforts to solve scientific problems.

As with the first edition, *Introduction to Computational Science: Modeling and Simulation for the Sciences (2nd ed.)* is designed to help the student to comprehend and exploit essential concepts of computational science, the modeling process, computer simulations, and scientific applications. Modeling and simulation are now crucial to the exploration of complex systems. These techniques enable scientists to conduct large numbers of experiments more quickly and cheaply than at the bench and to select promising research paths for their laboratories. Insights and discoveries made through modeling augment our understanding and can promote novel experimental approaches.

The **first edition** considered **three major approaches to computational science problems: system dynamics models, empirical modeling,** and **cellular automaton simulations**. The **second edition** includes **two additional computational approaches: agent-based simulations** and **modeling with matrices**.

System dynamics models afford global views of major systems that change with time. For example, such a model could include changes over time in the numbers of predator (e.g., wolf spider) and prey species (e.g., cricket). For system dynamics modeling, as with all approaches, the **text employs a nonspecific tool, or generic, approach**—to model such dynamic systems, students using the text can employ any one of several tools, such as *STELLA*®, *Vensim*®, Personal Learning Edition (PLE) (free for personal and educational use), *Berkeley Madonna*®, *NetLogo* (free), *Python* (free), and *R* (free). With many of these tools, the student can create visual representations of models, develop relationships, run simulations, and generate graphical results.

Unlike system dynamics, **cellular automaton simulations** present local views of individuals affecting individuals. Such a world is represented as a rectangular grid of cells, with each cell having a state that can change with time according to set rules. For example, the state of one cell could represent the presence of a cricket and the state of an adjacent cell could correspond to a wolf spider. One rule could be that, when adjacent, a spider captures a cricket with a probability of 25%. Thus, on the average at the next time step, a 25% chance exists that a particular cricket next to a spider will not chirp again. The text employs a generic approach for cellular automaton simulations and scientific visualizations of the results, so that students can employ any one of a variety of computational tools, such as *Maple*®, *Mathematica*®, *MATLAB*®, *Python*, and *R*.

Agent-based simulations are similar to cellular automaton simulations. Agents, such as spiders and crickets, often reside on a grid. Each autonomous, decision-making agent has a state represented by a set of state variables and behaviors that control its actions. Moreover, at each time step, the simulation sweeps through all the agents, each updating its state based on its location and what is nearby. By contrast, a cellular automaton simulation updates the states of all the cells in a grid at each time step. With both cellular automaton and agent-based simulations, individual actions and local interactions can help us to access their effects on the whole system. Both simulation techniques can be effective in modeling dynamic, spatially complex situations. With a generic approach in the text, tutorials and files associated with the modules are available on the text's website for various agent-based tools, such as *AgentSheets* and *NetLogo*.

Simulating with agents or a cellular automaton or modeling with system dynamics, we seek to develop a computational representation that explains a situation. On the other hand, with an **empirical model**, data, such as the prevalence of crickets in an area over several years, are our only source of information about the system. We find a function that captures the trend of the data and use this mathematical model to make predictions. The same computational tools employed in cellular automaton simulations can be used for empirical models and modeling with matrices.

Matrix models incorporate certain probabilities and averages, such as the probability that a cricket exhibits a certain behavior at the next time step or the probability that it matures to the next life cycle stage and its average number of offspring. Using matrix models, we can make long-term predictions about system behaviors and populations. Moreover, we can use matrices to represent contact networks that track the simulated behavior of individuals in a variety of situations from the spread of a disease to the strength of social connections.

Modules presented in Chapter 13 on modeling with matrices include sections that discuss the utility of **high-performance computing** (HPC) in computational science. Moreover, Chapter 12 provides an introduction to some of the hardware and algorithms for HPC, which is becoming increasingly important with the advent of "big data" and massive computational problems. The text's website includes additional related material and programs in C and MPI.

Tutorials, package-specific Quick Review Questions with answers, and files to accompany the text material are **available from the text's website in various system dynamics tools** (including *STELLA*®, *Vensim*®, *Berkeley Madonna*®, *NetLogo*, *Python*, and *R*), **in several computational tools** (including *Maple*®, *Mathematica*®, *MATLAB*®, *Python*, and *R*) and **in assorted agent-based tools** (including *AgentSheets* and *NetLogo*). Typically, an instructor selects one system dynamics tool, one computational tool, and one agent-based tool, depending on the types of models covered, for class use during the term. The variety of tools and examples of their use employed in this text provides an instructor with many options. Note that **at least one tool from each type is available as open source**.

Text Prerequisites

Prerequisites for *Introduction to Computational Science* are minimal. The text **does *not* require computer-programming experience**. Although the concept of rate of change, or derivative, from a first course in calculus is used throughout the text, the necessary background is contained in Module 2.2, "Unconstrained Growth and Decay." Otherwise, students do not need to know how to take derivatives in order to understand the material or develop the models. For those who would like additional discussions of the material, two modules on fundamental calculus concepts, "Rate of Change" and "Fundamental Concepts of Integral Calculus," are available on the text's website.

Learning Features

While the interdisciplinary nature of computational science is distinctive, it is also challenging, particularly for students with limited experience in computer science, mathematics, and various areas of the sciences. To mitigate this challenge, the text provides the background that is necessary for the student to understand the material, work the problems/projects, and confidently succeed in the course. Each module involving a scientific application **covers the prerequisite science without overwhelming the reader** with excessive detail. Furthermore, the text provides **a wide variety of application areas** for examples, exercises, and projects, including astronomy, biology, chemistry, economics, engineering, environmental science, finance, geology, medicine, physics, psychology, and the social sciences.

Introduction to Computational Science has chapters that consist of several modules each. The text's website contains **two tutorials on system dynamics tools**, **seven tutorials on computational tools**, and **two tutorials on agent-based tools**. The text presents the **tutorials in a just-in-time fashion**, covering the requisite information for the subsequent material.

Module 1.2 introduces the **modeling process**, and the text consistently uses the process to guide the user through numerous scientific examples. For instance, after covering the prerequisite scientific background, Module 4.4 develops a model of malaria by following the modeling process in a step-by-step fashion. Also, Module

9.2 introduces the **development of computer simulations** along with their utilities in the sciences. Thus, the text is structured to help students to learn how modelers model and to develop their own modeling skills.

The text presents material in a clear manner with generous use of examples and figures. Most sections of a module end with **Quick Review Questions** that provide an assessment of the student's comprehension of the material, and the text includes more than 230 such questions, averaging about three parts each. **Answers**, often with explanations, at the end of the module give immediate feedback and reinforcement to the student. In the case of system dynamics, computational, or agent-based tool-dependent questions, the questions and answers are on the text's website in PDF files for a variety of tools, as appropriate.

To further aid in understanding the material, most modules include a number of **exercises** (about 275 in the text) that correlate directly to the material and that the student is to complete for the most part with pencil and paper and occasionally with a computer. Answers to selected problems, whose exercise numbers are in color, appear in an appendix.

A subsequent "**Projects**" section provides numerous project assignments for students to develop individually or in teams. The text contains approximately 650 projects. While a module, such as "Modeling Malaria," might develop one model for an application area, the projects section suggests many other refinements, approaches, and applications. The ability to work well with an interdisciplinary team is important for a computational scientist. **Chapters 7 and 14 provide modules of additional, substantial projects** from a variety of scientific areas that are particularly appropriate for teams of students. These modules indicate prerequisite text material, and the projects sections of earlier modules refer to appropriate projects from Chapters 7 and 14 that the instructor can assign at that point in the term.

A "**References**" section occurs at the end of most modules. It provides a list of hyperlinks, books, and articles for further study.

A **glossary** of scientific, modeling, and simulation terms is provided for quick reference. The text's website provides links to **downloadable tutorials, models, PDF files, and datasets** for various tool-dependent quick review questions and answers, examples, and projects.

Using the Material

Because the area is emerging, a variety of departments offer introductory computational science courses, and instructors approach the material in diverse ways. The first edition has been used at various levels (undergraduate, graduate, and high school) and in different types of courses. The second edition provides two additional modeling approaches, agent-based and matrix modeling, and more than enough material so that an instructor can select one or more of the methodologies. Thus, *Introduction to Computational Science* provides **several pathways through the material**. For example, one could choose to start a course in any of a variety of places—with an overview of the modeling process (Chapter 1) and system dynamics modeling (Chapters 2–4); Chapter 8, "Data-Driven Models"; Chapter 9, "Monte Carlo Simulations"; Chapter 11, "Agent-Based Modeling"; or Chapter 13, "Matrix

Models." Moreover, Chapter 5, "Computational Error," can be covered or not covered any time during the course. The same options are true for Chapter 6, "Simulation Techniques," after consideration of Module 2.2. For those who wish to cover Chapter 13 before (or instead of) cellular automaton simulations, a tutorial for various computational tools, primarily covering matrix operations, is available on the text's website. One possible course outline is described shortly, and the website gives alternative syllabi and a prerequisite structure for the modules. Moreover, the text provides an abundance of discipline-specific applications, so that the text is suitable either for an **introductory course generally in computational science or**, with appropriate selection of applications, **specifically in computational applications for biology**.

The text begins with an introduction to computational science and the modeling process. Chapter 2 commences with a discussion of system dynamics and models where the rate of change of a quantity is proportional to that quantity. Two tutorials available in a choice of several tools lead the student step by step through the process of implementing a model with the software. "Unconstrained Growth and Decay" discusses models that exhibit exponential growth or decay and introduces concepts of time-driven simulations. The module also develops the analytical solution to unconstrained growth and decay problems for students who have had integral calculus and for those who have not. The module "Constrained Growth" considers situations in which the quantity under change, such as a population, has a maximum value, or carrying capacity. In this context, we introduce the concepts of equilibrium and stability. The module "Drug Dosage," which includes geometric series, provides other examples where rate is proportional to amount.

For those interested in physics models, Chapter 3, "Force and Motion," provides modules on falling and skydiving, bungee jumping (springs), pendulum clocks, and rocket motion. However, the instructor can choose to skip this chapter or to assign its modules and projects for work outside of class.

Numerous system dynamics models involve interactions, such as with population dynamics or chemical reactions. Chapter 4 considers such models with discussions of competition, predator-prey models, spread of SARS, modeling malaria, and enzyme kinetics.

With computational estimates, the modeler should always be aware of sources of computational error. Thus, after a beginning tutorial on a tool we can use for computation, empirical models, and cellular automaton simulations (tutorial versions on the text's website), Chapter 5, "Computational Error," contains a module "Errors." However, an instructor can delay coverage of the first two computational tool tutorials until considering material from Chapters 8–10.

After a second computational tool tutorial from the text's website, Chapter 6 covers three simulation techniques: Euler's, Runge-Kutta 2 (Euler's Predictor-Corrector), and Runge-Kutta 4. One or more of these techniques can be covered at any time after Chapter 2's module "Unconstrained Growth and Decay." For example, the instructor may choose to discuss Euler's Method immediately after that module and delay consideration of the other two techniques until later in the term.

Chapter 7 provides opportunities for students to learn system dynamics modeling by completing additional extensive projects. Unlike earlier chapters, the modules of this chapter do not include examples. Instead, each module contains sufficient background in a scientific application area for students to develop their own system dynamics models, as suggested by project descriptions. Each module lists the prerequisite material so that students can do Chapter 7's projects at any time after covering the earlier material. These projects and some of the more extensive projects in previous chapters provide excellent opportunities for teamwork. The chapter includes the following topics: radioactive chains, blood cell populations, scuba diving, carbon cycle, global warming, growth of a garden, cardiovascular system, electrical circuits, transmission of nerve impulses, antibiotic resistance, carbohydrate metabolism, mercury pollution, economics of commercial fishing, biochemical pathways, and colon cancer.

Chapter 8 shifts away from system dynamics modeling. After a third tutorial on a computational tool, an optional tutorial covers functions that often appear in modeling. With this background, empirical models, which are based only on data and are used to predict and not to explain a system, are considered.

Monte Carlo simulations of Chapter 9 form the basis for modules in this and the next chapter. After an appropriate tutorial, the chapter considers simulation development and area estimation using this technique. An instructor interested in doing so can also cover how to generate random numbers in other probability distributions for computer simulations and details of the multiplicative linear congruential method to generate uniformly distributed random numbers. After an additional computational tutorial, the chapter concludes with the random walk method that occurs in numerous computer simulations.

Chapter 10 considers many applications of cellular automata. A computational tool tutorial leads into a module involving applications, such as spreading heat through an iron bar, as well as fundamental concepts, such as periodic boundary conditions. An instructor can choose to cover one or more additional in-depth cellular automaton simulations on spreading of fire, movement of ants, or growth of biofilms.

On the text's website, two tutorials in various tools are associated with Chapter 11 on agent-based modeling. After the first tutorial, the student can consider the spread of disease in beef cattle as the animals reside on and travel to various locations. Movement of an invasive species, the cane toad, from one artificial watering point to another and attempts to contain their spread can be modeled in a subsequent module after completing a second tutorial.

Some modeling and simulation projects require massive computational power beyond the capabilities of present-day sequential computers. Thus, Chapter 12 provides an introduction to high-performance computing (HPC). The chapter covers the basic concepts and hardware configurations of HPC as well as some parallel-processing algorithms. With this background, the student can gain an appreciation of some of HPC's potential and challenges.

The utility of HPC is discussed in several sections of the following chapter, "Matrix Models." After a computational tool tutorial involving matrix operations, Module 13.2 covers various vector and matrix operations in the context of population applications. An instructor can then cover one or several applications in any order. When we can divide the life of an organism into ages/stages, Module 13.3 shows how we can often employ age-structured/stage-structured models to determine intrinsic growth rates (eigenvalues), stable distributions, threats of extinction, and how sensitive the long-term population growth rates and predicted times of extinction are to small changes in parameters. Covering the necessary background in probability theory, Module 13.4 develops a Markov chain model, which employs the probability of passing from one state to another and can solve a large variety of problems from predicting the behavior of animals to examining forest succession to locating genes in DNA. The last module covers individual-based (or network-based) epidemiology simulations that track the simulated behavior of individuals. Such a model involves a contact network, and the module covers the basics of graph theory, implementation of graphs with matrices, and characteristics of social networks.

As with Chapter 7, Chapter 14 provides opportunities for students, perhaps in teams, to enhance their computational science problem-solving abilities through completion of additional extensive projects that they can do at any time after covering prerequisite material. The modules do not have examples but do have sufficient scientific background for the projects. The applications of computational science empirical models, random walks, cellular automaton simulations, agent-based simulations, and modeling with matrices in Chapter 14 include the following: polymers, solidification, foraging, pit vipers and heat diffusion, mushroom fairy rings, spread of disease, HIV in the body, predator-prey relationships, clouds, fish schooling, invasive plants, numerous cellular automaton simulations originally considered with agent-based models, and vice versa, bioinformatics, and social science networks.

A Possible Course Outline

As with most courses, an instructor is likely to vary material coverage from term to term. Alterative approaches are on the text's website. With project assignments encompassing one to two weeks, the following is a possible schedule covering system dynamics and empirical modeling and cellular automaton simulations:

Week 1: Chapter 1, 2.1 (System Dynamics Tutorial 1) in a lab if possible, start 2.2 (project assignments from module)
Week 2: 2.2, 2.3, assign 2.4 (System Dynamics Tutorial 2)
Week 3: 2.5; project assignments from 2.5 or Chapter 3 and 7.1–7.6; 4.1, 4.2
Week 4: 4.3, project work in lab if possible, project assignments from 7.7–7.15
Week 5: 5.2 (exercises assigned to be graded), 6.2, 6.3
Week 6: In lab if possible—project work, 5.1 (Computational Tool Tutorial 1), and 6.2 (Computational Tool Tutorial 2)
Week 7: Project presentations, midterm
Week 8: Assign 8.1 (Computational Tool Tutorial 3), 8.3 (project assignments from module), review assignment on Computational Tool Tutorials 1–3
Week 9: Assign 9.1 and 9.5 (Computational Tool Tutorials 4 and 5), 9.2, 9.6, project assignments from 14.1 and 14.2
Week 10: Work on projects, assign 10.1 (Computational Tool Tutorial 6)
Week 11: 10.2, 10.3, project assignments from 14.3–14.12
Week 12: 10.3, work on projects
Week 13: 12.1, 12.2
Week 14: Project work and presentations
Week 15: Final exam

Students have found it helpful to have short quizzes taken directly from the Quick Review Questions. Other graded assignments include tutorials, selected exercises, projects, a midterm, and a final exam.

Supplementary Materials

Instructors and students can link to the text's website through the Princeton University Press' website (**http://pup.princeton.edu/**) or Wofford College's Computational Science website (**http://www.wofford.edu/ecs/**). The following **resources** are available on the text's site:

- Two system dynamics tool tutorials in several tools, such as *STELLA*, *Vensim* PLE, *Berkeley Madonna*, *Python*, and *R*
- Seven computational toolbox tutorials in several tools, such as *Maple®*, *Mathematica®*, *MATLAB®*, *Python*, and *R*
- For a variety of tools, PDF files that contain system-dependent Quick Review Questions and answers
- In a variety of tools, files that contain models, as indicated in the "Download" sections of modules
- Datasets
- References with links to other websites

The text's website also has an **online *Instructor's Manual***, which contains the following material:

- Solutions to all exercises in the text
- Solutions to the two system dynamics tool tutorials using several tools, such as *STELLA®*, *Vensim* PLE®, *Berkeley Madonna®*, *NetLogo*, *Python*, and *R*
- Solutions to the seven computational toolbox tutorials and a function tutorial using several tools, such as *Maple®*, *Mathematica®*, *MATLAB®*, *Python*, and *R*
- Solutions to the two agent-based tool tutorials using several tools, such as *NetLogo* and *AgentSheets*
- Test problems with answers
- Model solutions and suggested approaches for selected projects
- *PowerPoint* files of key figures and algorithms
- Suggested pathways through the material
- Prerequisite structure of the modules

Instructors who adopt the text may obtain a password from Princeton University Press to access the online *Instructor's Manual*.

Acknowledgments

The first edition of this text was published in 2006, and much in the world and our lives has changed. We are still eternally grateful for the support and encouragement of our colleagues and students, but we have also had enormously important input from many external users of the book. The development of the second edition has been no less a labor of love than the first, but we have written it with significantly more experience professionally. Colleagues at Wofford College have been especially accommodating in their willingness to class-test many of the new modules and to provide us with extraordinarily constructive suggestions for improvement. In particular, we would like to thank Drs. Anne J. Catllá, Joseph D. Sloan, and David A. Sykes. Drs. Catllá and Sloan used the first edition and new materials in their modeling and simulation classes, and Dr. Sykes meticulously reviewed most of the new modules and projects. Drs. Ted Monroe and Joseph Spivey also gave very helpful feedback on several of the new modules they utilized in their upper-level mathematics courses.

We are privileged to teach in an institution that attracts a remarkably talented group of science, computer science, and mathematics students, several of whom have contributed directly and indirectly to the development of various modules and associated files. We would like to recognize particularly Mayfield Reynolds, Shay Ellison, Jesse Hanley, and Whitney Sanders for their dedication, creativity, and diligence.

Additionally, we would like to acknowledge the generous comments and contributions, including R files, from Dr. Stephen Davies, computer scientist at Mary Washington University. Dr. Davies has not only shared excellent feedback for the first edition, but he has contributed directly to this edition by allowing us to incorporate an interesting module he developed (Module 7.6, "Plotting the Future—How Will Your Garden Grow") and a clever Project 10 on paratroopers in Module 3.1, "Modeling Falling and Skydiving."

This revised text has been significantly improved by experience we have gained from teaching with the material in various courses and through the sabbatical leave we were granted by Wofford College in 2010–11. Dr. David Wood, Senior Vice-President and Academic Dean at Wofford, has been very understanding of our work and the need for time to do research and writing. He and the college Committee on Non-Curricular Faculty Concerns were most munificent in accommodating this leave. The experiences we gained during the year changed our personal and professional lives immeasurably, and they continue to inspire us in our teaching, research, and writing.

We are unable to stress adequately the value of our time working abroad to the development of this work. For five months we worked with some extremely talented scholars of the Computational Biology Group, the University of Oxford, and we would like to acknowledge the graciousness of Professor David Gavaghan, who agreed to host us. Our focus at Oxford was the modeling of colon cancer, as part of a subgroup concentrating on projects in soft tissue mechanics and cancer. We worked most closely with a then-D.Phil. student in mathematics, Ornella Cominetti, who delighted us with her brilliance, her enthusiasm, and her wonderful sense of humor. The time there broadened our experience in modeling and programming (with *MATLAB*® and *Chaste*®).

This yearlong adventure was really enabled by a computer scientist we met at a meeting of the International Conference on Computational Science in Krakow, Poland, in 2008. Professor David Abramson, then of Monash University in Melbourne and more recently of the University of Queensland, Australia, was undoubtedly surprised by our queries about sabbatical positions in Australia and England. Thankfully, he agreed to help us, even providing the essential contact at Oxford. David gave us the opportunity to work within the eResearch and Grid Engineering Laboratory at Monash, where we had five months to do research and write. He encouraged us to present two workshops for graduate students and faculty in computational science at Monash, which provided us with invaluable experiences. Subsequently, one of the participants in the workshop, Dr. Valerie Maxville, who leads the Education Program at iVEC, Western Australia's supercomputing and eResearch facility, invited us to give a talk and workshop in Perth. Hopefully, our efforts have encouraged more computational applications at Monash and other Australian institutions.

We are indebted to the following reviewers, who offered many valuable constructive criticisms as we prepared the first edition:

Rob Cole, Evergreen State University
Richard Hull, Lenoir-Rhyne College
James Noyes, Wittenberg College
Bob Panoff, The Shodor Foundation
Sylvia Pulliam, Western Kentucky University
Joseph Sloan, Wofford College
Chuck Swanson, University of Minnesota
David Sykes, Wofford College
Peter Turner, Clarkson University
Ignatios Vakalis, Capital University

Second Edition Table of Contents and/or Module Reviewers:

Anne Catllá, Wofford College
Melanie Correll, University of Florida at Gainesville
Stephen Davies, Mary Washington University
Charles Epstein, University of Pennsylvania
David Joiner, Kean University
Hong Liu, Embry-Riddle
Richard Salter, Oberlin College
Joseph Sloan, Wofford College
David Sykes, Wofford College
Carlo Tomasi, Duke University

Vickie Kearn, Senior Editor at Princeton University Press, unfailingly supported us in the preparation of the first edition, which was a novel project in an emerging discipline. She has always had a clear understanding of the project and provided excellent guidance. In the same spirit, Vickie has been enthusiastic in the revision of that project. Her vision and trust have been remarkable, and we reiterate our gratitude for all that she has done to facilitate our latest efforts.

We are so grateful to the extraordinarily capable team at Princeton University Press who helped craft the second edition of this text. Quinn Fusting was always our "go to" person for answers and logistics. Debbie Tegarden very ably guided production, encouraging everyone else in the process. We were delighted to work again with Dimitri Karetnikov, who continues to transform mere figures into art, and Lorraine Doneker, who fashioned another attractive design. We were so fortunate to have Bytheway Publishing Services, directed by Lori Holland, working so carefully and capably again on the typesetting and composition of this very technical and complex project. Thanks also go to Linda Thompson, who skillfully copyedited this edition.

Linda Stoudt, George's cousin, has been so generous in allowing us to display some of her talent on the covers of both editions of our text. Everything that she does is done with loving care and powerful intellect.

There is no person more responsible for our writing this text than Dr. Robert M. Panoff of the Shodor Education Foundation/National Computational Science Institute. From the first Shodor workshop we attended to the present, Bob has been the source—ideas, encouragement and support. Bob has helped us in countless ways, including traveling to Australia to make the first workshop on *computational thinking* a resounding success. Moreover, the NSF-funded Blue Waters Undergraduate Petascale Education Program, for which he was a co-PI, funded development of six educational modules, which were preliminary versions of modules in this second edition, and internships for three students, who assisted us with HPC materials. He is passionate about computational science and computational science education, and he is enormously generous with his time and ideas. We are indebted to him for our professional transformation that has facilitated the development of the computational science program at Wofford and the creation of this book.

Our parents Isabell and Carroll Buzzett and Douglas and George Shiflet, Sr., worked so tirelessly to make our lives intense, joyful, and stimulating. We were blessed and inspired by their living efforts, and we miss them. Hopefully, they would be pleased by what we have been doing with our lives.

1

OVERVIEW

MODULE 1.1

Overview of Computational Science

> Scientific revolutions are "those non-cumulative developmental episodes in which in an older paradigm is replaced in whole or in part by an incompatible new one."
>
> —Thomas Kuhn, *The Structure of Scientific Revolutions*

Normally, "the scientific revolution" refers to the discoveries of the sixteenth and seventeenth centuries in Europe, which changed the western view of the natural world. This revolution began with the sun-centered universe of Copernicus and continued until Newton proposed universal gravitation and laws of motion. Nature was the object of much interest, and the exploration of the New World with all its discoveries continued to feed the desire to understand nature.

During the twentieth century, according to the eminent string-theory physicist Michio Kaku, there were three scientific revolutions—the quantum revolution, the biomolecular revolution, and the computer revolution (Kaku 1998). Few can doubt the rapidity with which recent scientific advances have been made with each new discovery or insight changing our view of our planet, its inhabitants, and often the universe. The accomplishments of that century augurs very well for the current one.

Early in the twenty-first century, Microsoft Research convened a workshop of international authorities to devise a "vision and roadmap of the evolution, challenges and potential of computer science and computing in scientific research during the next fifteen years." The outcome was "Towards 2020 Science." What they predicted marks the beginnings of a new scientific revolution, where computation will become more than an adjunct supporter of scientific research. Computational principles and tools will become integrated into science, changing the fundamental way that science is practiced. Computational science in both theoretical and experimental sciences will greatly augment the rates of scientific advances that will benefit the planet and our species (Microsoft Research 2006). For example, the results of the human genome project, which depended upon large-scale computational science, have encouraged a myriad of new research and development in government, university, and commercial laboratories. One significant outcome from these projects will be a far better understanding of molecular mechanisms that underlie human diseases and their more effective treatments.

In 2005, the President's Information Technology Advisory Committee released the report "Computational Science: Ensuring America's Competiveness" (Report to the President 2005). They concluded that computational science and high-performance computing could be integral to innovations in all of the sciences (biological/ biomedical, physical, and social), engineering, industry, and defense. Advances in computation allow us to acquire and analyze enormous streams of data, making it possible to consider and solve problems heretofore unapproachable. Computational science also allows us to build models, visualize phenomena, and conduct experiments difficult or impossible in the laboratory. We can now examine interactions in systems that involve more than one discipline, encouraging us to collaborate with specialists in other fields. Such collaboration should lead to solutions that are creative, synergistic, sustainable, and economically favorable.

Computational science, the fast-growing interdisciplinary field that is at the intersection of the sciences, computer science, and mathematics, will require scientists who are appropriately trained. The experts who produced "Towards 2020 Science" predicted that future scientists who are not computationally and mathematically literate will be unable to do science. Chemistry professor Robert Harrison, director of the Joint Institute for Computational Sciences at the University of Tennessee, states in the JICS Mission webpage, "To translate even the most elementary theories into useful tools for physical chemistry discovery, you have to do large-scale computation." He states further, "If you look at students coming into our graduate program from the undergraduate world, those that haven't already had some exposure to computation, such as thinking algorithmically, solving problems on the computer, and the little bits of applied math that you need to understand all of that, . . . have lost a year or two of productivity at the graduate level. But it's not only the undergraduate students coming into graduate school that have this issue; it's also our undergrads going off into the larger world. Industry and many other aspects of the commercial world use simulation and computation in diverse ways" (JICS).

Computational science, which combines computer simulation, scientific visualization, mathematical modeling, computer programming, data structures, networking, database design, symbolic computation, and high-performance computing, can transform practices in a diverse range of disciplines. Its computer models and simulations offer valuable approaches to problems in many areas, as the following examples indicate.

1. Scientists at Los Alamos National Laboratory and the University of Minnesota wrote, "Mathematical modeling has impacted our understanding of HIV pathogenesis. Before modeling was brought to bear in a serious manner, AIDS was thought to be a slow disease in which treatment could be delayed until symptoms appeared, and patients were not monitored very aggressively. In the large, multicenter AIDS cohort studies aimed at monitoring the natural history of the disease, blood typically was drawn every six months. There was a poor understanding of the biological processes that were responsible for the observed levels of virus in the blood and the rapidity at which the virus became drug resistant. Modeling, coupled with advances in technology, has changed all of this." Dynamic modeling has not only revealed important features of HIV pathogenesis but has advanced the drug treatment regime for AIDS patients (Perelson and Nelson 1999). Since then, Perelson and other researchers have applied modeling to enhance our understanding of the hepatitis C virus, which causes widespread infections and is the primary cause of liver cancer in the United States. Such models have already revealed much about the pathogenesis of the virus, the effectiveness of treatments (interferon/ribavirin and direct antiviral agents), and the influence of genetic variants in the kinetics of the virus (Dahari et al. 2011).

2. From the 1960s, numerical weather prediction has revolutionized forecasting. "Since then, forecasting has improved side by side with the evolution of computing technology, and advances in computing continue to drive better forecasting as weather researchers develop improved numerical models" (Pittsburgh Supercomputing Center 2001). A Weather Research and Forecasting (WRF) Model was released in 2000. The latest version of this model utilizes "multiple dynamical cores, a 3-dimensional variational data assimilation system, and a software architecture allowing for computational parallelism and system extensibility." This sophisticated, mesoscale [horizontal scale of 2 to 2000 kilometers (km)] numerical weather-prediction system is useful for forecasting and research. An array of partners, including the National Center for Atmospheric Research (NCAR), the National Oceanic and Atmospheric Administration (NOAA), other governmental and military organizations, universities, plus some international groups, continuously revise the WRF model. With this efficient, adaptable model for forecasting, researchers can conduct simulations using real data or idealized designs. In 2007, a specialized version of WRF was initiated for forecasting and research on hurricanes (WRF).

3. A multidisciplinary team at the University of Tennessee's Institute for Environmental Modeling is using computational ecology to study complex options for ecological management of the Everglades. Louis Gross, Director of the Institute, says that "computational technology, coupled with mathematics and ecology, will play an ever-increasing role in generating vital information society needs to make tough decisions about its surroundings" (Lymn 2003). South Florida has a well-known history of disruptions to normal water flow. The UT group has developed a parallelized landscape population model (ALFISH) to integrate with other models in a multiscale ecological multimodel (Across-Trophic Level System Simulation, ATLSS). ALFISH is used to model the effects on freshwater fish (planktivorous and piscivirous) of different water-management plans. These fish populations represent food resources for wading birds, and researchers can link the fish model with wading bird models to help sustain the higher-level multimodel (Wang et al. 2006).

4. Application of computer modeling has fueled the debate in another, rather unexpected area—linguistics. The origin of the Indo-European family of languages is rather hotly debated between proponents of two hypotheses—Eurasian steppes, 6000 years (yr) ago versus Anatolia (mostly in present-day Turkey), 8000 to 9500 yr ago. This family of languages has given rise to more than 400 modern languages, spoken by about three billion people. Recently, Quentin Atkinson and colleagues utilized evolutionary models, often employed to ascertain the origin of viruses that lead to epidemics, to analyze this problem. Based on common vocabulary words from various languages in the family, the model supports Anatolian origin, as agricultural techniques were broadcast. Although certainly not resolving the argument, the results have given experts in this field something to consider carefully. (Bouckaert et al. 2012)

Projects

1. Investigate three applications of computational science involving different scientific areas and write at least a paragraph on each. List references.
2. Investigate an application of computational science and write a three-page, typed, double-spaced paper on the topic. List references.

References

Bouckaert, Remco, Philippe Lemey, Michael Dunn, Simon J. Greenhill, Alexander V. Alekseyenko, Alexei J. Drummond, Russell D. Gray, Marc A. Suchard, and Quentin D. Atkinson. 2012. "Mapping the Origins and Expansion of the Indo-European Language Family." *Science* 337(6097): 957–960.

Dahari, H., J. Guedj, A. Perelson, and T. Layden. 2011. "Hepatitis C Viral Kinetics in the Era of Direct Acting Antiviral Agents and IL28B." *Curr Hepat Rep.* 10(3): 214–227. http://www.ncbi.nlm.nih.gov/pmc/articles/PMC3237049/pdf/nihms341709.pdf (accessed August 12, 2012)

JICS (Joint Institute for Computational Science). University of Tennessee and Oak Ridge National Laboratory. "Harrison Discusses JICS Mission." http://www.jics.tennessee.edu/jics-mission (accessed August 20, 2012)

Kaku, Michio. 1998. *Visions: How Science Will Revolutionize the 21st Century.* New York: Anchor.

Kuhn, Thomas. 1962. *The Structure of Scientific Revolutions.* Chicago: University of Chicago Press.

Lymn, Nadine. 2003. "Computational Ecology." Math & Bio 2010: Linking Undergraduate Disciplines, Kickoff Meeting—Articles and Reports. http://www.maa.org/mtc/Computational-Ecology.pdf (accessed January 7, 2013)

Microsoft Research. 2006. "Towards 2020 Science." http://research.microsoft.com/en-us/um/cambridge/projects/towards2020science/ (accessed August 12, 2012)

Perelson, Alan S., and Patrick W. Nelson. 1999. "Mathematical Analysis of HIV-1 Dynamics in Vivo." *SIAM REVIEW*, 41(1): 3–44.

Pittsburgh Supercomputing Center. 2001. "Pittsburgh System Marks a Watershed in Weather Prediction." News Release, December 4, 2001. http://www.psc.edu/publicinfo/news/2001/weather-12-04-01.html (accessed January 7, 2013)

Report to the President. June 2005. "Computational Science: Ensuring America's Competitiveness." President's Information Technology Advisory Committee. National Coordination Office for Information Technology Research and Development (NCO/IT R&D). http://www.nitrd.gov/pitac/reports/20050609_computational/computational.pdf (accessed January 7, 2013)

Wang, D., M. W. Berry, E. A. Carr, and L. J. Gross. 2006. "A Parallel Fish Landscape Model for Ecosystem Modeling." *SIMULATION* 2006, 82: 451.

WRF (The Weather Research and Forecasting Model). http://wrf-model.org/index.php (accessed August 23, 2012)

MODULE 1.2

The Modeling Process

Introduction

The process of making and testing hypotheses about models and then revising designs or theories has its foundation in the experimental sciences. Similarly, computational scientists use **modeling** to analyze complex, real-world problems in order to predict what might happen with some course of action. For instance, Professor Muneo Hori and colleagues from the Earthquake Research Institute, the University of Tokyo, Japan, use high-performance computation with sophisticated models to simulate earthquakes, making quantitative predictions of infrastructural damages, response, and recovery to help minimize damage, death, and injury (Lalith and Hori 2012). Professor Liming Liang, statistical geneticist at Harvard School of Public Health, uses computational and statistical tools to better understand the genetic variation in complex human diseases, such as dyslipidemia, cancer, and type 2 diabetes (Liang). Cognitive scientists, such as Professor Ken Koedinger at the Human-Computer Interaction Institute, Carnegie-Mellon University, develop computer models of student reasoning and learning to aid in the design educational software and to guide teaching practices (Koedinger). Both civilian and military organizations commonly employ drones (unmanned aerial vehicles, UAVs). Whether to monitor air quality or supervise combat forces, this technology is becoming more and more important, but the operation of the drones is quite complex. While a postdoctoral fellow at MIT, Dr. Luca Bertuccelli worked with a team using models to develop new decision support systems, enabling the operators of these craft to make better decisions (Bertuccelli). Arboviruses are arthropod-borne viruses that cause diseases, such as West Nile encephalitis, dengue fever, and yellow fever. A mathematical modeling team at the Universidad Nacional Autónoma de México has been modeling the dynamics of such infections. A better understanding of these viruses will improve outbreak predictions, interventions, and responses (Vargus and Cruz-Pacheco 2010).

Definition **Modeling** is the application of methods to analyze complex, real-world problems in order to make predictions about what might happen with various actions.

Model Classifications

Several classification categories for models exist. A system we are modeling exhibits **probabilistic**, or **stochastic**, **behavior** if it appears that an element of chance exists. For example, the path of a hurricane is probabilistic. In contrast, a behavior can be **deterministic**, such as the position of a falling object in a vacuum. Similarly, models can be deterministic or probabilistic. A **probabilistic**, or **stochastic**, **model** exhibits random effects, while a **deterministic model** does not. The results of a deterministic model depend on the initial conditions; and in the case of computer implementation with particular input, the output is the same for each program execution. As we see in Module 9.2, "Simulations," and other modules, we can have a probabilistic model for a deterministic situation, such as a model that uses random numbers to estimate the area under a curve.

Definitions A system exhibits **probabilistic**, or **stochastic**, **behavior** if an element of chance exists. Otherwise, the system exhibits **deterministic behavior**. A **probabilistic**, or **stochastic**, **model** exhibits random effects, while a **deterministic model** does not.

We can also classify models as static or dynamic. In a **static model**, we do not consider time, so that the model is comparable to a snapshot or a map. For example, a model of the weight of a salamander as being proportional to the cube of its length has variables for weight and length but not for time. By contrast, in a **dynamic model**, time changes, so that such a model is comparable to an animated cartoon or a movie. For example, the number of salamanders in an area undergoing development changes with time; hence, a model of such a population is dynamic. Many of the models we consider in this text are dynamic and employ a static component as part of the dynamic model.

Definitions A **static model** does not consider time, while a **dynamic model** changes with time.

When time changes continuously and smoothly, the model is **continuous**. If time changes in incremental steps, the model is **discrete**. A discrete model is analogous to a movie. A sequence of frames moves so quickly that the viewer perceives motion. However, in a live play, the action is continuous. Just as a discrete sequence of movie frames represents the continuous motion of actors, we often develop discrete computer models of continuous situations.

Definitions In a **continuous model**, time changes continuously, while in a **discrete model**, time changes in incremental steps.

Steps of the Modeling Process

The modeling process is cyclic and closely parallels the scientific method and the software life cycle for the development of a major software project. The process is cyclic because at any step we might return to an earlier stage to make revisions and continue the process from that point.

The steps of the modeling process are as follows:

1. **Analyze the problem.**
   We must first study the situation sufficiently to identify the problem precisely and understand its fundamental questions clearly. At this stage, we determine the problem's objective and decide on the problem's classification, such as deterministic or stochastic. Only with a clear, precise problem identification can we translate the problem into mathematical symbols and develop and solve the model.
2. **Formulate a model.**
   In this stage, we design the model, forming an abstraction of the system we are modeling. Some of the tasks of this step are as follows:
   a. **Gather data.**
      We collect relevant data to gain information about the system's behavior.
   b. **Make simplifying assumptions and document them.**
      In formulating a model, we should attempt to be as simple as reasonably possible. Thus, we frequently decide to simplify some of the factors and to ignore other factors that do not seem as important. Most problems are entirely too complex to consider every detail, and doing so would only make the model impossible to solve or to run in a reasonable amount of time on a computer. Moreover, factors often exist that do not appreciably affect outcomes. Besides simplifying factors, we may decide to return to Step 1 to restrict further the problem under investigation.
   c. **Determine variables and units.**
      We must determine and name the variables. An **independent variable** is the variable on which others depend. In many applications, time is an independent variable. The model will try to explain the **dependent variables**. For example, in simulating the trajectory of a ball, time is an independent variable; and the height and the horizontal distance from the initial position are dependent variables whose values depend on the time. To simplify the model, we may decide to neglect some variables (such as air resistance), treat certain variables as constants, or aggregate several variables into one. While deciding on the variables, we must also establish their units, such as days as the unit for time.
   d. **Establish relationships among variables and submodels.**
      If possible, we should draw a diagram of the model, breaking it into submodels and indicating relationships among variables. To simplify the model, we may assume that some of the relationships are simpler than they really are. For example, we might assume that two variables are related in a linear manner instead of in a more complex way.
   e. **Determine equations and functions.**

While establishing relationships between variables, we determine equations and functions for these variables. For example, we might decide that two variables are proportional to each other, or we might establish that a known scientific formula or equation applies to the model. Many computational science models involve differential equations, or equations involving a derivative.

3. **Solve the model.**
This stage implements the model. It is important not to jump to this step before thoroughly understanding the problem and designing the model. Otherwise, we might waste much time, which can be most frustrating. Some of the techniques and tools that the solution might employ are algebra, calculus, graphs, computer programs, and computer packages. Our solution might produce an exact answer or might simulate the situation. If the model is too complex to solve, we must return to Step 2 to make additional simplifying assumptions or to Step 1 to reformulate the problem.
4. **Verify and interpret the model's solution.**
Once we have a solution, we should carefully examine the results to make sure that they make sense (verification) and that the solution solves the original problem (validation) and is usable. The process of **verification** determines if the solution works correctly, while the process of **validation** establishes whether the system satisfies the problem's requirements. Thus, verification concerns "solving the problem right," and validation concerns "solving the right problem." Testing the solution to see if predictions agree with real data is important for verification. We must be careful to apply our model only in the appropriate ranges for the independent data. For example, our model might be accurate for time periods of a few days but grossly inaccurate when applied to time periods of several years. We should analyze the model's solution to determine its implications. If the model solution shows weaknesses, we should return to Step 1 or 2 to determine if it is feasible to refine the model. If so, we cycle back through the process. Hence, the cyclic modeling process is a trade-off between **simplification** and **refinement**. For refinement, we may need to extend the scope of the problem in Step 1. In Step 2, while refining, we often need to reconsider our simplifying assumptions, include more variables, assume more complex relationships among the variables and submodels, and use more sophisticated techniques.
5. **Report on the model.**
Reporting on a model is important for its utility. Perhaps the scientific report will be written for colleagues at a laboratory or will be presented at a scientific conference. A report contains the following components, which parallel the steps of the modeling process:
    a. **Analysis of the problem**
    Usually, assuming that the audience is intelligent but not aware of the situation, we need to describe the circumstances in which the problem arises. Then, we must clearly explain the problem and the objectives of the study.
    b. **Model design**
    The amount of detail with which we explain the model depends on the situation. In a comprehensive technical report, we can incorporate

much more detail than in a conference talk. In either case, we should state the simplifying assumptions and the rationale for employing them. Clearly labeled diagrams of the relationships among variables and submodels are usually very helpful in understanding the model.

c. **Model solution**
In this section, we describe the techniques for solving the problem and the solution. We should give as much detail as necessary for the audience to understand the material without becoming mired in technical minutia. For a written report, appendices may contain more detail, such as source code of programs and additional information about the solutions of equations.

d. **Results and conclusions**
Our report should include results, interpretations, implications, recommendations, and conclusions of the model's solution. Usually, we present some of the data and results in tables or graphs. Such figures should contain titles, sources, and labels for columns and axes. We may also include suggestions for future work.

6. **Maintain the model.**
As the model's solution is used, it may be necessary or desirable to make corrections, improvements, or enhancements. In this case, the modeler again cycles through the modeling process to develop a revised solution.

Definitions The process of **verification** determines if the solution works correctly, while the process of **validation** establishes if the system satisfies the problem's requirements.

Although we described the modeling process as a sequence or series of steps, we may be developing two or more steps simultaneously. For example, it is advisable to be compiling the report from the beginning. Otherwise, we can forget to mention significant points, such as reasons for making certain simplifying assumptions or for needing particular refinements. Moreover, within modeling teams, individuals or groups frequently work on different submodels simultaneously. Having completed a submodule, a team member might be verifying the submodule while others are still working on solving theirs.

The modeling process is a creative, scientific endeavor. As such, a problem we are modeling usually does not have one correct answer. The problems are complex, and many models provide good, although different, solutions. Thus, modeling is a challenging, open-ended, and exciting venture.

Exercises

1. Compare and contrast the modeling process with the scientific method: Make observations; formulate a hypothesis; develop a testing method for the hypothesis; collect data for the test; using the data, test the hypothesis; accept or reject the hypothesis.

2. Compare and contrast the modeling process with the software life cycle: analysis, design, implementation, testing, documentation, maintenance.

References

Bertuccelli, Luca F. Personal webpage. http://web.mit.edu/lucab/www/publications.html (accessed August 29, 2012)

Giordano, Frank R., Maurice D. Weir, and William P. Fox. 2003. *A First Course in Mathematical Modeling*. 3rd ed. Pacific Grove, CA: Brooks/Cole-Thompson Learning.

Koedinger. K. Personal webpage. http://www.hcii.cmu.edu/people/faculty/ken-koedinger (accessed August 28, 2012)

Lalith, Maddegedara and Muneo Hori. 2012. "Application of High Performance Computation for the Prediction of Urban Area Earthquake Disaster." Presented to the 8th International Symposium on Social Management Systems SSMS2012 (May), Kaohsiung, Taiwan. http://management.kochi-tech.ac.jp/ssms_papers/sms11-4706_4a089987c005e671dd603ce4162e6704.pdf (accessed August 28, 2012)

Liang, Liming. My Research, personal webpage. http://www.hsph.harvard.edu/faculty/liming-liang/index.html (accessed August 28, 2012)

Vargus, C., L. Esteva, and G. Cruz-Pacheco. 2010. "Mathematical Modelling of Arbovirus Diseases." Electrical Engineering Computing Science and Automatic Control (CCE), 2010 7th International Conference (September 8–10, 2010, Mexico City). pp. 205–209.

2

SYSTEM DYNAMICS PROBLEMS WITH RATE PROPORTIONAL TO AMOUNT

MODULE 2.1

System Dynamics Tool—Tutorial 1

Download

From the textbook's website, download Tutorial 1 in PDF format for your system dynamics tool. We recommend that you work through the tutorial and answer all Quick Review Questions using the corresponding software.

Introduction

Dynamic systems, which change with time, are usually very complex, having many components, with involved relationships. Two examples are systems involving competition among different species for limited resources and the kinetics of enzymatic reactions.

With a system dynamics tool, we can model complex systems using diagrams and equations. Thus, such a tool helps us perform Step 2 of the modeling process—formulate a model—by helping us document our simplifying assumptions, variables, and units; establish relationships among variables and submodels; and record equations and functions. Then, a system dynamics tool can help us solve the model—Step 3 of the modeling process—by performing simulations using the model and generating tables and graphs of the results. We use this output to perform Step 4 of the modeling process—verify and interpret the model's solution. Often such examination leads us to change a model. With its graphical view and built-in functions, a system dynamics tool facilitates cycling back to an earlier step of the modeling process to simplify or refine a model. Once we have verified and validated a model, the tool's diagrams and equations from the design and the results from the simulation should be part of our report, which we do in Step 5 of the modeling process. The tool can even help us as we maintain the model (Step 6) by making corrections, improvements, or enhancements.

This first tutorial is available for download from the textbook's website for several different system dynamics tools. Tutorial 1 in your system of choice prepares you to perform basic modeling with such a tool, including the following:

- Diagramming a model
- Entering equations and values
- Running a simulation
- Constructing graphs
- Producing tables

The module gives examples and Quick Review Questions for you to complete and execute with your desired tool.

MODULE 2.2

Unconstrained Growth and Decay

Introduction

Many situations exist where the rate at which an amount is changing is proportional to the amount present. Such might be the case for a population of people, deer, or bacteria, for example. When money is compounded continuously, the rate of change of the amount is also proportional to the amount present. For a radioactive element, the amount of radioactivity decays at a rate proportional to the amount present. Similarly, the concentration of a chemical pollutant decays at a rate proportional to the concentration of pollutant present.

Rate of Change

We deal with rate of change every time we drive a car. Suppose our position (y) is a function (s) of time (t), so we write $y = s(t)$. Suppose also that we start driving on a straight road at time $t = 0$ hours (h) at position marker $s(0) = 10$ miles (mi; about 16.1 km), and at time $t = 2$ h we are at position $s(2) = 116$ mi (about 186.7 km). Our **average velocity**, or average rate of change of position with respect to time, is the **change in position** (Δs) over the **change in time** (Δt) and incorporates average speed as well as direction by its sign:

$$\textbf{average velocity} = \frac{\Delta s}{\Delta t} = \frac{116 \text{ mi} - 10 \text{ mi}}{2 \text{ h} - 0 \text{ h}} = \frac{106 \text{ mi}}{2 \text{ h}} = 53 \text{ mi/h}$$

or

$$\textbf{average velocity} = \frac{\Delta s}{\Delta t} = \frac{186.7 \text{ km} - 16.1 \text{ km}}{2 \text{ h} - 0 \text{ h}} = \frac{170.6 \text{ km}}{2 \text{ h}} = 85.3 \text{ km/h}$$

We probably are not driving at a constant rate of 53 mi/h (85.3 km/h), but sometimes we are moving faster and other times, slower. To obtain a more accurate measure of our velocity at time $t = 1$ h, we can use a smaller interval. For instance, at time $t = 1$ h, our position might be at marker $s(1) = 51.2$ mi, while a short time before at $t = 0.98$ h, our position was $s(0.98) = 50.0$ mi. As the following calculation shows, over this interval of 0.02 h (1.2 min), our average velocity is faster, 60 mi/h:

$$\text{average velocity} = \frac{\Delta s}{\Delta t} = \frac{51.2\text{ mi} - 50\text{ mi}}{1.00\text{ h} - 0.98\text{ h}} = \frac{1.2\text{ mi}}{0.02\text{ h}} = 60\text{ mi/h}$$

or about 96.6 km/h.

Definition Suppose $s(t)$ is the position of an object at time t, where $a \le t \le b$. Then the **change in time**, Δt, is $\Delta t = b - a$; and the **change in position**, Δs, is $\Delta s = s(b) - s(a)$. Moreover, the **average velocity**, or the **average rate of change of s with respect to t**, of the object from time $a = b - \Delta t$ to time b is

$$\textbf{average velocity} = \frac{\textbf{change in position}}{\textbf{change in time}} = \frac{\Delta s}{\Delta t}$$

$$= \frac{s(b) - s(a)}{b - a} = \frac{s(b) - s(b - \Delta t)}{\Delta t}$$

Quick Review Question 1

Suppose on a windless day someone standing on a bridge holds a ball over the side and tosses the ball straight up into the air. After reaching its highest point, the ball falls, eventually landing in the water. The ball's height in meters (m) above the water (y) is a function (s) of time (t) in seconds (s), or $y = s(t)$.

- **a.** Determine the average velocity with units of the ball from $t = 1$ s to $t = 2$ s if $s(1) = 21.1$ m and $s(2) = 21.4$ m.
- **b.** Determine the average velocity with units of the ball from $t = 1$ s to $t = 3$ s if $s(1) = 21.1$ m and $s(3) = 11.9$ m.
- **c.** Using the notation of the definition of average velocity, for Part b determine the following, including units: b, $s(b)$, Δt, $b - \Delta t$, $s(b - \Delta t)$, Δs.

By making the interval smaller and smaller around the time $t = 1$ h, the average velocity calculation approaches our precise velocity at $t = 1$ h, or our **instantaneous rate of change of position with respect to time**, which is our odometer's reading. This instantaneous rate of change of s with respect to t is the **derivative** of s with respect to t, written as $\mathbf{s'(t)}$, or $\frac{dy}{dt}$, or dy/dt; and $\mathbf{s'(1)}$, or $\left.\frac{ds}{dt}\right|_{t=1}$, indicates the derivative at time $t = 1$ h.

Definition The **instantaneous velocity**, or the **instantaneous rate of change of s with respect to t**, at $t = b$ is the number the average velocity, $\frac{s(b) - s(b - \Delta t)}{\Delta t}$, approaches as Δt comes closer and closer to 0 (provided the ratio approaches a number). In this case, the **derivative of $y = s(t)$ with respect to t** at $t = b$, written $s'(b)$ or $\left.\frac{dy}{dt}\right|_{t=b}$, is the instantaneous velocity at $t = b$. In general, the **derivative of $y = s(t)$ with respect to t** is written as $s'(t)$, or $\frac{dy}{dt}$, or ***dy/dt***.

A function, such as $y = s(t)$, can represent many things other than position. Moreover, we are not restricted to using symbols, such as s. For example, $Q(t)$ might represent a quantity (mass) of radioactive carbon-14 at time t, and the instantaneous rate of change of Q with respect to t, $Q'(t) = dQ/dt$, is the instantaneous rate of decay. As another example, $P(t)$ might symbolize a population at time t, so that $P'(t) = dP/dt$, is the rate of change of the population with respect to t.

Differential Equation

Continuing with the population example, suppose we have a population in which no individuals arrive or depart; the only change in the population comes from births and deaths. No constraints, such as competition for food or a predator, exist on growth of the population. When no limiting factor exists, we have the **Malthusian model** for unconstrained population growth, where the rate of change of the population is **directly proportional** ($\propto$) to the number of individuals in the population. If P represents the population and t represents time, then we have the following proportion:

$$\frac{dP}{dt} \propto P$$

For a positive growth rate, the larger the population, the greater the change in the population. With the same positive growth rate in two cities, say New York City and Spartanburg, S.C., the population of the larger New York City increases more in magnitude in a year than that of Spartanburg. In a later section of this module, "Unconstrained Decay," we consider a situation in which the rate is negative.

We write the preceding proportion in equation form as follows:

$$\frac{dP}{dt} = rP$$

The constant r is the **growth rate**, or **instantaneous growth rate**, or **continuous growth rate**, while dP/dt is the **rate of change of the population**.

In "System Dynamics Tool—Tutorial 1" (Module 2.1), we started with a bacterial population of size 100, an instantaneous growth rate of 10% = 0.10, and time measured in hours. Thus, we had

$$\frac{dP}{dt} = 0.10P$$

with $P_0 = 100$. The equation $\frac{dP}{dt} = 0.10P$ with the **initial condition** $P_0 = 100$ is a **differential equation** because it contains a derivative. A **solution** to this differential equation is a function, $P(t)$, whose derivative is $0.10P(t)$, with $P(0) = 100$. We begin by reconsidering this example from Tutorial 1 for reinforcement and a more in-depth examination of the concepts.

Definitions A **differential equation** is an equation that contains one or more derivatives. An **initial condition** is the value of the dependent variable when the independent variable is zero. A **solution** to a differential equation is a function that satisfies the equation and initial condition(s).

Difference Equation

Each diagram in Figure 2.2.1, developed with a choice of modeling tools and with the generic format employed by the text, depicts the situation with *population* indicating P, *growth_rate* representing r, and *growth* meaning dP/dt. A **stock** (**box variable**, or **reservoir**), such as *population*, accumulates with time. By contrast, a **converter** (**variable-auxiliary/constant**, or **formula**), such as *growth_rate*, does not accumulate but stores an equation or a constant. The growth is the additional number of organisms that join the population. Thus, a **flow** (**rate**), such as *growth*, is an activity that changes the magnitude of a stock and represents a derivative. Because both population and growth rate are necessary to determine the growth, we have **arrows** (**connectors**, or **arcs**) from these quantities to the flow indicator.

For a simulation with a system dynamics tool or a program we write, we consider time advancing in small, incremental steps. For time, t, and length of a time step, Δt, the **previous time** is $t - \Delta t$. Thus, if t is 7.75 s and Δt is 0.25 s, the previous time is 7.50 s. A system dynamics tool might call the change in time dt, DT, or something else instead of Δt. As some tools do to avoid confusion, we replace each blank in a diagram component name with an underscore when using the name in equations and discussions. For example, we employ *growth rate* in the diagrams of Figure 2.2.1 and the corresponding *growth_rate* in the following discussion. Regardless of the notation, with initial *population* = 100, *growth_rate* = 0.1, and *growth* = *growth_rate* * *population*, as in Figure 2.2.1, a system dynamics tool generates an equation similar to the following, where *population*(t) is the population at time t and *population*($t - \Delta t$) is the population at time $t - \Delta t$:

$$population(t) = population(t - \Delta t) + (growth) * \Delta t$$

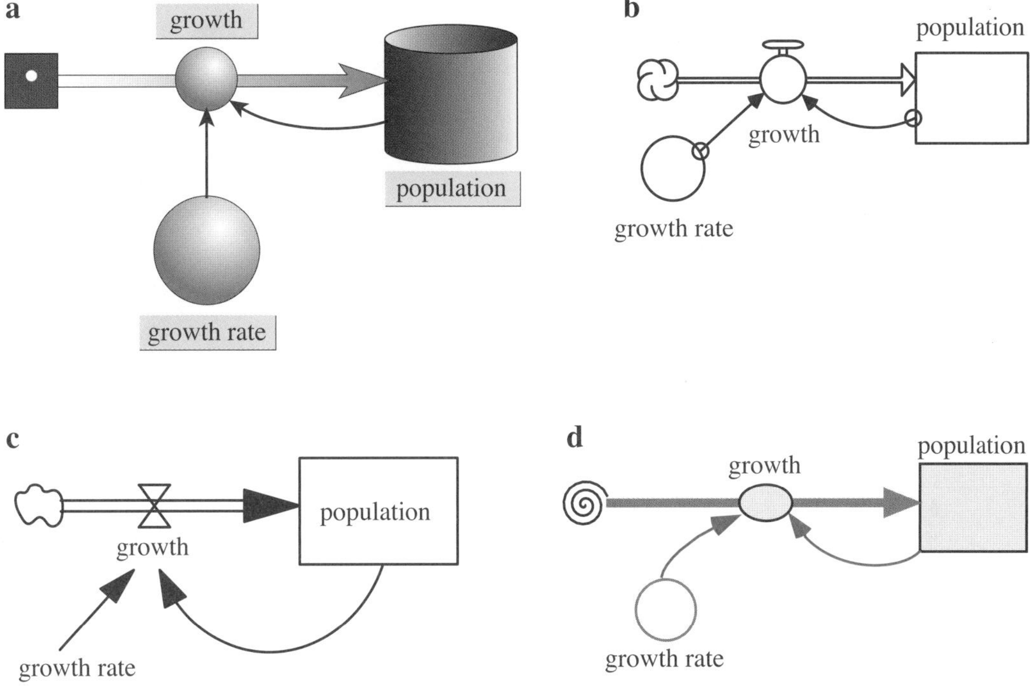

Figure 2.2.1 Diagrams of population models where growth rate is proportional to population: (a) *Berkeley Madonna®* (b) *STELLA®* (c) *Vensim PLE®* (d) Text's format

This equation, called a **finite difference equation**, indicates that the population at one time step is the population at the previous time step plus the change in population over that time interval:

$$\text{(new population)} = \text{(old population)} + \text{(change in population)}$$

or

$$population(t) = population(t - \Delta t) + \Delta population$$

where **Δpopulation** is a notation for the **change in population**. We approximate the change in the population over one time step, $\Delta population$ or $(growth) * \Delta t$, as the finite difference of the populations at one time step and at the previous time step, $population(t) - population(t - \Delta t)$. Thus, solving for *growth*, we have an approximation of the derivative dP/dt as follows:

$$growth = \frac{\Delta population}{\Delta t} = \frac{population(t) - population(t - \Delta t)}{\Delta t}$$

Computer programs and system dynamics tools employ such finite difference equations to solve differential equations.

> **Definition** A **finite difference equation** is of the following form:
>
> (new value) = (old value) + (change in value)
>
> Such an equation is a discrete approximation to a differential equation.

Quick Review Question 2

Consider the differential equation $dQ/dt = -0.0004Q$, with $Q_0 = 200$.

a. Using delta notation, give a finite difference equation corresponding to the differential equation.
b. At time $t = 9.0$ s, give the time at the previous time step, where $\Delta t = 0.5$ s.
c. If $Q(t - \Delta t) = 199.32$ and $Q(t) = 199.28$, give ΔQ.

The *growth* is the *growth_rate* (r previously) times the current *population* (P previously). For example, we can show that the population at time $t = 0.025$ h is approximately *population*(0.025) = 100.250250 bacteria, so that *growth* is about *growth_rate* * *population*(0.025) = 0.1 * 100.250250 = 10.025025 bacteria per hour at that instant. For $\Delta t = 0.005$ h, the change in the population of bacteria to the next time step, 0.025 + 0.005 = 0.030 h, is approximately *growth* * Δt = 10.025025 * 0.005 = 0.050125 bacteria[1]. We calculate the population at time 0.030 h as follows:

$$\begin{aligned} population(0.030) &= population(0.025) + (growth \text{ at time } 0.025 \text{ h}) * \Delta t \\ &= 100.250250 + 10.025025 * 0.005 \\ &= 100.250250 + 0.050125 \\ &= 100.300375 \end{aligned}$$

Thus, we compute the value at the line $t = 0.030$ h of Table 2.2.1 using the previous line.

Quick Review Question 3

Evaluate *population*(0.045), the population at the next time interval after the end of Table 2.2.1, to six decimal places.

Table 2.2.1
Table of Estimated Populations, Where the Initial Population is 100, the Continuous Growth Rate is 10% per Hour, and the Time Step is 0.005 h

t	*population*(t)	=	*population*($t - \Delta t$)	+	(*growth*)	*	Δt
0.000	100.000000						
0.005	100.050000	=	100.000000	+	10.000000	*	0.005
0.010	100.100025	=	100.050000	+	10.005000	*	0.005
0.015	100.150075	=	100.100025	+	10.010003	*	0.005
0.020	100.200150	=	100.150075	+	10.015008	*	0.005
0.025	100.250250	=	100.200150	+	10.020015	*	0.005
0.030	100.300375	=	100.250250	+	10.025025	*	0.005
0.035	100.350525	=	100.300375	+	10.030038	*	0.005
0.040	100.400701	=	100.350525	+	10.035053	*	0.005

[1] Computations in this model use **Euler's Method** for estimating values of a function. In Chapter 6, we examine this and two other techniques for numeric integration.

Because of compounding, the number of bacteria at $t = 1$ h is slightly more than 10% of 100, namely, 110.51. Table 2.2.2 lists the growth and the population on the hour for 20 h, and Figure 2.2.2 graphs the population versus time. The model states and the table and figure illustrate that as the population increases, the growth does, too.

The model gives an estimate of the population at various times. If the model is analytically correct, a simulation estimates the values for *growth* and *population*. Until computer round-off error (discussed in Module 5.2) causes the step size to be zero, it is usually the case that the smaller the step size, the more accurate will be the results. (In Exercise 9, we explore a situation where the smaller step size does not produce better results.) Because the additional computations resulting from a smaller step size cause the simulation to run longer, we often use a larger Δt during development and switch to a smaller Δt for more accurate results when the project is close to completion.

Table 2.2.2
Table of Estimated Growths and Populations, Reported on the Hour, Where the Initial Population is 100, the Growth Rate is 10%, and the Time Step is 0.005 h

t (h)	*growth*	*population*
0.000	10.00	100.00
1.000	11.05	110.51
2.000	12.21	122.13
3.000	13.50	134.98
4.000	14.92	149.17
5.000	16.49	164.85
6.000	18.22	182.18
7.000	20.13	201.34
8.000	22.25	222.51
9.000	24.59	245.90
10.000	27.18	271.76
11.000	30.03	300.33
12.000	33.19	331.91
13.000	36.68	366.81
14.000	40.54	405.38
15.000	44.80	448.00
16.000	49.51	495.11
17.000	54.72	547.16
18.000	60.47	604.69
19.000	66.83	668.27
20.000		738.54

Rule of Thumb Although the simulation takes longer because of more computation, it is usually more accurate to use a small step size (Δt), say, 0.01 or less.

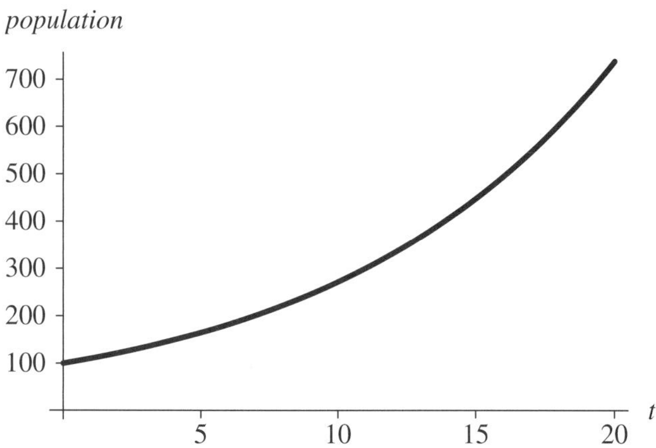

Figure 2.2.2 Graph of population versus time (hours) for the data in Table 2.2.2

Simulation Program

In developing a simulation program, we use statements similar to the preceding finite difference equations. We initialize constants, such as *growthRate*, *population*, Δt, and the length of time the simulation is to run (*simulationLength*), and we update the population repeatedly in a loop. The calculation for the total number of iterations (*numIterations*) of this loop is *simulationLength*/Δt. For example, if the simulation length is 10 h and Δt is 0.25 h, then the number of loop iterations is *numIterations* = 10/0.25 = 40. We have a loop index (i) go from 1 through *numIterations*. Inside the loop, we calculate time t as the product of i and Δt. For example, if Δt is 0.25 h, during the first iteration, the index i becomes 1 and the time is $1 * \Delta t = 0.25$ h. On loop iteration $i = 8$, the time gets the value $8 * \Delta t = 8 * 0.25$ h = 4.00 h.

Algorithm 1 contains **pseudocode**, or a structured English outline of the design, for generating and displaying the time, growth, and population at each time step. In the algorithm, a **left-facing arrow** (←) indicates assignment of the value of the expression on the right to the variable on the left. For example, *numIterations* ← *simulationLength*/Δt represents an assignment statement in which *numIterations* gets the value of *simulationLength*/Δt.

Algorithm 1 Algorithm for simulation of unconstrained growth

```
initialize simulationLength
initialize population
initialize growthRate
initialize length of time step Δt
numIterations ← simulationLength/Δt
for i going from 1 through numIterations do the following:
    growth ← growthRate * population
    population ← population + growth * Δt
    t ← i * Δt
    display t, growth, and population
```

If we do not need to display *growth* (derivative) at each step and the length of a step (Δt) is constant throughout the simulation, we can calculate the constant growth rate per step (*growthRatePerStep*) before the loop, as follows:

$$growthRatePerStep \leftarrow growthRate * \Delta t$$

Within the loop, we do not compute *growth* but estimate *population* as follows:

$$population \leftarrow population + growthRatePerStep * population$$

Thus, within the loop, we have two assignments instead of three and two multiplications instead of three, saving time in a lengthy simulation. The revised algorithm appears as Algorithm 2.

Algorithm 2 Alternative algorithm to Algorithm 1 for simulation of unconstrained growth that does not display *growth*

initialize *simulationLength*
initialize *population*
initialize *growthRate*
initialize Δt
growthRatePerStep* ← *growthRate* * Δ*t
numIterations ← *simulationLength*/Δt
for i going from 1 through *numIterations* do the following:
    population* ← *population* + *growthRatePerStep* * *population
    $t \leftarrow i * \Delta t$
    display t and *population*

Analytical Solution: Introduction

We can solve the preceding model analytically for unconstrained growth, which is the differential equation $\frac{dP}{dt} = 0.10P$ with initial condition $P_0 = 100$, as follows:

$$P = 100\, e^{0.10t}$$

The next three sections develop the analytical solution. The first section starts the explanation using indefinite integrals, while the second section begins the discussion using derivatives without using integrals. Thus, you may select the section that matches your calculus background. The third section completes the development of the analytical solution for both tracks. Those without calculus background may go immediately to the section "Completion of the Analytical Solution."

When it is possible to solve a problem analytically, we should usually do so. We have employed simulation of unconstrained growth with a system dynamic tool as an introduction to fundamental concepts and as a building block to more complex problems for which no analytical solutions exist.

Analytical Solution: Explanation with Indefinite Integrals (Optional)

We can solve the differential equation $\frac{dP}{dt} = 0.10P$ using a technique called **separation of variables**. First, we move all terms involving P to one side of the equation and all those involving t to the other. Leaving 0.10 on the right, we have the following:

$$\frac{1}{P}dP = 0.10\,dt$$

Then, we integrate both sides of the equation, as follows:

$$\int \frac{1}{P}dP = \int 0.10\,dt$$

$$\ln|P| = 0.10t + C \text{ for an arbitrary constant } C$$

We solve for $|P|$ by taking the exponential function of both sides and using the fact that the exponential and natural logarithmic functions are inverses of each other.

$$e^{\ln|P|} = e^{0.10t+C}$$

$$|P| = e^{0.10t}e^{C} = A\,e^{0.10t}$$

where $A = e^C$. Solving for P, we have

$$P = (\pm A)e^{0.10t}$$

or

$$P = ke^{0.10t}$$

where $k = (\pm A)$ is a constant.

Analytical Solution: Explanation with Derivatives (Optional)

We can solve the differential equation $\frac{dP}{dt} = 0.10P$ for P analytically by finding a function whose derivative is 0.10 times the function itself. The only functions that are their own derivative are exponential functions of the following form:

$$f(t) = ke^t, \text{ where } k \text{ is a constant}$$

For example, the derivative of $5e^t$ is $5e^t$. To obtain a factor of 0.10 through use of the chain rule, we have the general solution

$$P = ke^{0.10t}$$

For example, if $P = 5e^{0.10t}$, we have

$$\frac{dP}{dt} = \frac{d(5e^{0.10t})}{dt} = 5\frac{d(e^{0.10t})}{dt} = 5(0.10e^{0.10t}) = 0.10(5e^{0.10t}) = 0.10P$$

Completion of the Analytical Solution

Thus, the general solution to $\frac{dP}{dt} = 0.10P$ is $P = ke^{0.10t}$ for a constant k. Using the initial condition that $P_0 = 100$, we can determine a particular value of k and, thus, a particular solution of the form $P = ke^{0.10t}$. Substituting 0 for t and 100 for P, we have the following:

$$100 = ke^{0.10(0)} = ke^0 = k(1) = k$$

The constant is the initial population. For this example,

$$P = 100e^{0.10t}$$

Figure 2.2.3 displays the dramatic increase in the bacterial population as time advances.

In general, the solution to

$$\frac{dP}{dt} = rP \text{ with initial population } P_0$$

is

$$P = P_0e^{rt}$$

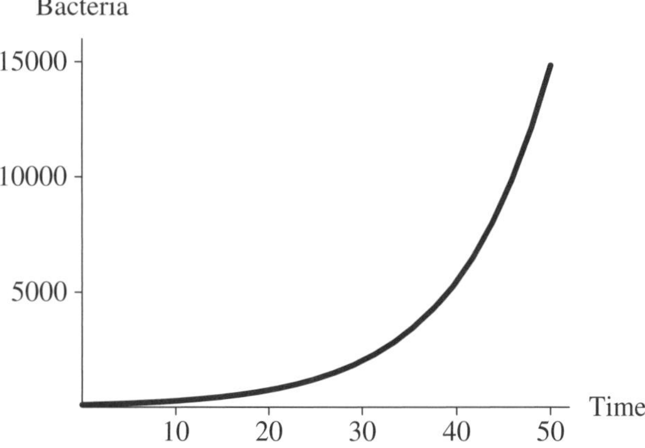

Figure 2.2.3 Bacterial population with a continuous growth rate of 10% per hour and an initial population of 100 bacteria

Quick Review Question 4

Give the solution of the differential equation

$$\frac{dP}{dt} = 0.03P \text{, where } P_0 = 57$$

The simulated values for the bacterial population are slightly less than those the model $P = 100e^{0.10t}$ determines. For example, after 20 h, a simulation may display, to two decimal places, a population of 738.54. However, $100e^{0.10(20)}$, expressed to two decimal places, is 738.91. The simulation compounds the population every step, and, in this case, the step size is $\Delta t = 0.005$ h. The analytic model compounds the population continuously; that is, as the step size goes to zero and the number of steps goes to infinity approaches, the simulated values approach the analytic solution.

Both the analytic model and simulation produce valid estimates of the population of bacteria. After 20 h, the number of bacteria will be an integer, not a decimal number, such as 738.54 or 738.91. Moreover, the population probably does not grow in an ideal fashion with a 10%-per-hour growth rate at every instant. Both the analytic model and the simulation produce estimates of the population at various times.

Further Refinement

We can refine the model further by having separate parameters for birth rate and death rate instead of the combined growth rate. Thus,

$$growth\_rate = birth\_rate - death\_rate$$

Unconstrained Decay

The rate of change of the mass of a radioactive substance is proportional to the mass of the substance, and the constant of proportionality is negative. Thus, the mass decays with time. For example, the constant of proportionality for radioactive carbon-14 is approximately –0.000120968. The continuous decay rate is about 0.0120968% per year, and the differential equation is as follows, where Q is the quantity (mass) of carbon-14:

$$\frac{dQ}{dt} = -0.000120968Q$$

As indicated in the section "Completion of the Analytical Solution," the analytical solution to this equation is

$$Q = Q_0 e^{-0.000120968t}$$

After 10,000 yr, only about 29.8% of the original quantity of carbon-14 remains, as the following shows:

$$Q = Q_0 e^{-0.000120968(10,000)} = 0.298292Q_0$$

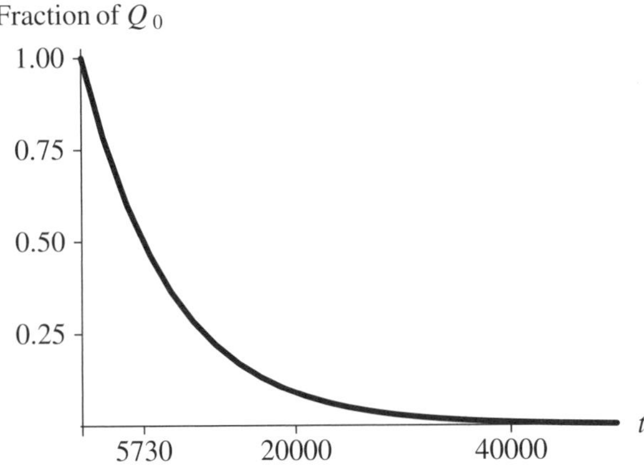

Figure 2.2.4 Exponential decay of radioactive carbon-14 as a fraction of the initial quantity Q_0, with time (t) in years

Figure 2.2.4 displays the decay of carbon-14 with time.

Carbon dating uses the amount of carbon-14 in an object to estimate the age of an object. All living organisms accumulate small quantities of carbon-14, but accumulation stops when the organism dies. For example, we can compare the proportion of carbon-14 in living bone to that in the bone of a mummy and estimate the age of the mummy using the model.

Example 1

Suppose the proportion of carbon-14 in a mummy is only about 20% of that in a living human. To estimate the age of the mummy, we use the preceding model with the information that $Q = 0.20Q_0$. Substituting into the analytical model, we have

$$0.20Q_0 = Q_0e^{-0.000120968t}$$

After canceling Q_0, we solve for t by taking the natural logarithm of both sides of the equation. Because the natural logarithm and the exponential functions are inverses of each other, we have the following:

$$\ln(0.20) = \ln(e^{-0.000120968t}) = -0.000120968t$$

$$t = \ln(0.20)/(-0.000120968) \approx 13{,}305 \text{ yr}$$

We often express the rate of decay in terms of the half-life of the radioactive substance. The **half-life** is the period of time that it takes for the substance to decay to half of its original amount. Figure 2.2.5 illustrates that the half-life of radioactive carbon-14 is about 5730 yr. We can determine this value analytically as we did in Example 1 using 50% instead of 20%; $Q = 0.50Q_0$.

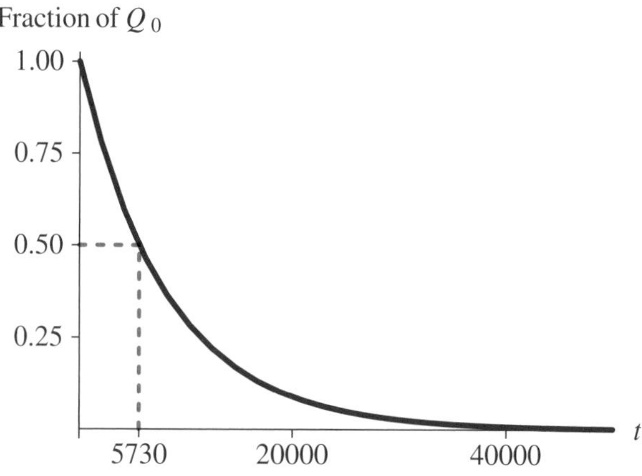

Figure 2.2.5 The half-life of radioactive carbon-14 indicated as 5730 yr

Definition The **half-life** is the period of time that it takes for a radioactive substance to decay to half of its original amount.

Quick Review Question 5

Radium-226 has a continuous decay rate of about 0.0427869% per year. Determine its half-life in whole years.

Reports for System Dynamics Models

The fifth step of the modeling process discussed in Module 1.2 is to "Report on the model." The following summarizes the items that would be included in a report for a system dynamics model:

a. **Analysis of the problem:** We begin by describing the problem, such as to model the growth of bacteria in media.
b. **Model design:** In this section, we should list simplifying assumptions, such as those in the section "Differential Equation"; equations, such as $\frac{dP}{dt} = 0.10P$ with $P_0 = 100$; reasoning for choices of constants, such as an instantaneous growth rate of 10%; the basic time step, such as hour; and other units. A diagram of the model, such as in Figure 2.2.1, is also appropriate to include.
c. **Model solution:** This part should contain the analytical solution or an algorithm, such as Algorithm 1.
d. **Results and conclusions:** Part d should include simulation tables, such as Table 2.2.2, and graphs, such as Figure 2.2.2. Moreover, the section should contain an explanation of verification accomplished by comparing the results to real data when available, descriptions of the outcomes of various scenar-

ios, a discussion of our conclusions with support from the results, and suggestions for model refinement.

e. **Appendices:** Usually, a copy of the file created with a system dynamics tool should be submitted with this report. Besides the model, this file should contain appropriate documentation, such as a text box with the authors' names, date, module and problem number, and problem description.

Exercises

Answers to marked exercises appear in the appendix "Answers to Selected Exercises."

1. a. For an initial population of 100 bacteria and a continuous growth rate of 10% per hour, determine the number of bacteria at the end of one week.
   b. How long will it take the population to double?

2. a. Suppose the initial population of a certain animal is 15,000 and its continuous growth rate is 2% per year. Determine the population at the end of 20 yr.
   b. Suppose we are performing a simulation of the population using a step size of 0.083 yr. Determine the growth and the population at the end of the first three time steps.

3. Adjust the model in Figure 2.2.1 to accommodate birth rate and death rate instead of just growth rate.

4. a. Newton's Law of Heating and Cooling states that the rate of change of the temperature (T) with respect to time (t) of an object is proportional to the difference between the temperatures of the object and of its surroundings. Suppose the temperature of the surroundings is 25 °C. Write the differential equation that models Newton's Law.
   b. Solve this equation for T as a function of time t.
   c. Suppose cold water at 6 °C is placed in a room that has temperature 25 °C. After 1 h, the temperature of the water is 20 °C. Determine all constants in the equation for T.
   d. What is the temperature of the water after 15 minutes (min)?
   e. How long will it take for the water to warm to room temperature?

5. a. Suppose someone, whose temperature is originally 37 °C, is murdered in a room that has constant temperature 25 °C. The temperature is measured as 28 °C when the body is found and at 27 °C 1 h later. How long ago was the murder committed from discovery of the body? See Exercise 4 for Newton's Law of Heating and Cooling.
   b. Suppose we are performing a simulation using a step size of 0.004 h. Using the decay rate from Part a, determine the temperature at the end of the first three time steps after discovery of the body.

6. a. What proportion of the original quantity of carbon-14 is left after 30,000 yr?
   b. If 60% is left, how old is the item?

7. a. The half-life of radioactive strontium-90 is 29 yr. Give the model for the quantity present as a function of time.
   b. What proportion of strontium-90 is present after 10 yr?

c. After 50 yr?
d. How long will it take for the quantity to be 15% of the original amount?

8. Suppose an investment has approximately a continuous growth rate of 9.3%. Calculate analytically the value of an initial investment of $500 after
a. 10 yr **b.** 20 yr **c.** 30 yr **d.** 40 yr
d. How long will it take for the value to double?
e. How long to quadruple?

9. Suppose the amount of deposited ash, A, in millimeters (mm) is a function of time t in days. Suppose the model states that the rate of change of ash with respect to time is 4 mm/day and the initial quantity is 3 mm.
a. Using a step size of 0.5 days (da), estimate the amount of ash when $t = 1$ da.
b. Repeat Part a using a step size of 0.25 da.
c. Does the smaller step size change the result?
d. Solve the model for A.
e. What kind of function do you obtain?

Projects

For additional projects, see Module 7.1, "Radioactive Chains—Never the Same Again"; Module 7.2, "Turnover and Turmoil—Blood Cell Populations"; Module 7.3, "Deep Trouble—Ideal Gas Laws and Scuba Diving"; Module 7.4, "What Goes Around Comes Around—The Carbon Cycle"; after completion of "System Dynamics Tool: Tutorial 2," Module 7.9, "Transmission of Nerve Impulses: Learning from the Action Potential Heroes"; Module 7.12 "Mercury Pollution—Getting on Our Nerves."

1. Develop a model for Newton's Law of Heating and Cooling (see Exercise 4). Using this model, answer the questions of Exercises 4 and 5.
2. In 1854, Dr. John Snow, the father of epidemiology, identified a particular London water pump as the point source of the Broad Street cholera epidemic, which spread in a radial fashion from the pump. Model such a spread of disease assuming that the rate of change of the number of cases of cholera is proportional to the square root of the number of cases.
3. Develop a model for Exercise 8.
4. A young professional would like to save enough money to pay cash for a new car. Develop a model to determine when such a purchase will be possible. Take into account the following issues: The price of a new car is rising due to inflation. The buyer plans to trade in a car, which is depreciating. This person already has some savings and plans to make regular monthly payments. Thus, use a Δt value of 1 mo. Assume appropriate rates and values.

Develop a spreadsheet for each of Projects 5–8.

5. Exercise 2
6. Exercise 4
7. Exercise 5
8. Exercise 8

Answers to Quick Review Questions

1. **a.** Average velocity from 1 to 2 s = $\frac{s(2)-s(1)}{2-1} = \frac{21.4-21.1}{1}$ = 0.3 m/s

   b. Average velocity from 1 to 3 s = $\frac{s(3)-s(1)}{3-1} = \frac{11.9-21.1}{2}$ = –4.6 m/s

   c. $b = 3$ s, $s(b) = 11.9$ m, $\Delta t = 2$ s, $b - \Delta t = 1$ s, $s(b - \Delta t) = 21.1$ m, $\Delta s = 11.9 - 21.1 = -9.2$ m
2. **a.** $Q(t) = Q(t - \Delta t) + \Delta Q$, where $\Delta Q = -0.0004Q(t - \Delta t)\Delta t$ and $Q(0) = 200$

   b. $t - \Delta t = 9.0 - 0.5 = 8.5$ s

   c. $\Delta Q = 199.28 - 199.32 = -0.04$
3. 100.450901

   growth = 100.400701 * 0.10 = 10.040070

   Thus, *population*(0.045) = 100.400701 + 10.040070 * 0.005 = 100.450901
4. $P = 57e^{0.03t}$
5. 1620. Reasoning:

$$Q = Q_0 e^{-0.000427869t}$$

$$\text{For } Q = 0.50Q_0,\ 0.50Q_0 = Q_0 e^{-0.000427869t} \text{ or } 0.50 = e^{-0.000427869t}$$

$$\ln(0.50) = -0.000427869t$$

$$t = \ln(0.50)/(-0.000427869) = 1620$$

Reference

Zill, Dennis G. 2013. *A First Course in Differential Equations with Modeling* Applications, 10th ed. Belmont, CA. Brooks-Cole Publishing (Cengage Learning).

MODULE 2.3

Constrained Growth

Introduction

An animal introduced to a new environment will often reproduce at a very high rate. That is what happened when the Eurasian perch, called the ruffe (*Gymnocephalus cernuus*), was introduced to Lake Superior from an ocean-going ship's ballast. A small fish, usually 4 to 6 inches (in.) long, with sharp, menacing spines on its gill covers and dorsal fin, ruffe is a meal of last resort for most predators. Moreover, the Eurasian perch has little or no value as a fishery and is a formidable competitor. Vying with other benthivorous fish (e.g., yellow perch), they have the advantage of being more adaptable in their dietary choices, including rotifers, microcrustaceans, immature insects, larval, and small, adult fish. Interestingly, they are preyed upon on by very few species of larger fish and then only if other prey are scarce. Ruffe not only tolerate wide ranges of temperature and pH, they are also prolific breeders (6000 to 200,000 eggs/batch), spawning on a variety of substrates.

People involved in fisheries in the Great Lakes have every right to be alarmed by this intruder. When introduced to Loch Lomond, Scotland, ruffe populations increased exponentially and decimated the eggs of local salmon (Adams 1998). Although there has been no established causal association, McLean (1993) found that populations of native North American fish like yellow perch, perch-trout, and emerald shiners have all declined since the ruffe were introduced. It is hypothesized that ruffe either predate on their competitors eggs or decrease their food resources.

Because births exceed the numbers maturing and reproducing, all populations, theoretically, have the potential for exponential growth. Endemic populations increase rapidly at first, but they eventually encounter resistance from the environment—competitors, predators, limited resources, and disease. Thus, the environment tends to limit the growth of populations, so that they usually increase only to a certain level and then do not increase or decrease drastically unless a change in the environment occurs. This maximum population size that the environment can support indefinitely is termed the **carrying capacity**. Many introduced species that become pests, such as the ruffe in the Great Lakes, have a very high reproductive potential in their new environments because they are very adaptable to habitat and food sources, they have few or less-fit competitors, and few to no predators.

Carrying Capacity

In Module 2.2, "Unconstrained Growth and Decay," we considered a population growing without constraints, such as competition for limited resources. For such a population, *P*, with instantaneous growth rate, *r*, the rate of change of the population has the following differential equation model:

$$\frac{dP}{dt} = rP$$

With initial population P_0, we saw that the analytical solution is $P = P_0e^{rt}$. In that module, we also developed the following finite difference equation for the change in *P* from one time to the next, which we used in simulations:

$$\begin{aligned}\Delta P &= P(t) - P(t - \Delta t)\\ &= (r\,P(t - \Delta t))\,\Delta t\end{aligned}$$

Simulation and analytical solution graphs in Figures 2.2.2 and 2.2.3, respectively, of Module 2.2 display the exponential growth of unconstrained growth.

After developing such a model in Step 2 of the modeling process and solving the model (Step 3) as before, we should verify that the solution (Step 4) agrees with real data. However, as the introduction indicates, no confined population can grow without bound. Competition for food, shelter, and other resources eventually limits the possible growth. For example, suppose a deer refuge can support at most 1000 deer. We say that the carrying capacity (*M*) for the deer in the refuge is 1000.

Definition The **carrying capacity** for an organism in an area is the maximum number of organisms that the area can support.

Quick Review Question 1

Cycling back to Step 2 of the modeling process, this question begins refinement of the population model to accommodate descriptions of population growth from the "Introduction" of this module.

a. Determine any additional variable and its units.

b. Consider the relationship between the number of individuals (*P*) and carrying capacity (*M*) as time (*t*) increases. List all the statements below that apply to the situation where the population is much smaller than the carrying capacity.

A. *P* appears to grow almost proportionally to *t*.

B. P appears to grow almost without bound.
C. P appears to grow faster and faster.
D. P appears to grow more and more slowly.
E. P appears to decline faster and faster.
F. P appears to decline more and more slowly.
G. P appears to grow almost linearly with slope M.
H. P is appears to be approaching M asymptotically.
I. P appears to grow exponentially.
J. dP/dt appears to be almost proportional to P.
K. dP/dt appears to be almost zero.
L. The birth rate is about the same as the death rate.
M. The birth rate is much greater than the death rate.
N. The birth rate is much less than the death rate.

c. List all the choices from Part b that apply to the situation where the population is close to but less than the carrying capacity.

d. List all the choices from Part b that apply to the situation where the population is close to but greater than the carrying capacity.

Revised Model

In the revised model, for an initial population much lower than the carrying capacity, we want the population to increase in approximately the same exponential fashion as in the earlier unconstrained model. However, as the population size gets closer and closer to the carrying capacity, we need to dampen the growth more and more. Near the carrying capacity, the number of deaths should be almost equal to the number of births, so that the population remains roughly constant. To accomplish this dampening of growth, we could compute the number of deaths as a changing fraction of the number of births, which we model as rP. When the population is very small, we want the fraction to be almost zero, indicating that few individuals are dying. When the population is close to the carrying capacity, the fraction should be almost $1 = 100\%$. For populations larger than the carrying capacity, the fraction should be even larger so that the population decreases in size through deaths. Such a fraction is P/M. For example, if the population P is 10 and the carrying capacity M is 1000, then $P/M = 10/1000 = 0.01 = 1\%$. For a population $P = 995$ close to the carrying capacity, $P/M = 995/1000 = 0.995 = 99.5\%$; and for the excessive $P = 1400$, $P/M = 1400/1000 = 1.400 = 140\%$.

Thus, we can model the instantaneous rate of change of the number of deaths (D) as the fraction P/M times the instantaneous rate of change of the number of births (r), as the following differential equation indicates:

$$\frac{dD}{dt} = \left(r\frac{P}{M}\right)P$$

The differential equation for the instantaneous rate of change of the population subtracts this value from the instantaneous rate of change of the number of births, as follows:

$$\frac{dP}{dt} = \underbrace{(rP)}_{births} - \underbrace{\left(r\frac{P}{M}\right)P}_{deaths}$$

or

$$\frac{dP}{dt} = r\left(1 - \frac{P}{M}\right)P \tag{1}$$

For the discrete simulation, where $P(t - 1)$ is the population estimate at time $t - 1$, the number of deaths from time $t - 1$ to time t is

$$\Delta D = \left(r\frac{P(t-1)}{M}\right)P(t-1) \quad \text{for } \Delta t = 1$$

In general, we approximate the number of deaths from time $(t - \Delta t)$ to time t by multiplying the corresponding value by Δt, as follows:

$$\Delta D = \left(r\frac{P(t-\Delta t)}{M}\right)P(t-\Delta t)\Delta t$$

where $P(t - \Delta t)$ is the population estimate at $(t - \Delta t)$. Thus, the change in population from time $(t - \Delta t)$ to time t is the difference of the number of births and the number of deaths over that period:

$$\Delta P = \text{births} - \text{deaths}$$

$$\Delta P = \underbrace{(rP(t-\Delta t))\Delta t}_{\text{births}} - \underbrace{\left(r\frac{P(t-\Delta t)}{M}\right)P(t-\Delta t)\Delta t}_{\text{deaths}}$$

$$= (r\Delta t)\left(1 - \frac{P(t-\Delta t)}{M}\right)P(t-\Delta t)$$

or

$$\Delta P = k\left(1 - \frac{P(t-\Delta t)}{M}\right)P(t-\Delta t), \text{ where } k = r\Delta t \tag{2}$$

Differential equation (1) and difference equation (2) are called **logistic equations**. Figure 2.3.1 displays the S-shaped curve characteristic of a logistic equation, where the initial population is less than the carrying capacity of 1000. Figure 2.3.2 shows how the population decreases to the carrying capacity when the initial population is 1500. Thus, the model appears to match observations from the "Introduction" qualitatively. To verify a particular model, we should estimate parameters, such as birth rate, and compare the results of the model to real data.

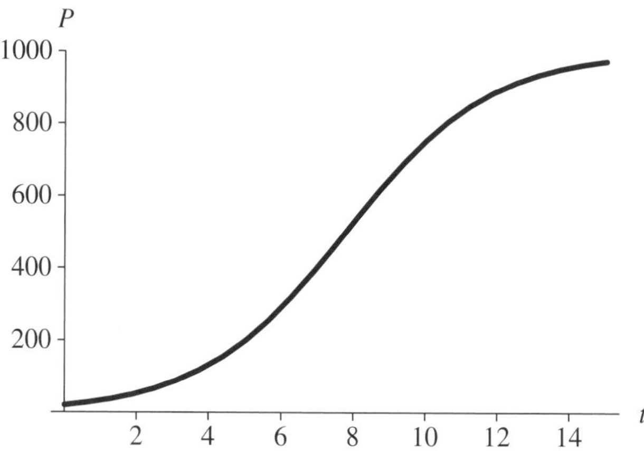

Figure 2.3.1 Graph of logistic equation, where initial population is 20, carrying capacity is 1000, and instantaneous rate of change of births is 50%, with time (t) in years

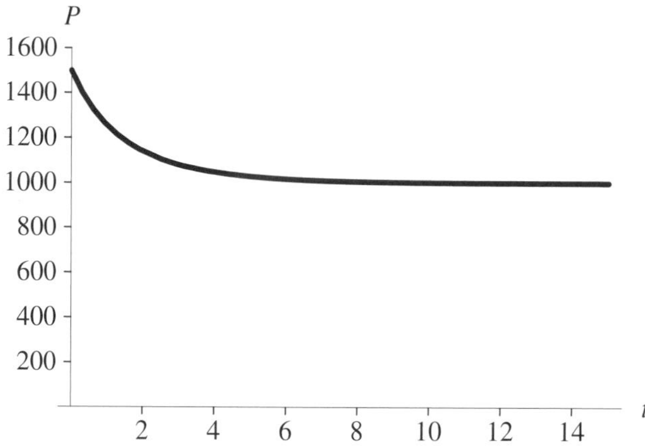

Figure 2.3.2 Graph of logistic equation, where initial population is 1500, carrying capacity is 1000, and instantaneous rate of change of births is 50%, with time (t) in years

Quick Review Question 2

a. Complete the difference equation to model constrained growth of a population P with respect to time t over a time step of 0.1 units, given that the population at time $t - \Delta t$ is $p \leq 1000$, the carrying capacity is 1000, the instantaneous rate of change of births is 105%, and the initial population is 20.
$\Delta P =$ ___(___ ___ _____)$(p)(0.1)$

b. What is the maximum population?

c. Suppose the population at time $t = 5$ yr is 600 individuals. What is the population, rounded to the nearest integer, at time 5.1 yr?

Equilibrium and Stability

The logistic equation with carrying capacity $M = 1000$ has an interesting property. If the initial population is less than 1000, as in Figure 2.3.1, the population increases to a limit of 1000. If the initial population is greater than 1000, as in Figure 2.3.2, the population decreases to the limit of 1000. Moreover, if the initial population is 1000, we see from Equation (1) that $P/M = 1000/1000 = 1$ and $dP/dt = r(1 - 1)P = 0$. In discrete terms, $\Delta P = 0$. A population starting at the carrying capacity remains there. We say that $M = 1000$ is an **equilibrium** size for the population because the population remains steady at that value or $P(t) = P(t - \Delta t) = 1000$ for all $t > 0$.

Definitions An **equilibrium solution** for a differential equation is a solution where the derivative is always zero. An **equilibrium solution** for a difference equation is a solution where the change is always zero.

Quick Review Question 3

Give another equilibrium size for the logistic differential equation (1) or logistic difference equation (2).

Even if an initial positive population does not equal the carrying capacity $M = 1000$, eventually, the population size tends to that value. We say that the solution $P = 1000$ to the logistic equation (1) or (2) is **stable**. By contrast, for a positive carrying capacity, the solution $P = 0$ is **unstable**. If the initial population is close to but not equal to zero, the population does not tend to that solution over time. For the logistic equation, any displacement of the initial population from the carrying capacity exhibits the limiting behavior of Figure 2.3.1 or 2.3.2. In general, we say that a solution is stable if for a small displacement from the solution, P tends to the solution.

Definition Suppose that q is an equilibrium solution for a differential equation dP/dt or a difference equation ΔP. The solution q is **stable** if there is an interval (a, b) containing q, such that if the initial population $P(0)$ is in that interval, then

1. $P(t)$ is finite for all $t > 0$;
2. As time, t, becomes larger and larger, $P(t)$ approaches q.

The solution q is **unstable** if no such interval exists.

Exercises

1. Using calculus, solve the following:
   a. The differential equation (1),

$$\frac{dP}{dt} = r\left(1 - \frac{P}{M}\right)P$$

where the carrying capacity, M, is 1000, $P_0 = 20$, and the instantaneous rate of change of the number of births, r, is 50%

b. The differential equation (1) in general

2. Consider $dy/dt = \cos(t)$.
   a. Give all the equilibrium solutions.
   b. Using calculus, find a function $y(t)$ that is a solution.
   c. Give the most general function y that is a solution.
3. It has been reported that a mallard must eat 3.2 ounces (oz) of rice each day to remain healthy. On the average, an acre of rice in a certain area yields 110 bushels (bu) per year; and a bushel of rice weighs 45 lb. Assuming that in the area 100 acres (ac) of rice are available for mallard consumption and mallards eat only rice, determine the carrying capacity for mallards in the area (Reinecke).
4. The **Gompertz differential equation**, which follows, is one of the best models for predicting the growth of cancer tumors:

$$\frac{dN}{dt} = kN \ln\left(\frac{M}{N}\right), \quad N(0) = N_0$$

   where N is the number of cancer cells and k and M are constants.
   a. As N approaches M, what does dN/dt approach?
   b. Make the substitution $u = \ln(M/N)$ in the Gompertz equation to eliminate N and convert the equation to be in terms of u.
   c. Using calculus, solve the transformed differential equation for u.
   d. Using the relationship between u and N from Part b, convert your answer from Part c to be in terms of N. The result is the solution to the Gompertz differential equation.
   e. Using calculus, verify that $N(t) = Me^{\ln\left(\frac{N_0}{M}\right)e^{-kt}}$ is the solution to the Gompertz differential equation.
   f. Using the solution in Part e, what does N approach as t goes to infinity?
5. a. Graph $y = e^{-t}$.

   Match each of the following scenarios to a differential equation that might model it.

   A. $dP/dt = 0.05P$ B. $dP/dt = 0.05P + e^{-t}$
   C. $dP/dt = 0.05(1 - e^{-t})P$ D. $dP/dt = 0.05P - 0.0003P^2 - 400$
   E. $dP/dt = 0.05e^{-t}P$ F. $dP/dt = 0.05P - 0.0003P^2$

   b. At first, a bacteria colony appears to grow without bound; but because of limited nutrients and space, the population eventually approaches a limit.
   c. Because of degradation of nutrients, the growth of a bacterial colony becomes dampened.
   d. A bacterial colony has unlimited nutrients and space and grows without bound.
   e. Because of adjustment to its new setting, a bacterial colony grows slowly at first before appearing to grow without bound.
   f. Each day, a scientist removes a constant amount from the colony.
6. Write an algorithm for simulation of constrained growth similar to Algorithm 1 for simulation of unconstrained growth in Module 2.2.

Projects

For additional projects, see Module 7.4, "What Goes Around Comes Around—The Carbon Cycle"; Module 7.5, "A Heated Debate—Global Warming"; and Module 7.6, "Plotting the Future: How Will the Garden Grow."

1. Develop a model for constrained growth.
2. Develop a model for the mallard population in Exercise 3. Have a converter or variable for the number of acres of rice available for mallard consumption, and from this value, have the model compute the carrying capacity. Report on the effect of decreasing the number of acres of rice available (Reinecke).
3. In some situations, the carrying capacity itself is dynamic. For example, the performance of airplanes had one carrying capacity with piston engines and a higher limit with the advent of jet engines. Many think that human population growth over a limited period of time follows such a pattern as technological changes enable more people to live on the available resources. In such cases, we might be able to model the carrying capacity itself as a logistic. Suppose M_1 is the first carrying capacity, and $M_1 + M_2$ is the second. The differential equation for the carrying capacity $M(t)$ as a function of time t would be as follows:

$$\frac{dM(t)}{dt} = a(M(t) - M_1)\left(1 - \frac{M(t) - M_1}{M_2}\right) \quad \text{for some constant } a > 0$$

By using $M(t)$, we have a logistic for the carrying capacity as well as a logistic for the population. Figure 2.3.3 displays population, $P(t)$, in black and $M(t)$ in color with the first carrying capacity $M_1 = 20$; the second, $M_1 + M_2 = 70$; and an inflection point for M at $t = 450$. Notice that we get a "bilogistic," or "doubly logistic," model for $P(t)$.

Develop a model for the following scenario. First, generate an appropriate logistic carrying capacity, $M(t)$. Then, use this dynamic carrying capacity to limit the population.

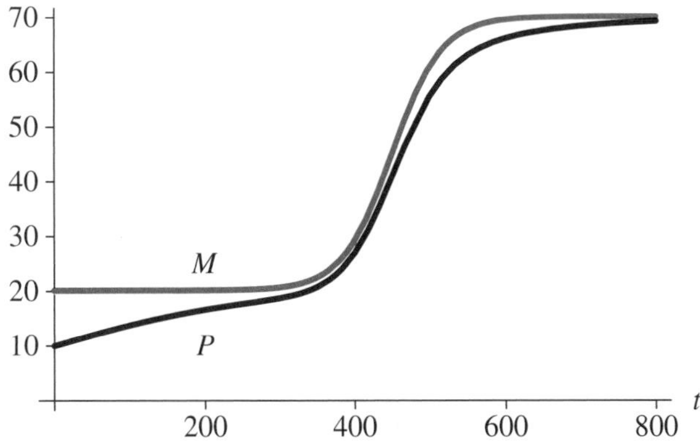

Figure 2.3.3 Graphs of functions for carrying capacity, $M(t)$, and population, $P(t)$, with time (t) in years

In a population study of England from 1541 to 1975, starting with a population of about 1 million, early islanders appear to have a carrying capacity of around 5 million people. However, beginning about 1800 with the advent of the Industrial Revolution, the carrying capacity appears to have increased to about 50 million people. The change in the concavity from concave up to concave down for this new logistic appears to occur in about 1850 (Meyer and Ausubel 1999).

4. Refer to Project 3 for a description of a logistic carrying-capacity function. Using that information, develop a model for the Japanese population from the year 1100 to 2000. With an initial population of 5 million, the island population was mainly a feudal society that leveled off to about 35 million. The industrial revolution came to Japan in the latter part of the nineteenth century, and the population rose rapidly over a 77-yr period, with the inflection point occurring about 1908 (Meyer and Ausubel 1999).
5. Develop a model for the number of trout in a lake initially stocked with 400 trout. These fish increase at a rate of 15%, and the lake has a carrying capacity of 5000 trout. However, vacationers catch trout at a rate of 8%.
6. It has been estimated that for the Antarctic fin whale, $r = 0.08$, $M = 400{,}000$, and $P_0 = 70{,}000$ in 1976. Model this population. Then, revise the model to consider harvesting the whales as a percentage of rM. Give various values for this percentage that lead to extinction and other values that lead to increases in the population. Estimate the **maximum sustainable yield**, or the percentage of rM that gives a constant population in the long term (Zill 2013).
7. Army ants on a 17-km^2 island forage at a rate of 1500 m^2/day, clearing the area almost completely of other insects. Once the ants have departed, it takes about 150 days for the number of other insects to recover in the area. Assume an initial number of 1million army ants and a growth rate of 3.6%, where the unit of time is a week. Model the population.

Answers to Quick Review Questions

1. **a.** carrying capacity, say M, in units of the population, such as deer or bacteria

   b. B. P appears to grow almost without bound.
   C. P appears to grow faster and faster.
   I. P appears to grow exponentially.
   J. dP/dt appears to be almost proportional to P.
   M. The birth rate is much greater than the death rate.

   c. D. P appears to grow more and more slowly.
   H. P is appears to be approaching M asymptotically.
   K. dP/dt appears to be almost zero.
   L. The birth rate is about the same as the death rate.

   d. F. P appears to decline more and more slowly.
   H. P is appears to be approaching M asymptotically.
   K. dP/dt appears to be almost zero.

L. The birth rate is about the same as the death rate.

2. **a.** $\Delta P = 1.05(1 - p/1000)(p)(0.1)$

b. 1000 individuals

c. 625 individuals because $P + \Delta P = 600 + 1.05(1 - 600/1000)\ 600(0.1) = 625.2$ individuals

3. 0 because $dP/dt = r(1 - P/M)P = r(1 - 0)0 = 0$

References

Adams, C. E., and P. S. Maitland. 1998. "The Ruffe Population in Loch Lomond, Scotland: Its Introduction, Population Expansion, and Interaction with Native Species." *Journal of Great Lakes Research*, 24: 249–262.

Fuller, P. G., J. Jacobs, J. Larson, and A. Fusaro. 2012. "*Gymnocephalus cernuus*." USGS Nonindigenous Aquatic Species Database. Gainesville, FL. http://nas.er.usgs.gov/queries/FactSheet.aspx?speciesID=7 (accessed November 11, 2012)

Hajjar, R. 2002. "Introduced Species Summary Project: Ruffe (*Gymnocephalus cernuus*)." Columbia University. http://www.columbia.edu/itc/cerc/danoffburg/invasion_bio/inv_spp_summ/Gymnocephalus cernuus.html (accessed November 11, 2012)

McLean, M. 1993. "Ruffe (*Gymnocephalus cernuus*) Fact Sheet." Minnesota Sea Grant Program, Great Lakes Sea Grant Network, Duluth, MN. http://www.dnr.state.mn.us/invasives/aquaticanimals/ruffe/index.html (accessed January 7, 2013)

Meyer, Perrin S., and Jesse H. Ausubel. 1999. "Carrying Capacity: A Model with Logistically Varying Limits." *Technological Forecasting and Social Change*, 61(3): 209–214.

Reinecke, Kenneth J. Personal communication. USGS, Pantuxent Wildlife Research Center. Beltsville Laboratory, 308-10300 Baltimore Avenue, Beltsville, MD 20705.

Sea Grant Pennsylvania. "Eurasian Ruffe: *Gymnocephalus cernuus*." http://www.paseagrant.org/wp-content/uploads/2012/09/Ruffe2012.pdf (accessed January 7, 2013)

Zill, Dennis G. 2013. *A First Course in Differential Equations with Modeling Applications*, 10th ed. Belmont, CA. Brooks-Cole Publishing (Cengage Learning).

MODULE 2.4

System Dynamics Tool: Tutorial 2

Prerequisite: Module 2.1, "System Dynamics Tool: Tutorial 1"

Download

From the textbook's website, download Tutorial 2 in PDF format and the *unconstrained* file for your system dynamics tool. We recommend that you work through the tutorial and answer all Quick Review Questions using the corresponding software.

Introduction

This tutorial introduces the following functions and concepts, which subsequent modules employ for model formulation and solution using your system dynamics tool:

- Built-in functions and constants, such as the *if-then-else* construct, absolute value, initial value, exponential function, sine, pulse function, time, time step, and π
- Relational and logical operators
- Comparative graphs
- Graphical input
- Conveyors, an optional topic useful for some of the later projects

MODULE 2.5

Drug Dosage

Downloads

The text's website has *OneCompartAspirin* and *OneCompartDilantin* files, which contain models for examples in this module, available for download in various system dynamics systems.

Introduction

Errors in the dispensing and administration of medications occur frequently. Although most do not result in great harm, some do. For instance, a Florida pharmacy dispensed 10 times the prescribed dose of a blood thinner to a mother of four, which resulted in her suffering a cerebral hemorrhage (Patel and Ross 2010). In other tragedies, a 10-mo-old infant died after receiving a 10-fold overdose of the chemotherapy agent Cisplatin (Fitzgerald and Wilson 1998), and three nurses were prosecuted for administering a 10-fold (fatal) overdose of penicillin to an infant (Ellis and Hartley 2004).

The National Quality Forum, a nonprofit whose mission involves enabling "private- and public-sector stakeholders to work together to craft and implement cross-cutting solutions to drive continuous quality improvement in the American healthcare system," has estimated that medication errors account for a conservative estimate of $21 billion in costs. This financial expenditure corresponds to serious preventable medication errors for 3.8 million hospital inpatients and 3.3 million outpatients per year (NQF 2010). These cases comprise an extraordinary amount of human suffering and, in some cases, death.

How do these errors occur? According to the Institute of Medicine, medication errors can be classified as errors in

ordering—incorrect drug or dosage;
transcribing—incorrect frequency of administration or missed dosages;
dispensing—incorrect drug, dosage, or timing;

administering—wrong dosage, technique;
monitoring—not observing effects of medication.

Whether these errors result from poor communication of orders, poor product labeling, or some other cause, the patients and their families suffer the consequences (IOM 2007).

It is not only health-care professionals who make mistakes in drug administration. On June 28, 2003, an Oklahoma teenager died from an overdose of Tylenol (acetaminophen). Suffering from a migraine headache, she took twenty 500-mg capsules, two and one-half times the maximum dosage recommended in 24 h. Apparently, the quantity was enough of the drug to cause liver and kidney failure. Assuming that an over-the-counter analgesic was safe, she apparently did not read the label and made a fatal dosage error (Robert 2004).

There are prescribed dosages for various drugs, but how do we determine what the correct/effective dosage is? There are quite a number of factors that are considered, including drug **absorption**, **distribution**, **metabolism**, and **elimination**. These factors are components of the quantitative science of **pharmacokinetics**.

One-Compartment Model of Single Dose

Metabolism of a drug in the human body is a complex system to represent in a model. Thus, in Step 2 of the modeling process, particularly for our first attempt, we should make simplifying assumptions about the drug and the body. A **one-compartment model** is a simplified representation of how a body processes a drug. In this model, we consider the body to be one homogeneous compartment, where distribution is instantaneous, the **concentration** of the drug in the system (amount of drug/volume of blood) is proportional to the drug dosage, and the rate of elimination is proportional to the amount of drug in the system. The concentration of a drug instead of the absolute quantity is important because a quantity that might be appropriate for a small child could be ineffective for a large adult. A drug has a **minimum effective concentration** (**MEC**), which is the least amount of drug that is helpful, and a **maximum therapeutic concentration**, or **minimum toxic concentration** (**MTC**), which is the largest amount that is helpful without having dangerous or intolerable side effects. The **therapeutic range** for a drug consists of concentrations between the MEC and MTC. A drug's **half-life**, or the amount of time for half the drug to be eliminated from the system, is useful for modeling as well as patient treatment. Often concentrations and half-life are expressed in relationship to the drug in the plasma or blood serum. The total amount of blood in an adult's body is approximately 5 liters (L), while the amount of **plasma**, or fluid that contains the blood cells, is about 3 L. Blood **serum** is the clear fluid that separates from blood when it clots, and an adult human has about 3 L of blood serum.

We begin by modeling the concentration in the body of aspirin (acetylsalicylic acid). For adults and children over the age of 12, the dosage for a headache is one or two 325-mg tablets every 4 h as necessary, up to 12 tablets/da. Analgesic effectiveness occurs at plasma levels of about 150 to 300 micrograms/milliliter (μg/mL), while toxicity may occur at plasma concentrations of 350 μg/mL. The plasma half-life of a dose from 300 to 650 mg is 3.1 to 3.2 h, with a larger dose having a longer half-life.

For simplicity, we assume a one-compartment model with the aspirin immediately available in the plasma. A stock (box variable), *aspirin_in_plasma*, represents the mass of aspirin in the compartment, which is the person's system, and has an initial value of the mass of two aspirin, (2)(325 mg)(1000 μg/mg), where 1 milligram (mg) is equivalent to 1000 μg.

The flow from *aspirin_in_plasma* (*elimination*) is proportional to the amount present in the system, *aspirin_in_plasma*. Thus, the rate of change of the drug leaving the system is proportional to the quantity of drug in the system (*aspirin_in_plasma*, or Q in the following equation):

$$dQ/dt = -KQ$$

As Module 2.2, "Unconstrained Growth and Decay," shows, the solution to this differential equation is as follows:

$$Q = Q_0 e^{-Kt}$$

Using this solution, as Exercise 1 shows, the constant of proportionality K given earlier and *elimination_constant* in the system dynamics software model have the following relationship to the drug's half-life ($t_{1/2}$):

$$\boldsymbol{K = -\ln(0.5)/t_{1/2}}$$

Pharmaceutical sources widely report a drug's half-life.

Quick Review Question 1

Determine the elimination constant with units for aspirin, assuming a half-life of 3.2 h.

To compute aspirin's plasma concentration (*plasma_concentration*) in a converter (variable), we have another converter for the volume of the system (*plasma_volume*) with a value of 3000 mL and appropriate connectors and equation. Figure 2.5.1 contains a one-compartment model for one dose of a drug, where the initial value of *plasma_concentration* is the dosage; and Equation Set 2.5.1 gives the corresponding equations and values explicitly entered for the model of aspirin.

Quick Review Question 2

In terms of the variables in the model of Figure 2.5.1, give the equation for *plasma_concentration*.

Equation Set 2.5.1

Explicitly entered equations and values for one-compartment model of aspirin:

half_life = 3.2 h
plasma_volume = 3000 mL
aspirin_in_plasma(0) = 2 * 325 * 1000 μg
elimination_constant = –ln(0.5)/*half_life*

*elimination = elimination_constant * aspirin_in_plasma*
plasma_concentration = aspirin_in_plasma/plasma_volume

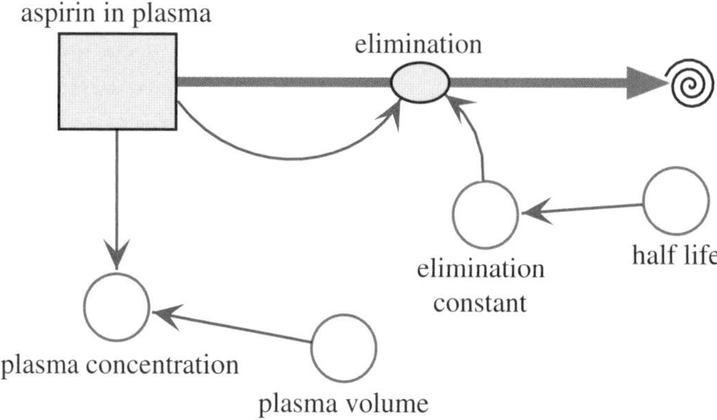

Figure 2.5.1 One-compartment model of aspirin

Running the simulation for 8 h and plotting *plasma_concentration*, the resulting graph in Figure 2.5.2 indicates that the concentration of the drug in the plasma is initially approximately 217 μg/mL, which is a safe, therapeutic dose. Subsequently, the concentration decreases exponentially.

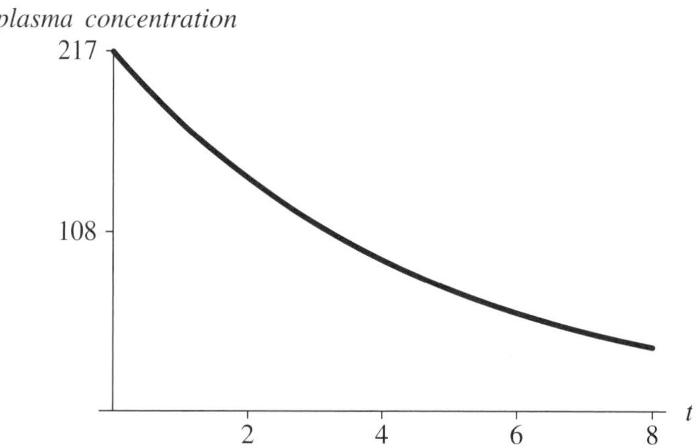

Figure 2.5.2 Graph of plasma_concentration (μg/mL) for aspirin versus time, t (h)

One-Compartment Model of Repeated Doses

As another example, we model the concentration in the body of the drug Dilantin, a treatment for epilepsy that the patient takes on a regular basis. Adult dosage is often one 100-mg capsule three times daily. The effective serum blood level is 10 to 20 μg/mL, which may take 7 to 10 da to achieve. Although individual variations occur, serious side effects can appear at a serum level of 20 μg/mL. The half-life of Dilantin ranges from 7 to 42 h but averages 22 h.

For simplicity, we assume a one-compartment model with instantaneous absorption. A stock (box variable), *drug_in_system*, represents the mass of Dilantin in the compartment, which is the person's blood serum. A flow, *ingested*, into *drug_in_system* is for the drug absorbed into the system. Because of the periodic nature of the dosage, we employ a pulse function with converters/variables for the dose (*dosage*), time of the initial dose (*start*), and time interval between doses (*interval*). Presuming that only a fraction (*absorption_fraction*) actually enters the system, we multiply this constant (say, 0.12, from experimental evidence) and the pulse value together for the equation of *entering*. We can estimate the value of *absorption_fraction* by plotting actual data of drug concentration versus time and employing techniques of curve fitting, which Module 8.3, "Empirical Models," discusses.

Quick Review Question 3

Give the equation for *entering*.

The flow from *drug_in_system* (*elimination*) is proportional to the amount present in the system, *drug_in_system*. Thus, between doses of a drug, the rate of change of the drug leaving the system is proportional to the quantity of drug in the system. As for the preceding aspirin example, we use a constant of proportionality (*elimination_constant*) of $-\ln(0.5)/t_{1/2}$, where $t_{1/2}$ is Dilantin's half-life.

For comparison purposes, we have converters (variables) for *MEC*, *MTC*, and the concentration of the drug in the system (*concentration*). To compute the latter, we have a converter (variable) for the volume of the blood serum (*volume*) with a possible value of 3000 mL and appropriate connectors and equation. Figure 2.5.3 contains a one-compartment model, and Equation Set 2.5.2 gives the corresponding explicitly entered equations and constants for Dilantin. Note that, except for name

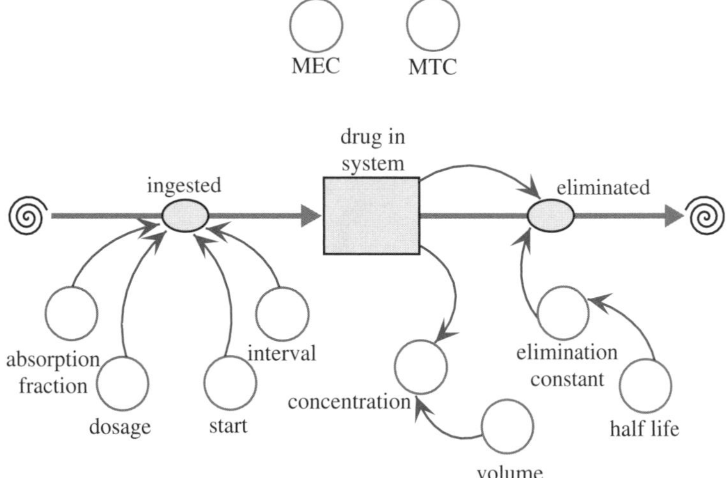

Figure 2.5.3 One-compartment model of Dilantin

changes, the middle and right side of the diagram agree with those of aspirin in Figure 2.5.1. The inflow for Figure 2.5.3 models the multiple doses of Dilantin, in contrast to no inflow for Figure 2.5.1 because of the assumption that exactly one dose of aspirin is immediately available in the plasma.

Equation Set 2.5.2

Explicitly entered equations and constants for one-compartment model of Dilantin:

half_life = 22 h; *interval* = 8 h; *MEC* = 10 μg/mL; *MTC* = 20 μg/mL; *start* = 0 h; *volume* = 3000 mL; *dosage* = 100 * 1000 μg; *absorption_fraction* = 0.12
elimination_constant = –ln(0.5)/*half_life*
drug_in_system(0) = 0
entering = *absorption_fraction* * (pulse of amount *dosage* beginning at *start* every *interval* hours)
elimination = *elimination_constant* * *drug_in_system*
concentration = *drug_in_system*/*volume*

Running the simulation and plotting the various concentrations that occur over 168 h (7 da), the resulting Figure 2.5.4 indicates that the concentration of the drug in the system between doses fluctuates. In less than 2 da, the concentration remains within the therapeutic range; and after about 5 da, the drug reaches a steady state.

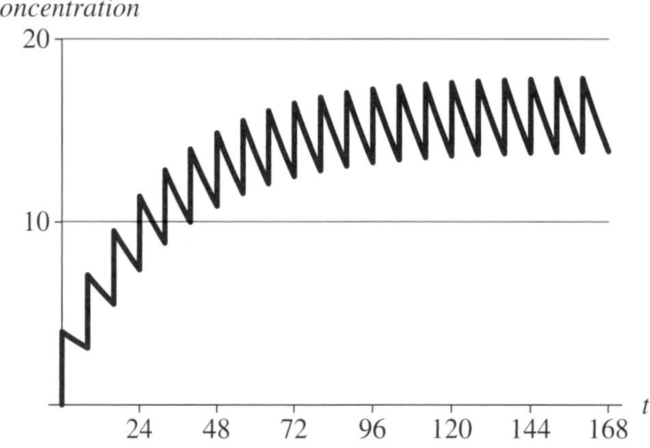

Figure 2.5.4 Graph of concentrations MEC = 10 μg/mL, MTC = 20 μg/mL, and concentration (μg/mL) versus time (h)

Mathematics of Repeated Doses

Let us show the mathematics of why the drug concentration in the Dilantin example tends to a fixed value, in this case about 12 μg/mL, immediately after a dose. Suppose that the patient takes a 100-mg tablet every 8 h. In the model, we assumed an absorption level of 0.12, so that the effective dosage is $Q_0 = (0.12)(100) = 12$ mg. With an elimination rate of –ln(0.5)/22, which is about 0.0315, the amount of drug in the system after 8 h is $Q = Q_0e^{-0.0315(8)} \approx (12)(0.7772) = 9.3264$ mg $= 9326.4$ μg. Thus, at the end of 8 h, about 77.72% of the drug remains in the system. The analytical value (9326.4 μg) for the mass of drug in the system is close to the simulated value (9327.91 μg) of *drug_in_system* at time 8.00 h (using a time step of 0.01 h and Runge-Kutta 4 numeric integration, which Module 6.4 discusses).

Suppose Q_n is the quantity (in mg) in the system immediately after the nth tablet. Thus, assuming 77.72% of the drug remains in the system at the end of an 8-h interval immediately before a dose, we have the following:

$$Q_1 = 12 \text{ mg}$$

$$Q_2 = \underbrace{(12 \text{ mg})(0.7772)}_{\text{remainder of tablet 1}} + \underbrace{12 \text{ mg}}_{\text{tablet 2}} = 21.3264 \text{ mg}$$

$$Q_3 = \underbrace{Q_2(0.7772)}_{\text{remainder of tablets 1 \& 2}} + \underbrace{12 \text{ mg}}_{\text{tablet 3}}$$

$$=(12(0.7772)+12)(0.7772)+12$$

$$= 12(0.7772)^2 + 12(0.7772) + 12 = 28.57488 \text{ mg}$$

$$Q_4 = \underbrace{Q_3(0.7772)}_{\text{remainder of tablets 1-3}} + \underbrace{12 \text{ mg}}_{\text{tablet 4}}$$

$$=(12(0.7772)^2+12(0.7772)+12)(0.7772)+12$$

$$= 12(0.7772)^3 + 12(0.7772)^2 + 12(0.7772) + 12 = 34.2084 \text{ mg}$$

Continuing in the same pattern, we determine that the general form of the quantity of the drug in the system immediately after the fifth tablet is as follows:

$$\begin{aligned} Q_5 &= 12(0.7772^4) + 12(0.7772^3) + 12(0.7772^2) + 12(0.7772) + 12 \\ &= 12(0.7772^4) + 12(0.7772^3) + 12(0.7772^2) + 12(0.7772^1) + 12(0.7772^0) \\ &= 12(0.7772^4 + 0.7772^3 + 0.7772^2 + 0.7772^1 + 0.7772^0) \end{aligned}$$

Similarly, the quantity of the drug immediately after the nth tablet, Q_n, follows:

$$Q_n = 12(0.7772^{n-1} + \cdots + 0.7772^2 + 0.7772^1 + 0.7772^0)$$

Quick Review Question 4

Suppose a patient takes a 200-mg tablet once a day, and within 24 h, 75% of the drug is eliminated from the body. With Q_n being the quantity of the drug in the body after the nth dose, determine the following:

a. Q_1

b. Q_2 expressed as a sum
c. Q_3 expressed as a sum
d. Q_4 expressed as a sum
e. Q_n expressed as a sum

We would like to determine what happens to the quantity of the drug in the system over a long period of time. To do so, we need a formula for the sum $0.7772^{n-1} + \cdots + 0.7772^2 + 0.7772^1 + 0.7772^0$ for positive integer n. This sum is a **finite geometric series**, and its general form is as follows:

$$a^{n-1} + \cdots + a^2 + a^1 + a^0 \text{ for } a \neq 1 \text{ and positive integer } n$$

As we verify in the next section, this sum is the following ratio:

$$a^{n-1} + \cdots + a^2 + a^1 + a^0 = \frac{(1-a^n)}{(1-a)} \text{ for } a \neq 1$$

Thus, for $a = 0.7772$ and $n = 5$, we can compute the value of Q_5:

$$\begin{aligned} Q_5 &= 12(0.7772^4 + 0.7772^3 + 0.7772^2 + 0.7772^1 + 0.7772^0) \\ &= 12 \cdot \frac{1-0.7772^5}{1-0.7772} = 38.5868 \text{ mg} = 38{,}586.8\ \mu\text{g} \end{aligned}$$

Within simulation error, this value agrees with *drug_in_system* (38,580.92) after the fifth dose, at time 32.01 h. In general, the quantity of the drug immediately after the nth tablet, Q_n, is as follows:

$$\begin{aligned} Q_n &= 12(0.7772^{n-1} + \cdots + 0.7772^2 + 0.7772^1 + 0.7772^0) \\ &= 12 \cdot \frac{1-(0.7772)^n}{1-0.7772} \end{aligned}$$

Definition $a^{n-1} + \cdots + a^2 + a^1 + a^0$ for $a \neq 1$ and positive integer n is a **finite geometric series** with **base** a.

Quick Review Question 5

Using the drug of Quick Review Question 4 and the formula for the sum of a finite geometric series, evaluate the following:

a. Q_{10}
b. Q_n

Using the formula for the sum of a finite geometric series, we can compute the quantity of drug after the nth tablet. To determine the long-range affect, we let n go to infinity and see that Q_n approaches 53.8599 mg, as follows:

$$Q_n = 12 \cdot \frac{1-(0.7772)^n}{1-0.7772} \to 12 \cdot \frac{1-0}{1-0.7772} \approx 53.8599 \text{ mg}$$

Thus, the serum concentration is about (53.8599 mg)/(3000 mL) = 0.0179533 mg/mL = 17.95 μg/mL, which agrees closely with the peak value of the concentration in Figure 2.5.4.

Quick Review Question 6

Using the drug of Quick Review Questions 4 and 5, determine the quantity of drug after the *n*th tablet when the patient has been taking the drug for a long time.

Sum of Finite Geometric Series

To derive the formula for the sum of a finite geometric series, we start by considering a particular example, Q_5 as before. Let s be equal to the sum of the powers from 0 through 4 of 0.7772, as follows:

$$s = 0.7772^4 + 0.7772^3 + 0.7772^2 + 0.7772^1 + 0.7772^0 \tag{1}$$

Multiplying both sides by 0.7772, we have the following:

$$0.7772s = (0.7772)\,(0.7772^4 + 0.7772^3 + 0.7772^2 + 0.7772^1 + 0.7772^0)$$
$$0.7772s = 0.7772^5 + 0.7772^4 + 0.7772^3 + 0.7772^2 + 0.7772^1 \tag{2}$$

Subtracting Equation 2 from Equation 1, we subtract off all but two terms on the right:

$$\begin{array}{rl} s = & \; 0.7772^4 + 0.7772^3 + 0.7772^2 + 0.7772^1 + 0.7772^0 \\ -0.7772s = & -0.7772^5 - 0.7772^4 - 0.7772^3 - 0.7772^2 - 0.7772^1 \\ \hline s - 0.7772s = & -0.7772^5 \qquad\qquad\qquad\qquad\qquad\qquad\qquad + 0.7772^0 \end{array}$$

With 0.7772^0 being 1, we factor out s on the left as follows:

$$s(1 - 0.7772) = -0.7772^5 + 1$$

or

$$s(1 - 0.7772) = 1 - 0.7772^5$$

Dividing both sides by the factor (1 – 0.7772), we obtain the following formula:

$$s = \frac{1-0.7772^5}{1-0.7772}$$

By the same reasoning, we have the general formula for the sum of a finite geometric series.

The formula for the sum of a finite geometric series is as follows:

$$a^{n-1} + \cdot\cdot\cdot + a^2 + a^1 + a^0 = \frac{(1-a^n)}{(1-a)} \text{ for } a \neq 1$$

Two-Compartment Model

The one-compartment model is more appropriate for an injection of a drug into the system than for a pill, which takes time to dissolve, be absorbed, and be distributed within the system. In such cases, a **two-compartment model** might yield better results. The first compartment represents the digestive system (stomach and/or intestines), while the second might indicate the blood, plasma, serum, or a particular organ that the drug targets. A flow pumps the drug from one compartment to the other in the model. One option for modeling the rate of change of absorption from the intestines to blood serum has the rate proportional to the amount of drug in the intestines. Probably a more accurate representation has the rate of change of absorption from the intestines to blood serum be proportional to the volume of the intestines and to the difference of the drug concentrations in the intestines and serum.

Although the one- or two-compartment model is appropriate for most situations, a drug dosage problem could benefit from more compartments in a **multicompartment model**. Various projects employ more than one compartment.

Quick Review Question 7

This question applies to the rate of change of absorption of a drug from the intestines to blood serum in a two-compartment model. Suppose k is a constant of proportionality; i and b are the masses of the drug in the intestines and blood serum, respectively; v_i and v_b are the volumes of the intestines and blood serum, respectively; c_i and c_b are the drug concentrations in the stomach and blood serum, respectively; and time t is in hours.

a. Give the differential equation for this rate if the rate of absorption is proportional to the mass of drug in the intestines.
b. In this case, give the units of k.
c. Give the differential equation for this rate if the rate of absorption is proportional to the volume of the intestines and to the difference of the drug concentrations in the intestines and blood serum.
d. In this case, give the units of k.

Exercises

1. Assuming that a quantity of a drug (Q) is $Q = Q_0e^{Kt}$, show that $K = -\ln(0.5)/t_{1/2}$, where $t_{1/2}$ is the drug's half-life.
2. **a.** In Figure 2.5.4, what are the units for *MEC* and *MTC*?

b. What are the units for *dosage*?
c. With a dosage of Dilantin being 100 mg, why is the value of *dosage* 100 * 1000?

3. Prove the general formula for the sum of a finite geometric series.
4. a. In Dilantin example, describe the effect a longer half-life has on *elimination_constant*.
b. Evaluate *elimination_constant* for $t_{1/2} = 7$ h.
c. Evaluate *elimination_constant* for $t_{1/2} = 22$ h.
d. Evaluate *elimination_constant* for $t_{1/2} = 42$ h.
5. a. Suppose a patient taking Dilantin decides for convenience to take 300 mg once a day instead of 100 mg every 8 h. Adjusting the model in *OneCompartDilantin*, determine the results of such a decision. Is the decision advisable?
b. Mathematically, determine the long-term value of Q_n, the quantity of Dilantin in the system immediately after the nth dose, assuming absorption of only (0.09)(300 mg).
6. a. Determine mathematically the quantity of Dilantin in the system immediately before the fifth dose. Use the same assumptions as in the section "Mathematics of Repeated Doses."
b. Determine mathematically the long-term value of the quantity of Dilantin in the system immediately before the nth dose.
c. Compare your answers to the values in *OneCompartDilantin*.
7. How should the one-dose aspirin example be adjusted to incorporate the weight of a male patient? About 65% to 70% of a male's body is liquid. Assume that 1 kilogram (kg) of body liquid has a volume of 1 L. Assume the patient has a mass of 90 kg (comparable to about 198 lb).

Projects

For additional projects, see Module 7.7, "Cardiovascular System—A Pressure-Filled Model."

1. Develop a two-compartment model for one dose of aspirin.
2. Develop a two-compartment model for aspirin, where someone with a headache takes three aspirin tablets and 2 h later takes two more aspirin tablets.
3. In attempt to raise the concentration of a drug in the system to the minimum effective concentration quickly, sometimes doctors give a patient a **loading dose**, which is an initial dosage that is much higher than the maintenance dosage. A loading dose for Dilantin is three doses—400 mg, 300 mg, and 300 mg 2 h apart. Twenty-four hours after the loading dose, normal dosage of 100 mg every 8 h begins. Develop a model for this dosage regime.
4. Develop a two-compartment model for Dilantin, where the rate of change of absorption from the stomach to the blood serum is proportional to the amount of drug in the stomach.
5. Develop a two-compartment model for Dilantin, where the rate of change of absorption from the stomach to the blood serum is proportional to the

volume of the stomach and to the difference of the drug concentrations in the stomach and serum. Assume the volume of the stomach is 500 mL.

6. Develop a two-compartment model for a pediatric dosage of Dilantin that includes the mass of the patient. The initial dose is 5 mg/kg per day in two or three equally divided doses. The maintenance dosage is usually 4 to 8 mg/kg per day.
7. Develop a model for vancomycin HCI, which is a treatment for serious infections by susceptible strains of methicillin-resistant staphylococci in penicillin-allergic patients. The drug is administered by IV infusion. The intravenous dose is usually 2 g divided either as 500 mg every 6 h or 1 g every 12 h, and the rate is no more than 10 mg/min or over a period of at least 50 min, whichever is longer. When kidney function is normal, multiple intravenous dosing of 1 g results in mean plasma concentrations of about 63 μg/mL immediately after infusion, 23 μg/mL in 2 h, and 8 μg/mL 11 h after infusion. In such patients, the mean elimination half-life from plasma is 4 to 6 h. The mean plasma clearance is approximately 0.058 L/kg/h (liter of drug per kilogram of patient mass each hour), while the mean renal clearance is about 0.048 L/kg/h (Hospira 2010). Thus, include the mass of the patient in the model.
8. Repeat Project 7 for patients with renal dysfunction in which the average half-life of elimination is 7.5 da (Hospira 2010).
9. Develop a model for Vancocin HCI in which the patient initially has normal kidney function (see Project 7). However, at the start of the third day, one of the patient's kidneys stops functioning; and the elimination rate becomes half its previous value. Consider using a step function.
10. Do Project 7 for children, where the dosage is 10 mg/kg every 6 h, and the rate of administration is over a period of at least 60 min (Hospira 2010).
11. Do Project 7 for neonates and young infants. The initial dose is 15 mg/kg. Thereafter, the dosage is 10 mg/kg every 12 h for neonates in their first week of life and afterward, up to age of 1 mo, every 8 h. Administration is more than 60 min (Hospira 2010).
12. Model drug dosage of aspirin for arthritis, where the initial dose is 3 g/da in divided doses. The dosage can be increased. Relief usually occurs at plasma levels of 20 to 30 mg per 100 mL. The plasma half-life of aspirin increases with dosage, so that a dose of 1 g has a half-life of about 5 h and a dose of 2 g has a half-life of about 9 h.
13. Considering the information about mass in Project 7, do any of the previous projects except one involving children or infants, accounting for the mass of a male patient.
14. By consulting a pharmacy reference or website, such as http://www.nlm.nih.gov/medlineplus/druginformation.html, obtain relevant information about some drug. Model the dosage of this drug.

Answers to Quick Review Questions

1. $K = -\ln(0.5)/3.2$ per hour $= 0.22$/h
2. *plasma_concentration = aspirin_in_plasma / plasma_volume*

3. *absorption_fraction* * (pulse of amount *dosage* beginning at *start* every *interval* hours), where the pulse function depends on the particular system dynamics tool
4. **a.** 200 mg
   b. (200 mg)(0.25) + 200 mg
   c. (200 mg)$(0.25)^2$ + (200 mg)(0.25) + 200 mg
   d. (200 mg)$(0.25)^3$ + (200 mg)$(0.25)^2$ + (200 mg)(0.25) + 200 mg
   e. (200 mg)$(0.25)^{n-1}$ + · · · + (200 mg)$(0.25)^2$ + (200 mg)(0.25) + 200 mg
5. **a.** (200 mg)(1 – $(0.25)^{10}$)/(1 – 0.25) = 266.67 mg
   b. (200 mg)(1 – $(0.25)^n$)/(1 – 0.25) = (200 mg)(1 – $(0.25)^n$)/(0.75)
6. (200 mg)(1 – 0)/(0.75) = 266.67 mg
7. **a.** $db/dt = ki$
   b. 1/h
   c. $db/dt = k(v_i)(c_i - c_b)$
   d. 1/h

References

Ellis, Janice Rider, and Celia Love Hartley. 2004. *Nursing in Today's World: Trends, Issues & Management*. Philadelphia: Lippincott, Williams and Wilkins.

Fitzgerald, Walter L., Jr., and Dennis B. Wilson. 1998. "Medication Errors: Lessons in Law." *Drug Topics*, January 19, 1998. *Hospital Pharmacy Reporter*. February 1998.

Hospira. 2010. "Vancomycin Hydrochloride for Injection, USP." Package insert. http://www.hospira.com/Images/EN-2613_32-5771_1.pdf (accessed December 30, 2012)

IOM (Institute of Medicine). 2007. Committee on Identifying and Preventing Medication Errors. Preventing Medication Errors: Quality Chasm Series. Philip Aspden, Julie Wolcott, J. Lyle Bootman, Linda R. Cronenwett, eds.

Khorasheh, Farhad, Amir Mahbod Ahmadi, and Abbas Gerayeli. 1999. "Application of Direct Search Optimization for Pharmacokinetic Parameter Estimation." *J Pharm Pharmaceutical Science* 2(3): 92–98. http://www.ualberta.ca/~csps/JPPS2(3)/F.Khorasheh/application.htm (accessed December 18, 2012)

NQF (National Quality Forum). 2010. "Preventing Medication Errors: A $21 Billion Opportunity." http://www.qualityforum.org/Publications/2010/12/Preventing_Medication_Errors_CAB.aspx

Patel, A., and B. Ross. 2010. "Walgreens Told to Pay $25.8 Million Over Teen Pharmacy Tech's Error." *The Blotter*. ABC News. http://abcnews.go.com/Blotter/walgreens-told-pay-285-mil-teen-pharmacy-techs/story?id=9977262#.UJXKrI7et38 (accessed October 20, 2012)

Pharmacy Network Group. "Pharmacy Network Group Home Page." http://www.pharmacynetworkgroup.com/ (accessed December 18, 2012)

Robert, Teri. 2004. "Teenager Dies from Acetaminophen Overdose." *Your Guide to Headaches/Migraine*, About, Inc. http://headaches.about.com/cs/medicationsusage/a/acet_death.htm. (accessed December 18, 2012) Source for this Web article: "Teenager Accidentally Overdoses on Over-the-Counter Analgesic," July 2, 2003. http://TheAssociatedPress.ChannelOklahoma.com

3

FORCE AND MOTION

MODULE 3.1

Modeling Falling and Skydiving

Downloads

The following files containing the models in this module are available for download on the text's website for various system dynamics tools: *Fall*, *FallFriction*, and *FallSkydive*.

Introduction

What is it like to skydive? Imagine ascending in a small plane to, say, 10,000 feet (ft), when the jumpmaster opens the door. The jumpmaster asks you if you are ready to jump. You head for the door and walk out onto a step under the wing, holding on to a strut. You experience lots of wind and noise. Your heart is pounding wildly. The jumpmaster yells, "Go!" You arch your body and release your grip on the strut. Your adrenalin levels have never been higher as you plunge toward earth at 120 mi/h. Nevertheless, you are in control. For the next 50 s, simple body movements can alter your speed, direction, and position. At 3000 ft, the landscape is fast approaching, and you pull your cord. As it deploys, your descent slows, and the mad rush of wind ceases, replaced by the rustling sounds of your canopy. Soon you gently settle to the ground.

The use of parachutes or parachute-like devices to slow the descent of jumpers from positions of considerable height may have begun with the twelfth-century Chinese. However, the first evidence of a parachute in the western world appeared in the late fifteenth-century drawings of Leonardo da Vinci. His pyramid-shaped design was to be constructed of linen and a wooden frame. There is no record of Leonardo experimenting with his invention, but late last century it was demonstrated successfully.

Not much development of parachutes took place until late in the eighteenth century, when hot-air balloons were being shown across Europe. Andres-Jacques Garnerin, a French balloonist of dubious reputation, was one of the first persons to demonstrate a parachute without a rigid frame. He successfully descended from his balloon (which exploded) at about 3000 ft using a gondola suspended by an umbrella-shaped parachute.

Jumps using parachutes from airplanes began in the early twentieth century but were primarily for rescuing observation balloon pilots. Barnstormers performed parachute-jumping demonstrations at air shows in the time between the world wars. During World War II, both sides exploited the capabilities of parachutes for dispersing men and supplies.

Sport parachuting (skydiving) probably has its roots in the first freefall conducted in 1914, but the sport really gained popularity only in the 1950s and 1960s (Bates; USPA 2008).

In this module, using a system dynamics tool we model the motion of someone skydiving. Such a jump has two phases, a free-fall stage followed by a parachute stage with greater air friction. In preparation, we develop a model for the motion of a ball thrown straight up from a bridge, first ignoring air friction and then refining the model to consider this additional force.

Acceleration, Velocity, and Position

As discussed in Module 2.2, "Unconstrained Growth and Decay," the instantaneous rate of change, or derivative, of position (s) with respect to time (t) is velocity (v). Moreover, the instantaneous rate of change of velocity with respect to time is **acceleration** (a). In derivative notation, we have the following:

$$v(t) = \frac{ds}{dt}$$

$$a(t) = \frac{dv}{dt}$$

In the first example, we use these derivatives in modeling the motion of a ball when, on a windless day, someone standing on a bridge holds a ball over the side and tosses the ball straight up into the air.

Quick Review Question 1

This question reflects on Step 2 of the modeling process—formulating a model—for developing a model for a falling object. We simplify this first attempt at a model by ignoring friction. After completing this question and before continuing in the text, we suggest that you develop a model for a falling object.

a. Determine four variables for the model and their units in the metric system.
b. Give a differential equation relating time (t), position (s), and velocity (v).
c. Give a differential equation relating time (t), velocity (v), and acceleration (a).
d. Ignoring friction, give any of the following that are constant in a fall: time,

distance, velocity, acceleration. In a model diagram, we will store such a value in a converter/variable.

e. In a model diagram, list the components that will be in stocks (box variables): *t*, *s*, *v*, *a*, *ds*/*dt*, *dv*/*dt*.

f. In a model diagram, give the value(s) that will flow into the position stock (box variable) for change in position: *t*, *s*, *v*, *a*, *ds*/*dt*, *dv*/*dt*.

g. In a model diagram, give the value(s) that will flow into the velocity stock (box variable) for change in velocity: *t*, *s*, *v*, *a*, *ds*/*dt*, *dv*/*dt*.

With a system dynamics tool to model the motion of a ball that someone throws straight up from a bridge, we have stocks (box variables) for the quantities that accumulate, the height (*position*) and velocity (*velocity*) of the ball. During the simulation, we can observe their changing values in a graph and table. A flow representing the change goes into velocity (*change_in_velocity*). Change in velocity is acceleration, and in this case, the acceleration is due to gravity. Therefore, a converter/variable (*acceleration_due_to_gravity*) contains the constant for **acceleration due to gravity**, which—with up being the positive direction—is approximately **–9.81 m/s²**. The converter connects to *change_in_velocity*, which has this constant as its equation. Also, the flow for the change in height (*change_in_position*) is identical to the current velocity, *velocity*. Thus, we have a connector from *velocity* to *change_in_position* and define the value of this flow to be *velocity*. Because velocity can be positive, zero, or negative, we specify that the flow can go into or out of *position*. For flexibility in models that we derive from this one, we also make *change_in_velocity* a biflow. Moreover, we specify that *velocity* and *position* can take on negative as well as positive values. For specificity, we initialize *velocity* to be 15 m/s and *position* to be 11 m, which is the height of the bridge. Figure 3.1.1 presents a diagram for a model of motion of the ball with a white arrowhead on each flow indicating the secondary biflow direction.

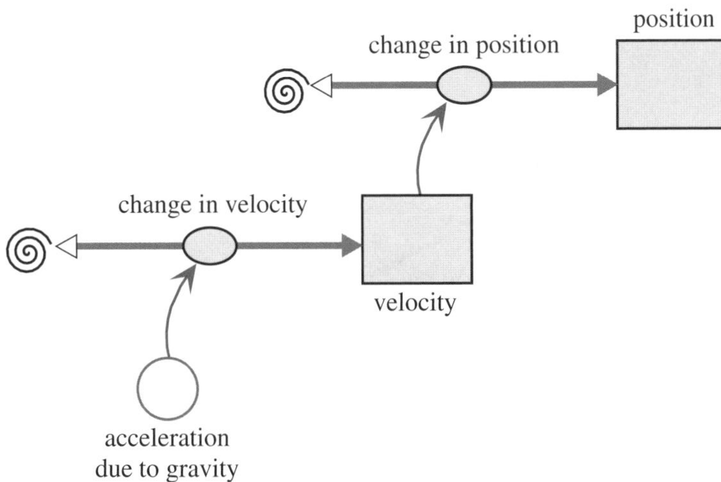

Figure 3.1.1 Diagram of motion of ball thrown straight up

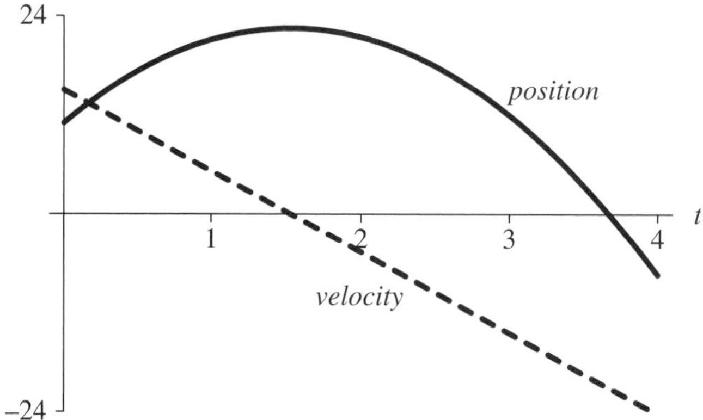

Figure 3.1.2 Graph of velocity (m/s) and position (m) of ball versus time (s)

Quick Review Question 2

Give the formula in metric units for each of the following components in Figure 3.1.1:

a. the converter *acceleration_due_to_gravity*
b. the flow *change_in_velocity*
c. the flow *change_in_position*

Output consists of a graph and a table of velocity and height versus time. With $\Delta t = 0.25$ s and the Runge-Kutta 4 integration technique, which Module 6.4 discusses, we obtain a graph of velocity, as in Figure 3.1.2. Moreover, the graph of velocity versus time, also in that figure, is the line $v(t) = 15 - 9.8t$.

For some of the models, it is more convenient to consider speed than velocity. The **speed** gives the magnitude of the change in position with respect to time, while

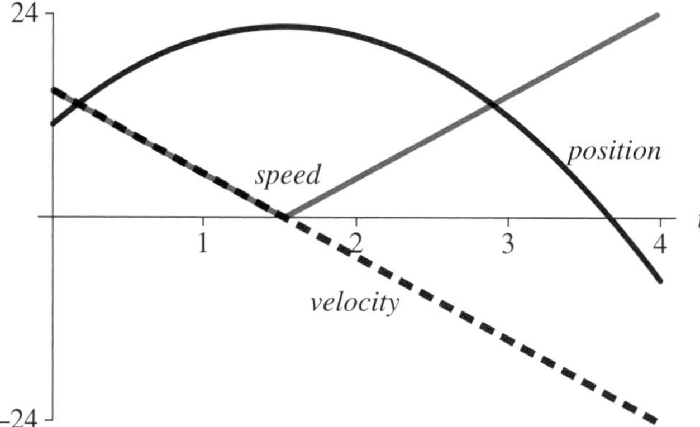

Figure 3.1.3 Graph of velocity (m/s), position (m), and speed (m/s) of ball versus time (s)

the velocity expresses the magnitude with the direction. Thus, speed is the absolute value of the velocity. To incorporate speed, we have a connector/arrow from the *velocity* stock (box variable) to a new converter/variable, *speed*, which stores the equation for the absolute value of velocity. The graph in Figure 3.1.3 shows speed and velocity decreasing in a linear fashion to 0 m/s at about time 1.5 s. Afterward, speed steadily increases.

Physics Background

Before developing additional examples of falling and skydiving, we need to consider some formulas from physics—Newton's second law and approximations of friction. Newton's second law concerns force applied to a mass imparting acceleration. So that we can refine models to account for air friction, we also consider several approximations of such a force.

Newton's second law has far-reaching significance. In this text, we employ the law in modeling situations from the motion of skydivers to the motion of the planets. The law states that a force F acting on a body of mass m gives the body acceleration a. Moreover, as the following models indicate, the acceleration is directly proportional to the force and inversely proportional to the mass:

$$a = F/m$$

or

$$F = ma$$

> **Newton's second law** A force F acting on a body of mass m gives the body acceleration a according to the following formula:
>
> $$\boldsymbol{F = ma}$$

We can apply this formula to obtain the relationship between weight and mass. **Weight** is a force and is not the same as mass. The acceleration involved is acceleration due to gravity, which at sea level is about -9.81 m/s^2 or -32 ft/s^2 for up being the positive direction. For example, an object that has mass of 20 kg has a weight of -196.2 newtons (N), as the following shows:

$$\text{weight} = F = (20\text{ kg})(-9.81\text{ m/s}^2) = -196.2\text{ kg m/s}^2 = -196.2\text{ N}$$

The metric unit for force is a **newton** (**N**), or kg m/s^2.

> **Definition** A **newton** (**N**) is a measure of force, and **1 N = 1 kg m/s^2**.

Quick Review Question 3

Determine the following, including units.

a. The mass of an object that weighs 981 N
b. The acceleration that results when a net force of 10 N is applied to an object with mass 5 kg

Kinetic friction, or **drag**, also is a force. This force between objects is in the opposite direction to a moving object and tends to slow motion. Thus, kinetic friction dampens motion of an object. When an object moves through a fluid, such as air or water, the fluid friction is a function of the object's velocity. For example, the faster we pedal a bicycle, the harder it is for us to do so. As our velocity increases, so does the friction of the air on our bodies.

Several models that estimate friction exist. In Module 8.3, "Empirical Models," we study how to derive our own model, such as a model for drag, from data. In this module, we consider two models for drag on a body traveling through a fluid.

For a small object traveling slowly, such as a dust particle floating through the air, we usually employ **Stokes's friction**, which states that friction on the particle is approximately proportional to its velocity,

$$F = kv$$

where k (kg/s) is a constant of proportionality for the particular object and fluid and v (m/s) is the velocity.

For a larger object moving faster through a fluid, we usually employ **Newtonian friction**, which states that the drag is approximately as follows:

$$F = 0.5CDAv^2$$

where C is a dimensionless constant of proportionality (the **coefficient of drag**, or **drag coefficient**) related to the shape of the object, D is the density of the fluid, and A is the object's projected area in the direction of movement. For a particular situation, C, D, and A are constants, so that the drag is approximately proportional to the velocity squared. At 0 °C, the density of air at sea level is 1.29 kg/m^3. For shapes that are hydrodynamically good, $C < 1$; for spheres, C is about 1; and for shapes that are hydrodynamically inefficient, $C > 1$. Many objects have a coefficient of drag of about 1. Thus, through air with $C = 1$, Newtonian friction is approximately the following:

$$F = 0.65Av^2$$

The **density of water at 3.98 °C**, where the fluid achieves its maximum density, is **1.00000 g/cm^3**, yielding a formula with a different coefficient. Table 3.1.1 summarizes the three models for fluid friction considered here.

The drag force is in the opposite direction of motion, and the sign of velocity indicates the direction. On the upward portion of a trajectory, drag and gravity both act downward; while on the downward part, drag is upward and gravity downward. Thus, for the general formula for Newtonian friction, we take the absolute value of

Table 3.1.1
Summary of Several Models for Magnitude of Fluid Friction

Name	*Formula*	*Meanings of Symbols*	*When to Use*
Stokes's friction	$F = kv$	k constant v velocity	Very small object moving slowly through fluid
Newtonian friction	$F = 0.5CDAv^2$	C coefficient of drag D density of fluid A object's projected area in direction of movement v velocity	Larger objects moving faster through fluid
Newtonian friction through air	$F = 0.65Av^2$	A object's projected area in direction of movement v velocity	Larger objects with $C = 1$ moving faster through sea-level air

only one of the velocity terms and multiply the entire formula by –1, yielding $-0.5CDAv|v|$. If ABS is the absolute value function, the translation of this formula into a system dynamics tool is as follows:

```
-0.5 * drag_coefficient * density * projected_area * velocity * ABS(velocity)
```

Quick Review Question 4

Calculate the following:

a. The density of 3.98 °C water in kg/m^3

b. The magnitude of friction in newtons of a ball falling through 3.98 °C water, where the coefficient of drag is 0.9, the cross-sectional area of the ball is 0.03 m^2, and its velocity is –20 m/s

c. Write the formula for Newtonian friction for a system dynamics tool, where the coefficient of drag is 1 and the air density is 1.29 kg/m^3, namely, $-0.65Av|v|$, A and v are appropriate variables, and ABS is the absolute value function.

Quick Review Question 5

This question reflects on refinement of the model of an object falling through sea-level air to account for friction. After completing this question and before continuing in the text, we suggest that you revise the model in the first example to account for drag friction for practice in model development.

a. Give the inputs to compute drag friction.

b. Give a formula for air friction in a system dynamics tool's model with v for velocity, A for projected area, and ABS for the absolute value function.

c. Give the force(s) acting on the object.

d. Give a formula for an object's weight in a system dynamics tool's model, where g is the acceleration due to gravity and m is the mass of the object.
e. Give a formula for an object's acceleration in a system dynamics tool's model, where F is the total force on the object (weight + air friction) and m is the mass of an object.

Friction during Fall

The previous example for modeling the motion of a ball thrown straight up does not account for air friction. To do so, we consider two forces on the ball, gravity and drag friction. The force due to gravity is its weight, which by Newton's second law is $F = ma$. Thus, adjusting the model diagram in Figure 3.1.1, we include a converter/variable for *weight* with connections from converters/variables for *mass* and *acceleration_due_to_gravity* (see Figure 3.1.4). Newtonian friction for the air friction including direction is $F = -0.65Av|v|$. In the diagram, connectors/arrows go

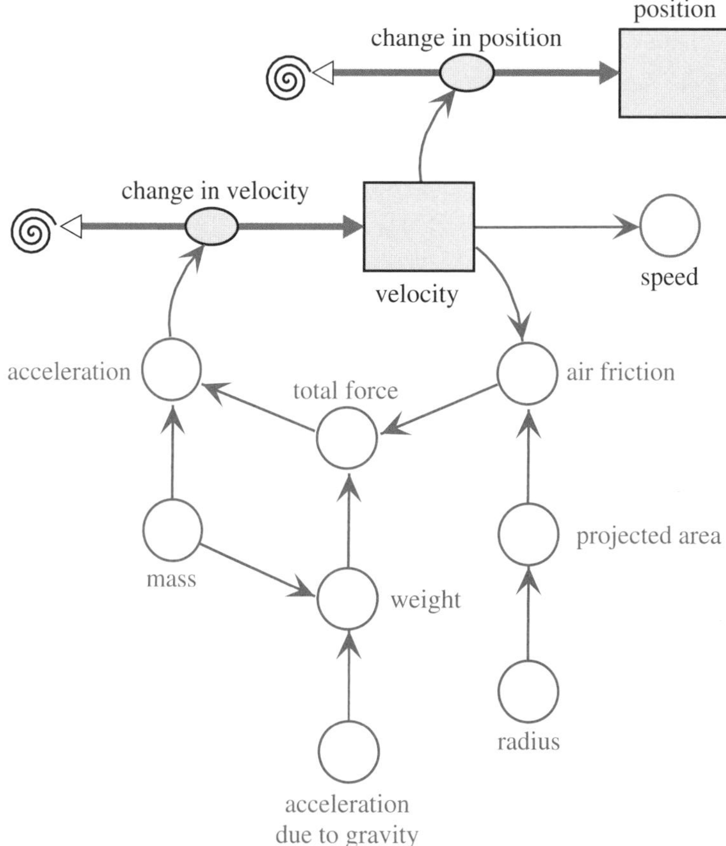

Figure 3.1.4 Diagram for motion of ball under influence of air friction; changes to converters/variables from Figure 3.1.1 in color

from *velocity* and from a new *area* converter to a new converter for *air_friction*. The *projected_area* converter/variable stores the cross-sectional, or projected, area of the object in the direction of motion. Assuming spherical objects, another converter/variable stores the *radius*; and the equation in *projected_area* is *pi* * *radius*^2, where *pi* is built in or is an approximated constant 3.15169, depending on the system dynamics tool. Both forces, *weight* and *air_friction*, connect to a new converter/variable for *total_force*, which is the sum of the individual forces. Employing Newton's second law again with $a = F/m$, *acceleration* is *total_force/mass*. This acceleration provides the change in velocity for the flow into *velocity*.

Figure 3.1.4 contains a **feedback loop**. The initial value of air friction employs the initial velocity, here 0 m/s; and *air_friction* contributes to the *total_force*, which *acceleration* uses. Acceleration is the *change_in_position*, which contributes to *velocity*. Then, the current value of *velocity* "feeds back" into *air_friction* for a new computation of that force.

To detect the influence of drag, we consider a ball of mass 0.5 kg and radius 0.05 m dropped (initial velocity = 0 m/s) from a height of 400 m. Equation Set 3.1.1 presents various underlying model equations with units for constants.

Equation Set 3.1.1

Various underlying equations to accompany diagram in Figure 3.1.4:

mass = 0.5 kg
acceleration_due_to_gravity = –9.81 m/s^2
radius = 0.05 m
weight = *mass* * *acceleration_due_to_gravity*
projected_area = 3.14159 * *radius*^2
air_friction = –0.65 * *projected_area* * *velocity* * ABS(*velocity*)
total_force = *weight* + *air_friction*
acceleration = *total_force/mass*
change_in_velocity = *acceleration*
change_in_position = *velocity*
speed = ABS(*velocity*)
velocity(0) = 0 m/s
velocity(t) = *velocity*($t - \Delta t$) + (*change_in_velocity*) * Δt
position(0) = 400 m
position(t) = *position*($t - \Delta t$) + (*change_in_position*) * Δt

Running the simulation for 15 s, we see in Figure 3.1.5 that the ball reaches a constant, or **terminal, speed**, of about 31 m/s. From about time 6 s on, the position graph is almost linear, so that acceleration is approximately 0 m/s^2.

Quick Review Question 6

At the terminal velocity, give the relationship between *weight* and *air_friction*:
A. *weight* < *air_friction* B. *weight* = *air_friction* C. *weight* > *air_friction*

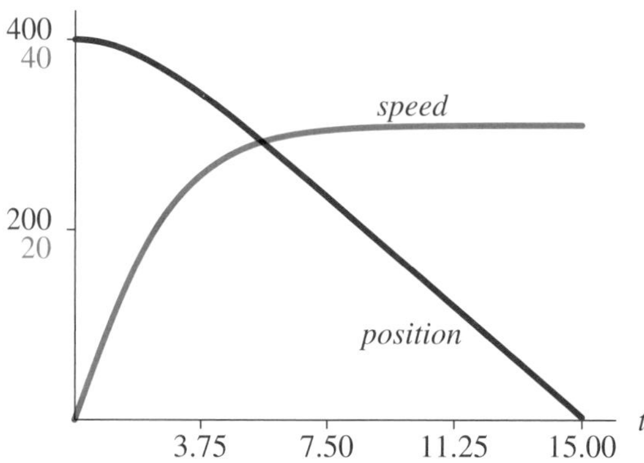

Figure 3.1.5 Graph of position (m) and speed (m/s) of object versus time (s) under influence of friction

Quick Review Question 7

This question reflects on refinement of the model to incorporate skydiving. After completing this question and before continuing in the text, we suggest that you revise the model.

a. Give the phases of the fall during the simulation.
b. Give the variable whose value we can use to trigger the change in phase: *acceleration, mass, position, velocity, weight*
c. Give the value(s) that change upon opening of the parachute: *acceleration_due_to_gravity, mass, projected_area, weight.*
d. Describe anticipated changes to the graphs in Figure 3.1.5 after deployment of a parachute.

Modeling a Skydive

To model a skydive, we build heavily on the example of a falling object under the influence of friction. For simplicity, we consider someone jumping out of a stationary helicopter at 2000 m (about 6562 ft), and we ignore changes in air density. Project 5 considers parachuting out of a moving plane, which imparts a horizontal velocity to the jumper. The model for a skydive out of a helicopter has two phases, one where the person is in a free fall and the other after the parachute opens, when the larger surface area results in more air resistance. For our model, the main difference in these two phases is the projected area in the direction of motion, down. The cross-sectional area of a jumper in the stable arch position with arms arched back and legs bent at the knees is approximately 0.4 m^2 (about 4.3 ft^2). Parachutes vary in their designs, but 28 m^2 (about 301 ft^2) is a reasonable value. We trigger the pull of the ripcord by the height (*position*) above the ground, say, 1000 m (about 3281 ft). Thus, the diagram contains a converter/variable (*position_open*) for this quantity and con-

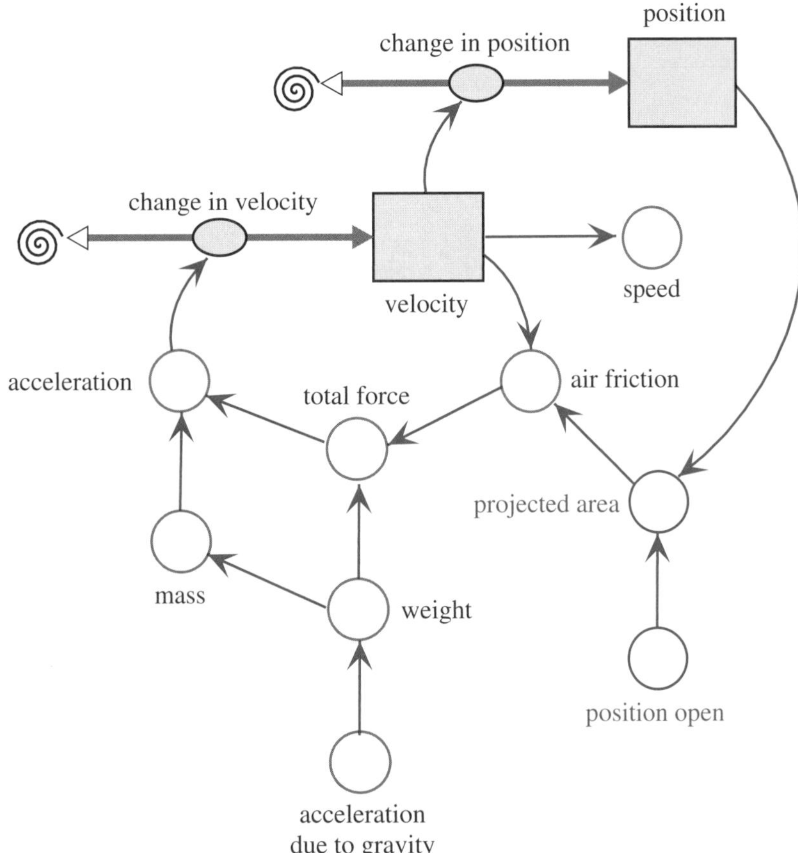

Figure 3.1.6 Diagram of skydiver's motion under influence of air friction

nectors/arrows from *position* to *position_open* and from *position_open* to *projected_area*. Figure 3.1.6 presents a model diagram for this example with changes in color from Figure 3.1.4 on a ball's fall. Assuming the parachute fully opens instantaneously, the equation in *projected_area* is no longer a constant but employs the following logic:

if (*position* > *position_open*)
  projected_area ← 0.4
else
  projected_area ← 28

Figure 3.1.7 shows graphs of the position and speed of a 90-kg (comparable to about 198 lb) skydiver versus time. Until a height of 1000 m, which occurs at about 21.3 s into the fall from 2000 m, the skydiver is in a free fall approaching a terminal speed of about 58.2 m/s (about 130.7 mi/h). At 1000 m, the person pulls the ripcord, and in a very short amount of time, the parachutist's speed slows to a new terminal speed of 6.96 m/s (about 15.6 mi/h).

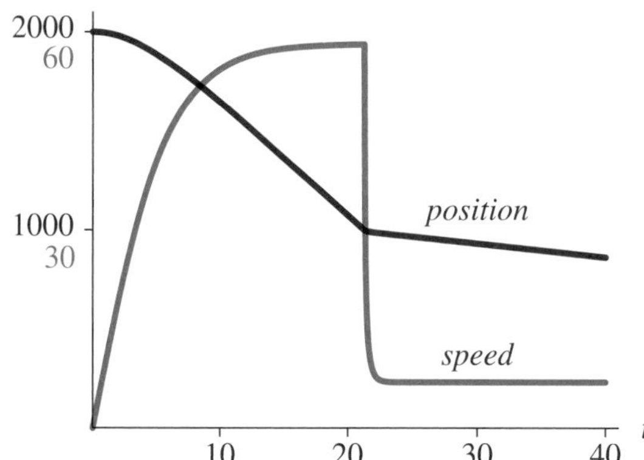

Figure 3.1.7 Position (m) and speed (m/s) versus time (s) of skydiver

Quick Review Question 8

a. How does the terminal speed of a skydiver who curls into a ball compare to that of the same skydiver who is in a stable arch position?

A. Less B. Equal C. Greater

b. Referring to Figure 3.1.7, approximately how long does it take for the skydiver in freefall to be close to terminal speed?

A. 13 s B. 21 s C. 40 s

c. Referring to Figure 3.1.7, at approximately what time does the skydiver pull the ripcord?

A. 13 s B. 21 s C. 40 s

Assessment of the Skydive Model

The shapes of the graphs of position and velocity in Figure 3.1.7 match the opening description of a skydive. However, our model exhibits a terminal speed of about 93.0 mi/h (about 41.4 m/s), while actual, measured speeds of 110 to 120 mi/h are common. The example employs the sea-level density of air, while the air density at 10,000 ft (about 3048 m) is about 73.8% (0.952 kg/m^3) of sea-level density. Adjusting the initial position to be 3048 m and using an air density of 0.952 kg/m^2 with the Newtonian friction of $F = 0.5CDAv^2$, the model indicates a terminal velocity of 48.15 m/s (about 108 mi/h) for the free fall for less than 50 s. However, the air density changes as the skydiver descends. Projects 4 and 7 explore refinements of the model to account for this variation.

Exercises

1. **a.** Using the equations and values for the example of a falling ball with no friction, write differential equations with initial conditions for acceleration and velocity.

   b. Using calculus, solve the differential equations of Part a to obtain velocity and position as functions of time.
2. Adjust *Fall* of the first example with no friction so that the object falls with an initial velocity of zero. Compare the results with those in *FallFriction* of the second example, which accounts for friction.
3. **a.** Using the equations and values of the second example with friction, write a differential equation involving the derivative of velocity for when an object reaches terminal velocity. At terminal velocity, the forces acting on the body are equal.
   b. Solve the equation of Part a using calculus.
4. Give the adjustments to the diagram in Figure 3.1.6 along with equations so that graphs of new converters/variables *adjusted_position* and *adjusted_speed* become horizontal lines at position 0 m after the parachutist lands.
5. Repeat Exercise 3 using Stokes's friction instead of Newtonian friction.
6. Suppose a raindrop evaporates as it falls but maintains its spherical shape. Assume that the rate at which the raindrop evaporates (that is, the rate at which it loses mass) is proportional to its surface area, where the constant of proportionality is –0.01. The density (mass per volume) of water at 3.98 °C is 1 g/cm^3. The surface area of a sphere is $4\pi r^2$, and its volume is $4\pi r^3/3$, where r is the radius. Assume no air resistance. (Project 8 models the motion of this raindrop under the influence of air resistance.)
   a. Assume that the initial radius is 0.3 cm. Determine the raindrop's initial mass.
   b. Write a differential equation for the rate of change of mass as a function of r.
   c. Write an equation for r as a function of mass.
7. Adjust the skydiving model so that the parachute opening depends on time, not height above the ground.
8. Write a system of differential equations for the skydiving model.
9. Using the models in your system dynamics tool's *Fall* and *FallFriction* files (see "Download"), compare position graphs for a dropped object with and without consideration of friction. Also, consider the velocity graphs. Discuss the results.

Projects

1. Using Table 3.1.2, which lists the velocities of a car at certain times, develop a model to estimate the total change in position of the car. Employ an input graph instead of an equation to record the table's values for the change in position. Compare your estimate to an exact value of 203$^1/_3$ m.

Table 3.1.2
The Velocities, v (m/s), of a Car at Certain Times, t (s)

t	0.0	0.5	1.0	1.5	2.0	2.5	3.0	3.5	4.0	4.5	5.0
v	24.00	28.75	33.00	36.75	40.00	42.75	45.00	46.75	48.00	48.75	49.00

2. Using Stokes's friction, develop a model for the motion of a dust particle floating down from a height of 50 m. Using comparative plots, determine its terminal speeds for various values of Stokes's constant of proportionality.
3. A bathysphere is a pressurized metal vessel in the shape of a sphere that allows people to explore the ocean to much greater depths than are possible by skin diving. A ship lowers and raises the sphere using a steel cable and communicates with its two occupants by telephone. In the 1930s, explorers William Beebe and Otis Barton developed the first bathysphere, which weighed 4500 lb and had a diameter of 4 ft 9 in. In a subsequent version, they descended to about 3000 ft in the ocean. Ignoring currents but not drag, model the sinking motion of a bathysphere. Assume that the boat reels out the steel cable fast enough so as not to affect the bathysphere's motion (Col 2010; Uscher 1999).
4. Table 3.1.3 contains air densities at various altitudes. Using these values on an input graph, refine the model for the skydiving example (Aber and Aber 2005).

Table 3.1.3
Approximate Air Densities at Various Altitudes

Altitude (m)	*Density (kg/m³)*
0	1.290
610	1.216
1219	1.146
1829	1.078
2438	1.014
3048	0.952
3658	0.894
4267	0.839
4877	0.786

5. Suppose an airplane is traveling in a straight line horizontally at 130 m/s at a height of 600 m when a parachutist jumps out of the plane at an angle of 30° with the horizon. Model the motion of the skydiver.
6. Model the motion of a meteor falling to the earth. Assume an initial height of 100,000 m, initial velocity of –10,000 m/s, coefficient of drag of 2, mass of 500 kg, and density of 8000 kg/m^3 for iron or 3500 kg/m^3 for stone (Schecker 1996). Give graphs for position, velocity, and acceleration versus time. Give comparison graphs for velocity versus height for meteors of various masses. Similarly, give comparison graphs for acceleration versus height. NASA's Glenn Research Center gives the following model for air density using variables D for density (slugs/ft^3), P for pressure (lb/ft^2), T for temperature (°F), and h for altitude (ft):

$$D = \frac{P}{1718(T + 459.7)}$$

where

for $h > 82{,}345$ ft, $T = -205.05 + 0.00164\,h$ and $P = 51.97\left(\dfrac{T + 459.7}{389.98}\right)^{-11.388}$

for $36{,}152 < h < 82{,}345$ ft, $T = -70$ and $P = 473.1\, e^{(1.73-0.000048h)}$

for $h < 36{,}152$ ft, $T = 59 - 0.00356h$, and $P = 2116\left(\dfrac{T+459.7}{518.6}\right)^{5.256}$

If you wish to use metric instead of English units, you can use the following: 1 slug = 14.5939 kg and 1 ft = 0.3048 m (Benson).

7. Using NASA's Glenn Research Center model for air density at heights less than 36,152 ft (see Project 6), refine the model in the skydiving example.

8. **a.** Model the change in mass of the raindrop that Exercise 6 describes.
b. Model the motion of this raindrop taking into account air resistance.

9. Develop a model to compare the terminal velocities of objects of different masses, such as a mouse, cat, human, horse, elephant, and so on. With the density of living protoplasm being almost constant across a wide variety of species, assume mass is proportional to the cube of a linear dimension, such as length or circumference; but surface area is proportional to the square of a linear dimension. How do the terminal velocities of more massive objects compare to those of less massive objects? Can a cat survive a fall from a tall building (Diamond 1989)?

10. (This project was contributed by and used with permission from Dr. Stephen Davies, University of Mary Washington (Davies 2012).) For a dangerous mission, a team of paratroopers must land on a 200-ft by 200-ft square roof of a 250-ft-tall building. The soldiers will be dropped at an altitude of 6400 ft from a transport jet traveling at a cruising speed of 200 knots. Certain restrictions apply to the jump: For safety, the paratroopers must not be airborne for more than 2 min and must hit the ground at less than 20 mi/h. Each paratrooper will be carrying a load of 40 lb of equipment, in addition to an assumed body weight of 180 lb. Before opening a chute, we can assume each soldier has a surface area of 0.4 m^2, in both horizontal and vertical directions. After opening, we can assume a 20-m^2 surface area projected vertically, and a 1-m^2 surface area projected horizontally. Assume good weather and negligible wind.

Develop a model for the jump. Plot altitude versus horizontal distance from the drop. Determine the following: The horizontal distance from the building the paratroopers should jump in order to land on the target area; the altitude at which they should pull their ripcords; the speed at which they will impact the rooftop; and the duration (in minutes and seconds) that they will be airborne. Use English units or metric units, but as with all models, be consistent throughout.

An important aspect of this problem that differs from the preceding skydiver model is that we must track and plot both x- and y-positions. In other words, in addition to following the trooper's altitude over time, we must also monitor the horizontal distance from the drop point, so that we can plot the trajectory of the falling trooper. "Time" is still the key independent variable. However, we now need a stock (box variable) for the horizontal (x) position in addition to one for the vertical (y) position. Moreover, we should graph the y-position versus the x-position, or altitude versus the horizontal distance, *not* altitude versus time.

The following physics facts are also useful: When the troopers make the jump, their initial horizontal (x) velocity will be the same as that of the plane (200 knots). The troopers' horizontal velocity can be considered separately from their vertical (y) velocity; in other words, neither one affects the other, so they can both be tracked independently. Air friction acts in the horizontal direction just as it does the vertical direction; that is, horizontal drag retards movement according to the formula, $-0.65Av|v|$, where v is the velocity in the horizontal direction.

Answers to Quick Review Questions

1. a. time, perhaps in s; distance, perhaps in ms; velocity, perhaps in m/s; acceleration, perhaps in m/s^2

b. $v(t) = \frac{ds}{dt}$

c. $a(t) = \frac{dv}{dt}$

d. Acceleration, which is acceleration due to gravity, –9.81 m/s^2

e. s and v

f. ds/dt

g. dv/dt or a, which is the constant acceleration due to gravity without friction

2. a. *acceleration_due_to_gravity* = –9.81 m/s^2

b. *change_in_velocity* = *acceleration_due_to_gravity*

c. *change_in_position* = *velocity*

3. a. $m = F/a = (981\text{ N})/(9.81\text{ m/s}^2) = 100\text{ kg}$

b. $a = F/m = (10\text{ N})/(5\text{ kg}) = 2\text{ m/s}^2$

4. a. $\frac{1\text{ g}}{\text{cm}^3} \times \frac{1\text{ kg}}{10^3\text{g}} \times \left(\frac{10^2\text{cm}}{1\text{ m}}\right)^3 = 10^3\frac{\text{kg}}{\text{m}^3}$

b. $F = -0.5CDAv|v| = -0.5(0.9)(10^3)(0.03)(-20)|-20| = 5400\text{ N}$

c. -0.65 * A * v * ABS(v)

5. a. velocity and projected area

b. –0.65 * A * v * ABS(v)

c. weight and air friction

d. m * g

e. F/m

6. B. *weight* = *air_friction*

7. a. before and after opening of the parachute

b. *position*

c. *projected_area*

d. The position curve should continue to decrease but not as steeply. The speed curve should suddenly drop and then level off to a new terminal velocity.

8. a. C. Greater because *projected_area* is less, causing *air_friction* to be less, making the absolute values of *total_force*, *acceleration*, *change_in_velocity*, *velocity*, and *speed* more.

b. A. 13 s

c. B. 21 s

References

Aber, James S., and Susan W. Aber. 2005. "High-Altitude Kite Aerial Photography," *Kite Aerial Photography*, April. http://www.geospectra.net/kite/weather/h_altit.htm (accessed December 30, 2012)

Bates, Jim. "The World of Parachutes, Parachuting, and Parachutists." Aero.com .http://www.aero.com/publications/parachutes/9511/pc1195.htm (accessed December 30, 2012)

Benson, Tom. "Earth Atmosphere Model." NASA Glenn Learning Technologies Project, Glenn Research Center. http://www.grc.nasa.gov/WWW/k-12/airplane/atmosmet.html (accessed December 30, 2012)

Col, Jeananda. 2010 "Undersea-Related Inventors and Inventions." EnchantedLearning.com. http://www.enchantedlearning.com/inventors/undersea.shtml (accessed December 30, 2012)

Davies, Stephen. 2012. "CPSC 370 Homework #3: Paratrooper Strike Force." University of Mary Washington. http://rosemary.umw.edu/~stephen/homework3.html. (accessed December 31, 2012)

Diamond, J. 1989. "How Cats Survive Falls from New York Skyscrapers." *Natural History* (August 1989): 20–26.

Elert, Glenn, ed. 2012. "Speed of a Skydiver (Terminal Velocity)." *The Physics Factbook*, 2002. http://hypertextbook.com/facts/JianHuang.shtml (accessed December 30, 2012)

Schecker, H. P. 1996. "Modeling Physics, System Dynamics in Physics Education." *The Creative Learning Exchange* (newsletter), 5(2): 1–8.

Uscher, Jennifer. 1999. "Deep-Sea Machines." NOVA Online, October. http://www.pbs.org/wgbh/nova/tech/deep-sea-submersibles.html (accessed December 30, 2012)

USPA (United States Parachuting Association). 2008. "Skydiving History." http://www.uspa.org/AboutSkydiving/SkydivingHistory/tabid/118/Default.aspx (accessed December 30, 2012)

Weisstein, Eric. 2012. "Eric Weisstein's World of Physics." Wolfram Research. http://scienceworld.wolfram.com/physics/ (accessed December 17, 2012)

Zill, Dennis G. 2013. *A First Course in Differential Equations with Modeling* Applications, 10th ed. Belmont, CA. Brooks-Cole Publishing (Cengage Learning).

MODULE 3.2

Modeling Bungee Jumping

Downloads

The text's website has the following files containing the models in this module available for download with various system dynamics tools: *VerticalSpring* and *Bungee*.

Introduction

On April Fool's Day in 1979, four members of Oxford University's Dangerous Sports Club, dressed in tails and top hats, climbed out onto the Clifton Suspension Bridge in Bristol, U.K. Each attached one end of a nylon-braided, rubber shock cord to himself and the other to the bridge. Then, they jumped off toward the 250-ft Avon Gorge. Voila! The sport of bungee jumping had begun in the western world.

What in the world possessed these men to do such a thing? The story goes that they watched a film on "land divers" from Pentecost Island in the South Pacific and became inspired to try diving themselves.

What are land divers? These divers are the male inhabitants of Pentecost who dive from platforms at various heights along a wooden tower. For these dives, lianas (vines) attached to the tower are tied to their ankles. Divers may be as young as 7 yr of age. Naturally, the lianas have to be selected very carefully. They must be just the right length and elasticity for the height of the platform and the weight of the diver. Consideration must be given to the length of the platform (which collapses and absorbs some shock), the slope of the land, and the swaying of the tower. A perfect dive will have the hair of the diver just brushing the ground. A miscalculation might be fatal. Land diving is part of ceremonies that ensure the yam harvest and fertility. Now, extreme-sports enthusiasts come from all over the world to experience land diving.

How did this practice get started? Why would men choose to jump from platforms with vines tied to their ankles? The annual land dives are based on local lore about a young girl betrothed to a much older man. The frightened young girl, attempting to escape her new husband, climbed high into a banyan tree. The angry husband pursued her up the tree. As he ascended, she tied vines to her ankles and jumped. The husband followed, but without the vines, and was killed. Today's young men may prove that they have learned the escape trick and will not be fooled again (Menz 1993).

Physics Background

The action of a bungee cord is similar to that of a spring. Thus, in modeling bungee jumping we employ a critical law of physics concerning springs, Hooke's law.

Hooke's law pertains to springs that are perfectly elastic, so that they can "spring back" fully. Thus, the law can be applied only as long as the spring has not been stretched too much. The law states that within the elastic limit of the spring, a restoring force (F) applied to a spring is in proportion to and in the opposite direction of the spring's displacement (s) from its equilibrium position. Thus, for a spring constant k, which varies depending on the spring, we have the following formula for the restoring force:

$$F = -ks$$

Figure 3.2.1 illustrates the situations for stretched and compressed springs. When a resting spring is pulled or pushed, a force is exerted to restore the spring. As long as we do not stretch it beyond its elastic limit, the further we pull or push the spring, the more force going in the opposite direction results. For example, as we tug farther on an exercise spring, we feel more resistance.

Hooke's Law Within the elastic limit of a spring, where F is the applied force, k is the spring constant, and s is the displacement (distance) from the spring's equilibrium position, the following formula holds:

$$F = -ks$$

Quick Review Question 1

a. If displacement is in meters, give the units of the spring constant.
b. For a displacement of 0.1 m and a spring constant of 5 kg/s^2, give the restoring force along with its units.

Quick Review Question 2

This question reflects on Step 2 of the modeling process—formulating a model—for a vertical spring's length. Suppose a spring has a weight on the end that we pull down or push up. Consider down as the positive direction. We simplify this first attempt at a model by ignoring friction and the weight of the spring. After completing this question and before continuing in the text, we suggest that you develop a model for the motion of the spring. The next section completes such a model.

a. Give the three lengths that sum to the total length of the spring.
b. Give the force(s).
c. Besides these lengths and forces, give other variables and constants for the model.

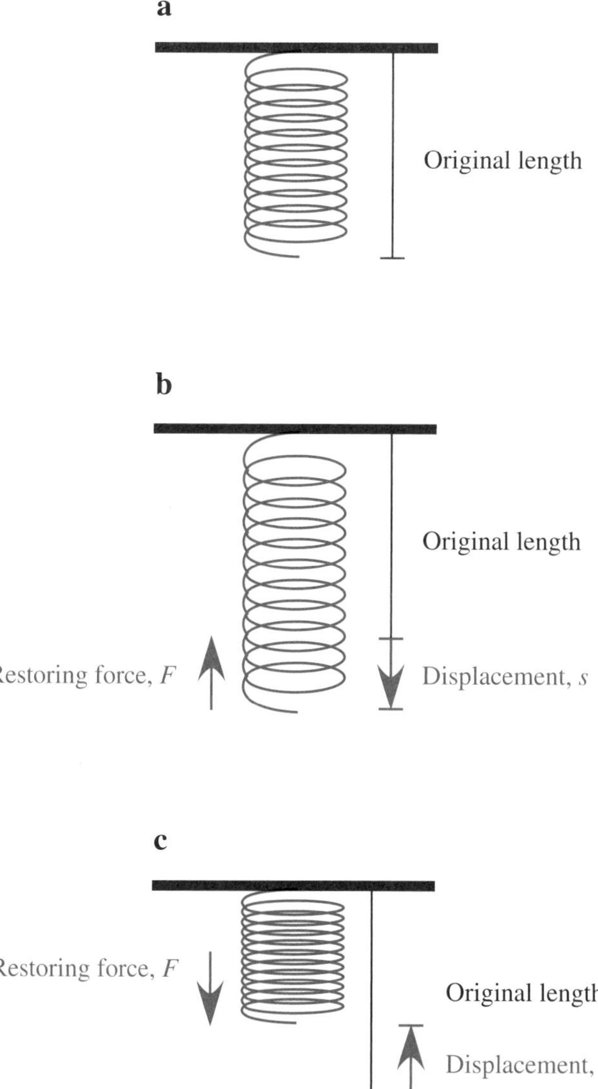

Figure 3.2.1 Original, stretched, and compressed spring

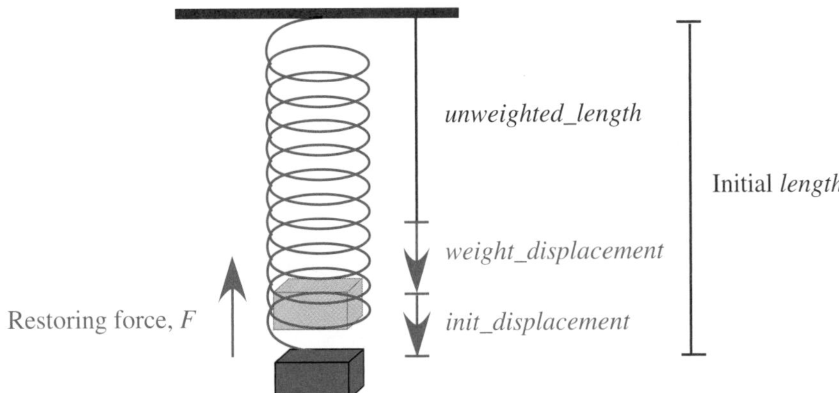

Figure 3.2.2 Vertical spring with attached weight

Vertical Springs

Before modeling bungee jumping, it is helpful to examine the action of a vertical spring, such as in Figure 3.2.2, hanging from a horizontal surface with an attached weight. Because we are considering lengths, having the positive direction be down is convenient. The initial length (*unweighted_length*) of the spring is augmented by a displacement due to the weight (*weight_displacement*) and an additional displacement due to stretching or compressing (*init_displacement*). Thus, we initialize the length of the spring (*length*) to be the following:

unweighted_length + weight_displacement + init_displacement

We enter *unweighted_length* and *init_displacement* as parameters, but a system dynamics tool can calculate *weight_displacement* because the displacement due to weight, which is a force, conforms to Hooke's law, $F = -ks$ or $s = -F/k$. Using the variables of the model, we have the following equation:

weight_displacement = -*weight*/*spring_constant*

A system dynamics diagram for the action of a vertical spring is similar to the diagram of the motion of a ball under the influence of air friction in Figure 3.1.4 of Module 3.1, "Modeling Falling and Skydiving." We change the name of the stock *position* to *length* and replace the section concerning air friction with one involving the force due to Hooke's law. Moreover, because down is positive, *acceleration_due_to_gravity* is +9.81 m/s^2 instead of –9.81 m/s^2, as in the skydiving module. Figure 3.2.3 presents a model diagram of the action of a vertical spring experiencing no friction, or an **undamped vertical spring**, with changes to converters/variables from Figure 3.1.4 in color. With the total displacement at any instant being

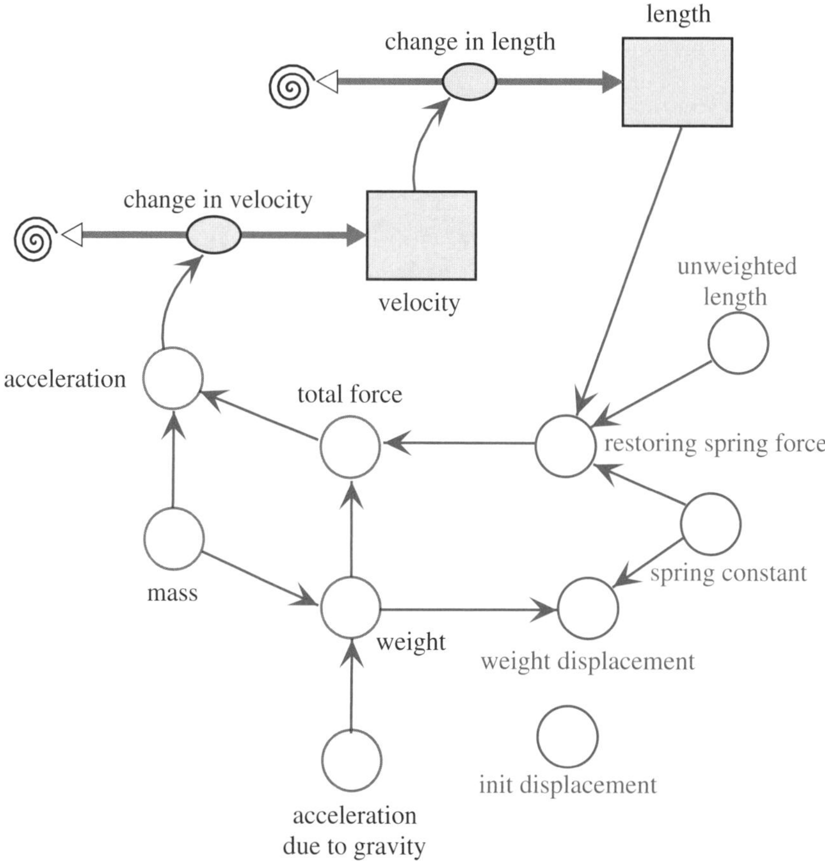

Figure 3.2.3 Model diagram of action of undamped vertical spring

(*length* – *unweighted_length*), the following Hooke's law equation yields the restoring force of the spring:

$$restoring\_spring\_force = -spring\_constant * (length - unweighted\_length)$$

For a simulation, suppose we consider hanging a 0.2-kg mass on the end of a 1-m spring that has a spring constant of 10 N/m. The 0.2-kg mass exerts a force of $F = mg = (0.2 \text{ kg})(9.81 \text{ m/s}^2) = 1.962$ N, its weight. The resulting displacement because of the weight is 1.962 N/(10 N/m) = 0.1962 m = 19.62 cm. Thus, the length of the resting spring with an attached 0.2-kg mass is 1 m + 0.1962 m = 1.1962 m. If we then consider pulling the weight an additional 0.3 m = 30 cm, the simulation with $\Delta t = 0.02$ s produces the graph of the length of the spring in Figure 3.2.4 with an initial length of 1 + 0.1962 + 0.3 = 1.4962 m. Because the spring is undamped, the simulation indicates perpetual oscillations. The length fluctuates from a maximum of 1.4962 m to a minimum of 0.8962 m. The equilibrium point, 1.1962 m, is the midway point between the two extremes and the length of the weighted motionless spring. The extremes of the oscillation are each *init_displacement* = 0.3 m from the

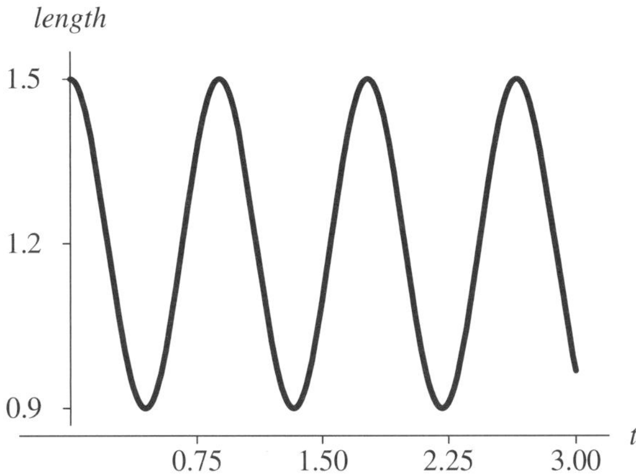

Figure 3.2.4 Graph of length (m) with respect to time (s) in undamped spring

equilibrium point. Such periodic oscillating motion is an example of **simple harmonic motion**, whose definition follows.

Definition A **simple harmonic oscillator** satisfies the following conditions:

1. The system oscillates around an equilibrium position.
2. The equilibrium position is the point at which no net force exists.
3. The restoring force is proportional to the displacement.
4. The restoring force is in the opposite direction of the displacement.
5. The motion is periodic.
6. All damping effects are neglected.

Quick Review Question 3

Suppose a weight of 8 N hangs from a spring with unstretched length of 2 m and spring constant of 100 N/m. Then, the spring is compressed 0.5 m. Using down as the positive direction and ignoring drag, give the following along with units:

a. The displacement caused by the weight
b. The equilibrium position
c. The maximum length
d. The sign of the initial restoring force
e. The restoring force when the length of the spring is 3.15 m

Before refining the model to include friction, we should point out that a small time step, $\Delta t = 0.02$ s, resulted in Figure 3.2.4, a smooth graph that has approximately the same amplitude (distance from midpoint to high point) at each cycle. However, had we picked a time step 10 times larger, $\Delta t = 0.2$ s, then the graph would be jagged with decreasing amplitudes, both of which are artifacts of the simulation instead of reflections of reality. Too large a time step can produce misleading results that can bring about faulty conclusions.

Rule of Thumb Although not always incorrect, we should be suspicious that a time step is too large if a resulting graph is not smooth but has sharp changes in direction with what appears like line segments glued together.

Returning to our model, a damped spring, which does experience friction, also is a simple harmonic oscillator. The diagram is as in Figure 3.2.3, with the addition of another force, **drag**, which becomes part of the total force. Exercise 5 and projects refine the model to include friction. The resulting graph in Figure 3.2.5 shows a damped oscillation with the same period as the corresponding spring with no friction.

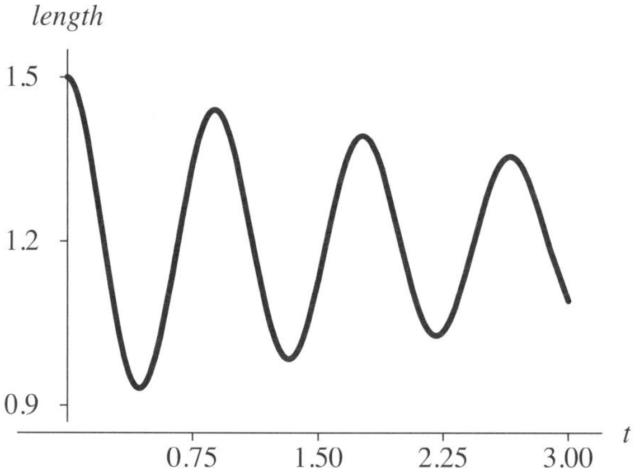

Figure 3.2.5 Graph of length (m) with respect to time (s) in damped spring

Modeling a Bungee Jump

A model of a bungee jump is very similar to that of the weighted spring under the influence of air friction. As a simplifying assumption, we ignore the weight of the bungee cord. Thus, the forces are the weight of the jumper, the restoring force of the cord, and air resistance. The only difference between this model and that of a weighted damped spring is in the equation for *restoring_spring_force*. A vertical spring is fairly rigid, so that when the weight is above the equilibrium point, the spring exerts a restoring force in the opposite direction. However, when the bungee jumper is above the unweighted length, the cord is slack and, we can assume, exerts no downward force. This unweighted length is the length of the original, unstretched cord without the jumper. Thus, the value of *restoring_spring_force* is a conditional expression. If the length is stretched beyond *unweighted_length*, then the restoring spring force obeys Hooke's law with the displacement being (*length – unweighted_length*). When the jumper flies above *unweighted_length*, the bungee cord does not exert such a force. Thus, the value of *restoring_spring_force* has the following logic:

if (*length* > *unweighted_length*) then
  restoring_spring_force ← –*spring_constant* * (*length* - *unweighted_length*)
else
  restoring_spring_force ← 0

Figure 3.2.6 presents a diagram for the motion of a bungee jumper with *total_force* summing *weight*, *air_friction*, and *restoring_spring_force* and with the additional converters/variables from Figure 3.2.3 in color. Figure 3.2.7 displays graphs for the bungee cord's length and the jumper's velocity for a cord with spring constant 6 N/m, unweighted length 30 m, and initial length, or distance from the top of the bridge, of 0 m. For this simulation, the jumper has mass 80 kg (equivalent to a

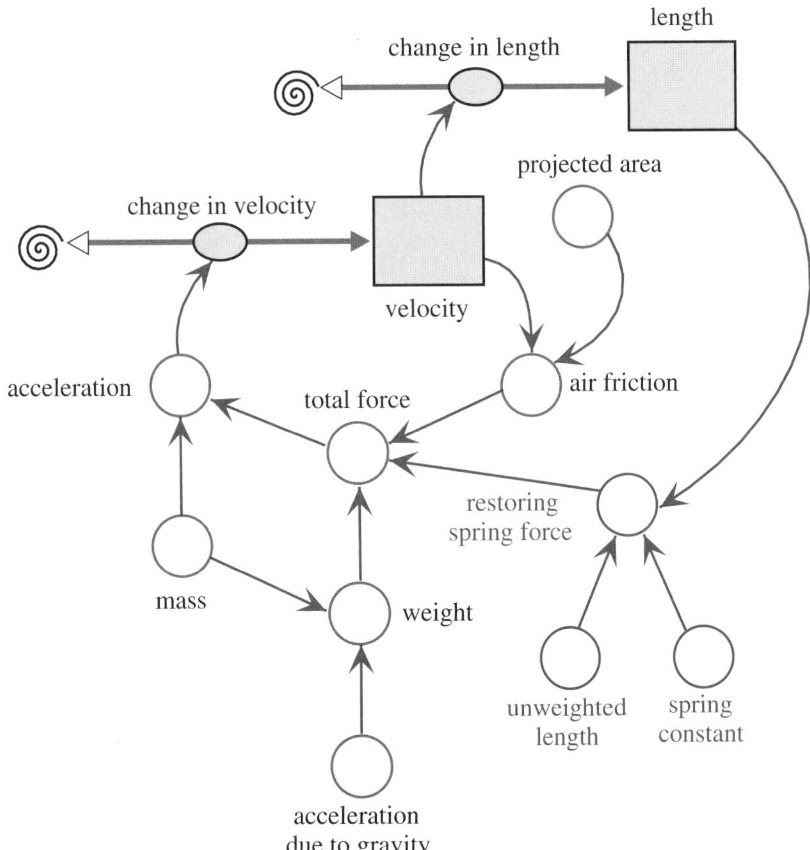

Figure 3.2.6 Diagram for motion of bungee jumper

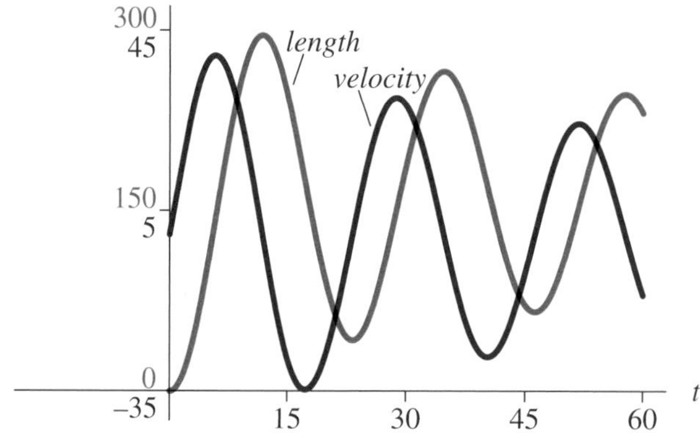

Figure 3.2.7 Graphs of length (m) and velocity (m/s) for bungee jump

weight of about 176.4 lb) and by jumping headfirst, has a small projected area of about 0.1 m^2. The graphs show the simple harmonic oscillation and the damping motion due to drag. One of the projects explores choosing an appropriate bungee cord for a jumper.

Quick Review Question 4

With down being the positive direction, give the formulas for each of the following components of the diagram for a bungee jump (Figure 3.2.6).

a. *total_force*
b. *acceleration*
c. *weight*
d. *acceleration_due_to_gravity* with two digits after the decimal
e. *air_friction* assuming Newtonian friction, sea-level air density, and absolute value function ABS
f. *restoring_spring_force* when *length* is greater than *unweighted_length*
g. *restoring_spring_force* when *length* is less than *unweighted_length*

Exercises

1. Give the changes in the diagram and equations for the model of the damped spring to account for air friction.
2. **a.** Write a differential equation in terms of displacement s to model the motion of an undamped vertical spring with mass m. Recall that weight should equal the spring's restoring force. Use the following variables: s, displacement; t, time; m, mass; k, spring constant.

   b. Using calculus, show that $s(t) = c_1 \cos\left(\sqrt{\frac{k}{m}}t\right) + c_2 \sin\left(\sqrt{\frac{k}{m}}t\right)$ is a solution to the differential equation.

c. From Part b, determine the period of the vibrations.
d. Using $k = 10$ N/m and $m = 0.2$ kg, determine the period of the vibrations. Does your answer agree with the graph in Figure 3.2.4?

3. Write a differential equation in terms of displacement s to model the motion of a damped vertical spring with mass assuming Newtonian friction.

4. Write a differential equation in terms of displacement s to model the motion of a damped vertical spring with mass assuming Stokes' friction.

5. Write differential equations in terms of displacement s to model the motion of a bungee jumper. The system should model two situations, $s \geq$ *unweighted_length* and $s <$ *unweighted_length*.

Projects

For additional projects, see Module 7.8, "Electrical Circuits—A Complete Story."

From the text's website, download a VerticalSpring *file with the model of the motion of an undamped vertical spring for Projects 1–7. Download a* Bungee *file with the model of the motion of a bungee jump for Project 8.*

1. Refine the model of the motion of an undamped vertical spring to account for air friction, and include a graph of *velocity*.
a. Determine if changing any of the following affects the period; and if it does, determine a relationship between the parameter and the period: *init_displacement*, *mass*, *spring_constant*.
b. How does changing *spring_constant* affect the graph of *length*?
c. Give the relationship between the length of the spring and its velocity. For example, when the velocity is zero, what is the position of the spring? At what stage(s) does the spring have a maximum velocity?

2. Refine the model of the motion of a spring to account for air friction. Run an experiment with a real spring and mass to determine the lengths at various times and the period. Estimate the spring constant using a system dynamics model.

3. Refine the model of the motion of a spring to account for drag using Stokes' friction. Suppose b is the constant of proportionality in Stokes' friction; define $u = b/(2m)$ for mass m. Also, define $w^2 = k/m$, where k is the spring constant. Show that when $u^2 > w^2$, with the damping coefficient large in comparison to the spring constant, the system is **overdamped** and displays no oscillation. However, when $u^2 < w^2$, the system is **underdamped** and does show oscillatory behavior.

4. The restoring force for a nonlinear hard spring is $ks(1 + a^2s^2)$, where k is the spring constant, s is the displacement, and a is a small constant. The restoring force for a nonlinear soft spring is $ks(1 - a^2s^2)$. Develop models for such springs. Discuss and compare their motions.

5. Model the motion of an aging spring by replacing the spring constant k with a decreasing function ke^{-at}, where a is a positive constant and t is time.

6. Adjust the equations of the spring model to compute the spring constant given a weight and a corresponding displacement.

7. Model a damped oscillator with parameters very close to those of the bungee jump. Discuss the results in comparison to those of the bungee jump.
8. A bungee jumper wants to have a "great ride," getting close to the ground without hitting it. Suppose the distance of the jumping bridge above a gorge is 80 m and the length of the cord is 30 m. Determine the most appropriate whole number spring constants for jumpers of masses 60 kg, 70 kg, and 80 kg. Employ comparison graphs. Discuss your results.
9. The **buoyancy** of a floating object is the restoring force to return the object to its normal floating layer after a vertical displacement. This force is equal to $-grA^s$, where g is the acceleration due to gravity, r is the fluid density, A is the cross-sectional area of the object, and s is the displacement from the normal floating layer. Design a model for the motion of a displaced object, and discuss the results of the simulation. Let the density of water be 1 g/cm^3.

Answers to Quick Review Questions

1. **a.** N/m or kg/s^2 because $k = -F/s$, s is in m, and F is in N or kg m/s^2
   b. –(5 kg/s^2)(0.1 m) = –0.5 N
2. **a.** resting length of spring with no weight, displacement from that length due to the weight, initial displacement due to pulling down or pushing up the weight
   b. weight and restoring force of spring
   c. Velocity, acceleration, mass, acceleration due to gravity, spring constant; a diagram of the model has the basic form of Figure 3.1.4 for an object falling with friction. Before continuing, we suggest that you revise this figure and include equations to model the motion of a spring, using *length* instead of *position* and ignoring friction.
3. **a.** 8 N/(100 N/m) = 0.8 m
   b. 2 m + 0.8 m = 2.8 m
   c. 2.8 m + 0.5 m = 3.3 m
   d. positive
   e. –(100 N/m)(3.15 m – 2.00 m) = –15 N
4. **a.** *weight* + *air_friction* + *restoring_spring_force*
   b. *total_force*/*mass*
   c. *mass* * *acceleration_due_to_gravity*
   d. 9.81
   e. –0.65 * *projected_area* * *velocity* * ABS(*velocity*)
   f. –*spring_constant* * (*length* – *unweighted_length*)
   g. 0

References

Danby, J.M.A. 1997. *Computer Modeling: From Sports to Spaceflight . . . From Order to Chaos*. Richmond, VA: Willmann-Bell.

Higdon, Don, Bud Rorison, Allen Skinner, and Charlotte Trout. 2001. "Modeling Oscillatory Systems: Physics Component." *Building the Other Models*, Teacher Notes, National Computational Science Leadership Program, July 2001. http://www.ncsec.org/team13/science/buildext.htm (accessed December 17, 2012)

Menz, Paul G. 1993. "The Physics of Bungee Jumping." Bungee.com. Originally from *The Physics Teacher* 31(8). http://www.bungee.com/bzapp/press/pt.html (accessed January 7, 2013)

Weisstein, Eric. 2012. *Eric Weisstein's World of Physics*. Wolfram Research. http://scienceworld.wolfram.com/physics/ (accessed December 17, 2012)

Zill, Dennis G. 2013. *A First Course in Differential Equations with Modeling Applications*, 10th ed. Belmont, CA: Brooks-Cole Publishing (Cengage Learning).

MODULE 3.3

Tick Tock—The Pendulum Clock

Download

The text's website has the file *simplePendulum*, which contains the model in this module, available for download for various system dynamics tools.

Introduction

When we think of a pendulum, many of us think about clocks—grandfather clocks, mantel clocks, kitchen clocks, and the like. We may even become nostalgic for the sound of such a clock in our grandparents' or parents' home. Such clocks had their origin in 1656, when a Dutch scientist, Christiaan Huygens, built a "pendulum" clock. Considering the early date, his first clock was incredibly accurate, losing less than a minute each day (Jespersen and Fitz-Randolph 1999).

Galileo is usually credited with the invention of the pendulum, studying its motion during the sixteenth century. Although a popular myth is that he studied gravity by dropping objects from the top of the Leaning Tower of Pisa, Galileo actually used the motion of a pendulum (Boyd 2002).

Physicists use the pendulum as a classic example of energy conservation. A mass (**bob**) on a string is attached to a pivot point (i.e., a pendulum) and is acted upon by both gravity and tension. The mechanical energy of the mass is affected only by external forces and, therefore, is influenced in this case only by tension. Mechanical energy is equal to the total of kinetic and potential energy and, assuming no friction with the air, is always constant during the oscillation of the pendulum. If you pull up the mass, you increase the potential energy. When you release it, the potential energy decreases as the mass falls, but the kinetic energy increases as the speed of the mass increases (Walker et al. 2010).

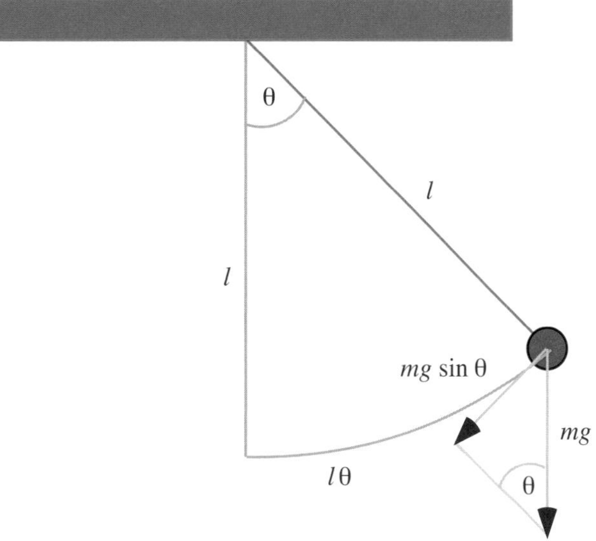

Figure 3.3.1 A simple pendulum

Simple Pendulum

We start our model of a pendulum's motion by considering a **simple pendulum**, which incorporates the following simplifying assumptions: All the mass for the bob is concentrated at a point; the stiff string has no mass; and friction does not exist. Figure 3.3.1 depicts such a simple pendulum with bob of mass m and string of length l being at an angle θ (in radians, rad) off the vertical (Elmer 2011; Weissteins 2003). We consider the angle to be positive when to the right of the vertical and negative when to the left.

As noted in the diagram, the weight of the bob is mg, where $g = 9.8$ m/s^2 is the acceleration due to gravity with down being the positive direction. The only force pulling the bob along an arc in this simple pendulum is the tangential component of the weight, whose magnitude is $mg \sin\theta$. Because the component points to the left when $\theta > 0$ and points to the right when $\theta < 0$, the force acting on the bob is $-mg \sin\theta$.

According to a geometric formula, the arc length from the bob to the vertical is the product of the string length and the angle in radians, $l\theta$. Thus, acceleration along the bob's path, the **angular acceleration**, is the second derivative of this arc length, or angular acceleration = $d^2(l\theta)/dt^2$. Because length is constant for a given pendulum, by calculus, $d^2(l\theta)/dt^2 = ld^2(\theta)/dt^2$. By Newton's second law of motion, the force is equal the mass times the acceleration. Thus, the force pulling the bob along the arc is as follows:

$$\text{force} = (\text{mass})(\text{angular acceleration}) = mld^2(\theta)/dt^2$$

Equating this expression to the negative of the earlier component of weight along the arc, $-mg \sin\theta$, we have the following:

$$ml\,d^2(\theta)/dt^2 = -mg \sin\theta$$

Canceling m and dividing by l, we have the following formula for angular acceleration:

$$\text{angular acceleration} = d^2(\theta)/dt^2 = -g(\sin\theta)/l$$

Initial conditions include specifying that at time $t = 0$, $\theta = \pi/4$ and the initial angular velocity is $d(\theta)/dt = 0$.

Because we cancel m, we see that the mass of the bob is irrelevant for the angular acceleration. According to the formula, for a given acceleration due to gravity at a particular location, the angular acceleration of a simple pendulum depends only on the length of the string and the angle off the vertical.

Quick Review Question 1

This question reflects on Step 2 of the modeling process—formulating a model—for developing a model for a pendulum. We simplify this first attempt at a model by ignoring friction. After completing this question and before continuing in the text, we suggest that you develop a model for a pendulum.

a. Determine variables for the model and their units in the metric system.
b. Which of these variables is the rate of change of angular velocity?
c. What is the flow into a stock (box variable) of angular velocity?
d. Give a differential equation relating angular acceleration ($d^2(\theta)/dt^2$), angle (θ), time (t), and pendulum length (l).
e. What is the value of the flow into a stock/box variable of angle?

Angular acceleration is the rate of change of angular velocity, and angular velocity is the rate of change of the angle, or $d\theta/dt$. With this information, we are in a position to develop a model diagram, which appears in Figure 3.3.2. The stocks (box variables) are *angle*, which has an initial value of *init_angle*, and *angular_velocity*, with an initial value of 0. Angular acceleration (*angular_acceleration*), whose formula developed previously is $-g(\sin\theta)/l$, is in a flow going into *angular_velocity*. The flow into *angle*, *angle_change*, has a value equal to *angular_velocity*. Both *angular_acceleration* and *angle_change* are bidirectional to allow for increasing and decreasing stocks as the pendulum swings back and forth. Moreover, *angular_velocity* and *angle* should be allowed to accommodate negative and positive values. Various underlying equations that accompany this diagram appear in Equation Set 3.3.1.

Equation Set 3.3.1

Various underlying equations to accompany Figure 3.3.2

init_angle = 3.14159/4
length = 1
g = 9.81

angle_change = *angular_velocity*
angle(0) = *init_angle*
angular_acceleration = –*g* * sin(*angle*) / *length*
angular_velocity(0) = 0

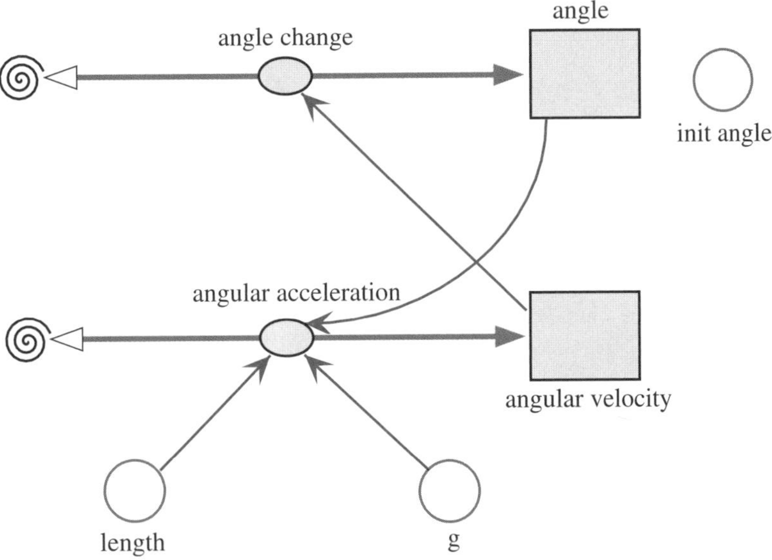

Figure 3.3.2 Model diagram for a simple pendulum

For a string length of 1 m and initial angle of $\pi/4$, Figure 3.3.3 presents a plot of the angle θ, angular velocity, and angular acceleration versus time. With only the force involving weight, the pendulum moves back and forth forever.

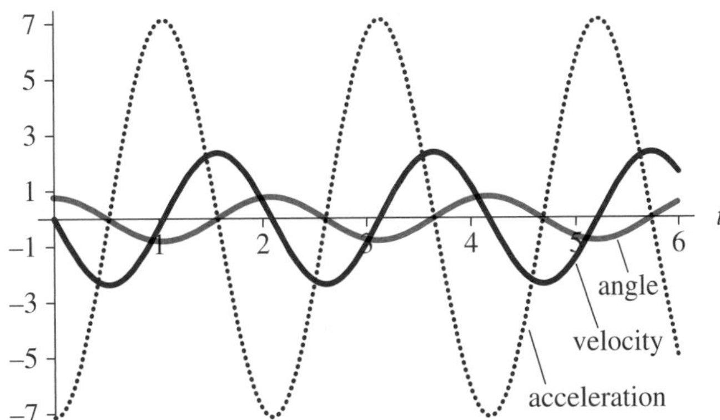

Figure 3.3.3 Plot of angle (rad), angular velocity (rad/s), and angular acceleration (rad/s^2) versus time (s)

Quick Review Question 2

The following questions relate to a simple pendulum with mass of 3 kg, length of 4 m, and initial angle of $\pi/6$.

a. Give the magnitude of the initial tangential component of weight along with its units.
b. Give the initial angular acceleration along with its units.

Linear Damping

In a real pendulum, the motion is damped by friction. As indicated in Module 3.1, a simple model of such effect is Stokes' friction ($F = kv$). Thus, we assume that the damping force is proportional to the angular velocity, $d(\theta)/dt$. In this case of **linear damping**, the model for the forces is as follows, where k is a positive constant:

$$ml\, d^2(\theta)/dt^2 = -mg \sin\theta - kd(\theta)/dt$$

The initial conditions continue to be $\theta = \theta_0$ and $d(\theta)/dt = 0$ when $t = 0$. Projects explore various models of a pendulum's motion that incorporate friction (Danby 1997).

Quick Review Question 3

a. Using linear damping, give the formula for angular acceleration.
b. With linear damping, give the effect (increases, decreases, or no effect) on the amplitude of the angular acceleration of increasing the mass.

Pendulum Clock

As the exercises and projects explore, the angle θ of a pendulum does not affect its **period**, or the length of time to complete a full cycle. Thus, we can use the device in construction of a clock.

We consider the construction of a 60-second clock with only one hand, as in Figure 3.3.4. Falling toward the ground, a **weight** attached to a rope wound around a **drum** provides potential energy to run the clock. The drum also has an attached clock hand so that the hand moves as the drum does. The weight is prevented from falling to the floor immediately by the action of the **pendulum** and a toothed wheel, or **escapement gear**. The shaft of the pendulum bob attaches rigidly to a shaft with an **anchor** that has an associated lever arm. As the pendulum swings in one direction, this attachment moves the anchor, and a tooth of the gear escapes the grasp of a **right** or **left stop** on the anchor, lowering the weight and producing a "tick" sound. Swinging in the opposite direction, the advancing gear hits the other stop with a "tock" sound. Besides regulating the lowering weight, the gear imparts enough of the falling weight's potential energy through the rigid shaft attachment to the pendulum for the latter to overcome friction that dampens its motion. Thus, the pendulum continues

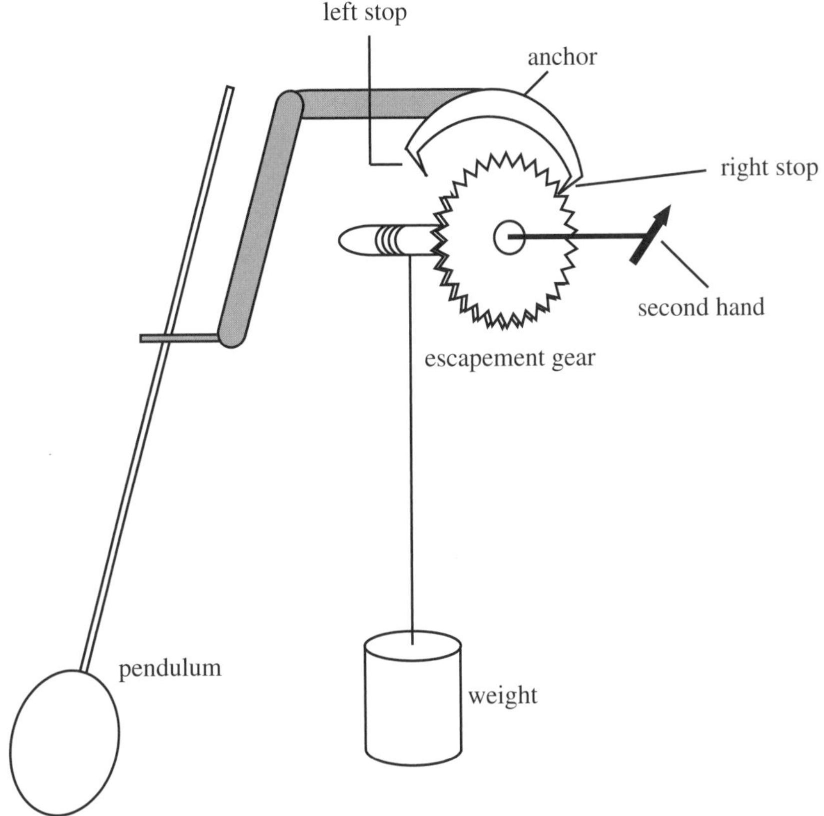

Figure 3.3.4 Pendulum clock with only a second hand

swinging for an extended period of time. Although we do not picture it here, through the interaction of gears, the mechanism also controls the hour and minute hands of the clock. A project models a pendulum clock with a second hand (Brain 2002).

Quick Review Question 4

For each of the following, match the clock part: escapement gear with anchor and stops, pendulum, second hand, stop, weight.

- **a.** Prevents weight from falling to floor
- **b.** Provides potential energy to run clock
- **c.** Regulates timing
- **d.** Transfers energy from weight to overcome friction

Exercises

1. a. Using a computational tool, evaluate $\sin(\theta)/\theta$ for values of θ getting closer and closer to 0.

b. Using your answer from Part a, give an approximation for sin(θ) for small angles θ.

c. Using your answer from Part b, give a formula that is an approximation for the restoring force due to gravity for a simple pendulum.

d. Using your answer from Part c and the formula for length of an arc, give a formula with independent variable of the arc length that is an approximation for the restoring force due to gravity for a simple pendulum.

e. Using your answer from Part a, give a differential equation that is an approximation for the angular acceleration for small angles θ.

f. Determine a, b, and c so that $\theta(t) = a\cos(ct) + b\sin(ct)$ is a solution to the differential equation of Part e with initial conditions $\theta(0) = \theta_0$ and $\theta'(0) = 0$.

2. a. Using the model for the simple pendulum in *simplePendulum* (see "Download") with length being 1, determine the period.

b. Change the length of the string to 4 and determine the period.

c. Change the length of the string to 9 and determine the period in relationship to your answer in Part a.

d. Determine a relationship between the period and the length of string for a given acceleration due to gravity.

3. a. Using the model for the simple pendulum in *simplePendulum* (see "Download") with length being 1, determine the period.

b. Change the acceleration due to gravity to 4×9.8 and determine the period.

c. Change the acceleration due to gravity to 9×9.8 and determine the period in relationship to your answer in Part a.

d. Determine a relationship between the period and the acceleration due to gravity for a given string length.

4. a. Using the model for the simple pendulum in *simplePendulum* (see "Download") with length being 1 and the acceleration due to gravity being 9.81 m/s^2, determine the period.

b. Determine a formula for period by using the answers from Part a, Exercise 2d, and Exercise 3d.

5. Using the model for the simple pendulum in *simplePendulum* (see "Download") with length being 1, determine the periods for different initial angles. Does the angle have any effect on the period?

6. Write a description of what the three graphs in Figure 3.3.3 show. Describe the relative phases of the three curves. Is it reasonable that the magnitude of the angular velocity is greatest when the angle is zero? Is it reasonable that the angular acceleration is greatest when the angle is least, and vice versa?

Projects

1. According to the **conservation of energy principle**, with only conservation forces in effect, the sum of a particle's potential and kinetic energies is constant throughout motion. The formula for potential energy is mgh and for kinetic energy is $0.5mv^2$, where m is mass, g is acceleration due to gravity, h is height, and v is velocity. Adjust the simple pendulum model in *simplePendu-*

lum (see "Download") to illustrate the Conservation of Energy Principle. Note that h is the y-value of the bob.

2. Develop a model of the pendulum assuming damping as described in the section "Linear Damping." Determine the period. With the model and with mathematics, show the effect of increasing mass on the resistance. Does the period change as the amplitude diminishes?
3. Develop a model of the pendulum assuming constant-magnitude dry friction whose sign is opposite that of $d(\theta)/dt$.
4. Develop a model of the pendulum assuming dry friction whose magnitude is greater at angular velocities closer to zero.
5. Develop a model of a pendulum clock that completes a cycle in 1 s. Assume linear damping as modeled in Project 2. Have an impulse change the angular velocity by an appropriate fixed amount at the bottom of a swing in one direction. If available in your system dynamics tool, a delay function that returns the value of an argument in the previous time step might be helpful in determining when *angle* changes sign. Approximately how long does your model run before the clock "runs down"?
6. Develop a model of a pendulum clock. Assume dry friction as modeled in Project 4. Have an impulse change the kinetic energy or equivalently, the angular acceleration, by an appropriate fixed amount (see Project 5). Approximately how long does your model run before the clock "runs down"?

Answers to Quick Review Questions

1. **a.** time in s; length of pendulum in m; angle in rad; angular velocity, perhaps in rad/s; angular acceleration in rad/s^2
   b. angular acceleration
   c. angular acceleration
   d. angular acceleration = $d^2(\theta)/dt^2 = -g(\sin\theta)/l$
   e. angular velocity
2. **a.** $|mg \sin(\theta)| = (3)(9.81)\sin(\pi/6)$ kg m/s^2 = $(3)(9.81)(0.5)$ N = 14.7 N
   b. $-g(\sin\theta)/l = -(9.81)\sin(\pi/6)/4$ m/s^2 = -1.23 m/s^2
3. **a.** $d^2(\theta)/dt^2 = -g(\sin\theta)/l - (k\cdot d(\theta)/dt)/(ml)$
   b. increases
4. **a.** escapement gear with anchor and stops
   b. weight
   c. pendulum
   d. escapement gear with anchor and stops

References

Boyd, Thomas M. 2002. "Pendulum Measurements." Colorado School of Mines, 2002. http://galitzin.mines.edu/INTROGP/notes_template.jsp?url=GRAV/NOTES/pend.html&page=Gravity: Notes: Pendulum Measurements (accessed December 17, 2012).

Brain, Marshall. 2002. "How Pendulum Clocks Work." HowStuffWorks, Inc. http://science.howstuffworks.com/clock.htm (accessed December 17, 2012)

Danby, J.M.A. 1997. *Computer Modeling: From Sports to Spaceflight . . . From Order to Chaos*. Richmond, VA: Willmann-Bell, p. 408.

Elmer, Franz-Josef. 2011. "The Pendulum Lab: The Lecture Room." http://www.elmer.unibas.ch/pendulum/lroom.htm (accessed December 17, 2012)

Jespersen, James, and Jane Fitz-Randolph, *From Sundials to Atomic Clocks: Understanding Time and Frequency*, 2nd (revi.) ed., Mineola, New York: Dover Publications, 1999. available at http://tf.nist.gov/general/pdf/1796.pdf (accessed December 17, 2012)

Walker, Jearl, David Halliday, and Robert Resnick. 2010. *Fundamentals of Physics*, 9th ed. Section 8.5, "Conservation of Mechanical Energy." New York: Wiley.

Weisstein, Eric. "Simple Pendulum." *Eric Weisstein's World of Physics*. Wolfram Research, Inc., 2003. http://scienceworld.wolfram.com/physics/SimplePendulum.html (accessed December 17, 2012)

MODULE 3.4

Up, Up, and Away—Rocket Motion

Download

The text's website has the following file containing a framework for the model in this module available for download for various system dynamics tools: *Rocket*.

Introduction

"Of course Peter had been trifling with them, for no one can fly unless the fairy dust has been blown on him."
—J. M. Barrie, *The Adventures of Peter Pan*

Human beings have long looked to the sky with a yearning to fly and have long experimented with various methods and contrivances to accomplish this goal. Around AD 1500, a mandarin named Wan-Hu attempted to fly to the moon in a wicker chair to which were attached 47 "rockets"—actually 47 bamboo tubes filled with black powder (Dvir 2003; Lethbridge 2000). Unfortunately, the innovator was unable to conduct other experiments with the potential of rockets to propel human beings into the sky. The successful launching of human beings into space would have to wait several centuries. Now, the launching of rockets with or without human beings is quite an ordinary event. The space above earth is littered with various types of satellites placed into orbit by rockets.

The Chinese generated a form of black powder, or "gun powder," during the first century AD from charcoal, saltpeter, and sulfur (Lethbridge 2000). Initially, they used this powder for fireworks; but sometime around the year AD 1000, they adapted this powder for use in fire arrows. These fire arrows were made by attaching powder-filled bamboo tubes to arrows and launching them with a bow. By 1232, they had modified these arrows by fixing tubes, open at one end and capped at the other end, to long sticks. They lit the powder, and the first true rockets were launched toward their Mongol attackers. The tips of these rockets were coated with either flammable materials or poison. How effective these rockets were as weapons is questionable, but the Chinese successfully warded off these invaders. Furthermore, the Mongols developed their own rockets and helped to spread their use to Europe. In fact, the word rocket probably originated from an Italian word "rochetta," coined by Muratori in his description of fire arrows used in medieval times (Lethbridge 2000).

From the time of the Chinese fire arrows, rockets have continued to play important military roles. During the last half of the twentieth century, however, rockets have taken on new roles in the exploration of the universe. Currently, satellites carried by rockets are providing us with three-dimensional views of polar ice sheets to give us insight into climate and its effects on life. Rockets have launched space telescopes and propelled probes to Mars, to the edge of our solar system, and beyond. More than 400 people have traveled into space borne by rockets since 1961 (NASA).

Physics Background

Before embarking on our development of a model of rocket motion, we need to consider some of the physics fundamentals. We have already worked extensively with Newton's second law, $F = ma$, where F is a force acting on an object of mass m and imparting an acceleration, a (see the section "Physics Background" from Module 3.1, "Modeling Falling and Skydiving"). In that same section, we discussed drag on an object.

With rockets, we consider another mechanical force, **thrust**. A rocket's engine generates thrust through the acceleration of a mass of gas through the bottom, propelling the rocket in the opposite direction. Thus, the forces obey **Newton's third law of motion**: "for every action, there is an equal and opposite reaction." The concept for a rocket is the same as that of a filled and released balloon, where expelled gas under pressure causes the balloon to fly around the room.

Definition **Thrust** is a mechanical force caused by the acceleration of a mass of fluid and in the opposite direction to flow.

Suppose c is the velocity of the gas relative to the rocket and v is the velocity of the rocket, so that $c + v$ is the velocity of the gas in space. If m is the mass and up is positive, then the **thrust** of the engine (T) is as follows:

$$T = c\frac{dm}{dt}$$

Over a period of time Δt, we have the following discrete version:

$$T = c\frac{\Delta m}{\Delta t}$$

or

$$T\,\Delta t = c\,\Delta m$$

Quick Review Question 1

a. Select all appropriate units of measure for thrust: kg, kg m/s^2, kg/s^2, m/s^2, mi/h, N, N s, lb, lb/s^2, s.
b. With up being positive, suppose a rocket is traveling up at a speed of 500 m/s, and the speed of the downward gas is 800 m/s. Give the value of c.
c. Suppose over a period of 0.1 s, 2 kg of propellant burns. Give the engine thrust.

As rocket fuel burns, the mass of the fueled rocket decreases. From time t to time $t + \Delta t$, the **impulse** is the product of the thrust and Δt, as follows:

$$\boldsymbol{I = T\,\Delta t}$$

During that period, the **specific impulse** (I_{sp}) is the impulse per newton (or pound) of fuel, or the quotient of impulse and the weight of the burned fuel, Δw.

$$\boldsymbol{I_{sp} = \frac{I}{\Delta w} = \frac{T\,\Delta t}{(\Delta m)g}}$$

Solving for T, we have the following value of thrust from time t to time $t + \Delta t$:

$$T = I_{sp}g\frac{\Delta m}{\Delta t}$$

Letting Δt approach 0, we have the equivalent derivative form:

$$T = I_{sp}g\frac{dm}{dt}$$

Definitions An **impulse** is the product of the thrust and the length of time. **Specific impulse** is the impulse per unit weight of burned fuel, or the quotient of impulse and the change in the fuel's weight.

As with earlier models of motion, our model of the motion of a rocket incorporates acceleration. In this case, we wish to have a formula for acceleration due to thrust. Because thrust is a force, for acceleration a we have the following equation by Newton's second law:

$$T = ma$$

Substituting for T and solving for acceleration, the following is true:

$$I_{sp}g\frac{dm}{dt} = ma$$

or

$$a = \frac{I_{sp}g\frac{dm}{dt}}{m}$$

Quick Review Question 2

a. Select all appropriate units of measure for impulse: kg, kg m/s^2, kg/s^2, m/s^2, mi/h, N, N s, lb, lb/s^2, s.
b. Suppose a fuel burning for 2 s imparts a thrust of 75 N to a rocket. Give the impulse.
c. Select all appropriate units of measure for specific impulse: kg, kg m/s^2, kg/s^2, m/s^2, mi/h, N, N s, lb, lb/s^2, s.
d. Suppose 0.5 kg of the fuel of Part b burns during 2 s. Give the specific impulse.

System Dynamics Model

The model of the motion of a ball tossed into the air in Figure 3.1.1 of Module 3.1, "Modeling Falling and Skydiving," serves as a basis for the rocket-motion model. For the extension, we make several assumptions:

- The only forces acting on the rocket are gravitation and thrust derived from burning fuel.
- Acceleration due to gravity is constant.
- The earth is flat.
- The rocket is vertical.
- The rocket has only one stage.

Quick Review Question 3

This question reflects on Step 2 of the modeling process—formulating a model—for developing a model for rocket motion. We employ the preceding simplifying assumptions. After completing this question and before continuing in the text, we suggest that you develop a model for rocket motion.

a. Building on the model in the section "Acceleration, Velocity, and Position" of Module 3.1, "Modeling Falling and Skydiving," determine additional variables for the rocket-motion model and their units in the metric system.
b. Give a differential equation for change in total mass (*dm/dt*) as a function of the constants mass of initial unburned fuel (*f*) and time to burn (*b*). Use the simplifying assumption that *dm/dt* is constant.
c. Give a differential equation for acceleration, or change in velocity (*dv/dt*), in terms of total mass (*m*), change in total mass (*dm/dt*), specific impulse (I_{sp}), and acceleration due to gravity (*g*).

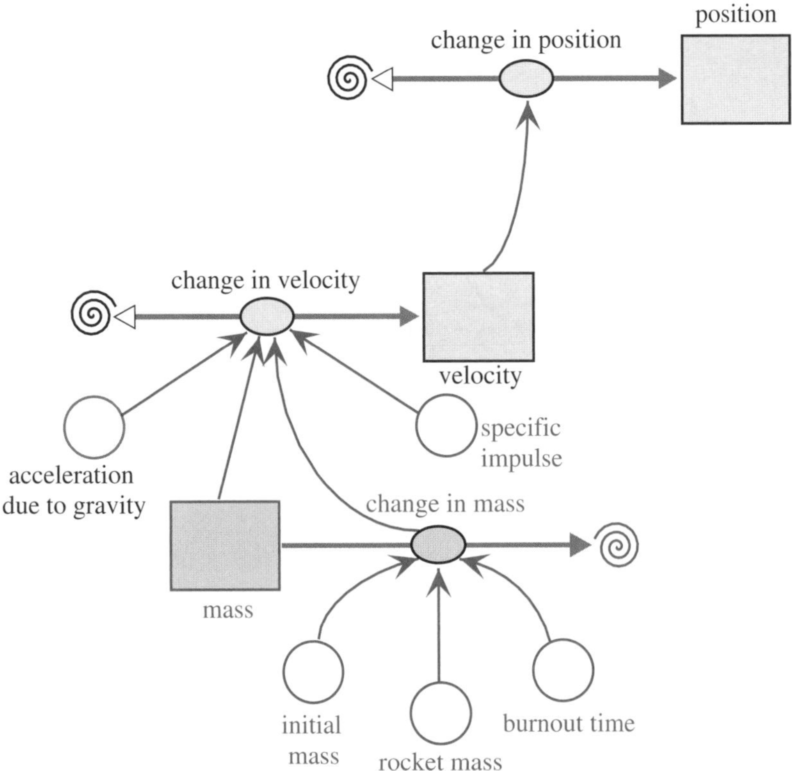

Figure 3.4.1 Model diagram of a rocket's motion

Extending the first model in "Modeling Falling and Skydiving," in which we assume no friction, thrust from burning fuel also impacts the motion of a rocket. The *change_in_velocity*, or acceleration, now involves acceleration due to this thrust as well as acceleration due to gravity. The extended model has an additional stock (box variable), *mass*, that contains the total mass of the fuel and rocket, which has mass *rocket_mass*. This stock has an initial value of *initial_mass*. We assume that while fuel is present, the flow out, *change_in_mass*, is constant and consists of the mass of the initial unburned fuel divided by the time for it to burn (*burnout_time*). After burnout, *change_in_mass* becomes zero. Figure 3.4.1, which is similar to Figure 3.1.1, contains a model diagram of a rocket's motion.

Quick Review Question 4

Refer to Figure 3.4.1 to give the formulas for the following:

a. The mass of the unburned fuel
b. The change in mass per unit of time of rocket with fuel

Figure 3.4.2 displays a graph of position (in color) and velocity versus time when *initial_mass* = 5000 kg, *rocket_mass* = 1000 kg, *burnout_time* = 60 s, and *specific_impulse* = 200 s. The graph shows the velocity increasing more and more until the

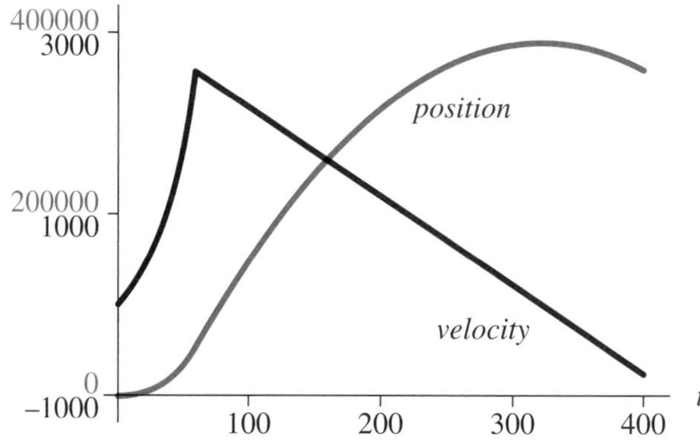

Figure 3.4.2 Graph of position (m) and velocity (m/s) versus time for a rocket

fuel completely burns. With a velocity of about 2567 m/s = 2.567 km/s at that instant, the rocket continues to rise to a height of about 388,500 m = 388.5 km before starting its descent. Various projects complete and expand upon the model in Figure 3.4.1.

Exercises

1. Write the differential equations in the model developed in this module.
2. Revise the model's differential equation for acceleration (dv/dt) to accommodate Newtonian friction (see the section "Physics Background" from Module 3.1, "Modeling Falling and Skydiving"). The following formula approximates the density of Earth's atmosphere:

$$D = 1.225e^{-0.1385y} \text{ kg/m}^3 \text{ for altitudes } y < 100 \text{ km}$$

The coefficient of drag for a very shiny rocket could be as low as 0.6, while rougher surfaces command higher values closer to 1.

3. Revise the model's differential equation for acceleration (dv/dt) where acceleration due to gravity is not constant but decreases with altitude according to the formula $g\frac{R^2}{(R+y)^2}$, where R is the radius of the earth, which is approximately 6.378×10^6 m, g is the acceleration due to gravity at sea level, and y is the distance of the rocket above the earth's surface. Continue to use $c = I_{sp}\, g$. The next exercise develops the formula for acceleration due to gravity at altitude y.
4. This exercise develops the formula for acceleration due to gravity at altitude y from the previous exercise.
   a. Newton's gravitation law states that the gravitation force between two objects of masses m_1 and m_2 is as follows:

$$F = G\frac{m_1 m_2}{d^2}$$

where G is the universal gravitational constant (6.672×10^{-11} N m^2 kg^{-2}) and d is the distance between the centers of mass of the objects. Let R be the radius of the earth, m_e its mass, and m the mass of an object. Write Newton's gravitation law for the weight of the object at the earth's surface using these masses.

b. Write the weight of the object at the earth's surface using Newton's second law and g.

c. Setting Parts a and b equal, solve for g, and simplify.

d. Consider the object at an altitude y above the surface of the earth. Write Newton's gravitation law for the weight of the object at height y.

e. Write the weight of the object using Newton's second law and g_y, the acceleration due to gravity at altitude y.

f. Setting Parts d and e equal, solve for g_y and simplify.

g. Evaluate the ratio g_y/g and simplify.

h. Solve for g_y.

5. National Association of Rocketry (NAR) codes, such as C6-3, appear on hobby rocket motors, as follows:

- The letter, which can be from 1/2A to K, specifies the total impulse, with C indicating 5.01 to 10.00 N s. A letter's range is double that of its predecessor, so that an impulse in the range of 1/2A is the smallest, and that in the range of K is the largest. Thus, the total impulse for rockets with letter B is from 2.51 to 5.00 N s.
- The subsequent number, such as 6, indicates **average thrust** (in newtons). The average thrust with total impulse indicates the length of time over which the motor releases its total energy.
- The number after the dash, such as 3, gives **time delay** (in seconds), or the time from motor burnout until activation of a recovery parachute. During that time, the rocket coasts to a higher altitude and slows.

For the following questions involving total impulse, use the higher range value, such as 10.00 N s for type-C motors.

a. Approximate the length of time for which a C6-5 motor delivers its energy.

b. Repeat Part a for a C4-5 motor.

c. Repeat Part a for a C10-5 motor.

d. Repeat Part a for a C5-3 motor.

e. Which of the preceding three engines is most powerful?

f. The C5-3 burns 12.7 g of propellant. Calculate its specific impulse.

Projects

1. Complete the model of rocket motion described in this module and begun in *Rocket* (see "Download"). Plot position (altitude) and velocity with respect to time. Obtain the maximum altitude and velocity. Try various parameter values, such as those for a hypothetical rocket with initial mass of 5000 kg, rocket mass of 1000 kg, burnout time of 60 s, and specific impulse of 200. Also, consider values of real engines, such as those in Table 3.4.1 with code

Table 3.4.1
Rocket Engine Specifications (also see Exercise 5; Culp)

Engine Type	*Maximum Lift (g)*	*Initial Mass (g)*	*Propellant Mass (g)*
A10-3T	141.5	7.9	3.78
C5-3	226.4	25.5	12.7
C6-3	113.2	24.9	12.48
D12-5	283.0	43.1	24.93

Table 3.4.2
Comparison of Chemical and Solar Electric Engines, Each with Total Impulse = 6×10^7 N s (Finke 1980)

Engine	I_{sp} *(s)*	*T (N)*	*Burn Time (s)*	*Propellant Mass (kg)*
Centaur (chemical)	440	66,000	880	13,600
Electrostatic Thruster	3,000	1	5×10^7	2,000

information (see Exercise 5). Write a discussion of the results. Augment your work by having a comparative plot of altitude and velocity versus time for various rocket masses. Discuss the impact of various rocket masses on the altitude and velocity at burnout. Similarly, investigate the impact of various specific impulses.

2. Complete the model of rocket motion described in this module and begun in *Rocket* (see "Download"). Use your model to compare a Centaur Upper Stage System, which is a chemical system, and the Electrostatic Thruster System, which utilizes the sun's nuclear energy as well as a propellant (see Table 3.4.2). Your comparison should include maximum velocity, propellant mass, thrust time, and the types of missions advisable for each (Finke 1980).
3. The first model assumption before was to ignore the effects of drag. In this project, refine the model to accommodate Newtonian friction (see Exercise 2). Investigate the impact of drag on altitude and speed at burnout. Discuss the results, including the reasonableness of the assumption to ignore drag.
4. One assumption before was that acceleration due to gravity is constant. In this project, refine the model to consider decreasing acceleration due to gravity as the rocket gains altitude according to the formula in Exercise 3. Investigate the impact on altitude and speed at burnout. Discuss the results, including the range of altitudes at which the assumption that acceleration due to gravity is constant seems reasonable.
5. Develop a model of a single-stage rocket in which after burnout and a time delay a parachute deploys so that the rocket falls safely to earth (see Module 3.1, "Modeling Falling and Skydiving").
6. Develop a model for a two-stage rocket. Each stage has an engine with propellant. After the initial burn, the rocket coasts for a few seconds before second stage ignition occurs. Discuss the advantages and disadvantages of such an arrangement.

Answers to Quick Review Questions

1. **a.** kg m/s^2, N, lb
   b. –300 m/s
   c. $c\ \Delta m/\Delta t = (-300)(2)/(0.1) = -6000$ N
2. **a.** N s
   b. $I = T\ \Delta t = (75)(2) = 150$ N s
   c. s
   d. $I_{sp} = I/\Delta w = I/(\Delta m\ g) = (150 \text{ N s}) / ((0.5 \text{ kg})(9.81 \text{ m/s}^2)) = 30.6$ s
3. **a.** rocket mass in kg, total mass of rocket and fuel in kg, change in total mass in kg, time for fuel to burn in s, specific impulse in s
   b. $dm/dt = f/b$
   c. Acceleration is the sum of acceleration due to gravity and acceleration due to thrust. Thus, $a = dv/dt = \dfrac{I_{sp} g \dfrac{dm}{dt}}{m} + g$
4. **a.** *initial_mass – rocket_mass*
   b. *(initial_mass – rocket_mass)/burnout_time*

References

Culp, Randy. "Rocket Simulations." http://my.execpc.com/~culp/rockets/rckt_sim.html (accessed December 17, 2012)

———. "East Coast Model Center Engine Specifications." *Estes Catalog*. http://my.execpc.com/~culp/rockets/estes_spec.html (accessed December 17, 2012)

Danby, J.M.A. 1997. *Computer Modeling: From Sports to Spaceflight . . . From Order to Chaos*. Richmond, VA: Willmann-Bell.

Dvir, Tal. 2003. "Four Decades After Gagarin, China Finally Reaches for the Stars." *Telegraph Newspaper Online*, May 10. http://www.telegraph.co.uk/news/worldnews/asia/china/1443319/Four-decades-after-Gagarin-China-finally-reaches-for-the-stars.html (accessed December 17, 2012)

Finke, Robert C. 1980. "Electric Propulsion Technology," NASA Lewis Research Center. http://ntrs.nasa.gov/archive/nasa/casi.ntrs.nasa.gov/19800022946_1980022946.pdf (accessed January 1, 2013)

Lethbridge, Cliff. 2000. "History of Rocketry Chapter 1: Ancient Times Through the 17th Century." Spaceline. http://www.spaceline.org/history/1.html (accessed December 17, 2012)

NASA Aeronautics Resources. "Brief History of Rockets." NASA. http://www.grc.nasa.gov/WWW/K-12/TRC/Rockets/history_of_rockets.html (accessed December 17, 2012)

National Association of Rocketry. "Standard Motor Codes." http://www.nar.org/NARmotors.html (accessed December 17, 2012)

Ratliff, Lisa, Larry Storm, James Blattman, and Gae McGibbon. "Rocketry: Modeling and Models." Computational Science in Education, National Computational Science Education Consortium. http://www.ncsec.org/cadre2/webview.cfm?menu=1.1&team=14&cadre=2 (accessed December 17, 2012)

Stillwater, Ryan A. 2003. "Spacecraft Propulsion—Electric" in "Visualization of a Micro-Electric Thruster." Class Project. ESM4714 Scientific Visual Analysis and Multimedia. College of Engineering. Virginia Tech. http://www.sv.vt.edu/classes/ESM4714/Student_Proj/class03/stillwater/prj/background/sp_electric.htm (accessed January 7, 2013)

Weisstein, Eric. 2003. *Eric Weisstein's World of Physics.* Wolfram Research. http://scienceworld.wolfram.com/physics/ (accessed December 17, 2012)

4

SYSTEM DYNAMICS MODELS WITH INTERACTIONS

MODULE 4.1

Competition

Download

The text's website has available for download for various system dynamics tools the file *sharkCompetition*, which contains a submodel for this module, available for download for various system dynamics tools.

Community Relations

In any population of organisms, the individuals are interacting with each other and with their environment. Populations, which are made up of only one species, are also interacting with other species in a particular area in what we term a **community**. These interactions influence the composition and dynamics of the community through time. Some of these interactions are robust, while others are not so robust or are even very weak. The magnitude of these interactions depends on the extent of their niche overlap. An **ecological niche** can be defined as the complete role that a species plays in an ecosystem. The more overlap two species have, the stronger the interaction will be. Two of these interactions between species are competition and predator-prey relationships.

Introduction to Competition

Everyone is familiar with competition. We compete for attention in families, for grades in school, for jobs and promotions, for parking spaces, and on and on. Competition is integral to most economic activity. Through competition in human societies, wages and prices are set; quantities and types of products manufactured are selected; businesses succeed or fail; and resources are distributed. Economic and social competition may occur even in noncapitalist systems.

More broadly, competition is a basic characteristic of all communities, human and nonhuman. It may occur within a population of the same species (**intraspecific**), like the human species, or it may occur between populations of different species (**interspecific**). Competitive interactions affect species distribution, community organization, and species evolution.

Simply speaking, **competition** is the struggle between individuals of a population or between species for the same limiting resource. If one individual (species) reduces the availability of the resource to the other, we term that type of competition **exploitative,** or **resource depletion**. This interaction is indirect and may involve removal of the resource or denial of living space. If there is direct interaction between individuals (species), where one interferes with or denies access to a resource, we term that competition **interference**. In this form, there may be physical contests for territory or resource. Interference may also, as in some plants, involve the production of toxic chemicals.

Modeling Competition

Sometimes two species are not eating each other but are competing for the same limited food source. For example, whitetip sharks (WTS) and blacktip sharks (BTS) in an area might feed on the same kinds of fish in a year when the fish supply is low. We anticipate that a large increase in one species, such as BTS, might have a detrimental effect on the ability of the other species, such as WTS, to obtain an adequate amount of food and, therefore, to thrive. Also, we expect that superior hunting skills of one species would diminish the food supply for the other species. As one species grows, the other shrinks, and vice versa.

In an unconstrained growth model (see Module 2.2, "Unconstrained Growth"), which ignores competition and limiting factors, we consider a population's (P) births to be proportional to the number of individuals in the population (r_1P) and its deaths to follow a similar proportionality (r_2P). Thus, in this model, the rate of change of the population is $dP/dt = r_1P - r_2P = (r_1 - r_2)P$, so that the solution is an exponential function, $P = P_0e^{(r_1 - r_2)t}$.

However, with competition, a competing species has a negative impact on the rate of change of a population. In this situation, we can model the number of deaths of each species as being proportional to its population size and the population size of the other species. Thus, for B being the population of BTS and W the population of WTS, the number of deaths of each species is proportional to the product BW. Moreover, the constant of proportionality associated with this proportionality for one species reflects competitive skills of the other species. (Projects explore various types of competition.) Consequently, we have the following equations for the change in the number of deaths of each species:

$$\Delta(\text{deaths of WTS}) = wBW, \text{ where } w \text{ is a WTS death proportionality constant}$$

$$\Delta(\text{deaths of BTS}) = bWB = bBW, \text{ where } b \text{ is a BTS death proportionality constant}$$

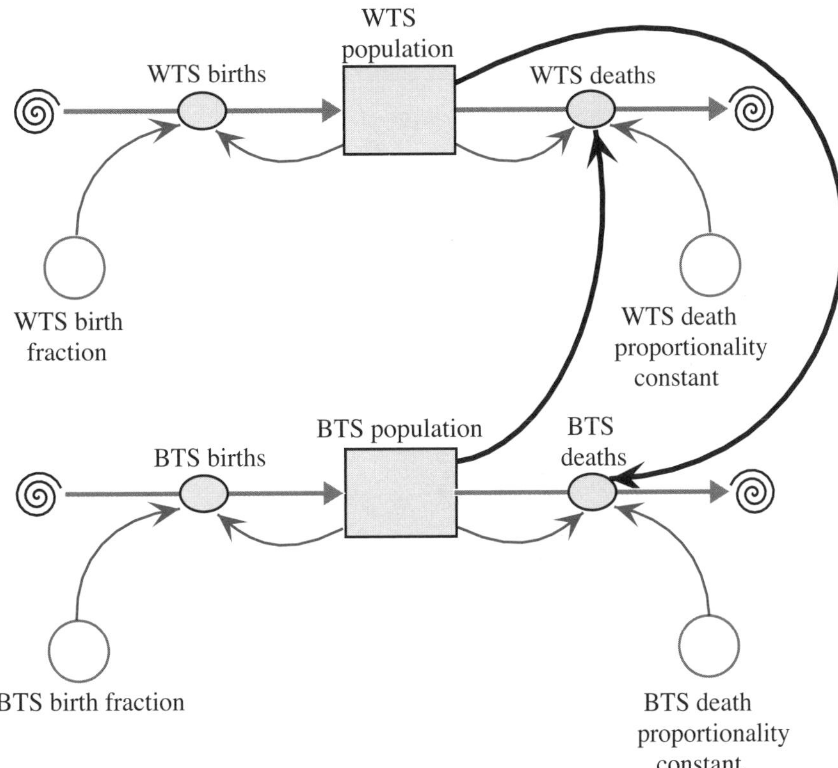

Figure 4.1.1 Model diagram of competition of species

Equation Set 4.1.1

Some equations to accompany Figure 4.1.1 with basic unit of time being 1 month

$BTS\_population(0) = 15$
$BTS\_birth\_fraction = 1$
$BTS\_births = BTS\_birth\_fraction * BTS\_population$
$BTS\_death\_proportionality\_constant = 0.20$
$BTS\_deaths = (BTS\_death\_proportionality\_constant * WTS\_population) * BTS\_population$
$WTS\_population(0) = 20$
$WTS\_birth\_fraction = 1$
$WTS\_births = WTS\_population * WTS\_birth\_fraction$
$WTS\_death\_proportionality\_constant = 0.27$
$WTS\_deaths = (WTS\_death\_proportionality\_constant * BTS\_population) * WTS\_population$

Figure 4.1.1 illustrates the interaction with the number of each species of shark affecting the deaths of the other species. With the basic unit of time being a month,

Equation Set 4.1.1 gives some of the equations and constants, which in this case models births as being unconstrained. The set of numbers serve as an example and, although realistic, do not represent any actual population. Typically, a computational scientist uses actual field data to establish reasonable parameters for a model.

Quick Review Question 1

This question reflects on Step 2 of the modeling process—formulating a model—for developing a model for competition. As before, let W be the number of WTS and B the number of BTS. We simplify this model by assuming unconstrained births. After completing this question and before continuing in the text, we suggest that you develop a model for competition.

a. Give an equation for WTS births.
b. Give an equation for WTS deaths.

Quick Review Question 2

If all other parameters are equal and the WTS death proportionality constant (w) is larger than the BTS death proportionality constant (b), which population should be larger after a few time steps?

A. WTS B. BTS C. Impossible to determine

Note that in this hypothetical example, the death proportionality constants (0.2 and 0.27) are much smaller than the birth fractions (1 and 1). The former constants are multiplied by product of the two populations, BW, potentially a very large number, while the later are multiplied by their respective populations, B or W. For birth fractions of 1, each type of shark gives birth to approximately one pup each month. With *BTS_population*(0) being 15 and *WTS_population*(0) being 20, initial predictions are for about 15 BTS and 20 WTS to be born in the first month. Should a death proportionality constant for BTS or WTS also be 1, the rate of change of deaths for that type of shark would initially be $1 \times 15 \times 20 = 300$ sharks/month; and the population would quickly become extinct. Thus, we have the following rule of thumb.

> **Rule of Thumb:** A constant of proportionality for a product of populations, such as BW, is frequently at least an order of magnitude (decimal point moved one place to the left) less than a constant of proportionality for one population, such as B or W.

With populations inhibited only by the competition for food, we might have a situation like the one illustrated in Figure 4.1.2 and Table 4.1.1. In this case, the WTS initially outnumber the BTS. However, the WTS death proportionality constant ($w = 0.27$) is larger than the BTS death proportionality constant ($b = 0.20$). Early in the simulation, the population of both species decreases. Eventually, the

WTS die out and the BTS thrive. The projects and exercises explore situations that have different initial populations and constants of proportionality and, consequently, different results.

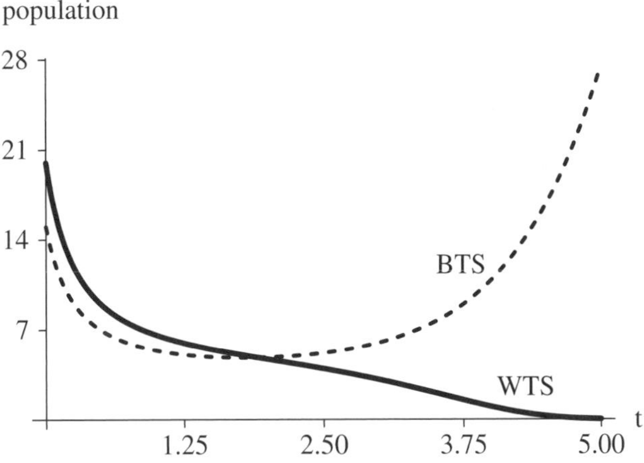

Figure 4.1.2 Graph of results of simulation from Figure 4.1.1, where the WTS death proportionality constant (w) is 0.27, the BTS death proportionality constant (b) is 0.20, and time (t) is in months

Table 4.1.1
Table of Results of Simulation from Figures 4.1.1 and 4.1.2 where $w = 0.27$ and $b = 0.20$

Time (months)	*WTS*	*BTS*
0	20.00	15.00
1	6.57	5.37
2	4.69	4.84
3	3.08	6.00
4	0.99	10.83
5	0.02	27.43

Exercises

1. **a.** Write the differential equations for modeling competition with unconstrained growth for both populations.
b. Find all equilibrium solutions to these equations.

2. **a.** Write the differential equations for modeling competition with constrained growth for both populations.
b. Find all equilibrium solutions to these equations.

3. What would be the effect on each of the following of increased intraspecific competition? *Hint:* Increased competition would be reflected in higher population densities.

a. Mortality in terms of number of pines/acre
b. Fertility in terms of number of seeds/plant/m^2
c. Average adult weight in terms of average adult bluegill weight per liter of water
d. Rate of growth in terms of increase in mallard duckling weight per unit of time

Projects

For additional projects, see Module 7.11, "Fueling Our Cells—Carbohydrate Metabolism."

1. **a.** Using your system dynamics tool's *sharkCompetition* file, which contains a model for competing species, find values for the initial populations and the constants of proportionality in which one population becomes extinct.
   b. Find values for which the two populations reach equilibrium.
   c. Discuss the results.
   d. Adjust the model to have the populations constrained by carrying capacities (see Module 2.3, "Constrained Growth").
   e. Adjust the parameters several times obtaining different results.
   f. Explain the models and discuss the results.
2. Argentine ants (*Linepithema humile*) are native to South America but have been invading the temperate zone of North America from the turn of the twentieth century. With its large and aggressive workers, Argentine ants are generally able competitively to exclude many native ant species. This success comes from the ant's ability to use exploitive as well as interference competitive mechanisms (Holway 1999).
   a. Develop a model of exploitive competition for the Argentine ant versus a native ant. The competitive factors include discovery time and rate of recruitment. The Argentine ant might discover a food source faster and attract other workers to the food source more quickly than the native ant.
   b. Develop a model of interference competition for the Argentine ant versus a native ant. The competitive factors include physical inhibition/removal and chemical repellents. Argentine ants might fight off or remove native ants from the food source, or they might use chemicals to repel them.
3. Model intraspecific competition. See Exercise 3 for examples. Discuss mortality and rate of growth in response to increasing intraspecific competition.
4. Plants can produce chemicals that, when released to the soil, inhibit the growth of other plants. These chemicals can act by inhibiting respiration, photosynthesis, cell division, protein synthesis, mineral uptake, or altering the function of membranes. For instance, sandhill rosemary (*Ceratiola ericoides*), an evergreen shrub found along the coastal plain of the southeastern United States, produces ceratiolin. This chemical washes from the leaves and degrades to hydrocinnamic acid, a compound that effectively inhibits seed germination of many competing species (Hunter and Menges 2002).

Assume that this chemical is increasingly effective at germination inhibition with increasing concentrations. Assume the highest concentration released to be 60 ppm (parts per million) and that concentration decreases linearly from the tips of the outermost leaves (for periods without rain).

a. Model inhibition of a competing plant species, where the effective concentrations of the toxin are between 20 and 60 ppm.

b. Model inhibition for this species with 2 cm rain per day. Set your own decrease in concentration per cm of rain for your model.

5. Model the interference competition of titmice versus other birds at feeders.

6. Model an environment with two competing species of flowering plants—species A and species B—and two essential resources—phosphorus and nitrogen. The constant renewal rate for each resource is 0.4 units/month. Initially, the availabilities of phosphorus and nitrogen are 12 units and 28 units, respectively. Each species has a starting population of 12 plants. At these levels, the maximum progeny produced per plant for species A and B are 1.2 plants/month and 1.0 plants/month, respectively; while their per plant deaths are 0.5 plants/month. Consider progeny production and deaths proportional to the number of species individuals. For maximum births, the phosphorus consumption amounts per plant for species A and B are 0.5/month and 0.25/month, respectively, and the nitrogen consumption amounts per plant are 0.25/month and 0.5/month, respectively. For fewer resources, the relative amounts of phosphorus and nitrogen consumption and the birth rates are proportionally smaller. Explain the model and discuss the results. Will this scenario result in equilibrium (Tilman 1980)?

Answers to Quick Review Questions

1. a. cW, where c is a birth rate

b. wBW or wWB, where w is a death proportionality constant

2. A. BTS, because a larger portion of the white tip sharks are dying

References

Holway, David A. 1999. "Competitive Mechanisms Underlying the Displacement of Native Ants by the Invasive Argentine Ant." *Ecology*, 80(1): 238–251.

Hunter, Molly E., and Eric S. Menges. 2002. "Allelopathic Effects and Root Distribution of *Ceratiola ericoides* (Empetraceae) on Seven Rosemary Scrub Species." *American Journal of Botany*, 89(7): 1113–1118.

Smith, Thomas M., and Robert Leo. Smith. 2012. *Elements of Ecology*. 8th ed. San Francisco: Benjamin Cummings.

Tilman, D. 1980. "Resources: A Graphical-Mechanistic Approach to Competition and Predation." *American Naturalist*, 116: 362–393.

MODULE 4.2

Predator-Prey Model

Download

The text's website has a *Predator-Prey* file, which contains the model of this module, available for download for various system dynamics tools.

Introduction

One of the interspecific interactions (see Module 4.1, "Competition") common to biological communities is the **predator-prey relationship**. When one species (**predator**) consumes another species (**prey**) while the latter is still living, the action is **predation**. Predation might involve the consumption of a young squirrel by a hawk, but examples also include tomato hornworms consuming tomato plant leaves and a tapeworm feeding off its mammalian host. Predator-prey interactions are important influences on population levels and ecosystem energy flow.

One of the most interesting characteristics of this type of relationship is that both predators and prey develop fascinating adaptations, which normally come about over long periods of time. Predator adaptations usually involve better prey detection and capture, whereas prey adaptations normally involve improved abilities to escape and avoid detection.

So, let's consider a 3/4-in. frog, commonly called a poison dart frog. We might expect that such a small animal would, to avoid predation, come out only at night or adopt some camouflaged coloration. However, this brazen creature forages for small invertebrates during the day (prey may also be predators) and is brilliantly colored (bright red, yellow, etc.). How might it manage then to avoid predation? The answer lies in the skin of the frog, which contains toxic, alkaloid chemicals that cause paralysis and/or death in the predator. Over time, predators associate the coloration with the toxic nature of the prey and, hence, avoid that prey. So the bright coloration is termed warning, or **aposematic, coloration**.

Lotka-Volterra Model

In the 1920s, mathematicians Vito Volterra and Alfred Lotka independently proposed a model for populations of a predator species and its prey, such as hawk and squirrel populations in a certain area. For simplicity, we assume that a hawk hunts only squirrels and that no other animal eats squirrels. If the hawk's only food source is squirrel and the number of squirrels diminishes significantly, then scarcity of food will result in starvation for some of the hawks. With reduced numbers of hawks, the squirrel population should increase.

Quick Review Question 1

This question reflects on the predator-prey situation before we begin the discussion.

a. Do predator-prey interactions have a direct impact on the births or deaths of the prey?
b. Based on other interaction model of Module 4.1, we can model the prey deaths as being directly proportional to what?
c. If we consider prey births as being unconstrained, we can model prey births as being directly proportional to what?
d. Are predator-prey interactions advantageous or disadvantageous for predators?
e. Based on other interaction models of Module 4.1, we can model predator births as being directly proportional to what?
f. If we consider predator deaths as being unconstrained, we can model the predator deaths as being directly proportional to what?

Let s be the number of squirrels in the area and h be the number of hawks. If no hawks are present, the change in s from time $t - \Delta t$ to time t is as in the unconstrained model (see Module 2.2, "Unconstrained Growth and Decay"):

$$\begin{aligned}\Delta s &= s(t) - s(t - \Delta t)\\ &= (\text{squirrel growth at time } t - \Delta t) * \Delta t\\ &= k_s * s(t - \Delta t) * \Delta t \text{ for constant } k_s\end{aligned}$$

However, this prey's population is reduced by an amount proportional to the product of the number of hawks and the number of squirrels, $h(t - \Delta t) * s(t - \Delta t)$. Thus, with a proportionality constant k_{hs} for this reduction, the change in the number of squirrels from time $t - \Delta t$ to time t is as follows:

$$\begin{aligned}\Delta s &= s(t) - s(t - \Delta t)\\ &= (\text{squirrel growth at time } t - \Delta t) * \Delta t\\ &= (k_s * s(t - \Delta t) - k_{hs} * h(t - \Delta t) * s(t - \Delta t)) * \Delta t\end{aligned}$$

for constants k_s and k_{hs}.

We can interpret the term $k_{hs} * h(t - \Delta t) * s(t - \Delta t)$ in a couple of ways. First, $h(t - \Delta t) * s(t - \Delta t)$ is the maximum number of distinct interactions of hawks with squirrels. For example, for $h(t - \Delta t) = 3$ hawks and $s(t - \Delta t) = 2$ squirrels, $(3)(2) = 6$

possible pairings exist. The decrease in the number of squirrels is proportional to this product, where the constant of proportionality, k_{hs}, is related to the hunting ability of the hawks and the survival ability of the squirrels. A second interpretation of $k_{hs} * h(t - \Delta t) * s(t - \Delta t) = (k_{hs} * h(t - \Delta t)) * s(t - \Delta t)$ is that the size of the squirrel population decreases in proportion to the size of the hawk population.

While the squirrel population decreases with more contacts between the predator and prey, the hawk population increases. Moreover, the death rate of hawks is proportional to the number of hawks. Thus, the change in the hawk population from time $t - \Delta t$ to time t is as follows:

$$\begin{aligned}\Delta h &= h(t) - h(t - \Delta t)\\ &= (\text{hawk growth at time } t - \Delta t) * \Delta t\\ &= (k_{sh} * s(t - \Delta t) * h(t - \Delta t) - k_h * h(t - \Delta t)) * \Delta t\end{aligned}$$

for constants k_{sh} and k_h. Although the deaths of the squirrels and the births of the hawks are both proportional to the product of the number of possible interactions of the two populations, their constants of proportionality, k_{hs} and k_{sh}, respectively, are probably different. For instance, 2% of the possible interactions might result in the death of a squirrel, while only 1% of the possible interactions might contribute to the birth of a hawk.

We can express the predator-prey model, known as the **Lotka-Volterra model**, as the following pair of difference equations for the change in prey (here, change in the squirrel population, Δs) and change in predator (here, change in the hawk population, Δh) from time $t - \Delta t$ to time t:

$$\begin{aligned}\Delta s &= (k_s * s(t - \Delta t) - k_{hs} * h(t - \Delta t) * s(t - \Delta t)) * \Delta t\\ \Delta h &= (k_{sh} * s(t - \Delta t) * h(t - \Delta t) - k_h * h(t - \Delta t)) * \Delta t\end{aligned} \qquad (1)$$

or as the following pair of differential equations:

$$\begin{aligned}\frac{ds}{dt} &= k_s s - k_{hs} h s\\ \frac{dh}{dt} &= k_{sh} s h - k_h h\end{aligned} \qquad (2)$$

Figure 4.2.1 contains a diagram for the predator-prey model with the prey population affecting the number of predator births and the predator population influencing the number of prey deaths.

Quick Review Question 2

Consider the following Lotka-Volterra difference equations:

$\Delta x = (2 * x(t - \Delta t) - 0.02 * y(t - \Delta t) * x(t - \Delta t)) * \Delta t$ with $x(0) = 100$

$\Delta y = (0.01 * x(t - \Delta t) * y(t - \Delta t) - 1.06 * y(t - \Delta t)) * \Delta t$ with $y(0) = 15$

a. Which equation (Δx, Δy, both, or neither) models the change in predator population?

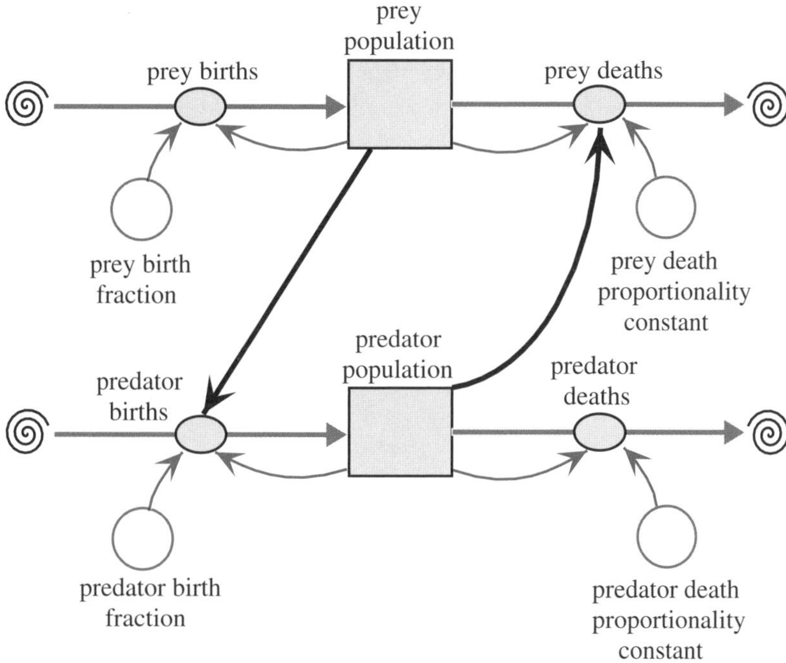

Figure 4.2.1 Predator-prey diagram

For each of the following questions, indicate the appropriate answer from the following choices:

A. 2	B. 0.02	C. –0.02	D. 0.01
E. 1.06	F. –1.06	G. 100	H. 15

b. Which number represents the predator birth fraction?
c. Which number represents the prey birth fraction?
d. Which number represents the predator death proportionality constant?
e. Which number represents the prey death proportionality constant?
f. What is the initial number of predators?
g. What is the initial number of prey?

Particular Situations

Historical Note During the Cultural Revolution in China (1958–1960), Chairman Mao Zedong decreed that all sparrows be killed because they ate too much of the crops and they seemed to be only for pleasure anyway. With reduction in its main predator, the insect population increased dramatically. The insects destroyed much more of the crops than the birds ever did. Consequently, the Chinese reversed the decision that caused the imbalance (PBS 2002).

Returning to the example of the hawks and squirrels, some of the model's equations and constants appear in Equation Set 4.2.1. In that example, *prey_birth_fraction* (k_s) = 2, *prey_death_proportionality_constant* (k_{hs}) = 0.01, *predator_birth_fraction* (k_{sh}) = 0.01, *predator_death_proportionality_constant* (k_h) = 1.06, the initial *prey_population* (s_0) = 100, and the initial *predator_population* (h_0) = 15. As suggested in the "Rule of Thumb" in Module 4.1, "Competition," the proportionality constants (0.01 and 0.01) for products, which involve interactions, are at least an order of magnitude less than the proportionality constants (2 and 1.06) for single populations.

Equation Set 4.2.1

Some of the equations and constants for model in Figure 4.2.1:

predator_population(0) = 15
predator_birth_fraction = 0.01
predator_births = (*predator_birth_fraction* * *prey_population*) * *predator_population*
predator_death_proportionality_constant = 1.06
predator_deaths = *predator_death_proportionality_constant* * *predator_population*
prey_population(0) = 100
prey_birth_fraction = 2
prey_births = *prey_birth_fraction* * *prey_population*
prey_death_proportionality_constant = 0.02
prey_deaths = (*prey_death_proportionality_constant* * *predator_population*) * *prey_population*

Table 4.2.1 and Figure 4.2.2 show the varying prey and predator populations as time advances through 12 months. Shortly after the squirrel, or prey, population

Table 4.2.1
Table of Prey and Predator Populations over 12-month period

Months	*Prey Population*	*Predator Population*
0.000	100.00	15.00
1.000	449.58	62.00
2.000	30.43	280.24
3.000	5.63	108.55
4.000	10.54	40.32
5.000	45.61	17.59
6.000	244.25	19.97
7.000	215.76	298.60
8.000	7.91	173.18
9.000	6.52	63.69
10.000	21.30	24.81
11.000	109.68	14.61
Final	470.44	74.28

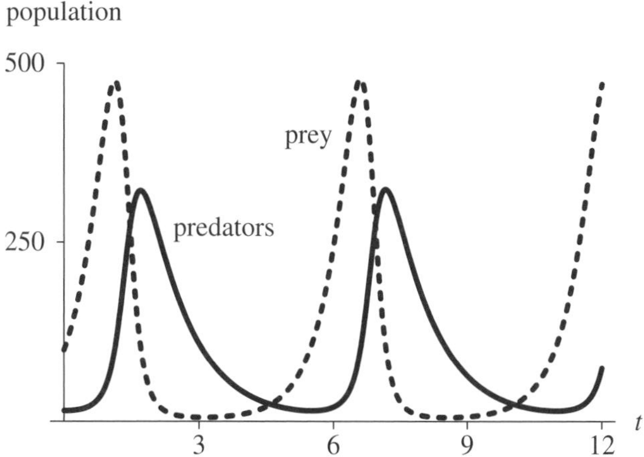

Figure 4.2.2 Graph of populations versus time in months

increases, the hawk, or predator, population does likewise. As the predators kill off their food supply, the number of predators decreases. Then, the cyclic process starts over.

Quick Review Question 3

To the nearest whole number, what is the period (in months) of the cyclic functions for population in Figure 4.2.2?

Figure 4.2.3 shows the graph of a solution to the difference or differential equations with the prey population along the horizontal axis and the predator population along the vertical axis. With the initial predator population being 15 and prey population being 100, the plot starts at the bottom toward the left and proceeds counterclockwise as time progresses. Initially, with few predators endangering them, the

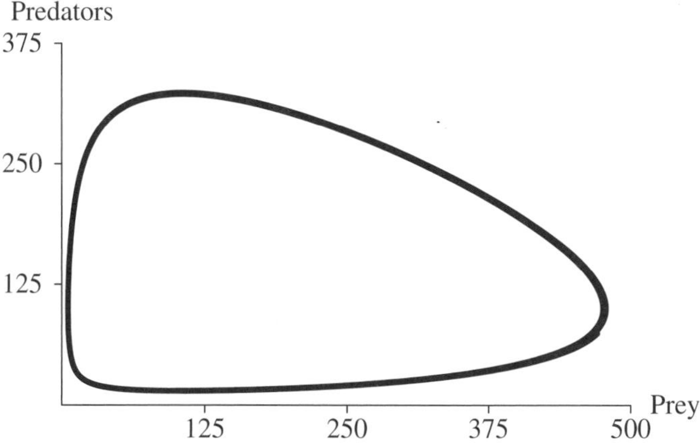

Figure 4.2.3 Graph of predator population versus prey population

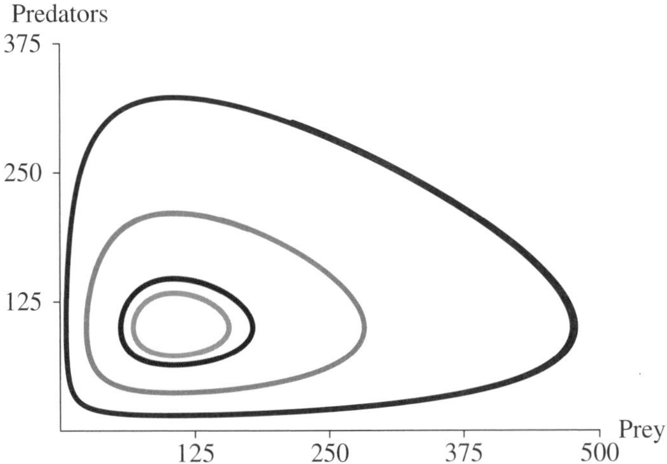

Figure 4.2.4 Several solutions to the predator-prey model using different initial conditions and the coloration shown

Predator	*Prey*	*Color of Graph*
15	100	black
75	125	gray
135	150	dark color
195	175	light color

prey population reaches a maximum of about 475 when the predator population is about 100. Then, with the graph developing to the left and up, we see that the prey population starts decreasing as the predator population continues to increase with the abundant supply of its food, the prey. At the graph's high point, about (107, 322), with approximately 107 prey, the predator population achieves a maximum of 322 individuals. That same number of predators, about 107, occurs toward the bottom of the graph when the prey only number about 15. After a maximum, the number of predators falls off rapidly because of the limited food supply, and the number of prey decreases as well. Eventually, on the bottom part of the graph, with the diminished number of predators, the prey are able to stage a comeback, and the cyclical process begins again. Figure 4.2.4 illustrates several such solutions employing different initial conditions.

Quick Review Question 4

The following are the Lotka-Volterra differential equations for the particular model we have been considering:

$$ds/dt = 2s - 0.02hs$$

$$dh/dt = 0.01sh - 1.06h$$

with $s(0) = 100$ and $h(0) = 15$.

a. Indicate all that must be true for the system to be in equilibrium: $ds/dt = 0$; $s = 0$; $dh/dt = 0$; $h = 0$; all of these; none of these.
b. A trivial solution for equilibrium is $s = 0$ and $h = 0$. Find a nontrivial solution, where $s \neq 0$ and $h \neq 0$.

Exercises

1. Give two sets of Lotka-Volterra equations with all coefficients being different that represent a system in equilibrium, such that the number of prey is always 3000 and the number of predators is always 500.
2. Write the differential or difference equations for a predator-prey model where there is a carrying capacity M for the predator. See differential equation 1 or difference equation 2 in Module 2.3, "Constrained Growth."
3. The blue whale, which can grow to 30 m in length, is a baleen whale whose favorite food is Antarctic krill, a small shrimp that is about 5 cm long. The difference equation for the change in the krill population is similar to that for Δs in (1), except the birth term must be logistic (see Equation 2 in Module 2.3, "Constrained Growth"). The difference equation for the change in the number of blue whales is a logistic equation, except that the carrying capacity is not a constant but is proportional to the krill population. Write the difference equations to model this system (Greenwood 1983)

Projects

For additional projects, see Module 7.11, "Fueling Our Cells—Carbohydrate Metabolism"; Module 7.12, "Mercury Pollution—Getting on Our Nerves"; Module 7.13, "Managing to Eat—What's the Catch?"; Module 7.14, "Control Issues: The Operon Model"; and Module 7.15, "Troubling Signals: Colon Cancer."

1. Develop a model where the prey birth fraction (k_s) is periodic, such as follows:

   $k_s = f + a\cos(p \times t)$, where f, a, and p are constants; $0 < a < f$; $0 < f$; and t is time.

   Note that a is the amplitude; the period is $2\pi/p$; and addition of f raises the graph of $a\cos(pt)$ by the amount f. (For a more detailed discussion, see the section "Trigonometric Functions" of Module 8.2, "Function Tutorial.") Have a table of population numbers, a graph of populations versus time, and a graph of one population versus the other. Determine values for the parameters so that the system is periodic, and then determine values where the system is chaotic. Discuss your results.
2. Using system dynamics software or a computer program, model the predator-prey example, including crop consumption discussed in the Historical Note about the Chinese Cultural Revolution.

3. Develop a predator-prey model where humans hunt both predator and prey equally. For example, the predator might be sharks and the prey tuna.
   a. Mathematically solve for the equilibrium point.
   b. Run the model for several situations assuming no fishing. Using those same situations, gradually increase the amount of fishing (the rate of catching shark and tuna), but keep this rate less than the prey birth rate. Because these simulations are run with a time scale of months, consider the human population to be constant, but vary the amount of fishing. Record all your results. What happens to the number of prey and predators? Why? Use well-written discussions with supporting work from your model.
   c. Continue increasing the amount of fishing. At what level of fishing do the predators all die? How does this level compare to the prey birth rate? Using the equilibrium point from Part a, discuss why.
   d. Alter your model to have seasonal fishing. Use a formula for the rate of fishing effort similar to that in Project 1. Run the model for several situations. Discuss your results.

Historical Note The situation in Project 3 was observed in the Mediterranean Sea between 1914 and 1923. Limited fishing occurred during World War I. Although fishing increased after the war, so did the number of tuna.

4. Implement the model of Exercise 2. Graph the populations as time progresses. Also, graph one population against the other. Run the model for several carrying capacities. Discuss the results. Compare this system to the one without a carrying capacity.
5. Are the two sets of equilibrium points in Quick Review Question 4 stable? Discuss your answers and give supporting evidence.
6. **a.** Implement the model in Exercise 3 for the Antarctic, assuming the following:

   krill carrying capacity = 4×10^8 tons
   krill birth rate = 5%
   blue whale growth rate = 10%
   yearly consumption of krill by a blue whale = 115 to 450 tons
   rate of blue whale consumption of krill = 1%

   Assume there are initially 5000 blue whales and 3×10^8 tons of krill and no other animal eats krill. Discuss your results (Mori and Butterworth 2005).
   b. Find the equilibrium point. Run the model assuming an ecological disaster kills 10% of the equilibrium level of whales and 80% of the equilibrium level of krill. Describe what happens.
   c. Run the model assuming whales are almost hunted to extinction, to 1% of their equilibrium level. Describe what happens after hunting stops. Develop a table of how long it takes for the whales to return to equilibrium level starting with different initial amounts: 1%, 2%, . . ., 10% of equilibrium level.

7. Suppose a rat population is growing logistically in an area of the city. Attempts to kill off the population through poisons or trapping are not 100% successful. Moreover, killing off half might cause the population to increase rapidly. Alternatively, people could attempt to decrease the carrying capacity by cleaning up garbage and sealing trash containers. Develop models for each of the proposed solutions to the rat problem. Compare and contrast the proposals, and discuss the circumstances under which each is best. Make recommendations to the city.

8. In predator-prey systems, predators make two basic types of responses to increasing prey density. Predators may react by taking more prey or taking them faster. This response is normally quick and termed by ecologists to be a **functional response**. On the other hand, predators may also respond by increased levels of immigration (movement in) or through producing more offspring. This type of response is typically going to take more time.

In 1959, the ecologist C. S. Holling presented a classification of three types of predator functional responses determined by the proportion of prey consumed (Holling 1959). This and the next two projects involve modeling these types.

For this project, model a **Type-I predator functional response**: The predator consumes a constant proportion of prey, regardless of prey density (density-independent). Predation rate increases in a linear fashion, until the current population of predators achieves satiation. Few good examples exist, but one is a spider feeding on insects they trapped in its web.

9. Model a **Type-II predator functional response** (see Project 8): the predator consumes less as it nears **satiation**, which determines the upper limit on consumption. In this type of response, prey handling time (T_H) and search time (T_S) are separated. Predators do not handle while they are searching and do not search while they are handling. Whatever time is necessary for handling decreases the time available for searching. Predation rate increases more slowly as prey populations increase than it does in Type I. A peak occurs when predators are consuming prey as fast as they can search and eat them. A praying mantis preys on insect prey and must process its prey before it can hunt for more food. This type of response is described by the following **disk equation** (Holling 1959):

$$\text{predation rate} = (N_a/P)/T = aN/(1 + aT_HN)$$

where N_a = prey attacked or killed, N = prey density, P = predator density, a = attack rate constant (= N_a/T), T = time predators and prey exposed, T_H = prey handling time, and ($T = T_S + T_HN_a$).

10. Model a **Type-III predator functional response** (see Projects 8 and 9): the predation increases slowly at low prey density, increases rapidly at higher densities, but levels off at satiation, even if prey density continues to increase. These predators also have separate handling and search times. This type of response is typical of predators that are **generalists**. They may use alternative prey as the prey densities of their primary prey decline. For instance, a hawk might switch to squirrels if smaller rodents became scarce (Holling 1959).

11. Where would an herbivorous animal (e.g., rabbit, deer, etc.) fit in the functional response schemes described in Projects 8, 9, and 10? Develop a model for this creature.

12. If you have visited the coast of northern California, Washington State, or southeastern Alaska and spent any time looking out to sea, you have probably seen a few captivating, furry animals swimming, diving, and floating on their backs. These creatures are sea otters. Sea otters have extremely dense fur to keep them warm in cold Pacific waters, because they lack the layer of insulating blubber possessed by other marine mammals of that area. For the fur, these creatures were hunted to near extinction during the eighteenth century. The sea otters have survived many challenges during the intervening centuries.

Sea otters are carnivorous and must also eat one-fourth to one-third of their body weight per day to maintain body temperature (adult males, 65 lb and adult females, 45 lb). Therefore, they spend much of their time (20% to 60%) hunting for and eating food. Sea otters eat a variety of grazing animals, such as sea urchins, snails, crabs, abalone, mussels, and clams that live in the rich kelp forests along the coast.

When sea otters are lost from a kelp forest, the grazing prey, particularly urchins, rapidly increase in numbers and feast on the kelp forests, sending them into decline. Kelp forests, some of the most highly productive communities in the world, are extremely important, especially for sheltering and feeding fish and shellfish communities. Loss of kelp results in the loss of many species. Where otters have been reintroduced, the healthy kelp forests return. For the key role otters play in maintaining the richness and diversity of the ecosystem, they are termed **keystone predators** (NOAA; Otter Project).

Model this situation and discuss the results.

13. This project allows you to model the changes in species diversity using the intertidal community of Washington State. The intertidal zone, in this case, is a rocky area covered by seawater at high tide, but uncovered at low tide. The community structure described for this project is based on the communities as reported by R. T. Paine during the 1960s (Paine 1966; Paine 1969). This community, like the kelp forest (see Project 12), has a keystone predator, the ochre sea star *Pisaster ochraceus*. This sea star can achieve a radius of 11 in. and is a voracious predator, preferring the delectable taste of mussels. The rest of the community is made up of various species of algae (primary producers), mussels, clams, chitons, barnacles, crustaceans, and snails. The organisms that live in this zone are specialists, adapted for the conditions that exist there. Thus, competition is intense. Following are the components you should consider in developing your model, in addition to *Pisaster*.

Molluscs (herbivores)	*Molluscs (carnivores)*	*Crustacea (filter feeders)*	*Algae*
Katherina tunicata (grazer)	*Nucella*	*Mitella*	*Porphyra*
Mytilus (filter feeder)		*Balanus*	*Neorhodamela*
			Corralina

The following are the feeding relationships in this community:

Pisaster: feeds on *Mytilus* preferentially but will also feed on *K. tunicata* and *Mitella*, depending on prey densities.
Nucella: feeds on *Mitella* and *Balanus*.
K. tunicata: feeds on all species of algae listed.
Mytilus, Mitella, and *Balanus*: filter food out of the ocean water.
All the algae photosynthesize for energy and organic matter production.

a. Before you try to model this community, generate a diagram that relates each of these organisms by feeding relationships. The following is the succession of changes that Paine found in the community, after excluding *Pisaster* from discrete areas of the intertidal zone:

Year 1: *Mitella* disappears, replaced by the other barnacle, *Balanus*.
Year 2: Both barnacles disappear. *Mytilus* (mussel) out-competes and replaces them. With no barnacles for food, the snail, *Nucella*, disappears. *Mytilus* begins crowding out the algae as well.
Years 3–6: Only *Mytilus* remains.

b. Given this scenario, what do you think the role of *Pisaster* is in this community?
c. Generate a model that describes community dynamics when *Pisaster* is present.
d. Generate a model that describes community dynamics when *Pisaster* is removed.

Answers to Quick Review Questions

1. a. deaths
b. the product of the number of predators and the number of prey
c. the number of prey
d. advantageous
e. the product of the number of predators and the number of prey
f. the number of predators
2. a. Δy
b. D. 0.01
c. A. 2
d. E. 1.06
e. B. 0.02
f. H. 15
g. G. 100
3. 6 mo
4. a. $ds/dt = 0$ and $dh/dt = 0$

b. $s = 106$ and $h = 100$. The following discussion derives the solution. We wish to solve the following system of equations:

$$0 = 2s - 0.02hs$$
$$0 = 0.01sh - 1.06h$$

Because $s \neq 0$, we can cancel out the factor s in the first equation to obtain $0 = 2 - 0.02\,h$. Thus, $h = 2/0.02 = 100$ hawks (predators). Similarly, because $h \neq 0$, we can cancel out the factor h in the second equation to obtain $0 = 0.01s - 1.06$. Thus, $s = 1.06/0.01 = 106$ squirrels (prey). See the section on "Equilibrium and Stability" in Module 2.3, "Constrained Growth" for a discussion of equilibrium.

References

Danby, J.M.A. 1997. *Computer Modeling: From Sports to Spaceflight . . . From Order to Chaos*. Richmond, VA: Willmann-Bell, pp. 99–119.

Greenwood, Raymond N. 1983. "Whales and Krill: A Mathematical Model." UMAP Module 610, COMAP.

Holling, C. S. 1959. "The Components of Predation as Revealed by a Study of Small-Mammal Predation of the European Pine Sawfly." *Canadian Entomologist* 91: 293–320.

Meerschaert, Mark M. 1993. *Mathematical Modeling*. Boston: Academic Press, pp. 115, 181.

Mori, M. and D. S. Butterworth. 2005. "Progress on Multi-species Modeling in the Antarctic." Institute for Cetacean Research, Tokyo, document: JA/J05/PJR23: 1–25.

NOAA. "Ecosystems: Kelp Forests." in National Marine Sanctuaries. http://sanctuaries.noaa.gov/about/ecosystems/kelpdesc.html (accessed November 17, 2012)

Otter Project. "The Otter Project." The Otter Project, Inc. http://www.otterproject.org/ (accessed December 17, 2012)

Paine, R. T. 1966. "Food Web Complexity and Species Diversity." *American Naturalist*, 100: 65–75.

———. 1969. "The Pisaster-Tegula Interaction: Prey Patches, Predator Food Preference and Intertidal Community Structure." *Ecology*, 50(6): 950–961.

PBS (Public Broadcasting Service). 2002. "Commanding Heights: The Battle for the World Economy, China" http://www.pbs.org/wgbh/commandingheights/lo/countries/ cn/cn_full.html

MODULE 4.3

Modeling the Spread of SARS—Containing Emerging Disease

Downloads

The text's website has *SIR* and *SARSRelationships* files, which contain models for the examples of this module, available for download for various system dynamics tools.

Introduction

Imagine being a college student in New York City and being told not to leave the city. That's what happened in 2003 in Beijing, when thousands of people were ordered to stay home and college students were told to stay in Beijing. Quarantine procedures were instituted for those who were thought to have had "intimate contact" with others who showed signs of a new rapidly spreading respiratory disease. More than 40 had died in the capital, and thousands of people in China were displaying symptoms of this pneumonia. Imagine the feelings of fear and panic that Beijing residents must have had—people in masks, disinfecting their homes, and hoarding of food and other necessities.

This new disease was called **SARS**, **severe acute respiratory syndrome**, with the first case occurring on November 16, 2002, in southern China. Chinese health officials reported the outbreak to the World Health Organization (WHO) on February 11, 2003. By April 2, the total reported cases of SARS were 2000; and by July, the count was over 8400 with more than 800 dead. In response to the initial report, WHO coordinated the investigation into the cause and implemented procedures to control the spread of this disease. The control measures were extremely effective, and the last new case was reported on June 12, 2003 (WHO).

By the third week in March several laboratories worldwide had identified the probable causative agent—*SARS-CoV*, the SARS coronavirus. Coronaviruses represent a large group of +-stranded RNA-containing viruses associated with various respiratory and gastrointestinal illnesses. Although the human diseases associated with these viruses have been mild previously, this coronavirus is quite different. Like many respiratory pathogens, SARS is spread by close personal contact and perhaps by airborne transmission.

The Centers for Disease Control and Prevention (CDC) in the United States uses clinical epidemiological and laboratory criteria to diagnose SARS. Severe cases exhibit a fever higher than 38 °C and one or more respiratory symptoms—difficulty breathing, cough, or shortness of breath. Additionally, the person must show radiographic evidence (lung infiltrates) of pneumonia, or **respiratory distress syndrome (RDS)**. RDS is an inflammatory disease of the lung, characterized by a sudden onset of edema and respiratory failure. A few others qualified if they exhibited an unexplained respiratory illness that resulted in death and an autopsy confirmed RDS with no identifiable cause. Epidemiological evidence might include close contact with a known SARS patient or travel to a region with documented transmission within 10 days of onset of symptoms. Today, laboratory tests confirm SARS if they reveal one of the following (CDC):

- Antibody to SARS virus in specimens obtained during acute illness or more than 28 days after onset of illness
- SARS viral RNA detected by RT-PCR
- SARS virus

On July 5, 2003, the World Health Organization declared that SARS had been contained. The outbreak resulted in 812 deaths, but the toll might have been much higher if WHO and other health agencies had not acted so quickly and effectively (WHO). Besides the direct effect on the victims and their families, SARS became a major drag on the economies of China, Taiwan, and Canada. Hong Kong's unemployment rate climbed to an unprecedented 8.3%, and travel warnings for Toronto cost Canada an estimated $30 million per day. One can only imagine the impact of this disease being spread into Africa, where there are poor healthcare systems and the astronomical HIV infection rates generate immunologically compromised populations.

SARS is an interesting disease for modeling, particularly because there is so much epidemiological information. We still have much to learn about SARS, and we still have no available, effective treatment.

SIR Model

Before developing a model for the spread of SARS, we consider the simpler situation of a disease in a closed environment in which there are no births, deaths, immigration, or emigration. A 1978 *British Medical Journal* article reported on such a situation—influenza at a boys' boarding school. On January 22, only one boy had the flu, which none of the other boys had ever had. By the end of the epidemic on February 4, 512 of the 763 boys in the school had contracted the disease (Murray 1989; NCSLIP).

To model this spread of influenza, we employ the **SIR Model**, which W. O. Kermack and A. G. McKendrick developed in 1927 (Kermack and McKendrick 1927).

Many systems models of the spread of disease, including the SARS model later in this module, are extensions of the SIR Model. The name derives from the following three populations considered:

Susceptibles (***S***) have no immunity from the disease.
Infecteds (***I***) have the disease and can spread it to others.
Recovereds ***(R)*** have recovered from the disease and are immune to further infection.

The model gives the differential equation for the rate of change for each of these populations. We assume that after a certain amount of time, an individual with the flu recovers. Thus, the rate of change of the number of recovereds is proportional to the number of infecteds.

Quick Review Question 1

With the constant of proportionality being the recovery rate (a), give the differential equation for the rate of change of the number of recovereds.

As the answer to Quick Review Question 1 states, the differential equation for the rate of change of the number of recovereds is $dR/dt = aI$ for recovery rate a. If the time unit is in days and d is the number of days that someone remains infected, we can consider a to be $1/d$. For example, if a boy is usually sick with the flu for 2 days, then $d = 2$ and $a = 0.5$/day, so that approximately half the infected boys get well in a day.

Model: In the SIR model, recovery rate = 1/(number of days infected).

A susceptible boy at the boarding school becomes infected with influenza by having contact with an infected boy. The number of such possible contacts is the product of the sizes of the two populations, SI. For example, suppose the set of susceptibles is S = {Joe, Lee, Orlando} and the set of infecteds is I = {Hondre, Leslee}. As Figure 4.3.1 illustrates, (3)(2) = 6 possible interactions exist between pairs of boys in different sets. The virus in Hondre can spread through contact to Joe, Lee, and Orlando. Similarly, Joe can become infected with the virus from Hondre or Leslee. With no new students entering the school, the number of susceptibles can only decrease, and the rate of change of the number of boys in this set is directly proportional to the number of possible contacts, SI, between susceptibles and infecteds.

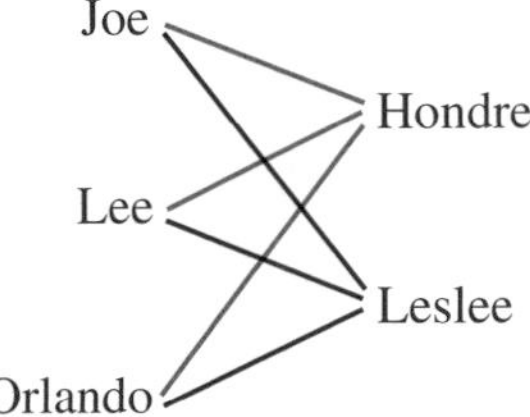

Figure 4.3.1 Possible contacts between S and I

Quick Review Question 2

a. Is the rate of change of S positive, zero, or negative?
b. With $r > 0$ being the constant of proportionality, give a differential equation for the rate of change of S.

In Module 4.1, "Competition," because of interactions, we modeled competitors' death rates of change using the same proportionality to a product of population sizes. Similarly, in Module 4.2, "Predator-Prey Models," considering contacts between predators and prey, we modeled predator births and prey deaths using the same type of product. Thus, three very different applications employ the same model for rates of change where interaction occurs—the product of a constant and the two interacting population sizes.

Model: One model for the rate of change involving the interaction of constituents with sizes A and B is cAB, where c is a constant.

As the answers to Quick Review Question 2 reveal, because of the interaction of susceptibles and infecteds in spreading the disease, we employ this model for the rate of change of susceptibles with respect to time: $dS/dt = -rSI$ for positive constant of proportionality r. The constant r, called the **transmission constant**, reflects the extent and the infectiousness of the disease and the interactions among the students. In the case of the boys' school, we use 0.00218 per day. Thus, 0.00218 = 0.218% of the total number of possible contacts, SI, results in the disease being spread from one child to another.

Notice how small the transmission constant (0.00218/day) is in comparison to the recovery rate (0.5/day). Also, recall in interactions for competition and predator-prey, where a rate-of-change model involves a product of populations, the constant of proportionality is small in comparison to constants multiplied by only one population. Breaking down dS/dt another way helps to explain why the constant of proportionality, here $r = 0.00218$ per day, is so small. For a sick child to pass the disease to someone else, the sick boy must come in contact with someone else, that person must be susceptible, and the interaction must result in the spread of the disease. Thus, the rate of change of S with respect to time (dS/dt) is minus the product of the mean number of contacts per day an infected has (k), the probability such a contact is with a susceptible, the probability that the disease is spread during such a contact (b), and the number of infecteds. Moreover, if N is the total population size (here 763) and the group is well mixed, then for an infected, the probability of that contact he has is with a susceptible is S/N, and the rate of change of S is as follows:

$$dS/dt = -k(S/N)bI = -(kb/N)SI = -rSI$$

Thus, the transmission constant r is (kb/N). For example, suppose on the average an infected child has 33.3 contacts per day and the probability that a contact results in the spread of the disease is 5% = 0.05. Then, for $N = 763$, the transmission constant is $r = (kb/N) = 0.00218$.

Note that this transmission constant, here 0.00218/day, is not the rate of infection. Suppose a report to the school's principal after all are well states that 80% of the boys had had the flu. The 80% is of the total population of $N = 763$ boys, not of the number of possible interactions, *SI*. Moreover, 80% of the susceptible boys do not become sick in one day. If flu lasted in the school for 3 weeks, as the following shows, on the average 3.81% of the boys get sick in 1 day:

$$\frac{0.80}{3\text{weeks}} \times \frac{1\text{week}}{7\text{days}} = \frac{0.0381}{\text{day}} = \frac{3.81\%}{\text{day}}$$

We must be careful to be consistent in units, such as not mixing days and weeks, and to understand of what we are taking a percentage, such as of *SI* instead of *S* or *N*.

Returning to our model, only susceptibles become infected, and infecteds eventually recover. What *I* gains comes from what *S* has lost; and what *I* loses, *R* acquires. Thus, the differential equation for the rate of change of the number of infecteds is the sum of the negatives of the other two rates of change:

$$dI/dt = -dS/dt - dR/dt$$

Quick Review Question 3

Give the differential equation for the rate of change of the number of infecteds in terms of *S*, *I*, *R*, the transmission constant (r), and the recovery rate (a).

Figure 4.3.2 presents a diagram for the SIR model with *susceptibles*, *infecteds*, and *recovereds* replacing the symbols *S*, *I*, and *R*, respectively, and with *transmission_constant* and *recovery_rate* representing the constants of proportionality r and a, respectively. Some of the corresponding equations and constants for a particular simulation appear in Equation Set 4.3.1.

Equation Set 4.3.1

With basic unit of time of 1 day, some equations and constants for SIR model in Figure 4.3.2

susceptibles(0) = 762
transmission_constant = 0.00218

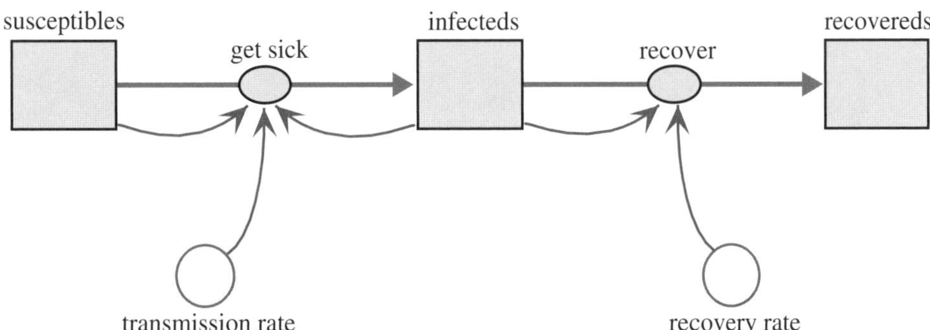

Figure 4.3.2 Diagram for the SIR model

get_sick = *transmission_constant* * *susceptibles* * *infecteds*
infecteds(0) = 1
recovery_rate = 0.5
recover = *recovery_rate* * *infecteds*
recovereds(0) = 0

The graphs of the three populations that result from running the simulation are in Figure 4.3.3. The number of *susceptibles* decreases slowly at first before experiencing a rapid decline and subsequent leveling. In contrast, the number of *recovereds*, which is initially 0, has a graph that appears similar to the logistic curve. When the number of *susceptibles* decreases sharply, the *infecteds* increase to their maximum. Afterwards, as the number of *infecteds* decreases, the number of *recovereds* rises. Although not mimicking the final numbers exactly, this model does capture the trend of the data along with the epidemic increase and decrease and the progress towards a steady state.

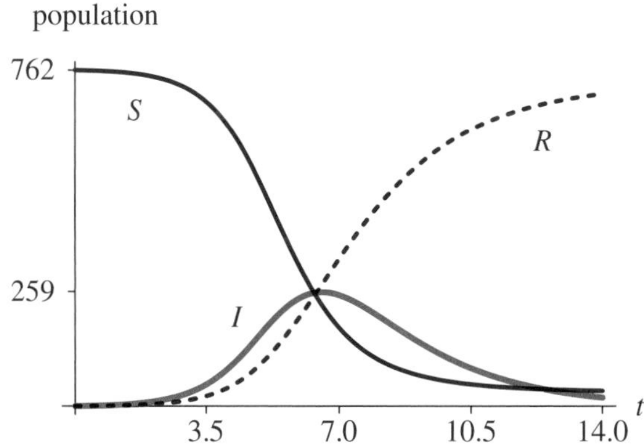

Figure 4.3.3 Graphs of *susceptibles* (*S*), *infecteds* (*I*), and *recovereds* (*R*) versus time (*t*) in days

Quick Review Question 4

Answer the following questions referring to Figure 4.3.3.

a. On what day was the number of cases the largest?
b. On what day were most of the boys sick or recovered?
c. On what day were most of the boys recovered?

SARS Model

Marc Lipsitch in collaboration with others developed a model for the spread of **severe acute respiratory syndrome** (**SARS**) and used the model to make predictions on the impact of public health efforts to reduce disease transmission (Lipsitch et al. 2003). Such efforts included **quarantine** of exposed individuals to separate them from the susceptible population, perhaps by confinement to their homes, and **isolation** of those who had SARS to remove them to strictly supervised hospital areas with no contacts other than by healthcare personnel. The Lipsitch model is an extension of the SEIR model, which is a refinement of the SIR model. Besides the populations considered by SIR, the **SEIR Model (susceptible-exposeds-infecteds-recovereds)** has an intermediate **exposed (*E*)** population of individuals who have the disease but are not yet infectious. The Lipsitch model modifies SEIR to allow for quarantine, isolation, and death. The modelers make the following simplifying assumptions:

1. There are no births.
2. The only deaths are because of SARS.
3. The number of contacts of an infected individual with a susceptible person is constant and does not depend on the population density.
4. For susceptible individuals with exposure to the disease, the quarantine proportion (q) is the same for non-infected as for infected people.
5. Quarantine and isolation are completely effective. Someone in quarantine or isolation cannot spread disease or, in the case of a susceptible, cannot catch the disease.

The populations considered are as follows:

susceptible (S) do not have but can catch SARS from infectious individuals.
susceptible_quarantined (S_Q) do not have SARS, quarantined because of exposure, so cannot catch SARS.
exposed (E) have SARS, no symptoms, not yet infectious.
exposed_quarantined (E_Q) have SARS, no symptoms, not yet infectious, quarantined because of exposure.
infectious_undetected (I_U) have undetected SARS, infectious.
infectious_quarantined (I_Q) have SARS, infectious, quarantined, cannot transmit.
infectious_isolated (I_D) have SARS, infectious, isolated, cannot transmit.
SARS_death (D) are dead due to SARS.
recovered_immune have recovered from SARS, immune to further infection.

Because we are assuming that quarantine is completely effective, only someone in the *susceptible* (S) category can catch SARS, and transmission to a susceptible can occur only through exposure to an individual in the *infectious_undetected* (I_U) category. Those with SARS in other categories are under quarantine or isolation or are not yet infectious.

Quick Review Question 5

After completing this question and before continuing in the text, we suggest that you make a diagram with stocks (box variables) and flows only to represent possible transitions between categories. For each of the following, give the possible category(ies):

a. Flows out of S into what categories?
b. Flows into S from what categories?
c. Flows into D from what categories?

Without inclusion of converters and connectors, Figure 4.3.4 displays a diagram with the stocks that represent these populations along with the flows between them. As illustrated, a susceptible individual who has had contact with someone having SARS and has moved from the *susceptible* group can be quarantined with or without the disease (to *exposed_quarantined* or *susceptible_quarantined*, respectively) or can be infected and not quarantined (to *exposed*). A susceptible, quarantined person who does not have SARS (in *susceptible_quarantined*) eventually is released from quarantine (to *susceptible*). An exposed but not yet infectious individual who does have SARS, whether quarantined or not (in *exposed_quarantined* or *exposed*, respectively), eventually becomes infectious (to *infectious_quarantined* or *infectious_undetected*, respectively). Regardless of quarantine status, an infectious individual can recover (to *recovered_immune*), go into isolation after discovery (to *infectious_isolated*), or die (to *SARS_death*). Isolated patients who are sick with SARS can recover or die.

Quick Review Question 6

Using this model, indicate if each of the following situations is possible or not:

- **a.** A susceptible person dies of SARS.
- **b.** A person who has undetected SARS in the early stages recovers without ever becoming infectious.
- **c.** Someone in quarantine diagnosed with SARS recovers without going into isolation.
- **d.** Someone who has recovered from SARS becomes infected with the disease again.
- **e.** Someone is transferred from isolation to quarantine.

The model employs the following parameters:

- b probability that a contact between person in *infectious_undetected* (I_U) and someone in *susceptible* (S) results in transmission of SARS
- k mean number of contacts per day someone from *infectious_undetected* (I_U) has. By assumption, the value does not depend on population density.
- m per capita death rate
- N_0 initial number of people in the population
- p fraction per day of exposed people who become infectious; this fraction applies to the transitions from *exposed* (E) to *infectious_undetected* (I_U) and from *exposed_quarantined* (E_Q) to *infectious_quarantined* (I_Q). Thus, $1/p$ is the number of days in the early stages of SARS for a person to be infected but not infectious.
- q fraction per day of individuals in *susceptible* (S) who have had exposure to SARS that go into quarantine, either to category *susceptible_quarantined* (S_Q) or to *exposed_quarantined* (E_Q)
- u fraction per day of those in *susceptible_quarantined* (S_Q) who are allowed to leave quarantine, returning to the *susceptible* (S) category; thus, $1/u$ is the number of days for a susceptible person to be in quarantine.
- v per capita recovery rate; this rate is the same for the transition from category *infectious_undetected* (I_U), *infectious_isolated* (I_D), or *infectious_quarantined* (I_Q) to category *recovered_immune*.

w fraction per day of those in *infectious_undetected* (I_U) who are detected and isolated and thus transferred to category *infectious_isolated* (I_D)

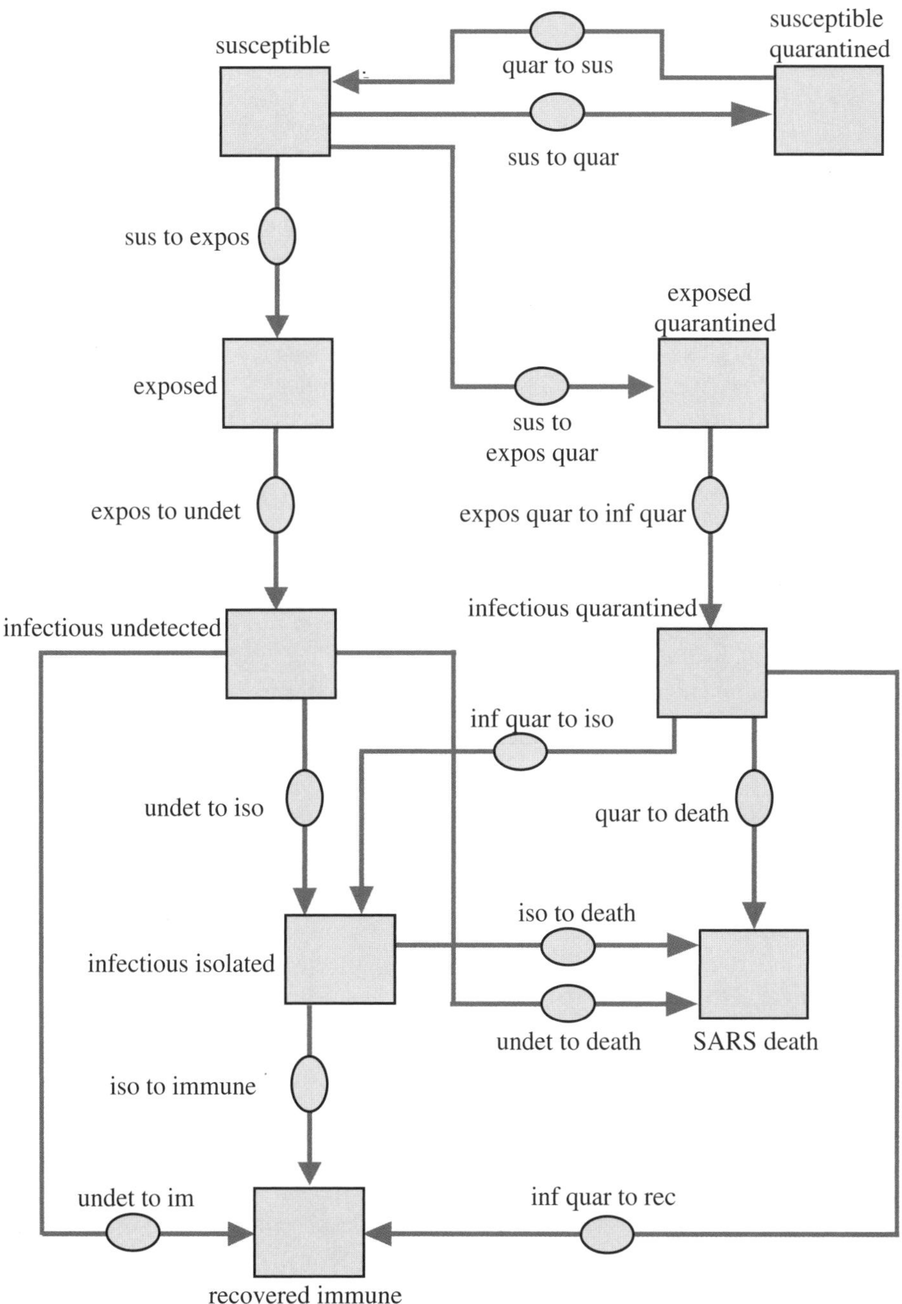

Figure 4.3.4 Initial diagram of relationships for SARS

Quick Review Question 7

a. Suppose it takes an average of 5 days for someone who has SARS but is not infectious to progress to the infectious stage. Give the value of p along with its units.

b. Give the formula for the rate of change of exposed individuals who are not quarantined to move into the phase of being infectious and undetected, from E to I_U.

c. Give the formula for the rate of change of exposed individuals who are quarantined to move into the phase of being infectious and quarantined, from E_Q to I_Q.

d. Suppose 10% of the people who have been in quarantine but who do not have SARS are allowed to leave quarantine each day. Give u and the average number of days for a susceptible person to be in quarantine.

e. Suppose the duration of quarantine is 16 days. If someone has not developed symptoms of SARS during that time period, he or she may leave quarantine. Give the corresponding parameter and its value.

f. Give the formula for the rate of change of susceptible, quarantined individuals leaving quarantine, from S_Q to S.

As illustrated in Figure 4.3.4, three paths exist for someone to leave *infectious_undetected* (I_U)—to *recovered_immune* at a rate of v, to *SARS_death* at a rate of m, or to *infectious_isolated* (I_D) at a rate of w. Thus, the total rate of change to leave *infectious_undetected* (I_U) is $(v + m + w)$/day. For example, if $v = 0.04$, $m = 0.0975$, and $w = 0.0625$, $v + m + w = 0.2$/day. In this case, $1/(v + m + w) = 5$ day is the average duration of infectiousness.

By assumption, k is the number of contacts an undetected infectious person has, regardless of population density. Thus, with N_0 being the initial population size, k/N_0 is the fraction per day of such contacts. Because b is the probability of transmitting the disease, the product $(k/N_0)b$ is the transmission constant. As in the SIR model, the product I_US gives the total number of possible interactions. Thus, $(k/N_0)b\ I_US = kbI_US / N_0$ is the number of new cases of SARS each day. Of these new cases, a fraction (q) go into category *exposed_quarantined* (E_Q), while the remainder, the fraction $(1 - q)$, go into *exposed* (E).

Quick Review Question 8

a. Suppose $k = 10$ contacts/day, and $N_0 = 10{,}000{,}000$ people. Give the percentage of contacts per day.

b. Suppose 6% of contacts between an infectious and a susceptible person result in transmission of the disease. Give the corresponding parameter, its value, and units.

c. Using your answers to Parts a and b, what percentage of all possible contacts results in transmission of SARS each day?

d. If the sizes of *infectious_undetected* (I_U) and *susceptible* (S) are 5000 and 9,000,000, respectively, give the total number of possible contacts.

e. Using your answers to Parts c and d, give the number of contacts per day that result in transmission of SARS.
f. Suppose $q = 0.1 = 10\%$ of the individuals who have had contact with an infectious person go into quarantine. Give the number of those from Part e who go into *exposed_quarantined* (E_Q).
g. Give the formula for the rate of change from *susceptible* (S) to *exposed_quarantined* (E_Q).
h. Assuming $q = 0.1$, give the number of those from Part e who go into *exposed* (E).
i. Give the formula for the rate of change from *susceptible* (S) to *exposed* (E).

For those transferring from *susceptible* (S) to *susceptible_quarantined* (S_Q), although they have been exposed to an infectious person, the disease was not transmitted to them. The fraction of total possible contacts, I_US, is (k/N_0), and the probability of nontransmittal is $(1 - b)$. Thus, the total number of nontransmission contacts is $(k/N_0)(1 - b)I_US = k(1 - b)I_US / N_0$. However, only a fraction (q) of those go into quarantine. Thus, the rate of change of those going from *susceptible* (S) to *susceptible_quarantined* (S_Q) is $qk(1 - b)I_US / N_0$.

Quick Review Question 9

Using the values from Quick Review Question 8, determine the rate of change of those going from *susceptible* (S) to *susceptible_quarantined* (S_Q).

Reproductive Number

Several exercises deal with the differential equations for this SARS model, and a project completes the model. In this model, an important value in evaluating the effectiveness of quarantine and isolation is the **reproductive number R**, which is the expected number of secondary infectious cases resulting from an average infectious case once the epidemic is in progress. The **basic reproductive number, R_0**, is the initial reproductive number with one infectious individual and all others being susceptible. For example, if at the start of a disease in an area the infectious individual transmits SARS to a mean of three other people who eventually become infectious, then the basic reproductive number is $R_0 = 3$. Such a number results in the alarming prospect of exponential growth of the disease. On the average, one person transmits infectiousness to three other people, who each cause three other people to become infectious, and so forth. In such a situation, at stage n of transmission, 3^n new people would eventually become infectious. For example, at stage $n = 13$, 3^{13}, or more than 1.5 million, new people, would get sick. Because of such exponential growth, it very important that R be less than 1. With $R < 1$, there is no epidemic. For $R > 1$, there is an epidemic. The larger the reproductive number, the more virulent the epidemic.

For this SARS model, on the average, an undetected infectious person has k contacts per day. At the beginning of the disease with all individuals except one being

> **Definitions** The **reproductive number *R*** is the expected number of secondary infectious cases resulting from an average infectious case once the disease has started to spread. The **basic reproductive number R_0** is the expected number of secondary infectious cases resulting from one infectious individual in a completely susceptible population.

susceptible, each such contact can result in the disease spreading. Thus, with a probability *b* of transmission, approximately *kb* secondary cases of SARS per day derive from the first infections individual. Thus, for mean disease duration of *D* days, the **basic reproductive number**, R_0, is *kbD*. Because the average duration of infectiousness is $1/(v + m + w)$ da (see explanation after Quick Review Question 7), without quarantine being a factor, one infectious person eventually gives rise to $R_0 = kb/(v + m + w)$ secondary infectious cases of SARS. However, when a fraction, *q*, go into quarantine so that a fraction (1 - *q*) do not, the reproductive number is $R_0 = \dfrac{kb}{v+m+w}(1-q)$. The larger *q* is, the smaller R_0 is, and the less severe the impact of the disease is.

> **Model:** A model of the basic reproductive number is as follows:
>
> $$R_0 = kbD$$
>
> where *k* is the mean number of contacts an undetected infectious person has per time unit (such as day), *b* is the probability of disease transmission, and *D* is the mean duration of the disease.

Quick Review Question 10

Evaluate the basic reproduction number, R_0, using the values of Quick Review Question 8 and text material: $k = 10$ contacts/da, $b = 0.06$, $v = 0.04$, $m = 0.0975$, $w = 0.0625$, and $q = 0.1$.

Examining R_0, the death rate, and other factors, WHO and other health organizations realized that they must act quickly with bold measures involving quarantine and isolation to avoid a major, worldwide epidemic of SARS. Computer simulations with scenario analyses verified the seriousness of the disease. Thanks to aggressive actions, a terrible catastrophe was averted.

Exercises

1. Write the system of differential equations for the SIR model using a transmission constant of 0.0058 and a recovery rate of 0.04.

In the SARS model, give the differential equation for each rate of change in Exercises 2–10.

2. dS_Q/dt **3.** dE/dt **4.** dE_Q/dt **5.** dS/dt **6.** dI_U/dt
7. dI_D/dt **8.** dI_Q/dt **9.** $d(recovered\_immune)/dt$ **10.** dD/dt

11. a. For basic reproductive number of $R_0 = 3$, give the number of new people that will eventually become infectious at stage $n = 10$ of transmission of the disease.
b. Give the total number of people who will eventually become infectious.
c. Repeat Part a for $n = 15$.
d. Repeat Part b for $n = 15$.

Projects

For additional projects, see Module 7.11, "Fueling Our Cells—Carbohydrate Metabolism"; Module 7.14, "Control Issues: The Operon Model"; and Module 7.15, "Troubling Signals: Colon Cancer."

1. Adjust the SIR model to allow for vaccination of susceptible boys. Assume that 15% are vaccinated each day, and make a simplifying assumption that immunization begins immediately. Discuss the effect on the duration and intensity of the epidemic. Consider the impact of other vaccination rates.
2. Adjust the SIR model to allow for vaccination of susceptible boys. Assume that 15% are vaccinated each day and that immunization begins after 3 days. Discuss the effect on the duration and intensity of the epidemic. Consider the impact of other vaccination rates.
3. Adjust the SIR model to allow for vaccination of susceptible boys. Assume that all children are vaccinated 2 days before a boy comes down with the flu and that immunization begins after 4 days. Discuss the effect on the duration and intensity of the epidemic. Consider the impact of other vaccination rates.
4. Develop an SEIR model of disease.
5. Complete the Lipsitch SARS model introduced in the text. Have the model evaluate R. Produce graphs and a table of appropriate populations, including *susceptible*, *recovered_immune*, *SARS_death*, and the total of the five categories of infecteds. Employ the following parameters: $k = 10$/day; $b = 0.06$; $1/p = 5$ days; $v = 0.04$, $m = 0.0975$, and $w = 0.0625$, so that $v + m + w = 0.2$/day and $1/(v + m + w) = 5$ days; $1/u = 10$ days; $N_0 = 10{,}000{,}000$ people. Vary q from 0 upward. Note that in each case, the graph of the number of susceptibles appears logistic and the solution eventually reaches equilibrium. Describe the shapes of the graphs and discuss the results.
6. After developing the model of Project 5, with a fixed value of q, test other ranges of k from 5 to 20 per day. Discuss the results.
7. After developing the model of Project 5, with a fixed value of q, test other ranges of $1/(v + m + w)$ from 1 to 5 days.
8. Adjust the model of Project 5 so that the simulation is allowed to run for a while before quarantine and isolation measures that reduce R to below 1 are instituted. Discuss the implications on the number of people quarantined and on the health care system of not taking aggressive measures initially.

9. Complete the Lipsitch SARS model introduced in the text. Run the simulation for various values of R_0. Produce graphs and a table of appropriate populations, including *susceptible*, *recovered_immune*, *SARS_death*, and the total of the five categories of infecteds. Describe the shift of the steady state as R_0 becomes larger, and discuss the implications.

10. Develop a model of strep throat. Bacterium Group A *Streptococcus* causes strep throat, which occurs most frequently in school-aged children. The bacterium spreads through direct or airborne contact with the mucus from an infected person. Symptoms start from 1 to 5 days after exposure and include fever, sore throat, and tender and swollen neck glands. If untreated, people with strep throat are infectious for 10 to 21 days. Usually, 24 h after antibiotic treatment, those who are ill are no longer contagious. The spread of strep throat can be minimized by infectious people covering their mouths when sneezing or coughing and by washing their hands frequently (IDEHA).

11. Develop a model of the viral infection mumps. Symptoms include painful and swollen salivary glands, painful swallowing, fever, weakness, fatigue, and a tender, swollen testicle. Infection is spread through breathing of infected saliva droplets. About one-third of those with mumps experience no symptoms. If present, symptoms usually start 2 to 3 weeks after infection. The person is contagious from approximately 1 day before salivary gland swelling occurs and remains contagious for at least another 3 days. As the swelling diminishes, so does the degree of the contagion. Before licensing of the mumps vaccine in 1967, the United States had more than 200,000 cases per year. Since then, the country has had fewer than 1000 cases per year (Mayo Clinic Staff 2012).

12. Diphtheria has been virtually eradicated in the United States because of a vaccine, which was introduced in the 1920s. Before that time, the United States had 100 to 200 cases per 100,000 people. The disease is still a problem in developing countries. Two types of diphtheria exist, respiratory and cutaneous. The former is more serious, and death results in about 10% of those cases. The disease is spread through respiratory droplets and from contaminated objects or food. The incubation period for the disease is usually 2 to 5 days. Develop a model for respiratory diphtheria (NCBI).

13. Using data and mathematical models implemented in spreadsheets, the Dutch Ministry of Health, Welfare and Sports developed "a national plan to minimize effects of pandemic influenza." Through scenario analysis, scientists examined various intervention options and estimated the number of hospitalizations and deaths. In the base case, in which no intervention was possible, they assumed 30% of the population would become ill with influenza. In the Influenza Vaccination Scenario, they considered two strategies:

1. Vaccinate two risk groups, persons 65 years of age or older ($N = 2.78$ million (M)) and healthcare workers ($N = 0.80$ M)
2. Vaccinate the total population ($N = 15.6$ M)

They assumed the vaccine to be 56% effective in preventing hospitalizations and deaths for the older at-risk group and 80% effective for those younger than 65. Develop a model for the first strategy. With no intervention, assume a hospitalization rate (per 100,000) for influenza and influenza-related illnesses of 125 (per 100,000) for persons 65 years of age or older and a rate (per 100,000) of 50 for the younger age group; and assume death rates (per 100,000) of 56 and15, respectively, for the two age groups. (In the actual study, scientists considered three age groups and a more involved set of input variables; van Genugten et al. 2003)

14. Develop a model for the second strategy in Project 13.

15. Develop models for the two strategies in Project 13, discuss the results, and make recommendations.

16. Adjust a SIR model to have seasonal changes in infectiousness by having a periodic function for a transmission coefficient (see Project 1 in Module 4.2, "Predator-Prey Model"). Discuss the results.

17. Repeat Project 16 for a SEIR model.

18. Obtain information and data about another infectious disease, where the disease spreads from one individual to another. Model at least one aspect of the spread of the disease, starting with one infected individual in a particular area. Run the model for various scenarios, produce graphs and tables, and discuss the results. The following are some suggested diseases: pinkeye in cattle (see "Introduction" in Module 11.2, "Agents of Interaction: Steering a Dangerous Course"), rotavirus, pertussis, meningitis, bacterial/viral pneumonia, cold (rhinovirus), tuberculosis, various STD's, impetigo, herpes (cold sores).

Answers to Quick Review Questions

1. $dR/dt = aI$

2. **a.** negative while people are getting sick because the number of susceptibles is decreasing
b. $dS/dt = -rSI$

3. $dI/dt = rSI - aI$

4. **a.** day 7
b. day 6
c. day 8

5. **a.** E, E_Q, S_Q
b. S_Q
c. I_U, I_Q, I_D

6. **a.** no
b. no
c. yes
d. no
e. no

7. **a.** 0.2/day
b. pE
c. pE_Q
d. 0.1/day, 10 days
e. u = 1/16 per day = 0.0625/day
f. $u\, S_Q$

8. **a.** $k/N_0 = 10/10{,}000{,}000 = 0.000001 = 0.0001\%/\text{day}$
b. $b = 0.06/\text{day}$
c. $(0.000001)(0.06) = 0.00000006 = 0.000006\%/\text{day}$
d. $(5000)(9{,}000{,}000) = 45{,}000{,}000{,}000$
e. $(0.00000006)(45{,}000{,}000{,}000) = 2700$
f. $(0.1)(2700) = 270$ people
g. $qkbI_US / N_0$
h. $(1 - 0.1)(2700) = (0.9)(2700) = 2430$ people; or $2700 - 270 = 2430$ people
i. $(1 - q)kbI_US / N_0$

9. $qk(1 - b)I_US / N_0 = (0.1)(10)(1 - 0.06)(5000)(9{,}000{,}000)/(10{,}000{,}000) = 4230$ people

10. $R_0 = (1 - q)kb/(v + m + w) = (1 - 0.1)(10)(0.06)/(0.04 + 0.0975 + 0.0625) = 2.7$

References

CDC (Centers for Disease Control and Prevention). "Severe Acute Respiratory Syndrome. Supplement F: Laboratory Guidance." Department of Health and Human Services. http://www.cdc.gov/sars/guidance/F-lab/app7.html (accessed November 17, 2012)

IDEHA (Infectious Disease and Environmental Health Administration). "Strep Throat Fact Sheet." Maryland Department of Health and Mental Hygiene. http://ideha.dhmh.maryland.gov/SitePages/Strep-Throat.aspx (accessed November 17, 2012)

Kermack, W. O., and A. G. McKendrick. 1927. "A Contribution to the Mathematical Theory of Epidemics." *Proceedings of the Royal Society of London*, Series A, 115(772) (August 1): 700–721.

Lipsitch, Marc et al. 2003. "Transmission Dynamics and Control of Sever Acute Respiratory Syndrome." *Science* 300(5627): 1966–1970.

Mayo Clinic Staff. 2012. "Mumps." Mayo Foundation for Medical Education and Research, August 5. http://www.mayoclinic.com/health/mumps/DS00125

Murray, J. D. 1989. *Mathematical Biology*. New York: Springer-Verlag.

NCBI (National Center for Biotechnology Information). "Diphtheria." National Library of Medicne. National Institutes of Health http://www.ncbi.nlm.nih.gov/pubmedhealth/PMH0002575/ (accessed November 17, 2012)

NCSLP (National Computational Science Leadership Program). "Influenza Epidemic in a Boarding School." The Shodor Educational Foundation, Inc. Original from J. D. Murray, *Mathematical Biology*, New York: Springer-Verlag, 1989. http://www.shodor.org/ncslp/talks/basicstella/sld009.htm (accessed November 17, 2012)

van Genugten, Marianne L. L., Marie-Louise A. Heijnen, and Johannes C. Jager. 2003. "Pandemic Influenza and Healthcare Demand in the Netherlands: Scenario Analysis." *Emerging Infectious Diseases* 9(5), Centers for Disease Control and Prevention (CDC). http://www.medscape.com/viewarticle/453679 (accessed November 17, 2012)

WHO (World Health Organization). "Severe Acute Respiratory Syndrome." The United Nations. http://www.who.int/topics/sars/en/ (accessed November 17, 2012)

MODULE 4.4

Modeling a Persistent Plague: Malaria

Download

The text's website has a *malaria* file, which contains the model of this module, available for download for various system dynamics tools.

Introduction

How important to civilization can a mosquito-borne protozoan be? Unfortunately, we cannot ask the ancient Romans, but we do have some recent evidence that implicates this tiny parasite in the fall of one of the mightiest empires of all time. Excavations in a cemetery near Lugano, Italy, have uncovered at least one infant from AD 450 that yielded the DNA of *Plasmodium falciparum*—the deadliest of all the human malarias. Nearby, 50 other infants, who also showed fingerprints of this parasite, were buried in a relatively short period of time.

The death of many infants would be expected during a malaria epidemic, partially because *falciparum* induces high rates of miscarriages and infant death. The mosquitoes that transmit malaria flourish in marshy areas found in the Tiber River valley; and if malaria swept through Rome, the disease may indeed have contributed to its downfall. Even if Roman troops were not directly affected by disease, disruptions to the production and supply of food and war materials could have drastically impaired the military's ability to protect Rome.

Interestingly, around the time the infant lived, Attila's Huns were pillaging in the north of Italy en route to Rome. Although legend credits Pope Leo the Great with persuading Attila to withdraw, it is more likely that the presence of malaria in the city was even more convincing (Carroll 2001).

Malaria is a very old disease (probably prehistoric), originating in Africa, spreading as humankind migrated to other lands. The disease gets its name from an Italian word for "bad air" (RPH).

After more than 1500 yr, we still have mosquitoes and malaria. In fact, the WHO estimates that malaria sickens 216 million people, killing more than 655,000 of them, each year—that is almost 1800 per day. Most of these fatalities are among African children (WHO, World Malaria Report). In fact, 1 out of 20 African children die of malaria before the age of 5. (NetMark).

So long as Woman has walked the earth,
malaria may have stalked her.

—Duffy et al.

Pregnant women and their developing young seem to be at special risk from infection, which seems to have been true from the earliest human records (Duffy et al. 2001). *P. falciparum* malaria in pregnant women is associated with high levels of maternal and fetal morbidity and mortality (Desai et al. 2007). The anemia associated with malaria infections results in approximately 10,000 maternal deaths and thousands of low-birth-weight infants. Up to 200,000 of such infants in Africa may die annually (ter Kuile and Rogerson 2008).

Background Information

A **vector** is an animal that transmits a pathogen, or something that causes a disease, to another animal. Mosquitoes are the only vectors for malaria, but only 60 out of the 380 species of ***Anopheles*** mosquitoes can host malaria-causing ***Plasmodium*** (Ryan 2008).

Three-fifths of the female *Anopheles* mosquitoes, like their sisters of other lines, are dependent on blood meals to feed their maturing eggs. While sipping blood, a *Plasmodium*-infected female mosquito injects thread-shaped, infectious agents called **sporozoites** into her human **host**. Sporozoites circulate for a time and then enter the parenchymal cells of the liver to hide out from the immune system. Here, they live for 1 to 2 weeks, multiplying asexually to produce thousands of offspring, which mature into other invasive cells, **merozoites**. Eventually, all this activity causes the parenchymal cell to break open and release merozoites into the blood. In other malaria-causing parasites, *Plasmodium vivax* and *ovale*, some of the sporozoites become dormant **hypnozoites**. Later, these mature to reinvade other liver cells, where they continue to produce more merozoites, causing recurring bouts with malaria. Interestingly, the most deadly species, *Plasmodium falciparum*, does not produce these hypnozoites (Despommier et al. 2005; NIAID; Wiser).

Merozoites enter red blood cells to feed on the blood. They reproduce asexually to form more merozoites, which invade other red blood cells. This cycle continues unless stopped by the body's defenses or medicine (NIAID).

While in the red blood cells, some merozoites mature into **male** and **female gametocytes**. Upon release, these do not enter the red blood cells, but circulate, awaiting transfer to the mosquito host. The female mosquito takes her blood meal from an infected host and simultaneously sucks up some of the gametocytes.

In the mosquito's stomach, the male gametocyte (sperm) and the female gametocyte (egg) fuse. The resulting **oocyst** divides to produce thousands of sporozoites. The sporozoites migrate to the salivary glands of the mosquito awaiting their journey into a vertebrate host. Figure 4.4.1 diagrams the life cycle of *Plasmodium falciparum*.

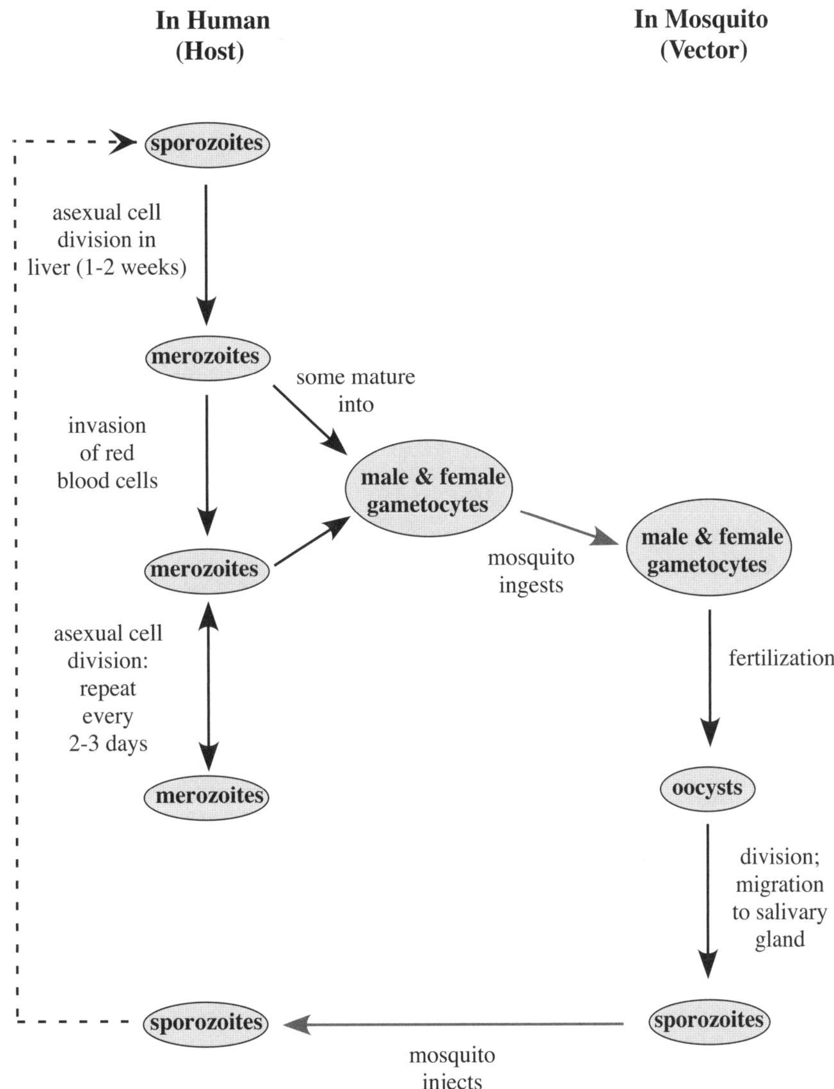

Figure 4.4.1 Life cycle of *Plasmodium falciparum*

Analysis of Problem

In the discussion that follows in this module, we consider the modeling process involving malaria (see Module 1.2, "The Modeling Process"). We begin by analyzing the situation to identify the problem and understand its primary questions.

In this problem, we wish to investigate the progress of malaria. In particular, we consider the relationships between human and *Anopheles* mosquito populations, both of which are necessary for the life cycle of *Plasmodium*. Thus, with a system dynamics model, we wish to study the changing numbers of various categories of humans and mosquitoes as time progresses.

Formulating a Model: Gather Data

Data on malaria are often difficult to find and may be undependable. Countries with high rates of malaria are also often desperately poor, and the effectiveness of data collection can vary dramatically from year to year. Moreover, climate can play a significant role, with the number of cases of malaria correlating to periods of high rainfall. Also, the values associated with mosquitoes, such as numbers in each category, birth and death rates, bite probabilities, and constants of proportionality, are usually not available.

In a computational science study, such as of malaria, an interdisciplinary team approaches a problem from many directions. "Wet-lab" team members conduct initial experiments and, with their "dry-lab" counterparts, pose questions for the latter to consider in modeling. In formulating a model, the group may uncover the need for additional parameters, such as birth and death rates of anopheline mosquitoes. If not available from other sources, the team may decide to conduct additional experiments to collect data for empirical computations of such values.

Websites for the WHO, the U.S. Central Intelligence Agency (CIA), and other organizations do provide some enlightening and startling data concerning people. For example, the entire population of Malawi lives in malarious areas. In 2012, the population of Malawi was more than 16 million, with a birth rate of 40.42 births/1000 population, a death rate of 12.84 deaths/1000 population, and a life expectancy of only 52.31 years (CIA 2012). According to the Kaiser Family Foundation, in 2010, there were 6,851,108 reported cases of malaria, thousands of which will result in death (KFF 2010). Pregnant women and children under the age of 5 are at particular risk. Infected mothers are more likely to miscarry or to experience intrauterine demise, premature delivery, low-birth-weight neonates, and neonatal death. They are also more likely to develop anemia and/or die during delivery (Schantz-Dunn and Nour 2009).

Formulating a Model: Make Simplifying Assumptions

For our first model of malaria, we make several simplifying assumptions. We model the serious form of malaria that *Plasmodium falciparum* causes, in which relapses do not occur. In the model, we primarily consider the number of individuals in several categories of humans and mosquitoes and ignore *Plasmodium*.

Quick Review Question 1

Considering the simplifying assumption in the preceding paragraph, give the major submodels of the malaria model.

Because the life expectancy of a human is much greater than that of a mosquito, we assume that the population of humans is closed with no births, no immigration, and no deaths except from malaria. We presume that as soon as a vector bites a human, the individual becomes a host. No immunity exists for uninfected individuals, and no incubation period occurs. Some human hosts eventually become immune and others die, while still others recover and become susceptible again. We ignore the chance of relapse. Deceased individuals pass from consideration in the model.

Because of their relatively short life expectancy, we do consider mosquito births and deaths. We have the assumption that the death rates for infected and uninfected mosquitoes are identical. Similarly, we assume that all mosquitoes reproduce at the same rate. At birth, a mosquito is uninfected. As a simplification for this first version of the model, we suppose that an infected mosquito immediately becomes a host that can infect humans.

Quick Review Question 2

Based on these assumptions, list major categories of organisms for a model. In a system dynamics diagram, we represent these categories, which can accumulate individuals, as stocks (box variables).

Also, for simplification in this version of the model, except where relevant interactions between mosquitoes and humans occur, let us assume that the number of organisms in each category (uninfected humans, human hosts, immune humans, uninfected mosquitoes, mosquito vectors) expands or contracts in an unconstrained manner. In such situations, constraints, such as competition for food or predators, do not exist.

Formulating a Model: Determine Variables and Units

Based on these simplifying assumptions, we monitor three categories of humans, employing the following variables with the basic unit being one person:

uninfected_humans, who are susceptible to the disease
human_hosts, who have malaria and can infect mosquitoes that bite them
immune, who cannot get the disease again

For the mosquito submodel, as with the human submodel, we do not count the number of dead individuals. Consequently, assuming no incubation period for *Plasmodium*, we consider the following two categories of mosquitoes:

uninfected_mosquitoes, which do not carry *Plasmodium*
vectors, which carry *Plasmodium*

We employ a day as the basic unit of time, ***t***.

Quick Review Question 3

This question considers the relationships among these categories of humans and mosquitoes. After completing the question, we recommend that you develop a relationship diagram with stocks (box variables) representing the categories and with appropriate flows. Making the foregoing simplifying assumptions, give the requested flow information by selecting from the categories *uninfected_humans; human_hosts; immune; uninfected_mosquitoes; vectors;* and undesignated "clouds," such as for deceased humans, dead mosquitoes, and unborn mosquitoes:

a. Destination(s) of flow(s) from *uninfected_humans*
b. Source(s) of flow(s) to *uninfected_humans*
c. Destination(s) of flow(s) from *human_hosts*
d. Source(s) of flow(s) to *immune*
e. Source(s) of flow(s) to undesignated "cloud(s)" for deceased humans
f. Destination(s) of flow(s) from *uninfected_mosquitoes*
g. Source(s) of flow(s) to *uninfected_mosquitoes*
h. Source(s) of flow(s) to *vectors*
i. Destination(s) of flow(s) from *vectors*
j. Source(s) of flow(s) to undesignated "cloud(s)" for dead mosquitoes

Formulating a Model: Establish Relationships

Based on biology of the organisms and the simplifying assumptions, Figure 4.4.2 presents a relationship diagram with the two major submodels for humans and mosquitoes, stocks (box variables) representing the three human and two mosquito categories, and appropriate flows between stocks. Arrows (connectors) represent the

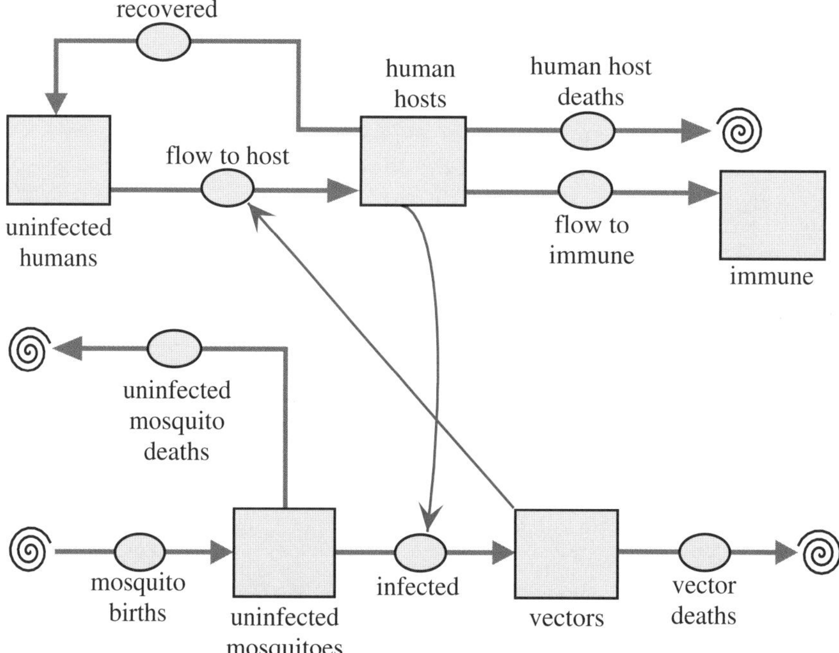

Figure 4.4.2 Relationship diagram

impact of one population on the other. For example, the infected mosquito category (*vectors*) is a necessary component in an uninfected human becoming infected, a member of *human_hosts*. Similarly, for an uninfected mosquito to become to a vector, the former must bite a person in *human_hosts*.

Quick Review Question 4

One of the simplifying assumptions is that, except where interactions between mosquitoes and humans occur, the number of organisms in a category (stock or box variable in Figure 4.4.2) exhibits unconstrained expansion or contraction. Give requested differential equations that utilize this assumption. Incorporate additional proportionality constants as needed.

a. *d*(*immune*)/*dt*
b. *d*(*deceased_humans*)/*dt*
c. rate of change from *human_hosts* to *uninfected_humans*
d. *d*(deceased_vectors)/*dt*

Formulating a Model: Determine Equations and Functions

Because we assume unconstrained growth or decay except where interactions between mosquitoes and humans occur, the following flow equations with constants of proportionalities in boldface correspond to proportionalities:

flow_to_immune = ***immunity_rate*** * *human_hosts*
human_host_deaths = ***malaria_induced_death_rate*** * *human_hosts*
recovered = ***recovery_rate*** * *human_hosts*
mosquito_births = ***mosquito_birth_rate*** * *mosquitoes*, where *mosquitoes* = *uninfected_mosquitoes* + *vectors*
uninfected_mosquito_deaths = ***mosquito_death_rate*** * *uninfected_mosquitoes*
vector_deaths = ***mosquito_death_rate*** * *vectors*

Because only uninfected mosquitoes are born and both categories of mosquitoes reproduce, *mosquito_births* is proportional to the total number of mosquitoes. Moreover, we assume that the death rate is the same for uninfected mosquitoes and vectors, so the last two equations have the same constant of proportionality, *mosquito _death_rate*.

Quick Review Question 5

Suppose for a simulation that the change in time (Δt) from one time step to another is 0.1 da.

a. Give the unit of measure for *d*(*immune*)/*dt*.
b. If at day 6 *immunity_rate* is 0.2, *human_hosts* is 500, and *immune* is 400, using the technique discussed in the section on "Difference Equation" from Module 2.2, "Unconstrained Growth and Decay," estimate the number of im-

mune people at day 6.1. (This technique is Euler's method, which Module 6.2 considers in greater detail.)

For uninfected humans, we have individuals entering from and exiting to the population of human hosts. The differential equation contains a positive term for a growth component with constant *recovery_rate* while subtracting a decay term, as follows:

$$d(uninfected\_humans)/dt = (recovery\_rate)(human\_hosts) - (transmission\_constant)(uninfected\_humans) \quad (1)$$

We can break *transmission_constant* into two factors, the probability that a human is bitten by a mosquito (***prob_bit***) and the probability that a mosquito is a vector (***prob_vector***). The product of these two probabilities forms the transmission constant. For example, if the probability that someone is bitten by a mosquito is 60% = 0.60 and the probability that a mosquito is a vector is 20% = 0.20, then the probability that a human is bitten by a vector is (0.60)(0.20) = 0.12 = 12%. With no presumed immunity, the transmission constant is equal to this probability. Thus, the following differential equation reflects a refinement of Equation 1:

$$d(uninfected\_humans)/dt = (recovery\_rate)(human\_hosts) - (prob\_bit)(prob\_vector)(uninfected\_humans) \quad (2)$$

The probability of a vector is the quotient of the number of vectors (*vectors*) and the total number of mosquitoes (*mosquitoes*). Thus, we have the following equation:

$$prob\_vector = vectors/mosquitoes$$

Substituting into Equation 2, the rate of change of *uninfected_humans* is as follows:

$$d(uninfected\_humans)/dt = (recovery\_rate)(human\_hosts) - (prob\_bit)(vectors)(uninfected\_humans)/mosquitoes$$

Similar to the situation for humans, the rate of change from uninfected mosquito to vector is the product of a rate and *uninfected_mosquitoes*. Assuming that an uninfected mosquito that bites a human host always becomes infected, we again break the rate into two factors, the probability that the mosquito bites a human (***prob_bite_human***) and the probability that a human is a host (***prob_host***). Thus, the differential equation for the rate of change from uninfected to infected mosquito (*vector_formation*) is as follows:

$$d(vector\_formation)/dt = flow\_to\_host = (prob\_bite\_human)(prob\_host)(uninfected\_mosquitoes) \quad (3)$$

The probability that a mosquito is a vector (*prob_vector*) and the probability that a human is a host (*prob_host*) are the connections between the models for humans and mosquitoes. For ***humans***, being the total number of humans (*uninfected_humans* + *human_hosts* + *immune*), we have the following identities:

$$prob\_host = human\_hosts/humans$$
$$= human\_hosts/(uninfected\_humans + human\_hosts + immune)$$

Solving the Model

We can use a system dynamic tool to help model the spread of malaria, perform a simulation, and generate graphs and tables of the results. Figure 4.4.3 pictures a human submodel of the malaria model. A converter/variable, *humans*, stores the sum of the quantities in the three stocks (box variables) for humans. Other converters/variables store constants of proportionality and probabilities. For example, *immunity_rate* might store the constant 0.01 to indicate that the rate at which human hosts become immune from malaria is 1% a day. An initial value 1 for *human_hosts* would indicate that the human population has one human host at the start of the simulation.

Quick Review Question 6

Consider Figure 4.4.3.

a. Give the number of terms in the differential equation for *d*(*human_hosts*)/*dt*.
b. Give the number of these terms that contribute to an increase in *human_hosts*.

The mosquito submodel in Figure 4.4.4 also has four flows, with associated converters/variables for constants of proportionality and probabilities. Similar to the human submodel, a converter/variable, *mosquitoes*, contains the sum of the populations for *uninfected_mosquitoes* and *vectors*.

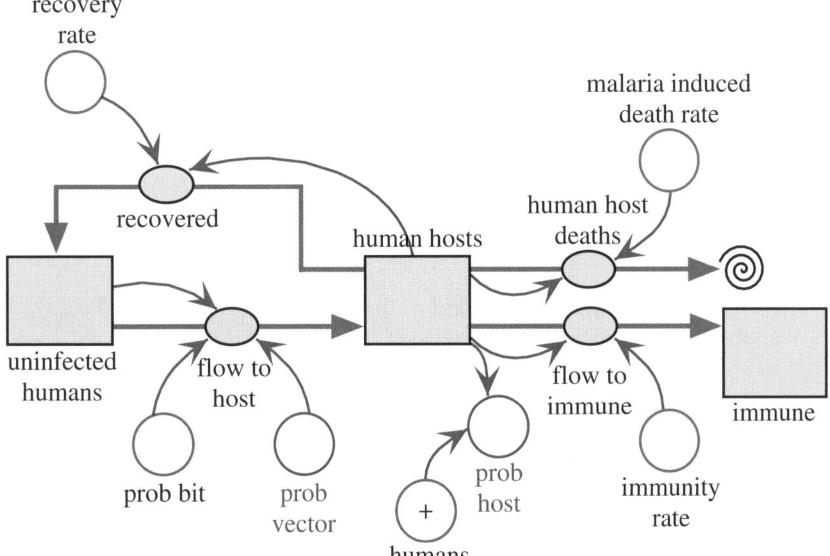

Figure 4.4.3 Human submodel for a closed system

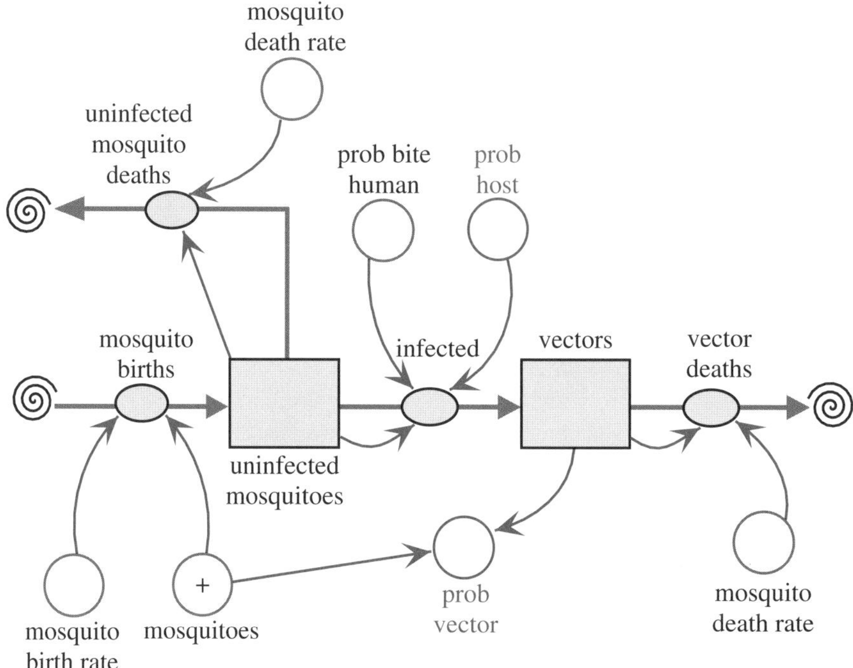

Figure 4.4.4 Mosquito submodel

Converters/variables for the probability of a vector (*prob_vector*) and the probability of a host (*prob_host*) appear in both submodels and in color in the figures. We calculate *prob_vector* in the mosquito submodel and use it in the human submodel. Symmetrically, the calculation for *prob_host* is in the human submodel, while the mosquito submodel employs the result.

Quick Review Question 7

Give the difference equation to estimate *vectors*(t) using the technique discussed in the section "Difference Equation" from Module 2.2, "Unconstrained Growth and Decay."

We specify a simulation length of 200 days and $\Delta t = 0.0625$ days with Euler's method for the integration technique. Equation Set 4.4.1 shows parameters for one run of the simulation. We begin with equal numbers of uninfected mosquitoes and humans (300), no vectors or immune humans, and one human host. (However, we could change the units and consider, for example, the number of humans in the thousands and the number of mosquitoes in the millions.) From such a small incidence, we hope to observe the dramatic spread of the disease to become an epidemic. In this run of the simulation, for humans we make the rates of immunity, recovery to being susceptible once more, and malaria death be 1%, 30%, and 0.5% per day, respectively. We give the probability that a human is bitten by a mosquito or that a mosquito bites a human as 30%/day. For a constant number of mosquitoes, we make their birth and death rates equal, in this case 1%/day.

Equation Set 4.4.1

Parameters for one run of the simulation:

$uninfected\_humans(0) = 300$
$human\_hosts(0) = 1$
$immune(0) = 0$
$prob\_bit = 0.3$
$recovery\_rate = 0.3$
$immunity\_rate = 0.01$
$malaria\_induced\_death\_rate = 0.005$
$mosquito\_birth\_rate = 0.01$
$mosquito\_death\_rate = 0.01$
$vectors(0) = 0$
$uninfected\_mosquitoes(0) = 300$
$prob\_bite\_human = 0.3$

Verifying and Interpreting the Model's Solution

Figure 4.4.5 presents the graphs of the five stocks, and a table of values with a reporting interval of 1 day appears in Table 4.4.1. Over 80 days, we observe a dramatic drop from 300 to a minimum of about 24 uninfected mosquitoes and a corresponding rise from 0 to a maximum of about 276 vectors. Afterward, the number of uninfected mosquitoes begins to increase, while the number of vectors drops. With equal birth and death rates, the population of mosquitoes holds constant at 300, which helps to verify the solution.

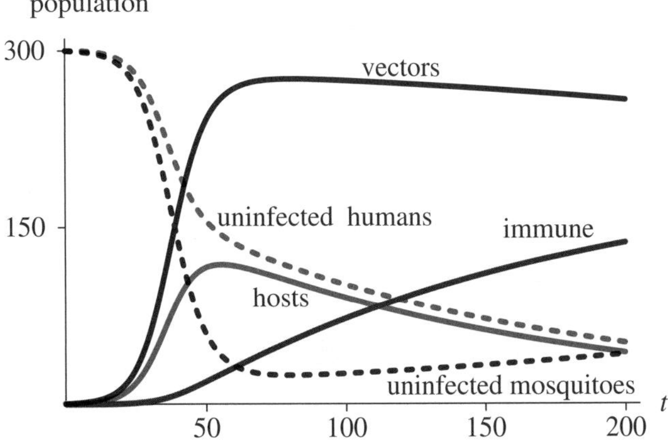

Figure 4.4.5 Graphs resulting from simulation with parameters of Equation Set 4.4.1 and time, t (da)

Table 4.4.1
Table of Values Corresponding to Graphs of Figure 4.4.5

Time	*Uninfected Humans*	*Human Hosts*	*Immune*	*Uninfected Mosquitoes*	*Vectors*
0	300.00	1.00	0.00	300.00	0.00
10	299.34	1.53	0.09	297.52	2.48
20	292.08	8.20	0.48	286.58	13.42
30	261.42	35.96	2.41	239.30	60.70
40	199.40	88.69	8.61	136.33	163.67
50	156.09	116.17	19.16	59.11	240.89
60	136.89	117.66	30.97	32.17	267.83
70	125.57	111.74	42.46	25.29	274.71
80	116.63	104.46	53.28	24.15	275.85
90	108.70	97.25	63.36	24.65	275.35
100	101.44	90.44	72.75	25.65	274.35
110	94.72	84.08	81.47	26.86	273.14
120	88.48	78.16	89.58	28.19	271.81
130	82.67	72.65	97.12	29.62	270.38
140	77.28	67.53	104.13	31.13	268.87
150	72.27	62.77	110.64	32.73	267.27
160	67.61	58.35	116.69	34.41	265.59
170	63.27	54.25	122.32	36.19	263.81
180	59.24	50.43	127.56	38.06	261.94
190	55.49	46.88	132.42	40.03	259.97
200	52.00	43.58	136.94	42.10	257.90

Trailing the rapid increase in the number of vectors is a fast decrease in the number of uninfected humans and quick increase in the number of human hosts over about a 25-day period. The number of hosts reaches a maximum of about 119 about day 55. Then, the number of hosts gradually falls, while the number of uninfected humans continues declining, but not as rapidly. Eventually, the two graphs appear almost parallel, while the graph of the number of immune humans increases in a concave-down fashion.

Extending the length of the simulation to 1500 days, we obtain the graphs of Figure 4.4.6. The numbers of uninfected humans and human hosts approach zero, and most of the surviving humans are immune. The total number of humans is reduced by about one-third to around 200. With the number of vectors tending to zero, the vast majority of the mosquitoes are uninfected. Because most humans are immune and almost no mosquito carries *Plasmodium*, malaria is virtually eradicated.

The results seem reasonable under the assumptions and simplifications. However, considerations suggest several stages of refinement.

We have extended the length of the simulation to more than 4 years and presumed no births or deaths from causes other than malaria for humans. Also, we have not considered the incubation period for *Plasmodium*.

For another model we might consider a different form of malaria, in which an individual could have a relapse of the disease. Moreover, expectant mothers tend to have lower-birth-weight children and more miscarriages, which affects the birth and death rates for humans. Also, data show that children have a higher death rate from

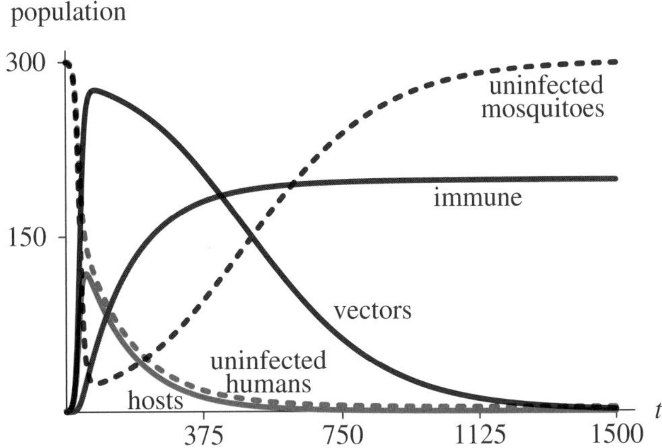

Figure 4.4.6 Simulation run for 1500 da

malaria than older humans. Consequently, other model refinements could involve expanding the number of human categories with varying death rates. Various projects consider such refinements.

Exercises

1. Discuss possible factors that could contribute to the situation in which 90% of malaria cases occur in Africa, south of the Sahara.
2. Give a differential equation for the rate of change of *human_hosts*.
3. Give a differential equation for the rate of change of *uninfected_mosquitoes*.
4. Give a differential equation for the rate of change of *vectors*.

Projects

For additional projects, see Module 7.11, "Fueling Our Cells—Carbohydrate Metabolism"; Module 7.14, "Control Issues: The Operon Model"; and Module 7.15, "Troubling Signals: Colon Cancer."

1. Run the simulation for the following situations. Describe and explain the long-term results.
   a. Various initial values of stocks (box variables)
   b. Slightly higher birth rate than death rate for mosquitoes
   c. No human host and one vector
   d. Zero death rate for humans
   e. Probability that a human is bitten reduced by a factor of 10 to 3%
   f. Probability that a mosquito bites a human reduced by a factor of 10 to 3%
2. Refine the malaria model of this module to accommodate human births and deaths.

3. Refine the malaria model of Project 2 of this module to include a stock (box variable) of tainted mosquitoes, who are infected but not yet vectors. In these mosquitoes, the *Plasmodium* protozoans are in incubation.
4. Develop an alternative implementation of the model for Project 3 that employs a conveyor for the tainted mosquitoes. Because we are considering births and deaths for mosquitoes, the inflow multiplier from the conveyor to the stock of vectors is not 1. However, the value is not 1 – *mosquito_death_rate*, which is 0.99 for a death rate of 0.01, because the inflow multiplier applies only to the number exiting the conveyor at that time step, not over the period of incubation. Consequently, you must employ the actual, accumulated survival rate over the period of incubation as a multiplier. You can use mathematics or your system dynamics tool to compute the actual survival rate for the number exiting the conveyor.

   Because we only remove mosquitoes from the conveyor at the end of the incubation period, using the number in the conveyor to calculate the total number of mosquitoes, *mosquitoes*, results in an overestimate for *mosquitoes*. Thus, for this project, assume a constant number of mosquitoes, so *mosquitoes* is the sum of the initial values of the various mosquito stocks.
5. Develop an alternative implementation of the model for Project 3 that employs a separate stock (block variable) for each day of incubation. A portion (*mosquito_death_rate*) of the mosquitoes is siphoned off each day, and the remainder is transferred to the next day's stock or, after the final day of incubation, to the stock *vector*.
6. Model malaria caused by *Plasmodium vivax* or *ovale* in which a human host can go into remission and have relapses.
7. Refine one of the previous models to reflect a seasonal increase in the number of mosquitoes, such as in a rainy season (see Project 1 in Module 4.2, "Predator-Prey Model").
8. Using one of the previous models, consider the effect on the epidemic of distribution of a prophylactic drug, such as Malazone, which travelers take to prevent malaria. Suppose everyone in the population takes the drug. Investigate varying degrees of effectiveness. Such drugs are expensive, especially relative to the economy of populations in which malaria thrives. Discuss the practicality of such treatments.
9. Using one of the previous models, consider the effect on the epidemic of using insecticides to control the mosquito population. Investigate varying degrees of insecticide effectiveness. Discuss the practicality of such an approach relative to the ecosystem.
10. Starting with the model from Project 8, consider the effect on the epidemic of a combined approach consisting of distribution of a prophylactic drug, mosquito netting, and use of insect repellant and insecticides.
11. Refine a malaria model of Projects 2–5 to consider that a person does not obtain permanent, complete immunity from malaria, but only temporary, partial immunity.
12. Refine a malaria model by expanding the number of human categories with varying contraction and death rates. In particular, data shows that *falciparum* is lethal to children under 5 years old. Each day, approximately 3000

children under the age of 5 die from malaria. In some of the worst areas, it is estimated that more than 40% of the toddlers die from the disease. More than 30% of the children in Africa get malaria by time they are 3 months old. However, approximately, one-eighth of the children in some countries of sub-Saharan Africa are born with sickle cell anemia, which makes it more difficult for them to contract malaria.

13. Because mothers are more likely to suffer malarial relapses during pregnancy, malaria is an important cause of low weight births and stillbirths. More than half of the miscarriages in endemic areas are caused by malaria. Adjust the model for Project 12 to reflect this information.

14. Adjust one of the earlier models to have constrained growth with a carrying capacity for humans and a carrying capacity for mosquitoes. Examine and discuss the effects of these changes. A logistic equation can model such constrained growth (see Module 2.3, "Constrained Growth").

15. Obtain information and data about another infections disease, where the disease spreads through a vector. Model at least one aspect of the spread of the disease, starting with one infected individual in a particular area. Run the model for various scenarios, produce graphs and tables, and discuss the results. The following are some suggested diseases: West Nile virus, Lyme disease, Chagas' disease, bubonic plague, typhus.

16. Repeat Project 16 for a zoonotic disease, which can spread from one animal species to another. The following are some suggested diseases: Nipah virus, hantavirus, avian flu, rabies, tularemia.

Answers to Quick Review Questions

1. humans and mosquitoes

2. uninfected humans, human hosts, immune humans, uninfected mosquitoes, mosquito vectors

3. **a.** *human_hosts*
   b. *human_hosts*
   c. *uninfected_humans*, *immune*, and undesignated "cloud" for deceased humans
   d. *human_hosts*
   e. *human_hosts*
   f. *vectors* and undesignated "cloud" for dead mosquitoes
   g. undesignated "cloud" for unborn mosquitoes
   h. *uninfected_mosquitoes*
   i. undesignated "cloud" for dead mosquitoes
   j. *uninfected_mosquitoes* and *vectors*

4. **a.** *d*(*immune*)/*dt* = *immunity_rate* * *human_hosts*
   b. *d*(deceased_humans)/*dt* = *malaria_induced_death_rate* * *human_hosts*
   c. rate of change from *human_hosts* to *uninfected_humans* = *recovery_rate* * *human_hosts*
   d. *d*(deceased_vectors)/*dt* = *mosquito_death_rate* * *vectors*

5. **a.** people per day

b. 410 because *immune*(6.1) = *immune*(6) + *immunity_rate* * *human_hosts*(6) * Δt = 400 + (0.2)(500)(0.1) = 400 + 10 = 410

6. a. 4, one for each flow entering or leaving the stock *human_hosts*

b. 1, because only one flow enters the stock *human_hosts*.

7. *vectors*(t) = *vectors*($t - \Delta t$) + (*infected* – *vector_deaths*) * Δt

References

Carroll, Rory. 2001. "Skeleton Find Links Malaria to Fall of Rome." *The Guardian*, February 21. http://news.nationalgeographic.com/news/2001/02/0221_malaria rome.html (accessed December 17, 2012)

CIA (U.S. Central Intelligence Agency). 2012. "The World Fact Book. Africa: Malawi." https://www.cia.gov/library/publications/the-world-factbook/geos/mi .html (accessed December 18, 2012)

Desai, Meghna, Feiko O. ter Kuile, François Nosten, Rose McGready, Kwame Asamoa, Bernard Brabin, and Robert D. Newman. 2007. "Epidemiology and Burden of Malaria in Pregnancy." *The Lancet Infectious Diseases* 7(2): 93–104.

Despommier, D. D., Robert W. Gwadz, Peter J. Hotez and Charles A. Knirsch. 2005. *Parasitic Diseases*. 5th ed.. IV. The Protozoas. 9. The Malarias. pp. 50–68. New York: Apple Trees Productions.

Duffy, Patrick E., Robert S. Desowitz, P. Duffy, and M. Fried. 2001. "Pregnancy Malaria Throughout History: Dangerous Labors." *Malaria in Pregnancy: Deadly Parasite, Susceptible Host*, 1–25.

KFF (Henry J. Kaiser Family Foundation). 2010, U. S. Global Health Policy. Malaria. http://www.globalhealthfacts.org/data/topic/map.aspx?ind=30 (accessed December 17, 2012)

NetMark. "About Malaria." NetMark Plus. http://www.netmarkafrica.org/Malaria/ (accessed December 17, 2012)

NIAID (National Institute of Allergy and Infectious Diseases). "Life Cycle." National Institutes of Health. http://www.niaid.nih.gov/topics/malaria/pages/life cycle.aspx (accessed December 17, 2012)

RPH (Royal Perth Hospital) "History of Malaria." Malaria On-Line Resource. http://www.rph.wa.gov.au/malaria/history.html (accessed December 17, 2012)

Ryan, S. J. 2008. "Malaria. Lecture 13." Anthropological Sciences 178A/278A. University of California, Santa Barbara. http://www.nceas.ucsb.edu/~sjryan/PPP/ (accessed December 17, 2012)

Schantz-Dunn, J., and N. M. Nour. 2009. "Malaria and Pregnancy: A Global Health Perspective." *Rev Obstet Gynecol*. 2(3): 186–92.

ter Kuile, Feiko O., and Stephen J. Rogerson. 2008. "Plasmodium Vivax Infection during Pregnancy: An Important Problem in Need of New Solutions." *Clinical Infectious Diseases* 46(9): 1382–1384.

WHO (World Health Organization). The United Nations. http://www.who.int (accessed December 18, 2012)

———. "Malaria—Fact Sheet" http://www.who.int/mediacentre/factsheets/fs094 /en/index.html (accessed December 18, 2012)

———. "World Malaria Report." 2011. http://www.who.int/malaria/world_malaria_report_2011/en/ (accessed December 18, 2012)

Wiser, Mark F. "Cellular and Molecular Biology of *Plasmodium.*" Tulane University. http://www.tulane.edu/~wiser/malaria/cmb.html (accessed November 20, 2012)

MODULE 4.5

Enzyme Kinetics: A Model of Control

Download

The text's website has the file *substrate*, which contains a submodel for this module, available for download for various system dynamics tools.

Introduction

Enzymes catalyze, or hasten, chemical reactions for biological systems. Although most enzymes are proteins, there are some RNA enzymes as well. They are remarkably adapted for this role, because in minute quantities they are very specific and can be quite active. Enzymes do not influence the direction of the reaction. Unchanged by the reaction, they can be used over and over again. Without them, even spontaneous reactions would not proceed fast enough to support living cells. Enzymes can increase the rate of reaction by a factor of up to 10^{20}. Additionally, they are "regulatable" by both physical and chemical factors. Many biochemists study the activities of enzymes and the factors that regulate enzyme activity.

Archibald Garrod (1902), studying the disease **alkaptonuria**, proposed that the instructions for producing specific enzymes in the cell were inherited. He elaborated on his work in his *Inborn Errors of Metabolism*, published in 1923 (Garrod 1923). Essentially, his hypothesis was that diseased individuals lacked a normal enzyme in the catabolism of proteins. This enzyme deficiency resulted from receiving one recessive gene from each parent. Later investigation confirmed that alkaptonurics lack the activity of *homogentisate dioxygenase*. This enzyme normally converts homogentisate, one of the intermediate compounds in the breakdown of the amino acid **tyrosine**, into maleylacetoacetate. Hence, homogentisate accumulates in and darkens various body tissues (e.g., bone, skin, prostate), causing symptoms of arthritis. Some is eliminated in the urine, which turns dark if allowed to stand. We now understand the bases of many metabolic diseases, which are consequences of defective enzymes.

All the attention to metabolic diseases has catalyzed great interest in enzymes and how they work. One area of focus has been on the rate of enzyme activity and its control. The quantitative study of enzyme behavior is **enzyme kinetics**. Using mathematical approaches we can examine the factors that influence the rate of an enzymatic reaction. Commonly, we consider things like substrate concentration, enzyme concentrations, cofactors, inhibitors, pH, and temperature.

Enzymatic Reactions

We begin by considering a simple reaction in which a **substrate**, **S**, in the presence of an **enzyme**, **E**, converts to one **product**, **P**. (Not all reactions involve breaking one substrate into parts; some combine or rearrange.) In their seminal work, biochemist Leonor Michaelis and physician Maud Menten hypothesized that the enzyme catalyzes the reaction by interacting with the substrate to form an intermediate **enzyme-substrate complex**, **ES** (Michaelis and Menten 1913). This complex undergoes a catalytic reaction to form the enzyme E and the product P. The following diagram represents the situation where k_1, k_2, k_3, and k_4 are rate constants, and Figure 4.5.1 depicts the enzyme reaction in the forward direction:

$$\mathrm{E} + \mathrm{S} \underset{k_2}{\overset{k_1}{\rightleftarrows}} \mathrm{ES} \underset{k_4}{\overset{k_3}{\rightleftarrows}} \mathrm{E} + \mathrm{P}$$

Thus, Reaction 1 for the chemical reactions indicates the following about the rate constants:

- k_1—rate of change of substrate S in the presence of enzyme E to intermediate enzyme-substrate complex ES
- k_2—reverse reaction rate of change of intermediate enzyme-substrate complex ES back to substrate S and enzyme E
- k_3—rate of change of intermediate enzyme-substrate complex ES to product P and enzyme E
- k_4—reverse reaction rate of change of product P and enzyme E back to intermediate enzyme-substrate complex ES

As Figure 4.5.1 illustrates with the **forward reaction**, one molecule of substrate (S) combines with one molecule of enzyme (E) to form a molecule of the enzyme-substrate complex (ES), which then **dissociates**, or breaks apart, into a molecule of enzyme and a molecule of product (P). Reaction 1 moves forward and backward, so that in the **backward**, or **reverse, reaction** a molecule of E and a molecule of P combine to create a molecule of ES, which can dissociate into a molecule each of E and S.

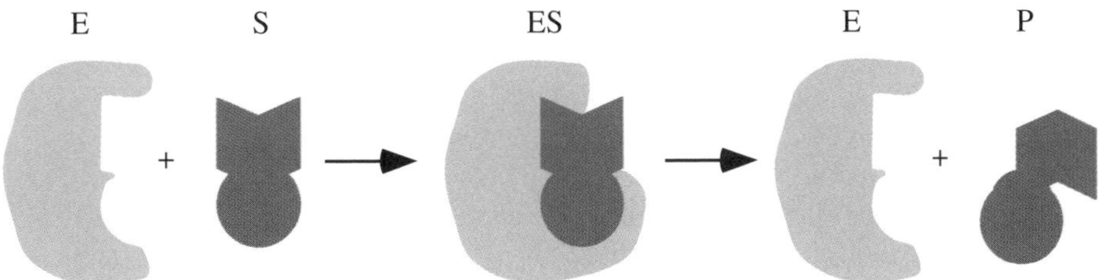

Figure 4.5.1 Depiction of a simple enzyme reaction in the forward direction

We wish to model Reaction 1 so that given rate constants and initial **concentrations of enzyme and substrate**, written **[E]** and **[S]**, respectively, we can determine the eventual **product concentration, [P]**.

Differential Equations

As with interactions of competing species (see Module 4.1, "Competition"), of predator and prey (see Module 4.2, "Predator-Prey Model"), of susceptibles and infecteds (see Module 4.3, "Modeling the Spread of SARS—Containing Emerging Disease"), and humans and mosquitoes (see Module 4.4, "Modeling a Persistent Plague—Malaria"), the rate of change of removal of [S] is proportional to the product of the concentrations of E and S, k_1[E][S], which obeys the **law of mass action**. Because some of the enzyme-substrate complex ES reverts to form a molecule of E and a molecule of S, the rate of change of the formation of [S] is proportional to the concentration of ES, k_2[ES]. Thus, as the following differential equation indicates, the rate of change of the concentration of S is equal to the rate of change of formation minus the rate of change of removal:

$$d[\mathrm{S}]/dt = k_2[\mathrm{ES}] - k_1[\mathrm{E}][\mathrm{S}]$$

Because in this reaction a molecule of enzyme reacts with a molecule of substrate, the rate of change of removal of [E] in the forward direction is the same as the rate of change of removal of [S], namely, k_1[E][S]. Similarly, in the backward direction, the rate of change of removal of [E] is proportional to the product of [E] and [P], k_4[E][P]. However, [E] can be formed from [ES] by forward and backward reactions. The rate of change of each of these reactions is proportional to [ES]. With rate constants of k_3 and k_2 for the forward and backward reactions, respectively, the rate of change of the forward reaction for the formation of [E] is k_3[ES], while the rate of change of the backward reaction equals k_2[ES]. Thus, the total rate of change of the formation of [E] is the sum of these two values.

Quick Review Question 1

Considering the simple reaction shown in (1), give the formula for each of the following quantities:

a. the rate of change of the formation of [E]
b. $d[\mathrm{E}]/dt$
c. $d[\mathrm{P}]/dt$
d. $d[\mathrm{ES}]/dt$

Quick Review Question 2

In the model diagram for the simple reaction shown in (1), what do the stocks (box variables) represent?

Model

The model mimics the differential equations of the last section. A stock (box variable) exists for each of the four concentrations, [E], [S], [ES], and [P]. Recall that the rate of change of [S] is as follows:

$$d[\mathrm{S}]/dt = k_2[\mathrm{ES}] - k_1[\mathrm{E}][\mathrm{S}]$$

Thus, the flow into [S] has connectors/arrows from the stock for [ES] and from a converter storing the value of the rate constant k_2, and its equation is the product of these two values. *Flows do not connect stocks for different concentrations*, such as [ES] and [S], because the concentration for one substance, such as ES, does not become the concentration for another substance, such as S. With connectors/arrows from the converter for k_1 and the stocks for [E] and [S], the flow out of [S] is the product of these values. Figure 4.5.2 depicts a submodel for [S] with $d[\mathrm{S}]/dt = k_2[\mathrm{ES}] - k_1[\mathrm{E}][\mathrm{S}]$. The figure does not include some diagram elements, such as flows, associated with [E], [ES], and [P].

Quick Review Question 3

In a submodel for [E] with differential equation $d[\mathrm{E}]/dt = (k_2 + k_3)[\mathrm{ES}] - k_1[\mathrm{E}][\mathrm{S}] - k_4[\mathrm{E}][\mathrm{P}]$, give the number of connectors/arrows that

a. go to the flow into [E]'s stock,
b. go to the flow that leaves [E]'s stock,
c. come out of [E]'s stock.

Moles vs. Molar

Before observing the graphs resulting from running the simulation, we should clarify units for amounts and concentrations of chemicals. The **mole (mol)** is widely used by chemists to measure the amount of a substance. They might use it in the same way that farmer or grocer would use a dozen to describe a group of 12 eggs. By for-

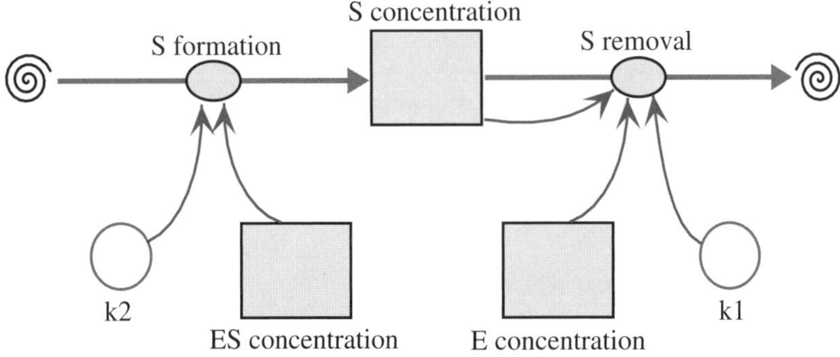

Figure 4.5.2 Submodel for [S]

mal definition, a mole is the number of carbon (C) atoms in exactly 12 g of carbon-12 (^{12}C), which is 6.02214×10^{23}, or **Avogadro's number**. We can apply the term to other substances, defining a mole of a substance as the quantity of that substance containing an equal number of units, such as atoms or molecules, as there are in exactly 12.0 g of ^{12}C. A mole of any compound is its formula mass in grams. Thus, 1 mol of sodium chloride (NaCl) would equal 58.45 g, or the sum of the atomic weights of sodium (23) and chlorine (35.45).

Chemists often express the concentration of a solute in a solvent, such as water, as **molarity** (**M**), which is the number of moles of solute per liter of solution. For a 0.05-M NaCl solution, we have 0.05 mol of sodium chloride in every liter of water. A **molar**, then, is equivalent to 1 mol/L. For example, suppose we add 50 g of NaCl to a liter of water. We know from above that the mass of 1 mol of NaCl is 58.45 g. Thus, using cancellation of units as follows, we determine that 0.855 mol of NaCl is in the liter of water:

$$50 \text{ g NaCl} \times \frac{1 \text{ mol NaCl}}{58.45 \text{ g NaCl}} = 0.855 \text{ mol NaCl}$$

Because we have dissolved 0.855 mol of NaCl in 1 L of solvent, the concentration of NaCl ([NaCl]) in the solution is 0.855 M.

Definition A **mole (mol) of a substance** is the quantity of that substance containing 6.02214×10^{23} units (atoms, molecules, or some other unit). A mole of any compound is its formula weight in grams. The concentration of a solute in a solvent is often express as **molarity** (**M**), or the number of moles of solute per liter of solution. One **molar** is 1 mol/L.

Quick Review Question 4

The following questions relate to magnesium chloride, $MgCl_2$, where each molecule contains one atom of magnesium and two atoms of chlorine. The atomic weight of magnesium is 24.31, while that of chlorine is 35.45.

a. Find the formula mass of $MgCl_2$.
b. Find the number of grams in a mole of $MgCl_2$.
c. Give the number of molecules in a mole of $MgCl_2$.
d. Determine the moles in 50 g of $MgCl_2$.
e. If 50 g of $MgCl_2$ are dissolved in 700 mL of solution, calculate the molarity.
f. Determine [$MgCl_2$] for Part e.

Results

Returning to our model, the **velocity of the reaction** is the rate of change of [P], $d[P]/dt$. This derivative (*rate_of_P*) in a model is the rate of change of formation minus the rate of change of removal of [P], k_3[ES] – k_4[E][P], or the flow into [P] minus the flow out of [P]. For use in graphing, we have a converter/variable evaluating this difference. In an actual chemical reaction, we consider the initial velocity of the reaction because of restrictive factors, such as the amount of substrate. In this model, for a particular substrate concentration, the rate of change of [P] approaches a limit. To observe the impact of substrate concentration on velocity, we have a system dynamics tool generate a comparative graph of *rate_of_P* versus *S_concentration* with *S_concentration* taking on values 0.0, 1.5, 3.0, . . ., 30.0 millimolar (mM). For each value, we run the simulation using the Runge-Kutta 4 method for 3 s with $\Delta t = 0.05$ s. Figure 4.5.3 displays a resulting graph for parameters of $k_1 = 0.05\ s^{-1}$, $k_2 = 0.1\ s^{-1}$, $k_3 = 0.02\ s^{-1}$, and $k_4 = 0.0\ s^{-1}$. Tabular output gives the top of the last column as about 0.0001837 mM/s.

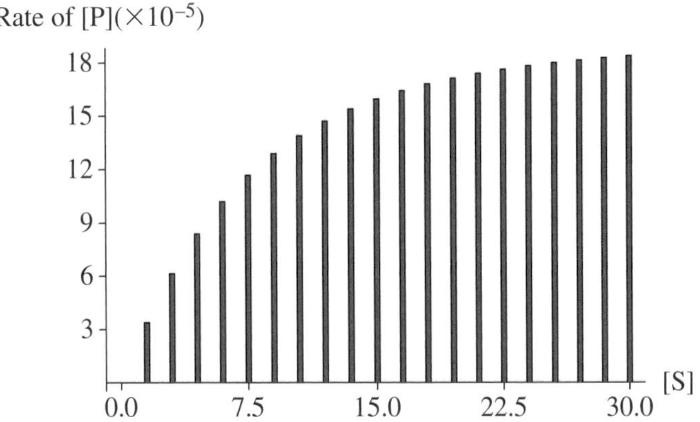

Figure 4.5.3 Comparative graph of *rate_of_P* (× 10^{-5} mM/s) versus *S_Concentration* (mM)

Michaelis-Menten Equation

For most enzymes, if we increase substrate concentration and hold enzyme concentration constant, the resulting **initial velocities** (initial d[P]/dt), or **reaction rates, of the reaction** (***v***) produce an asymptotic curve, which is the situation we observe by following the tops of the columns in Figure 4.5.3. In other words, v increases rapidly at first as we increase [S]. Then, with the constant number of enzyme molecules, the rate of increase in v decreases, and v approaches a **limit of the reaction rate**, called $\boldsymbol{V_{max}}$. No further increases in [S] will increase velocity. With some assumptions, the **Michaelis-Menten equation** describes this relationship between [S] and v, as follows:

$$v = \frac{V_{max}[S]}{K_m + [S]}$$

The graph of this model appears in Figure 4.5.4. $\boldsymbol{K_m}$ is the **Michaelis-Menten constant** and is equal to the [S] where $v = \frac{V_{max}}{2}$. K_m is an indicator of the enzyme's affinity for the substrate. The lower the K_m value, the higher the affinity, so it takes less substrate to reach half of V_{max} and the enzyme is a better catalyst for the reaction. Table 4.5.1 provides K_m values for several enzyme-substrate combinations.

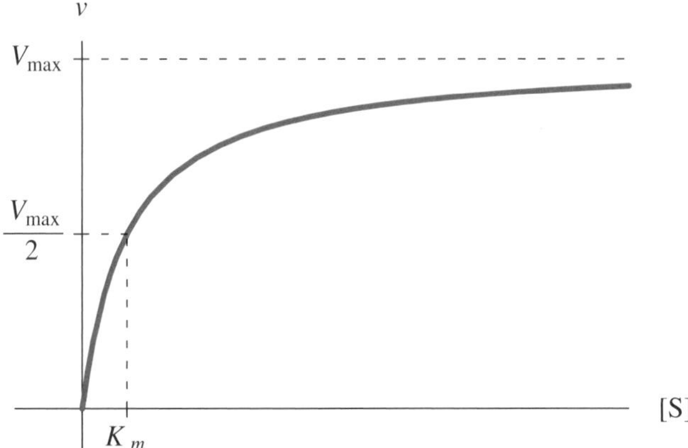

Figure 4.5.4 Graph of initial reaction velocity versus substrate concentration for the Michaelis-Menten equation

Table 4.5.1
Michaelis-Menton Constants (Hardin et al. 2012; Kimball 2003)

Enzyme	*Substrate*	K_m *(mMol/L)*
Acetylcholinesterase	Acetylcholine	0.09
Carbonic anhydrase	CO_2	12
Catalase	H_2O_2	1,100
Chymotrypsin	Gly-Tyr-Gly	108
Fumarase	Fumarate	0.005
Triose phosphate isomerase	Glyceraldehyde-3-phosphate	0.5
Beta-lactamase	Benzylpenicillin	0.02

Quick Review Question 5

The following questions refer to Figure 4.5.3.

a. Select the best estimate of V_{max} from the tick values.
b. Using your answer from Part a, estimate K_m to the nearest whole number.

To derive their model, Michaelis and Menten made the following simplifying assumptions:

1. The reaction rate is determined before very much product is formed. Consequently, the reverse reaction from E + P to ES is negligible, so that k_4 is zero.
2. k_3 is small in comparison to k_1 and k_2; that is, the rate of product formation is slow in comparison to the rate of ES formation and the rate of ES dissociation to E + S.
3. [S] is much greater than [E], so that [S] is virtually constant.
4. [E] + [ES] is constant.

Under these assumptions, the Michaelis-Menten equation models Reaction 1 as follows (Danby 1997):

$$v = \frac{V_{max}[S]}{K_m + [S]}$$

Exercise 3 derives this equation. Note that our choices of parameters adhere to these conditions and that our graph in Figure 4.5.3 corresponds to the graph of the Michaelis-Menten equation in Figure 4.5.4.

Quick Review Question 6

a. Using Assumption 1, give an approximate value for k_4.
b. Using Assumption 3, give an approximate value for $d[S]/dt$.
c. Using the preceding assumptions, give the approximate relationship between the initial velocity of the reaction, v, and the rate of change of the product, $d[P]/dt$.

Although the Michelis-Menten equation captures the relationship of reaction velocity to substrate concentration, K_m and V_{max} are difficult to ascertain from its graph, such as in Figure 4.5.4. Hans Lineweaver and Dean Burk reorganized the equation into a form that is more helpful for determination of these constants (Danby 1997). As Exercise 5 develops, taking the reciprocal of both sides, they solved for $1/v$ in terms of $1/[S]$, as follows:

$$\frac{1}{v} = \frac{K_m}{V_{max}}\left(\frac{1}{[S]}\right) + \frac{1}{V_{max}}$$

$\frac{K_m}{V_{max}}$ and $\frac{1}{V_{max}}$ are constants. Moreover, considering $x = 1/[S]$ to be an independent variable and $y = 1/v$ to be a dependent variable, the equation has the form of a line, $y = mx + b$. Thus, in the graph of $1/v$ versus $1/[S]$, the slope is $\frac{K_m}{V_{max}}$, and the vertical intercept is $\frac{1}{V_{max}}$. Setting $1/v$ equal to zero, we find that the horizontal intercept is $-\frac{1}{K_m}$. Figure 4.5.5 presents the graph of $1/v$ versus $1/[S]$ for the Michaelis-Menten equation in Figure 4.5.4. The following Quick Review Question determines V_{max} and K_m from this graph.

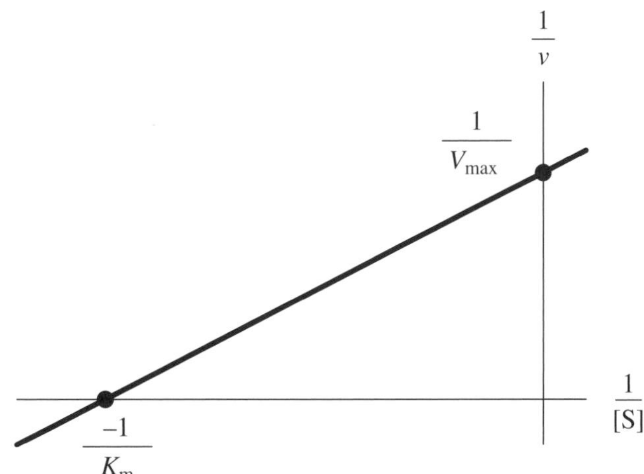

Figure 4.5.5 Graph of 1/v versus 1/[S] for Figure 4.5.4

Quick Review Question 7

Determine the following for Figure 4.5.5, where the vertical intercept is 0.2 and the horizontal intercept is -4:

a. V_{max}
b. K_m
c. K_m/V_{max}
d. The slope

Modeling Inhibition

The form of the Michelis-Menten equation, $v = \frac{V_{max}[S]}{K_m + [S]}$, parallels one way to model the process involving **inhibition**, which we employ in various project modules. Suppose without inhibition the rate of change of some quantity, p, is u, as follows:

$$dp/dt = u$$

However, suppose x inhibits this process. The larger that the value of x is, the smaller dp/dt is. A preliminary refinement of the model to account for this inhibition is to divide the right-hand side of the equation by x, as follows:

$$\frac{d(p)}{dt} = \frac{u}{x}$$

However, we have a division-by-zero error if x is not present. One possible solution is to add a small number, say 0.01 to the denominator, such as follows:

$$\frac{d(p)}{dt} = \frac{u}{x + 0.01}$$

However, if x is zero, division by 0.01 multiplies u by 100. Thus, we complete the inhibition model by multiplying the numerator by 0.01, as follows:

$$\frac{d(p)}{dt} = \frac{u \cdot 0.01}{x + 0.01} = \frac{0.01u}{x + 0.01}$$

If x is zero, the 0.01 in the numerator and the 0.01 in the denominator cancel out each other. As x becomes larger and larger, the fraction, and hence the instantaneous rate of change of p, approaches zero. In the Michelis-Menten equation, $v = \frac{V_{max}[S]}{K_m + [S]}$, K_m plays the role of the inhibitor of v, the initial velocity of the reaction: The larger K_m is, the smaller v is.

Model A model for **inhibition** of $dp/dt = u$ by x is as follows:

$$\frac{dp}{dt} = \frac{cu}{x+c}, \text{ where } c \text{ is a small constant}$$

Exercises

1. For the simple model of enzyme kinetics in (1), give the relationship between $d[\mathrm{E}]/dt$ and $d[\mathrm{ES}]/dt$.
2. In the Michaelis-Menten equation, by considering the instant when $[\mathrm{S}] = K_m$, show that K_m is equal to the substrate concentration when $v = \frac{V_{\max}}{2}$.
3. In this exercise, we derive the Michaelis-Menten equation (Danby 1997).
   a. The first assumption is that the reverse reaction from E + P to ES is negligible, so that k_4 is zero or very small. Write the differential equation $d[\mathrm{E}]/dt$ for assuming $k_4 = 0$.
   b. Write the differential equation $d[\mathrm{ES}]/dt$ for assuming $k_4 = 0$.
   c. Write the differential equation $d[\mathrm{P}]/dt$ for assuming $k_4 = 0$.
   d. The first assumption also states that the reaction rate is determined very early. Give the derivative that the initial velocity v of the reaction equals.
   e. In the first assumption, $d[\mathrm{S}]/dt$ is zero. Using this assumption, solve for [E][S] in terms of [ES].
   f. Using your answer to Part e, [E][S] is proportional to [ES]. Give the constant of proportionality and call it K_m.
   g. Using your answers to Parts e and f, solve for K_m in terms of [ES], [E], and [S].
   h. By Assumption 4, [E] + [ES] is constant. Call this constant $[\mathrm{E}_0]$. Using this equation, solve for [E] in terms of [ES] and $[\mathrm{E}_0]$.
   i. Substitute your solution of [E] from Part h into the answer to Part g, and solve for [ES].
   j. Substitute your solution of [ES] from Part i into the differential equation for $d[\mathrm{P}]/dt$ in Part c.
   k. As [S] increases, the initial velocity of the reaction, $v = d[\mathrm{P}]/dt$, approaches its maximum, $V_{\max}$. Moreover, as [S] increases, K_m is small in comparison to [S], so that [S] and K_m + [S] are approximately the same. Using your answer to Part j, we can also say that $d[\mathrm{P}]/dt$ approaches what value?
   l. Using Parts j and k, solve for $v = d[\mathrm{P}]/dt$, and obtain the Michaelis-Menten equation.
4. The **Briggs-Haldane model** assumes that very soon the rate of change of [ES] is small in comparison to [E] and [S], so that $d[\mathrm{ES}]/dt$ is almost zero. Using this assumption and the differential equation from your answer to Part b of Exercise 3, solve for [ES] in terms of [E][S]. Define K as an appropriate constant, and in a similar fashion to Exercise 3 complete the solution of

$v = d[P]/dt$ as $k_3[E_0][S]/(K + [S])$, where $[E_0]$ is the constant $[E] + [ES]$ (Danby 1997).

5. From the Michaelis-Menten equation, derive the solution of $1/v$ in terms of 1/[S], as follows:

$$\frac{1}{v} = \frac{K_m}{V_{\max}}\left(\frac{1}{[S]}\right) + \frac{1}{V_{\max}}$$

6. In Reaction 1, one molecule of the enzyme reacts with one molecule of the substrate. In the following reaction, one molecule of the enzyme reacts with n molecules of the substrate, and we assume that the reverse reaction from E + P to ES is negligible:

$$\mathrm{E} + n\mathrm{S} \underset{k_2}{\overset{k_1}{\rightleftarrows}} \mathrm{ES} \xrightarrow{k_3} \mathrm{E} + \mathrm{P} \qquad (2)$$

Write a differential equation for each of the following parts. Because the reaction involves one molecule of enzyme for n molecules of substrate, we take the product of [E] and n copies of [S], that is $[E][S]^n$, where appropriate (Danby 1997).

a. $d[S]/dt$
b. $d[E]/dt$
c. $d[P]/dt$
d. $d[ES]/dt$

7. Give the Michaelis-Menten approximation for the reaction in Exercise 6.

Projects

For additional projects, see Module 7.11, "Fueling Our Cells—Carbohydrate Metabolism"; Module 7.14, "Control Issues: The Operon Model"; and Module 7.15, "Troubling Signals: Colon Cancer."

1. Model the following simplified reaction:

$$\mathrm{S} \underset{k_2}{\overset{k_1}{\rightleftarrows}} \mathrm{P}$$

Plot substrate concentration [S] and product concentration [P] versus time. Describe the growth of [P] and decline of [S]. With your simulator, determine experimentally $[P]_{eq}$, $[S]_{eq}$, when equilibrium occurs, and the equilibrium constant K_{eq} for several values of k_1 and k_2. Do the experimental results agree with the analytical ones?

2. Model the complete Reaction 1. Generate tables with time, the values of all concentrations, and the rate of change of [P]. Generate the following graphs: [S] and [P] versus time, the rate of change of [P] versus time, and the rate of change of [P] versus [S]. Generate a comparative graph as in Figure 4.5.3. Estimate $V_{\max}$ and K_m from this graph. Using the generated data, plot $1/v$ versus 1/[S] with a computational tool, such as a spreadsheet. From the graph, estimate $V_{\max}$ and K_m.

3. Model the complete Reaction 1. Also, compute the Briggs-Haldane and Mi-

chaelis-Menten approximations. The Briggs-Haldane approximation assumes that after an initial period, the rate of formation of [ES] is small in comparison with the change in [S] and change in [P] (see Exercise 4). This approximation for the rate of change of the concentration of [P] is as follows:

$$\frac{d[\mathrm{P}]}{dt} = \frac{k_3([\mathrm{E}]+[\mathrm{ES}])[\mathrm{S}]}{\frac{(k_2+k_3)}{k_1}+[\mathrm{S}]}$$

The Michaelis-Menten approximation is as follows:

$$\frac{d[\mathrm{P}]}{dt} = \frac{k_3([\mathrm{E}]+[\mathrm{ES}])[\mathrm{S}]}{\frac{k_2}{k_1}+[\mathrm{S}]}$$

Find parameter values that satisfy the assumptions of the approximations. Graph and compare the rate of change of [P] for your simulation and the approximations versus time (Danby 1997).

4. Complete Project 2 and explore the situation of having a very low substrate concentration, where $[\mathrm{S}] \ll K_m$. In this case, [S] is much lower than K_m, so that $K_m + [\mathrm{S}]$ is approximately K_m. Simplify the Michaelis-Menten equation in this situation. Have your model compute this v. Compare your model with the computed value. How does v vary with [S] (Danby 1997)?
5. Complete Project 2 and explore the situation of having very high substrate concentration, where $[\mathrm{S}] \gg K_m$. In this case, [S] is much higher than K_m, so that $K_m + [\mathrm{S}]$ is approximately [S]. Simplify the Michaelis-Menten equation in this situation. Have your model compute this v. Compare your model with the computed value. How does v vary with [S]? How does $V_{\max}$ vary with [E] (Danby 1997)?
6. Complete Project 2 and explore the situation of $[\mathrm{S}] = K_m$. Simplify the Michaelis-Menten equation in this situation. Have your model compute this v. Compare your model with the computed value. What is the meaning of K_m (Danby 1997)?
7. Model the complete Reaction 2 of Exercise 6. Generate tables with time, the values of all concentrations, and the rate of change of [P]. Generate the following graphs: [S] and [P] versus time, the rate of change of [P] versus time, and the rate of change of [P] versus [S]. Generate a comparative graph as in Figure 4.5.3. Estimate $V_{\max}$ and K_m from this graph. Using the generated data, plot $1/v$ versus $1/[\mathrm{S}]^n$ with an appropriate computational tool, such as a spreadsheet. From the graph, estimate $V_{\max}$ and K_m.
8. We have observed oscillations in several contexts, including predator-prey and pendulum problems. However, most chemists were disbelieving when Boris P. Belousov claimed in the 1950s to have found a chemical reaction that oscillated between yellow and clear. In 1964, Anatol M. Zhabotinsky developed what became known as the **Belousov-Zhabotinsky (BZ) reaction**, a refinement of Belousov's work that oscillated in time and space with changing geometric patterns.

This project involves modeling the following theoretical chemical reaction, called the **Brusselator**, which exhibits oscillation of the concentrations of X and Y:

$$A \xrightarrow{k_1} X + B$$
$$B + X \xrightarrow{k_2} Y + C$$
$$2X + Y \xrightarrow{k_3} 3X + D$$
$$X \xrightarrow{k_4} E + F$$

Assume that $[A]$ is constant, say, 1 M, and that B, C, D, E, and F are inactive. Graph $[X]$ and $[Y]$ versus time and $[X]$ versus $[Y]$ for increasing values of $[B]$. Verify that $[X] = [A]$ and $[Y] = [B] / [A]$ is an equilibrium point. Is the point stable or unstable for $[B] \le 1 + [A]^2$? For $[B] > 1 + [A]^2$? Describe the graphs close to the equilibrium if $1 - [A] + [A]^2 < [B] < 1 + [A] + [A]^2$ (Danby 1997).

9. Model the following chemical reaction, called the **Oregonator**, which oscillates in time (see the first paragraph of Project 8):

$$A + Y \xrightarrow{k_1} X$$
$$X + Y \xrightarrow{k_2} P$$
$$B + X \xrightarrow{k_3} 2X + Z$$
$$2X \xrightarrow{k_4} Q$$
$$Z \xrightarrow{k_5} Y$$

where X is bromous acid ($HBrO^2$), Y is bromide ion (BR^-), and Z is cerium ion (Ce(IV)). Assume $[A]$ and $[B]$ are constant and P and Q are inert. The original paper gave the following rate constants: $k_1 = 1.34\ M^{-1}\ s^{-1}$, $k_2 = 1.6 \times 10^9\ M^{-1}\ s^{-1}$, $k_3 = 8 \times 10^3\ M^{-1}\ s^{-1}$, $k_4 = 4 \times 10^7\ M^{-1}\ s^{-1}$, and $k_5 = 1\ M^{-1}\ s^{-1}$ (Field and Noyes 1974). For a variety of initial concentrations, graph the logarithms of $[X]$, $[Y]$, and $[Z]$ versus time for at least 2000 s. Verify that an equilibrium occurs at $[X] = 2.45562 \times 10^{-8}\ M^{-1}\ s^{-1}$, $[Y] = 2.99388 \times 10^{-7}\ M^{-1}\ s^{-1}$, and $[Z] = 1.1787 \times 10^{-5}\ M^{-1}\ s^{-1}$. Explore the situations with initial conditions near the equilibrium. Discuss the results (Danby 1997; Field and Noyes 1974).

Answers to Quick Review Questions

1. a. rate of change of formation of [E] = k_2[ES] + k_3[ES] = $(k_2 + k_3)$[ES]
b. $d[E]/dt = (k_2 + k_3)[ES] - k_1[E][S] - k_4[E][P]$
c. $d[P]/dt = k_3[ES] - k_4[E][P]$
d. $d[ES]/dt = k_1[E][S] + k_4[E][P] - (k_2 + k_3)[ES]$

2. the four concentrations, [E], [S], [ES], and [P]

3. a. 3, from k_2, k_3, and [ES]
b. 5, from k_1, [E], [S], k_4, and [P]
c. 4, one to each flow

4. a. 95.21 g = (24.31 + 2 × 35.45) g

b. 95.21 g
c. 6.02214×10^{23}
d. 0.525 mol = 50 g × 1 mol/(95.21 g)
e. 0.75 M = 0.525 mol/(0.700 L) = 0.75 mol/L
f. 0.75 M

5. a. $V_{max} \approx 18 \times 10^{-5}$ mM/s = 0.00018 mM/s
b. $K_m \approx 5.0$ mM. If V_{max} = 0.00018 mM/s, V_{max} / 2 = 0.00009 mM/s, which occurs at approximately *S_Concentration* = 5.0 mM.

6. a. $k_4 \approx 0$ because the reverse reaction from E + P to ES is negligible
b. $d[S]/dt \approx 0$ because [S] is virtually constant, giving an almost 0 value for the rate of change of [S] with respect to time
c. $v = d[P]/dt$. By Assumption 1, the reverse reaction from E + P to ES is negligible, so the rate of change in [P] comes from formation, not from deformation of P. By Assumption 4, [E] + [ES] is constant, so that the total amount of the enzyme, which is in the free form of E and ES, is constant. Moreover, $d[S]/dt \approx 0$ (see Part b). Thus, the initial velocity of the reaction is the rate of change of [P].

7. a. 1/0.2 = 5
b. $K_m = -1/(-4) = 0.25$
c. $K_m/V_{max} = 0.25/5 = 0.05$
d. 0.05

References

Danby, J.M.A. 1997. *Computer Modeling: From Sports to Spaceflight . . . From Order to Chaos*. Richmond, VA: Willmann-Bell, pp. 351–367.

Field, R. J., and R. M. Noyes. 1974. "Oscillations in Chemical Systems IV. Limit Cycle Behavior in a Model of a Real Chemical Reaction." *J. Chem. Phys.* 60(1974): 1877–1884.

Garrod, Archibald E. 1902. "The Incidence of Alkaptonuria: A Study in Chemical Individuality." *Lancet*, ii: 1616–1620.

———. 1923. *Inborn Errors of Metabolism*, 2nd ed.. London: Henry Frowde and Hodder & Stoughton.

Hardin, Jeff, Gregory Bertoni, and Lewis J. Kleinsmith. 2012. *Becker's World of the Cell*. 8th ed. San Francisco: Benjamin Cummings, pp. 138–150.

Harvey, David. 2009. *Analytical Chemistry 2.0*. http://acad.depauw.edu/harvey_web/eText Project/AnalyticalChemistry2.0.html (accessed September 23, 2012)

Kimball, John W. 2003. "Enzyme Kinetics." Kimball's Biology Pages. http://users.rcn.com/jkimball.ma.ultranet/BiologyPages/E/EnzymeKinetics.html (accessed November 17, 2012)

Michaelis, L., and M. L. Menten. 1913. "Die Kinetik der Invertinwirkung." *Biochem. Z.*: 333–369

Rogers, A., and Y. Gibon. 2009. "Enzyme Kinetics: Theory and Practice." In *Plant Metabolic Networks* (J. Schwender, Ed.), Berlin: Springer; New York: Heidelberg, pp. 71–103. http://www.bnl.gov/pubweb/alistairrogers/linkable_files/pdf/Rogers_&_Gibon_2009.pdf (accessed November 17, 2012)

5

COMPUTATIONAL ERROR

MODULE 5.1

Computational Toolbox—Tools of the Trade: Tutorial 1

Download

From the textbook's website, download Tutorial 1 in the format of your computational tool or in PDF format. We recommend that you work through the tutorial and answer all Quick Review Questions using the corresponding software.

Introduction

Various computer software tools are useful for graphing, numeric computation, and symbolic manipulation. This first computational toolbox tutorial is available for download from the textbook's website for several different software systems. Tutorial 1 in your system of choice gives an introduction to that software and prepares you to use the tool to complete various projects in the next few chapters. The tutorial introduces concepts and functions, such as the following:

- Getting started
- Evaluation
- Saving
- File organization, such as cells
- Styles
- Numbers
- Arithmetic operators
- Built-in functions, such as the natural logarithm, sine, and exponential functions
- Variables
- Assignments
- User-defined functions
- Online documentation
- Printing
- Looping
- Plotting
- Differentiation (optional)
- Solving differential equations (optional)
- Integration (optional)

The module gives computational examples and Quick Review Questions for you to complete and execute in your desired software system.

MODULE 5.2

Errors

Introduction

Errors can occur in the solution of a computational science problem at any stage, from the earlier steps of data collection and simplifying assumptions to the later time of computer implementation. The modeler must be aware of possible errors to minimize their occurrence and to avoid drawing incorrect conclusions from flawed solutions. In this module, we discuss various concepts surrounding and the sources of errors.

Data Errors

Unfortunately, the sources for errors in the data upon which we base and verify our models can be numerous. For example, a sensor measuring barometric pressure might malfunction, giving incorrect values or values that are valid in one range but not in another. Moreover, the accuracy of the sensor might not be sufficient. In addition to equipment error, someone can fail to calibrate an instrument properly, misread measurements, or record results incorrectly.

Modeling Errors

Humans can also make errors in formulation of a model. Perhaps the modeler makes simplifying assumptions or determines incorrect equations that cause the model's results to deviate drastically from reality. He or she may not even be aware of crucial factors. For example, Lord Kelvin (William Thomson, Baron Kelvin of Largs), the accomplished nineteenth-century scientist who proposed the absolute temperature scale that bears his name, developed, in the mid to late part of that century, a mathematical model to calculate the age of the earth. Kelvin based his model on the assumption that the earth was cooling from a molten mass, with the sun being its only source of energy. Using only conduction, he calculated the time it would have taken for the earth, as it cooled, to reach the surface temperature at the time, and from that he made his estimates. It was difficult to determine the temperature of the earth's innermost region; but assuming that the earth was solid, he calculated that the temperature could not be high enough to melt the rock. His model estimated the age of the earth to be between 20 and 400 million years old (although later in life he reduced the upper value to 40 million years). His computed range is drastically different from its currently accepted age, which is about 4.5 *billion* years. However, his underlying assumptions were faulty. The earth is not solid, but is composed of a very hot core, surrounded by a viscous, but plastic, mantle. So, under the earth that we know, called the crust, is another source of heat. Convection from the interior of the earth warms the surface, and that makes Kelvin's calculations too low. With Becquerel's work with radioactivity, scientists proposed the heat provided from decaying radioactive elements in the earth's crust as another source of error. However, most scientists today think that this heat source is likely irrelevant, despite the pervasiveness of this hypothesis throughout the Internet. Kelvin was a fine scientist, flawed in some ways perhaps; but at the time, he could not have included subsequent knowledge in is calculations (England et al. 2007; Mareschal and Jaupart 2009).

Implementation Errors

In a computer program implementing a model, computational scientists can make logical errors that produce disastrous results. For example, in 1999, NASA's Mars Climate Orbiter spacecraft was lost because the builder of the spacecraft, Lockheed Martin Corp., programmed the system to use English units, such as pounds and feet, and NASA's Jet Propulsion scientists employed metric units, such as newtons and meters.

Precision

Other errors we consider in this module also involve computer calculations. In this section, we discuss some basic terms involved with such computations; and in the next section, we define two metrics of error. Then, we return to a discussion of other errors encountered in modeling and simulation.

Many computer languages allow floating-point numbers to be printed in exponential form as a decimal fraction times a power of 10. For instance, output of 9.843600e02 means $9.843600 \times 10^2 = 984.36 = 0.98436 \times 10^3$. Usually, **floating-point numbers**, or numbers with a decimal expansion, are stored in the computer in three parts: 0 or 1 representing the sign + or –, respectively; a **significand, fractional part**, or **mantissa**, such as 98436; and an exponent, such as 3. Every day, we use the **decimal**, or **base 10**, number system with digits, 0, 1, 2, . . ., 9, A computer usually employs the **binary**, or **base 2**, number system with only 2 binary digits, or **bits**, 0 and 1, but the concepts are the same regardless of base.

Definitions A **floating-point number** is expressed with a decimal expansion. **Exponential notation** represents a floating-point number as a decimal fraction times a power of 10. With *a* being a decimal fraction and *n* an integer, the exponential notation ***aen*** represents $a \times 10^n$. The integer formed by dropping the decimal point from *a* is the **significand**, **fractional part**, or **mantissa**, and *n* is the **exponent**.

A **normalized** number in exponential notation has the decimal point immediately preceding the first nonzero digit, as in 0.98436×10^3. This notation is similar to **scientific notation**, which places the decimal point immediately after the first nonzero digit, such as 9.8436×10^2. When a number is expressed in normalized exponential notation, as with 0.98436×10^3, all the digits of the significand, such as 98436, are what we call **significant digits**. For integers written without a decimal, all the digits except leading and trailing zeros are **significant digits**; for other numbers, all digits are significant except leading zeros. For example, there are 4 significant digits in $003{,}704{,}000 = 0.3704 \times 10^7$: 3, 7, 0, and 4. The **most significant digit** is the leftmost one, 3. The most significant digit of $0.09200 = 0.9200 \times 10^{-1}$ is 9 because the leading zero is not significant. All other digits after the decimal point (9, 2, 0, 0), however, are significant in this number.

Definitions A **normalized** number in exponential notation has the decimal point immediately preceding the first nonzero digit. The **significant digits** of a floating-point number are all the digits except the leading zeros. The **significant digits** of an integer are all the digits except the leading and trailing zeros.

Precision is the number of significant digits. Thus, 003,704,000 and 0.09200 each have a precision of 4. **Magnitude** is an indication of the relative size of a number and is 10 to the power when the number is expressed in normalized exponential notation. Therefore, 0.3704×10^7 has a magnitude of 10^7. In C and C++, the precision of a floating-point number of type *float*, which we call a **single-precision number**, is about 6 or 7 decimal digits, whereas the magnitude ranges from about 10^{-38} to 10^{38}. Taking up twice as much computer storage space as a *float* variable, a variable of type *double*, which stores a **double-precision number**, has 14 or 15 significant digits and magnitude from about 10^{-308} to 10^{308}.

Definitions The **precision** of a number is the number of significant digits. **Magnitude** is 10 to the power when the number is expressed in normalized exponential notation.

Quick Review Question 1

Use 0.0004500 for the following problems.

a. In normalized exponential notation, give the significand.
b. In normalized exponential notation, give the exponent of 10.
c. Give the precision.

Absolute and Relative Errors

To understand the size of a problem, it is helpful to have ways of measuring error. The **absolute error** is the absolute value of the difference between the exact answer and the computed answer. The **relative error** is this difference divided by the absolute value of the exact answer, provided the exact answer is not zero. We often express relative error as a percentage. For example, we can write a relative error of 0.03 as 3%.

Definitions If *correct* is the correct answer and *result* is the result obtained, then

$$\textbf{absolute error} = |\,correct - result\,|$$

$$\textbf{relative error} = \frac{(\text{absolute error})}{|\,correct\,|} = \frac{|\,correct - result\,|}{|\,correct\,|}$$

$$= \frac{(\text{absolute error})}{|\,correct\,|} \times 100\% = \frac{|\,correct - result\,|}{|\,correct\,|} \times 100\%,$$

provided $correct \neq 0$.

Example 1

Suppose a computer has a precision of 3, allowing only 3 digits in the significand, and **truncates**, or chops off, the significand to 3 digits. No computer has such limited precision, but limiting the precision to 3 simplifies our computations and still illustrates the problem. We evaluate the absolute and relative errors in the computation $(0.356 \times 10^8)(0.228 \times 10^{-3})$. The exact answer is as follows:

$$(0.356 \times 10^8)(0.228 \times 10^{-3}) = (0.356)(0.228)(10^8)(10^{-3})$$
$$= 0.081168 \times 10^5$$

Normalizing, we obtain $correct = 0.81168 \times 10^4$.

For a computer with a precision of 3, the result of this computation is $result = 0.811 \times 10^4$. Thus, an error has been introduced. The absolute error is as follows:

$$|correct - result| = |0.81168 \times 10^4 - 0.811 \times 10^4| = 0.00068 \times 10^4 = 6.8$$

The relative error is the ratio of the absolute error and the positive correct answer, as shown:

$$(0.00068 \times 10^4)/(0.81168 \times 10^4) = 0.0008378 = 0.08378\%$$

The error is about eight-hundredths of a percent of the exact answer.

Definition To **truncate** a normalized number to k significant digits, eliminate all digits of the significand beyond the kth digit.

Quick Review Question 2

Using the number 6.239, find the following:

a. The absolute error as it is truncated to 2 decimal places.
b. The relative error for Part a. Express your answer as a percentage.

Round-off Error

Instead of truncating a number to fit in storage, a computer might **round**. To round the significand of 0.81168×10^4 to a precision of 3, we consider the fourth significant digit, 6. If that digit is less than 5, we **round down**; but if the digit is greater than or equal to 5, we **round up**. Thus, 0.811**6**8 × 10^4 and 0.811**5**8 × 10^4 round up to 0.81**2** × 10^4, while 0.811**3**8 × 10^4 rounds down to 0.81**1** × 10^4.

Definitions To **round** a normalized number to precision k, consider the $(k + 1)^{th}$ significant digit, d. If d is less than 5, **round down** the normalized number by truncating the significand to k significant digits. If d is greater than or equal to 5, **round up** the normalized number by truncating the significand to k significant digits and then adding 1 to the kth significant digit of the significand, carrying as necessary to digits on the left.

Quick Review Question 3

Round each of the following so that the significand has a precision of 2:

a. 0.93742×10^{-5} **b.** 0.93472×10^{-5} **c.** 0.93572×10^{-5}

Example 2

This example illustrates the difficulty of expressing exact decimal floating-point numbers in the computer. Suppose we enter a computation for 1/3 into a cell of a spreadsheet, as follows:

```
=1/3
```

Alternatively, suppose we make the following **assignment statement** in the programming language C, C++, or Java that gives the value of the expression on the right to the variable, *x*, on the left:

```
x = 1.0/3.0;
```

The computer stores the floating-point representation of 1/3 = 0.333. . . in the spreadsheet cell or in the location for *x*. If our computer can store only 3 digits of the significand, the machine rounds or truncates the value to 0.333.

Definitions An **assignment statement** causes the computer to store the value of an expression in a memory location associated with a variable. In most programming languages, the assignment statement has a format similar to the following, with the expression always appearing on the right and the variable getting the value always being on the left of an **assignment operator**, here an equal sign:

variable = expression

Round-off error is the problem of not having enough bits to store an entire floating-point number. We have round-off error whether the computer rounds or truncates the number to fit in a location. If the computer uses a greater number of bits to store the number, the round-off error will not be as serious. For example, if we store the significand with 7 digits, the value of *x* will be 0.3333333; storage for 15 significant digits will yield the even more accurate 0.333333333333333.

Definition **Round-off error** is the problem of not having enough bits to store an entire floating point number and approximating the result to the nearest number that can be represented.

Overflow and Underflow

The problems of overflow and underflow can also ensue from finite storage and binary representation of numbers in a computer. Suppose we are working with a very small computer that uses 16 bits to store an integer. If we ask the computer to perform the sum 20480 + 16384, the result, surprisingly, will be a negative number, –28672. The problem arises when the leftmost bit, the sign bit, gets a carry from the addition on the right, converting the result to a negative number. There simply are not enough bits to express the answer, so the final answer has the wrong sign. Overflow also occurs when we add two negative numbers and get a positive result.

Definition **Overflow** is an error condition that occurs when there are not enough bits to express a value in a computer.

An overflow error caused the European Space Agency's Ariane 5 rocket to explode in 1996. Less than 37 seconds into the launch, the guidance system's computer attempted to convert the rocket's sideways velocity from a 64-bit floating-point number to a 16-bit integer. However, because the number was too large, overflow resulted; the guidance system attempted a severe correction for a wrong turn that had not occurred; and very quickly the rocket had to self-destruct. The overflow of a few bits caused the loss of a rocket that took 10 years and $7 billion to develop (Gleick 1996).

Problems can also arise when the result of a computation is too small for a computer to represent, in a situation called **underflow**. For example, suppose the smallest floating-point number a computer can express has magnitude 10^{-39}. If the correct value for an arithmetic expression, such as 10^{-48}, is smaller than the smallest positive floating-point value a computer can represent, then underflow occurs, and the computer evaluates the expression as zero.

Definition **Underflow** is an error condition that occurs when the result of a computation is too small for a computer to represent.

Arithmetic Errors

Errors can arise in addition. Consider $(0.684 \times 10^3) + (0.950 \times 10^{-2})$. Unlike in multiplication, the decimal points must be aligned for addition, so we have the following:

$$\begin{aligned} 0.684 \times 10^3 &= \ 684.0000 \\ 0.950 \times 10^{-2} &= \underline{+\ \ \ \ 0.0095} \\ & \ \ \ \ 684.0095 \end{aligned}$$

If our computer allows for only 3 significant digits, the normalized result is 0.684×10^3, and the effect of the 0.950×10^{-2} is lost.

Quick Review Question 4

Suppose a particular computer rounds stored floating-point numbers to 4 significant digits. Calculate $(0.1235 \times 10^2) + (0.2499 \times 10^{-1})$. Do not use exponential notation in the answer.

Because of such problems, when adding numbers whose magnitudes are drastically different, we should accumulate smaller numbers first before combining them with larger ones. Thus, the sum of the smaller numbers has a chance of being large enough to make a difference in the final answer.

Similarly, when multiplying and dividing in a term, to avoid loss of precision, we usually should perform all multiplications in the numerator before dividing by the denominator. For example, in our computer that rounds to 3 significant digits, suppose we are to calculate $(x/y)z$, where $x = 2.41$, $y = 9.75$, and $z = 1.54$. The quotient $x/y = 0.247179$ rounds to 0.247, so that the product is $(x/y)z = (0.247)(1.54) = 0.38038$, or 0.380 after rounding. However, algebraically $(x/y)z = (xz)/y$. Performing multiplication first, we have $xz = 3.7114$, or 3.71 in rounded form. Dividing, we have $(xz)/y = 3.71/1.54 = 0.380513$. The rounded 0.381 is closer to the exact answer of 0.380656.

Of course, whether or not we should perform all numerator multiplications before division depends on the numbers. In the example $(xz)/y$, if the product xz would cause overflow, we should first divide x by y to obtain a smaller result before multiplying by z.

Rule of Thumb In an expression involving multiplication and division on a computer, it is generally best to perform the division last.

Quick Review Question 5

Suppose variables r, u, x, y, and z store floating-point numbers. Write the following expression to minimize round-off error upon evaluation, assuming no problems with overflow or underflow:

$$\left(\frac{3}{z}\right)(r)\left(\frac{x+y}{u}\right)$$

Error Propagation

Looping enables the computer to execute a segment of code several times. Such a segment that is executed repeatedly is called a **loop**. Performing floating-point operations within loops can compound round-off error.

Definition A **loop** is a segment of code that is executed repeatedly.

An accumulation error in a loop had disastrous consequences during the First Gulf War in Dharan, Saudi Arabia, when an American Patriot missile battery failed to intercept a Scud missile. The Scud hit an American army barracks, killed 28 soldiers, and injured more than 100 others. The Patriot's internal computer clock measured time in tenths of a second and multiplied the number of ticks by 1/10 to obtain the actual time in seconds. For example, 15 ticks indicated an elapsed time of (15 ticks)(0.1 s/tick) = 1.5 s. The missile's computer used 24 bits to store numbers. However, because 1/10 has an infinite expansion in binary representation (similar to the expansion 1/3 has in decimal representation), the system could not hold all the number's bits. Each 1/10 increment produced an error of about 0.000000095 s. At the time of the disaster, the Patriot Missile had been operating for 100 h, causing an error of about (100 h)(60 min/h)(60 s/min)(10 ticks/s)(0.000000095 s/tick) = 0.34 s. During that third of a second, a Scud flew about 1676 m, so that the intercepting Patriot Missile missed its target (Arnold 2000).

Example 3

In many simulations of scientific phenomena, such as pollution in a stream, we have the computer calculate the values of various quantities as time advances in small, discrete time steps. Suppose time, *t*, starts at 0.0 s and is to end at 10 min = 600 s. The length of a time step is *dt*. Perhaps we wish to designate *dt* to be 0.1 s, so that the number of time steps (*numberOfTimeSteps*) would be 6000. However, because of conversion to base 2 and finite storage in our exaggeratedly small computer, suppose the actual stored value of *dt* is $0.09961 = 0.9961 \times 10^{-1}$ s. Inside a loop, we compute new values for time and other quantities, such as a simulated amount of mercury in a stream. A **loop variable** or **index,** *i*, takes on values 1, 2, 3, . . ., 6000. Thus, this simulation has the following general algorithm:

```
numberOfTimeSteps = number of time steps for simulation
dt = 0.09961, the length of a time step in seconds
t = 0.0, the starting time in seconds
initialize other quantities as necessary
for i going from 1 through numberOfTimeSteps do the following:
    compute a new value for t
    compute quantities being simulated
```

One method of updating time can lead to a serious accumulation of round-off error. Suppose that each time through the loop, we calculate the new value of time (*t* on the left-hand side of the assignment statement) as the old value of time (*t* on the right-hand side of the assignment statement) plus the change in time (*dt*):

$$t = t + dt$$

Suppose the machine for this example uses the decimal system and rounds to 4 significant digits. Ignoring problems of having 0.09961 instead of 0.1 for *dt* and conversion between the decimal and binary number systems, Table 5.2.1 enumerates the absolute and relative errors of *t* for several iterations of the loop.

As the table illustrates, round-off error increases with the number of loop executions. After the eleventh iteration ($i = 11$), the new value of *t* should be (11)(0.09961) = 1.09571. For a computer with a precision of 4, however, the rounded value is 1.096. The subsequent absolute error is 0.00029, and the relative error is about 0.026%. After iteration $i = 51$, the relative error is about 0.3128% and is more than 30 times the relative error in the loop with $i = 2$.

Table 5.2.1
Accumulation of Error in Time ($t = t + dt$) for Example 3

Value of i	*Correct New Value of t*	*Accumulated New Value of t = t + dt*	*Absolute Error*	*Relative Error*
1	0.09961 +0.00000 0.09961	0.09961 +0.00000 0.09961 is 0.09961 rounded with precision 4	0	0
2	0.09961 +0.09961 0.19922	0.09961 +0.09961 0.19922 is 0.1992 rounded with precision 4	0.19922 −0.1992 0.00002	0.00002/0.19922 ≈ 0.0001 = 0.01%
3	0.09961 +0.19922 0.29883	0.09961 +0.1992 0.29881 is 0.2988 rounded with precision 4	0.29883 −0.2988 0.00003	0.00003/0.29883 ≈ 0.0001 = 0.01%
4	0.09961 +0.29883 0.39844	0.09961 +0.2988 0.39841 is 0.3984 rounded with precision 4	0.39844 −0.3984 0.00004	0.00004/0.39844 ≈ 0.0001 = 0.01%
5	0.09961 +0.39844 0.49805	0.09961 +0.3984 0.49801 is 0.4980 rounded with precision 4	0.49805 −0.4980 0.00005	0.00005/0.49805 ≈ 0.0001 = 0.01%
11	1.09571	1.096	1.096 −1.09571 0.00029	0.00029/1.09571 ≈ 0.00026 = 0.026%
51	5.08011	5.096	5.096 −5.08011 0.01589	0.01589/5.08011 ≈ 0.003128 = 0.3128%

To avoid the cumulative error of this loop, we should compute t as the index, i, times $dt = 0.09961$, as follows with * indicating multiplication:

$$t = i * dt$$

We still have round-off error, but its effect is minimized because there is no accumulation. For example, in the evaluation with $i = 51$ of $(51)(0.09961) = 5.08011$, the computer stores 4 significant digits, 5.080. The absolute error is $5.08011 - 5.080 =$

0.00011 = 0.011%, while the relative error is 0.00011/5.08011 = 0.000022 = 0.0022%. In the iteration $i = 51$, the relative error using an accumulated $t = t + dt = t + 0.09961$ is about 140 times greater than the corresponding relative error using $t = i * dt = (i)(0.09961)$. In some simulations, it is not possible to avoid such repeated additions; but where possible, we should.

Rule of Thumb In looping, whenever possible, avoid accumulating floating-point values through repeated addition or subtraction.

Quick Review Question 6

Which assignment statement is better (if there is a difference) in a loop with index k whose initial value is 1? Assume that *sum* is initialized to 0 before the loop.

A. $sum = sum + 0.00492$ B. $sum = 0.00492 * k$ C. It doesn't matter.

Violation of Numeric Properties

The section "Arithmetic Errors" hints at a problem that can lead to errors—numeric properties do not necessarily hold in computer arithmetic. Expressions that are numerically equivalent might not evaluate to equal values on the computer. Mathematically, $(x/y)z = (xz)/y$, but for $x = 2.41$, $y = 9.75$, and $z = 1.54$ on a 3-significant-digit machine with rounding, we showed $(x/y)z = 0.380$, while $(xz)/y = 0.381$. Specifically, the following are some of the properties that can be violated using computer arithmetic:

Associative Properties: $(a + b) + c = a + (b + c)$ and $(ab)c = a(bc)$

These properties indicate that grouping of numbers in addition or multiplication does not matter. However, a computer can yield different results, for example, if a and b are small numbers and c is very large.

Distributive Property: $a(b + c) = ab + ac$

This property involves the distribution of multiplication over addition. Under certain circumstances, this property may not hold in a computer, such as in the case where a and b are very small in comparison to c.

In the exercises and projects, we explore these and other properties that can fail using computer arithmetic.

Quick Review Question 7

By evaluating $x(y + z)$ and $x(y) + x(z)$, show that the distributive property does not hold for $x = 2.48$, $y = 9.34$, and $z = 1.55$ on a machine that truncates intermediate and final results to 3 significant digits.

Comparison of Floating-Point Numbers

Conversion back and forth between the decimal system that people usually employ and the binary system of computers can result in the loss of information. For example, with a subscript indicating the base, the decimal number 0.6_{10} is equivalent to the binary number $0.1\overline{001}_2$, where the line over 1001 indicates that this pattern continues forever. However, a computer cannot store an infinite expansion. If the particular computer truncates to 20 bits to store the significand, the stored value would be 0.10011001100110011001_2, which is equivalent to $0.59999942779541015625_{10}$. Thus, if we want to print this value, the computer might display 0.599999 instead of the expected 0.600000.

This conversion between bases, a finite amount of computer storage, and error propagation are the reasons we should *not* test if floating-point numbers are exactly equal in a computer. Thus, in a spreadsheet, we should not test if the floating-point contents of cells *B2* and *B3* are equal as follows:

```
=If(B2 = B3, . . .) # Do NOT test floating-point numbers this way
```

Similarly, in a programming language, we should not test if floating-point variables x and z are identical, as in the following C/C++/Java statement, which employs two adjacent equal signs, ==, to check for equality:

```
if (x == z) . . . // Do NOT test floating-point numbers this way
```

Example 4

For ease of computation, let us consider a computer that uses the decimal system but truncates to 3 digits for the significand. Suppose x has the value 0.536. Now, suppose we multiply x by 7, using * for multiplication, to obtain y and divide this value by 7, assigning the final result to z, as follows:

```
x = 0.536
y = 7 * x
z = y / 7
```

In mathematics x and z are identical; but when we multiply $x = 0.536$ by 7, we obtain $3.752 = 0.3752 \times 10^1$, which truncates to $y = 0.375 \times 10^1$ in a system that has 3-digit significands. Division by 7 yields an infinite decimal expansion, $0.53\overline{571428}$, which truncates in 3 decimal places to 0.535 for z. Thus, $x = 0.536$ and $z = 0.535$ are not identical. We have exaggerated the problem by using only 3 digits for the significand, but the idea is the same for a larger significand.

To avoid the problem, we should instead test if the difference between two floating-point values is "close enough" to zero. For our limited system that uses only 3 digits for the significand, we might decide that if the difference is within 0.001 of zero, then we will consider the values to be equal. Because the difference could be positive or negative, we take the absolute value of the difference and make the following test:

$$\text{If } |x - z| < 0.001, \text{ consider } x \text{ and } z \text{ to be the same.} \tag{1}$$

In a more-realistic computer system, to determine equality we might test if x and z are within some very small number of each other, say, 0.000001 of each other.

Rule To determine if floating-point numbers x and z are equal on a computer, test if x and z are within some small value of each other, that is, test if $|x - z| < c$ for some small constant, c, such as 0.000001.

Quick Review Question 8

Write a statement similar to (1) to test if two floating-point variables are *not* equal to within 0.000001 of each other.

Truncation Error

Just as a computer cannot exactly store numbers with infinite precision, a computer cannot perform an infinite number of calculations. Let us consider a result of calculus that says e^x has the following infinite series expansion:

$$e^x = 1 + x + \frac{x^2}{1\cdot 2} + \frac{x^3}{1\cdot 2\cdot 3} + \frac{x^4}{1\cdot 2\cdot 3\cdot 4} + \cdots + \frac{x^n}{n!} + \cdots$$

Therefore, $e^1 = e$, which is Euler's number, 2.718281828459045 . . ., is exactly equal to the following infinite series, with 1 replacing x in the last equation:

$$e = 1 + 1 + \frac{1}{1\cdot 2} + \frac{1}{1\cdot 2\cdot 3} + \frac{1}{1\cdot 2\cdot 3\cdot 4} + \cdots + \frac{1}{n!} + \cdots$$

However, a finite machine is unable to perform such an infinite number of additions. If we wish to use this series to evaluate e, we must truncate the sum. For example, if we perform the additions to $n = 20$ to obtain a partial sum, we have the following close approximation of e:

$$e \approx 1 + 1 + \frac{1}{1\cdot 2} + \frac{1}{1\cdot 2\cdot 3} + \frac{1}{1\cdot 2\cdot 3\cdot 4} + \cdots + \frac{1}{20!} = \frac{6{,}613{,}313{,}319{,}248{,}080{,}001}{2{,}432{,}902{,}008{,}176{,}640{,}000}$$

This finite sum does not include terms from 1/(21!) on and results in the following **truncation error**:

$$\frac{1}{21!} + \frac{1}{22!} + \frac{1}{23!} + \cdots \approx 2.05\times 10^{-20}$$

> **Definition** A **truncation error** is an error that occurs when a truncated, or finite, sum is used as an approximation for the sum of an infinite series.

Figure 5.2.1 displays graphs of e^x (in color) and the following approximating partial-sum functions (in black): $f_n(x) = 1 + x + \frac{x^2}{1\cdot 2} + \frac{x^3}{1\cdot 2\cdot 3} + \frac{x^4}{1\cdot 2\cdot 3\cdot 4} + \cdots + \frac{x^n}{n!}$ for $n = 1$ through $n = 4$:

$$f_1(x) = 1 + x$$

$$f_2(x) = 1 + x + \frac{x^2}{1\cdot 2} = 1 + x + \frac{x^2}{2}$$

$$f_3(x) = 1 + x + \frac{x^2}{1\cdot 2} + \frac{x^3}{1\cdot 2\cdot 3} = 1 + x + \frac{x^2}{2} + \frac{x^3}{6}$$

$$f_4(x) = 1 + x + \frac{x^2}{1\cdot 2} + \frac{x^3}{1\cdot 2\cdot 3} + \frac{x^4}{1\cdot 2\cdot 3\cdot 4} = 1 + x + \frac{x^2}{2} + \frac{x^3}{6} + \frac{x^4}{24}$$

As n becomes larger, the graphs of the partial-sum functions $f_n(x)$ approach, or converge to, the graph of $f(x) = e^x$.

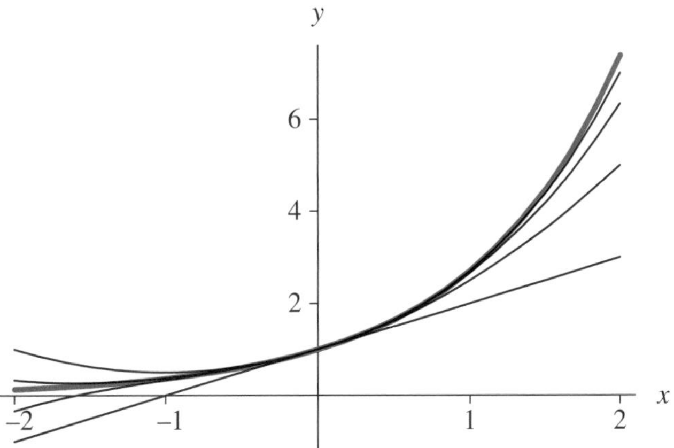

Figure 5.2.1 Graphs of e^x (in color) and four finite series approximations

Quick Review Question 9

Calculus shows that sin(x) equals the following infinite series:

$$\sin(x) = x - \frac{x^3}{3!} + \frac{x^5}{5!} - \frac{x^7}{7!} + \cdots$$

a. Obtain an approximation for sin(2), truncating the infinite series to 2 terms.
b. Give the error for this approximation.

Exercises

Write the numbers of Exercises 1–12 in normalized exponential notation.

1. 63,850	**2.** 29.748	**3.** 0.00032	**4.** 53.7×10^3
5. 0.496	**6.** 0.0000017	**7.** 0.009×10^{-5}	**8.** 0.009×10^5
9. –0.82	**10.** –82	**11.** –0.00082	**12.** 4.4

13. Give the magnitude and precision of 0.743621×10^{25}.

14. Give the magnitude and precision of 93.6×10^7.

15. Give the precision and the largest magnitude of numbers in a computer where the significand has 8 digits and the largest power of 10 is 125.

Give the number of significant digits and the most significant digit for the numbers in Exercises 16–21.

16. 63,850	**17.** 29.004	**18.** 0.00074
19. 10^3	**20.** 4×10^{-5}	**21.** 0.300500

22. Give the range of the normalized positive numbers where the significand has three digits and the exponent of 10 is from –5 to +5.

23. Give the range of the normalized positive numbers that can be expressed with 7 digits in the significand and an exponent of 10 from –78 to +73.

For exercise 24–26, find the following: **(a)** *the absolute error and* **(b)** *the relative error of each number as it is rounded to 2 decimal places. Then compute* **(c)** *the absolute error and* **(d)** *the relative error of each number as it is truncated to 2 decimal places.*

24. 6.239 **25.** 6.231 **26.** 1.0/3.0 stored with 5 significant digits

27. **a.** Perform the arithmetic $(9.4 \times 10^{-5}) + (3.6 \times 10^4)$, expressing the answer in normalized exponential notation.

b. Give the final answer if the representation allows only 5 significant digits, rounded.

c. Give the absolute error of Part b.

d. Give the relative error of Part b.

28. Repeat Exercise 27 for $(0.7 \times 10^3) - (0.825 \times 10^2)$. Use 3 significant digits instead of 5 significant digits for Parts b–d.

29. Suppose the following sequence is executed: $x = 6.239$; $x = x + x$. For a machine that truncates to 3 significant digits, give the values stored for x after execution of each statement and the relative error for the value of x after the last statement. Compare this error with your answer in Exercise 24.

30. Consider a machine that rounds to 4 significant digits. Suppose initially $y = 9.649$ and $x = 7.834$. The following assignment statement, which calculates the expression on the right, $(y + x)$, and then replaces the value of y on the left with the result, is in a loop that executes four times: $y = y + x$. After each iteration of the loop, give the value stored in y and the absolute and relative errors:

31. Using the computer described in Example 3 and $t = i * dt$, evaluate the computer's value for t, the absolute error, and the relative error for the values of i in Table 5.2.1. Compare your results with those of Table 5.2.1.

32. Mathematics can prove that $1.0 = 0.9\overline{9} = 0.99\overline{9}$. Suppose this value is assigned to x and to y as a series of 9s truncated to 4 significant digits.

a. If $x = x + y$ is executed four times in a loop, each time replacing the old value of x with the result of the sum on the right, give the values of x and the absolute and relative errors for the original assignment and after each iteration of the loop.

b. By observing the results of Part a, give the value of x and the absolute and relative errors after the tenth iteration of the loop.

33. Refer to Example 4 for the proper testing of equality of floating-point numbers.

a. Write an *if* statement that puts 1 in the current cell of a spreadsheet if the floating-point values in cells *B2* and *B3* are equal and otherwise puts 0 in the current cell.

b. In another computational tool, write an *if* statement that displays 1 if the floating point values in x and z are equal and otherwise displays 0.

34. a. Calculus shows that cos(x) equals the following infinite series:

$$\cos(x) = 1 - \frac{x^2}{2!} + \frac{x^4}{4!} - \frac{x^6}{6!} + \cdots$$

Obtain a decimal expansion approximation for cos(2), truncating the infinite series to three terms.

b. Find finite series approximations for cos(2) until successive approximations are the same for the first 4 significant digits and give that approximation.

35. a. With a computational tool, evaluate the following:

10000000000000000. + 1. – 10000000000000000.

that is, $1.0 \times 10^{16} + 1. - 1.0 \times 10^{16}$.

b. What should the result be?

c. Explain what happened.

36. For a machine that rounds to 3 significant digits, give an example where a floating-point number, x, does not have a multiplicative inverse, y, where $xy = 1$.

37. For a machine that rounds to 3 significant digits, give an example where the associative property for addition does not hold.

Projects

1. Using a computational tool, evaluate each of the following expressions for $t = 355/113$, $r = 101/113$, and $s = 52/113$: $t - s - r - r - r$, $t - r - r - r - s$, $t - r - r - s - r$, $t - r - 2r - s$, $t - r - s - r - r$, $t - 2r - r - s$, $t - 3r - s$, $t - r - s - 2r$. Using mathematics, what are the values of the expressions? What numeric property or properties are violated (Panoff 2004)?

2. Using a computational tool, evaluate e^x for a given value of x using the series expansion in the section "Truncation Error" and enough terms so that successive approximations differ by no more than 1 hundred-thousandth. Show all steps of the developing expansion on separate rows. Also, evaluate e^x using the built-in exponential function.
3. Do Project 2 for $\sin(x)$ using the series expansion in Quick Review Question 9.
4. Do Project 2 for a function and its Taylor series expansion. Do not use e^x or $\sin(x)$. Refer to a calculus text for such an expansion.
5. Using a computational tool and the infinite series expansions of $\sin(x)$ (see Quick Review Question 9), define five partial-sum approximation functions. Graph $\sin(x)$ and the approximating functions. Evaluate $\sin(x)$ and the approximating functions at several values of x, and compute the absolute and relative errors of the approximations.
6. Do Project 5 for a function and its Taylor Series expansion. Do not use e^x or $\sin(x)$. Refer to a calculus text for such an expansion.
7. Develop a system dynamics model for Exercise 8 in Module 2.2, "Unconstrained Growth and Decay." Have your system dynamics tool calculate the absolute and relative errors of the simulated values in comparison to the analytical values.
8. This project requires the use of programming in a computer language or a computational tool. A chemical is added a drop at a time to a container in which a reaction is occurring. Each drop is measured as precisely 0.xxxx mL. The total amount of the chemical must be computed after each drop is added. Write a program to perform the calculation in two ways: by incrementing the previous total by 0.xxxx (*accumulated*) and by multiplying 0.xxxx (*multiplied*) by the number of drops so far. At the beginning of the program, ask the user for the number of iterations and the reporting interval. Print the iteration number and both ongoing totals at the requested intervals. Use the last 4 digits of your phone number as the nonzero digits in the measurement of the drop of the chemical. Replace zeros with different odd digits. For example, with a phone number of 555-9389, use 0.9389; and with 555-8090, possibly use 8193. Sample output for 0.xxxx = 0.9389 is as follows:

```
Give the number of iterations: 1000
Give the reporting interval: 200

Iteration 200
Accumulated = 187.781
Multiplied  = 187.78

Iteration 400
Accumulated = 375.561
Multiplied  = 375.56

Iteration 600
Accumulated = 563.342
Multiplied  = 563.34

Iteration 800
Accumulated = 751.123
Multiplied  = 751.12
```

```
Iteration 1000
Accumulated = 938.904
Multiplied  = 938.9
```

Print the output and report for each of the following:

a. Run the program with a reporting interval of 1 to determine the first iteration in which there is a difference between the two computations. What are the absolute and relative errors for *accumulated* at that point? Is there an error for *multiplied*?

b. Run the program with a reporting interval of 1,000,000. What are the absolute and relative errors for both totals?

c. Run the program with a reporting interval of 999,999. What are the absolute and relative errors for both totals? *Note:* To compute the correct answer, take the correct answer for the millionth iteration and subtract 0.xxxx.

d. Run the program with a large enough number of iterations and a large reporting interval so that eventually the value of *accumulated* does not change from one report to the next. What explanation do you have for this phenomenon?

Answers to Quick Review Questions

1. a. 4500
b. –3
c. 4 because the 4 digits of 4500 are significant

2. a. 6.239 – 6.23 = 0.009
b. (6.239 – 6.23)/6.239 = 0.144%

3. a. 0.94×10^{-5}
b. 0.93×10^{-5}
c. 0.94×10^{-5}

4. 12.37 because the sum is 12.35 + 0.02499 = 12.37499, rounded to 4 significant digits.

5. $\frac{3r(x+y)}{zu}$, or $3r(x + y)/(zu)$. *Note:* In the second form, parentheses around the denominator product are essential because of priority of operations. Without the parentheses around zu, z would be divided into $3r(x + y)$ and—incorrectly—the result would be multiplied by u, which is equivalent to $\left(\frac{3r(x+y)}{z}\right)u = \frac{3r(x+y)u}{z}$.

6. B. $sum = 0.00492 * k$

7. $x(y + z) = 2.48(9.34 + 1.55) = 2.48(10.89)$. However, 10.89 truncates to 10.8. Thus, 2.48(10.8) = 26.784, which truncates to 26.7.
$x(y) + x(z) = 2.48(9.34) + 2.48(1.55) = 23.1632 + 3.844$, which after truncation is 23.1 + 3.84 = 26.94, or 26.9.

$$x(y + z) = 26.7 \neq x(y) + x(z) = 26.9$$

8. If $|x - z| \geq 0.000001$, consider x and z not to be the same.

9. a. $2 - 2^3/(3!) = 2 - 8/6 = 2/3$

b. $\frac{2^5}{5!} - \frac{2^7}{7!} + \cdots = \sin(2) - 2/3$

References

Arnold, Douglas N. 2000. "The Patriot Missile Failure." University of Minnesota, August 23. http://www.ima.umn.edu/~arnold/disasters/patriot.html (accessed December 18, 2012)

Darden, Lindley. 1998. "The Nature of Scientific Inquiry." University of Maryland, College Park. http://www.philosophy.umd.edu/Faculty/LDarden/sciinq/ (accessed December 18, 2012)

England, Phillip, Peter Molnar, and Frank Richter. 2007. "John Perry's Neglected Critique of Kelvin's Age for the Earth: A Missed Opportunity in Geodynamics." *GSA TODAY* 17(1): 4. http://www.colorado.edu/geolsci/faculty/molnarpdf/2007GSAT.England.PerryKelvinBlownOpportunity.pdf (accessed January 8, 2013)

Gleick, James. 1996. "Sometimes a Bug Is More Than a Nuisance." http://www.around.com/ariane.html (accessed December 18, 2012)

Mareschal, J., and C. Jaupart. 2009. "Heat Loss of the Earth and Energy Budget of the Mantle." Meeting American Geophysical Union. Meeting Abstract #U24A-01

Panoff, Robert. 2004. National Computational Science Institute Workshop on Computational Science at Wofford College, Spartanburg, South Carolina.

6

SIMULATION TECHNIQUES

MODULE 6.1

Computational Toolbox—Tools of the Trade: Tutorial 2

Prerequisite: Module 5.1, "Computational Toolbox—Tools of the Trade: Tutorial 1."

Download

From the textbook's website, download Tutorial 2 in the format of your computational tool or in PDF format. We recommend that you work through the tutorial and answer all Quick Review Questions using the corresponding software.

Introduction

Various computer software tools are useful for graphing, numeric computation, and symbolic manipulation. This second computational toolbox tutorial, which is available from the textbook's website in your system of choice, prepares you to use the tool to complete projects for this and subsequent chapters. The tutorial introduces the following functions and concepts:

- Lists/arrays
- Plotting data
- Comments
- Appending

The module gives computational examples and Quick Review Questions for you to complete and execute in the desired software system.

MODULE 6.2

Euler's Method

Download

The text's website has the file *unconstrainedError*, which contains a model for the "Error" section in this module, available for download for various system dynamics tools.

Introduction

With system dynamics tools, we often have the choice of simulation techniques, such as Euler's Method, Runge-Kutta 2, Runge-Kutta 4, and others. These numerical methods are for solving ordinary differential equations, as we have done in Chapters 2–4, and estimating definite integrals for which the indefinite integral does not exist. In this module, we discuss the most straightforward of these, Euler's method.

Reasoning behind Euler's Method

In Module 2.2, "Unconstrained Growth and Decay," to simulate with $P_0 = 100$, we employ the following underlying equations with *INIT* meaning "initial" and *dt* representing a small change in time, Δt:

```
growth_rate = 0.10
INIT population = 100
growth = growth_rate * population
population = population + growth * dt
```

These equations, which we enter explicitly or implicitly with a model diagram, represent the following finite difference equations using **Euler's method**:

```
growth_rate = 0.10
population(0) = 100
growth(t) = growth_rate * population(t - Δt)
population(t) = population(t - Δt) + growth(t) * Δt
```

The population at one time step, *population*(t), is the population at the previous time step, *population*($t - \Delta t$), plus the estimated change in population, *growth*(t) * Δt. The derivative of population with respect to time is *growth*, and

$$\begin{aligned} growth(t) &= growth\_rate * population(t - \Delta t) \\ &= 0.10 * population(t - \Delta t) \end{aligned}$$

or $dP/dt = 0.10P$. The change in the population is the flow (in this case, *growth*) times the change in time, Δt; and the flow (*growth*) is the derivative of the function at the previous time step, $t - \Delta t$.

Figure 6.2.1 illustrates Euler's method used to estimate $P_1 = P(8)$ for the preceding differential equation by starting with $P_0 = P(0) = 100$ and using $\Delta t = 8$. In this situation, $t = 8$, $t - \Delta t = 0$, and the derivative at that time is $P'(0) = 0.1(100) = 10$, which, as Figure 6.2.1 depicts, is the slope of the tangent line to the curve $P(t)$ at (0, 100). We multiply Δt, 8, by this derivative at the previous time step, 10, to obtain the estimated change in P, 80. Consequently, the estimate for P_1 is as follows:

$$\begin{aligned} \text{estimate for } P_1 &= \text{previous value of } P + \text{estimated change in } P \\ &= P_0 + P'(0)\Delta t \\ &= 100 + 10(8) \\ &= 180 \end{aligned}$$

In Module 2.2, "Unconstrained Growth and Decay," we discovered analytically that the solution to the preceding differential equation is $P = 100\ e^{0.1t}$. Because the graph of the actual function is concave up, this estimated value, 180, is lower than the actual value at $t = 8$, $100e^{0.1(8)} \approx 223$.

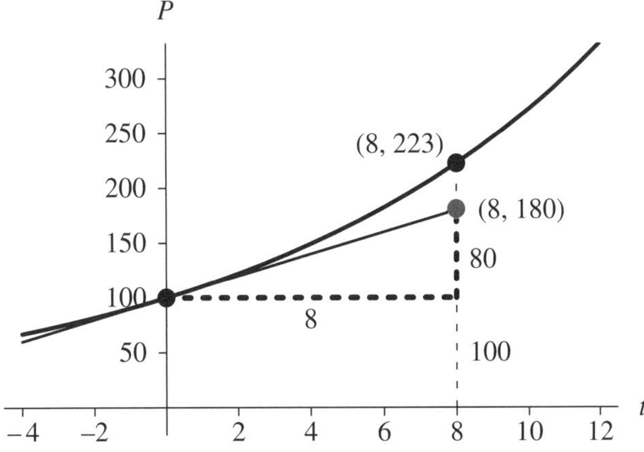

Figure 6.2.1 Actual point, (8, 223), and point obtained by Euler's method, (8, 180)

Quick Review Question 1

For $dP/dt = 10 + P/5$, with $P_0 = 500$ and $\Delta t = 0.1$, calculate the following:

a. dP/dt at $t = 0$
b. Euler's estimate of P_1

Algorithm for Euler's Method

Following the preceding description, Algorithm 1 presents Euler's method.

Algorithm 1: Euler's Method

$t \leftarrow t_0$
$P(t_0) \leftarrow P_0$
Initialize *SimulationLength*
while $t <$ *SimulationLength* do the following:

$t \leftarrow t + \Delta t$
$P(t) = P(t - \Delta t) + P'(t - \Delta t)\,\Delta t$

Thus, this simulation uses a sequence of times—t_0, t_1, t_2, ...—and calculates a corresponding sequence of estimated populations—P_0, P_1, P_2, ... In Algorithm 1, $t_n = t_{n-1} + \Delta t$ or $t_{n-1} = t_n - \Delta t$, and $P_n = P_{n-1} + P'(t_{n-1})\Delta t$.

However, as illustrated in Module 5.2, "Errors," repeatedly accumulating Δt into t usually produces an accumulation error. To minimize error, we calculate the time as the sum of the initial time and an integer multiple of Δt. Using the functional notation $f(t_{n-1}, P_{n-1})$ to indicate the derivative dP/dt at Step $n - 1$, Algorithm 2 presents a revised Euler's Method that minimizes accumulation of error.

Algorithm 2 Revised Euler's Method to minimize error accumulation of time with $f(t_{n-1}, P_{n-1})$ indicating the derivative dP/dt at step $n - 1$

Initialize t_0 and P_0
Initialize *NumberOfSteps*
for n going from 1 to *NumberOfSteps* do the following:

$t_n = t_0 + n\,\Delta t$
$P_n = P_{n-1} + f(t_{n-1}, P_{n-1})\,\Delta t$

Quick Review Question 2

Match each of the following symbols to its meaning in Algorithm 2 for Euler's method. Here, previous means immediately previous.

A. Change in time between time steps
B. Derivative of function at estimated value of function for current time step

C. Derivative of function at estimated point for previous time step
D. Estimated value of function at current time step
E. Estimated value of function at previous time step
F. Initial time
G. Initial value of function
H. Number of current time step
I. Time at current time step
J. Time at previous time step

a. t_n
b. t_0
c. n
d. Δt
e. P_n
f. P_{n-1}
g. $f(t_{n-1}, P_{n-1})$
h. t_{n-1}

Error

As we saw in Module 2.2, "Unconstrained Growth and Decay," the analytical solution to $dP/dt = 0.10P$ with $P_0 = 100$ is $P = 100e^{0.10t}$. Even with Algorithm 2, comparison of the results of Euler's method with the analytical solution reveals an accumulation error. As Figure 6.2.2 and an *unconstrainedError* file illustrate (see "Download"), we can adjust the unconstrained growth model to demonstrate the variation.

The converter/variable *actual_population* evaluates P_0e^{rt}, or $100e^{0.10t}$. The formula does not use the changing value of *population*, but the initial population, *initial_population*. With *Time* as the current value of time and *EXP* as the exponential function, the equation in the converter *actual_population* is as follows:

```
initial_population * EXP(growth_rate * Time)
```

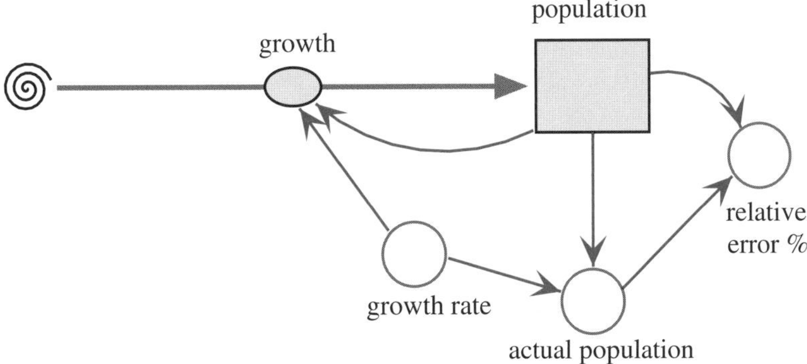

Figure 6.2.2 Unconstrained growth model with monitoring

Similarly, the converter *relative_error_%* computes the percent relative error as the product of 100 and the absolute value (ABS) of the difference in Euler's estimate and the analytical population with the result divided by the latter, as follows:

```
ABS(population - actual_population) * 100 / actual_population
```

Figure 6.2.3 presents graphs of the analytical solution and the solution using Euler's Method, with Δt being 1. As demonstrated, the simulated solution is below the analytical one. At the end of the run, at time 100, the analytical value for the population is 2,202,647, while the simulated solution using Euler's method produces 1,378,061, so that the relative error is more than 37.4%. For Δt being cut in half, the relative error is almost cut in half to 21.5% at time 100. The new simulated population is 1,729,258, which is considerably closer to the analytical solution of 2,202,647. If we cut the time step in half again so that Δt is 0.25, the relative also reduces by about half to 11.6% at time 100. Thus, the relative error is proportional to Δt. We say that the relative error is **on the order of Δt, $O(\Delta t)$**.

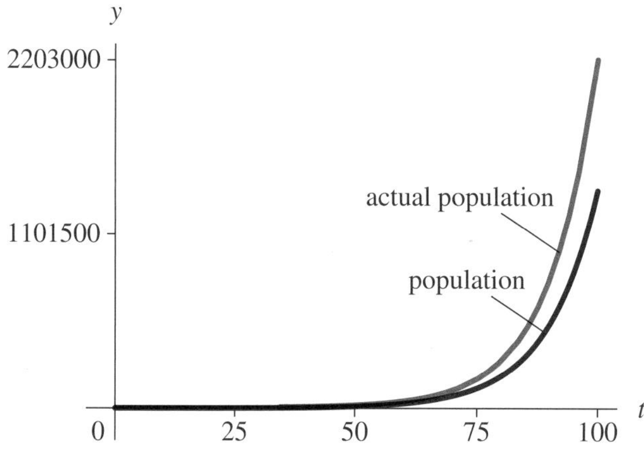

Figure 6.2.3 Graphs of analytical solution and Euler's Method solution with $\Delta t = 1$

Quick Review Question 3

The analytical solution to $dP/dt = 10 + P/5$, with $P_0 = 500$, of Quick Review Question 1 is $P = 550e^{t/5} - 50$. Part b of that question showed that for $\Delta t = 0.1$, the Euler's Method estimate of P_1 is 511. Calculate the relative error as a percentage with 4 decimal places.

Of the three simulation techniques in this chapter, Euler's method is the easiest to understand and has the fastest execution time but is the least accurate. We usually can reduce the error of the Euler method by employing a smaller Δt, which unfortunately slows the simulation. Despite its shortcomings, the reasoning behind Euler's method serves as an excellent introduction to the other techniques, Runge-Kutta 2 and Runge-Kutta 4, because each of these has Euler's method embedded as the first step in its simulation.

Exercises

1. Use $dP/dt = 0.10P$, with $P_0 = 100$, and Euler's method to calculate P_2 starting with $P_1 = 180$ at $t = 8$ h in Figure 6.2.1.

In Exercises 2–5, for each differential equation with initial condition and Δt, calculate the following using Euler's method and any other requested computation:

a. *The first estimated point, such as P_1, where the differential equation is in terms of dP/dt*

b. *The second estimated point, such as P_2*

2. $dP/dt = 0.10P$, with $P_0 = 100$ and $\Delta t = 2$

   c. The relative error for P_1

3. The logistic equation $\dfrac{dP}{dt} = 0.5\left(1 - \dfrac{P}{1000}\right)P$ with $P_0 = 20$ and $\Delta t = 2$

   c. The relative error for P_1, where the exact solution is $P(t) = \dfrac{10}{0.01 + 0.49e^{-0.5t}}$

4. The rate of change of the number of particles of radioactive carbon-14 in a dead tree $dA/dt = -2.783e^{-0.000121t}$ with $A_0 = 23{,}000$ particles and $\Delta t = 0.2$ yr

   c. The relative error for A_2, where the exact solution is $A(t) = 23{,}000e^{-0.000121t}$

5. The Gompertz differential equation $dN/dt = kN\ln(M/N)$, with $N(0) = 200$, $k = 0.1$, $M = 1000$, and $\Delta t = 0.5$

   c. The relative error for N_2, where the exact solution is $N(t) = Me^{\ln(N_0/M)e^{-kt}}$

Projects

Using Algorithm 2 for Euler's method, develop a computational tool file to perform the simulations of Projects 1–7. Run the simulation for the indicated length of time and perform any other requested tasks.

1. Calculate P from $t = 0$ through $t = 100$, where $dP/dt = 0.10P$, with $P_0 = 100$ and $\Delta t = 2$. Calculate the relative error at each time step using the solution $P = 100e^{0.1t}$. Repeat the computation with $\Delta t = 0.25$. Check your results using a system dynamics tool. Use your results in a discussion of relative error.
2. Calculate P from $t = 0$ through $t = 100$ for logistic equation $\dfrac{dP}{dt} = 0.5\left(1 - \dfrac{P}{1000}\right)P$, with $P_0 = 20$ and $\Delta t = 2$. Calculate the relative error at each time step using the solution $P(t) = \dfrac{10}{0.01 + 0.49e^{-0.5t}}$. Repeat the computation with $\Delta t = 0.25$. Check your results using a system dynamics tool. Use your results in a discussion of relative error.
3. Suppose the instantaneous rate of change of the number of particles, A, of radioactive carbon-14 in a gram of a dead tree is $dA/dt = -2.783e^{-0.000121t}$ particles/year from the time, t, the tree dies with $A_0 = 23{,}000$ particles. Use Euler's method to estimate the total change in the number of particles of carbon-14 between years 10 and 20. Calculate the exact value of the definite integral with calculus or an appropriate computational tool and compute the relative error.

4. The Gompertz differential equation, which is one of the best models for predicting the growth of cancer tumors, follows:

$$dN/dt = kN \ln(M/N)$$

where N is the number of cancer cells. Calculate N from $t = 0$ through $t = 20$, where $k = 0.1$, $M = 1000$, and $\Delta t = 0.5$. Using the solution $N(t) = Me^{\ln(N_0/M)e^{-kt}}$, calculate the relative error at each time step. Repeat the computation with $\Delta t = 0.25$. Check your results using a system dynamics tool. Use your results in a discussion of relative error.
5. Estimate $\int_1^5 (-t^2 + 10t + 24)dt$ using $\Delta t = 0.25$. Calculate the exact value using calculus or an appropriate computational tool and compute the relative error of the simulated result.
6. Estimate $\frac{1}{\sqrt{2\pi}}\int_0^2 e^{-t^2}dt$ using $\Delta t = 0.1$. The corresponding indefinite integral does not exist. The function being integrated is the normal distribution with mean 0 and standard deviation 1.
7. Calculate $h(t)$ and $s(t)$ from $t = 0$ through $t = 50$ using $\Delta t = 0.25$ for the following system of differential equations:

$$\begin{cases} \dfrac{ds}{dt} = 2s - 0.02hs, & s_0 = 100 \\ \dfrac{dh}{dt} = 0.01sh - 1.06h, & h_0 = 15 \end{cases}$$

As Module 4.2, "Predator-Prey Model," discusses, this system is a model for predator (h) and prey (s) populations.

Answers to Quick Review Questions

1. **a.** 110 because 10 + (500)/5 = 110
   b. 511 because 500 + 0.1(110) = 511
2. **a.** t_n — I. Time at current time step
   b. t_0 — F. Initial time
   c. n — H. Number of current time step
   d. Δt — A. Change in time between time steps
   e. P_n — D. Estimated value of function at current time step
   f. P_{n-1} — E. Estimated value of function at previous time step
   g. $f(t_{n-1}, P_{n-1})$ — C. Derivative of function at estimated point for previous time step
   h. t_{n-1} — J. Time at previous time step
3. 0.0217% because

$$550e^{0.1/5} - 50 = 511.111 \text{ and}$$
$$|(511 - 511.111)|/511.111 = 0.000217 = 0.0217\%$$

References

Burden, Richard L., and J. Douglas Faires. 2011. *Numerical Analysis,* 9th ed. Belmont, CA: Brooks/Cole (Cengage Learning).

Danby, J. M. A. 1997. *Computer Modeling: From Sports to Spaceflight . . . From Order to Chaos.* Richmond, VA: Willmann-Bell.

Woolfson, M. M., and G. J. Pert. 1999. *An Introduction to Computer Simulation.* Oxford, UK: Oxford University Press.

Zill, Dennis G. 2013. *A First Course in Differential Equations with Modeling Applications*, 10th ed. Belmont, CA: Brooks/Cole (Cengage Learning).

MODULE 6.3

Runge-Kutta 2 Method

Introduction

In Module 6.2, "Euler's Method," which is a prerequisite to the current module, we discuss the simplest of this chapter's simulation techniques for solving differential equations and computing definite integrals numerically. In this section, we consider a second and better simulation technique, **Euler's predictor-corrector (EPC) method**, also called **Runge-Kutta 2**.

Euler's Estimate as a Predictor

In the current module, we consider the example of Module 6.2, "Euler's Method," $dP/dt = 0.10P$, with $P_0 = 100$. As in that section, $f(t_n, P_n)$ is sometimes a more convenient notation for the derivative dP/dt at Step n. Thus, at $(t, P) = (0, 100)$, the derivative is $f(0, 100) = 0.1(100) = 10$. According to that technique, using the derivative at (t_{n-1}, P_{n-1}), which is always equal to the slope of the tangent line there, we have the following computation for t_n and estimation of P_n:

$$t_n = t_0 + n\,\Delta t$$
$$P_n = P_{n-1} + f(t_{n-1}, P_{n-1})\,\Delta t$$

As Figure 6.2.1 of "Euler's Method" and Figure 6.3.1 illustrate for $t_0 = 0$ and $\Delta t = 8$, the estimate at $t_1 = 8$ is the vertical coordinate of the point on the tangent line, $100 + 8(10) = 180$.

Corrector

To estimate (t_n, P_n), we would really like to use the slope of the chord from (t_{n-1}, P_{n-1}) to (t_n, P_n) instead of the slope of the tangent line at (t_{n-1}, P_{n-1}). As in Figure 6.3.2, if

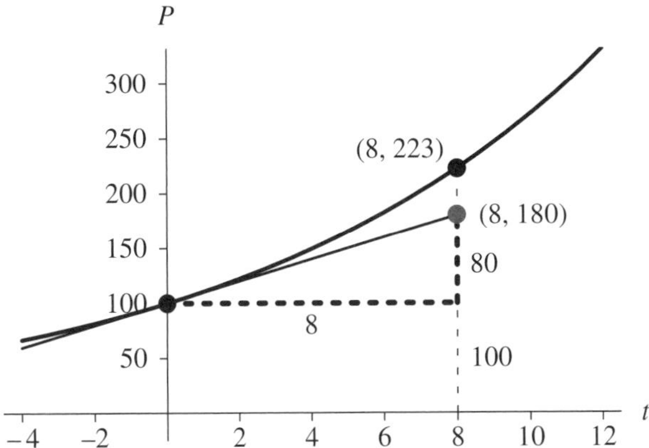

Figure 6.3.1 Actual point, $(8, P(8)) \approx (8, 223)$, and point obtained by Euler's method, (8, 180)

we know the slope of the chord between $(0, P(0)) = (0, 100)$ and $(8, P(8)) = (8, 100e^{0.10(8)}) \approx (8, 223)$ is approximately $\dfrac{223-100}{8-0} = \dfrac{123}{8}$, we can evaluate $P(8) = P(0) + slope\_of\_chord * \Delta t = 100 + \left(\dfrac{123}{8}\right)8 = 223$. However, to evaluate the slope of the chord, we must know $P(8) \approx 223$, which is the value we are trying to estimate. If we know the actual value, there is no need to estimate.

Although we do not know the slope of the chord between $(0, P(0))$ and $(8, P(8))$, we can estimate it as approximately the average of the slopes of the tangent lines at $P(0)$ and $P(8)$:

$$\left(\begin{array}{c}\text{slope of the chord}\\ \text{between } (0, P(0)) \text{ and } (8, P(8))\end{array}\right) \approx \frac{(\text{slope of tan at } P(0)) + ((\text{slope of tan at } P(8))}{2}$$

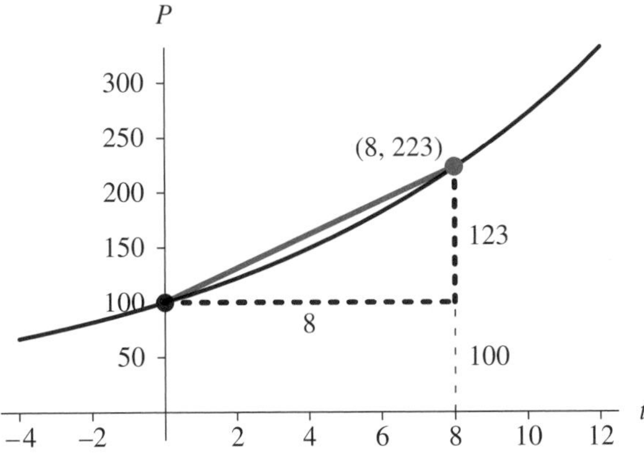

Figure 6.3.2 Actual point, $(8, P(8)) \approx (8, 223)$, along the chord between (0, 100) and (8, 223)

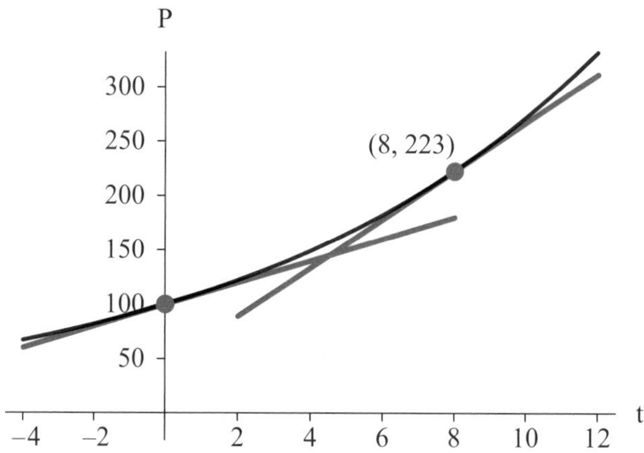

Figure 6.3.3 Tangent lines at (0, $P(0)$) and (8, $P(8)$)

Figure 6.3.3 depicts these two tangent lines.

How can we find the slope of the tangent line at $P(8)$ when we do not know $P(8)$? Instead of using the exact value, which we do not know, we predict $P(8)$ as in Euler's method. As the computation in the first section, "Euler's Estimate as a Predictor," shows, in this case, the prediction is $Y = 180$. We use the point (8, 180) in derivative formula to obtain an estimate of slope at $t = 8$. In this case, the slope of the tangent line at (8, 180), or the derivative, is $f(8, 180) = 0.1(180) = 18$. Using 18 as the approximate slope of the tangent line at (8, $P(8)$), we estimate the slope of chord between (0, $P(0)$) and (8, $P(8)$) as the following average of tangent line slopes:

$$\text{slope of chord} \approx (10 + 18)/2 = 0.5(10 + 18) = 14$$

As Figure 6.3.4 shows, using 14, the corrected estimate is $P_1 = 100 + 14(8) = 212$, which is closer to the actual value of 223.

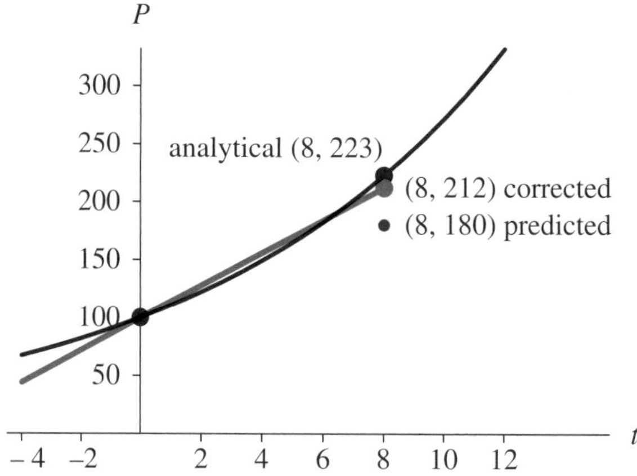

Figure 6.3.4 Predicted and corrected estimation of (8, $P(8)$)

Quick Review Question 1

Quick Review Question 1 of Module 6.2, "Euler's Method," considered $dP/dt = 10 + P/5$, with $P_0 = 500$ and $\Delta t = 0.1$. Part a calculated the derivative at $t = 0$ to be 110, and Part b evaluated Euler's estimate of P_1 to be 511. Calculate the corrected estimate for P_1 using the technique of this section and decimal notation for your answer.

Runge-Kutta 2 Algorithm

Computations of the previous section illustrate Euler's predictor-corrector method for estimating P numerically given a differential equation involving dP/dt. The algorithm for **Euler's predictor-corrector (EPC) method**, or **Runge-Kutta 2**, is the same as Euler's, with only one more statement in the loop to obtain the corrected value.

Algorithm for Euler's Predictor-Corrector (EPC) Method, or **Runge-Kutta 2,** with $f(t_{n-1}, P_{n-1})$ indicating the derivative dP/dt at step $n - 1$

Initialize t_0 and P_0
Initialize *NumberOfSteps*
for n going from 1 to *NumberOfSteps* do the following:

$t_n = t_0 + n\,\Delta t$
$Y_n = P_{n-1} + f(t_{n-1}, P_{n-1})\Delta t$, which is the Euler's method estimate
$P_n = P_{n-1} + 0.5\,(f(t_{n-1}, P_{n-1}) + f(t_n, Y_n))\Delta t$

Quick Review Question 2

Match each of the following symbols to its meaning in the algorithm for Euler's predictor-corrector (EPC) method. Here, previous means immediately previous; EPC estimate means estimated value of function using the EPC method; and Euler estimate means estimated value of function using Euler's method.

A. Average of derivatives of function at previous EPC estimate and current Euler estimate
B. Average of derivatives of function at previous Euler estimate and current EPC estimate
C. Derivative of function at EPC estimate for current time step
D. Derivative of function at EPC estimate for previous time step
E. Derivative of function at Euler estimate for current time step
F. Derivative of function at Euler estimate for previous time step
G. EPC estimate at current time step
H. EPC estimate at previous time step

I. Euler estimate at current time step
J. Euler estimate at previous time step
   a. Y_n
   b. P_{n-1}
   c. $f(t_{n-1}, P_{n-1})$
   d. P_n
   e. $f(t_n, Y_n)$
   f. $0.5\ (f(t_{n-1}, P_{n-1}) + f(t_n, Y_n))$

Error

As noted previously, the actual slope of the chord is $(P(8) - 100)/8 \approx (222.6 - 100)/8 \approx 15.3$, but 14 is certainly a better slope to use than 10 from Euler's method. With Euler's method, $P_1 = 180$, giving a relative error of $(180 - P(8))/P(8) \approx |180 - 222.6|/\ 222.6 \approx 0.191 \approx 19.1\%$. We get a much better estimate with Euler's predicator-corrector (Runge-Kutta 2) method, $P_1 = 212$, which has a relative error of $|212 - P(8)|/\ P(8) \approx |212 - 222.6|/222.6 \approx 0.047 \approx 4.7\%$.

As we saw in the "Error" section of Module 6.2, "Euler's Method," the relative error of Euler's method is on the order of Δt, $O(\Delta t)$. If we cut the time interval Δt in half, the relative error is halved as well. Using the same model with the Runge-Kutta 2 simulation method, Table 6.3.1 shows estimates of $P(100)$, whose actual value to 0 decimal places is 2,202,647, for $\Delta t = 1$, 0.5, and 0.25. As the time interval is cut by 1/2, the relative error is cut by about $(1/2)^2 = (1/4)$. Thus, the relative error of the EPC method is $O((\Delta t)^2)$, or on the order of $(\Delta t)^2$. Thus, although we must compute a correction in each EPC algorithm iteration, the EPC method is more accurate than Euler's method.

Table 6.3.1
Estimates of $P(100)$ and Relative Errors for Various Changes in Time Using Runge-Kutta 2 Simulation Method, where $dP/dt = 0.10P$ with $P_0 = 100$

	EPC Estimates at Time 100	
Δt	*Estimated* $P(100)$	*Relative Error*
1.00	2,168,841	1.53%
0.50	2,193,824	0.40%
0.25	2,200,396	0.10%

Quick Review Question 3

The analytical solution to $dP/dt = 10 + P/5$, with $P_0 = 500$, of Quick Review Question 1 is $P = 550e^{t/5} - 50$. That question showed that the Euler's predictor-corrector method estimate of P_1 is 511.11 for $\Delta t = 0.1$. Calculate the relative error as a percentage rounded to 4 decimal places.

Exercises

Repeat the exercises of Module 6.2. "Euler's Method," using the Runge-Kutta 2 Method. Compare the relative errors with those of the corresponding exercises from Module 6.2.

Projects

Repeat the projects of Module 6.2, "Euler's Method," using the Runge-Kutta 2 method.

Answers to Quick Review Questions

1. 511.11 because $Y_1 = 511$; $f(0.1, 511) = 10 + 511/5 = 112.2$; $500 + (0.5)(110 + 112.2)(0.1) = 511.11$

2. **a.** Y_n — I. Euler estimate at current time step
b. P_{n-1} — H. EPC estimate at previous time step
c. $f(t_{n-1}, P_{n-1})$ — D. Derivative of function at EPC estimate for previous time step
d. P_n — G. EPC estimate at current time step
e. $f(t_n, Y_n)$ — E. Derivative of function at Euler estimate for current time step
f. $0.5\,(f(t_{n-1}, P_{n-1}) + f(t_n, Y_n))$ — A. Average of derivatives of function at previous EPC estimate and current Euler estimate

3. 0.0002% because $550e^{0.1/5} - 50 = 511.111$ and $|511.11 - 511.111|/511.111 = 0.000002 = 0.0002\%$

References

Burden, Richard L. and J. Douglas Faires. 2011. *Numerical Analysis,* 9th ed. Belmont, CA: Brooks/Cole (Cengage Learning).

Danby, J.M.A. 1997. *Computer Modeling: From Sports to Spaceflight... From Order to Chaos.* Richmond, VA: Willmann-Bell.

Woolfson, M. M., and G. J. Pert. 1999. *An Introduction to Computer Simulation.* Oxford, UK: Oxford University Press.

Zill, Dennis G. 2013. *A First Course in Differential Equations with Modeling* Applications, 10th ed. Belmont, CA: Brooks/Cole (Cengage Learning).

MODULE 6.4

Runge-Kutta 4 Method

Introduction

Of the three integration techniques of this chapter—Euler's, Runge-Kutta 2, and Runge-Kutta 4 methods—the last is the most involved but the most accurate. The relative errors of the techniques are $O(\Delta t)$, $O(\Delta t^2)$, and $O(\Delta t^4)$, respectively, with the names Runge-Kutta 2 and 4 indicating the exponents of Δt. Thus, the latter technique improves the most as Δt gets smaller.

To illustrate Runge-Kutta 4 method, we again use the example $f(t, P) = dP/dt = 0.10P$, with $P_0 = 100$ and $\Delta t = 8$, to show the derivation of P_1 from P_0. To estimate P_n, the technique adds to P_{n-1} a weighted average of four estimates—∂_1, ∂_2, ∂_3, and ∂_4—of the change in P.

First Estimate, ∂_1, Using Euler's Method

As with the Runge-Kutta 2 method, in the Runge-Kutta 4 method we employ the estimated value of the function from Euler's method for the first predicted change in P. As the section "Reasoning behind Euler's Method" from Module 6.2 explains, we multiply the derivative of the function at (t_0, P_0) times Δt for the change in the value of the function from the initial value, P_0, to the new estimate, P_1. In our example, $f(0, 100) = 0.1 \times 100 = 10$, so the first estimate of the change in P is $\partial_1 = f(0, 100)\Delta t = 10 \times 8 = 80$. Figure 6.4.1 illustrates this change with a dashed line in color to the estimated point that is also in color.

In general, the **first estimate** of $\Delta P = P_n - P_{n-1}$ is as follows:

$$\partial_1 = f(t_{n-1}, P_{n-1})\Delta t$$

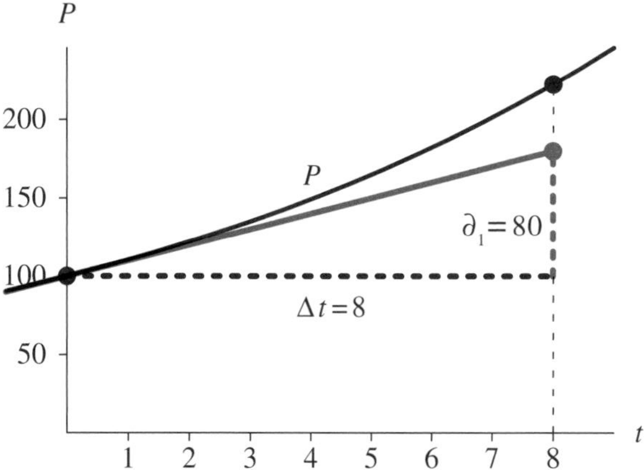

Figure 6.4.1 First estimate of change in P, $\partial_1 = 80$

Quick Review Question 1

Suppose $dP/dt = -P^2/1000$, $t_{30} = 1$, and $P_{30} = 500$. Evaluate ∂_1 for $\Delta t = 6$.

Second Estimate, ∂_2

To calculate the second estimate of ΔP for the previous example, we use the point halfway between the initial point (t_0, P_0), and point from Euler's estimate, $(t_0 + \Delta t, P_0 + \partial_1)$, in Figure 6.4.1. The midpoint is on the tangent line to the graph of the function P at $(t_0, P_0) = (0, 100)$. Its first coordinate is $t_0 + 0.5\Delta t = 0 + 0.5(8) = 4$, and its second coordinate is $P_0 + 0.5\partial_1 = 100 + 0.5(80) = 140$. Figure 6.4.2 depicts this point, (4, 140), in color.

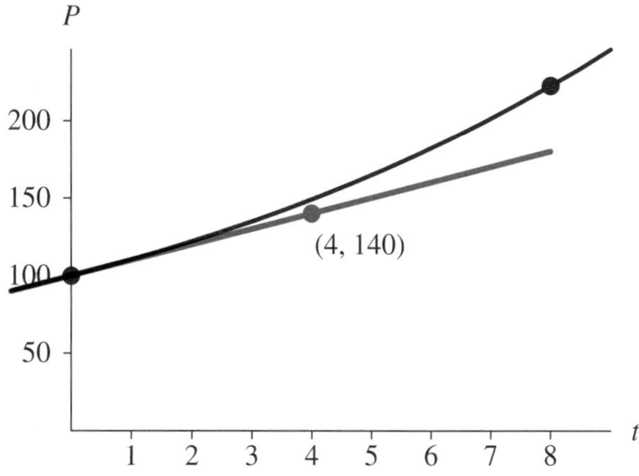

Figure 6.4.2 Midpoint (4, 140) between $(t_0, P_0) = (0, 100)$ and $(t_0 + \Delta t, P_0 + \partial_1) = (8, 180)$

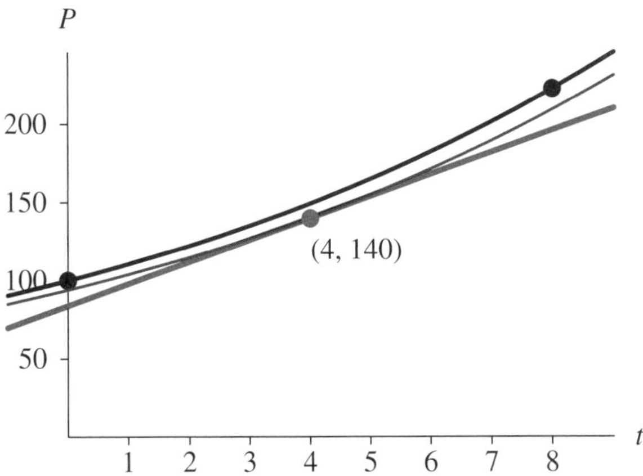

Figure 6.4.3 Estimate slope at midpoint between (0, 100) and (8, 180)

We calculate the derivative, f, for this midpoint using the derivative formula $f(t, P) = 0.1P$, as follows:

$$f(4, 140) = 0.1(140) = 14$$

Figure 6.4.3 shows, with less thickness, the exponential function through (4, 140) that has derivative 14 at $t = 4$. Thus, at $t = 4$ the curve's tangent line, which is in color, has slope 14.

For the second estimate of the change in P, ∂_2, we determine the change in the vertical direction for this line for $\Delta t = 8$, as follows:

$$\partial_2 = ((0.1)(140))\ (8) = 14\ (8) = 112$$

Figure 6.4.4 pictures a line of the same slope (14) that passes through the initial point (0, 100). After a change in t of 8 units, P increases by 112. Thus, the second estimate

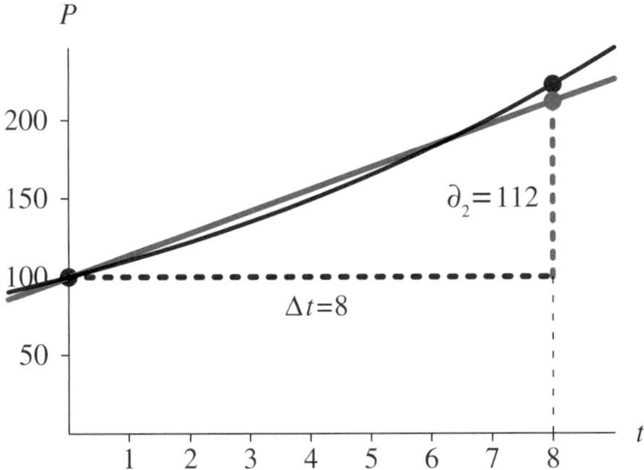

Figure 6.4.4 Second estimate of change in P, $\partial_2 = 112$

of P_1 is 100 + 112 = 212. The actual value of P at $t = 8$ is about 222.6, so 212 is an improvement over the estimate using Euler's method, 100 + 80 = 180. Figure 6.4.4 depicts the new estimated point in color as being significantly closer to the actual point than Euler's estimate in Figure 6.4.1. The improvement comes from making a midway correction.

The **second estimate** for the change in P employs the estimated slope at the point $(t_{n-1} + 0.5\Delta t, P_{n-1} + 0.5\partial_1)$, as follows:

$$\partial_2 = f(t_{n-1} + 0.5\Delta t, P_{n-1} + 0.5\partial_1)\Delta t$$

Quick Review Question 2

Suppose $dP/dt = -P^2/1000$, $t_{30} = 1$, and $P_{30} = 500$. Quick Review Question 1 showed that $\partial_1 = -1500$ for $\Delta t = 6$.

a. Give the t-coordinate of the point at which to calculate the derivative for ∂_2.
b. Give the P-coordinate of the point at which to calculate the derivative for ∂_2.
c. Evaluate ∂_2.

Third Estimate, ∂_3

For the third estimate, ∂_3, we use the same process as for the second estimate on the line in Figure 6.4.4 that passes through the initial point (0, 100) and the second estimate point, $(t + \Delta t, P_0 + \partial_2) = (8, 212)$. First, we find the midpoint, (4, 156), between the endpoints (see Figure 6.4.5).

Using the derivative formula, $f(t, P) = 0.1P$, we estimate the slope of the curve at $t = 4$ as $f(4, 156) = 0.1(156) = 15.6$. The line through (4, 156) with slope 15.6 appears in color in Figure 6.4.6.

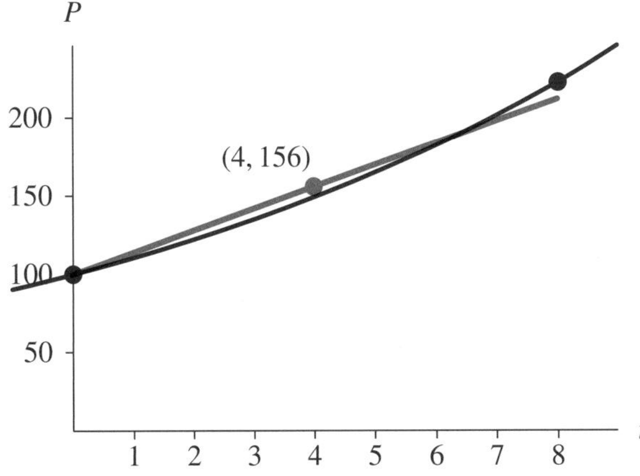

Figure 6.4.5 Midpoint (4, 156) between $(t_0, P_0) = (0, 100)$ and $(t_0 + \Delta t, P_0 + \partial_2) = (8, 212)$

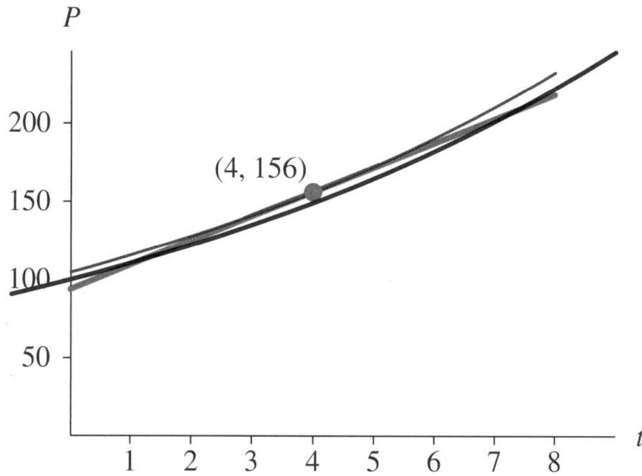

Figure 6.4.6 Estimate slope at midpoint between (0, 100) and (8, 212)

Using the slope of this line, we determine the third estimated change in P over $\Delta t = 8$, ∂_3, as follows:

$$\partial_3 = ((0.1)(156))\ (8) = 15.6\ (8) = 124.8$$

Figure 6.4.7 displays this third estimate of ΔP, ∂_3, as the length of the colored, vertical dashed line to the point, (8, 224.8), both of which are in color.

The **third estimate** for the change in P employs the estimated slope at the point $(t_{n-1} + 0.5\Delta t,\ P_{n-1} + 0.5\partial_2)$, as follows:

$$\boldsymbol{\partial_3 = f(t_{n-1} + 0.5\Delta t,\ P_{n-1} + 0.5\partial_2)\Delta t}$$

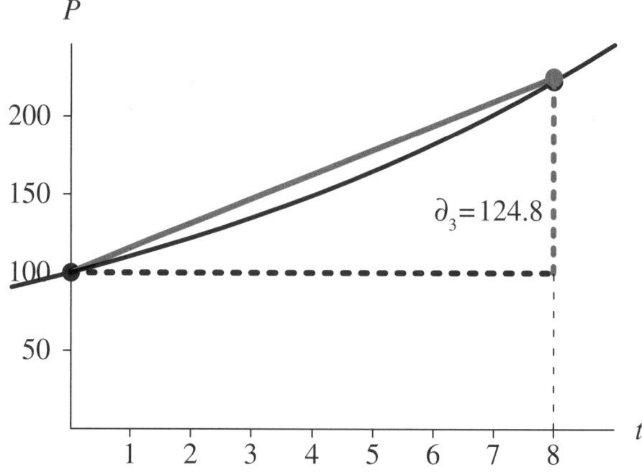

Figure 6.4.7 Third estimate of change in P, $\partial_3 = 124.8$

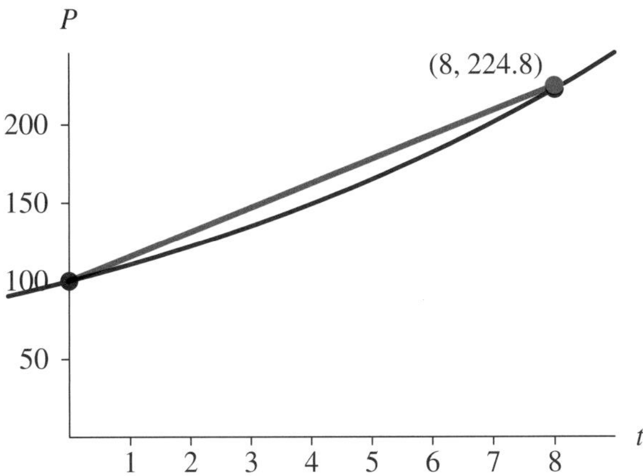

Figure 6.4.8 Endpoint $(t_0 + \Delta t, P_0 + \partial_3) = (8, 224.8)$

Quick Review Question 3

Suppose $dP/dt = -P^2/1000$, $t_{30} = 1$, and $P_{30} = 500$. Quick Review Question 2 showed that $\partial_2 = -375$ for $\Delta t = 6$. Express your answers to 1 decimal place.

a. Give the t-coordinate of the point at which to calculate the derivative for ∂_3.
b. Give the P-coordinate of the point at which to calculate the derivative for ∂_3.
c. Evaluate ∂_3.

Fourth Estimate, ∂_4

The fourth estimate, ∂_4, of the change in P over the interval of length Δt occurs at the end of the interval. As Figure 6.4.8 illustrates, using the third estimate ∂_3, the endpoint is $(t_0 + \Delta t, P_0 + \partial_3) = (8, 224.8)$ for the example under discussion.

With $dP/dt = f(t, P) = 0.1P$, The following computation estimates the slope at the endpoint:

$$f(8, 224.8) = 0.1(224.8) = 22.48$$

Figure 6.4.9 shows the endpoint along with the exponential function and tangent line of slope 22.48 through that point.

With this slope, we estimate ∂_4, the increase in P as t increases, by $\Delta t = 8$, as follows:

$$\partial_4 = ((0.1)(224.8))(8) = 22.48(8) = 179.84$$

This fourth estimate of ΔP is the length of the boldfaced, vertical dashed line to the point (8, 279.84), both of which are in color in Figure 6.4.10. Using ∂_4, 279.84 is the new estimate of P_1.

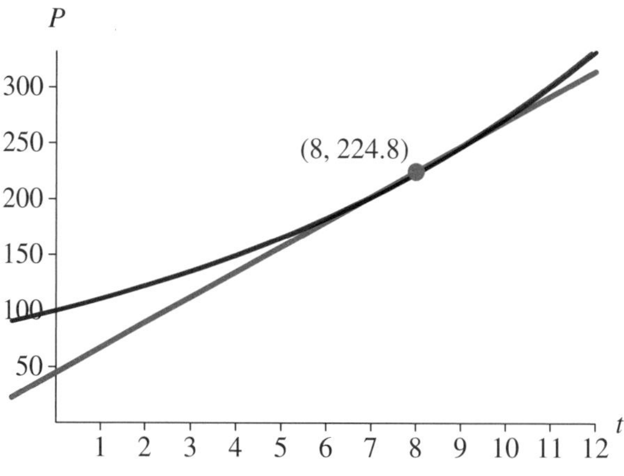

Figure 6.4.9 Estimate slope at (8, 224.8)

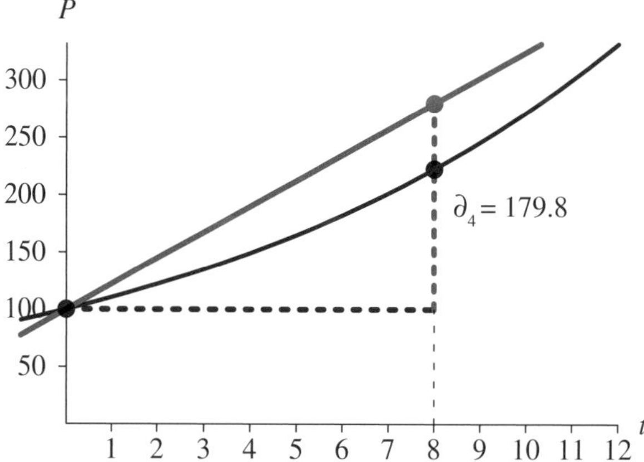

Figure 6.4.10 Fourth estimate of change in P, $\partial_4 = 179.84$

The **fourth estimate** for the change in P employs the estimated slope at the point $(t_{n-1} + \Delta t, P_{n-1} + \partial_3)$, as follows:

$$\boldsymbol{\partial_4 = f(t_{n-1} + \Delta t, P_{n-1} + \partial_3)\Delta t}$$

Quick Review Question 4

Suppose $dP/dt = -P^2/1000$, $t_{30} = 1$, and $P_{30} = 500$. Quick Review Question 3 showed that $\partial_3 = -585.9$ for $\Delta t = 6$. Express your answers to 1 decimal place.

a. Give the t coordinate of the point at which to calculate the derivative for ∂_4.

b. Give the P coordinate of the point at which to calculate the derivative for ∂_4.
c. Evaluate ∂_4.

Using the Four Estimates

We have obtained estimates of the rate of change of P with respect to t, $dP/dt = f(t, P)$, at four places on the interval of length Δt, at each end and twice at the midpoint. Using these computations, we derived four estimates (∂_1, ∂_2, ∂_3, and ∂_4) of the change in P over the interval from $t_0 = 0$ to $t_0 + \Delta t = 8$. Figure 6.4.11 shows, corresponding to the black points, the estimates at the left and right endpoints (for $\partial_1 = 80$ and $\partial_4 = 179.84$, respectively) and, corresponding to the points in color, the two estimates at the midpoint (for $\partial_2 = 112$ and $\partial_3 = 124.8$). Each value indicates a length of vertical dashed line in color from a height of $P_0 = 100$ to a point whose second coordinate is the corresponding estimate of P_1.

To determine the Runge-Kutta 4 estimate of P_1, we add to $P_0 = 100$ a weighted average of ∂_1, ∂_2, ∂_3, and ∂_4. Giving twice the weight to the estimates at the midpoint, the computation is as follows:

$$\begin{aligned} P_1 &= P_0 + (\partial_1 + 2\partial_2 + 2\partial_3 + \partial_4)/6 \\ &= 100 + (80 + 2 \cdot 112 + 2 \cdot 124.8 + 179.84)/6 \\ &= 100 + 122.24 \\ &= 222.24 \end{aligned}$$

The **Runge-Kutta 4 estimate** of P_n is as follows:

$$\boldsymbol{P_n = P_{n-1} + (\partial_1 + 2\partial_2 + 2\partial_3 + \partial_4)/6}$$

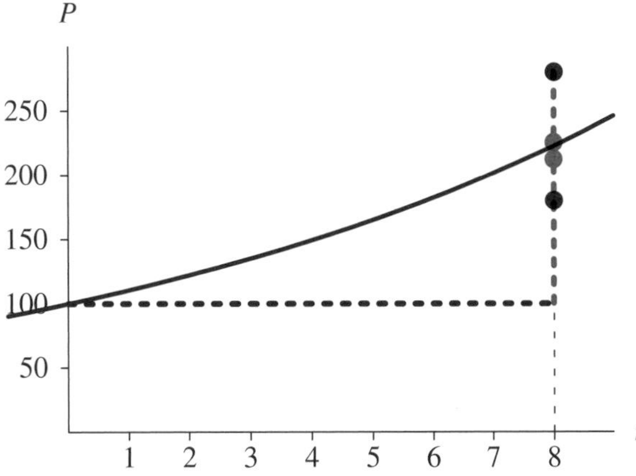

Figure 6.4.11 Four estimates of ΔP

Quick Review Question 5

Suppose $dP/dt = -P^2/1000$, $t_{30} = 1$, and $P_{30} = 500$. For $\Delta t = 6$, we found the following estimations of ΔP:

∂_i	*Quick Review Question*
$\partial_1 = -1500$	1
$\partial_2 = -375$	2
$\partial_3 = -585.9$	3
$\partial_4 = -44.3$	4

Evaluate Runge-Kutta 4's estimate of P_{31} to 1 decimal place.

Runge-Kutta 4 Algorithm

The following Runge-Kutta 4 algorithm combines the computation of the four estimates of ΔP, the weighted average, and the final estimate of P_n into a loop.

Runge-Kutta 4 Algorithm with $f(t_{n-1}, P_{n-1})$ indicating the derivative dP/dt at step $n - 1$

Initialize t_0 and P_0
Initialize *NumberOfSteps*
for n going from 1 to *NumberOfSteps* do the following:
$t_n = t_0 + n\,\Delta t$
$\partial_1 = f(t_{n-1}, P_{n-1})\,\Delta t$
$\partial_2 = f(t_{n-1} + 0.5\Delta t, P_{n-1} + 0.5\partial_1)\Delta t$
$\partial_3 = f(t_{n-1} + 0.5\Delta t, P_{n-1} + 0.5\partial_2)\Delta t$
$\partial_4 = f(t_{n-1} + \Delta t, P_{n-1} + \partial_3)\Delta t$
$P_n = P_{n-1} + (\partial_1 + 2\partial_2 + 2\partial_3 + \partial_4)/6$

Quick Review Question 6

Match each of the following symbols to its meaning in the Runge-Kutta 4 algorithm. Here, previous means immediately previous and estimate means estimated value of function using the indicated method.

A. Derivative at midpoint between (t_{n-1}, P_{n-1}) and point with Euler estimate as second coordinate
B. Derivative at midpoint between (t_{n-1}, P_{n-1}) and point with EPC estimate as second coordinate
C. Estimate of ΔP at a midpoint
D. Estimate of ΔP at right endpoint

E. Euler estimate at current time step
F. Euler estimate at previous time step
G. Runge-Kutta 2 estimate
H. Runge-Kutta 4 estimate
I. Weighted average of intermediate estimates of ΔP
  a. ∂_1
  b. $f(t_{n-1} + 0.5\Delta t, P_{n-1} + 0.5\partial_1)$
  c. ∂_3
  d. ∂_4
  e. $(\partial_1 + 2\partial_2 + 2\partial_3 + \partial_4)/6$
  f. P_n

Error

For the preceding example, the analytical solution of P_1 to 2 decimal places is 222.55, while the Runge-Kutta 4 estimate is 222.24. Thus, as the following shows, the relative error is small:

$$|222.24 - 222.55|/222.55 = 0.0014 = 0.14\%$$

An even more dramatic illustration of the improvement in accuracy of the Runge-Kutta 4 method over Euler's and Runge-Kutta 2 (Euler's predictor-corrector, EPC) methods occurs at the estimate of $P(100)$. The analytical solution of $P(100)$ to 0 decimal places is 2,202,647. Table 6.4.1 lists the relative errors for the three techniques using $\Delta t = 1$, 0.5, and 0.25. With the Runge-Kutta 4 method at $\Delta t = 0.25$, the relative error is extremely small, and the rounded estimate and analytical solutions are identical.

Producing such small errors, simulations can usually have larger step sizes, or Δt values, with the Runge-Kutta 4 method than with the other two techniques. However, the computation is slower because on each step, the Runge-Kutta 4 algorithm computes the derivative, f, four times instead of one or two times. Thus, a trade-off of time for accuracy exists.

Table 6.4.1
Relative Errors of $P(100)$ for Various Time Changes and Simulation Methods, where $dP/dt = 0.10P$ with $P_0 = 100$

	Relative Errors at Time 100		
Δt	*Euler's*	*EPC*	*Runge-Kutta* 4
1.00	37.4%	1.53%	0.000767%
0.50	21.5%	0.40%	0.000050%
0.25	11.6%	0.10%	0.000003%

Exercises

Repeat the exercises of Module 6.2, "Euler's Method," using the Runge-Kutta 4 method. Compare the relative errors with those of the corresponding exercises from Module 6.2, "Euler's Method," and Module 6.3, "Runge-Kutta 2 Method."

6. Download the file *simplePendulum* from the text's website in one of the system dynamics tools. Figure 3.3.3 of Module 3.3, "Tick Tock—The Pendulum Clock," shows a plot of a simple pendulum's angle, angular velocity, and angular acceleration versus time that is the result of the simulation with $\Delta t = 0.01$ and Runga-Kutta 4 integration. Run the simulation with $\Delta t = 0.01$ using, in turn, Runga-Kutta 4, Runga-Kutta 2, and Euler's methods or whatever methods are available with your system dynamics tool. Describe any anomalies in the graphs. Repeat the simulations and description for $\Delta t = 0.1$. Discuss the implications of your findings.

Projects

Repeat the projects of Module 6.2, "Euler's Method," using the Runge-Kutta 4 Method.

Answers to Quick Review Questions

1. –1500 because the product of the derivative and Δt at (t_{30}, P_{30}) is $f(1, 500)\Delta t = (-(500^2)/1000)(6) = -1500$

2. a. 4 because $t_{30} + 0.5\Delta t = 1 + 6/2 = 4$

b. –250 because $P_{30} + 0.5\partial_1 = 500 + (-1500/2) = 500 - 750 = -250$

c. –375 because $f(t_{30} + 0.5\Delta t, P_{30} + 0.5\partial_1)\Delta t = f(4, -250)\Delta t = (-(-250)^2/1000)\ 6 = -375$

3. a. 4 because $t_{30} + 0.5\Delta t = 1 + 6/2 = 4$

b. 312.5 because $P_{30} + 0.5\partial_2 = 500 + (-375/2) = 500 - 187.5 = 312.5$

c. –585.9 because $f(t_{30} + 0.5\Delta t, P_{30} + 0.5\partial_2)\Delta t = f(4, 312.5)\Delta t = (-(312.5)^2/1000)\ 6 = -585.9$

4. a. 7.0 because $t_{30} + \Delta t = 1 + 6 = 7$

b. –85.9 because $P_{30} + \partial_3 = 500 - 585.9 = -85.9$

c. –44.3 because $f(t_{30} + \Delta t, P_{30} + \partial_3)\Delta t = f(7, -85.9)\Delta t = (-(-85.9)^2/1000)\ 6 = -44.3$

5. –77.7 because $500 + (-1500 + 2(-375) + 2(-585.9) + (-44.3))/6 = -77.7$

6.

a. ∂_1	F.	Euler estimate at previous time step
b. $f(t_{n-1} + 0.5\Delta t, P_{n-1} + 0.5\partial_1)$	A.	Derivative at midpoint between (t_{n-1}, P_{n-1}) and point with Euler estimate as second coordinate
c. ∂_3	C.	Estimate of ΔP at a midpoint
d. ∂_4	D.	Estimate of ΔP at right endpoint

e. $(\partial_1 + 2\partial_2 + 2\partial_3 + \partial_4)/6$	I. Weighted average of intermediate estimates of ΔP
f. P_n	H. Runge-Kutta 4 estimate

References

Burden, Richard L. and J. Douglas Faires. 2011. *Numerical Analysis,* 9th ed.. Belmont, CA: Brooks/Cole (Cengage Learning).

Danby, J.M.A. 1997. *Computer Modeling: From Sports to Spaceflight. . . From Order to Chaos.* Richmond, VA: Willmann-Bell.

Woolfson, M. M., and G. J. Pert. 1999. *An Introduction to Computer Simulation.* Oxford, UK: Oxford University Press.

Zill, Dennis G. 2013. *A First Course in Differential Equations with Modeling* Applications, 10th ed. Belmont, CA: Brooks/Cole (Cengage Learning).

7

ADDITIONAL SYSTEM DYNAMICS PROJECTS

Overview

In the previous chapters, we have studied techniques, issues, and applications of computational science system dynamics models. Projects often extended the examples discussed and developed in the modules. We cannot overemphasize the importance of developing solutions to such projects, for it is in doing modeling that we learn computational science problem-solving abilities.

This chapter provides opportunities to further this learning through additional extensive projects. Unlike earlier chapters, these modules do not include examples. Instead, each module contains sufficient background in a scientific application area for you to complete the projects. Chapter 7's modules list the prerequisite modules. Moreover, the project sections of those previous modules cross-reference the material in Chapter 7. Thus, students can work with projects in the current chapter as soon as they have covered the appropriate prerequisites or at a later time.

As with earlier projects, the projects in this chapter are well suited for teamwork. In computational science, most research and development are done using interdisciplinary teams. Thus, experiences developing models with teams, perhaps on applications out of an area of major study, are important for a student studying computational science.

Chapter 7's applications with projects involving system dynamics models are in a variety of scientific areas, including the following: radioactive chains, blood cell populations, scuba diving, the carbon cycle, global warming, growth in a garden, the cardiovascular system, electrical circuits, transmission of nerve impulses, carbohydrate metabolism, mercury pollution, the economics of commercial fishing, lac operon, and colon cancer.

MODULE 7.1

Radioactive Chains—Never the Same Again

Prerequisite: Module 2.2, "Unconstrained Growth and Decay."

Introduction

The mass $Q(t)$ of a radioactive substance decays at a rate proportional to the mass of the substance (see the section "Unconstrained Decay" in Module 2.2, "Unconstrained Growth and Decay"). Thus, for positive **disintegration constant**, or **decay constant**, r, we have the following differential equation:

$$dQ/dt = -rQ(t)$$

and its difference equation counterpart:

$$\Delta Q = -rQ(t - \Delta t)\Delta t$$

In this module, we model the situation where one radioactive substance decays into another radioactive substance, forming a chain of such substances. For example, radioactive bismuth-210 decays to radioactive polonium-210, which in turn decays to lead-206. We consider the amounts of each substance as time progresses.

Modeling the Radioactive Chain

If a radioactive substance, *substanceA*, decays into substance *substanceB*, we say that *substanceA* is the **parent** of *substanceB* and that *substanceB* is the **child** of *substanceA*. If *substanceB* is also radioactive, *substanceB* is the parent of another substance, *substanceC*, and we have a **chain** of substances. Figure 7.1.1 depicts the situation where A, B, and C are the masses of radioactive substances, *substanceA*, *substanceB*, and *substanceC*, respectively; and different disintegration constants, *decay_rate_of_A* (a) and *decay_rate_of_B* (b), exist for each decay.

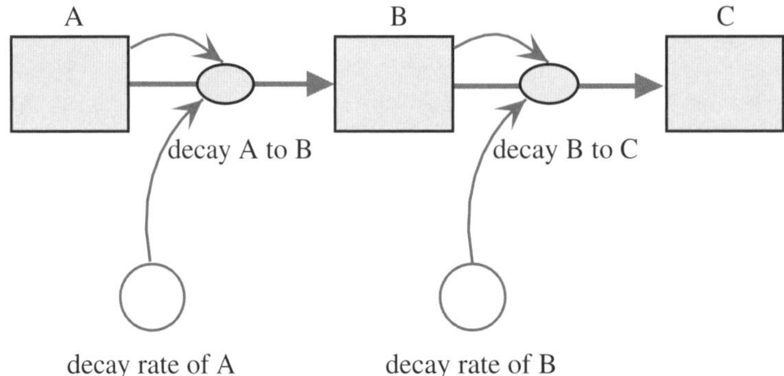

Figure 7.1.1 Chain of decays

Quick Review Question 1

Suppose A and B are the masses of *substanceA* and *substanceB*, respectively, at time t; ΔA and ΔB are the changes in these masses; and a and b are the positive disintegration constants.

a. Using these constants and variables along with arithmetic operators, such as minus and plus, give the difference equation for the change in the mass of *substanceA*, ΔA.

b. Through disintegration of *substanceA*, *substanceB*'s mass increases, while some of *substanceB* decays to *substanceC*. Give the difference equation for the change in the mass of *substanceB*, ΔB.

c. In Figure 7.1.1, where A, B, and C are the masses of three radioactive substances, give the formula as it appears in a system dynamics tool's equation for the flow *decay_A_to_B*.

d. Give the formula as it appears in a systems dynamics tool's equation for the flow *decay_B_to_C*.

The mass of *substanceA* that decays to *substanceB* is aA. Thus, in Figure 7.1.1, the flow *decay_A_to_B* contains the formula *decay_rate_of_A* * *A*. What *substanceA* loses, *substanceB* gains. However, *substanceB* decays to *substanceC* at a rate proportional to the mass of *substanceB*, bB. Consequently, in Figure 7.1.1, the flow *decay_B_to_C* contains the mass that flows from one stock to another, *decay_rate_of_B* * *B*. The total change in the mass of *substanceB* consists of the gain from *substanceA* minus the loss to *substanceC* with the result multiplied by the change in time, Δt:

$$\Delta B = (aA - bB)\,\Delta t$$

We consider the initial amounts of *substanceB* and *substanceC* to be zero.

Projects

1. **a.** With a system dynamics tool or a computer program, develop a model for a radioactive chain of three elements, from *substanceA* to *substanceB* to *substanceC*. Allow the user to designate constants. Generate a graph and a table for the amounts of *substanceA*, *substanceB*, and *substanceC* versus time. Answer the following questions using this model.

   b. Explain the shapes of the graphs.

   c. As the decay rate of A, a, increases from 0.1 to 1, describe how the time of the maximum total radioactivity changes. The total radioactivity is the sum of the change from *substanceA* to *substanceB* and the change from *substanceB* to *substanceC*, or the total number of disintegrations. Why?

   d. (The verification in Part d requires calculus.) With b being the decay rate of B, in several cases where $a < b$, observe that eventually we have the following approximation:

$$\frac{B}{A} \approx \frac{a}{b-a}$$

   With the ratio of the mass of *substanceB* (B) to the mass of *substanceA* (A) being almost constant, $a/(b - a)$, we say the system is in **transient equilibrium**. Eventually, *substanceA* and *substanceB* appear to decay at the same rate. Using the following material, verify this approximation:

   Find the exact solution to the differential equation for the rate of change of A with respect to time, $dA/dt = -aA$ (see the section "Analytic Solution" in Module 2.2, "Unconstrained Growth and Decay").

   Verify that $B = \frac{aA_0}{b-a}(e^{-at} - b^{-bt})$, where A_0 is the initial mass of *substanceA*, is a solution to the differential equation for the rate of change of B with respect to time (see the difference equation for ΔB). What number does e^{-at} approach as t goes to infinity? For $a < b$, which is smaller, e^{-at} or e^{-bt}? Thus, for large t, B is approximately equal to what?

   e. Using your model from Part a, observe in several cases where $a > b$ that the ratio of the mass of *substanceB* to the mass of *substanceA* does not approach a number. Thus, transient equilibrium (see Part d) does not occur in this case.

   f. (Requires calculus) Verify the observation from Part e analytically using work similar to that in Part d.

   g. If a is much smaller than b, we have $A \approx A_0$ and $B \approx \frac{aA_0}{b-a}$. With the two amounts being almost constant, we have a situation called **secular equilibrium**. Observe this phenomenon for the radioactive chain from radium-226 to radon-222 to polonium-218: $Ra^{226} \rightarrow Rn^{222} \rightarrow Po^{218}$, where the decay rate of Ra^{226}, a, is 0.00000117/da and the decay rate of Rn^{222}, b, is 0.181/da. Using your work from Part a, run the simulation for at least one year.

h. (Requires calculus) Show analytically that the approximations from Part g hold.

i. In the radioactive chain $Bi^{210} \rightarrow Po^{210} \rightarrow Pb^{206}$ (bismuth-210 to polonium-210 to lead-206), the decay rate of Bi^{210}, a, is 0.0137/da and the decay rate of Po^{210}, b, is 0.0051/da. Assuming the initial mass of Bi^{210} is 10^{-8} g and using your model from Part a, find, approximately, the maximum mass of Po^{210} and when the maximum occurs.

j. (Requires calculus) In Part d, we verified that $B = \dfrac{aA_0}{b-a}(e^{-at} - b^{-bt})$. Using this result, find analytically the maximum of mass of *substanceB* and when this maximum occurs.

k. Check your approximations of Part i using your solution to Part j.

l. For the chain in Part g, use your solution to Part j to find when the largest mass of Rn^{222} occurs.

m. For the chain in Part g, use your simulation of Part a to approximate the time when the largest mass of Rn^{222} occurs. How does your approximation compare with the analytical solution of Part l?

2. Develop a model for a chain of four elements. Perform simulations, observations, and analyses similar to those before. Discuss your results

Answers to Quick Review Question

1. a. $\Delta A = -aA\,\Delta t$

b. $\Delta B = (aA - bB)\Delta t$

c. *decay_A_to_B = decay_rate_of_A * A*

d. *decay_B_to_C = decay_rate_of_B * B*

Reference

Horelick, Brindell, and Sinan Koont. 1979, 1989. "Radioactive Chains: Parents and Children." *UMAP Module 234*. COMAP, Inc.

MODULE 7.2

Turnover and Turmoil—Blood Cell Populations

Prerequisite: Module 2.2, "Unconstrained Growth and Decay."

Introduction

In a healthy individual, the count of blood cells is usually constant. However, for certain diseases, blood cell counts may oscillate, perhaps in an involved or chaotic manner. Such disorders are in a category of **dynamical diseases**, which include HIV, forms of leukemia, and anemia. We can use modeling to study the origins, behaviors, and treatments of these diseases.

Formation and Destruction of Blood Cells

Blood is composed of fluid, called **plasma**, and **blood cells**. The following are major types of blood cells:

- **Red blood cells**, which are for oxygen transport from the lungs to tissues
- **White blood cells**, which are part of the body's defense mechanism against infections
- **Platelets**, which help the blood to clot

In a healthy individual, if a deficit for a particular type of blood cell occurs, physiological mechanisms in the body cause an increase in production of that type of cell. Similarly, if an oversupply exists, the production rate decreases. Thus, the production of new blood cells of a particular type depends on the number of blood cells of that type.

Aging, disease, or infection causes the eventual death of any cell. As with production, the number of blood cells destroyed also depends on the number of blood cells of that type.

Basic Model

Suppose x is the **number of blood cells** of a particular type, and x_i is the **number of such blood cells at time t_i**. As indicated before, the **number of such blood cells produced (p)**, or the **production rate**, and the **number of such blood cells destroyed (d)**, or the **destruction rate**, are functions of the number in existence, x. Thus, $p(x_i)$ and $d(x_i)$ are the production and destruction rates, respectively, for the given number of cells at time t_i.

Quick Review Question 1

Suppose the number of blood cells of a particular type at time t_i is below the normal range. For each part, give the relationship that is desirable.

a.	$x_{i+1} > x_i$	$x_{i+1} < x_i$	$x_{i+1} = x_i$
b.	$p(x_{i+1}) > p(x_i)$	$p(x_{i+1}) < p(x_i)$	$p(x_{i+1}) = p(x_i)$
c.	$d(x_{i+1}) > d(x_i)$	$d(x_{i+1}) < d(x_i)$	$d(x_{i+1}) = d(x_i)$

For a healthy mammal, one widely accepted model for d is that the number of blood cells destroyed is directly proportional to the number of blood cells existing. Thus, for a constant of proportionality, c, called the **destruction coefficient**, we have the following model for the number of blood cells destroyed:

$$d(x_i) = cx_i \tag{1}$$

Quick Review Question 2

Indicate which of the following are true about the destruction function d and the destruction coefficient c: $c > 0$; $c < 0$; $c = 0$; $c \geq 1$; $c \leq 1$; $d(0) = 0$; the graph of d is increasing; the graph of d is decreasing; the graph of d is a line; the graph of d cannot be a line.

A model for the number of blood cells of a particular type produced is more complicated. As noted earlier, the production is a function of the number of blood cells of that type. Certainly, if there were no blood cells ($x = 0$), we would expect no production because the animal would be dead. As x increases, p increases rapidly to some maximum and then production decreases, tailing off to zero. One possible such graph of production versus population for a particular type of blood cell appears in Figure 7.2.1.

For a healthy person, we expect no change in the number of blood cells from one time step to another, so that $x_{i+1} = x_i = v$, a constant called the **steady-state level**. In this case, $x_{i+1} - x_i = 0$. Thus, the production rate $p(v)$ and the destruction rate $d(v)$ are equal at v, or $p(v) = d(v)$.

Model Parameters

As noted in Gearhart and Martelli (1990), it takes about 6 days for red blood cells to reach maturity, while in a healthy person about 2.3% of the cells are destroyed per

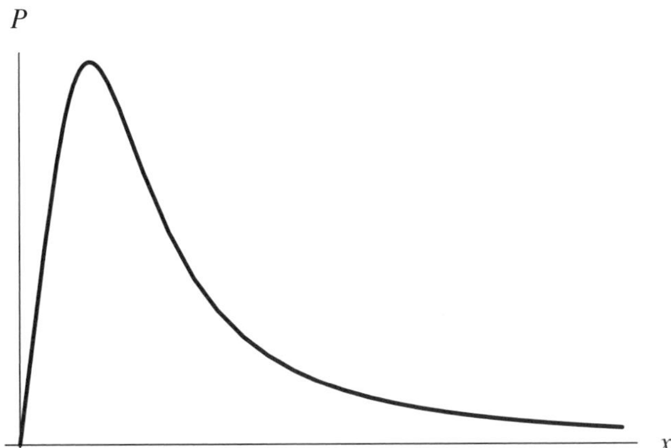

Figure 7.2.1 General graph of blood cell production (p) versus blood cell population (x)

day. Thus, for a change in time of 6 days, an approximation for the destruction coefficient is $c = (6)(0.023) = 0.14$; and the destruction function, or the number of blood cells destroyed as a function of the number of blood cells at time $t_{i,}$ is as follows:

$$d(x_i) = 0.14x_i \tag{2}$$

In 1977 Lasota used the following production function, which has the shape of Figure 7.2.1 (Lasota 1977):

$$p(x) = bx^s e^{-sx/r} \tag{3}$$

where p is the number of blood cells produced as a function of the number of blood cells, x, and b, r, and s are positive constants. To determine these constants, use the following data (Gearhart and Martelli 1990):

- Normal red-cell count $\approx$ 3.3 × 10^{11} cells/kg
- Maximum red-cell production $\approx$ 10 times the steady-state production rate
- For a red-cell population of 75% of the steady-state level (count),
- red-cell production $\approx$ five times steady-state level

Because the normal red-cell count is about 3.3 × 10^{11} cells/kg, we can use this figure as the steady-state level or

$$v = 3.3 \times 10^{11} \text{ cells/kg}$$

At v, production equals destruction or $p(v) = d(v)$. From Equation 2

$$d(v) = 0.14v$$

so we know the following:

$$p(v) = 0.14v$$
$$p(3.3 \times 10^{11}) = 0.14 \times 3.3 \times 10^{11} = 4.62 \times 10^{10} \text{ cells/kg}$$

At $v = 3.3 \times 10^{11}$ cells/kg, production and destruction equal 4.62×10^{10} cells/kg, which is the steady-state production rate.

Quick Review Question 3

Using this information, indicate which of the following points are on the graph of the production function p: $(3.3 \times 10^{11}, 4.62 \times 10^{10})$, $(4.62 \times 10^{10}, 3.3 \times 10^{11})$, $(3.3 \times 10^{11}, 0.14)$, $(0.14, 3.3 \times 10^{11})$, $(v, 0.14v)$, $(0.14v, v)$.

According to the second bullet, the maximum value of p is 10 times the steady-state rate, or $10 \times 4.62 \times 10^{10}$ cells/kg $= 4.62 \times 10^{11}$ cells/kg. To determine where this maximum occurs, we take the first derivative of Equation 3 and set it equal to 0, solving for x (see Project 1). Keeping in mind that b, r, and s are constants, we find that the maximum occurs at $x = r$; and the maximum is as follows:

$$p(r) = br^s e^{-s} = 4.62 \times 10^{11}$$

Quick Review Question 4

Interpret the third bullet given earlier as one or more equations:

A. $p(5v) = 0.75v$
B. $p(0.75v) = 5v$
C. $p(\,5(0.14v)\,) = 0.75v$
D. $p(0.75 \times 3.3 \times 10^{11}) = 5 \times 4.62 \times 10^{10}$
E. $p(0.75v) = 5(0.14v)$
F. $p(0.75 \times 3.3 \times 10^{11}) = 5 \times 3.3 \times 10^{11}$
G. $p(5 \times 4.62 \times 10^{10}) = 0.75 \times 3.3 \times 10^{11}$

Projects

For all model development, use an appropriate system dynamics tool.

1. Develop a model for a red blood cell population using the following production function, which is a variation of Equation 3: $p(x) = bv(x/v)^s e^{-sx/(vr)}$, where v is the normal red-cell count, $b = 1.1 \times 10^6$, $s = 8$, and $r = 0.5$ (Gearhart and Martelli 1990). Account for the (approximate) 6-day maturation of cells, perhaps using a conveyor; and determine reasonable constants by referring to the "Model Parameters" section. Graph blood cells per kilogram, blood cells produced per kilogram, and blood cells destroyed per kilogram versus time. Discuss your results

   Find analytically where the maximum occurs and the maximum for the production function in Equation 3, and verify that your model approximately agrees with this value.

   For $r = 3$, $b = 50$, $s = 5$, and $c = 0.5$, what is the period?
2. Develop a model for a type of white blood cell (granulocyte) population using a destruction coefficient of 0.6 for a time step of 6 days and a production function of $p(x) = bxe^{-x/(vr)}$, where the steady state is about $v = 8.2\times10^9$ granulocytes/kg (Gearhart and Martelli 1990). Considering steady-state and

maximum productions, determine reasonable constants. Alternatively, for p, use graphical input where $p(0) = 0$, p increases initially to a maximum and then decreases to 0. Graph blood cells per kilogram, blood cells produced per kilogram, and blood cells destroyed per kilogram versus time.

3. Develop a model for granulocytes, a type of white blood cell. Use the following production function by Mackey and Glass (1977), where b, a, and m are positive constants (Gearhart and Martelli 1990):

$$p(x) = \frac{ba^m x}{a^m + x^m} \tag{4}$$

The units of a are cells/kilogram, while b and m are unitless. Determine reasonable constants by referring to the "Model Parameters" section. Graph blood cells per kilogram, blood cells produced per kilogram, and blood cells destroyed per kilogram versus time.

4. Complete Project 3. Show that we can use this work to model chronic myelogenous leukemia (CML), a cancer resulting in an overproduction of the white blood cells. In CML, the white cell count may be 150 times the normal, and counts can oscillate around the elevated level with a period of 30 to 70 da. As Gearhart and Martelli (1990) indicates, the following are parameters for a normal person:

$$c = \mu\partial \quad \text{with } \mu = 0.16/\text{day}$$
$$b = \beta\partial \quad \text{with } \beta = 1.43/\text{day}$$
$$a = 3.22 \times 10^8 \text{ cells/kg}$$
$$m = 3$$

and the delay in production, ∂, is 0.68 day.

Show that increasing a can result in a gain in white blood cell count, and increasing ∂ can cause the indicated periodicity. Find appropriate values for a and ∂ that match the abnormal variations in white blood cell counts of CML.

Answers to Quick Review Questions

1. **a.** $x_{i+1} > x_i$
   b. $p(x_{i+1}) > p(x_i)$
   c. $d(x_{i+1}) < d(x_i)$
2. $c > 0$ because as the number of cells increase, the number destroyed increases. $d(0) = 0$ because if there are no cells, none can be destroyed. The graph of d is increasing because c and x are positive, and as the number of cells increases, the number destroyed increases. The graph of d is a line.
3. $(v, 0.14v)$ and $(3.3 \times 10^{11}, 4.62 \times 10^{10})$ because $p(v) = 0.14v = p(3.3 \times 10^{11}) = 4.62 \times 10^{10}$ cells/kg
4. Equations B and F, $p(0.75v) = 5v$ and $p(0.75 \times 3.3 \times 10^{11}) = 5 \times 3.3 \times 10^{11}$, because $v = 3.3 \times 10^{11}$ is the steady-state level.

References

Gearhart, William B., and Mario Martelli. 1990. "A Blood Cell Population Model, Dynamical Diseases, and Chaos." *UMAP Module 709*. COMAP, Inc.

Lasota, A. 1977. "Ergodic Problems in Biology." *Astérisque* (Société Mathématique de France) 50: 239–250.

Mackey, M.C., and L. Glass. 1977. "Oscillation and Chaos in Physiological Control Systems." *Science*, 197: 287–89.

MODULE 7.3

Deep Trouble—Ideal Gas Laws and Scuba Diving

Prerequisite: *Module 2.2, "Unconstrained Growth and Decay."*

Pressure

Scuba divers are often under great amounts of pressure and, therefore, should be very concerned with it. **Pressure**, which is the weight of matter per unit area, increases rapidly with increasing depths. For divers, the total pressure is a combination of the weight of air and water per square centimeter. **Air pressure at sea level** is about **10.1 N/cm²**, meaning that a square centimeter column of air as tall as the atmosphere (about 80 km, or 50 mi) weighs about 10.1 N (**10.1 N/cm² = 760 mm of mercury (mm Hg) = 760 torr = 14.7 lb/in.²**). The atmospheric pressure at sea level is by definition equal to **1 atmosphere (atm)**. Water pressure derives from the weight of water, which is considerably greater than air. As indicated in Module 3.1, "Modeling Falling and Skydiving," the density of water at 3.98 °C is 1.00000 g/cm³. Thus, at this temperature, a column of water 10 m high with a base of area cm² weighs 9.81 N, as the following calculations show:

$$\text{weight} = F = ma = \left(\frac{1.00000\,\text{g}}{\text{cm}^3}\right)(10\,\text{m})(1\,\text{cm}^2)\left(\frac{9.81\,\text{m}}{\text{s}^2}\right)$$

$$= \left(\frac{1.00000\,\text{g}}{\text{cm}^3}\right)\left(\frac{\text{kg}}{1000\,\text{g}}\right)(10\,\text{m})\left(\frac{100\,\text{cm}}{\text{m}}\right)(1\,\text{cm}^2)\left(\frac{9.81\,\text{m}}{\text{s}^2}\right)$$

$$= (1.00000\text{ kg})(9.81\text{ m/s}^2) = 9.81\text{ N}$$

This weight of a 10-m by 1-cm² column of water approximately matches the weight of the entire column of air above it. Hence, a diver at a depth of about **10 m (33 ft)** experiences **2 atm of pressure**, resulting from the pressure of 1 atm of air plus the equivalent of approximately an additional atmosphere of pressure from the water. Each 10 m of depth adds approximately 1 atm of pressure to the diver.

Definitions **Pressure** is the weight of matter per unit area**. One atmosphere (atm)** is the atmospheric pressure at sea level.

Quick Review Question 1

Determine the water pressure at 15 m in terms of each of the following units:

a. atm
b. N/cm^2
c. torr

Ideal Gas

In this module, we employ several **ideal gas laws**, which describe the behaviors of an ideal gas, in models related to scuba diving. An ideal gas is one in which the volume of its atoms is insignificant in comparison to the total volume of the gas and in which atom interactions are negligible except for the energy and momentum exchanged during collisions. Remarkably, under most circumstances, the ideal gas laws model well the behaviors of real gases because of the great distances between the atoms and molecules of a gas.

Definitions The **ideal gas laws** describe the behaviors of an ideal gas. An **ideal gas** is one in which the volume of its atoms is insignificant in comparison to the total volume of the gas and in which atom interactions are negligible except for the energy and momentum exchanged during collisions.

Dalton's Law

Scuba equipment provides air to divers at the higher pressures of deeper waters. Air normally contains 21% oxygen (O), 78% nitrogen (N), and 1% various inert gases. We must consider nitrogen and the inert gases in calculating the speeds and rest schedules for a scuba diver. Thus, scuba computations frequently group nitrogen and the inert gases together and **assume N_2 is 79% of the air**. Because the partial pressure of a gas is determined by its fractional portion in a mixture, at sea level 1 atm of pressure is composed of 0.21 atm of oxygen and 0.79 atm of nitrogen (assuming 79% nitrogen). This relationship follows one of the ideal gas laws as proposed by **Dalton**. Although air for divers is compressed, the percentage of the gas components is the same. So, as pressure on a diver increases during a dive, the partial pressure exerted by each gas in the diver's body and tank increases. For example, a diver reaching 20 m (66 ft) experiences 3 atm of pressure, with partial pressures of O_2 and N_2 equaling 0.63 and 2.37 atm, respectively.

Dalton's law states that the partial pressure of a gas (P_g) is the product of the fraction of the gas in the mixture (F_g) and the total pressure (P) of all gases, excluding water vapor:

$$\boldsymbol{P_g = F_g P}$$

Quick Review Question 2

Determine the partial pressure in atmospheres (atm) of nitrogen in a mixture of air at 15 m.

Boyle's Law

Probably the most important gas law to divers is that of Robert Boyle, who discovered that at a particular temperature, the volume of a gas is inversely proportional to pressure. Hence, the product of pressure (P) and volume (V) yields a constant (K).

$$\boldsymbol{PV = K}$$

When pressure increases at constant temperature, gas volume decreases, and vice versa. For example, assuming constant temperature, **Boyle's law** means that if we take an air-filled balloon, which is 3 m^3 in volume at the surface of the ocean to a depth of 20 m, the balloon would shrink to a volume of 1 m^3. We can obtain this result by using the pressure at the surface, P_1 = 1 atm, with volume V_1 = 3 m^3. At 20 m, pressure is P_2 = 3 atm. Thus, assuming constant temperature, by Boyle's law we have the following relationship:

$$P_1V_1 = P_2V_2$$

Substituting the values, we have (1)(3) = 3V_2, or V_2 = 1 m^3.

Boyle's law for gas at a particular temperature is as follows:

$$\boldsymbol{PV = K}$$

where P is pressure, V is volume, and K is a constant.

Quick Review Question 3

Suppose an air-filled balloon is 3 m^3 in volume at the ocean's surface. Assuming constant temperature, determine its size to three decimal places at a depth of 32 m.

Skin divers hold their breath when they dive. During descent to, say, 10 m, the air-filled lungs are reduced to one-half their surface volume. As they ascend, the lungs again expand to their normal volume. Of course, there are limits of depth for diving without the aid of scuba equipment.

Scuba divers breathe air from tanks through regulators that deliver the air at **ambient pressure**, the pressure of the surrounding water pressure. As divers at 20 m (3 atm pressure) inhale, they take in the equivalent of 3 breaths of air from the surface (1 atm pressure). Accordingly, it is important for divers to determine their **surface air consumption** (**SAC**) **rate** so that they can calculate how long their air tanks will last at the depths they are diving.

Another important consideration that comes from breathing air at these pressures is that as divers rise from the deep, the gases in their lungs obey Boyle's law and expand with the decreasing pressure. Rapid and extensive expansion of the equivalent of three times the normal lung volume could cause the lungs to burst. Consequently, scuba divers are always cautioned never to hold their breath, but return to the surface slowly, exhaling.

Charles's Law

Charles's law takes into consideration pressure (P), volume (V), and temperature (T). The relationship is as follows:

$$\boldsymbol{PV = nRT}$$

where T is temperature in kelvins (K), n is the number of moles, and the constant R = 0.0821 atm/(mol K). The conversion from a Celsius temperature (T_C) to a Kelvin temperature (T_K) is the following:

$$\boldsymbol{T_K = T_C + 273.15}$$

The number of moles, n, in a mass m of molecular weight M is the following:

$$\boldsymbol{n = m/M}$$

The molecular mass of dry air is 29.0 g/mol, of nitrogen is 28.0 g/mol, and of oxygen is 32.0 g/mol. However, frequently in scuba diving examples, the form of the law we employ relates the pressures, volumes, and temperatures at two depths, as follows:

$$\frac{P_1V_1}{T_1} = \frac{P_2V_2}{T_2}$$

Charles's law states that

$$PV = nRT$$

where P is pressure, V is volume, T is temperature in kelvins, n is the number of moles, and the constant $R = 0.0821$ atm/(mol K). Thus, if an object has volume V_1 at pressure P_1 and Kelvin temperature T_1, then at pressure P_2 and Kelvin temperature T_2 its volume is V_2, according to the following:

$$\frac{P_1V_1}{T_1} = \frac{P_2V_2}{T_2}$$

The conversion from a Celsius temperature (T_C) to a **Kelvin temperature** (T_K) is the following:

$$T_K = T_C + 273.15$$

The **number of moles**, n, in a mass m of molecular weight M is the following:

$$n = m/M$$

Quick Review Question 4

Suppose a balloon is 1 m^3 in 30 °C weather on the ocean's surface. Determine the balloon's volume at 15 m in 10 °C seawater.

Henry's Law

Divers who dive deeply for long periods of time also face another problem that follows from **Henry's law**—the amount of any gas in a liquid at a particular temperature is a function of the partial pressure of the gas and its solubility coefficient in that liquid. Thus, divers must be concerned with the amount of nitrogen gas in their blood. For V_g being gas volume, V_L being liquid volume, s being the solubility coefficient for the gas in that liquid, and P_g being the pressure of gas, Henry's law is as follows:

$$V_g/V_L = sP_g$$

The **solubility coefficient for nitrogen in blood** is **0.012**, and the total **volume of blood** in an adult's body is approximately **5 L**. With the greater pressure, blood can absorb a greater volume of nitrogen, which the body cannot use.

Henry's law for the amount of any gas in a liquid at a particular temperature is as follows:

$$V_g/V_L = sP_g$$

where V_g is gas volume, V_L is liquid volume, s is the solubility coefficient for the gas in that liquid, and P_g is the pressure of gas.

Quick Review Question 5

Determine the volume of nitrogen that can go into solution in the blood at the following pressures:

a. 1 atm
b. 2 atm

Rate of Absorption

The rate at which tissue (a compartment) takes up an inert gas, such as nitrogen, is proportional to the difference in the partial pressures of the gas in the lungs and the gas in the tissues:

$$dP_{\text{tissue}}/dt = k(P_{\text{lungs}} - P_{\text{tissue}})$$

where P_{lungs} is the partial pressure of the gas in the lungs, P_{tissue} is the partial pressure of the gas in the tissue, and k depends on the type of tissue. (This formula is similar to Newton's Law of Heating and Cooling in Exercise 4 of Module 2.2, "Unconstrained Growth and Decay.") We can show that

$$k = (\ln 2)/t_{\text{half}}$$

where t_{half} is the **half-time**, or the time for the tissue to absorb or release half of the partial difference of the gas. Thus, if it takes 20 min for such absorption, then we have the following calculation of the constant k:

$$k = (\ln 2)/t_{\text{half}} = (\ln 2)/20 = 0.0346574$$

The **rate of absorption** at which tissue (a compartment) takes up an inert gas is as follows:

$$dP_{\text{tissue}}/dt = k(P_{\text{lungs}} - P_{\text{tissue}})$$

where P_{lungs} is the partial pressure of the gas in the lungs, P_{tissue} is the partial pressure of the gas in the tissue, and k is a proportionality constant, which depends on the type of tissue. The value of k is as follows:

$$k = (\ln 2)/t_{half}$$

where t_{half} is the time for the tissue to absorb or release half of the partial difference of the gas.

Quick Review Question 6

Suppose the half-time for nitrogen absorption into a certain tissue is 4 min, the partial pressure in the lungs is 1.58, and the partial pressure in the tissue is 1.22. Compute the following to 3 decimal places:

- **a.** The constant of proportionality, k, in the rate of absorption equation
- **b.** The rate at which the tissue takes up nitrogen

Decompression Sickness

Remember that the partial pressure of nitrogen in air under higher pressures is proportionately higher. Divers at 20 m, for instance, receive nitrogen pressures of 2.37 atm, increasing its solubility in body fluids. Nitrogen gas has no role in cellular metabolism of the diver, so it accumulates in solution. The total amount of residual nitrogen is dependent on the depth and duration of the dive. Deep divers may experience **nitrogen narcosis**, a sudden feeling of euphoria that can impair judgment and lead to serious or even fatal consequences. Furthermore, such divers returning to the surface too rapidly risk **decompression sickness** ("bends"), which is not only painful, but also potentially lethal.

Decompression sickness results from the reduced solubility of nitrogen gas in the blood, causing the release of nitrogen bubbles as pressure decreases. As the diver ascends, these bubbles continue to form and to expand. Bubbles may cause joint pain and block blood vessels. Such blockages may lead to heart attack, stroke, or ruptured blood vessels in the lungs.

Professional diving organizations publish dive tables to use for calculating how much nitrogen is absorbed during dives at varying depths and durations. Generally speaking, the deeper the dive, the shorter the duration should be.

Projects

1. Develop a scuba diving model, including pressure and volume of air in the lungs. Assume that the temperature is constant, the descent rate is less than or equal to 23 m/min, and the ascent rate is no more than 12 m/min = 0.2 m/s (U.S. Navy 2008).
2. Repeat Project 1 with comparison graphs for seawater at various locations,

such as Juneau, Alaska, at 10 °C (50 °F); Paita, Peru, or Santa Monica, Calif., at 20 °C (68 °F); Savannah Beach, Ga., at 30 °C (86 °F; NOAA 2008).

3. Develop a model for the pressure and volume of the air in a diver's suit. Suppose initially the volume of the gas is 8 L = 0.008 m^3.
4. Develop a model for the duration of scuba tank usage. Suppose the surface air consumption rate is 3.3×10^{-4} m^3/s, and the tank initially holds 12 L (de Lara 2002). A scuba tank delivers air to a diver at ambient pressure. Consider the depth consumption rate, or consumption rate at a depth, as part of your model.
5. Develop a model for the amount of nitrogen in tissue. At about 1.5 times the partial pressure of nitrogen at surface, nitrogen is not able to go into solution in blood.
6. Repeat Project 5 and have comparative graphs with half times of 5, 10, 20, 40 and 75 min.
7. The DECOM dive tables give the list of decompression stops for someone who dives to 39.6 m (130 ft) of seawater for the indicated amount of time (Table 7.3.1) (NAUI). For example, if someone dives to 39.6 m (130 ft) of seawater and stays there 25 min, on return the diver should stop for 4 min at a depth of 9.1 m (30 ft), 9 min at 6.1 m (20 ft), and 14 min at 3.0 m (10 ft). A rate of ascent is assumed to be about 9.1 m/min (30 ft/min). Develop a model for the amount of nitrogen in the body using each of these scenarios. Run the model long enough to determine the length of time for the amount of nitrogen in the blood to return to normal.

Table 7.3.1
DECOM Dive Table for Dive to 39.6 m (130 ft) of Seawater

		Time at		
39.6 m (130 ft) Bottom	*12.2 m (40 ft)*	*9.1 m (30 ft)*	*6.1 m (20 ft)*	*3.0 m (10 ft)*
15			3	6
20		1	7	9
25		4	9	14
30	2	6	11	19

Answers to Quick Review Questions

1. **a.** 2.5 atm
   b. 25.25 N/cm^2 = 2.5(10.1) N/cm^2
   c. 1900 torr = 2.5(760) torr
2. 1.975 = (0.79)(2.5 atm)
3. 0.714 m^3 because at 32 m, P_2 = 1 atm + (32 m)(1 atm /10 m) = 4.2 atm; and $P_1V_1 = P_2V_2$, or (1)(3) = 4.2V_2, or $V_2 = 3/4.2 \approx 0.714$
4. 0.374 m^3: at 15 m, the pressure is 2.5 atm; 30 °C = 303.15 K, and 10 °C = 283.15 K; (1)(1)/(303.15) = (2.5)(V_2)/283.15, or V_2 = 283.15/((2.5) (303.15)) = 0.374

5. **a.** 0.0474 L because $V_g = V_L s P_g = (5 \text{ L})(0.012)(0.79)$
   b. 0.0948 L because $V_g = V_L s P_g = (5 \text{ L})(0.012)(2 \times 0.79)$
6. **a.** $0.173 = k = (\ln 2)/4$
   b. $0.062 = dP_{\text{tissue}}/dt = k(P_{\text{lungs}} - P_{\text{tissue}}) = 0.173(1.58 - 1.22)$

References

Baker, Erik C. "Some Introductory 'Lessons' About Dissolved Gas Decompression Modeling." http://www.shearwaterresearch.com/wp-content/uploads/2012/08/Introductory-Deco-Lessons.pdf (accessed December 11, 2012)

Deep Ocean Diving. 2005. "Decompression Theory—Neo-Haldane Models." *Diving Science.* http://www.deepocean.net/deepocean/index.php?science05.php (accessed December 11, 2012)

de Lara, Michel. 2002. "Diving into Mathematics or Some Mathematics of Scuba Diving." École Nationale des Ponts et Chaussés. http://cermics.enpc.fr/~delara/plongee/math_diving/math_diving.html (accessed December 11, 2012)

Martin, Lawrence. 1997. "An Explanation of Pressure and the Laws of Boyle, Charles, Dalton, and Henry." *Scuba Diving Explained,* Section D. http://www.lakesidepress.com/pulmonary/books/scuba/sectiond.htm (accessed December 12, 2012)

NAUI (National Association of Underwater Instructors). "NAUI Dive Tables." http://www.naui.org/tables.aspx (accessed December 11, 2012)

NOAA Satellite and Information Service. 2008. "Coastal Water Temperature Guide." NOAA-NESDIS-National Oceanographic Data Center. http://www.nodc.noaa.gov/dsdt/cwtg/ (accessed 12/11/12)

U.S. Navy Diving Manual. 2008. "SCUBA Diving: Water Entry and Descent." Ch. 7-6, pp.7-26–7-29. http://www.usu.edu/scuba/navy_manual6.pdf (accessed December 12, 2012)

MODULE 7.4

What Goes Around Comes Around—The Carbon Cycle

Prerequisites: Module 2.2, "Unconstrained Growth and Decay" for Project 1; Module 2.3, "Constrained Growth" for Projects 2 and 3.

Introduction

Most of us are familiar with carbon in the form of the gas carbon dioxide (CO_2). However, carbon dioxide is only one form of carbon (C), an element with very wide distribution on the earth. Carbon combines with elements like calcium, iron, and magnesium to form rocks. Carbon-containing compounds are dissolved in the oceans and other bodies of water. All things living are made up of organic molecules, which are all carbon based. Carbon moves in varying forms among the four major **environmental subsystems** (interdependent parts of the earth's system) of the earth: **lithosphere** (ground and inside the earth), **atmosphere** (air surrounding the earth), **hydrosphere** (lakes, rivers, and oceans), and **biosphere** (all living things). Because of the importance of carbon dioxide accumulation in the atmosphere and its effect on climate, scientists are particularly interested in carbon and its movement. This movement of carbon is described as the **carbon cycle**, and as carbon is transferred from one subsystem to another, it is often transformed from one form of carbon to another.

Flow between Subsystems

Most estimates of atmospheric CO_2 fall somewhere near 2745 gigatons (1 gigaton (Gt) = 10^{15} g). Some CO_2 is taken up by plants and converted to various organic compounds through **photosynthesis**. Most plants, animals, and some other life forms **oxidize** organic molecules to release CO_2 back into the atmosphere. Other CO_2 dissolves in seawater. Some CO_2 is also released from solution back into the atmosphere. The atmosphere, photosynthetic organisms (part of the biosphere), and

Table 7.4.1
Major Reservoirs in Carbon Cycle

Reservoir	*Initial Amount of Carbon (Gt)*
atmosphere	750
terrestrial biosphere	600
ocean surface	800
deep ocean	38,000
soil	1,500

Table 7.4.2
Major Fluxes in Carbon Cycle

Flux	*Rate (Gt C/yr)*	*Source*	*Sink*
terrestrial photosynthesis	110	atmosphere	terrestrial
marine photosynthesis	40	atmosphere	ocean surface
terrestrial respiration	55	terrestrial biosphere	atmosphere
marine respiration	40	ocean surface	atmosphere
carbon dissolving	100	atmosphere	ocean surface
evaporation	100	ocean surface	atmosphere
upwelling	27	deep ocean	ocean surface
downwelling	23	ocean surface	deep ocean
marine death	4	ocean surface	deep ocean
plant death	55	terrestrial biosphere	soil
plant decay	55	soil	atmosphere

the ocean (hydrosphere) all represent **reservoirs** for carbon. Soil, sediments, and various rock formations represent other reservoirs of carbon. The transfer of carbon from one reservoir to another is usually termed a **flux** and is given as gigatons of carbon transferred per year. (1 Gt carbon is equivalent to 3.66 Gt CO_2.) Tables 7.4.1 and 7.4.2 list the major reservoirs and fluxes actively involved in the carbon cycle (Allmon et al. 2006). In Table 7.4.2, the **source** is the origin, and the **sink** is the destination of the carbon flow. Carbon dioxide **gas exchange** between the atmosphere and the ocean surface, which involves gas dissolving into and evaporating from water, is in the direction of greater to lesser carbon concentration. **Upwelling** occurs when deep currents bring cool, nutrient-rich bottom ocean water to the surface. By contrast, with **downwelling** currents move ocean surface water to lower depths.

Fossil Fuels

Human activity, such as deforestation and combustion of fossil fuels, especially the latter, has greatly accelerated the release of carbon dioxide from more static reservoirs into the atmosphere. Fossil fuels are essentially combinations of carbon and hydrogen (**hydrocarbons**), which are oxidized into CO_2 upon burning. In 2011, 9.5 Pg (Pg = Petagram = 1×10^{15}g) of carbon (= 3.47×10^{10} tonnes = 3.8502×10^{10} tons, where 1 tonne (t), or metric ton, is 1000 kg and 1 ton is 2000 lb) were released into the atmosphere, a 3% increase from the previous year. Not surprisingly, 56% of these emissions came from just four political entities—China, United States, European Union, and India (GCP 2012). Other gases are released as well—carbon monoxide, various hydrocarbons, nitrogen oxides, and sulfur dioxide. Some of these gases, like carbon dioxide, contribute further to global warming, while others contribute to severe respiratory problems, smog, and acid rain (EPA 2011).

Why all the concern about the spiraling levels of atmospheric CO_2 and other so-called greenhouse gases? Each greenhouse gas gets this designation because it is capable of trapping some of the outgoing radiant energy from the earth's surface, increasing the surface temperatures. Even though the increase of gas concentrations each year is seemingly small, over time they may lead to increases in the average surface temperature sufficiently to alter precipitation patterns, raise sea level, and decrease the pH of the oceans. Should these changes be effected, they will threaten our food and water supplies, wreck ecological networks, and threaten the health of all living organisms (Shah 2012).

Projects

1. Prepare a model of the carbon cycle using the reservoirs and fluxes in Tables 7.4.1 and 7.4.2. Assume that the rate of change of marine materials sinking to the deep ocean is constant, but in all other cases, assume that the rate of carbon transfer from source to sink is proportional to the amount of carbon in the source. Have a separate flow corresponding to each flux with a converter for its rate constant. Running the simulation for one simulation year using a Runge-Kutta 4 technique and a time step of 0.01 yr, determine proportionality constants to obtain the indicated fluxes. Then, running the simulation for a longer period of time, produce appropriate graphs, such as the quantities in the reservoirs versus time. Vary these values, and discuss the results.

2. **a.** Modify the model you developed in Project 1 to include the effects of deforestation and fossil fuel combustion. Suppose a fossil-fuel-deposits reservoir has an initial value of 4000 Gt and fluxes for combustion and deforestation have values of 5 and 1.15 Gt C/year, respectively. Assume that the rate of change of fossil fuel emissions has constrained growth with a carrying capacity of 15 Gt C/year and growth rate of 0.03/year (see Module 2.3, "Constrained Growth"). Produce a similar model for deforestation (Houghton et al. 1999).

b. What is the effect on carbon reservoirs of various atmospheric carbon dioxide concentrations?

c. What is the effect of doubling the rate of deforestation or fossil fuel combustion?

d. What is the effect of doubling both?

3. In a 1999 article from *Science*, Houghton et al. estimated that appropriate land management (e.g., reforestation, fire suppression) might offset some of the CO_2 emissions from fossil fuel consumption by 10% to 30%. Factor land management into your evolving model from Project 2 of the carbon cycle (Houghton et al. 1999; Mersereau and Zareba-Kowalska 1997).

References

Allmon, Warren, Bryan Isacks, and William White. 2006. "Modeling the Carbon Cycle." Dept. of Earth and Atmospheric Sciences, Cornell University. http://www.geo.cornell.edu/eas/education/course/descr/EAS302/302_06Lab11.pdf (accessed December 19, 2012)

EPA (Environmental Protection Agency). 2011. "Climate Change Indicators in the United States. Atmospheric Concentrations of Greenhouse Gases." http://www.epa.gov/climatechange/science/indicators/ghg/ghg-concentrations.html (accessed December 19, 2012)

GCP (Global Carbon Project). 2012. "Global Carbon Budget Highlights." http://www.globalcarbonproject.org/carbonbudget/12/hl-full.htm (accessed December 19, 2012)

Houghton, R. A., J. L. Hackler, and K. T. Lawrence. 1999. "The U.S. Carbon Budget: Contributions from Land-Use Change." *Science* 285: 574-578.

Mersereau, Martha, and Anna Zareba-Kowalska. 1997. "Changes in the Carbon Cycle Due to Deforestation and Fossil Fuel Consumption in the U.S. Using Stella Modeling." The Woodrow Wilson National Fellowship Foundation. http://www.woodrow.org/teachers/environment/institutes/1997/26/ (accessed December 19, 2012)

Shah, Anup. 2012. "Climate Change and Global Warming Introduction." Global Issues. http://www.globalissues.org/article/233/climate-change-and-global-warming-introduction. (accessed December 19, 2012)

MODULE 7.5

Heated Debate—Global Warming

Prerequisites: Module 2.3, "Constrained Growth"; Module 7.4, "What Goes Around Comes Around—The Carbon Cycle."

Greenhouse Effect

If you drive your car into an uncovered parking lot on a clear, sunny day, one of the first things you do is to look for some shade. You know from experience that if you leave your car parked in the sunlight for even a short amount of time, the temperature inside will become exceptionally high. What is happening in your automobile is what happens in a greenhouse. The visible light waves enter your car, passing through the glass. The light is absorbed by the interior of the car and is emitted as heat (infrared). This heat warms the air contained in the car so that the interior temperature is far greater than the air temperature outside the car.

The heating of the earth by sunlight is often described as a result of the **greenhouse effect**. Visible light from the sun passes through the atmosphere, with more than 50% of the original solar energy reaching the earth's surface (NASA 2000; Bothun 1998). Clouds and various gases and particles absorb 23% of the incident solar energy, while clouds and particles reflect another 25% (Bothun 1998). Most of the solar energy that reaches the earth's surface is absorbed, increasing the temperature of the ground or water. Energy is then radiated from the surface as heat or infrared radiation. Atmospheric gases absorb most of this radiation. In fact, 70% of atmospheric heating is realized from this energy, with the rest from the incoming light energy (Heywood 1998). The gases of the atmosphere, now warmer, begin to radiate infrared energy themselves, much of it toward the earth, which absorbs the infrared. This "natural" greenhouse effect is responsible for increasing the annual, all-latitude average temperature to 15 °C. Without such effect the earth would average a chilly –18 °C (Pidwirny and Jones 2010).

Global Warming

If this so-called greenhouse effect makes the earth a hospitable place for human beings, why is there so much concern about it? This effect, without our help, helps to sustain life, but with our help, it might become "too much of a good thing." Human activity (e.g., combustion of fossil fuel, deforestation, etc.) has gradually increased the atmospheric concentration of the greenhouse gases. From the development of James Watts' steam engine in the mid-eighteenth century, carbon dioxide (CO_2) had risen from 280 parts per million (ppm = one part in one million = mg/L) to 392 ppm in 2012 (NOAA 2012). With more absorptive gases in the atmosphere, there is greater potential for heat absorption, which can then lead to a gradual heating of the earth—**global warming**. There is substantial evidence for this global warming. Between 1900 and 2009, the average surface temperature for the earth increased approximately 0.7 °C overall, but was increasing during the last 50 years of that time by at almost double the average rate (0.13 °C per decade; Dahlman 2009). According to the EPA, during the thermometer-based temperature-recording period until that time, the decade 2001–2010 was the warmest. Mean surface temperature has increased at 0.078 °C, on average, for each decade since 1901. Alarmingly, during the last two decades of the twentieth and the first decade of the twenty-first centuries, the United States warmed more rapidly than the rest of the world (EPA 2012).

The vast majority of evidence leads to the conclusion that the alarming trend of global warming is largely a result of **anthropogenic** (human influence on nature) activities. Models, using various emission scenarios, predict CO_2 concentrations will reach 540 to 970 ppm by the end of the twenty-first century. Increases of these magnitudes may boost the earth's average temperature by 1.4 to 5.8 °C (IPCC 2001).

Greenhouse Gases

Carbon dioxide is only one of several **greenhouse gases** (**GHGs**). Greenhouse gases are atmospheric gases that absorb infrared radiation (IR), preventing its loss to space. The most common of these gases is actually water vapor, but other naturally occurring examples include methane and nitrous oxide. Human activities add to the increase of these gases and synthetic GHGs, such as chlorofluorocarbons (CFCs) and hydrofluorocarbons (HFCs). Some of these gases have much higher absorptive power (hence, greater warming potential) than carbon dioxide. Nevertheless, carbon dioxide is the focus of most studies and concern, because CO_2 will contribute more than 50% of the increase in **radiative forcing** (increased IR absorption and warming) for the next century (IPCC 2001; Schlesinger 2001).

Consequences

Let us accept for the purpose of modeling that the accumulation of certain gases in the atmosphere results in greater absorption of heat from the earth's surface and leads to rising global temperatures. We also accept that human activities lead to an increase in the concentration of these gases. Some people might think that a 1 °C increase is not very great over a century, and we would hardly live long enough to notice. Why are so many people concerned about this problem? The answer is quite complicated because temperature affects so many processes on earth, most especially climate.

Climate changes of even minor amplitude may have drastic and dramatic consequences of life on this planet. These effects will be seen globally and regionally. Global warming will result in thermal expansion of the oceans and in glacial and ice cap melting that will raise the sea level. Projections based on various emissions/warming models predict a rise of 0.09 to 0.88 m by the end of this century (IPCC 2001). Thousands of coastal miles would be inundated by seawater. Parts of lower Manhattan, for instance, might return to the sea (Claussen 2002).

Projects

1. Assume that there is a relationship between increasing concentrations of atmospheric carbon dioxide ([CO_2]) and average global temperature (*T*). Add these components to your carbon cycle model of Project 2 from Module 7.4, "What Goes Around Comes Around—The Carbon Cycle," using the following relationships for CO_2 concentration and change in temperature from the start to the end of the simulation, respectively:

$$[CO_2] \text{ in ppm} = 350 \times (\text{mass of } CO_2 \text{ in the atmosphere})/750$$
(Allmon et al. 2006)

$$\text{temperature change (°C) over entire period} = 0.01([CO_2] - 350)$$
(Ward and Johnson 2004)

2. a. Methane (CH_4) is produced naturally by some anaerobic bacteria, termites, and domestic grazing animals. Human activities, such as landfills, burning, rice cultivation, coal mining, and oil/gas extraction, have drastically increased the release of this gas into the atmosphere. Although not nearly as prevalent as CO_2 in the atmosphere, methane is an important and powerful greenhouse gas with the ability to absorb 21 times as much heat per molecule as CO_2. In 1978, the concentration was 1.52 ppm, and methane concentration increased by about 1% per year until 1990. In 2011, the concentration was about 1.818 ppm (EPA 2011). Factor the effect of methane into your global warming model of Project 1. Experiment with increases and decreases in emissions.

b. Because much of the increase in methane levels is related to the production of food, we can tie the levels of atmospheric methane to population changes. According to Kremer (1993), world populations have increased as follows (with the years in parentheses): 720 million (1750), 1.2 billion (1850), 1.8 billion (1900), 2.5 billion (1950), and over 7 billion today. Corresponding methane levels were approximately 0.70 ppm (1750), 0.85

Table 7.5.1
Population and Methane Concentrations

Country	*Year*	*Population*	[*Methane*] in *ppm*)
United States	1750	2,059,000	0.70
	2001	278,059,000	310
Denmark	1769	797,600	0.70
	2001	5,355,000	310
Malawi	1890	543,000	0.70
	2001	14,696,000	310

ppm (1850), 0.90 ppm (1900), 1.1 (1950), and 1.818 ppm (20011; Etheridge et al. 2001; EPA 2011). Factor in population changes to methane concentrations.

c. Growth rates in various parts of the world differ significantly. For instance, Table 7.5.1 compares the population and methane concentration figures for Malawi, Denmark, and the United States. If world population had followed the overall growth rates of each of these countries, how might that change the model and results from Part b?

3. A third greenhouse gas is **nitrous oxide** (**N_2O**). Although production is much smaller, nitrous oxide absorbs 270 times as much heat as CO_2 per molecule and resides in the atmosphere for about 150 year. This gas is released with land-use conversion, combustion of fossil fuels, burning, and nitrogen fertilization of agricultural lands. The atmospheric concentration of nitrous oxide in 2011 was 324 ppb (parts per billion; EPA 2011). Factor the effect of nitrous oxide into your global warming model of Project 1 or 2. Experiment with increases and decreases in emissions. Discuss the results.

References

Allmon, Warren, Bryan Isacks, and William White. 2006. "Modeling the Carbon Cycle." Dept. of Earth and Atmospheric Sciences, Cornell University. http://www.geo.cornell.edu/eas/education/course/descr/EAS302/302_06Lab11.pdf (accessed December 19, 2012)

Bothun, Greg. 1998. "Greenhouse Effect." University of Oregon. http://zebu.uoregon.edu/1998/es202/l13.html (accessed December 19, 2012)

Claussen, Eileen. 2002. "Climate Change: Myths and Realities." Swiss Re Conference, *Emissions Reductions: Main Street to Wall Street*. New York (July 17). http://www.c2es.org/newsroom/speeches/climate-change-myths-and-realities (accessed December 19, 2012)

Dahlman, LuAnn. 2009. "Climate Change: Global Temperature." *ClimateWatch Magazine*. National Oceanic and Atmospheric Administration. http://www.climatewatch.noaa.gov/article/2009/climate-change-global-temperature (accessed December 19, 2012)

Etheridge, D. M., L. P. Steele, R. J. Francey, and R. J. Langenfelds. 1998. "Atmospheric Methane Between 1000 A.D. and Present: Evidence of Anthropogenic Emissions and Climatic Variability." *Journal of Geophysical Research*, 103, (D13)m: 15979–15993.

EPA (Environmental Protection Agency). 2011. Climate Change Indicators in the United States. Atmospheric Concentrations of Greenhouse Gases. http://www.epa.gov/climatechange/science/indicators/ghg/ghg-concentrations.html (accessed December 19, 2012)

———. 2012. Climate Change Indicators in the United States. http://www.epa.gov/climatechange/science/indicators/weather-climate/temperature.html (accessed December 19, 2012)

Heywood, N. C. 1998. "Energy & Heat. " University of Wisconsin, Stephens Point. http://www4.uwsp.edu/geo/faculty/heywood/GEOG101/energyhe/ (accessed December 19, 2012)

IPCC (Intergovernmental Panel on Climate Change). 2001. "Climate Change 2001: Synthesis Report." R. T. Watson and the Core Writing Team (eds.). Cambridge, UK, and New York: Cambridge University Press. http://www.grida.no/publications/other/ipcc_tar/ (accessed December 19, 2012)

———. 2007. "Climate Change 2007: Synthesis Report. Fourth Assessment Report." Core Writing Team, R. K. Pachauri and A. Reisinger. (eds.). Geneva, Switzerland: IPCC. http://www.ipcc.ch/publications_and_data/ar4/syr/en/contents.html (accessed December 19, 2012)

Kremer, Michael. 1993. "Population Growth and Technical Change, One Million B.C. to 1990." *Quarterly Journal of Economics,* 108(3: August): 681–716.

NASA (National Aeronautics and Space Administration). 2000. *Studying Earth's Environment From Space.* Ch. 2.2.1, "Energy from the Sun to the Earth's Surface." http://www.ccpo.odu.edu/~lizsmith/SEES/veget/class/Chap_2/2_1.htm (accessed December 19, 2012)

NOAA (National Oceanic & Atmospheric Administration). 2012. "Trends in Atmospheric Carbon Dioxide." Earth System Research Laboratory, Global Monitoring Division. http://www.esrl.noaa.gov/gmd/ccgg/trends/ (accessed December 19, 2012)

———. 2010. "Greenhouse Gases." National Climatic Data Center. http://www.ncdc.noaa.gov/oa/climate/gases.html (accessed December 19, 2012)

Pidwirny, Michael, and Scott Jones. 2010. *Fundamentals of Physical Geography*, 2nd ed. Ch. 7, "Introduction to the Atmosphere: The Greenhouse Effect." http://www.physicalgeography.net/fundamentals/7h.html (accessed December 19, 2012)

Schlesinger, William H. 2001. "The Carbon Cycle: Human Perturbations and Potential Management Options." *Global Climate Change: The Science, Economics and Policy Symposium*, Bush School of Government and Public Service, Texas A&M University (April 6). http://www.soc.duke.edu/~pmorgan/Schlesinger.htm (accessed December 19, 2012)

Ward, Dennis, and Roberta Johnson. 2004. "Modeling the Fast Carbon Cycle: The Terrestrial Loop." MGW 2004 Fall Workshop. http://eo.ucar.edu/staff/dward/carbon/modeling_the_fast_carbon_cycle.doc (accessed December 19, 2012)

MODULE 7.6

Plotting the Future—How Will the Garden Grow?

Prerequisite: Module 2.3, "Constrained Growth."

The Problem

This problem and project were contributed by and used with permission from Dr. Stephen Davies, University of Mary Washington (Davies 2012).

The caretaker for a wealthy landowner in southern Austria is planning an ornamental garden with a meditative footpath. A 40-ft by 40-ft plot for the garden will be entirely cleared before seeding. The gardener will plant five types of foliage: coneflowers, hostas, sedums, ferns, and ornamental trees. The caretaker wants to predict how the garden will grow and look in the future in order to plan an aesthetically pleasing arrangement.

The climate in this area has two annual seasons, rainy and dry. Each period lasts about half the year—January through June is the rainy season, and July through December is the dry season.

We can employ a logistic growth model for each of the plant species (not including the trees). For example, the number of new coneflowers that grow in a given period is related to the carrying capacity and the number of coneflowers already in the garden. With the plot able to support only a limited amount of plant life, we assume that the soil's carrying capacity is one plant per square foot. We use a logistic equation for all population calculations, so that the total plant population asymptotically approaches that carrying capacity. (Note that this carrying capacity applies only to the four plant types, not to the trees.)

Also, the different plants thrive in different moisture conditions. Hostas and ferns do well when moist, whereas coneflowers and sedums prefer drier soil. For this problem, we assume the following growth rates for each of the four species:

- Coneflowers and ferns: 0.02 in rainy season, –0.01 in dry season
- Hostas: –0.01 in rainy season, 0.02 in dry season
- Sedums: –0.01 in rainy season, 0.03 in dry season

Note carefully that some of the preceding numbers have a negative sign in front of them, indicating that the plant does not do well in that season.

The ornamental trees will grow in size but not in number. Planted as mere saplings, after 4 yr the trees will be large enough to produce shade. The shade a tree provides is a function of the tree's **canopy**, or the uppermost branches and leaves that block sunlight. We assume that each tree effectively has zero canopy until the plant is 4 years old. Starting at age 4, we assume the radius of the canopy grows at a rate of 0.8 ft/year until age 14, when the tree ceases to grow. Thus, a 5-year-old tree has a canopy radius of 0.8 ft; a 5.5-year-old tree has a radius of 1.2 ft; a 6-year-old tree's radius is 1.6 ft; and a tree's maximum canopy radius (from year 14 onward) is 8 ft. To calculate the area of sun blockage for each tree, use the formula for the area of a circle, πr^2, where r is the radius. For simplicity, we assume that no two trees' canopies overlap each other.

The amount of sun shining on the plants also can affect their growth rate. Specifically, both coneflowers and sedums require a great deal of sunlight, and we should adjust their growth rates as follows:

Coneflowers: If the percentage of the garden's total area that is sunny falls below 70%, coneflowers' growth rate should be 0.1 lower than otherwise. If the percentage of sunny area falls below 50%, the rate should be 0.2 lower. For example, in rainy season, if the garden had only 60% sunlight, the coneflowers' growth rate would be –0.08.

Sedums: If the sunny area percentage falls below 50%, sedums' growth rate should be 0.05 lower than otherwise.

Project

1. Simulate the garden's growth over a period of 20 yr, so that we can predict the plot's contents. Use 1 da as the value of Δt, where t represents time. Assume the gardener initially populates the garden as follows: 12 coneflowers, 19 hostas, 14 sedums, 7 ferns, and 12 ornamental trees (saplings). Keep track of the time of year, so that in rainy season the simulation uses one set of growth rates and in the dry season, the other set. Also, calculate the fraction of the garden's sunlight that the trees block at each time step, and adjust the coneflowers and sedums growth rates as necessary. Remember also to calculate the total number of plants at each clock tick in order to apply the logistic factor (1 – *totalPlants* / *carryingCapacity*).

Plot the growth of the four plants over time. Prepare a brief written narrative about what you discovered when trying different simulation parameters. In particular, experiment with the initial numbers of different plant and tree types, the amount of shade provided by the trees, and the climate (e.g., the rainy season shortened or lengthened by a month or two). Discuss which of the preceding three changes (or others you might try) have large effects and which have negligible effects (Davies 2012).

Reference

Davies, Stephen. 2012. "CPSC 370 Homework #3: Ornamental Garden." University of Mary Washington. http://rosemary.umw.edu/~stephen/homework3.html (accessed 06/28/2012)

MODULE 7.7

Cardiovascular System—A Pressure-Filled Model

Prerequisites: Module 2.4, "System Dynamics Software Tutorial 2," or Module 2.5, "Drug Dosage"; and for Projects 1 and 2, also the section on "Modeling Inhibition" from Module 4.5, "Enzyme Kinetics—A Model of Control."

Circulation

As organisms assume larger dimensions, they are confronted with a number of problems resulting from this increase in size. Cells become separated from one another, sometimes by great distances; and they take on specialized functions. The functional integration of these specialized cells becomes paramount for the success of the organism. One necessary adaptation is the acquisition of effective and efficient transport systems, made up of interconnected spaces and tubes that transport fluids. In multicellular animals, we refer to these systems as **circulatory systems**. Even vascular plants have such systems. As animals grew and evolved, tubular systems **(cardiovascular systems)** that included a muscular pump **(heart)** replaced primitive circulatory systems. In these animals, a fluid (**blood**) delivers oxygen, nutrients, hormones, and wastes to their proper destinations.

The human cardiovascular system is made up of a heart and two circulatory loops—the **pulmonary** (lungs) and **systemic** (rest of body) circulations. The heart consists of four chambers—right and left **atria** and right and left **ventricles**. In both circulations, blood is pumped from the ventricles through a series of tubes—arteries, arterioles, capillaries, venules, veins—and is returned to the opposite atrium of the heart. Blood leaving the left ventricle enters the arteries of the systemic circulation, which perfuses through capillaries in muscles, digestive tract, brain, and vital organs, such as the kidneys and liver. Blood is returned through venules and veins to the right atrium of the heart. This blood, low in oxygen and high in carbon dioxide, is squeezed into the right ventricle, which pumps it into the pulmonary arteries and on to the capillaries surrounding the tiny sacks in the lungs, where oxygen is exchanged for carbon dioxide. Pulmonary veins return this blood to the left atrium. From there, the blood enters the left ventricle to be sent out in the systemic circulation.

Blood Pressure

Blood pressure is the hydrostatic (fluid) pressure that moves the blood through the circulation. We monitor the pressure exerted on the arteries of the systemic circuit. Blood pressure is pulsatile, not continuous, because the cardiac cycle is intermittent. We measure the highest pressure exerted as the left ventricle contracts (**systolic pressure**). Traditionally, this pressure in a healthy adult is approximately 120 mm of mercury (mm Hg). Likewise, the pressure in the arteries as the left ventricle relaxes (**diastolic pressure**) is approximately 80 mm Hg.

Mean arterial pressure (***MAP***) is the average pressure during an aortic pulse cycle. For a normal resting person, *MAP* approximately obeys the following model:

$$MAP = \text{(diastolic pressure)} + \frac{\text{(systolic pressure)} - \text{(diastolic pressure)}}{3}$$

Cardiac output (***CO***) and **systemic vascular resistance** (***SVR***) regulate this pressure according to the following model:

$$MAP = CO \times SVR$$

Cardiac output is the product of the **stroke volume** (***SV***) and the **heart rate** (***HR***):

$$CO = SV \times HR$$

Stroke volume, the volume of blood that the left ventricle ejects, ranges from 50 to 100 mL in a healthy adult. A number of factors influence the stroke volume, including the volume of blood returned to the ventricle and **contractility**, or the ability of heart muscle to shorten. Heart rate in a resting, healthy adult is normally between 60 and 80 **beats/minute** (**bpm**). Consequently, cardiac output, or the volume of blood that the left ventricle ejects over a period of time, ranges from 4 to 8 L/min.

Nervous Systems

Controlling various involuntary activities of the body, including heart rate, the **autonomic nervous system** consists of the **sympathetic** and the **parasympathetic nervous systems**. The **pacemaker** of the heart, which is in the right atrium, fires at an intrinsic rate of 100 to 115 bpm. The vagus nerve, which is part of the parasympathetic nervous system, can inhibit the pacemaker's normal beat to 50 bpm. The stimulatory influence of the sympathetic nerves counters the inhibitory effects of the vagus and under certain conditions can increase heartbeat to as much as 200 bpm.

Stroke Volume

As well as heart rate, the sympathetic nervous system controls stroke volume (*SV*) directly and indirectly. Directly, sympathetic stimulation causes greater contraction of the heart and, consequently, larger *SV*. The nervous system achieves this increase with an influx of calcium ions into the cardiac muscle cells. Calcium ions promote the formation of cross-bridges between the muscle fibers, increasing the strength of contraction. Likewise, epinephrine release promotes increased contractility of heart muscle. Indirectly, active sympathetic stimulation promotes **vasoconstriction** (decreasing the vessel diameter) of the veins, leading to greater **venous return** (flow of blood to the heart). Greater venous return increases end-diastolic volume and the contractile tension of the heart muscle to a more optimal length for contraction, resulting in the heart pumping out more blood.

Venous Return

Factors other than the sympathetic nervous system influence venous return. Skeletal muscle activity, respiration, and increases in blood volume also amplify venous return. In particular, salt-water balance and the vasopressin-angiotensin system (hormones that are important to fluid balance and are vasoconstrictors) influence blood volume.

Systemic Vascular Resistance

Systemic vascular resistance (***SVR***) is the resistance or impediment of the blood vessels in the systemic circulation to the flow of blood. Although there are a number of factors that regulate *SVR*, one of the most important is the diameter of the perfusing blood vessels. Increases in *SVR* are caused by numerous factors that promote **vasoconstriction**, whereas decreases are triggered by factors that encourage **vasodilation**. These factors include those that are neurohumoral (e.g., epinephrine and vasopressin promote vasoconstriction), endothelial (e.g., nitric oxide promotes vasodilation; endothelin promotes vasoconstriction), local hormones (e.g., arachidonic acid metabolites, which may promote vasoconstriction or vasodilation) and myogenic (usually promotes vasoconstriction).

Normal ranges for *SVR* are 800 to 1200 dyn s/cm^5, where dyn represents dynes. From the preceding section, "Blood Pressure," we see that systemic vascular resistance is the quotient of mean arterial pressure and cardiac output, or $SVR = MAP/CO$. For *MAP* in mm Hg and *CO* in L/min, we multiply the result by 79.9 to obtain the value with units of dyn s/cm^5. We can determine this conversion factor (79.9) with the facts that 1 mm Hg = 1333.22 dyn/cm^2 and 1 mL = 1 cm^3. The definition of the force 1 **dyn** is 10^{-5} N.

Blood Flow

Blood flow through tissues is vital for the delivery of nutrients, oxygen, and chemical messages and for the removal of carbon dioxide and wastes. Regulation of blood flow is, therefore, vital to the proper functioning of those tissues. **Blood flow** (***Q***) through a vessel over time is equal to the **mean velocity of the blood flow** (***v***) times the **cross-sectional area of the vessel**. With the cross-sectional area being the area of a circle, πr^2, where $\boldsymbol{r}$ is the **radius of the vessel**, we have the following equation for blood flow:

$$Q = v\pi r^2$$

A model for the mean velocity (v) is the **pressure gradient** ($\boldsymbol{\Delta P}$) times the square of the radius (r^2) divided by the product of 8, the **viscosity of the blood** (η), and the **vessel length** ($\boldsymbol{L}$):

$$v = \frac{\Delta P r^2}{8\eta L}$$

Viscosity indicates the degree to which the fluid resists flow. If we substitute for v in our blood flow equation, we determine blood flow in an arteriole using **Poiseuille's equation** (Klabunde 2010):

$$Q = v\pi r^2 = \frac{\Delta P r^2}{8\eta L} \cdot \pi r^2 = \frac{\pi r^4 \Delta P}{8\eta L}$$

Projects

For each of the following projects, discuss your simplifying assumptions and results and how closely those results match reality.

1. **a.** Model the regulation of heart rate by the sympathetic nervous system, holding the parasympathetic constant. Assume heart rate is linearly dependent on this system. Employ converters/variables.
   b. Refine your model to include inhibition of the heart rate by the parasympathetic nervous system.
   c. Extend your model to include the regulation of heart rate by **epinephrine** (adrenalin), a chemical messenger of the sympathetic nervous system. At times of stress, such as exercise, excitement, or excessive bleeding, the adrenal medulla releases this epinephrine to increase heart rate. Illustrate its action by having the adrenal medulla release the epinephrine quickly and having the epinephrine stay in the body for a few minutes before gradually diminishing at a rate proportional to the concentration of epinephrine. Have a stock (box variable) for this concentration.
   d. Modify your model to include the influence of stroke volume and heart rate on cardiac output. Consider two versions of the model. One models cardiac output as the product of stroke volume and heart beat. The other has the flow out of the heart being a pulse of a volume of blood every heartbeat. Some typical parameters for a normal person are as follows: volume of blood in the body = 5000 mL; stroke volume = 70 mL; immediately after beat, volume in ventricle = 60 mL.
   e. Investigate factors controlling blood volume and modify your model for the control of cardiac output. A normal value for fluid intake and urine output is 1 mL/min. Blood volume is approximately 70 mL/kg. Consult other sources as necessary to complete the project.

2. a. Develop a simple model of mean arterial pressure, where cardiac output (*CO*) and systemic vascular resistance (*SVR*) govern mean arterial pressure (*MAP*).
   b. Modify your simple model to include the regulation of *SVR* by vasoconstriction and vasodilation, which increase (promote) and decrease (inhibit) resistance, respectively.
3. Using the components of Poiseuille's Equation (see the section "Blood Flow") as well as cardiac output, develop a model for regulation of blood flow in the systemic loop of the cardiovascular system. Have heart rate and stroke volume regulate blood pumping (pulsing) from the heart. Consider both arterial and venous flow with stocks (box variables) for the heart, artery system, and body tissues. In arterial flow, systolic and diastolic pressures determine the pressure gradient. With P being *MAP*, some possible arterial parameter values are as follows: $r = 6$ mm, $\eta = 0.04$ g/(cm s), $L = 1000$ mm. Some possible venous parameter values are as follows: $r = 3.5$ mm, $\eta = 0.04$ g/(cm s), $L = 10$ mm. Venous pressure, which varies over a large range, averages 17 mm Hg (Romstedt 2003).
4. Refine Project 3 to include pulmonary circulation.
5. Refine Project 4 so that each heart chamber is a separate stock (box variable).

References

Cheatham, Michael L. 2009. "Hemodynamic Calculations." Orlando Regional Medical Center. http://www.surgicalcriticalcare.net/Lectures/PDF/hemodynamic calculations I.pdf (accessed December 21, 2012)

Gannon, Patrick. J. 2012. "Introduction to Cardiovascular Physiology." https://profgannon.wikispaces.com/Introduction+to+Cardiovascular+Pharmacology+ (accessed December 20, 2012)

Klabunde, Richard E. 2009. "Cardiovascular Pharmacology Concepts." Cardiovascular Pharmacology website. http://www.cvpharmacology.com/ (accessed December 21, 2012)

———. 2010. "Cardiovascular Physiology Concepts." Cardiovascular Physiology Concepts website. http://www.cvphysiology.com/ (accessed December 21, 2012)

MacLeod, Rob. "Bioengineering 3202: An Integrated Approach to Human Physiology II." University of Utah. http://www.sci.utah.edu/~macleod/bioen/be3202/ (accessed December 20, 2012)

Rogers, James. 1999. "Cardiovascular Physiology." *World Anaesthesia Online*, Issue 10, Article 2. http://www.nda.ox.ac.uk/wfsa/html/u10/u1002_01.htm (accessed December 20, 2012)

Romstedt, Karl. 2003. "Modeling the Cardiovascular System using STELLA: A Module for Computational Biology." *Biology Modules*. Capital University. http://www.capital.edu/uploadedFiles/Capital/Academics/Schools_and_Departments/Natural_Sciences,_Nursing_and_Health/Computational_Studies/Educational_Materials/Content/Modeling the Cardiovascular System using STELLA®.pdf (accessed December 21, 2012)

Sherwood, Lauralee. 2010. *Human Physiology: From Cells to Systems*. 7th ed. Belmont, CA: Brooks-Cole/Thompson Learning.

Walton, D. Brian. 1999. "Physics of Blood Flow in Small Arteries." Mathematics and Biology, Mathematics Awareness Month. University of Arizona. http://math.arizona.edu/~maw1999/blood/poiseuille/ (accessed December 21, 2012).

MODULE 7.8

Electrical Circuits—A Complete Story

Prerequisite: Module 3.2, "Modeling Bungee Jumping."

Defibrillators

When someone has a heart attack—what doctors call a **myocardial infarction**—insufficient blood flows to a specific portion of the heart. Interruption of blood flow may occur when the coronary arteries supplying heart muscle cells with blood are obstructed by a blood clot. The muscle soon suffers from lack of oxygen and nutrients; and if the blood supply is not restored immediately, the muscle dies. The patient may experience various symptoms, including pain, as the heart is thrown into disarray. The distress of heart muscle is accompanied by electrical instability, which may lead to **ventricular fibrillation**, or chaotic electrical disturbance (Kulick 2012).

To restore orderly electrical signals during fibrillation, medical personnel may use an instrument called a **defibrillator**. This device causes a predetermined amount of current to flow across the heart. Paddles are positioned properly on a patient's chest. Flipping of a switch forms a bridge, or, we say, completes an **electrical circuit**, and stored electrons can then flow from a negatively charged plate of the defibrillator's capacitor, through the patient's heart, and back to the capacitor's positive plate. This current synchronizes the depolarization of the heart muscle and helps to restore normal electrical rhythm and normal, coordinated beating.

The use of the defibrillator provides us with but one example of the utility of electrical circuits. Circuits may be simple, such as in the flashlight we keep in an automobile, or very complex, such as those in sophisticated computers. In any application, completed circuits make functions possible.

Current and Potential

In an atom, **electrons** orbit a nucleus, which contains **neutrons** and **protons**. With opposite charges, electrons and protons are attracted to each other. A **coulomb (C)** is a unit of **electric charge**, Q. The charge on an electron is $-1.6\ 10^{-19}$ C, while a proton has a charge of $+1.6\ 10^{-19}$ C.

When a charge flows through a region, we say that a current exists. The **current** I is the rate of change of the charge with respect to time, or

$$I = \frac{dQ}{dt}$$

One **ampere**, or **amp** (**A**), is the unit of current for a charge of 1 C to pass through a region in 1 s, or **1 A = 1 C/s**. A charge is analogous to water, while current corresponds to the movement of the water. Similarly, a ball and a ball falling form analogies to charge and current, respectively.

Definitions A **coulomb** (**C**) is a unit of **electric charge**, **Q**. **Current**, I, is the rate of change of the charge with respect to time, or $I = \frac{dQ}{dt}$. One **ampere**, or **amp** (**A**), is the unit of current for a charge of 1 C to pass through a region in 1 s, or **1 A = 1 C/s**.

A metal wire is a good **conductor** of current, and an electrical **circuit** usually consists of wires and other components. By convention, we say that the **direction of the current** is opposite to the direction in which the electrons flow.

Energy must be employed to pull opposite charges apart. When together, they have potential energy. The **electronic potential**, or **potential**, V, at a point is the potential energy per unit charge, or the work per unit charge to bring a positive charge from infinity to the point. The **potential difference**, or **voltage difference**, between two points A and B is the difference in potential between the points. A unit of measure of potential difference is a **volt** (**V**). Figure 7.8.1 presents the circuit symbol for an **imposed voltage** E, such as a battery. We define the **voltage** at a point A in the circuit as the voltage difference between A and a circuit reference point, the **ground** (often the negative terminal of a battery). In a circuit, current flows from a region of high voltage to one of low voltage. Electronic potential is analogous to mechanical potential energy. For example, water flows from the top of a waterfall, where it has high potential energy, to the bottom, where its potential energy is lower.

Definitions The **electronic potential**, or **potential**, V, at a point is the potential energy per unit charge, or the work per unit charge to bring a positive charge from infinity to the point. The **potential difference**, or **voltage difference**, between two points A and B is the difference in potential between the points. A unit of measure of potential difference is a **volt** (**V**). The **voltage** at a point A in the circuit is the voltage difference between A and a circuit reference point, the **ground**.

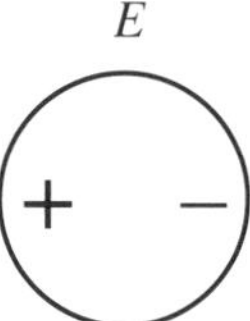

Figure 7.8.1 Electrical circuit symbol for imposed voltage, *E*

Resistance

Voltage is virtually constant along a wire. However, other components in the circuit cause voltage to drop. For example, a **resistor** slows the current flow and, thus, controls the current level. A resistor is analogous to a constriction in a hose that slows the water flow. Figure 7.8.2 displays the electrical circuit symbol for a resistor. A constant **resistance *R*** measures the ability of a resistor to reduce the flow of charges. According to **Ohm's law**, the voltage drop or potential change across a resistor is as follows:

$$V = IR = R\frac{dQ}{dt}$$

or

$$I = V/R$$

or

$$R = V/I$$

We measure the resistance of a resistor in **ohms** (Ω), and 1 Ω = 1 V/A. A good wire has resistance much less than 1 Ω. With an incandescent lightbulb, the dissipated potential from resistance appears as light and heat. For a toaster, resistance dissipates potential that results mostly in heat.

> **Definitions** A **resistor** slows the current flow. A constant **resistance *R*** measures the ability of a resistor to reduce the flow of charges. **Ohm's law** states that $V = IR = R\frac{dQ}{dt}$. A measure of resistance is 1 **ohm** (Ω) = 1 V/A.

Quick Review Question 1

Suppose a circuit has a battery with voltage = 4.5 V and a resistor with resistance =100 Ω. Calculate the current through the circuit and give its units.

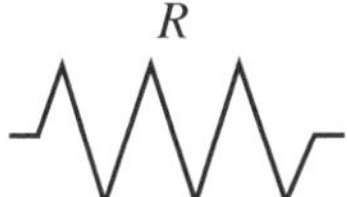

Figure 7.8.2 Electrical circuit symbol for a resistor

Figure 7.8.3 Electrical circuit symbol for a capacitor

Capacitance

A **capacitor**, whose symbol appears in Figure 7.8.3, is a component for storing charge. A simple capacitor consists of two conductors, such as metal plates, one with a positive charge and one with an equal negative charge, with an insulator between them. The potential difference can build between the two conductors. Just as a dam can hold water from a river, a capacitor can hold charge. The ability to store charge is **capacitance (*C*)**, which we can measure in **farads (F)**. One farad of capacitance is equivalent to having a capacitor hold a charge of 1 C for a potential difference of 1 V across its conductors, or 1 F = 1 C/V. We have the following relationship among capacitance, charge, and voltage drop or change in potential across a capacitor:

$$C = Q/V$$

or

$$Q = CV$$

> **Definitions** A **capacitor** is a component for storing charge. The ability to store charge is **capacitance (*C*)**. One **farad (F)** of capacitance for a capacitor is equivalent to having a capacitor hold a charge of 1 C for a potential difference of 1 V across its conductors, or 1 F = 1 C/V.

Quick Review Question 2

Suppose a capacitor has a capacitance of 32 μF, and the voltage across the capacitor is 5000 V. Calculate the amount of charge that the capacitor stores in millicoulombs (mC).

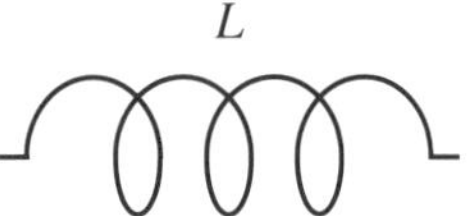

Figure 7.8.4 Electrical circuit symbol for an inductor

Inductance

A third component that reduces current is an **inductor**, a coil of wire that dampens sudden changes in current. As Figure 7.8.4 shows, the electrical symbol for an inductor suggests a coil. An inductor prevents the instantaneous increase in current and prolongs current flow. The constant **inductance *L*** of the coil measures the opposition to a change in current and has the following formula:

$$L = \frac{V}{dI/dt}$$

Because $I = dQ/dt$, dI/dt is the second derivative of charge with respect to time, d^2Q/dt^2, or the rate of change of the rate of change of Q with respect to time, and

$$L = \frac{V}{d^2Q/dt^2}$$

A unit of measure for inductance is a **henry** (**H**), which is 1 V s/A. Table 7.8.1 summarizes some of the terms associated with electrical circuits along with their symbols, units, and formulas.

> **Definitions** An **inductor** is a device that dampens sudden changes in current. A constant **inductance *L*** of a coil measures the opposition to a change in current and has the formula $L = \frac{V}{dI/dt}$. A unit of measure for inductance is a **henry** (**H**), which is 1 V s/A.

Quick Review Question 3

Suppose a large inductor has 1 H inductance and a current of 10 A flows through the inductor. Estimate the voltage difference in volts if we cut off the current in 1.0 ms.

Circuit for Defibrillator

Figure 7.8.5 contains a circuit diagram for a defibrillator. Initially, a switch is set so that the battery can charge the capacitor. When the switch is set in the other direction, the capacitor discharges sending a surge of electricity through the heart, which

Table 7.8.1
Electrical Circuit Terms

Term	*Symbol*	*SI Unit*	*Formula*	*Formula*
Capacitance	C	Farad (F)	$C = Q/V$	
Charge	Q	Coulomb (C)	$Q = CV$	
Current	I	Ampere (A)	$I = \frac{dQ}{dt}$	$I = V/R$
Inductance	L	Henry (H)	$L = \frac{V}{dI/dt}$	$\mathrm{L} = \frac{V}{d^2Q/dt^2}$
Resistance	R	Ohm (Ω)	$R = V/I$	
Voltage	V	Volt (V)	$V = IR$	$V = R\frac{dQ}{dt}$

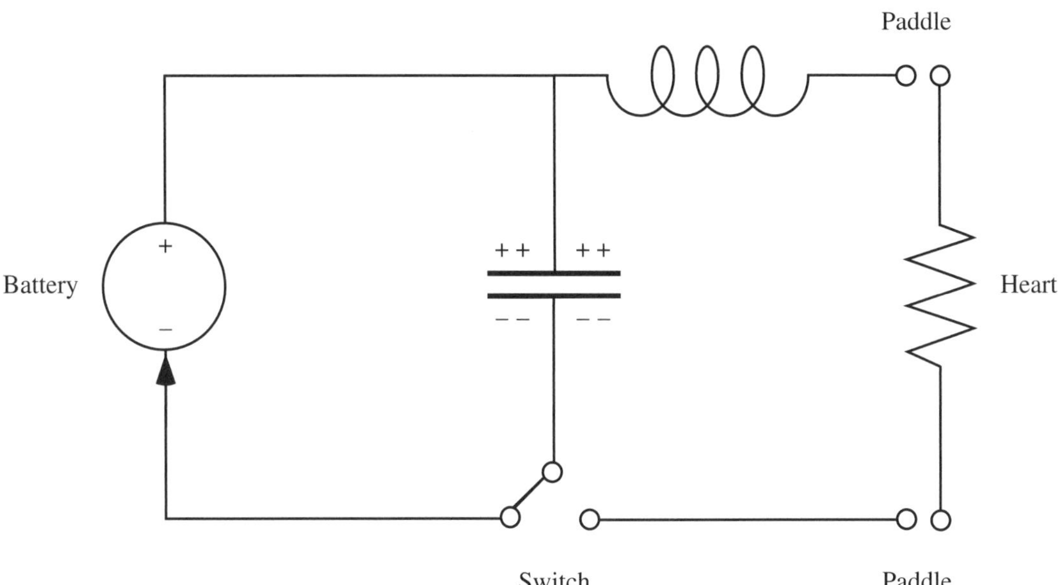

Figure 7.8.5 Circuit diagram for a defibrillator (Williams et al. 2003)

is a resistor. An inductor dampens sudden increase in current and prolongs current flow.

Kirchhoff's Voltage Law

An important connection among the components of a circuit is **Kirchhoff's voltage law**, which states that in a closed loop, the sum of the changes in voltage is zero. For example, consider the ***RLC* circuit** (circuit with a resistor, a inductor, and a capacitor) in Figure 7.8.6, with a battery providing voltage, $E(t)$. Resistance causes a voltage drop of $IR = R\frac{dQ}{dt}$; the voltage drop due to the capacitor is Q/C; while induc-

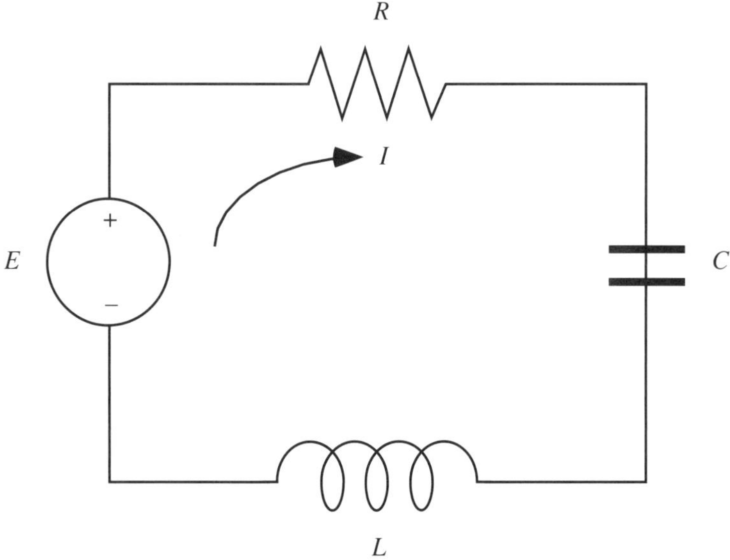

Figure 7.8.6 An *RLC* circuit

tance causes a voltage drop of $L\dfrac{dI}{dt} = L\dfrac{d^2Q}{dt^2}$. Thus, by Kirchhoff's voltage law, the following equation holds:

$$E(t) - R\frac{dQ}{dt} - \frac{Q}{C} - L\frac{d^2Q}{dt^2} = 0$$

or

$$E(t) = R\frac{dQ}{dt} + \frac{Q}{C} + L\frac{d^2Q}{dt^2} \tag{1}$$

Sometimes it is convenient to express this equation using current instead of charge. Recalling that $I = dQ/dt$, we differentiate the above equation and substitute appropriately to obtain the following:

$$E'(t) - R\frac{dI}{dt} - \frac{I}{C} - L\frac{d^2I}{dt^2} = 0$$

or

$$E'(t) = R\frac{dI}{dt} + \frac{I}{C} + L\frac{d^2I}{dt^2}$$

Kirchhoff's Voltage Law In a closed loop, the sum of the changes in voltage is zero.

Quick Review Question 4

Using Kirchhoff's voltage law on the defibrillator circuit diagram in Figure 7.8.5, give the equations for the following:

a. The left loop using Q
b. The left loop using I
c. The right loop after the switch is thrown to complete that circuit using Q
d. The right loop after the switch is thrown to complete that circuit using I

Kirchhoff's Current Law

Many circuits, such as the one in Figure 7.8.7, consist of several loops. **Kirchhoff's current law** states that the sum of the currents into a junction, such as node J_1, equals the sum of the currents out of that junction. Thus, $I_1 = I_2 + I_3$.

Quick Review Question 5

Give Kirchhoff's current law as it applies to the following junctions in Figure 7.8.7:

a. Junction J_2
b. E

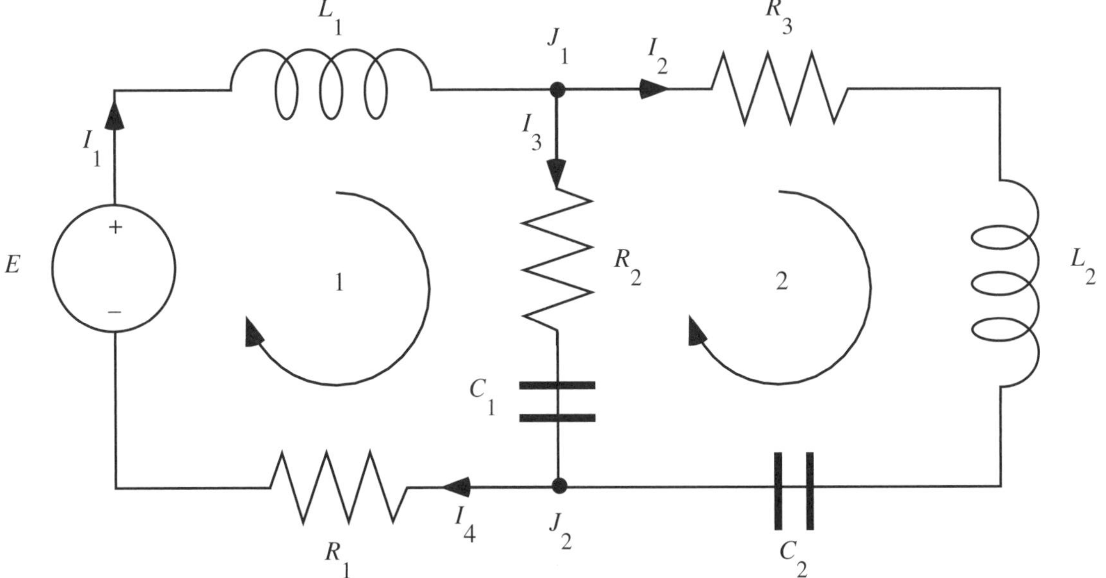

Figure 7.8.7 Circuit with more than one loop

Using Kirchhoff's voltage and current laws, we obtain a system of differential equations that models the circuit. Applying Kirchhoff's voltage law to Loop 2 of Figure 7.8.7, we obtain the following differential equation involving current:

$$R_3 \frac{dI_2}{dt} + L_2 \frac{d^2 I_2}{dt^2} + \frac{I_2}{C_2} - \frac{I_3}{C_3} - R_2 \frac{dI_4}{dt} = 0 \qquad (2)$$

Because the assumed direction of the current through the components for C_1 and R_2 is opposite to that of the current through the other components in the loop, the terms involving C_1 and R_2 are negative.

Quick Review Question 6

Apply Kirchhoff's voltage law to Loop 1 of Figure 7.8.7 to obtain a differential equation involving current.

By applying Kirchhoff's current law, we can simplify the differential equations to involve fewer currents. For example, we know that I_4 and I_1 are the same and that $I_1 = I_2 + I_3$ or $I_3 = I_1$ - I_2. Taking the derivative of the latter, we have the following relationship:

$$\frac{dI_3}{dt} = \frac{dI_1}{dt} - \frac{dI_2}{dt}$$

Thus, substituting in Equation 2 and the answer to Quick Review Question 6, we have the following system of differential equations involving currents I_1 and I_2:

$$\text{Loop 1: } \quad E'(t) = L_1 \frac{d^2 I_1}{dt^2} + R_2 \frac{dI_1}{dt} - R_2 \frac{dI_2}{dt} + \frac{I_1}{C_1} - \frac{I_2}{C_1} + R_1 \frac{dI_1}{dt}$$

$$\text{Loop 2: } \quad R_3 \frac{dI_2}{dt} + L_2 \frac{d^2 I_2}{dt^2} + \frac{I_2}{C_2} - \frac{I_1}{C_1} + \frac{I_2}{C_1} - R_2 \frac{dI_1}{dt} + R_2 \frac{dI_2}{dt} = 0$$

Projects

For the projects, use an appropriate system dynamics tool with small time steps.

1. **a.** Develop a model for the *RLC* circuit in Figure 7.8.6. Assume $L = 0.05$ H, $R = 20\ \Omega$, $C = 100\ \mu\text{F}$, $E(t) = 100$ V, and $Q(0) = Q'(0) = 0$ C. Produce appropriate graphs, such as current and charge versus time.
   b. Model and discuss the impact of having zero inductance.
   c. Model and discuss the impact of having zero resistance.
   d. Model and discuss the impact of having zero capacitance.
   e. Vary the values for the constants in Part a. Observe and discuss the results.
   f. Referring to Module 3.2, "Modeling Bungee Jumping," develop an analogy between an RLC circuit modeled by Equation 1 and a forced, damped spring-mass system (Davis 1992).

2. Repeat Project 1 assuming $L = 0.2$ H, $R = 50\ \Omega$, $C = 10\ \mu$F, $E(t) = 120 \cos(120\pi t)$, $Q(0) = 10^{-6}$ C, and $Q'(0) = 0$ A.
3. **a.** Write a differential equation for the voltage applied to the heart by a defibrillator. This equation is piecewise and consists of an equation for $E'(t)$ during the time when the capacitor is charging and sending no current to the heart and an equation when the capacitor is discharging and conveying an electrical impulse to the heart.
   b. Develop a model for a defibrillator circuit. Suppose the defibrillator has a 5000-V battery and a 32-μF capacitor. The resistance of a patient is between 50 Ω and 150 Ω. Plot voltage applied to the heart versus time as well as other appropriate graphs.
4. Model a heart pacemaker, which is similar to a defibrillator. The pacemaker alternates between a time, such as 4 s, in which the capacitor is charging and a time, such as 2 s, in which it is discharging and sending an electrical impulse to the heart. Suppose the pacemaker has a 12-V battery.
5. Develop a model for the circuit in Figure 7.8.7. Assume $L_1 = 0.2$ H, $L_2 = 1.0$ H, $R_1 = 10\ \Omega$, $R_2 = 220\ \Omega$, $R_3 = 330\ \Omega$, $C_1 = 0.1\ \mu$F, $C_2 = 1.0\ \mu$F, $E(t) = 117$ V, and $Q(0) = Q'(0) = 0$. Produce appropriate graphs. Discuss the results.
6. Repeat Project 5 with $E(t) = 3 \cos(20\pi t)$.
7. Develop a model for a circuit of your choosing.

Answers to Quick Review Questions

1. 0.045 A because the current is $I = V/R = 4.5$ V/100 Ω = 0.045 A
2. 160 mC because $Q = CV = (32\ \mu\text{F})(5000\ \text{V})(1\ \text{mF}/(1000\ \mu\text{F})) = 160$ mC
3. $V = L\,(dI/dt) \approx L\,(\Delta I / \Delta t) = (1\ \text{H})(10\ \text{A})/(1\ \text{ms}) = 10/0.001 = 10{,}000$ V
4. **a.** $E(t) = Q/C$
   b. $E'(t) = I/C$
   c. $\dfrac{Q}{C} + L\dfrac{d^2Q}{dt^2} + R\dfrac{dQ}{dt} = 0$
   d. $\dfrac{I}{C} + L\dfrac{d^2I}{dt^2} + R\dfrac{dI}{dt} = 0$
5. **a.** $I_2 + I_3 = I_4$
   b. $I_4 = I_1$
6. $E'(t) = L_1\dfrac{d^2I_1}{dt^2} + R_2\dfrac{dI_3}{dt} + \dfrac{I_3}{C_1} + R_1\dfrac{dI_4}{dt}$

References

Bolooki, H. Michael, and Arman Askari. 2011. "Acute Myocardial Infarction." Cleveland Clinic. http://www.clevelandclinicmeded.com/medicalpubs/diseasemanagement/cardiology/acute-myocardial-infarction/ (accessed December 21, 2012)

Burghes, D. N., and M. S. Borrie. 1981. *Modeling with Differential Equations.* Chichester, England: Ellis Hordwood Limited, p. 172.

Davis, Paul W. 1992. *Differential Equations for Mathematics, Science, and Engineering*. Englewood Cliffs, NJ: Prentice Hall, p. 565.

Kulick, Daniel Lee. 2012. "Heart Attack (Myocardial Infarction)." *MedicineNet*. http://www.medicinenet.com/Heart_Attack/article.htm

Ross, Clay C. 1995. *Differential Equations, An Introduction with Mathematica*. New York: Springer-Verlag, p. 503.

Urone, Paul Peter. 2001. *College Physics*. 2nd ed. Sacramento, CA: Brooks/Cole, p. 893.

Williams, David J., Fiona J. McGill, and Hywel M. Jones. 2003. "Principles of Physical Defibrillators." *Anaesthesia and Intensive Care Medicine, Physics*. Abingdon Oxon, UK: The Medicine Publishing Company, pp. 23–31.

Zill, Dennis G. 2013. *A First Course in Differential Equations with Modeling* Applications, 10th ed. Belmont, CA: Brooks/Cole (Cengage Learning).

MODULE 7.9

Transmission of Nerve Impulses—Learning from the Action Potential Heroes

Prerequisites: Modules 2.2, "Unconstrained Growth and Decay," and 2.4, "System Dynamics Tool: Tutorial 2."

Background: Module 7.8, "Electrical Circuits—A Complete Story," sections "Current and Potential"," "Resistance," and "Capacitance."

Introduction

You may be familiar with a group of mollusks called the cephalopods. These animals include octopus, squid, cuttlefish, and the chambered nautilus. What sets these animals apart from other mollusks is their well-defined head region. You likely have dined on an appetizer or pasta that included calamari, which means you were eating squid. The cephalopods are an ancient group (~500 million years of evolution) found in marine environments all over the earth. They are not only old, but they are considered the "smartest" of the invertebrates, because of their well-developed brain and nervous system. Their remarkably keen eyes look much like mammalian eyes, although their actual structures are quite different (Wood 2012).

One species of squid, *Loligo pealei*, common in the coastal waters of the east coast of the United States, has been of enormous help to neurophysiologists. Their nervous system contains some of the largest neurons, especially axons, in the animal world—up to 10 cm long and 1 mm in diameter (Namnezia 2011). On the other hand, although some are quite long, mammalian axons are only about one-tenth the diameter of the squid axon and are more delicate. Two physiologists, Alan Hodgkin and Andrew Huxley, using the squid's giant axon, uncovered much about how nerve impulses are transmitted (the action potential); and, as it turns out, the squid's axon functions pretty much like that of mammals. Hodgkin and Huxley were awarded the 1963 Nobel Prize in Physiology or Medicine, and their model still serves as the basis of our understanding of neuronal functions (Nelson and Rinzel 2003). We might call these two scientists the "Action Potential Heroes."

The squid giant axon has been a rich experimental model for neuronal function for decades. Today, scientists are still using the model, but now to study neuronal malfunction. For instance, NIH researchers have used the calculations to explore the accumulation of intracellular filamentous tangles that form from hyperphosphorylated cytoskeletal proteins in the nerve cell body. The squid neuron's two principle compartments (cell body and axon) can be isolated for study and comparison. This ability is important, because diseases like Alzheimer's and amyotrophic lateral sclerosis (Lou Gehrig's disease) are both characterized by such tangles (involving different proteins) in the cell bodies of the human brain or spinal cord neurons. So, we still have lots to learn from these remarkable creatures. Next time you have a nice serving of calamari, be sure you give thanks to the intrepid physiologists and the squid, all of whom contributed so much to our understanding of the mammalian nervous system.

The Neuron—Basic Structure and Function

The human nervous system is a highly complex network of cells that are essential for receipt and integration of sensory information, communication, and coordination of body activities. **Neurons**, the functional nerve cells (those that conduct signals), are excitable cells, capable of conducting an electrical impulse, that can receive and integrate signals and transmit them to target cells (other neurons or effector cells). There are probably thousands of types of neurons, but here we will concentrate on a **motor neuron**. Motor neurons are also called "efferent" neurons, because they carry signals away from the central nervous system (brain and spinal cord). Their role is to transmit that signal to **effector cells** (e.g., muscle cells, glands) and "effect" a response (e.g., muscle contraction, secretion). A typical motor neuron is diagramed in Figure 7.9.1. Each cell is composed of a **cell body** (or **soma**) and two types of cytoplasmic extensions: the **axon** and the **dendrite**. There are many dendrites that can transmit signals to the cell body along the plasma membrane. The signal passes along the membrane of the cell body, which narrows and forms a single axon. The axon then transmits the signal from the cell body toward the effector. **Terminal buttons**, produced at the branched ends of the axon, interface with the effector cells by way of a junction called the **synapse**.

The plasma membrane that encloses the neuron is surrounded by a sea of ions that is quite different quantitatively and qualitatively from the ionic solution in its cytoplasm. Overall, the cell maintains this difference so that the inside is much more negative than outside. Therefore, along the inner surface is an excess of negative charges, which are provided primarily by large, organic anions and phosphates that cannot pass through the membrane. On the outside, positive ions (primarily **sodium ions, Na^+**) accumulate. The difference in concentration yields an ***electrical* (membrane) potential** (potential = voltage difference across a membrane). At rest, this potential (***V***) is negative. So, the gradient across the membrane is both *chemical* and *electrical*. The plasma membrane would be totally impermeable to charged particles, except that it is riddled with **ion channels**. These channels are specialized, permitting only certain ions through when they are open. Each ion moves through open channels by diffusion until the **equilibrium potential** (**EP**) is achieved. The EP for any ion is

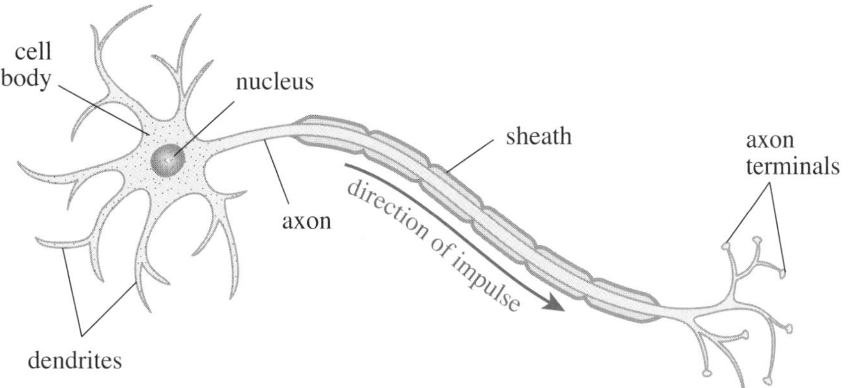

Figure 7.9.1 Diagram of typical motor neuron (NIH 2007)

the membrane potential where there is no net diffusion of the ion across the membrane. EP occurs because the chemical concentration and electrical forces are acting equally on that ion. Without these channels, the ions could not diffuse, and the cells could not function for signal transmittance. Some of the channels are **gated**, requiring specific stimuli to open the gates. These stimuli may be mechanical, electrical (voltage), or chemical (ligand), yielding three principle types of **gated channels**. Others are sometimes called **leak channels**, because they are essentially open all the time.

Resting nerve cells maintain an electrical difference across the membrane, and we refer to this electrical potential as the **resting potential** (**RP**). The value for this voltage varies, but we will assume that it is –65 millivolts (mV). RP is maintained primarily by regulating the sodium ion (Na^+) and the **potassium ion** (**K^+**) concentrations on either side of the membrane, with more Na^+ ions outside the cell and more K^+ ions inside the cell. There is some contribution to the RP from the diffusion of ions through leak channels, much more K^+ out than Na^+ in. Also, note that although there are negative ions (e.g., **chloride ions**, **Cl^-**) outside the cell, there are far more negative charges inside the cell. What largely reduces the positive charges inside the cell, however, is the exchange of *three* Na^+ ions for *two* K^+ ions that the **Na^+-K^+-ATPase pump** in the plasma membrane constantly performs. Both ions are moved against their concentration gradients by this pump, so much ATP energy produced in the cell is used to maintain this essential imbalance (Guyton and Hall 2011; Byrne 2012).

Initiating an Action Potential

Nerve cells transmit signals via a very rapid change in the membrane potential along the plasma membrane, called the **action potential** (**AP**; Figure 7.9.2). When resting, the neuron is **polarized**, but when ions are able to diffuse across the membrane, changes will occur in the membrane potential. For instance, if gated channels for Na^+ are opened, Na^+ ions flood into the nerve cell, temporarily making the inside more positive. We then say that the cell is **depolarized**. What would make these gates open? The Na^+ channels in the axon are **voltage-gated**. Therefore, something that

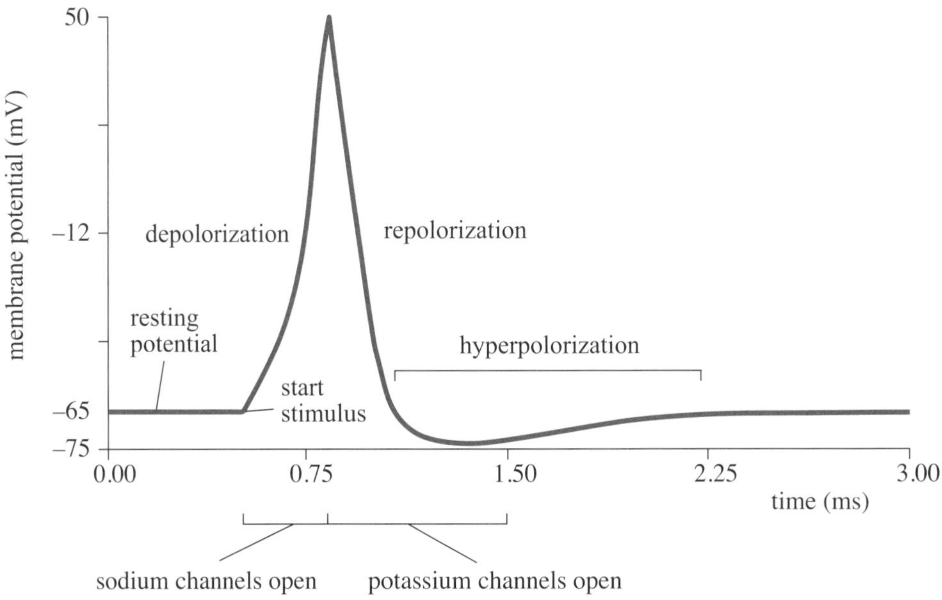

Figure 7.9.2 Action potential with membrane potential (mV) versus time (ms)

causes depolarization of the axon membrane, if great enough, could change the voltage enough to cause these gates to open.

An AP follows the arrival of a signal at a dendrite or soma of the neuron. These signals, for the most part, are chemical (although there are mechanically and thermally gated channels, also). On the surface of the dendrite are **ligand-gated channels** that bind specifically to a chemical signal (**ligand**). Some chemical signals are **excitatory**, causing a localized depolarization; and some signals are **inhibitory**, causing a **hyperpolarization** (making the membrane potential even more negative). How might this happen?

Signals are released from terminal buttons from other neurons that synapse with the dendrites/soma of the motor neuron. There will be hundreds or thousands of these synapses, both excitatory and inhibitory. Let's assume that our motor neuron is receiving excitatory chemical signals, and these signals are binding to ligand-gated channels, which open and allow Na^+s into the cell body. This influx changes the membrane potential locally (depolarizes), but the depolarization along the membrane decreases as the depolarization moves away from the site of stimulation. The potential spreads passively, because there are fewer or no voltage-gated Na^+ channels or the voltage threshold is higher than in the axon. So, the size of the stimulus is proportional to the number of gated channels opened. This proportionality results in a graded potential that can be summated spatially (lots of small stimuli from different sites) or temporally (high frequency of stimulus) near the junction of the cell body and the axon. If this summation results in depolarization exceeding the thresholds for the voltage-gated channels in this area, an action potential commences.

Inhibitory neurotransmitters open different channels—Cl^- or K^+. Either the Cl^-s enter, making the potential more negative, or the K^+s exit the cell, with a similar effect. In either case, the membrane becomes hyperpolarized, making it less likely that another AP will begin. In the end, whether or not an AP occurs depends on the ratio of excitatory and inhibitory stimuli.

If the membrane in the initiating area is depolarized sufficiently (to about –55 mV in our example), an AP will be initiated. The initiation event opens voltage-gated channels in this area, and sodium ions diffuse through the channels into the cytoplasm of the axon. This depolarizes the membrane, which in turn opens other voltage-gated Na^+ channels in the adjacent area. The open channels are now **activated**. This process is perpetuated down the entire axon in an all-or-none response. Although the sodium ions spread out in all directions, the AP is directional and generally proceeds from the initiating zone to the terminal buttons of the axon. The reason for this will be explained shortly.

While the sodium channels are activated to their maximum levels, the voltage-gated potassium channels remain closed. Maximum depolarization occurs with this influx of positive ions, and the membrane depolarizes to about +50 mV in our example. Then, after about a millisecond, the sodium gates close and become **inactive**. About this time, voltage-gated potassium channels open, and potassium ions diffuse out of the cell, helping to make the internal potential to become more negative. This process is called **repolarization**, and the membrane may actually become hyperpolarized with this efflux. At this time, the potassium channels begin to close, and the Na^+-K^+-ATPase pump exchanges enough sodium and potassium ions to restore the membrane potential to its normal resting state. With the reestablishment of the RP, the sodium channels return to their resting (closed) state and are receptive to depolarization signals.

During an action potential, when all the sodium gates are open, an ensuing stimulus, no matter how strong, cannot initiate another AP. This **absolute refractory period** helps to ensure the unidirectionality of the impulse. Because it cannot respond to depolarization of gated channels in front of it, the impulse does not flow back up the axon but moves quickly toward the terminal end of the axon. During the time that the sodium channels become inactivated and the potassium channels are beginning to close, the membrane enters a **relative refractory period**. An action potential can be initiated, if a sufficiently strong stimulus is applied. A larger stimulus is needed to counteract the hyperpolarization following the efflux of K^+. As the repolarization succeeds, the stimulus needed to initiate an AP decreases. In this way, the neuron can vary the rate of impulse conduction in response to different strengths of stimuli (Guyton and Hall 2011; Byrne 2012).

Hodgkin and Huxley Model

Before continuing, please read the sections on "Current and Potential," "Resistance," and "Capacitance" from Module 7.8, "Electrical Circuits—A Complete Story."

In 1952, Alan Lloyd Hodgkin and Andrew Huxley published the following model of the propagation of action potentials in a squid giant axon:

$$dV/dt = (\underbrace{I}_{\text{applied current}} - \underbrace{g_K n^4 (V - V_K)}_{\text{K current}} - \underbrace{g_{Na} m^3 h (V - V_{Na})}_{\text{Na current}} - \underbrace{g_L (V - V_L)}_{\text{leakage current}})/C_M$$

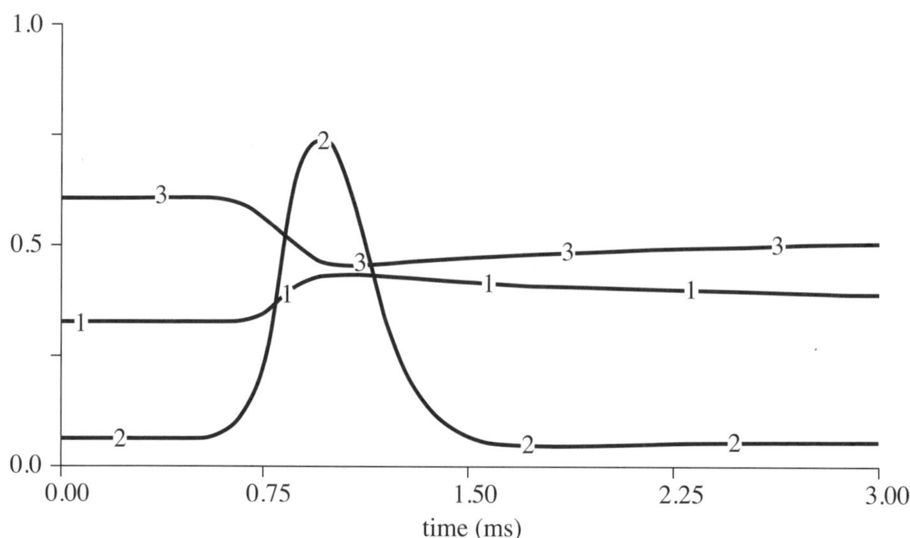

Figure 7.9.3 Graphs for n (1), m (2), and h (3) versus time

where dV/dt is the rate of change of the action potential, V; I is a current applied to a patch of the axon; $g_K n^4(V - V_K) = I_K$ is the potassium current; $g_{Na} m^3 h(V - V_{Na}) = I_{Na}$ is the sodium current; $g_L(V - V_L) = I_L$ is the leakage current; and C_M is the capacitance. Table 7.9.1 gives the meanings of each symbol along with values for Projects 2 and 3. Hodgkin and Huxley (1952) used empirical modeling, which we cover later in the text, to determine functions $g_K n^4$ and $g_{Na} m^3 h$, for the coefficients of $(V - V_K)$ and $(V - V_{Na})$, respectively, that captured the trends of their experimental data. Using the values in Table 7.9.1, which are based on Ermentrout (1998), the graph for the membrane potential is in Figure 7.9.2, and the corresponding graphs for n, m, and h appear in Figure 7.9.3.

Projects

1. Suppose that a membrane is permeable to just one ion so that the membrane potential (V) is equal to the equilibrium potential (E_{ion}) for that ion. The Nernst equation provides a model for the ion's equilibrium potential:

$$E_{ion} = \frac{RT}{zF} \ln\left(\frac{[\text{ion}]_o}{[\text{ion}]_i}\right),$$

where E_{ion} is the equilibrium potential for the ion, R is the gas constant, T is the temperature in kelvin, z is the valence of the ion, and F is the number of faradays. At 25 °C, RT/F is 25; while at 6 °C, RT/F is 24 (Moore 2007). Plot initial membrane potential (E_{ion}) versus $[\text{ion}]_o$ for each of the following ions at 25 °C and describe the results:

a. Potassium (K^+), where $z = 1$, $[K]_o = 5$ mM, $[K]_i = 124$ mM

Table 7.9.1
Hodgkin-Huxley Model Symbols with Values for Project 2 (Ermentrout 1998)

Symbol	*Meaning (units)*	*Formula/Value*
C_M	capacitance (μF/cm^2)	0.1 μF/cm^2
I	applied current (nA)	15 nA
I_K	potassium channel current (nA)	$g_K n^4(V - V_K)$
I_L	leakage current (nA)	$g_L(V - V_L)$
I_{Na}	sodium channel current (nA)	$g_{Na}m^3h(V - V_{Na})$
V	action potential (mV)	Initially –65 mV
V_K	displacement from the equilibrium potential for K^+ (mV)	–77 mV
$[K^+]_i$	potassium ion concentration inside (mM/L)	150 mM/L
$[K]_o$	potassium ion concentration outside (mM/L)	5.5 mM/L
V_{Na}	displacement from the equilibrium potential for Na^+ (mV)	50 mV
$[Na^+]_i$	sodium ion concentration inside (mM/L)	15 mM/L
$[Na^+]_o$	sodium ion concentration outside (mM/L)	150 mM/L
V_L	displacement from the equilibrium potential for leakage (mV)	–54.4 mV
g_K	maximum K conductance (mS/cm^2)	36 mS/cm^2
g_{Na}	maximum Na conductance (mS/cm^2)	120 mS/cm^2
g_L	maximum leakage conductance (mS/cm^2)	0.3 mS/cm^2
n	potassium activation gating variable; probability of K gate being open	initially 0.317
dn/dt	rate of change of n (ms^{-1})	$\alpha_n(1 - n) - \beta_n n$
m	sodium activation gating variable; probability of Na gate being open	initially 0.05
dm/dt	rate of change of m (ms^{-1})	$\alpha_m(1 - m) - \beta_m m$
h	sodium inactivation gating variable; probability of Na gate being inactivated	initially 0.6
dh/dt	rate of change of h (ms^{-1})	$\alpha_h(1 - h) - \beta_h h$
α_n	opening rate constant (ms^{-1})	$\phi\left(0.01(V+10)/\left(\exp\left(\frac{V+10}{10}\right)-1\right)\right)$
α_m	opening rate constant (ms^{-1})	$\phi\left(0.01(V+25)/\left(\exp\left(\frac{V+25}{10}\right)-1\right)\right)$
α_h	opening rate constant (ms^{-1})	$\phi(0.07 \exp(V/20))$
β_n	closing rate constant (ms^{-1})	$\phi(0.125 \exp(V/80))$
β_m	closing rate constant (ms^{-1})	$\phi(4 \exp(V/18))$
β_h	closing rate constant (ms^{-1})	$\phi\left(1/\left(\exp\left(\frac{V+30}{10}\right)+1\right)\right)$
T	temperature (°C)	6.3 °C
ϕ	factor for temperature correction	$3^{(T-6.3)/10}$, 1 in projects

b. Sodium (Na^+), where $z = 1$, $[Na]_o = 140$ mM, $[Na]_i = 14$ mM
c. Calcium (Ca^{++}), where $z = 2$, $[Ca]_o = 2$ mM, $[Ca]_i = 2.4 \times 10^{-4}$ mM

2. Develop a system dynamics model for the Hodgkin-Huxley model, plotting the action potential versus time for at least 3 ms. Use the Runga-Kutta 4 integration technique if available and a small t, such as 0.001 ms. Starting at time 0.5 ms, apply a stimulus current of 15 mV for 0.5 ms.

In the model, have stocks (box variables) for V, $[Na^+]_i$, $[Na^+]_o$, $[K^+]_i$, $[K^+]_o$, n, m, and h. The potassium channel current flows from inside the cell (stock for $[K^+]_i$) to the outside (stock for $[K^+]_o$). Assume that only the potassium ions leak into the cell. Because initially V is –65 mV and V_K is –54.4 mV, $I_L < 0$, so we must multiply its formula by a unary minus to obtain the leakage flow from $[K^+]_o$ to $[K^+]_i$. Similarly, I_{Na} is negative, so for the flow from the stock for $[Na^+]_o$ to the one for $[Na^+]_i$, we multiply the formula by a unary minus. We can model the Na^+-K^+-ATPase pump with two flows, one that transports a constant, c, times 3 Na^+ ions to the exterior (from the stock for $[Na^+]_i$ to the stock for $[Na^+]_o$) and another that moves c * 2 K^+ ions inside, for a net change of c positive ions to the outside. This pump counteracts the effect of the leakage current, so that at rest the membrane potential is –65 mV. Thus, to maintain the resting potential until application of the stimulus current, c should be the initial value of I_L for the net change of charges to be zero. Note that the pump should operate only if the sodium concentration on the inside and the potassium concentration on the outside are both positive. The flow into V adds flows going into the neuron (sodium channel, potassium leaked and pumped, stimulus), subtracts those going to the outside (potassium channel, sodium pumped), and divides the result by the membrane capacitance.

Although the preceding flows are unidirectional, the ones into n, m, and h are bidirectional, because, as Figure 7.9.3 shows, the graphs of n, m, and h increase and decrease.

Start the voltage-sensitive sodium channel current (gates are opening) when the membrane potential becomes greater than or equal to $V_{Na} = -55$ mV. When the membrane potential grows to be greater than or equal to $V_{Na} = 50$ mV, stop this flow and start the potassium channel current. End this potassium channel current during hyperpolorization, immediately after the membrane potential achieves its minimum and starts increasing.

3. Refine Project 2 to account for the fact that potassium gates open and close much slower than those for sodium.

References

Byrne, John H., Project Developer. 2012. *Neuroscience Online*. http://neuroscience.uth.tmc.edu/ (accessed June 29, 2012)

Ermentrout, G. Bard. 1998. "Channeling with Bard, Comp Neuroscience." http://www.math.pitt.edu/~bard/classes/channel/channel.html (accessed June 30, 2012)

Guyton, Arthur C., and John E. Hall. 2011. *Textbook of Medical Physiology*, 12th ed. Philadephia, PA: Sanders Elsevier.

Hodgkin, A. L., and A. F. Huxley. 1952. "A Quantitative Description of Membrane Current and Its Application to Conduction and Excitation in Nerve." *J. Physiol.* 117:500–544.

Lee, Chuang-Chung (Justin). 2008. "Appendix C: Hodgkin-Huxley Equations," in "Investigation of Synaptic Plasticity as Memory Formation Mechanism and Pathological Amyloid Fibrillation Caused by β-amyloids Aggregation." Ph.D. Diss., MIT.

Moore, John W., and Anne E. Stuart. 2007. *Neurons in Action 2: Tutorials and Simulations using NEURON*. Sunderland, MA: Sinauer Assoc., Inc.

Namnezia. 2011. "A Neuroscience Field Guide: The Squid Giant Axon." http://scientopia.org/blogs/bridgeblog/2011/09/30/a-neuroscience-field-guide-the-squid-giant-axon/ (accessed June 29, 2012)

Nelson, Mark, and John Rinzel. 2003. "Chapter 4: The Hodgkin-Huxley Model," *The Book of Genesis, Exploring Realistic Neural Models with the GEneral NEural SImulation System*, James M. Bower and David Beeman (eds.). http://www.genesis-sim.org/GENESIS/bog/bog.html (accessed June 29, 2012)

NIH. 2007. Figure 9, "Information about Alcohol." http://www.ncbi.nlm.nih.gov/books/NBK20360/ (accessed June 29, 2012)

Wood, James B. 2012. "The Cephalopod Page." http://www.thecephalopodpage.org/ (accessed June 29, 2012)

MODULE 7.10

Feeding the Problem—Antibiotic Resistance

Prerequisite: Module 4.3, "Modeling the Spread of SARS—Containing Emerging Disease."

Introduction

Cephalosporins, a class of antibiotics first isolated in 1948, have been used clinically since 1964. Drug companies have derived newer versions of these antimicrobials, used to treat various human respiratory infections, including pneumonia, certain skin and urinary tract infections, and uncomplicated gonorrhea (FDA 2009; Lemke 2008). Unfortunately, bacteria have developed various ways to elude these drugs, so that physicians must try to find new drugs, which may have other, undesirable side effects. In January 2012, the U.S. Food and Drug Administration (FDA) banned certain cephalosporins, which doctors may use to treat bacterial infections in humans, for farm animals (FDA 2012; Gilbert 2012). So, why has the FDA taken this step?

Antibiotics have been very effective in fighting infections in other animals, as well as in human beings. Not only employed to treat diseases in farm animals, the drugs have also been used for years to prevent disease or to promote growth. The use of antibiotic growth promoters (AGPs)—antibiotics added to animal food and water to promote growth—has now come under serious scrutiny. Arguably, these AGPs have been at least partially responsible for the incredible increases in world livestock production over the past 40 years. Between 1960 and 2010, the Food and Agriculture Organization of the United Nations estimates that pig production doubled and chicken production almost quadrupled (FAOSTAT 2010). Although there may be some debate about how AGPs promote growth, there is no doubt that they work, which farmers like (Ministry 2011; Turndige 2004).

The FDA and others in public health began to worry in the 1970s about the possibility that the gut microbial flora of livestock might contaminate human consumers' food supply and that because of the animals' previous exposure to antibiotics, this transfer might include antimicrobial-resistant strains. At that time, the FDA tried to ban the use of some antimicrobials in animal feed and water, but under pressure from agricultural interests, Congress passed resolutions against such a ban, and the FDA backed off. Since then, much research has examined the possibility of transfer of resistant microbes to human beings through consumption or handling (Harris 2012).

According to researchers, zoonotic, food-borne pathogens carrying resistance genes have already been found for several bacterial genera, including *Salmonella*, toxic strains of *E. coli*, *Campylobacter*, and *Listeria* (White et al. 2002). Even some nonpathogenic (**commensal**) gut bacteria can carry resistance genes into the human intestines (Tollefson and Karp 2004). If pathogens with such genes become established, they may then resist treatment by antibiotics. Whether pathogenic or commensal, the resistant bacteria may transfer their resistance genes to human microbes via mobile DNA elements, such as plasmids, transposons (Frankel 1994). Although there is some debate about how frequently this might cause serious infections in human beings, most public health officials are unwilling to risk it. Certainly, there is increased dissemination of antimicrobial resistance, and that is "bad news" for treating human and animal diseases. Infections caused by MDR strains of bacteria are more complicated (and often more expensive) to treat, cause longer-lasting illnesses with extended hospital stays, and more often lead to mortality (Tollefson and Karp).

So, these concerns may explain why the FDA has recently banned the use of antibiotics like cephalosporins from animal feed and water. Furthermore, as consumers have become more aware of the dangers of antibiotic resistance, pressure is placed on the food industry to reduce/eliminate the use of antibiotics in food animals. McDonald's Corporation, a major purchaser of chicken, beef, and pork, has made a policy to phase out the use of growth-promoting antibiotics. In fact, the company prohibits its suppliers from employing such antibiotics, if health professionals use the drugs to treat human disease (McDonald's Corporation 2003).

Projects

1. Develop a system dynamics model of the impact of animal antibiotic use on antibiotic resistance in human commensal bacteria. Assume that the population is constant with no births or deaths.

A susceptible human can become exposed to or infected with antibiotic resistant (AR) bacteria at a fairly low level. In some of these individuals, the bacteria colonize at a rate θ, but the number of AR bacteria is still small. However, because of medical antibiotic use (MAU), at a prescription rate of ρ, an exposed or a colonized person can become an amplified individual, carrying high loads of the bacteria and being very contagious. Amplified individuals can get better, becoming a colonized person, at a recolonization rate of φ. A susceptible human can become infected by exposure to an amplified person or, to a lesser degree, to a colonized person with contact rates β and η, respectively. Moreover, infection can occur through contaminated animal food products caused by agricultural antibiotic use (AAU) or from such bacteria that are otherwise in the environment, the background. In the latter two cases, the rates of change are proportional to the number of susceptibles with

Table 7.10.1
Parameter Symbols and Meanings for Project 1 (Smith et al. 2002)

Symbol	*Meaning*
θ	Colonization rate
ρ	Medical antibiotic use prescription rate
φ	Recolonization rate
β	Contact rate between susceptible and amplified
η	Contact rate between susceptible and colonized
λ	Exposure rate to contaminated animal food products
μ	Background exposure rate
α	Transient loss rate
σ	Colonization loss rate
γ	Amplified loss rate

proportionality constants λ and μ, respectively. We assume that AR bacteria in exposed, colonized, and amplified individuals are lost at rates α, σ, and γ, respectively. Table 7.10.1 summarizes the parameters.

a. Assume that transient bacteria populations in exposed humans last on the average 10 days, while colonized populations last about a year and amplified populations last about 100 days. Using these assumptions, calculate α, σ, and γ using units of "per day." Humans who become well from any of these categories are again susceptible. Write a differential equation for the rate of change of people becoming susceptible.

b. Suppose each day 0.1% of the susceptibles become exposed through contaminated animal food products and 0.0001% become exposed through other nonhuman-to-human (background) sources. Determine λ and μ. Suppose per day that 50% of the contacts are between susceptibles and amplified individuals result in the susceptible person becoming exposed, but 0.001% of the possible contacts between susceptibles and colonized cause exposure. Determine β and η. Write a differential equation for the rate of change of people becoming exposed.

c. Suppose each day 0.1% of the exposed humans have colonization occur and 0.3% of the amplified individuals change to being colonized. Determine θ and φ. Write a differential equation for the rate of change of people becoming colonized.

d. Suppose each day 0.3% of the exposed and 0.3% of the colonized humans become amplified because of medical antibiotic use. Determine ρ. Write a differential equation for the rate of change of people becoming amplified.

e. Complete the model and run the simulation for 9 years. Plot the prevalence of AR bacteria, or the number of exposed, colonized, and amplified individuals, versus time.

f. Run the simulation several times with a sequence of increasing medical antibiotic use prescription rates, ρ. Describe and discuss the results.

g. Run the simulation several times with a sequence of decreasing contaminated animal food products caused by agricultural antibiotic use, λ. Describe and discuss the results.

h. Using your model, support or refute the following statements and conclusions by Smith et al. (2002): "After AR is common in humans, infection control and prudent MAU are more likely to reduce the prevalence of AR in hospitals than eliminating AAU." "Restricting AAU is most effective when AR bacteria remain rare." "We conclude that agricultural use of antibiotics in new resistance classes should be delayed until the period of maximum medical utility has passed."

2. Develop a system dynamics model of the spread of antibiotic resistant bacteria in a hospital. Such bacteria might reside in respiratory passages, on the skin, or in digestive tracts. We suppose that two drugs, drug 1 and drug 2, are available to treat an infection by these bacteria; a strain of the bacteria exists that is resistant to drug 1; but no resistance to drug 2 exists. Another model assumption we make is that admitted patients do not have the drug resistant strain of the bacteria. Also, assume that the rate of patients entering and leaving the hospital each day is the same, so that the total population of the hospital is constant. Because of the wide use of antibiotics in hospitals, we assume that patients receive drug 1 or drug 2 at rates independent of whether they are infected with the bacteria under consideration or not.

For the model consider three interrelated systems: patients who are uncolonized by bacteria under consideration (X), patients who are colonized by a strain of the bacteria that are sensitive to treatment by drugs 1 and 2 (S), and patients who are colonized with a strain of the bacteria resistant to drug 1 but sensitive to drug 2 (R). The transmission rate from X to S is β per day, so that each day a fraction, β, of the possible interactions between patients in category X and those in category S results in patients moving from X to S. A fraction, c, of the X population is not susceptible to the antibiotic resistant strain. Thus, the transmission rate from X to R is $\beta(1 - c)$ per day, and each day a fraction, β, of the possible interactions between those in category R and a fraction, $(1 - c)$, of the patients in category X results in patients moving from X to R. Assume that patients cannot change directly from category S to category R, or vice versa (Lipsitch et al. 2000).

Table 7.10.2 lists symbols and meanings for project parameters. Researches have determined several ranges for various parameters. The proportion, m, of entering patients in category S is between 20% and 100%; the remainder of admissions falls into category X. The average duration of a

Table 7.10.2
Parameter Symbols and Meanings for Project 2 (Lipsitch et al. 2000)

Symbol	*Meaning*
m	Proportion of entering patients in category S
μ	Rate (per day) of patients entering and leaving the hospital
γ	Rate (per day) at which patients spontaneously (without drug treatment) become clear of bacterial colonization
τ_1	Rate (per day) at which patients are treated with drug 1
τ_2	Rate (per day) at which patients are treated with drug 2
β	Base transmission rate (per day) from S to X
c	Fitness "cost" of resistance to drug 1; proportional reduction in transmission rate from X to R

hospital stay is 7 to 20 days, while the average time from hospital admission to colonization is 6 to 50 days. The mean time from hospital admission or colonization by the bacteria until a spontaneous clearance without drug treatment is 30 to 60 days (Lipsitch et al. 2000).

For Parts a–f, plot the frequencies of X, S, *and* R *versus time for 60 days. For Parts a–g employ the following parameters except as noted*: $\beta = 1.0/day$, $c = 0.05$, $\mu = 1/(10\ days)$, $\gamma = 1/(30\ days)$, $m = 0.75$, $\tau_1 = 1/(5\ days)$, *and* $\tau_2 = 1/(10\ days)$.

a. Produce plots of X, S, and R versus time for within-hospital transmission rates, β, varying from 0 to 0.5 per day. Discuss the impact on the prevalence of sensitive and of resistant bacteria. Based on your findings, should transmission reduction interventions, such as better hand washing and quarantines, have a greater effect on sensitive or resistant carriage?

b. Produce plots for levels of treatment with drug 1, τ_1, varying from 0 to 0.5 per day. Based on your results, discuss the impact of increased levels of drug 1 treatment on the prevalence of bacteria resistant to and sensitive to the drug.

c. Repeat Part b for levels of treatment with drug 2, τ_2, varying from 0 to 0.5 per day.

d. Produce plots of X, S, and R versus time for $\tau_1 = 0.05$, 0.2, and 0.4 per day with $\beta = 0.05$, 0.2, and 0.4 per day. Thus, you will generate nine plots. For which combinations of parameter values does the resistant bacteria persist in the hospital? Discuss the implications including the related intervention.

e. Repeat Part d for $\tau_2 = 0.05$, 0.2, and 0.4 per day with $\beta = 0.05$, 0.2, and 0.4 per day.

f. Repeat Part d for $\mu = 0.05$, 0.15, and 0.2 per day with $\beta = 0.05$, 0.2, and 0.4 per day.

g. Define **the prevalence of carriage of bacteria resistant to drug 1** as $\rho = R/S$ (Lipsitch et al. 2000). Describe how the change of ρ is a way to measure the effectiveness of an intervention. Add the calculation of ρ to your model; and with $c = 0$ and $\tau_2 = 0$, generate a plot of ρ versus time for 60 days. Show separately the effect on ρ of reducing β by 50%; reducing τ_1 by 50%; reducing τ_1 by 100%; replacing 50% of use of drug 1 by use of drug 2; and replacing 100% of use of drug 1 by use of drug 2. Discuss the results and the most effective approach.

h. Consider the case where resistant and sensitive bacteria are equally transmissible ($c = 0$) and all patients enter the hospital uncolonized. Determine the formula for the basic reproductive number R_0. Calculate R_0 for three situations, or sets of parameters, and discuss the results.

3. Refine the model for Project 2 to track patients by their histories of treatment by drug 2. Thus, each compartment from Project 2 (S, X, and R) is separated into two compartments, those who have not been treated with drug 2 (S_U, X_U, and R_U) and those who have (S_T, X_T, and R_T). Once treated by drug 2, an appropriate proportion of untreated individuals move to X_T (Lipsitch et al. 2000).

a–h. Repeat the corresponding part from Project 2.

i. Define the **prevalence odds ratio** as the probability of an individual treated with drug 2 carrying resistant bacteria over the probability of an individual who has not been so treated carrying the resistant bacteria (Lipsitch et al. 2000):

$$OR = \frac{R^T}{S^T + R^T} \bigg/ \frac{R^U}{S^U + R^U}$$

If the ratio is greater than 1, is a treated individual more or less likely to carry bacteria resistant to drug 1 than a person who has not been so treated? Add the calculation of *OR* to your model. If the use of drug 2 increases with a corresponding reduction in use of drug 1, what happens to the prevalence of resistance to drug 1 in the hospital and to an individual's odds of carrying the resistant bacteria? Discuss your results.

References

FAOSTAT (Food and Agricultural Organization of the United States). 2010. http://faostat.fao.org/ (accessed June 19, 2012)

FDA (U.S. Food and Drug Administration). 2009. "Use and Importance of Cephalosporins in Human Medicine." http://www.fda.gov/AdvisoryCommittees/CommitteesMeetingMaterials/VeterinaryMedicineAdvisoryCommittee/ucm129875.htm (accessed June 19, 2012)

———. 2012. "Cephalosporin Order of Prohibition Questions and Answers." http://www.fda.gov/AnimalVeterinary/NewsEvents/CVMUpdates/ucm054434.htm (accessed June 19, 2012)

Frankel, W. L., W. Zhang, A. Singh, D. M. Klurfeld, S. Don, T. Sakata, I. Modlin, and J. L. Rombeau. 1994. "Mediation of the Trophic Effects of Short-Chain Fatty Acids on the Rat Jejunum and Colon." *Gastroenterology*, 106: 375–380.

Gilbert, Natasha. 2012. "Rules Tighten on Use of Antibiotics on Farms." *Nature*, 481: 125.

Harris, Gardiner. 2012. "U.S. Tightens Rules on Antibiotics Use for Livestock," *New York Times*, April 11.

Lemke, Thomas. 2008. *Foye's Principles of Medicinal Chemistry*. Philadelphia: Lippincott Williams & Wilkins, pp. 1028–1082.

Lipsitch, M., C. T. Bergstrom, and B. R. Levin. 2000. "The Epidemiology of Antibiotic Resistance in Hospitals: Paradoxes and Prescriptions." *PNAS* 97: 1938–1943.

McDonald's Corporation. 2003. "McDonald's Global Policy on Antibiotic Use in Food Animals." http://www.aboutmcdonalds.com/content/dam/AboutMcDonalds/Sustainability/Sustainability%20Library/antibiotics_policy.pdf (accessed June 19, 2012)

Ministry of Agriculture, Food, and Rural Affairs, Canada. 2011. "Antibiotic Use For Growth Improvement—Controversy And Resolution." http://www.omafra.gov.on.ca/english/livestock/animalcare/amr/facts/05-041.htm (accessed June 19, 2012)

Smith D. L., A. D. Harris, J. A. Johnson, E. K. Silbergeld, and J. G. Morris, Jr. 2002.

"Animal Antibiotic Use Has an Early but Important Impact on the Emergence of Antibiotic Resistance in Human Commensal Bacteria." *PNAS*, 99: 6434–6439.

Tollefson, L., and B. E. Karp. 2004. "Human Health Impact from Antimicrobial Use in Food Animals." *Médecine et Maladies Infectieuses*, 34: 514–521.

Turndige, J. 2004. "Antibiotic Use in Animals—Prejudices, Perceptions and Realities." *J. of Antimicrobial Chemotherapy*, 53: 26–27.

White, David G., Shaohua Zhao, Shabbir Simjee, David D. Wagner, and Patrick F. McDermott. 2002. "Antimicrobial Resistance of Foodborne Pathogens." *Microbes and Infection* 4(4): 405–412.

MODULE 7.11

Fueling Our Cells—Carbohydrate Metabolism

Prerequisite: Module 4.5, "Enzyme Kinetics—A Model of Control."

Glycolysis

Carbohydrates are organic molecules composed of the elements carbon (C), hydrogen (H), and oxygen (O). Many organic molecules include these elements, but what sets carbohydrates apart is the general ratio of these elements—usually 1:2:1 (C:H:O). Carbohydrates serve as primary sources of energy for most living organisms. In fact, certain cells of many tissues, including the brain, prefer to use carbohydrates to any other energy source.

An animal may consume carbohydrates as simple molecules (sugars) or as long chains of sugars, such as starch. Once consumed, animals, using enzymes of the digestive tract, break down the larger carbohydrates to produce even more sugars. By the time these sugars (**monosaccharides**, mostly **glucose**) reach the small intestine, they are small enough to be absorbed into the blood stream and distributed to the liver and other organs of the body, where they are taken up by the cells. Depending on the cell type and the metabolic state of the animal's body, these monosaccharides may be converted into other organic constituents of the cell (e.g., fatty acids, amino acids, animal starch (**glycogen**)), or they may be broken down to produce energy.

Energy from carbohydrates and other organic food sources is obtained through gradual chemical degradation, or **oxidation**. Let's define oxidation as the removal of electrons or hydrogens from a molecule. In a cell, enzymes, called **dehydrogenases**, catalyze oxidation reactions; and the molecules within a pathway that provide the electrons or hydrogens we call the **substrates.**

Most cellular monosaccharides are **glucose**, which has 6 carbons, 12 hydrogens, and 6 oxygens. The more highly **reduced** (with lots of hydrogens) a molecule is, the better source of energy it is for the cell; and we consider glucose to be highly reduced. So in this section, we are examining the cell's sequential oxidation of glucose for energy. The complete oxidation of glucose yields carbon dioxide, water, and energy (**ATP**).

If we oxidize glucose by combustion, the reaction yields 686 kcal/mol, all released as heat. Combustion in the cell is not very practical, so the cell oxidizes glucose step by step, ensuring that the cell does not burn up and that some of the energy is in a form of energy the cell can use. We term the energy made available for cellular work **free energy**, and the cell is able to garner 275 kcal/mol of free energy from the 686 kcal available in one glucose molecule.

Glucose oxidation begins in the cytoplasm of the cell, and the initial sequence of chemical reactions is collectively referred to as **glycolysis**. The first few reactions of glycolysis essentially "prime" the molecule and require a total of 2 molecules of **ATP** (adenosine triphosphate). ATP is referred to as the "universal coupling agent," because its synthesis from ADP + P_i (inorganic phosphate) "captures" some of the free energy from oxidation (ADP + P_i + energy → ATP). This captured energy, released through hydrolysis of ATP (ATP → ADP + P_i + energy), can then be used to power other cellular reactions.

At the end of the priming steps, the enzymes have converted glucose to another 6-carbon molecule, called ***fructose 1,6-bisphosphate*** (**F1,6BP**). F1,6BP is then processed through the oxidizing steps of glycolysis to yield 2 pyruvates, 4 ATPs, and 2 reduced coenzyme NADHs (from NAD^+ + H). **Pyruvates** are 3-carbon products of glycolysis, resulting from the splitting and oxidation of glucose. Note that the cell has only netted 2 ATPs from glycolysis. Although 4 were synthesized, 2 were consumed in priming glucose for oxidation.

You probably remember from basic chemistry that when something is oxidized, something else is reduced. **Coenzymes** are organic cofactors that associate with enzymes and help them catalyze. One of the most common coenzymes, often associated with dehydrogenases, is **NAD^+ (nicotinamide adenine dinucleotide)**. The enzyme removes two hydrogens (H = 1 electron and 1 proton) from the glycolytic substrate but adds 2 electrons and 1 proton from oxidation to NAD^+, converting NAD^+ to **NADH** (reduced coenzyme). These reduced coenzymes are particularly important for aerobic cells, because electron transport systems can reoxidize them to yield more ATP.

Recycling NAD^+s

With Project 1, we can find that increasing the numbers of NAD^+s available increases the ATP yield. For most cells, suddenly increasing the normal pool of cytoplasmic NAD^+s is not really feasible, so they employ another solution—they recycle. The NADHs shed their electrons (hydrogens). The resulting NAD^+s are reused to sustain glycolysis, allowing ATP production to continue. The recycling of coenzymes demands electron acceptors, which vary from organism to organism. Under anaerobic conditions, cells like your over-exercised muscle cells enzymatically remove the electrons (hydrogens) from the reduced coenzymes and return them to **pyruvate**, converting pyruvate to **lactate**. The enzyme, called **lactate dehydrogenase**, that catalyzes this reaction at the same time reoxidizes NADH to NAD^+. This process is referred to as **lactate fermentation** and is common to many types of cells.

Aerobic Respiration

It might have occurred to you that 2 net ATPs is not a tremendous amount of ATP produced from glucose. At the end of this pathway, without fermentation, we have 2 three-carbon molecules (pyruvates) and 2 molecules of NADH. In fact, the oxidation of glucose is quite unfinished, and many cells possess more elaborate pathways to complete the job. Aerobic cells, when supplied with adequate quantities of oxygen, transform pyruvate into CO_2 and H_2O. In doing so, they also generate a considerable amount of ATP. In a typical aerobic cell, pyruvate is transported into membrane-bound compartments, called **mitochondria**. These organelles contain sets of enzymes organized into a pathway often referred to as the **Krebs' Cycle**. In this pathway, the remaining electrons are removed from pyruvate and placed onto oxidized coenzymes. Very little ATP is produced directly in the Krebs' cycle. Therefore, the cell needs a way of getting the prospective energy found in all these reduced coenzymes, even those from the cytoplasm.

Within the inner membrane of the mitochondrion are sets of electron carriers, arranged into precisely structured complexes. These complexes represent **electron transport systems**, which remove and pass along the electrons and protons from all these reduced coenzymes. During the passage of the electrons, protons are actually pumped into the space outside the inner membrane, which is not very permeable to protons. This process establishes an electrochemical gradient that represents a potent force. There are other protein complexes in the membrane that form channels in that membrane for protons. Attached to these channels are **ATP synthase** particles. When protons pass through the channels, they interact with these particles in such a way that they generate enough conformational changes to promote ATP synthesis. Thus, the proton gradient set up by the electron transport system has essentially powered the transfer of energy from glucose into a high-energy bond of ATP. The electrons are eventually passed on to oxygen, which serves as the terminal electron acceptor for aerobic cells.

ATP synthesis in the cytoplasm and in the Krebs' cycle occurs by what is called **substrate-level phosphorylation**. In this type of phosphorylation, the P (phosphate) used to make ATP has come from an organic compound that has a higher energy level than does ATP. From one molecule of glucose, an aerobic cell produces 2 net molecules of ATP during glycolysis and 2 net molecules during the Krebs' Cycle.

Oxidative phosphorylation involves the production of ATP using the proton gradient established by the electron transport system. Pairs of electrons from NADH help to establish sufficient electrochemical gradient to power the synthesis of 3 ATPs. Because the electrons from the other major coenzyme (**$FADH_2$**) enter the electron transport system at a lower level, they establish enough gradient to power the synthesis of only 2 ATPs.

Projects

1. Design a simple model of glycolysis that includes the following components: glucose, NADH, ATP, pyruvate.

a. Use the following values for your model, and test to see how many ATPs, NADH's, and pyruvates the model produces.

Materials	*Number of Molecules*
glucose	2000
NAD^+	500
ADP	1000

b. What is the maximum number of ATPs obtained with these starting values in your model? What is the maximum number of pyruvates?

c. How many ATPs and pyruvates should glycolysis be able to produce from 2000 molecules of glucose? Why does your model yield less? Suggest ways to increase ATP production here.

2. Extend the model you developed in Project 1 to include recycling NAD^+s.

3. Extend the model you developed in Project 1 to include **aerobic respiration**, which uses oxygen.

References

Farabee, M. J. 2010. "Cellular Metabolism and Fermentation." http://www.emc.maricopa.edu/faculty/farabee/biobk/biobookglyc.html (accessed December 21, 2012)

Hardin, Jeff, Gregory Bertoni, and Lewis J. Kleinsmith, 2012. *Becker's World of the Cell*. 8th ed. San Francisco, CA: Benjamin Cummings, pp. 138–150.

King, Michael W. 2012. "Glycolysis." From *The Medical Biochemistry Page*, Indiana University School of Medicine. http://themedicalbiochemistrypage.org/glycolysis.php (accessed December 21, 2012)

MODULE 7.12

Mercury Pollution—Getting on Our Nerves

Prerequisites: Module 2.2, "Unconstrained Growth and Decay" for Project 1a and b; Module 2.4, "System Dynamics Tool: Tutorial 2" for Project 1c; Modules 2.5, "Drug Dosage," and 4.2, "Predator-Prey Model," for Projects 2–4.

Introduction

Many people think of rock bands when they hear the term *heavy metal*. However, the "real" **heavy metals** are highly toxic elements, which generally lack any known biological function. **Mercury** (**Hg**) is one of these elements. Mercury's distinctive chemical and physical characteristics have been put to use in numerous commercial, industrial, and medical applications—thermometers, barometers, batteries, antiseptics, pesticides, dental restorations, fluorescent lamps, and the like. This element is also a common trace component in fossil fuels. Through all these uses, mercury has been widely dispersed in various ecosystems; and mercury pollution has become a serious problem (NJTF 2002; Riley and Thomas 1999).

The most significant threat to human health is through the consumption of fish contaminated with methylmercury (NJTF 2002). **Methylmercury**, more available and more toxic than other chemical forms, is produced by the addition of a methyl group to mercury. Much of this **methylation** is accomplished by sulfate-reducing bacteria, which live at the sediment-water interface or among algal mats (Riley and Thomas 1999). The bacteria are consumed by plankton, which are consumed by larger planktonic or nektonic predators. At each level of this chain of consumption, the mercury accumulates in higher and higher concentrations, increasing by up to tenfold at each level (USGS 1996). Fish may accumulate up to one million times the concentration of mercury in their aquatic environment. This accumulation is an excellent example of **biomagnification** (NJTF 2002). For those who consume fish, the presence of this neurotoxin and possible carcinogen (Riley and Thomas) is of great importance and concern.

Mercury exists in the atmosphere primarily as **elemental Hg** (**Hg^0**) and as **oxidized Hg** (**Hg^{2+}**). The form Hg^0 is easily emitted into the atmosphere from the earth.

Moreover, this elemental mercury tends to remain in the atmosphere for a year or more. Hg^{2+} is far more reactive and soluble, dissolving easily in rainwater. Some will be adsorbed to particles and aerosols. **Particulate mercury** (**dry**) and **oxidized mercury** (**wet**) are deposited on various surfaces of the earth. A portion of the mercury in the atmosphere originates from naturally occurring emissions from the earth's surface, and the remainder is **anthropogenic**, or of human origin (Mason et al. 1994). The oxidized mercury entering the terrestrial/marine environments tends to form inorganic and organic complexes, with methylmercury being one of the most common organic forms. Because methylmercury is the most bioavailable and the most toxic, we concentrate on that form of mercury in some of our modeling. Additionally, we focus on the aquatic environment, where much of the mercury accumulates for transfer to human beings.

Projects

1. Table 7.12.1 has the pools and fluxes for the estimated "mercury budget" for the preindustrial earth, while Table 7.12.2 contains the current, estimated values.

a. Model the global mercury cycle for preindustrial and current times. Have a separate flow corresponding to each flux with a converter for its rate constant. Running the simulation for one simulation year using a Runge-Kutta 4 technique and a time step of 0.01 yr, determine proportionality constants to obtain the indicated fluxes.

b. In the atmosphere, elemental mercury is converted into oxidized mercury, which is either deposited dissolved in precipitation or adsorbed to particles (2%). Once in the terrestrial pool, oxidized mercury may be reduced to elemental mercury and reemitted, be combined to form a variety of complexes (e.g., HgS) in the soil, or be methylated. Similar possibilities exist for mercury deposited in marine environments. We do not know much about the rates of these chemical conversions. Modify your model to include these transformations.

c. We know that methylation of mercury is more likely to occur under certain conditions. Methylation is favored under low pH, low oxygen, high levels of organic matter, higher temperatures, and high sulfate concentrations. Modify your model to include some of these factors.

2. a. Mercury movement through the food chain is often given as an example of biomagnification. Because elimination is not easy, mercury tends to accumulate. Sulfate-reducing bacteria are thought to be responsible for much of the methylation of mercury. These organisms are consumed by plankton, which are consumed by insect larvae, which may be consumed by fish fry, which are consumed by minnows, which may be consumed by small fish, which are, in turn, eaten by still-bigger fish. These bigger fish are the ones usually consumed by human predators. Assume that the half-life of mercury in an organism is 50 da. Start with a mercury concentration of 0.6 ng/L (nanograms of mercury per liter of water) of mercury in the water and generate a model of biomagnification.

Table 7.12.1
Pools and Fluxes for the Estimated "Mercury Budget" for Preindustrial Earth (Seigel and Seigel 1997)

Pools	(× 10^3 *kg*)	*Fluxes*	(× 10^3 *kg/yr*)
Atmosphere	1600	terrestrial deposition	1000
Mixed layer (marine and terrestrial)	3600	marine deposition	600
		evasion[1]	600
		natural emission(terr)[2]	1000
		riverine flow[3]	60
		particulate removal[4]	60

[1] **Evasion**—elemental mercury entering the atmosphere from the ocean
[2] **Natural emission (terr)**—mercury from natural and anthropogenic sources transferred from the terrestrial pool to the atmosphere
[3] **Riverine flow**—mercury transfer from the terrestrial pool to the ocean through runoff of streams and rivers
[4] **Particulate removal**—particles containing mercury settling to deep ocean sediments that are essentially removed from active cycling

Table 7.12.2
Current Pools and Fluxes for the Estimated Mercury Budget (See the notes for Table 7.12.1; Seigel and Seigel 1997)

Pools	(× 10^3 *kg*)	*Fluxes*	(× 10^3 *kg/yr*)
Atmosphere	5000	terrestrial deposition	3000
Mixed layer (marine and terrestrial)	10800	marine deposition	2000
		evasion[1]	2000
		natural emission(terr)[2]	1000
		riverine flow[3]	200
		particulate removal[4]	200
		anthropogenic (total)	4000
		atmosphere	2000
		terrestrial deposition	2000

b. In some parts of the world, fish is a primary part of the diet and an important source of protein. Modify your model in Part a to calculate the accumulation of mercury in the bodies of adults with varying percentages of fish in their diets. Assume that all adults weigh 65 kg and that they eat the same amount of food (in kg).

c. Not all fish species accumulate the same amount of mercury. Table 7.12.3 contains maxima of mercury concentrations in fish species from a report by the Environmental Protection Agency (EPA) to Congress (EPA 1997a). Consider that all numbers are in mg/kg dry weight. Modify your model to predict the accumulation of mercury in human adults consuming diets of different amounts of different kinds of fish.

Table 7.12.3
Maxima of Mercury Concentrations in Fish Species (EPA 1997a)

Fish Species	*Dry Weight (mg/kg)*	*Fish Species*	*Dry Weight (mg/kg)*
carp	0.250	largemouth bass	1.369
brown trout	0.418	catfish	0.890
northern pike	0.531	walleye	1.383

Table 7.12.4
Body Mass Index (BMI) Categories for Adults (CDC 2011)

Category	*BMI*
Underweight	< 18.5
Normal weight	18.5–24.9
Overweight	25–29.9
Obese	≥ 30

d. Methylmercury is **lipophilic**, which means it likes fat and tends to accumulate there. Fat insulates our bodies, pads organs, and serves as energy storage depots. One might hypothesize that someone with more fat also tends to accumulate more mercury. Assuming this association, develop a model based on the **body mass index (BMI)**. BMI, based on a mathematical relationship between height and weight, is a commonly used method to determine the fat content of our bodies. To calculate BMI, divide weight in pounds by height in inches. Divide the results by height in inches again and then multiply by 703. The outcome may be evaluated using the common BMI categories for adults from Table 7.12.4.

e. The U.S. EPA publishes **reference doses (RfDs)** for methylmercury. This value represents the amount of methylmercury that may be ingested on a daily basis for a lifetime with no adverse effects on health. Methylmercury is rapidly and efficiently absorbed through the gastrointestinal tract. Moreover, methylmercury passes through the blood-brain and placental barriers. With a biological half-life in human beings of up to 80 da, we can acquire toxic amounts in small doses over a long time or through massive doses at one time. Many of the toxic effects are in the nervous system, and some of these are fatal. The RfD for methylmercury is 0.1 μg/kg body weight per day. Using a dose-conversion equation, this translates into a 1.1-μg methylmercury/kg body weight/day ingested by a 60-kg adult. Monitoring is usually done from blood or hair concentrations. Blood with 44 μg/(L of blood) or hair with 11 μg/(g of hair) corresponds to the RfD. Incorporate this information into the model you developed in Part c. How much fish is too much?

f. Modify your latest model to include the effects of mercury toxicity in Table 7.12.5.

3. In a food chain that includes bivalve mollusks (e.g., clams, mussels, oysters), these filter feeders take in many small bits of organic matter and small organisms. Some of these organisms include bacteria and plankton. Assume that

Table 7.12.5
Effects of Mercury Toxicity (Bakir et al. 1973)

Symptom	*Total Body Burden (mg)*
Abnormal skin sensations (paresthesia)	25–40
Difficulties in coordinated movement (ataxia)	55
Difficulty speaking (dysarthria)	90
Blindness	170
Death (LD_{50})	200

Table 7.12.6
Average mercury accumulation in prey (EPA 1997a)

Prey Organism	*Average Mercury Accumulation Wet Weight (mg/kg)*
shrimp	0.047
clam	0.023
crab	0.117
scallop	0.042

the half-life of mercury in an organism is 50 da. Model the accumulation of mercury in these animals. Table 7.12.6 presents the average mercury accumulation in prey (EPA 1997a).

4. In a diet for shorebirds, which includes lots of shellfish, model mercury accumulation in the birds from varying mixtures of the prey in Table 7.12.6 (EPA 1997a). Assume that the half-life of mercury in an organism is 50 da.

References

Bakir, F., S. F. Damluji, L. Amin-Zaki, M. Murtadha, A. Khalidi, N. Y. Al-Rawi, S. Tikriti et al. 1973. "Methylmercury Poisoning in Iraq." *Science*, 181(96): 230–241.

CDC (Center for Disease Control and Prevention). 2011. "Healthy Weight. About BMI for Adults." http://www.cdc.gov/healthyweight/assessing/bmi/adult_BMI/index.html (accessed December 20, 2012)

EPA (United States Environmental Protection Agency). 1997a. "Mercury Study Report to Congress. Vol. 1 Executive Summary."

———.1997b. "Mercury Study Report to Congress. Vol. VI. An Ecological Assessment for Anthropogenic Mercury Emissions in the United States." http://www.epa.gov/ttn/atw/112nmerc/volume6.pdf (accessed December 19, 2012)

Mason, R.P., W. F. Fitzgerald, and F. M. M. Morel. 1994. "Biogeochemical Cycling of Elemental Mercury: Anthropogenic Influences." *Geochemica et Cosmochimica Acta*,, 58: 3191–3198.

NJTF (New Jersey Mercury Task Force). 2002. *New Jersey Mercury Task Force Report*. State of New Jersey. http://www.state.nj.us/dep/dsr/mercury_task_force.htm (accessed December 20, 2012)

Riley, D. M., and V. M. Thomas. 1999. "Mercury Pollution: Sources, Consequences, and Remedies." *PU/CEES Working Paper* No. 140. Princeton: Princeton University.

Seigel, A., and H. Seigel, eds. 1997. *Metal Ions and Biological Systems: Mercury and It Effects on Environment and Biology*, Vol. 34. New York: Marcel Dekker.

Selin, N. E. 2009. "Global Biogeochemical Cycling of Mercury: A Review." *Annu. Rev. Environ. Resour.*, 34: 43–63.

USGS (U.S. Geological Survey). 1996. Mercury Transfer Through an Everglades Aquatic Food Web. http://sofia.usgs.gov/projects/index.php?project_url=merc_foodweb (accessed December 20, 2012)

MODULE 7.13

Managing to Eat—What's the Catch?

Prerequisite: Module 4.2, "Predator-Prey Model."

Introduction

In 1970, fishing and aquaculture employed about 13 million people worldwide, producing 65 million tons of seafood. By 2010, the numbers had increased to 54.8 million workers, hauling in almost 148 million tons (FAO 2012). Most of these fishers are in Asia. With burgeoning populations (and accompanying demand), destruction of habitat, and improved technology, many fisheries face tremendous pressures. The United Nations Food and Agriculture Organization (FAO) claims that worldwide almost 70% of the marine fish species are overfished or nearly so (FAO 2005). The National Marine Fisheries Service indicates that about 21% of the fish species in U.S. waters are overfished, and another 14% are being fished at a rate that is unsustainable (NMFS 2011). If fisheries are to remain sustainable and profitable, proper stewardship is essential. Already, according to the World Wildlife Fund, the world fishing fleet is 2½ times bigger than necessary (Porter 1998). We need to implement management systems that ensure that a gainful harvest does not exceed nature's capacity to maintain the resource.

Like many other species, Alaskan halibut began facing tremendous stresses from the fishing industry during the 1970s and 1980s. Even fishers supported the catch limits that were imposed then. By 1995, the season ran for only 2 days. At that time, there was open entry, and the fishing season was little more than a contest where each boat attempted to gather as large a share as possible. With shortened access times, fishing crews braved long hours and dangerous weather conditions. They cut loose tangled long lines and left them to lure and kill fish—fish that would never be harvested. Consequently, this situation resulted in lost fish, lost equipment, lost boats, and lost lives. Adding insult to injury, because all participants brought in their catches at the same time, boats had to sell the fish in a glutted market to large processors at depressed prices. For consumers, the situation meant no real market for fresh fish, so they consumed only frozen fish (Hartley and Fina 2001; PBS 2002).

Today, commercial halibut fishing operations are working with a closed fishery. Participant fishers must own part of the total allowable catch, called an **individual fishing quota** (**IFQ**). IFQs are property that fishers actually buy and sell. This system has helped to replace the "derby fishery" that existed prior to the mid-1990s. The season is 8 months long. Fishers no longer have to risk themselves and their boats during treacherous conditions. Their major income is no longer dependent on a 1- or 2-day venture. Although the IFQs may have saved the halibut fishery in Alaska, fishers in other parts of the United States worry that the system favors people with more money and may lead to aggregation of quota shares into monopolies. On the other hand, most fishers have become better caretakers of this resource, if for no other reason than the IFQs make them "owners" (Hartley and Fina 2001; PBS 2002).

Economics Background

Business decisions frequently involve maximizing profit. Profit in the fishing industry depends on the **cost**, or expense, of fishing and the **revenue**, or income, from fish sales. However, for an industry such as fishing, conservation of the product, the fish, for the ecosystem and future profits should be an essential part of the decision-making process.

For a **quantity** of product (***q***), such as metric tons of fish, the **cost function *C*(*q*)** returns the total cost, or expense, of producing a quantity of q items, such as catching q metric tons of fish. Figure 7.13.1 presents a particular cost function, $C(q) = 0.01q^3 - 0.6q^2 + 13q + 35$. In this example from a very small company, the cost of producing 10 items is the corresponding value, \$115, on the vertical axis. (Instead of dollars, $C(q)$ could indicate cost in thousands of dollars, millions of dollars, etc.) As the quantity produced increases, so does the total cost of production. Starting at about $q = 30$, this cost rises rapidly, perhaps with the company requiring new machinery to keep pace with the rising production demands or having to pay overtime to workers. Even if the company does not manufacture any product, the initial value of this cost function is a **fixed cost** of $C(0) = \$35$. Perhaps a fixed cost represents rental for warehouse space or workers' wages, even when no production occurs.

> **Definitions** For a **quantity** of product (***q***), the **cost function *C*(*q*)** returns the total cost, or expense, of producing a quantity of q items. The **fixed cost** is $C(0)$.

Quick Review Question 1

Consider the cost function $C(q) = 2000 + 50q$ for a scientific equipment company to manufacture q number of barometers.

a. Give the cost for manufacturing 100 barometers.
b. Give the fixed cost.

While cost is the total money going out, the **revenue function *R*(*q*)** gives the total amount of money coming in, or income, from selling q items. Figure 7.13.2a pres-

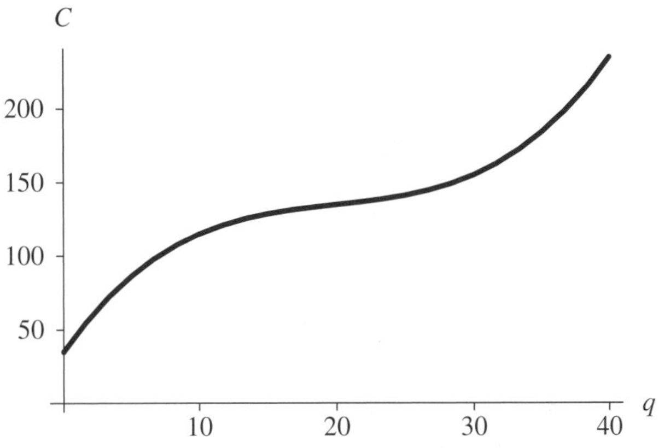

Figure 7.13.1 Example of a cost function $C(q)$

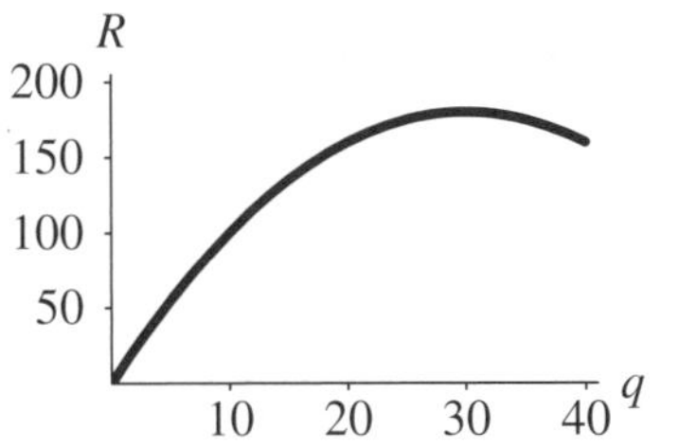

a. Revenue for price decreasing with oversupply

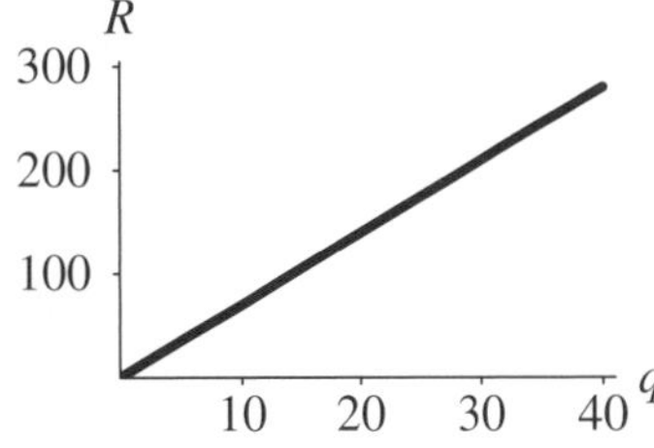

b. Revenue for constant price

Figure 7.13.2 Examples of a revenue function $R(q)$

ents an example of a revenue function, $R(q) = 12q - 0.2q^2$. Typically, with no product, no revenue exists, or the initial value is $R(0) = \$0$. In the figure's example, the revenue rises to a maximum of \$180 for selling $q = 30$ items, or $R(30) = \$180$. Afterward, revenue decreases. Perhaps oversupply results in a drop of the **price per item**, or charge for one item. Thus, the price per item, $\boldsymbol{p(q)}$, is a function of quantity. In general, the revenue for producing quantity q items is the product of the price per item $p(q)$ and the number of items, q, as follows:

$$\boldsymbol{R(q) = p(q) \times q}$$

If the price per item is constant regardless of the production quantity, the revenue function is linear, as in Figure 7.13.2b. In this example, the price of an item is $p(q) = p = \$7$; so that the revenue function is $R(q) = 7q$, a line with slope $p = 7$. Thus, the income for 10 items is $R(10) = 7 \times 10 = \$70$.

Definition The **revenue function** $\boldsymbol{R(q)}$ is the total amount of income from selling q items. The **price per item**, $\boldsymbol{p(q)}$, is the charge for one item when selling q items. Thus, the following equality holds:

$$\boldsymbol{R(q) = p(q) \times q}$$

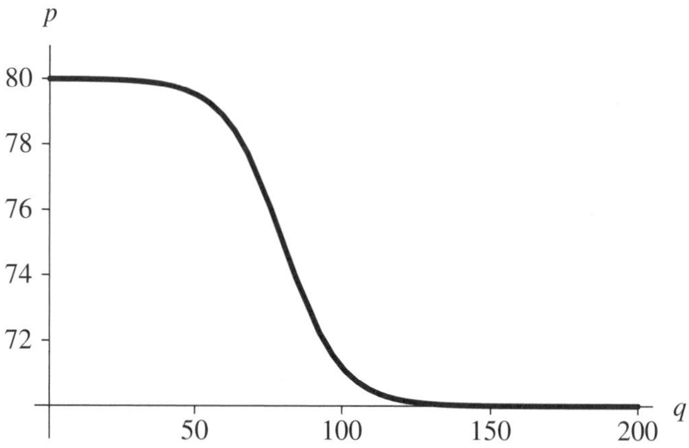

Figure 7.13.3 Graph of a price function $p(q)$

Quick Review Question 2

a. Suppose a barometer sells for \$73, regardless of the number sold. Give the revenue for 100 barometers sold.

b. Give revenue as a function of quantity q for a barometer that sells for \$73, regardless of the number sold.

c. Suppose the price of the barometer depends on quantity, according to the function whose graph is in Figure 7.13.3. To the nearest dollar, determine the revenue from selling 10 barometers.

d. To the nearest dollar, determine the revenue from selling 200 barometers.

Companies or fishers are ultimately concerned with their **profit**. The **profit function $\pi(q)$** is the profit, or the total gain, from producing and selling q items. (Economists usually employ i for the name of the profit function. This symbol is not the number $\pi \approx 3.14$.) Thus, the profit is the difference of the amount of money coming in and the amount of money going out, as indicated by the following equation:

$$\text{profit} = \text{revenue} - \text{cost}$$

or

$$\pi(q) = R(q) - C(q)$$

Thus, for revenue $R(q) = 12q - 0.2q^2$ and cost $C(q) = 0.01q^3 - 0.6q^2 + 13q + 35$, the profit function is $\pi(q) = (12q - 0.2q^2) - (0.01q^3 - 0.6q^2 + 13q + 35) = -0.01q^3 + 0.4q^2 - q - 35$. A company is working at a profit when revenue exceeds cost. Figure 7.13.4 displays in color shading the region where the company is profitable. Revenue and cost are equal, or $R(q) = C(q)$, where the graphs intersect at $q \approx 13.5$ and

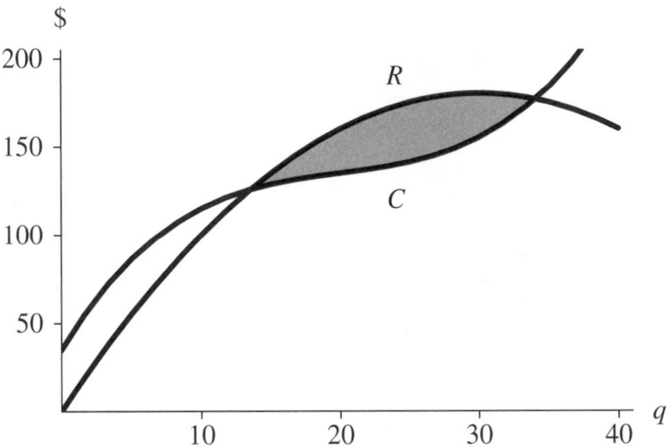

Figure 7.13.4 Region of profitability shown with color shading

34.0. For production and sales of fewer than 13 items or more than 34 items, cost is more than revenue, and the company is operating at a deficit.

Definition The **profit function $\pi(q)$** is the profit, or the total gain, from producing and selling q items. Thus, profit for selling q items is the difference in revenue and cost, so that the following equation holds:

$$\boldsymbol{\pi(q) = R(q) - C(q)}$$

Quick Review Question 3

For the cost function $C(q) = 2000 + 50q$ and the revenue function $R(q) = 73q$, determine the profit function.

Certainly, a company wishes to maximize profits. From calculus, we know that to maximize (or minimize) a profit function, we set the function's derivative equal to 0; solve for the independent variable, q; and determine if a maximum does indeed occur at that q. Thus, we have the following identities when a maximum occurs:

$$\pi'(q) = R'(q) - C'(q) = 0$$

or

$$\boldsymbol{R'(q) = C'(q)}$$

Economists call the derivative of the revenue function, or the instantaneous rate of change of revenue with respect to quantity, the **marginal revenue**. Similarly, the derivative of the cost function, or the instantaneous rate of change of cost with respect to quantity, is **marginal cost**. Thus, a maximum profit occurs at the quantity q

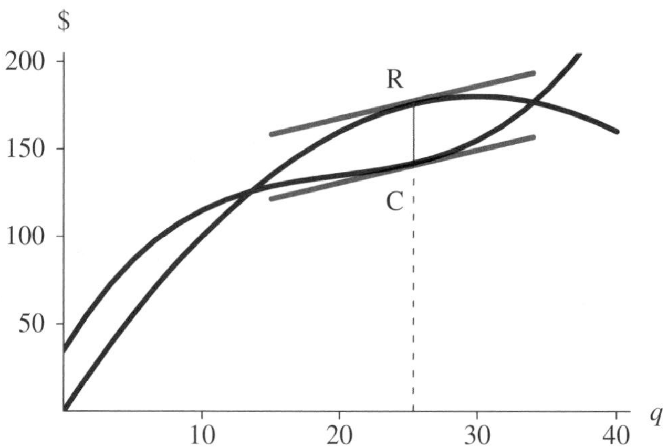

Figure 7.13.5 Maximum profit where tangent lines are parallel, $q \approx 25.35$

where marginal revenue equals marginal cost, $R'(q) = C'(q)$, and revenue exceeds cost. Because a derivative at a point is the slope of the tangent line to the curve at that point, the tangent lines to the curves are parallel at a quantity that yields a maximum profit. Figure 7.13.5 displays the revenue and cost functions with parallel tangent lines at $q \approx 25.35$. For that quantity, the profit has a maximum value of $\pi(25.35) \approx \$33.79$.

Definitions **Marginal revenue** is the derivative of the revenue function, $R'(q)$. **Marginal cost** is the derivative of the cost function, $C'(q)$.

Quick Review Question 4

For the sale of barometers, consider cost function $C(q) = 200 + 72q$ and revenue function $R(q) = 21q^2 - q^3$, where q is the quantity in thousands of barometers and $C(q)$ and $R(q)$ are in thousands of dollars. The graphs appear in Figure 7.13.6.

- **a.** Give the marginal cost.
- **b.** Give the marginal revenue.
- **c.** Determine the quantity of barometers sold for maximum profit.
- **d.** Give the profit function.
- **e.** Determine the maximum profit.

Gordon-Schaefer Fishery Production Function

In the 1950s, fishery scientist M. B. Schaeffer developed a model of biological yield, and H. Scott Gordon enhanced the model to include economics. A version of this **Gordon-Schaefer Fishery production curve** appears in Figure 7.13.7. The model assumes a quadratic yield function and linear cost-of-effort function. Effort involves such items as numbers of boats, traps, and days fishing. Initially, as effort increases,

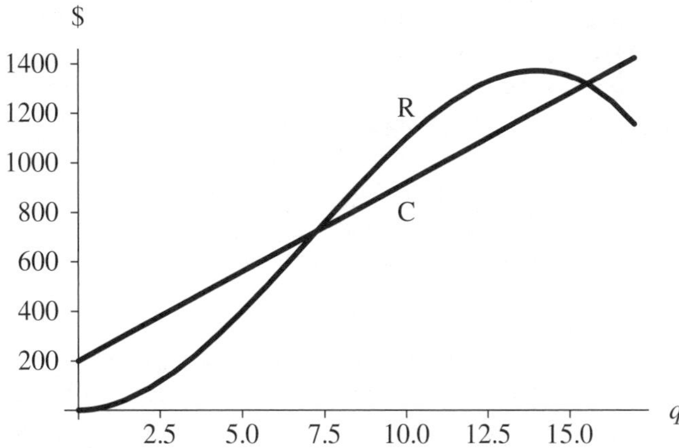

Figure 7.13.6 Cost function $C(q) = 200 + 72q$ and revenue function $R(q) = 21q^2 - q^3$

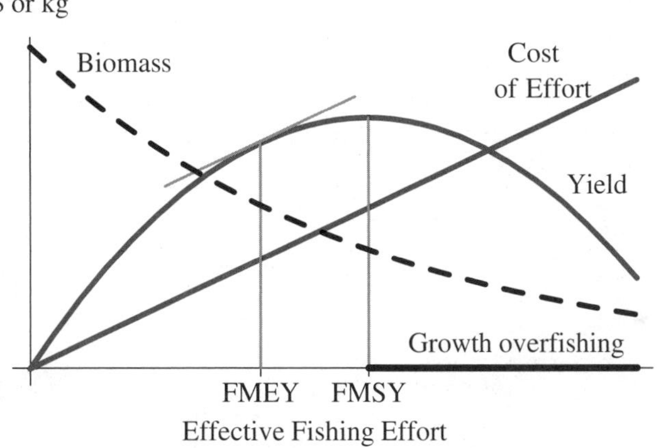

Figure 7.13.7 Gordon-Schaefer fishery production curve (Seijo et al. 1998)

so does yield in dollars. As discussed earlier, maximum profit occurs where the derivatives of these two functions are equal, or where the tangent line to the yield function is parallel to the cost-of-effort function. We call this quantity the **fishing maximum economic yield** (**FMEY**). At the high point of the yield curve, the cost of effort is the **fishing maximum sustainable yield** (**FMSY**). After this value, increased effort results in decreased yield (Seijo et al. 1998).

Projects

For all model development, use an appropriate system dynamics tool.

1. This project concerns the economics of fishing one species.

a. Consider the growth rate of a species of fish to be logistic when no fishing occurs (see Module 2.3, "Constrained Growth"). Let $E(t)$ be an effort function with respect to time t that is some fixed amount at the beginning and end of a 90-day season and maximum midseason. Suppose the rate of catching fish is proportional to the product of this effort and the population of fish. Determine a differential equation for the rate of change of the population of fish with respect to time (Danby 1997).

b. Let the price per unit catch be p and the cost per unit effort be c. Write an equation for profit per unit catch.

c. Develop a model involving the economics of fishing.

d. Form another version of this model with the cost per unit effort being periodic due to seasonal changes.

e. Form another version of this model with abrupt changes to the cost per unit effort. Discuss situations that could cause such changes.

f. Form another version of this model with demand and, thus, price, suddenly increasing. Discuss situations that could cause such increases.

2. Augment Project 1 to include IFQs.

3. Table 7.13.1 contains a list of the lobster commercial catch and effort from 1942 through 1979. Adjust Project 1 to accommodate these data.

Table 7.13.1

U.S. Commercial Lobster Catch and Effort, Territorial Sea and Fishery Conservation Zone (now called the U.S. Exclusive Economics Zone) Combined (1 metric ton (t) = 1.102311 tons) (NEFMC 1983)

Year	*Total Catch (t)*	*Total Effort (10^3 Traps)*	*Year*	*Total Catch (t)*	*Total Effort (10^3 Traps)*
1942	5,577	279	1961	12,700	978
1943	7,450	305	1962	13,378	1,003
1944	8,130	327	1963	13,731	964
1945	10,307	480	1964	14,043	1,043
1946	11,012	589	1965	13,719	1,163
1947	10,850	677	1966	13,399	1,096
1948	9,519	625	1967	12,131	1,099
1949	11,183	615	1968	14,769	1,168
1950	10,521	586	1969	15,327	1,333
1951	11,767	517	1970	15,489	1,851
1952	11,351	553	1971	15,279	1,905
1953	12,749	581	1972	14,626	1,858
1954	12,465	648	1973	13,152	2,307
1955	13,132	701	1974	12,945	2,303
1956	12,028	697	1975	13,698	2,334
1957	13,679	708	1976	14,293	2,305
1958	12,349	785	1977	14,434	2,302
1959	13,193	898	1978	15,653	2,302
1960	14,136	896	1979	16,870	2,255

4. This project concerns the economics of fishing two species.

a. Consider the growth rates of tuna and shark to follow Lotka-Volterra's predator-prey model when no fishing occurs (see Module 4.2, "Predator-Prey Model"). Let $E(t)$ be an effort function with respect to time t, and suppose the rate of catching each kind of animal is proportional to the product of this effort and that population. Suppose catching tuna involves equal effort to catching shark. Determine differential equations for the rates of change of the populations with respect to time (Danby 1997).

b. Let p_T and p_S be the price per unit catch of tuna and shark, respectively, with p_T being much greater than p_S. Let c be the cost per unit effort. Write an equation for profit per unit catch.

c. Assuming that the rate of change of effort is proportional to profit, determine a differential equation for the rate of change of E with respect to time.

d. Determine the equilibrium points.

e. Develop a model involving the economics of fishing tuna and shark.

f. Form another version of this model with the cost per unit effort being periodic due to seasonal changes.

g. Form another version of this model with abrupt changes to the cost per unit effort. Discuss situations that could cause such changes.

h. Form another version of this model with demand, and, thus, price, suddenly increasing. Discuss situations that could cause such increases.

4. Repeat Project 3 using two species of competing fish (see Module 4.1, "Competition").

Answers to Quick Review Questions

1. a. $\$7000 = C(50) = 2000 + 50(100)$
b. $\$2000 = C(0)$

2. a. $\$7300 = R(100) = 73(100)$
b. $R(q) = 73q$
c. $\$800 = (80)(10)$
d. $\$14,000 = (70)(200)$

3. $\pi(q) = 23q - 2000 = 73q - (2000 + 50q)$

4. a. $C'(q) = 72$
b. $R'(q) = 42q - 3q^2$
c. 12,000 because $72 = 42q - 3q^2$, or $-3q^2 + 42q - 72 = 0$, or $-3(q - 2)(q - 12) = 0$. A minimum occurs at $q = 2$, while a maximum occurs at $q = 12$.
d. $\pi(q) = 21q^2 - q^3 - (200 + 72q) = -200 - 72q + 21q^2 - q^3$
e. $\pi(12) = 232$, so that the profit is \$232,000.

References

ADFG (Alaska Department of Fish and Game). 2002. "Marine Protected Areas in Alaska: Recommendations for a Public Process." Report to the Alaska Board of Fisheries. The Alaska Department of Fish and Game Marine Protected Areas Task Force. http://www.adfg.alaska.gov/static/lands/protectedareas/pdfs/5j02-08_p1.pdf (accessed January 7, 2013)

Danby, J.M.A. 1997. *Computer Modeling: From Sports to Spaceflight . . . From Order to Chaos*. Richmond, VA: Willmann-Bell, p. 408.

FAO (Food and Agriculture Organization of the United Nations). 2005. "Fisheries and Aquaculture Topics. Trends in Capture Fisheries Development. Topics Fact Sheets." Text by Peter Manning. In: *FAO Fisheries and Aquaculture Department* [online]. Rome. Updated 27 May 2005. http://www.fao.org/fishery/topic/13838/en (accessed December 27, 2012)

———. 2012. "The State of the World Fisheries and Aquaculture." http://www.fao.org/docrep/016/i2727e/i2727e00.htm (accessed December 27, 2012)

Hartley, M., and M. Fina. 2001. "Allocation of Individual Vessel Quota in the Alaskan Pacific Halibut and Sablefish Fisheries. Case studies on the allocation of transferable quota rights in fisheries." FAO Fisheries Technical Paper No. 411. Rome, FAO, pp. 251–265.

NEFMC (New England Fisher Management Council). 1983. American Lobster Fishery Management Plan. Saugus, MA.

NMFS (National Marine Fisheries Service). 2011. "Status of Stocks 2011." Report to Congress on the Status of U.S. Fisheries. http://www.nmfs.noaa.gov/sfa/statusoffisheries/2011/RTC/2011_RTC_FactSheet.pdf

PBS (Public Broadcasting Service). 2002. "Management, Overfishing, & Alaskan Halibut." http://www.pbs.org/emptyoceans/eoen/halibut/ (accessed December 27, 2012)

Porter, Gareth. 1998. "Estimating Overcapacity in the Global Fishing Fleet." World Wildlife Fund.

Seijo, J.C., O. Defeo, and S. Salas. 1998. "Fisheries bioeconomics. Theory, modelling and management." *FAO Fisheries Technical Paper*. No.368. Rome, FAO, 108p.

MODULE 7.14

Control Issues: The Operon Model

Prerequisite: One of the following modules: 4.1, "Competition," 4.2, "Predator-Prey Model," 4.3, "Spread of SARS," or 4.5, "Enzyme Kinetics."

Proteins

Proteins are basic molecules of life, performing many critical functions. Some proteins are the fundamental, structural components of cells and tissue, while others (**enzymes**) are catalysts for chemical reactions. A simple protein is a linear polymer or chain of **amino acids**. Table 7.14.1 lists the 20 amino acids common to proteins along with their one-letter and three-letter codes. Each amino acid contains a central carbon (C), which bonds with 4 chemical groups—an **amino group** (NH_3^+), a **carboxyl group** (COO^-), a hydrogen (H), and a variable side-chain (**R-group**; Figure 7.14.1). The R-group determines the chemical nature (acidic, nonpolar, etc.) of each amino acid in the chain. Chains of amino acids are linked by **peptide bonds**, which form through the interaction of an amino group of one amino acid with the carboxyl group of another (Figure 7.14.2). This interaction results in condensation, or release of water. Because one end (**N-terminal**) of a protein has a free amino group and the other (**C-terminal**) has a free carboxyl group, we can assign an orientation to the chain and list the amino acids from the "beginning" (N-terminal) of the chain to the "end" (C-terminal).

Nucleic Acids

In the cell, the nucleic acid **DNA** (**deoxyribonucleic acid**) contains the encoded information for the manufacture of all the proteins a cell needs. However, DNA does not oversee protein synthesis directly but acts through an intermediary nucleic acid,

Table 7.14.1
The 20 Commonly Occurring Amino Acids Along with Their One-Letter and Three-Letter Codes. (Note: B is used when one cannot distinguish between D and N because of amino acid analytical processing. Similarly, Z is used when it is ambiguous whether the amino acid is E or Q. X represents an unknown or nonstandard amino acid.)

One-Letter Code	*Three-Letter Code*	*Name*
A	Ala	Alanine
R	Arg	Arginine
N	Asn	Asparagine
D	Asp	Aspartic acid
C	Cys	Cysteine
Q	Gln	Glutamine
E	Glu	Glutamic acid
G	Gly	Glycine
H	His	Histidine
I	Ile	Isoleucine
L	Leu	Leucine
K	Lys	Lysine
M	Met	Methionine
F	Phe	Phenylalanine
P	Pro	Proline
S	Ser	Serine
T	Thr	Threonine
W	Trp	Tryptophan
Y	Tyr	Tyrosine
V	Val	Valine

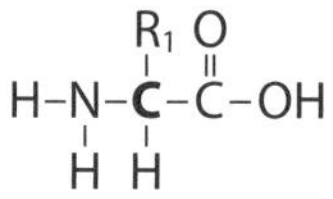

```
    R1 O
    |  ||
H-N-C-C-OH
  | |
  H H
```

Figure 7.14.1 Structure of an amino acid

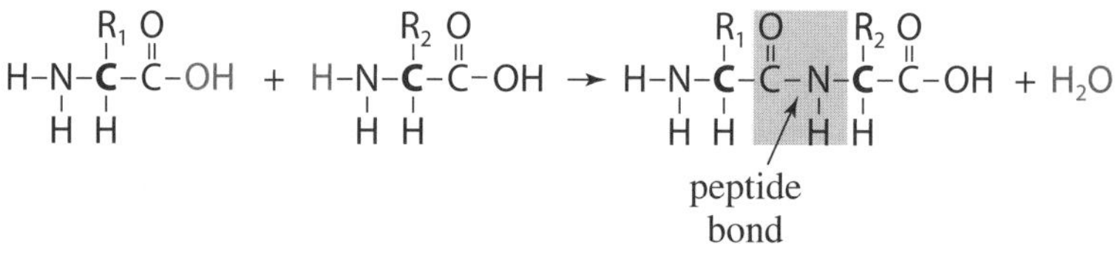

```
    R1 O           R2 O             R1 O   R2 O
    |  ||          |  ||            |  ||  |  ||
H-N-C-C-OH  +  H-N-C-C-OH  →  H-N-C-C-N-C-C-OH  +  H2O
  | |            | |            | |   | |
  H H            H H            H H   H H
                                   peptide
                                    bond
```

Figure 7.14.2 Formation of peptide bond.

RNA (**ribonucleic acid**). The RNA sequences subsequently specify the amino acid order of proteins. Both DNA and RNA are polymers, or long chains, of molecules called **nucleotides**. A nucleotide is a compound molecule made up of a sugar (either **deoxyribose** for DNA or **ribose** for RNA), a phosphate, and a nitrogen base (**adenine** (A), **guanine** (G), **cytosine** (C), and **thymine** (T) in DNA or **uracil (U)** in

deoxyadenosine-5′monophosphate adenosine-5′monophosphate

Figure 7.14.3 Deoxyribo- and ribonucleotide structures.

RNA). A and G are **purines**, while C, T, and U are **pyrimidines** (Figure 7.14.3). DNA is composed of a double strand of nucleotides, whereas RNA is composed of a single strand. Each nonterminal nucleotide is linked to its neighbors via phosphodiester linkages (Figure 7.14.4). Each phosphate is attached to the #5 carbon of the ribose or deoxyribose and is designated the **5′ carbon**. The phosphate then joins the hydroxyl group attached to the #3 carbon of its neighbor, designated the **3′ carbon**. So, each nucleotide of a chain is linked to its neighbors by 5′ and 3′ ends, and that gives the nucleotide chain a specific 5′–3′ orientation. This consistent organization allows us canonically to give direction to the sequence of nucleotides (or bases) in a strand.

In DNA, bases in one strand may bond with bases in another. Because of their structure, A and T always bond together, and C and G always bond together. Each pair is said to be made up of **complementary bases** and is referred to as a **base pair (bp).** The number of such **base pairs** is used to describe the **length** of a DNA molecule. Because of pairing consistency, by knowing the sequence of bases in one strand, we can deduce the sequence of bases in the other strand through **reverse complementation**. For example, suppose one sequence is s = ATGAC. Because of the required pairing, A–T and C– G, we know the base pairs must appear as follows:

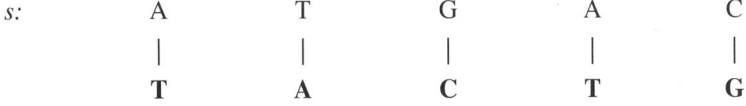

In contrast to DNA, RNA is a single strand of nucleotides made up of ribose sugars and bases A, C, G, and U instead of the nitrogen base thymine (T; Table 7.14.2). Several types of RNA with different functions exist in the cell. Also, RNA strands may base-pair with complementary regions within the same molecule, with strands of other RNA molecules, or with single-stranded portions of DNA molecules.

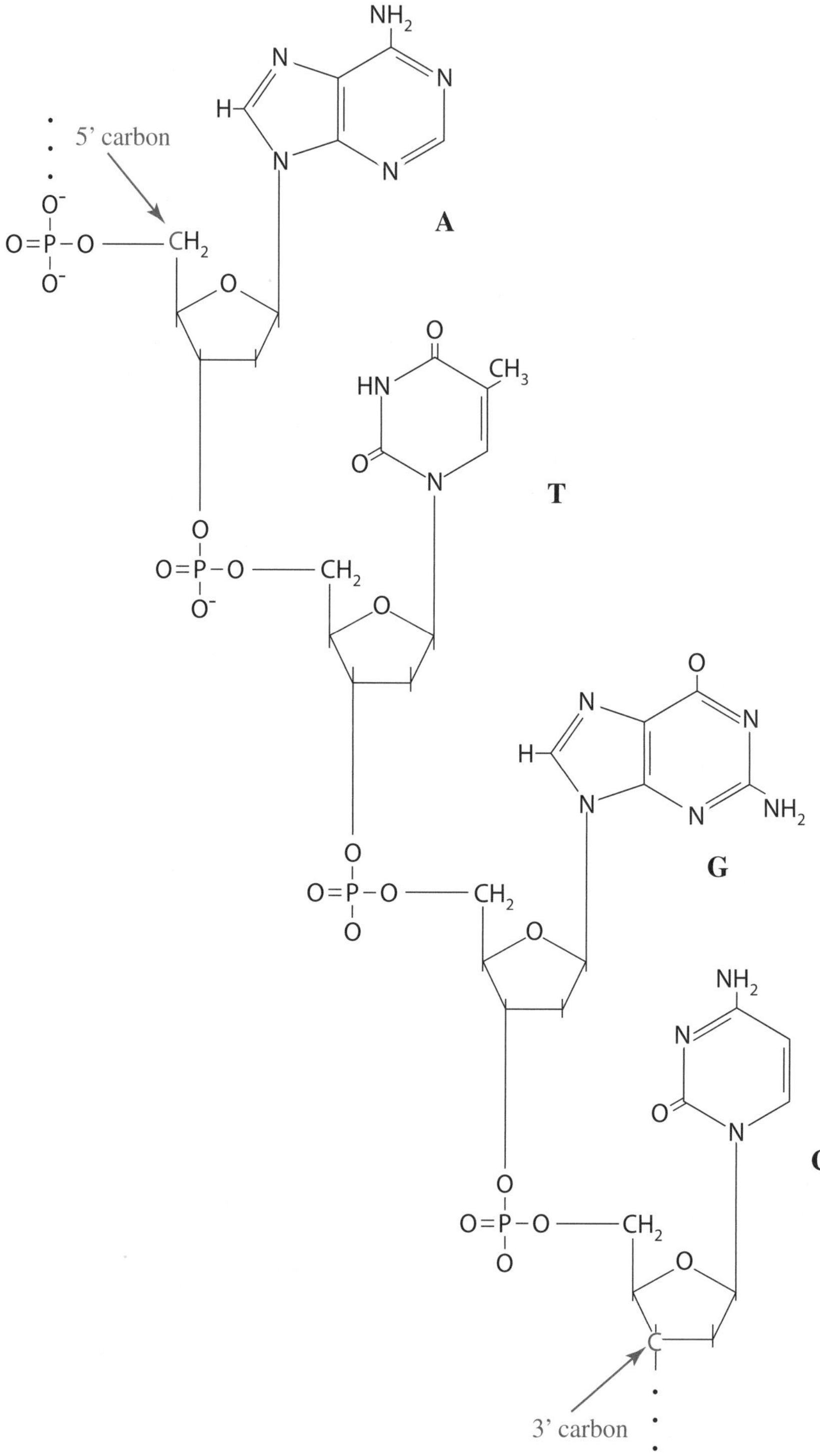

Figure 7.14.4 Single-strand nucleotide chain.

Table 7.14.2
Bases in DNA and RNA

Base	*Abbreviation*	*Complement*	*In DNA*	*In RNA*	*Group*
Adenine	A	T in DNA, U in RNA	yes	yes	Purine
Cytosine	C	G	yes	yes	Pyrimidine
Guanine	G	C	yes	yes	Purine
Thymine	T	A	yes	no	Pyrimidine
Uracil	U	A	no	yes	Pyrimidine

From Genes to Proteins

Each cell contains **chromosomes**, which are very long DNA molecules. A **gene** is a contiguous section of a chromosome that encodes information to build a protein or an RNA molecule. In humans, the average gene is composed of about 3,000 bp. A chromosome contains genes and contiguous sections that are not part of any gene. Some scientists believe that genes (coding sequences) comprise only a small percentage of a human chromosome. The function of these nongene bits of DNA is still debated. Some are known to be important for regulation of gene expression and others are important for matching homologues and structure. A complete set of chromosomes in a cell contains the organism's hereditary information and is called the **genome**. For example, a human genome has 46 chromosomes in 23 pairs.

For simplicity, we assume that a particular protein in an organism corresponds to exactly one gene. In a gene, a sequence of three nucleotides (**triplet**) specifies an amino acid. For example, the sequence ACG or ACA encodes the information for the amino acid Threonine (Thr). The **genetic code** represents such a correspondence between these triplets and the amino acids they specify. With four base choices, a pair of bases could encode information for only (4)(4) = 16 amino acids. With three bases, (4)(4)(4) = 64 possible triplets exist. Several, such as ACG and ACA, encode the same amino acid; and three sequences do not encode for any amino acid.

Protein synthesis uses the genetic code to direct the building of proteins. Synthesis begins in the nucleus, where enzymes catalyze the production of a molecule of RNA, termed **messenger RNA**, or **mRNA**. Each DNA triplet specifies a complementary sequence of three nucleotides, which we call a **codon**, in the RNA. The synthesis of RNA is called **transcription**. During transcription, base-pairing ensures formation of a strand of RNA that is complementary to the gene sequence, with U replacing T. Once transcribed, the mRNA will bind to a **ribosome** that will **translate** the mRNA sequence into a sequence of amino acids for the protein. Ribosomes contain the molecular apparatus to translate groups of three nucleotides (codons) into specific amino acids, thereby converting a specific sequence of nucleotides into a specific sequence of amino acids.

In bacteria, genes that code for functionally related proteins are sometimes positioned contiguously on the single chromosome and are transcribed into one elongated piece of mRNA. This mRNA frequently contains the instructions for a set of enzymes that function in the same biochemical pathway. Control of this transcription is coordinated, and if one of the genes is transcribed, so are all. Or, the genes may all be inactive, awaiting some extrinsic signal that will stir them from their quiescence. This organization and regulation of bacterial genes is widely known as the **operon model** of gene regulation, proposed in the early 1960s by two French scientists—François Jacob and Jacque Monod (1961).

One of the earliest examples of the operon model was the **lac operon** in *Escherichia coli* (*E. coli*). Normally, this bacterium preferentially utilizes glucose if that sugar is available in the medium. Metabolism of other sugars, like lactose, is carried out only when that sugar is supplied in place of glucose in the growth medium. Consequently, the cell produces the enzymes to metabolize lactose only when that sugar is present. Lactose is a disaccharide (two sugars—glucose and galactose—linked by a glycosidic bond). Therefore, lactose must first be split into the two constituent sugars before the sugars are metabolized in the process of glycolysis. Two enzymes are known to be necessary for processing lactose:

- A **permease**, coded for by the *lacY* gene, helps to transport lactose into the cell.
- ***β*-galactosidase**, coded for by the *lacZ* gene, splits the disaccharide into glucose and galactose.

As it happens, the genes for these enzymes and a transacetylase enzyme (gene *lacA*) of unknown function are located next to each other on the chromosome and are regulated by the same factors. Let's look at the organization of this operon (Figure 7.14.5).

The three genes that code for enzymes are termed **structural genes**. Adjacent and upstream from the structural genes are two regulatory sequences—the **promoter** and the **operator**. The promoter ($\mathbf{P_{lac}}$), in this case, is the portion of the DNA where the **RNA polymerase**, an enzyme that makes RNA from DNA, will bind to begin transcription of the structural genes. Another regulatory region made up of the **repressor gene** (lacI) and its promoter (P_I) lie in a more distant part of the chromosome. The repressor gene, when active, transcribes RNA that is translated into the **repressor protein.** The **repressor protein** recognizes and binds to the operator (**O**) sequence, which will act as a type of molecular switch. When the operator is unoccupied by the repressor protein, the RNA polymerase can bind to the promoter and proceed with transcription of the mRNAs for the structural genes. When the operator is bound to the repressor protein, the RNA polymerase is blocked from transcription and the mRNAs for the enzymes are not produced.

The lactose-repressor gene is essentially always expressed, but at a relatively low level. We say that this gene is not regulated (constitutive), and its transcription is

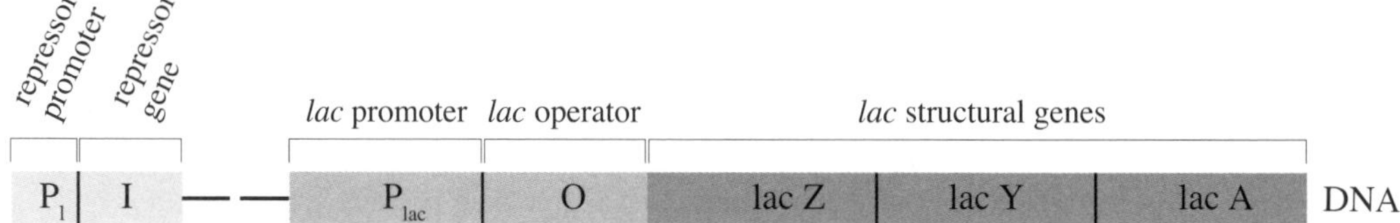

Figure 7.14.5 Structural organization of the lac operon in *E. coli*

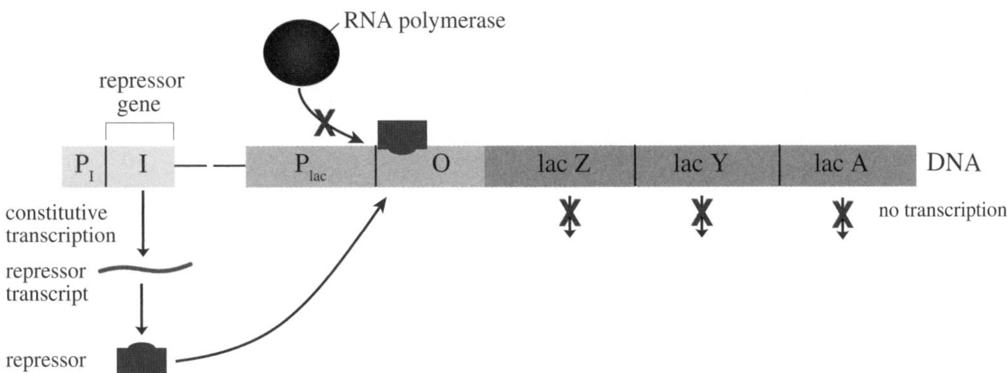

Figure 7.14.6 Lac operon function while glucose is present. The repressor binds to the operator, inhibiting the transcription of enzymes that are needed for lactose utilization.

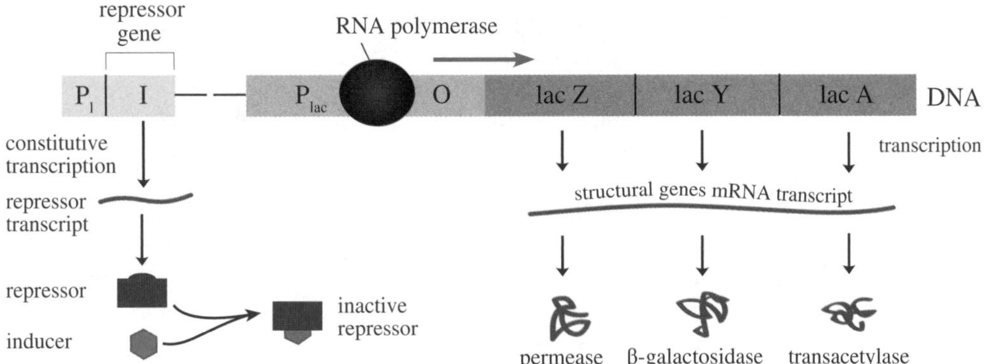

Figure 7.14.7 Lac operon function while lactose is present. The repressor binds to the lactose inducer, inactivating it. RNA polymerase binds to the lac promoter and stimulates active transcription of enzymes that are needed for lactose utilization.

dependent only on the dynamics of its promoter. Because the promoter for the repressor is only weakly active, only small amounts of repressor are produced. When glucose is present, the active repressor binds to the operator for the lac structural genes and inhibits their transcription (Figure 7.14.6). This mechanism represents efficiency in the expression of *E. coli*'s genes because the cell saves materials and energy needed for synthesis of unneeded proteins.

When lactose replaces glucose in the medium, lactose can bind to the **repressor** molecule, inactivating it (Figure 7.14.7). The **inactive repressor** will not bind to the operator; and when RNA polymerase binds to the lac operon promoter, which is very powerful, it will be able to transcribe many copies of the genes for utilization of lactose for cellular energy. Lactose, in this system, acts as an **inducer** for the expression of the lac genes.

Projects

1. Assuming a linear relationship between enzyme activity and enzyme concentrations and assuming that the induction of the lac operon results in an increase in transcription/translation for enzymes required for metabolizing lactose, model the following and discuss the results for varying amounts of glucose and/or lactose:
   a. Lac operon with glucose present
   b. Lac operon with both glucose and lactose present
   c. Lac operon without glucose and with lactose present
2. In an *E. coli* cell, we have a pool of ATP (energy) of 100,000 molecules, and the cell hydrolyzes about 1000 molecules/min of ATP to sustain basic life functions. If we assume that the cell gains 32 ATP/molecule of glucose metabolized, model the lac operon and the effect of glucose on ATP levels with time. Consider the situation with fixed concentration of glucose and then with glucose influx. Discuss the results.
3. Lactose is a second choice for an energy source in *E. coli*; but when there is no glucose available or when concentrations of glucose are low, the cell responds by importing and processing lactose. As glucose concentrations decline, the cell transcribes the mRNA for permease and β-galactosidase and translates the mRNA into these two enzymes. As more enters the cell, lactose is converted to an inducer (allolactose). This allolactose binds to the repressor protein, changing its conformation so that the repressor protein no longer has an affinity for the operator. Subsequently, the RNA polymerase can freely bind to the promoter, and production of the two enzymes escalates. Assuming that induction of the lac operon requires at least 100 molecules of lactose, model the lac operon and the changes in ATP (energy) concentrations of a cell as glucose declines and lactose increases. Each lactose molecule may be oxidized to yield 30 ATP. (This model should include the transcription and translation of the lac enzymes.)
4. In Project 3, the transcription of lac genes is very slow, even when the repressor is inactive. To speed things up considerably, another regulatory molecule, which binds in another promoter site, is required. When associated with a molecule of cAMP, this molecule, called cAMP-receptor protein (CRP) or catabolite-activator protein (CAP), binds to the DNA of the promoter. cAMP levels in the cell are inversely related to the intracellular glucose concentration. When bound, CRP opens up the promoter and enhances the binding of the RNA polymerase for transcription. Model this system to include the effects of glucose concentration on levels of cAMP and subsequent CRP binding on transcription rates of the lac operon genes. Relate these parameters to the production of ATP (energy).

References

Beckwith, J. 1987. "The Operon: An Historical Account," in F. C. Neidhardt, J. L. Ingraham, K. B. Low, B. Magasanik, M. Schaechter, and H. E. Umbarger (eds.).

Escherichia coli and Salmonella typhimurium: Cellular and Molecular Biology. Washington, DC: ASM Press.

Jacob, F., and Monod, J. 1961. "Genetic Regulatory Mechanisms in the Synthesis of Proteins." *J. Mol. Biol.* 3: 318–356

McGill, Clint. "Gene Regulation—The Lac Operon." Class Notes. Genetics 310 – Principles of Heredity. McGill University. http://www.tamu.edu/faculty/magill/gene310/PDF files/Gene Regulatio1.pdf (accessed January 7, 2013)

Reznikoff, William S. 1992. "The Lactose Operon—Controlling Elements: A Complex Paradigm." *Molecular Microbiology*. 6(17): 2419–2422.

Tajbakhsh, Shahragim, Giacomo Cavalli, and Evelyne Richet. 2011. "Integrated Gene Regulatory Circuits: Celebrating the 50th Anniversary of the Operon Model." *Molecular Cell*. 43(4): 505–514.

MODULE 7.15

Troubling Signals: Colon Cancer

Prerequisites: Section on "Modeling Inhibition" from Module 4.5, "Enzyme Kinetics," and one of the following modules: 4.1, "Competition, " 4.2,"; Predator-Prey Model"; 4.3, "Spread of SARS"; or 4.5, "Enzyme Kinetics."

I don't want to achieve immortality through my work. . . . I want to achieve it through not dying.

—Woody Allen (Lax 1975)

Introduction

Howard, a 56-year-old teacher, waits nervously on a gurney following his first colonoscopy. His physician arrives and tells him that he removed several polyps and that one of the polyps was of a precancerous type. Howard can't follow anything else the physician is saying. All he can think about is his father, who died at the age of 52 of colon cancer.

Cancer is one of the most terrifying diagnoses a patient can receive. Just about everyone knows someone who has died of the disease, and early in the twentieth century, such a diagnosis was essentially a death sentence. Physicians have battled for decades with surgery, radiation, and various chemotherapies to cure their cancer patients—or at least to prolong their lives. We have come far in our understanding of cancer and its prevention, as well as in our ability to detect it; and yet, the U.S. National Cancer Institute (NCI) predicted more than 1.6 million new cases of cancer (excluding nonmelanoma skin cancers) with more than a half million deaths in 2012 (NCI 2013).

Cancer is really more than one disease, although we tend to think of it as only one. Cancers may originate in the body covering or organ linings, blood-forming and immune tissues, or various connective tissues, but all types of cancers have one common characteristic: *abnormal* and *unrestrained growth*. Nevertheless, the 100 or so types of cancer often behave quite differently and require different treatments (ACS 2013; NCI 2013).

Colon Cancer

In the United States, the NCI predicted that there would be about 150,000 new cases of colorectal cancer and about 50,000 deaths in 2012 (NCI 2013). Worldwide, about 600,000 deaths per year are attributable to this type of cancer (WHO 2012). Actually, there is more than one form of colon cancer, but the vast majority of these tumors are **adenocarcinomas** (cancers of the glandular endothelium). Understanding how colon cancer develops is important if we want to advance in its prevention or cure.

The normal colon is lined with cells that are continually dying, sloughing off, and being replaced, somewhat like an internal skin. The cells that line the lumen (cavity) of the intestine are replaced weekly (Medema and Vermeulen 2011). The replacement cells arise within millions of invaginations (**infoldings**) in the lining called **crypts**. At the base of these crypts are stem cells that divide persistently and produce new cells that gradually mature (**differentiate**) and migrate to the surface to replace their more seasoned predecessors (Figure 7.15.1). This vigorous renewal system is under very stringent control by chemical signals, many of which are produced by nearby mesenchymal cells of the **stem-cell niche** and by more mature epithelial cells (Medema and Vermeulen 2011).

Particular gene mutations in the stem cells or early progeny of stem cells may cause aberrations in some of the developing cells. Such changes can disrupt the normal division and maturation process. If an accumulation of mutations (**hits**) takes place, the combination of changes may make these cells cancerous (Stanford Medicine Cancer Institute 2013). In fact, colorectal cancer often develops through accumulation of so-called genetic hits (Fearon and Vogelstein 1990).

Modeling Crypt Dynamics

To better understand how colon cancer develops, we need to better understand the normal growth and development of the cells lining the colonic crypts. A powerful approach, supplementing bench experiments, is mathematical modeling. For example, the Computational Biology Group (CBG) at the University of Oxford is developing and maintaining **Chaste** (Cancer, Heart and Soft-Tissue Environment), an open source simulation package that models the dynamics of the colonic crypt (Chaste 2013).

To build our model of the complex processes of crypt dynamics, we will start simply. Therefore, we begin with a sequence of molecular signals, called the **Wnt pathway**, which scientists know plays an essential role in tissue homeostasis. This pathway helps to control cell division and maturation in the crypt, as well as to sustain the stem cells (Nusse 2013).

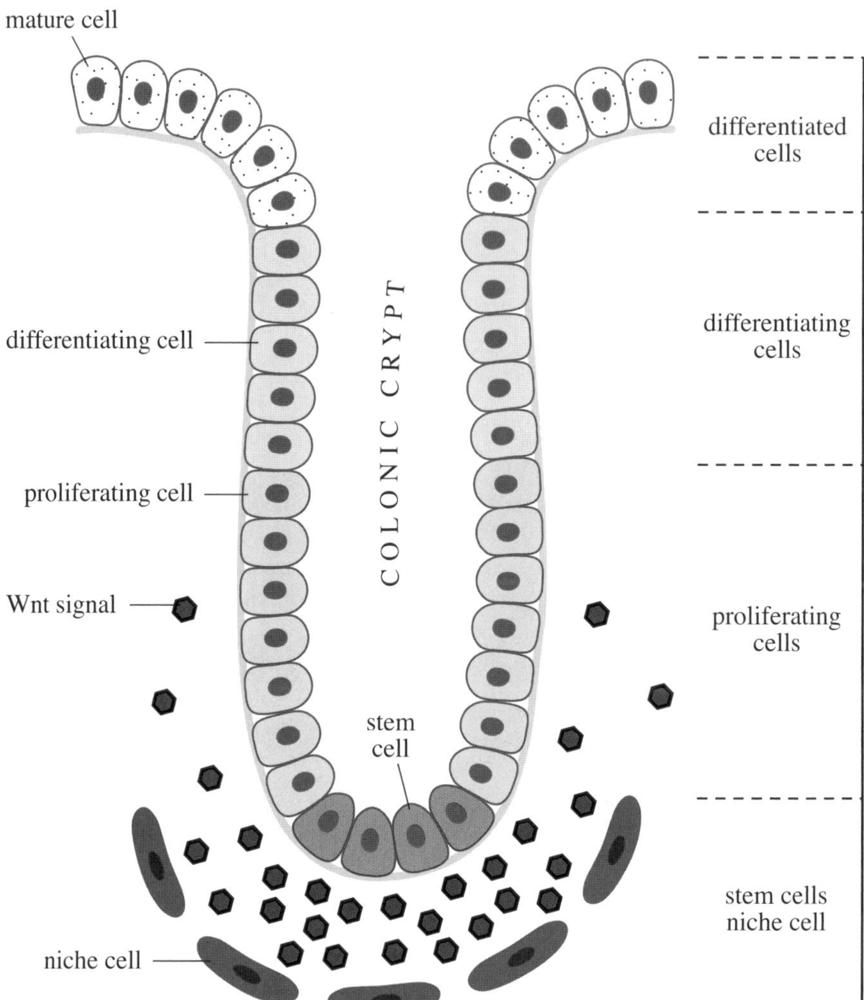

Figure 7.15.1 Diagram of colonic crypt. New cells are produced by stem cells at the base of the crypt. The cells of the stem-cell niche secrete the Wnt signal molecules, which forms a gradient of signal molecules, where the concentration decreases as the cells move away from the bottom of the crypt. Wnt signal molecules bind to receptors on the crypt cells, promoting cell division. This gives rise to the region of **proliferating cells**. As the concentration of Wnt declines, proliferation slows, and the cells begin to differentiate. Cells in the differentiating region gradually move toward the colonic lumen and mature into differentiated cells.

Wnt signals are part of a family of chemicals (glycolipoproteins) that serve as **morphogens** in Wnt pathways. These locally produced molecules form a concentration gradient as they diffuse away from the source and interact with nearby cells to bring about some response (e.g., cell division or differentiation). The response is concentration dependent. Mesenchymal cells at the base of the crypt (stem-cell niche) secrete **Wnt signals**. The highly processed signal molecules are not very diffusible, so their strength diminishes as the cells move away from the crypt base. Cells divide and differentiate at rates determined, at least partially, by their interaction with this signal gradient.

Wnt signals influence the cytoplasmic concentration of the protein **β-catenin**, which has at least two roles in epithelial cells: (1) binding to membrane-localized cadherin proteins to form **adherens junctions** (**AJ**s) that help maintain the structural integrity of the epithelial layer and (2) acting as a **transcription factor**. In a target cell without Wnt signals, there is a basal rate of synthesis that provides more than enough β-catenin for the AJs. Any excess β-catenin is broken down by **proteasomes** (protein complexes that degrade excess or damaged proteins in the cell). For the proteasomes to degrade a protein, the protein is often chemically marked by the attachment of a short chain of small proteins called **ubiquitins**. Ubiquitin molecules act as recognition signals for the proteasomes. So, in a cell sensitive to Wnt signals, but where there are none, the excess β-catenin will be **ubiquitinated** (ubiquitin attached) and then degraded.

Crypt cells have receptors on their membranes that are sensitive to Wnt signals. The signals bind to these receptors, initiating specific cellular responses. When there are no Wnt signals, the cell produces β-catenin, uses what is needed for AJs, and degrades the rest. A complex of proteins that combine to form a sort of "doomsday machine" initiates this degradation sequence. Two of these proteins are **axin** and **APC**, scaffolding proteins that bind to β-catenin. The complex also includes two enzymes that add phosphates to specific sites on the β-catenin when the β-catenin is bound to the scaffolding. Phosphorylated β-catenin is now more recognizable to **ubiquitinating enzymes**. Once the ubiquitin is attached, the β-catenin is doomed for destruction by proteasomes (Figure 7.15.2).

When present, the Wnt signal binds to a pair of receptors, **Fzl** and **LRP6/5** (Figure 7.15.3). The binding of signal to receptor induces the recruitment of another protein, **dvl**, which promotes the phosphorylation of **LRP6**, which, in turn, binds to **axin**. With insufficient axin available, the "doomsday machine" does not form, and β-catenin does not undergo phosphorylation. Consequently, the ubiquitination and subsequent degradation do not occur, and β-catenin concentrations build up in the cell.

Some of the excess β-catenin is transported into the nucleus, where it interacts with other protein factors (e.g., TCF) to become **transcription factors** (**TF**s). These TFs bind to specific control regions on the DNA to activate (or in some cases inactivate) the expression of a restricted set of genes. Many of the products coded for by these genes are involved in the control of growth and development of the cell. For instance, the β-catenin-associated TFs may induce the production of factors that promote cell division and others that inhibit cell differentiation. It is easy to see why defects in the components or steps of this signaling pathway might lead to abnormal growth and even cancer.

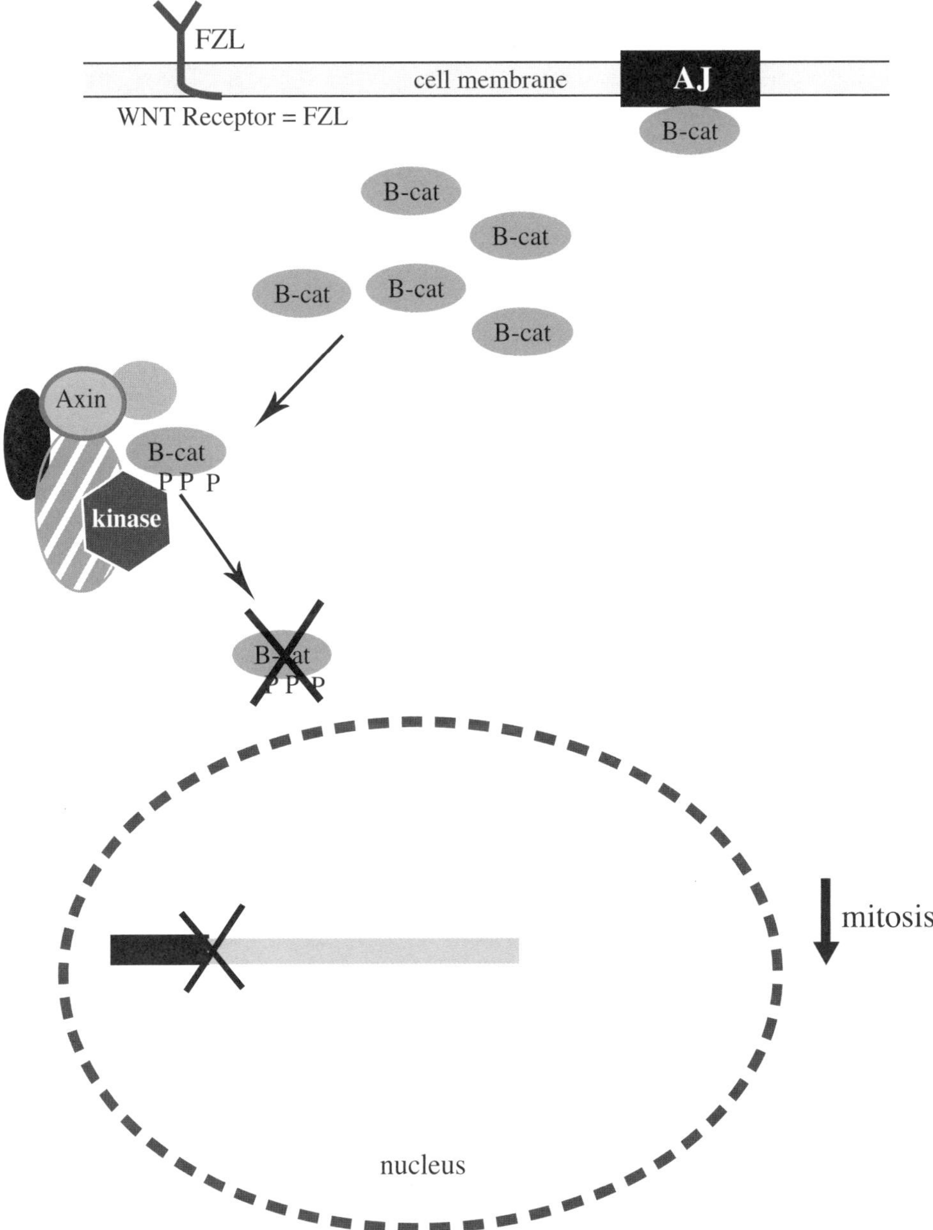

Figure 7.15.2 Cell-division activity with no or low levels of the Wnt signal

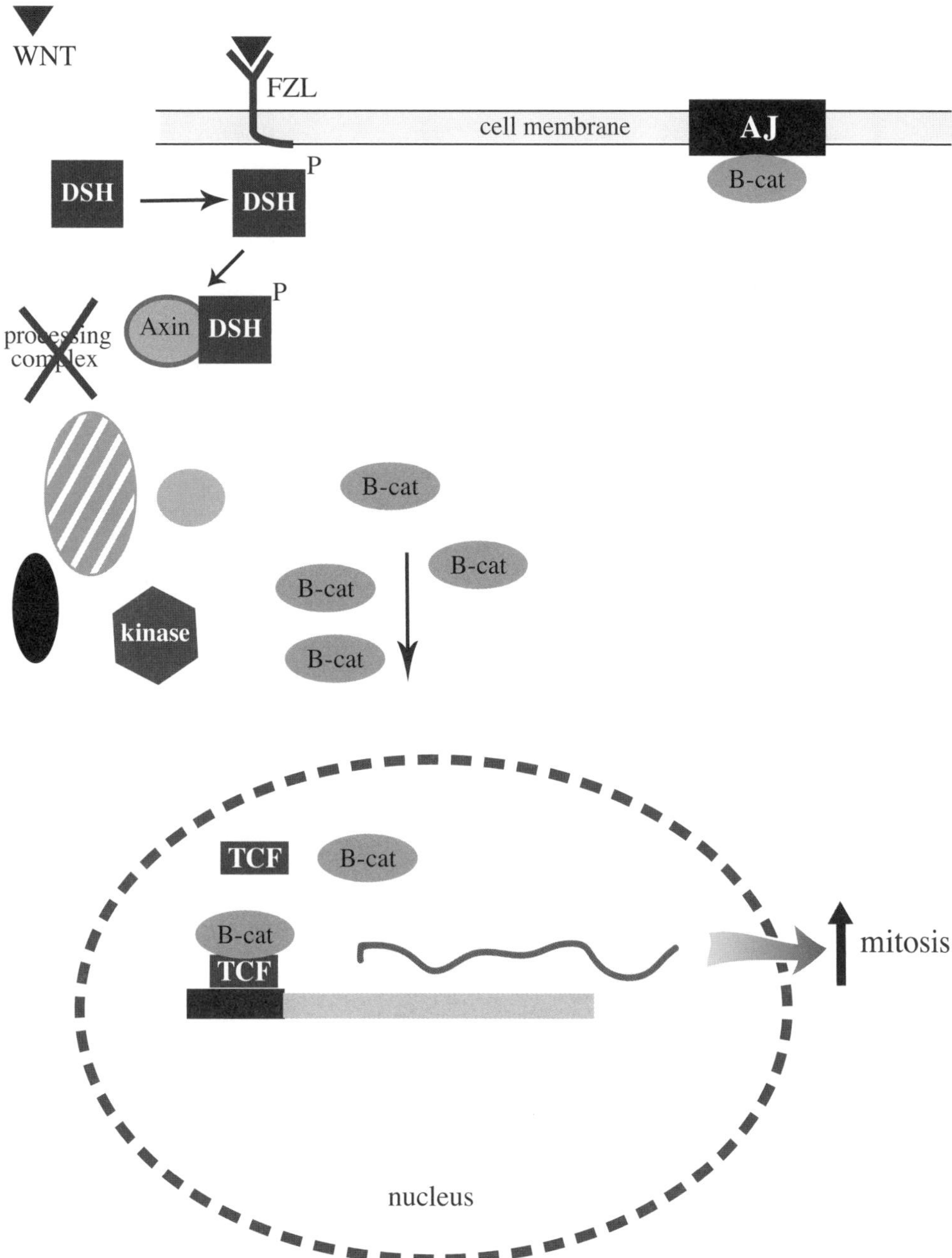

Figure 7.15.3 Cell-division activity under conditions of high levels of WNT signal

Table 7.15.1
Substances for Simplified Wnt-Signaling Model Diagram

Name	*Initial Quantity*
Wnt	5.616
TCF	10.000
β-catenin	20.000
FZL	10.000
Active receptor	0.000
Target genes	10.000
β-catenin/TCF	6.000
Cell-division promoters	20.000
Cell-division inhibitors	20.000

Table 7.15.2
Parameters for Simplified Wnt-Signaling Model

Name	*Value*
Constant for interaction of WNT and FZL	0.00475
Catenin activator, constant for interaction of β-catenin and active receptor	0.01
Transcription factor, constant for interaction of TCF and β-catenin (t)	5.9×10^{-4}
Degradation constant	0.01
Transcription factor1, constant for cell-division promotion	0.00500
Transcription factor2, constant for cell-division inhibition	0.00500

Projects

1. Develop a system dynamics model of the Wnt pathway, graphing important values such as the amounts of Wnt, cell-division inhibitors, and cell-division promoters. Possible initial quantities and parameter values appear in Tables 7.15.1 and 7.15.2, respectively. The section "Modeling Inhibition" from Module 4.5, "Enzyme Kinetics—A Model of Control," describes one way to model the process involving inhibition, such as β-catenin/TCF inhibiting the rate of change of division inhibitors. Run the model for a variety of Wnt amounts. Discuss the results.

References

ACS (American Cancer Society). 2013. "Learn about Cancer." http://www.cancer.org/Cancer/index (accessed January 6, 2013)

Chaste (Cancer, Heart and Soft Tissue Environment). 2012. http://www.cs.ox.ac.uk/chaste/about.html (accessed December 30, 2012)

Clevers, Hans. 2006. "Wnt/β-catenin Signaling in Development and Disease." *Cell*, 127: 469-480.

Clevers, Hans, and Roel Nusse. 2012. "Wnt/β-catenin Signaling and Disease." *Cell*, 149 (June 8).

Fearon E. R., and B. Vogelstein. 1990. "A Genetic Model for Colorectal Tumorigenesis." *Cell*. 1990, 61: 759–767.

Lax, Eric. 1975. *On Being Funny: Woody Allen and Comedy. New York: Charterhouse.*

Logan, C .Y., and R. Nusse. 2004. "The Wnt-Signaling Pathway in Development and Disease." *Annual Review of Cell and Developmental Biology*, 20: 781–810.

MacDonald, B. T., K. Tamai, and X. He. 2009. "Wnt/β-catenin Signaling: Components, Mechanisms and Disease." *Dev Cell* 17(1): 9–26.

Medema, J. P., and L. Vermeulen. 2011. "Microenvironmental Regulation of Stem Cells in Intestinal Homeostasis and Cancer," *Nature* 474: 318–326

Melo, F., and J. P. Medema. 2012. "Axing Wnt Signals." *Cell Res*, 22: 9–11.

Nusse Lab, Stanford University. 2013. "The Wnt Homepage." http://www.stanford.edu/group/nusselab/cgi-bin/wnt/ (accessed January 6, 2013)

NCI (U.S. National Cancer Institute). 2013. http://www.cancer.gov/ (accessed January 6, 2013)

Stanford Medicine Cancer Institute. 2013. "How Genes Cause Cancer." http://cancer.stanford.edu/information/geneticsAndCancer/genesCause.html (accessed January 6, 2013)

WHO (World Health Organization). 2012. Cancer Fact Sheet, No. 297.http://www.who.int/mediacentre/factsheets/fs297/en/ (accessed January 6, 2013)

8

DATA-DRIVEN MODELS

MODULE 8.1

Computational Toolbox—Tools of the Trade: Tutorial 3

Prerequisite: Module 6.1, "Computational Toolbox—Tools of the Trade: Tutorial 2."

Download

From the textbook's website, download Tutorial 3 in the format of your computational tool or in PDF format. We recommend that you work through the tutorial and answer all Quick Review Questions using the corresponding software.

Introduction

Various computer software tools are useful for graphing, numeric computation, and symbolic manipulation. This third computational toolbox tutorial, which is available from the textbook's website in your system of choice, prepares you to use the tool to complete projects for this and subsequent chapters. The tutorial introduces the following functions and concepts:

- List/array operations
- Additional graphics options
- Showing several graphics together
- Fitting curves to data
- Rules
- Reading from a file

The module gives computational examples and Quick Review Questions for you to complete and execute in the desired software system.

MODULE 8.2

Function Tutorial

Download

We recommend that you download the function tutorial in the format of your desired computational tool from the textbook's website and work through the tutorial using the software. Alternatively, you can download the corresponding tutorial in PDF format and answer the Quick Review Questions using a new file in the appropriate computational software. For the questions that do not involve using a computational tool, type the answers into the tutorial file or write the answers on a separate sheet of paper. When plotting several functions together, distinguish between the curves, such as by color, line thickness, or dashing. As with other software-dependent tutorials, answers to the Quick Review Questions are not available at the end of the module. Material in the printed text, which does not depend on a particular computational tool, contains important generic information about functions.

Introduction

In this chapter, we deal with models that are driven by the data. In such a situation, we have data measurements and wish to obtain a function that roughly goes through a plot of the data points capturing the trend of the data, or **fitting the data**. Subsequently, we can use the function to find estimates at places where data do not exist or to perform further computations. Moreover, determination of an appropriate fitting function can sometimes deepen our understanding of the reasons for the pattern of the data.

In this module, we consider several important functions, some of which we have already used. By being familiar with basic functions and function transformations, the modeler can sometimes more readily fit a function to the data.

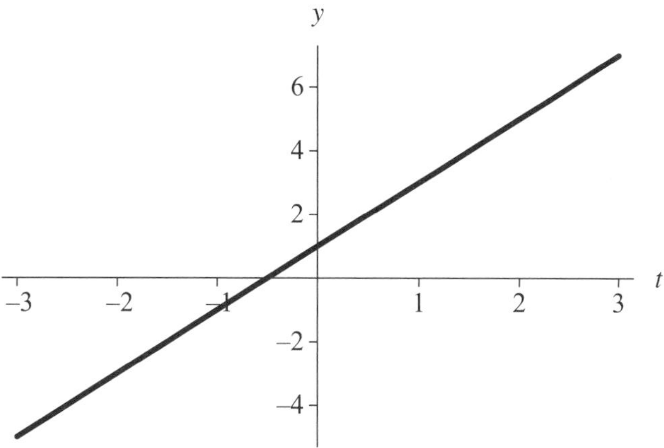

Figure 8.2.1 Graph of the linear function $y = 2t + 1$

Linear Function

The concept of a linear function was essential in our discussions of simulation techniques, such as Euler's method. Here, we review some of the characteristics of functions whose graphs are lines.

Figure 8.2.1 presents the graph of the linear function $y = 2t + 1$. This line has y-intercept 1, because $y = 1$ when $t = 0$. Thus, the graph crosses the y-axis, which occurs when $t = 0$. With data measurements where t represents time starting at 0, the y-intercept indicates the initial data value. The slope of this particular line is 2, which is the coefficient of t. Consequently, when we go over 1 unit to the right, the graph rises by 2 units.

Definitions A **linear function**, whose graph is a straight line, has the following form:

$$y = mx + b$$

The **y-intercept**, which is b, is the value of y when $x = 0$, or at the place where the line crosses the y-axis. The **slope** m is the change in y over the change in x. Thus, if the line goes through points (x_1, y_1) and (x_2, y_2), the slope is as follows:

$$m = \frac{\Delta y}{\Delta x} = \frac{y_2 - y_1}{x_2 - x_1}$$

Quick Review Question 1

For this and every Quick Review Question in this module, use an appropriate computational tool to complete the question.

a. Plot the preceding function, $f(t) = 2t + 1$, from $t = -3$ to 3.
b. Plot f along with the equation of the line with the same slope as f but with y-intercept 3. Distinguish between the graphs of f and the new function, such as by color, line thickness, or dashing.
c. Copy the command from Part b, and change the second function to have a y-intercept of –3.
d. Describe the effect that changing the y-intercept has on the graph of the line.
e. Copy the command from Part b, and change the second function to have the same y-intercept as f but slope 3.
f. Copy the command from Part b, and change the second function to have the same y-intercept as f but slope –3.
g. Describe the effect that changing the slope has on the graph of the line.

Quadratic Function

In Quick Review Question 1 of Module 2.2, "Unconstrained Growth and Decay," we considered a ball thrown upward off a bridge. If the bridge is 11 m high and the initial velocity is 15 m/s, then the function for height of the ball with respect to time is the following quadratic function:

$$s(t) = -4.9t^2 + 15t + 11$$

The general form of a **quadratic function** is as follows:

$$f(x) = a_2x^2 + a_1x + a_0$$

where a_2, a_1, and a_0 are real numbers. The graph of the ball's height $s(t)$ in Figure 8.2.2 is a **parabola** that is concave down. The next two Quick Review Questions develop some of the characteristics of quadratic functions.

Definitions A **quadratic function** has the following form:

$$f(x) = a_2x^2 + a_1x + a_0$$

where a_2, a_1, and a_0 are real numbers. Its graph is a **parabola**.

Quick Review Question 2

a. Plot the preceding function, $s(t) = -4.9t^2 + 15t + 11$, from $t = -1$ to 4.
b. Give the command to plot $s(t)$ and another function with the same shape that

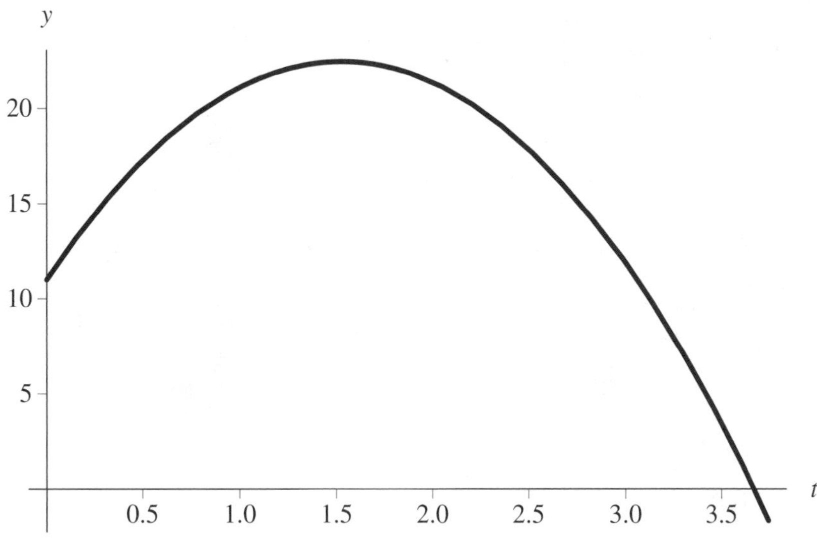

Figure 8.2.2 Height (y) in m versus time (t) in s of a ball thrown straight up from a bridge

crosses the y-axis at 2. Distinguish between the graphs, such as by color, line thickness, or dashing.

c. Using calculus, determine the time t at which the ball reaches its highest point. Verify your answer by referring to the graph.

d. What effect does changing the sign of the coefficient of t^2 have on the graph?

Quick Review Question 3

For this and every Quick Review Question, when plotting several functions together, distinguish between the curves, such as by color, line thickness, or dashing.

a. Plot t^2, $t^2 + 3$, and $t^2 - 3$ on the same graph.

b. Describe the effect of adding a positive number to a function.

c. Describe the effect of subtracting a positive number from a function.

d. Plot t^2, $(t + 3)^2$, and $(t - 3)^2$ on the same graph.

e. Describe the effect of adding a positive number to the independent variable, such as t, in a function.

f. Describe the effect of subtracting a positive number from the independent variable in a function.

g. Plot t^2 and $-t^2$ on the same graph.

h. Describe the effect of multiplying a function by -1.

i. Plot t^2, $5t^2$, and $0.2t^2$ on the same graph.

j. Describe the effect of multiplying the function by number greater than 1.

k. Describe the effect of multiplying the function by positive number less than 1.

Summary of graphical impacts of several operations on $y = f(t)$ for positive constant c:

$f(t) + c$ adds c to each y-value, so addition of $c > 0$ moves the graph of f up c units.
$f(t) - c$ moves the graph of f down c units.
$c\,f(t)$ multiplies each y value by c, so multiplication by $c > 1$ stretches the graph of f. Multiplication by c for $0 < c < 1$ shrinks the graph of f.
$-f(t)$ rotates the graph of $f(t)$ around the t-axis.
$f(t + c)$ moves the graph of f to the left c units.
$f(t - c)$ moves the graph of f to the right c units.

Polynomial Function

Linear and quadratic functions are polynomial functions of degree 1 and 2, respectively. The general form of a **polynomial function of degree *n*** is as follows:

$$f(x) = a_n x^n + \cdots + a_1 x + a_0$$

where $a_n, \ldots, a_1$, and a_0 are real numbers and n is a nonnegative integer. The graph of such a function with degree greater than 1 consists of alternating hills and valleys. The quadratic function of degree 2 has one hill or valley. In general, a polynomial of degree n has at most $n - 1$ hills and valleys.

Definition A **polynomial function of degree *n*** has the following form:

$$f(x) = a_n x^n + \cdots + a_2 x^2 + a_1 x + a_0$$

where $a_n, \ldots, a_1$, and a_0 are real numbers and n is a nonnegative integer.

Quick Review Question 4

a. Plot the polynomial function $p(t) = t^3 - 4t^2 - t + 4$ from $t = -2$ to 5 to obtain a graph similar to Figure 8.2.3.
b. To what value does $p(t)$ go as t goes to infinity?
c. To what value does $p(t)$ go as t goes to minus infinity?
d. Plot $p(t)$ and another function with each coefficient of t having the opposite sign as in $p(t)$. Distinguish between the curves, such as by color, line thickness, or dashing.
e. To what does the new function from Part d go as t goes to infinity?
f. To what does the new function from Part d go as t goes to minus infinity?

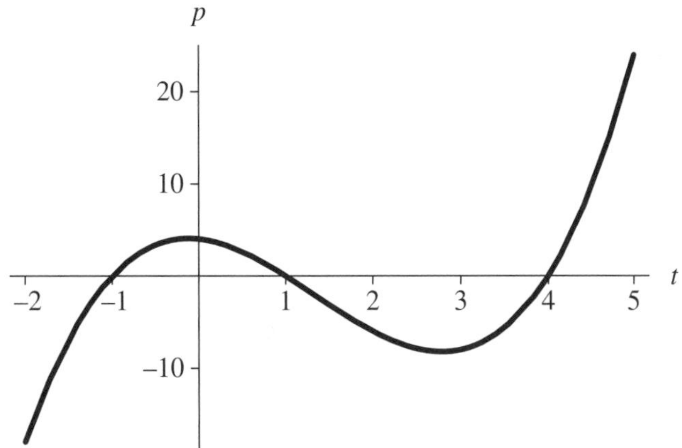

Figure 8.2.3 Graph of polynomial function $p(t) = t^3 - 4t^2 - t + 4$ from $t = -2$ to 5

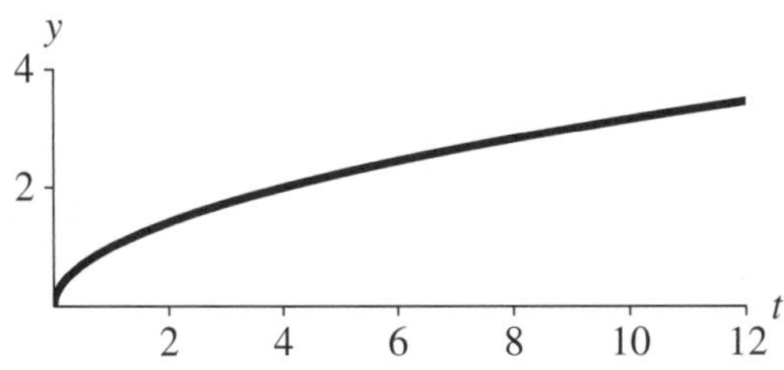

Figure 8.2.4 Square root function

Square Root Function

The square root function, whose graph is in Figure 8.2.4, is increasing and concave down. Its domain and range are the set of nonnegative real numbers.

Quick Review Question 5

Plot each of the following transformations of the square root function.

a. Move the graph to the right 5 units.
b. Move the graph up 3 units.
c. Rotate the graph around the x-axis
d. Double the height of each point.

Exponential Function

In Module 2.2, "Unconstrained Growth and Decay," we considered situations where the rate of change of a quantity, such as the size of a population, is directly proportional to the size of the population, such as $dP/dt = 0.1P$, with initial population $P_0 = 100$. As we saw, the solution to this differential equation is the exponential function $P = 100e^{0.1t}$, whose graph is in Figure 2.2.3 of that module. Similarly, the solution to the differential equation $dQ/dt = -0.000120968Q$ for radioactive decay is $Q = Q_0e^{-0.000120968t}$, with graph in that module's Figure 2.2.4. As indicated in both solutions, the coefficient is the initial amount and the coefficient of t is the continuous rate. For a positive initial amount and a positive rate, the function increases and is concave up; while a negative rate results in a decreasing, concave-up function.

The base can be any positive real number, not just e, which is approximately 2.71828. For example, we can express $P = 100e^{0.1t}$ as an exponential function with base 2. Setting $100(2^{rt})$ equal to $100e^{0.1t}$, we cancel the 100's, take the natural logarithm of both sides, and solve for r, as follows:

$$\begin{aligned} 100e^{0.1t} &= 100(2^{rt}) \\ 0.1t &= \ln(2^{rt}) \\ 0.1t &= rt\ln(2) \\ r &= 0.1/\ln(2) = 0.14427, \text{ when } t \neq 0 \end{aligned}$$

Thus, $P = 100e^{0.1t} = 100(2^{0.14427t})$.

Definition An **exponential function** has the following general form:

$$P(t) = P_0a^{rt}$$

where P_0, a, and r are real numbers.

Quick Review Question 6

a. Define an exponential function $u(t)$ with initial value 500 and continuous rate 12%.

b. Plot this function.

c. On the same graph, plot exponential functions with initial value 500 and continuous rates of 12%, 13%, and 14%. Which rises the fastest?

d. Express the function $u(t)$ as an exponential function with base 4.

Quick Review Question 7

a. Define an exponential function $v(t)$ with initial value 5 and continuous rate –82%.

b. Plot this function.

c. Plot $v(t)$ and $v(t) + 7$ on the same graph. Distinguish between the curves, such as by color, line thickness, or dashing.

d. What effect does adding 7 have on the graph?

e. As t goes to infinity, what does $v(t)$ approach?

f. As t goes to infinity, what does $v(t) + 7$ approach?
g. Copy the answer to Part b. In the copy, plot $v(t)$ and $-v(t)$.
h. What effect does negation (multiplying by –1) have on the graph?
i. Copy the answer to Part g. In the copy, plot $v(t)$ and $7 - v(t)$.
j. As t goes to infinity, what does $7 - v(t)$ approach?
k. Give the value of $7 - v(t)$ when $t = 0$.

Quick Review Question 8

a. From $t = 0$ to $t = 5$, plot $12te^{-2t}$, a function that has an independent variable t as a factor and as an exponent.
b. Initially, with values of t close to 0, give the factor that has the most impact, t or e^{-2t}.
c. As t gets larger, give the factor that has the most impact, t or e^{-2t}.

Logarithmic Functions

In Module 2.2, "Unconstrained Growth and Decay," we employed the logarithmic function to obtain an analytical solution to the differential equation $dP/dt = 0.1P$, with initial population $P_0 = 100$. In that same module, the logarithmic function was useful in solving a problem to estimate the age of a mummy.

John Napier, a Scottish baron who considered mathematics a hobby, published his invention of logarithms in 1614. Unlike most other scientific achievements, his work was not built on that of others. His highly original invention was welcomed enthusiastically, because problems of multiplication and division could be reduced to much simpler problems of addition and subtraction using logarithms.

By definition, m is the **logarithm to the base 10**, or **common logarithm**, **of** $\boldsymbol{n}$, written as $\mathbf{log_{10}}\boldsymbol{n} = m$ or $\mathbf{log}\,\boldsymbol{n} = m$, provided m is the exponent of 10 such that 10^m is n, or

$$\mathbf{log_{10}}\boldsymbol{n} = \boldsymbol{m} \textbf{ if and only if } \boldsymbol{n} = \mathbf{10}^{\boldsymbol{m}}$$

A logarithm is an exponent, in this case, an exponent of 10. Thus,

$\log_{10}1000 = 3$	because	$1000 = 10^3$
$\log_{10}1{,}000{,}000 = 6$	because	$1{,}000{,}000 = 10^6$
$\log_{10}0.01 = -2$	because	$0.01 = 10^{-2}$

Because 10^m is always positive, we can take the logarithm only of positive numbers, so that the domain of a logarithmic function is the set of positive real numbers. However, the exponent m, which is the logarithm, can take on values that are positive, negative, or zero. Thus, the range of a logarithmic function is the set of all real numbers. Figure 8.2.5 shows the graph of the common logarithm. Because the logarithm is an exponent, the logarithmic function increases very slowly, and the graph is concave down.

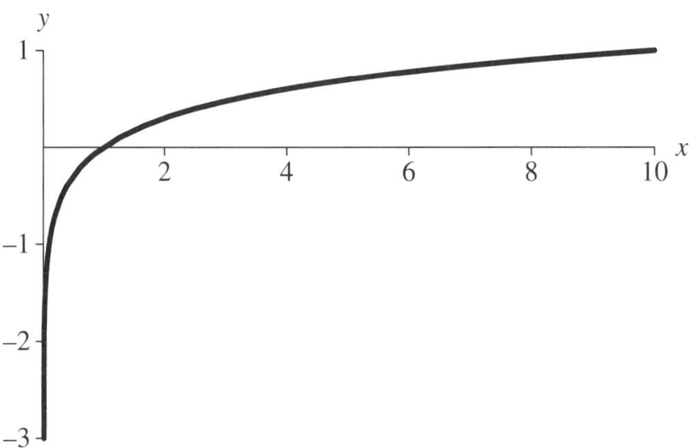

Figure 8.2.5 log x

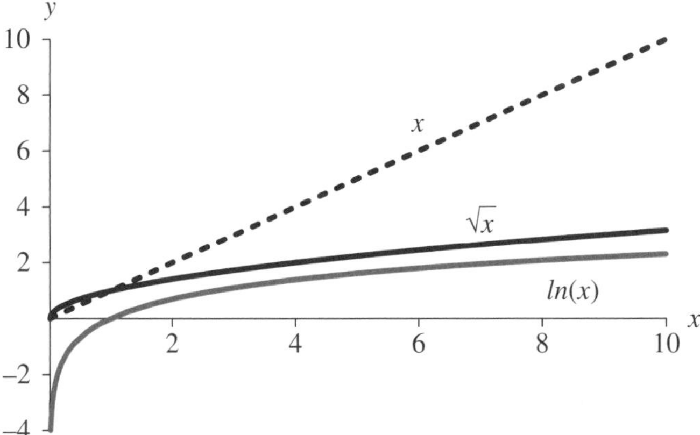

Figure 8.2.6 x and $\sqrt{x}$ dominate ln x

In scientific applications, we frequently employ the **logarithm to the base *e***, or the **natural logarithm**. The notation for $\log_e n$ is $\ln n$. Similar to the common logarithm, we have the following equivalence:

$$\ln n = m \text{ if and only if } n = e^m$$

Moreover, the graph of the natural logarithm has a similar shape to that of the common logarithm in Figure 8.2.5.

Definitions The **logarithm to the base *b* of *n***, written $\log_b n$, is m if and only if b^m is n. That is, $\log_b n = m$ is equivalent to $n = b^m$. The **common logarithm** of n, usually written $\log n$, has base 10; the **natural logarithm** of n, usually written $\ln n$, has base e.

In comparing the graph of ln x to that of x and $\sqrt{x}$ in Figure 8.2.6, we see that the linear and square root functions dominate the logarithmic function, which is in color.

Quick Review Question 9

a. Evaluate $\log_2 8$.
b. Write $y = \log\ 7$ as a corresponding equation involving an exponential function.
c. Evaluate $\ln(e^{5.3})$.
d. Evaluate $10^{\log(6.1)}$.

Logistic Function

In Module 2.3, "Constrained Growth," we modeled the rate of change of a population with a carrying capacity that limited its size. The model incorporated the following differential equation with carrying capacity M, continuous growth rate r, and initial population P_0:

$$\frac{dP}{dt} = r\left(1 - \frac{P}{M}\right)P \quad \text{where } P = P_0 \text{ when } t = 0$$

The resulting analytical solution, which is a **logistic function**, is as follows:

$$\boldsymbol{P(t) = \frac{MP_0}{(M - P_0)e^{-rt} + P_0}}$$

Figure 2.3.1 of Module 2.3, "Constrained Growth," depicts the characteristic *S*-curve of this function.

Quick Review Question 10

a. Plot the logistic function with initial population $P_0 = 20$, carrying capacity $M = 1000$, and instantaneous rate of change of births $r = 50\% = 0.5$ from $t = 0$ to 16 to obtain a graph as in Figure 2.3.1 of Module 2.3, "Constrained Growth."
b. On the same graph, plot three logistic functions that each have $M = 1000$ and $r = 0.5$ but P_0 values of 20, 100, and 200.
c. What effect does P_0 have on a logistic graph?
d. On the same graph, plot three logistic functions that each have $M = 1000$ and $P_0 = 20$ but r-values of 0.2, 0.5, and 0.8.
e. What effect does r have on a logistic graph?
f. On the same graph, plot three logistic functions that each have $P_0 = 20$ and $r = 0.5$ but M-values of 1000, 1300, and 2000.
g. What effect does M have on a logistic graph?

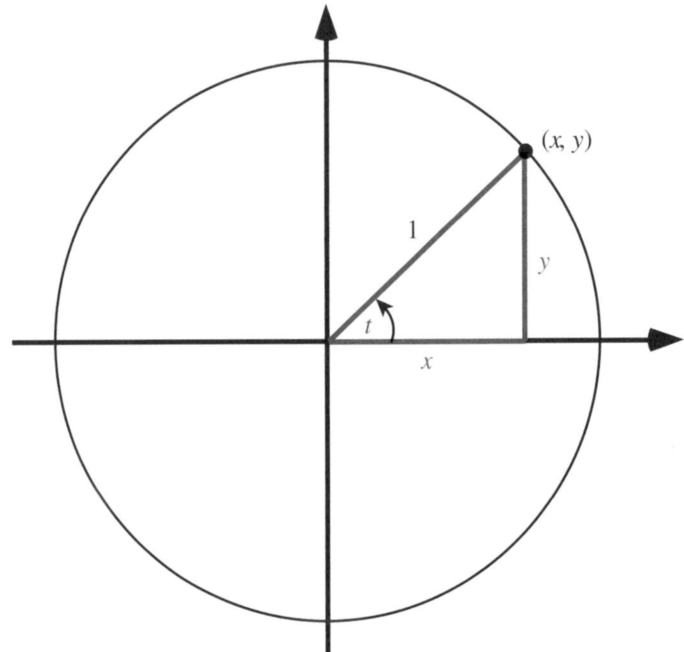

Figure 8.2.7 Triangle for evaluation of trigonometric functions

Trigonometric Functions

The sine and cosine functions are employed in many models where oscillations are involved. For example, two projects in Module 4.2, "Predator-Prey Model," considered seasonal birth rates and fishing and employed the cosine or sine function to achieve periodicity.

To define the trigonometric functions sine, cosine, and tangent, we consider the point (x, y) on the unit circle of Figure 8.2.7. For the angle t off the positive x-axis, with t being positive in the counterclockwise direction and negative in the clockwise direction, the definitions of these trigonometric functions are as follows:

$$\begin{aligned}\sin t &= y\\ \cos t &= x\\ \tan t &= y/x\end{aligned}$$

For example, if $x = 0.6$ and $y = 0.8$, then in radians (rad) t is approximately $0.9273 = 53.13°$, so that the following hold:

$$\begin{aligned}\sin(0.9273) &= 0.8\\ \cos(0.9273) &= 0.6\\ \tan(0.9273) &= 0.8/0.6 \approx 1.33\end{aligned}$$

For an angle of 0 rad, the opposite side, y, is zero, so that $\sin(0) = 0$. An angle of $\pi/2$ results in $(1, 0)$ being the point on the unit circle and the sine function achieving

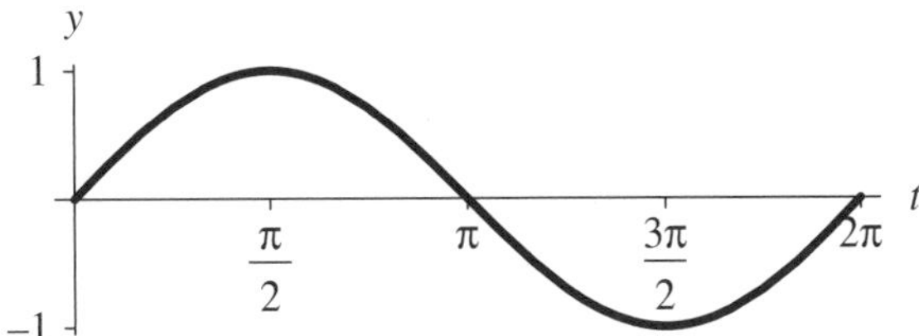

Figure 8.2.8 One cycle of the sine function

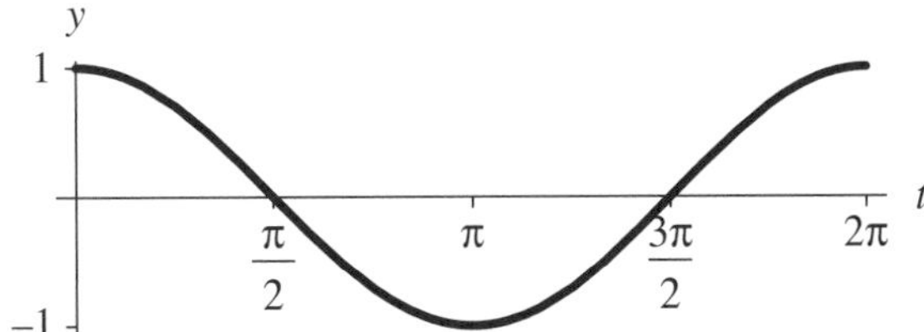

Figure 8.2.9 One cycle of the cosine function

its maximum value of 1. The sine returns to 0 for the angle $\pi = 180°$. Then, sin(t) obtains its minimum, namely, −1, at $3\pi/2$, where the point on the unit circle is (0, −1). At $t = 2\pi = 360°$, the sine function starts cycling through the same values again. Figure 8.2.8 presents one cycle of the sine function, and Figure 8.2.9 gives a cycle of the cosine function.

Quick Review Question 11

a. Evaluate sin t where $x = 0.6$ and $y = 0.8$ for angle t.
b. Evaluate $\sin(\pi/6)$ where the corresponding point on the unit circle is $(1/2, \sqrt{3}/2)$.
c. Give the domain of the sine function.
d. Give the range of the sine function.
e. Give the sine's **period**, or length of time before the function starts repeating.
f. Is sin t positive or negative for values of t in the first quadrant?
g. Is sin t positive or negative for values of t in the second quadrant?
h. Is sin t positive or negative for values of t in the third quadrant?
i. Is sin t positive or negative for values of t in the fourth quadrant?

Quick Review Question 12

a. Evaluate cos(0).
b. Evaluate $\cos(\pi/2)$.
c. Evaluate $\cos(\pi)$.
d. Evaluate $\cos(3\pi/2)$.

e. Evaluate $\cos(\pi/6)$ where the corresponding point on the unit circle is $(1/2, \sqrt{3}/2)$.
f. Give the maximum value of $\cos t$.
g. Give the minimum value of $\cos t$.
h. Give the domain of the cosine function.
i. Give the period of the cosine function.
j. Is $\cos t$ positive or negative for values of t in the first quadrant?
k. Is $\cos t$ positive or negative for values of t in the second quadrant?
l. Is $\cos t$ positive or negative for values of t in the third quadrant?
m. Is $\cos t$ positive or negative for values of t in the fourth quadrant?

For a function of the form $f(t) = A \sin(Bt)$ or $g(t) = A \cos(Bt)$, where A and B are positive numbers, A is the **amplitude**, or maximum value of the function from the horizontal line going through the middle of the function. For example, $h(t) = 3 \sin(7t)$ has amplitude 3; the function oscillates between y values of –3 and 3. Because the period of the sine and cosine functions is 2π, the period of f and g is $2\pi/B$. When $t = 0$, $Bt = 0$. When $t = 2\pi/B$, $Bt = B(2\pi/B) = 2\pi$. Thus, the period of $h(t) = 3 \sin(7t)$ is $2\pi/7$.

Definitions The **amplitude** of an oscillating function is the maximum value of the function from the horizontal line going through the middle of the function. A **periodic function** is one function whose values repeat at regular intervals, and the **period** of a periodic function is the length of such an interval.

A function of the form $\boldsymbol{f(t) = A \sin(Bt)}$ or $\boldsymbol{g(t) = A \cos(Bt)}$, where A and B are positive numbers, has amplitude A and period $2\pi/B$.

Quick Review Question 13

Plot the following functions.

a. $\sin t$ and $2 \sin(7t)$
b. $\sin t$ and a function involving sine that has amplitude 5 and period 6π
c. $\sin t$ and a function involving sine that has minimum value –2 and maximum value 4
d. $\sin t$ and a function involving sine that has amplitude 4 and crosses the t-axis at each of the following values of t: . . ., $-\pi/6$, $\pi/3$, $5\pi/6$, . . .
e. $\cos t$ and a function involving cosine that has amplitude 3, period π, and maximum value 2 at $t = \pi/5$
f. $\sin(5t)$ and $e^{-t} \sin(5t)$ (The latter is a function of decaying oscillations. The general form of such a function is $Ae^{-Ct}\sin(Bt)$, where A, B, and C are constants.)

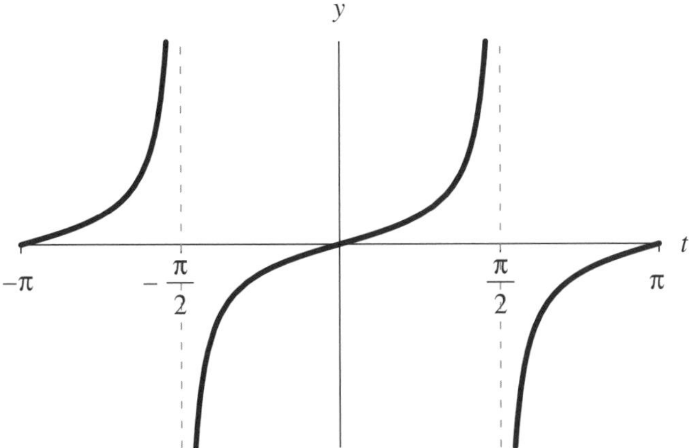

Figure 8.2.10 Tangent function

The tangent function is also periodic. Because tan $t = y/x$, for a corresponding point (x, y) on the unit circle (see Figure 8.2.7), tan t = sin t/cos t. The graph of this function appears in Figure 8.2.10, and the next Quick Review Question explores some of its properties.

Quick Review Question 14

- **a.** Evaluate $\tan(\pi/6)$ where the corresponding point on the unit circle is $(1/2, \sqrt{3}/2)$.
- **b.** Evaluate tan(0).
- **c.** Evaluate $\tan(\pi)$.
- **d.** Evaluate $\tan(\pi/2)$.
- **e.** As t approaches $\pi/2$ from values less than $\pi/2$, what does tan t approach?
- **f.** As t approaches $\pi/2$ from values greater than $\pi/2$, what does tan t approach?
- **g.** Evaluate $\tan(-\pi/2)$.
- **h.** As t approaches $-\pi/2$ from values less than $-\pi/2$, what does tan t approach?
- **i.** As t approaches $-\pi/2$ from values greater than $-\pi/2$, what does tan t approach?
- **j.** Give the range of the tangent function.
- **k.** Give all the values between -2π and 2π for which tan t is not defined.
- **l.** Give an angle in the third quadrant that has the same value of tan t, where t is in the first quadrant.
- **m.** Give an angle in the fourth quadrant that has the same value of tan t, where t is in the second quadrant.
- **n.** Give the period of the tangent function.

MODULE 8.3

Empirical Models

Downloads

For several computational tools, the text's website has an *8_3QRQ.pdf* file available for download, which contains system-dependent Quick Review Questions and answers for this module.

Moreover, the text's website has an *EmpiricalModels* file, which contains the models of this module, available for download for various computational tools, Table 8.3.1 lists data files that are also available on the website and where they are employed in the text. The data files are based on files from the National Institute of Standards and Technology (NIST) website, "Statistical Reference Datasets," as indicated in the references. The name of each data file is as on the NIST site, except that

Table 8.3.1
Data Files on Textbook's Website

Description File	*Data File*	*Where Used*
BoxBODEM.txt	*BoxBODEM.dat*	Project 9
DanWoodEM.txt	*DanWoodEM.dat*	"Nonlinear One-Term Model"
FilipEM.txt	*FilipEM.dat*	"Multiterm Models"
Gauss1EM.txt	*Gauss1EM.dat*	Project 5
Lanczos1EM.txt	*Lanczos1EM.dat*	Project 7
Lanczos3EM.txt	*Lanczos3EM.dat*	Project 4
MGH10EM.txt	*MGH10EM.dat*	Project 8
MGH17EM.txt	*MGH17EM.dat*	Project 6
Misra1aEM.txt	*Misra1aEM.dat*	"Solving for y in a One-Term Model"
NoInt1EM.txt	*NoInt1EM.dat*	Project 1
NorrisEM.txt	*NorrisEM.dat*	"Linear Empirical Model," Exercise 1
PontiusEM.txt	*PontiusEM.dat*	Project 2
Wampler1EM.txt	*Wampler1EM.dat*	Project 3

EM appears before the extension *.txt* or *.dat*. File names with the extension *.txt* give the file name for the data file, URL reference, original dataset name, description, reference, data in column format (*y*, then *x*), and statements in several computational tools assigning appropriate data lists to *xLst* and *yLst*. The corresponding files with the extension *.dat* store only the data in column format (*x*, then *y*), which most computational tools can read.

Introduction

Sometimes it is difficult or impossible to develop a mathematical model that explains a situation. However, if data exist, we can often use these data as the sole basis for an **empirical model**. The empirical model consists of a function that fits the data. The graph of the function goes through the data points approximately. Thus, although we cannot employ an empirical model to explain a system, we can use such a model to predict behavior where data do not exist. Data are crucial for an empirical model. We utilize data to suggest the model, to estimate its parameters, and to test the model.

Definition An **empirical model** is based only on data and is used to predict, not explain, a system. An empirical model consists of a function that captures the trend of the data.

When we derive a mathematical model through analysis of a system, we may accept a model that does not fit the data as closely as we would wish because the model explains the situation well. However, with an empirical model, the data are our only source of information about the system.

Sometimes with a derived model, which helps to explain the science, it may be difficult or impossible to differentiate or integrate a function to perform further analysis. In this case, too, we can derive an empirical model, such as a polynomial function, that is differentiable and integrable. For example, a step function might accurately model a pulsing signal, but we cannot differentiate such a function where it is discontinuous, or jumps from one step to the next. In this case, we might use trigonometric functions, which we can differentiate and integrate, in an empirical model that captures the trend of the data.

Linear Empirical Model

We begin studying empirical models by considering a National Institute of Standards and Technology (NIST) study involving calibration of ozone monitors, where x is NIST's measurement of ozone concentration and y is the customer's measurement. For the purpose of this example, we take the subset of the data shown in Table 8.3.2.

Table 8.3.2
Subset of NIST *Norris* Dataset, Where x is "NIST's Measurement of Ozone Concentration" and y is "the Customer's Measurement"

x	y
0.2	0.1
0.4	0.3
0.3	0.3
0.3	0.6

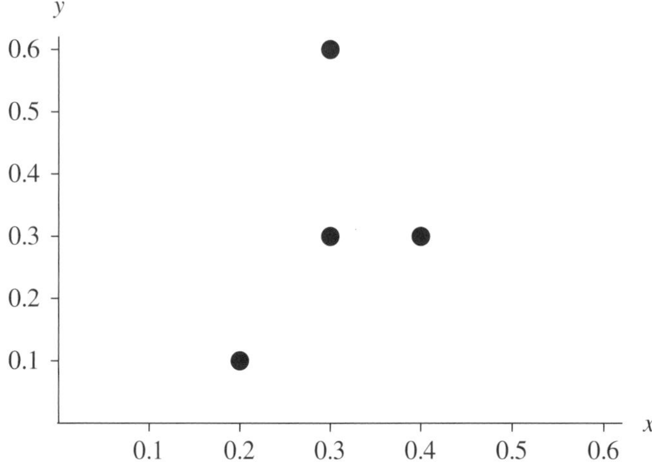

Figure 8.3.1 Plot of data in Table 8.3.2

Using an appropriate computational tool, we define a set of ordered pairs, assigning the result to a variable $pts = \{(0.2, 0.1), (0.4, 0.3), (0.3, 0.3), (0.3, 0.6)\}$. Figure 8.3.1 shows a plot of these data.

To explain obtaining a line that fits the data well, we need the definition of **linear combination**. The linear combination of p and q is a sum of the form $ap + bq$, where a and b are constants. For example, $3p + 7q$ is a linear combination of p and q. A linear combination of 1 and x has the form $b \cdot 1 + m \cdot x = b + mx = mx + b$, for constants b and m. For example, $-2.2x + 9.3$ is a linear combination of x and 1. We can extend the definition of linear combination to any number of terms. Thus, $4 - 3x + 19x^2$ is a linear combination of 1, x, and x^2, while $5x$ is a linear combination of just x.

Definition For positive integer n, a **linear combination** of $x_1, x_2, \ldots, x_n$ is a sum

$$a_1x_1 + a_2x_2 + \cdots + a_nx_n$$

where $a_1, a_2, \ldots, a_n$ are constants.

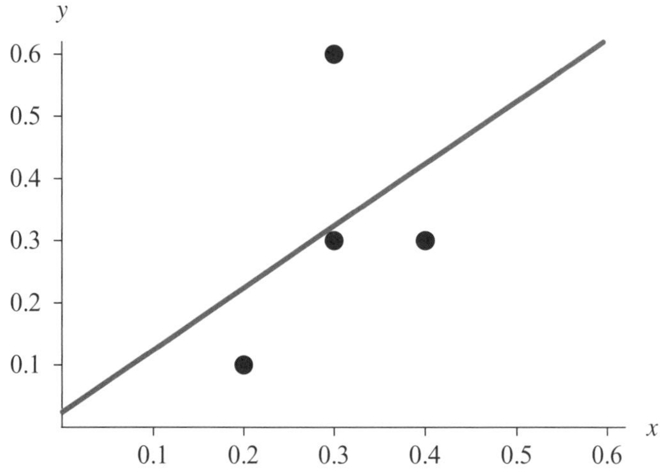

Figure 8.3.2 Plot of Figure 8.3.1, along with the best-fit line

Quick Review Question 1

List the expressions that are linear combinations of u and v.

A. $5u - 18v$ B. $-18v + 5u$ C. $7u$
D. $15uv$ E. $u/5 + v/3$ F. $5/u + 3/v$

A computational tool usually has a function that can return an equation that is a **least-squares** fit to a list of points. In the section "Linear Regression," we discuss the algorithm, but we can use such a fit function without knowing the formulas involved. The equation $y = 0.025 + 1.0x$, which is a linear combination of 1 and x, is the least-squares linear function that best fits the data in Table 8.3.2. We can plot this line, along with the original data, to obtain a graph similar to Figure 8.3.2.

Quick Review Question 2

From the text's website, obtain your computational tool's *8_3QRQ.pdf* file for this system-dependent question concerning a command to obtain a least-squares line that best fits a set of points.

Predictions

We can use the result of the least-squares linear fit, $y = 0.025 + 1.0x$, to predict y-values where no data value exists as long as those values are within the range of values used to determine the formula. For example, for NIST's measurement of ozone concentration of $x = 0.34$, the predicted customer's ozone concentration measurement is $y = 0.025 + 1.0(0.34) = 0.365$. Figure 8.3.3 displays the point (0.34, 0.365), which is larger than the other points, on the curve.

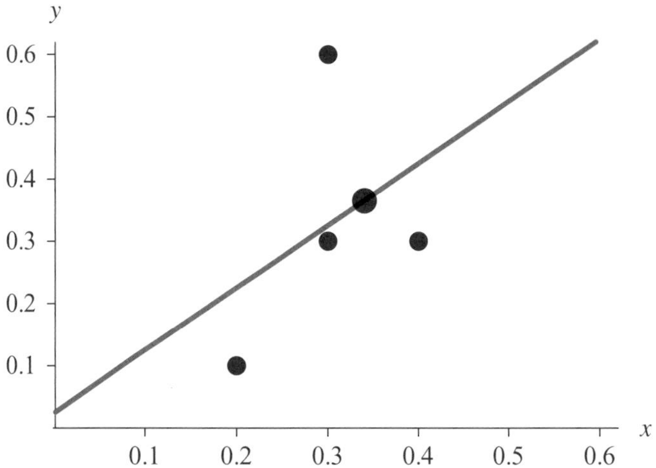

Figure 8.3.3 Predicted point (0.34, 0.365), which is larger

We must be careful not to employ this predictive function beyond the range of the data. With an empirical model, the data drive the model. Outside the range of the data, we cannot depend on the data behaving in a similar manner to observations within the range. For example, the ozone monitor being calibrated could be fairly accurate when measuring small concentrations, but completely unreliable for large concentrations.

Linear Regression

A fit function in an appropriate computational tool returns a least-squares fit to the data. In the preceding example, a fit function determined that $y = 0.025 + 1.0x$ is the line that best captures the trend of the data using a technique called **linear least-squares regression,** or **linear regression**. We call x the **predictor variable** and y the **response variable**. The method can find the line $y = mx + b$ that minimizes the sum of the squares of the vertical distances from the data points to the line. For example, the point that is directly above or below (0.2, 0.1) on a line $y = mx + b$ is $(0.2, m \cdot 0.2 + b)$. We obtain the y-value, $m \cdot 0.2 + b$, on the line by substituting the x-value, 0.2, into the linear function. The difference in the y-values of the point on the line and the point (0.2, 0.1) is $m \cdot 0.2 + b - 0.1$. The lengths of the dotted lines in Figure 8.3.4 are the absolute values of such differences. Linear regression finds m and b so that the sum of the squares of the vertical distances is as small as possible. Thus, for n points, $(x_1, y_1), (x_2, y_2), \ldots, (x_n, y_n)$, the method does the following, where the summation ($\sum_{i=1}^{n}$) indicates summing the squared terms for $i = 1, 2, \ldots, n$:

$$\text{minimize} \quad \sum_{i=1}^{n}(mx_i + b - y_i)^2 = (mx_1 + b - y_1)^2 + (mx_2 + b - y_2)^2 + \cdots + (mx_n + b - y_n)^2$$

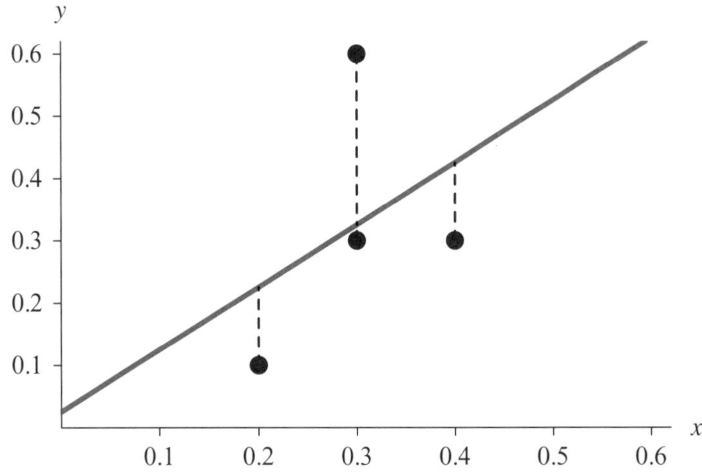

Figure 8.3.4 Data points with dashed vertical lines to the least-squares regression line

Using calculus, minimization techniques yield m and b, as follows:

$$m = \frac{n\sum x_i y_i - \sum x_i \sum y_i}{n\sum x_i^2 - \left(\sum x_i\right)^2}$$

$$b = \frac{\sum x_i^2 \sum y_i - \sum x_i y_i \sum x_i}{n\sum x_i^2 - \left(\sum x_i\right)^2}$$

Fortunately, a fit function in an appropriate computational tool performs these calculations for us.

In a fashion similar to this computation, a fit function can return the equation that is a linear combination of given functions and that yields the minimum of the sum of the squares of the vertical distances from the points to the corresponding curve. Consequently, using a fit function we can obtain nonlinear functions with multiple terms that model the data empirically.

Nonlinear One-Term Model

Table 8.3.3 presents data from NIST's *DanWood* dataset for the next example (see the section "Downloads"). The predictor variable x is the absolute temperature of the filament in 1000 K, while the response variable y is the energy radiated from a carbon filament lamp per square centimeter per second.

Figure 8.3.5 shows a plot of the data from Table 8.3.3. Although not the regression line, a faint line through the first and last points helps us to see that the configu-

Table 8.3.3
Data from NIST's *DanWood* Dataset

x	y
1.309	2.138
1.471	3.421
1.490	3.597
1.565	4.340
1.611	4.882
1.680	5.660

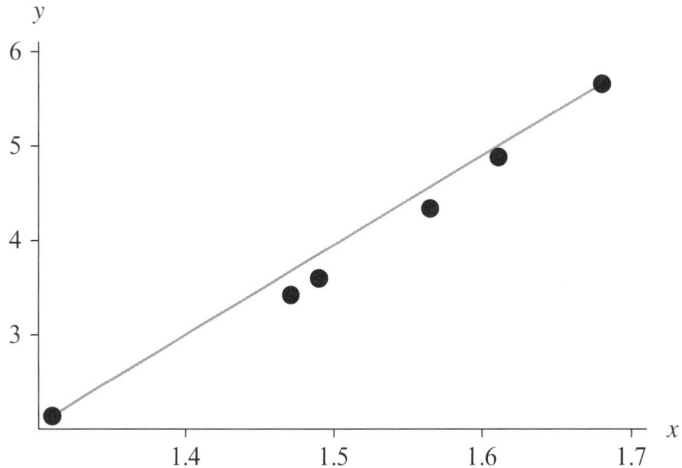

Figure 8.3.5 Plot of points from Table 8.3.3 with a line through the first and last points

ration of points is slightly concave up. Such a data set, whose plot is concave up or down throughout, can usually be modeled effectively with a function in which only one term has the single dependent variable. To effectively determine a mathematical **one-term model** for this data using linear regression, we use a transformation of the data that appears linear.

Quick Review Question 3

From the text's website, obtain your computational tool's *8_3QRQ.pdf* file for this system-dependent question concerning commands to plot data and a line together.

We can accomplish the transformation on this data from being concave up to straight in one of two ways:

1. Extend the points to the right, stretching the distance from the y-axis to the rightmost points more than to those on the left. Thus, perform an operation

on x that results in greater values and has more effect on the larger x-values than on the smaller ones.

2. Pull points down, shrinking the distance between the x-axis and the higher points more than to the lower ones. Thus, perform an operation on y that results in lesser values and has more effect on the larger y-values than on the smaller ones.

For the first alternative with the preceding data, the operation might be to raise all x values to a power greater than 1, such as 2. To see the effect, consider the data points (1.309, 2.138) and (1.680, 5.660), where the former is to the left of the latter. Squaring both x-coordinates, we find that $1.309^2 = 1.713$, while $1.680^2 = 2.822$. The difference $x^2 - x$ for the first point is $1.309^2 - 1.309 = 0.404$, but the effect on the second, rightmost point is much greater with a difference of $1.680^2 - 1.680 = 1.142$. Thus, the transformation of squaring the x-coordinate, where $x > 1$, stretches the rightmost points to the right even more than the points that are further to the left.

Similarly, for the second alternative, taking the square root of the y-coordinates, which are all greater than 1, gives smaller values. However, the effect is more pronounced on the larger y-values. The point $(1.309, \sqrt{2.138})$ is 0.676 units lower than (1.309, 2.138), but $(1.680, \sqrt{5.660})$ is 3.28 units lower than (1.680, 5.660).

We should note that these operations perform as indicated because the coordinates are all greater than 1. Recall that for a number c between 0 and 1, c^2 is smaller than c, while $\sqrt{c}$ is larger. Moreover, when the values are negative, we cannot perform certain operations, such as taking the square root or logarithm. Also, raising a negative value to an even exponent gives a positive number. To obtain the desired results, we should reason carefully and not apply operations randomly.

Table 8.3.4 gives a sequence of transformations that have an increasingly greater impact on larger values, where the numbers are greater than 1. The transformations, such as $-1/z$, that involve a unary minus do so to maintain the same ordering of the data points. For example, (1.309, 2.138) is to the left of (1.680, 5.660). Using the transformation $1/x$, (1/1.309, 2.138) = (0.7639, 2.138) is to the right of (1/1.680, 5.660) = (0.5952, 5.660). However, with $-1/x$, the point (–1/1.309, 2.138) = (–0.7639, 2.138) remains to the left of (–1/1.680, 5.660) = (–0.5952, 5.660).

Taking the first alternative given before, which performs an operation on x, we pair various powers of x with the corresponding y. We plot the resulting ordered pairs in an attempt to find a graph that appears approximately linear. As Figures 8.3.6 and 8.3.7 show, with the assistance of lines through the first and last points, squaring and cubing the x-coordinates seem still to result in plots that are concave up. However, raising the x-values to the fourth power, as in Figure 8.3.8, appears to

Table 8.3.4
Sequence of Transformations for $z > 1$

$$\ldots, -\frac{1}{z^2}, -\frac{1}{z} - \frac{1}{\sqrt{z}}, \ln(z), \sqrt{z}, z, z^2, z^3, \ldots$$

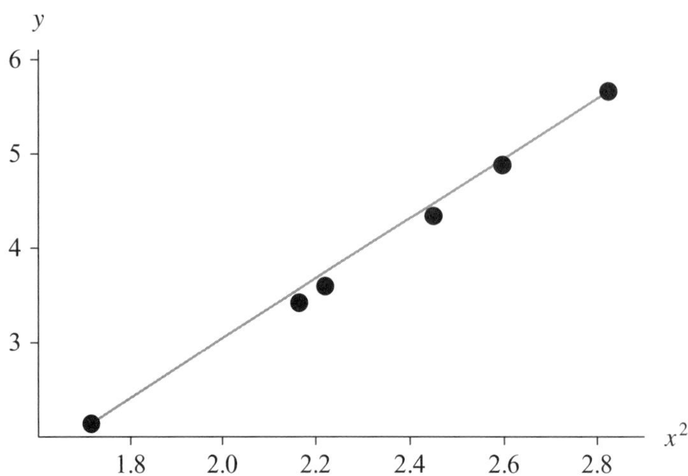

Figure 8.3.6 Plot of points (x^2, y) for data in Table 8.3.3, with a line through the first and last points

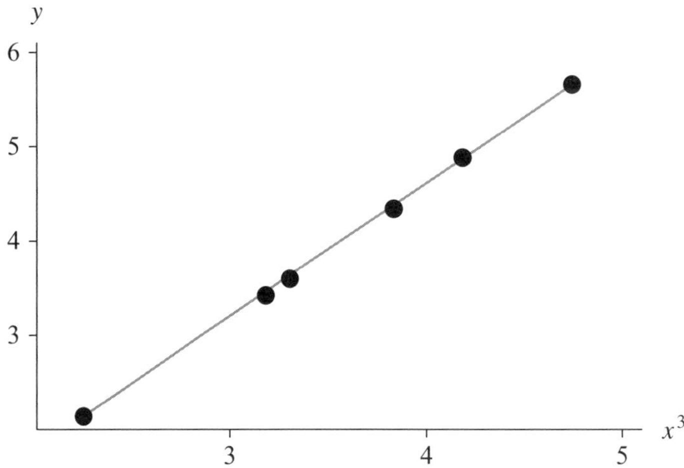

Figure 8.3.7 Plot of points (x^3, y) for data in Table 8.3.3, with a line through the first and last points

cause a graph that is slightly concave down. If we are not satisfied with the exponent 3 or 4, we might try powers between these values. Figure 8.3.9 shows a plot of the points $(x^{3.5}, y)$.

Thus, using a fit function in an appropriate computational tool, we employ linear regression on the transformed set of points $(x^{3.5}, y)$ and obtain the following best-fit line: $y = -0.393131 + 0.988186z$. Figure 8.3.10 shows the graph of this line with the points of Figure 8.3.9.

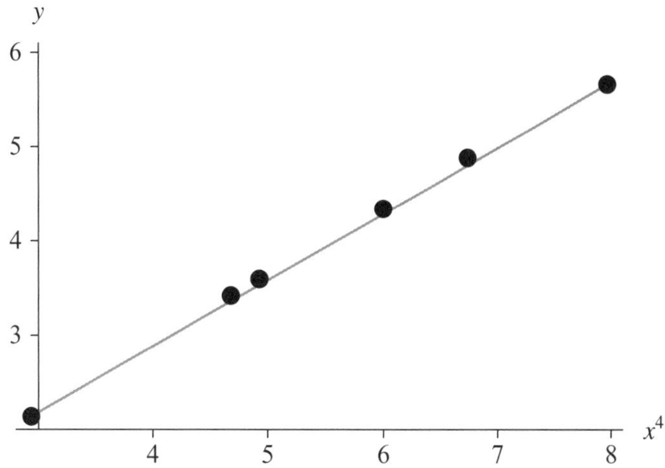

Figure 8.3.8 Plot of points (x^4, y) for data in Table 8.3.3, with a line through the first and last points

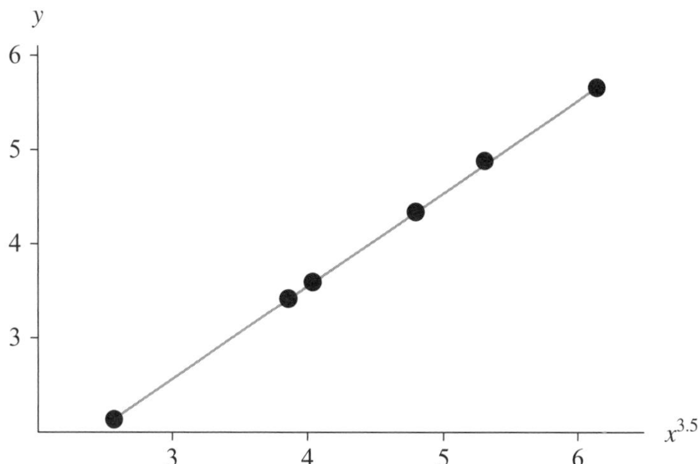

Figure 8.3.9 Plot of points ($x^{3.5}$, y) for data in Table 8.3.3, with a line through the first and last points

Satisfied with this result, we still must determine the function through the original set of points and view its curve through those points. Because we fit a line ($y = -0.393131 + 0.988186z$) to the transformed points of the form ($x^{3.5}$, y), we now substitute $x^{3.5}$ for z to obtain our empirical model, which is as follows:

$$f(x) = -0.393131 + 0.988186x^{3.5}$$

Figures 8.3.11 and 8.3.12 present two graphs of this function, along with the original data from Table 8.3.3 for different ranges of x.

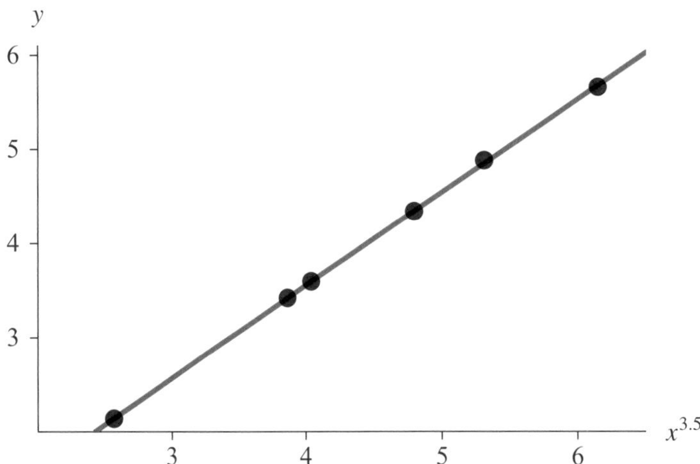

Figure 8.3.10 Graph of linear regression line with points of Figure 8.3.9

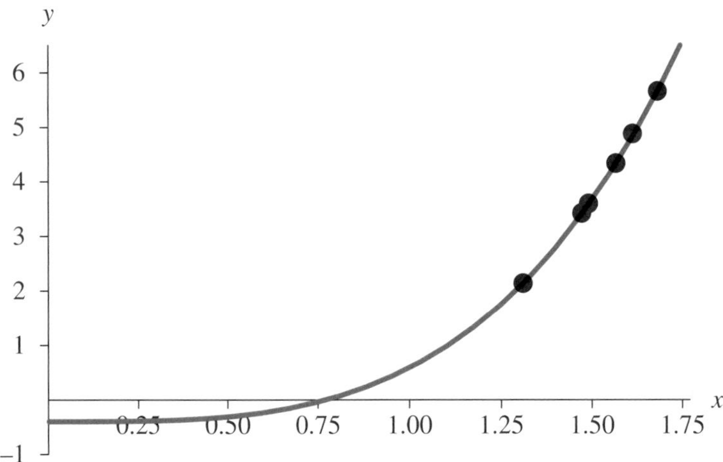

Figure 8.3.11 Graph of $f(x) = -0.393131 + 0.988186x^{3.5}$ and the data of Table 8.3.3

The graphs indicate that our empirical model f seems to be satisfactory for x from 1.3 to 1.7. To verify the model, we should collect additional data in this range and plot all the data with the graph of the model to observe how they agree. Moreover, for each newly observed x value, we should determine how closely the observed and predicted y values agree.

Clearly, other empirical models than $y = -0.393131 + 0.988186x^{3.5}$ approximate the data. Empirical modeling is an art as well as a science. Several metrics aid in determination of which model to use but are beyond the scope of this text.

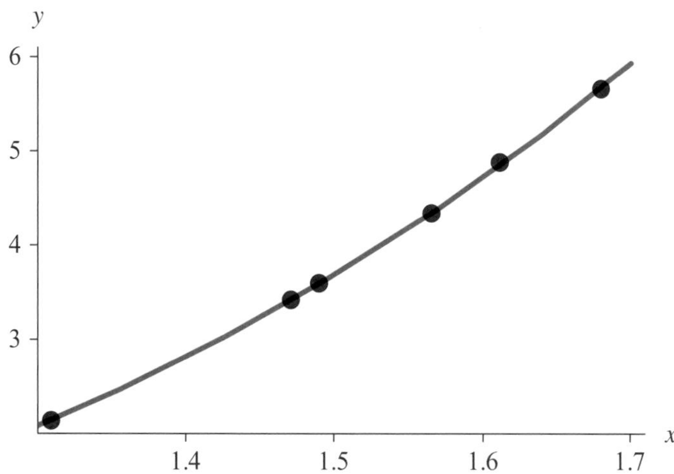

Figure 8.3.12 Graph of Figure 8.3.11 for x between 1.3 and 1.7

Solving for y in a One-Term Model

In this section, we consider empirical model development in which we make a transformation on y and perhaps on x, too, instead of just on x, as in the previous section. The data, which Table 8.3.5 lists, are from NIST dental research in monomolecular adsorption, where x represents pressure and y volume (NIST *Misra1a* Dataset). Figure 8.3.13 displays a plot of these data, with a faint line between the first and last points to emphasize concavity.

Table 8.3.5
Data from *Misra1aEM.dat*, Available on the Textbook's Website and in the NIST *Misra1a* Dataset

x	y
77.6	10.07
114.9	14.73
141.1	17.94
190.8	23.93
239.9	29.61
289.0	35.18
332.8	40.02
378.4	44.82
434.8	50.76
477.3	55.05
536.8	61.01
593.1	66.40
689.1	75.47
760.0	81.78

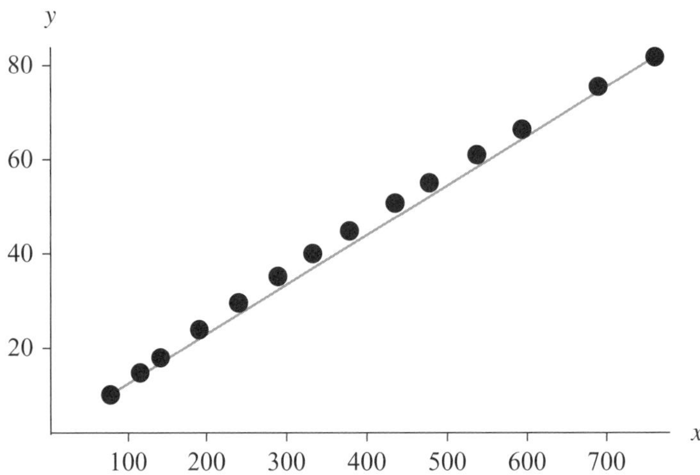

Figure 8.3.13 Plot of data from Table 8.3.5 with a line through the first and last points

Quick Review Question 4

Suppose *xLst* and *yLst* are lists of x- and y-values, respectively. Give the command in an appropriate computational tool to plot these data with large points, as in Figure 8.3.13. If necessary in your tool, assign to *pts* the list of ordered pairs of corresponding x- and y-values, and then use *pts* in a plot command.

The dots of Figure 8.3.13 are in a concave-down pattern. As with concave-up graphs having coordinates greater than 1, we can transform these data from concave down to straight in one of two ways:

1. Pull points left, shrinking the distance from the y-axis to the rightmost points more than to those on the left. Thus, perform an operation on x that results in smaller values and has more effect on the larger x-values than on the smaller ones.
2. Extend points up, stretching the distance from the x-axis to the higher points more than to the lower ones. Thus, perform an operation on y that results in greater values and has more effect on the larger y-values than on the smaller ones.

Using transformations from the sequence in Table 8.3.4, we can transform x or y. As Figure 8.3.14 shows, the plot of points $(x, y^{6/5})$ is close to being linear.

With a fit function, we can obtain the following equation of a line that captures the trend of these points:

$$u = -5.46747 + 0.267834z$$

Substituting $y^{6/5}$ for u and x for z, we obtain the following equation:

$$y^{6/5} = -5.46747 + 0.267834x$$

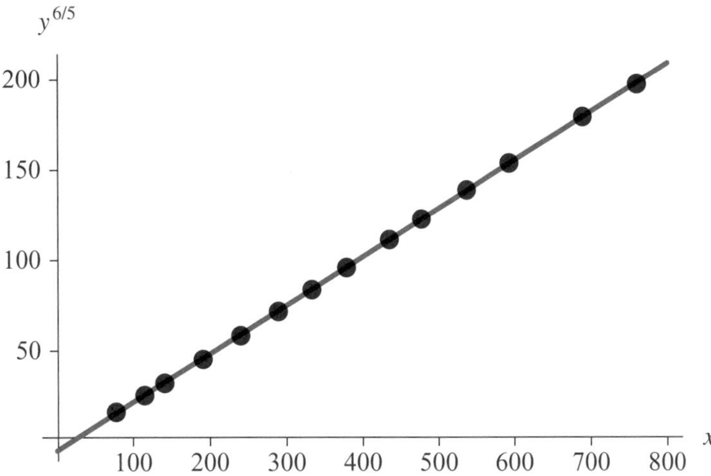

Figure 8.3.14 Plot of $(x, y^{6/5})$ for data from Table 8.3.5, with a line through the first and last points

After solving for y by raising each side to the 5/6 power, we can define our model, as follows:

$$f(x) = (-5.48629 + 0.267869x)^{5/6}$$

Figure 8.3.15 shows the graph of this function, along with the original data. We should always plot our model with the original data to verify that the function really does capture the trend of the untransformed data.

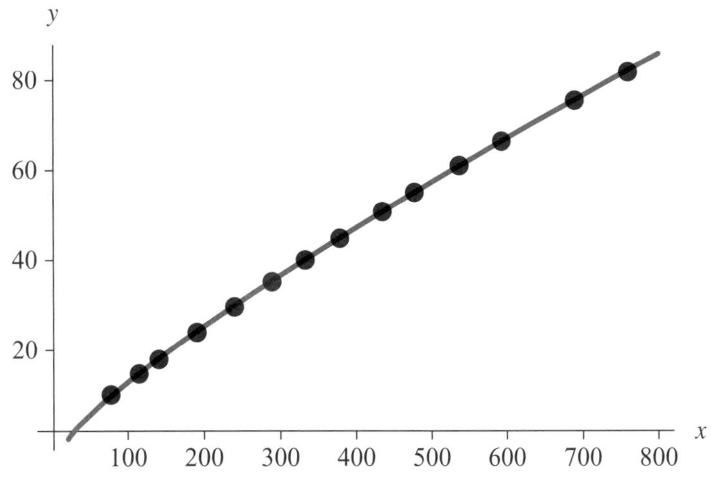

Figure 8.3.15 Graph of $y = (-5.48629 + 0.267869x)^{5/6}$ and data from Table 8.3.5

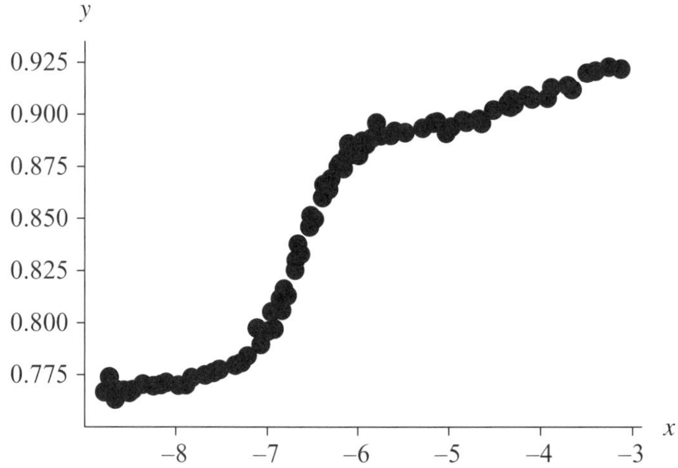

Figure 8.3.16 Plot of data from NIST's *Filip* Dataset

Multiterm Models

In a NIST study, A. Filippelli collected the data that Figure 8.3.16 displays (see the section "Downloads"). The curve almost has the look of a logistic function. However, the right tail does not trend asymptotically to a value but keeps increasing in a wavy fashion. Thus, one possibility is to develop a polynomial empirical model. Such models are particularly useful because we can readily differentiate and integrate a polynomial for further analyses.

We can fit to the data a fourth-degree polynomial, which is a linear combination of 1, z, z^2, z^3, and z^4 with the form $y = b_0 + b_1z + b_2z^2 + b_3z^3 + b_4z^4$. Without the use of a fit function, in a process called **interpolation**, we can determine the coefficients b_i by solving five equations simultaneously. For each equation, a different data point is substituted into the general fourth-degree polynomial, with z being replaced by the first coordinate and y by the second. Then, we solve the five equations simultaneously for b_i, $i = 0, 1, 2, 3, 4$. Often, as in this example, we have many more data than interpolation requires. A fit function obtains a least-squares fit of the general fourth-degree polynomial to all the data instead of interpolating through a limited number of specific points. Thus, for the remainder of this section, we continue to use a fit function. Figure 8.3.17, which is a plot of the data and the fourth-degree polynomial that a fit function returns, reveals shortcomings of the model.

Because this fourth-degree polynomial has an inadequate number of "hills" and "valleys" to represent the data, we fit higher-degree polynomials to the points. For a linear combination of expressions, 1, z, z^2,. . ., z^{10}, a fit function returns the following tenth-degree polynomial: $8.45174 + 1.36940z - 5.35707z^2 - 0.34983z^3 - 0.410472z^4 +$

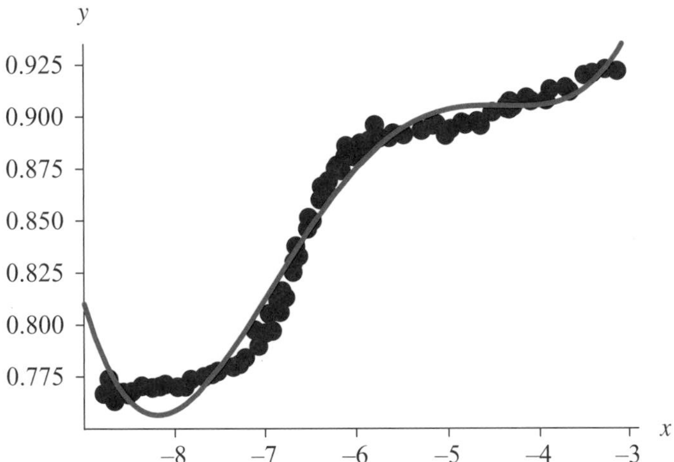

Figure 8.3.17 Plot of the fitted fourth-degree polynomial and the data

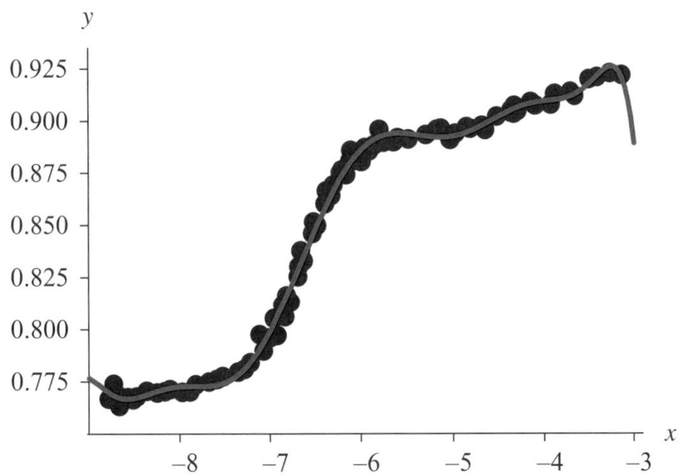

Figure 8.3.18 Plot of tenth-degree polynomial fitted to the data in NIST's *Filip* Dataset

$0.256553z^5 + 0.119554z^6 + 0.0231150z^7 + 0.00240206z^8 + 0.000131536z^9 + 2.98847 \times 10^{-6}z^{10}$. A graph of this polynomial with the data reveals that this empirical model captures the trend of the data better than smaller-degree polynomials (see Figure 8.3.18).

As indicated previously, we must be very careful not to apply this model outside the range of the data. The danger is particularly striking when we consider polynomials. For example, Figure 8.3.19 shows the dramatic slant of this polynomial just one unit to the right and left of the graph in Figure 8.3.18.

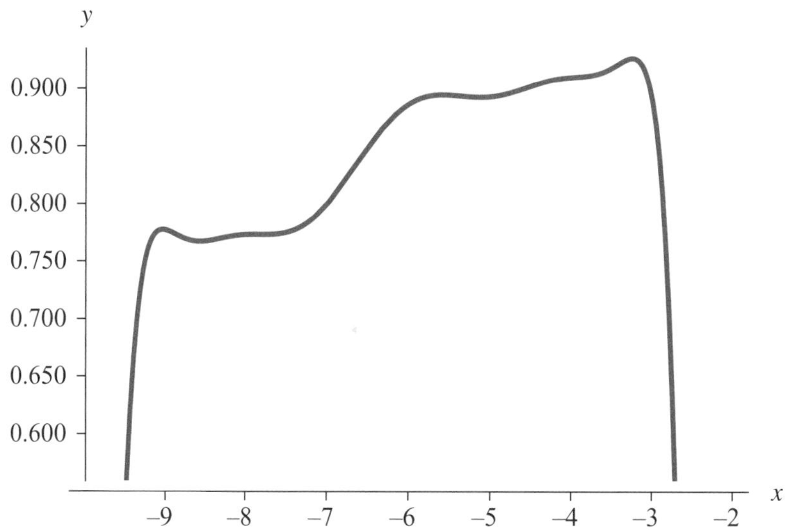

Figure 8.3.19 Plot of tenth-degree polynomial model inside and outside the range of the data

Advanced Fitting with Computational Tools

Many computational tools have resources for fitting polynomial, logistic, sine, exponential, and general functions to data. Tutorial 3 for your computational tool explores some of these techniques.

Exercise

1. Using all the data in *NorrisEM.dat*, which is available on the textbook's website, construct an empirical model using a computational tool. Compare your results to the model that was developed with a subset of the data in the section "Linear Empirical Model."

Projects

For Projects 1–9, develop a model for each of the following data sets (see the section "Downloads").

1. *NoInt1EM.dat*
2. *PontiusEM.dat*
3. *Wampler1EM.dat*
4. *Lanczos3EM.dat*
5. *Gauss1EM.dat*
6. *MGH17EM.dat*
7. *Lanczos1EM.dat*
8. *MGH10EM.dat*
9. *BoxBODEM.dat*

10. With a system dynamics tool or a computational tool, solve $\frac{dP}{dt} = 0.10P$ with $P_0 = 100$ using Euler's, Runge-Kutta 2 (EPC), and Runge-Kutta 4 techniques (see Chapter 6). For each method, compute the relative errors at $t = 100$ for several values of Δt less than or equal to 1.0, such as $\Delta t = 0.1$, 0.2, 0.3, . . ., 1.0; plot relative error at time 100 versus Δt; and fit a function to the data. Do your results agree with the statement in Module 6.4, "Runge-Kutta 4 Method," that relative errors of the techniques are $O(\Delta t)$, $O(\Delta t^2)$, and $O(\Delta t^4)$, respectively?

Using an advanced data-fitting function from your computational tool, develop an empirical model for each of the following data sets.

11. With the *SIR* file in your system dynamics tool, for a particular situation, generate data for the simulated number of *susceptibles* (*S*), *infecteds* (*I*), and *recovereds* (*R*) (see Module 4.3, "Modeling the Spread of SARS—Containing Emerging Disease," and Figure 4.3.3). Use a small time step, but take data over larger intervals, such as $t = 1$ da. Discover functions that fit each of the simulated datasets.
12. With the *VerticalSpring* file in your system dynamics tool, for a particular situation, generate data for length (m) of an undamped spring with respect to time in seconds (see Module 3.2, "Modeling Bungee Jumping," and Figure 3.2.4). Use a small time step, but take data over larger intervals. Discover a sine function that fits the simulated data.
13. With the *Bungee* file in your system dynamics tool, for a particular situation, generate data for length (m) of a damped spring with respect to time in seconds (see Module 3.2, "Modeling Bungee Jumping," and Figure 3.2.5). Use a small time step, but take data over larger intervals. Discover a polynomial or other function that fits the simulated data.
14. With the *simplePendulum* file in your system dynamics tool, for a particular situation, generate data for angle in radians, angular velocity in radians per second, and angular acceleration in radians per second squared versus time in seconds (see Module 3.3, "Tick Tock—The Pendulum Clock," and Figure 3.3.3). Use a small time step, but take data over larger intervals. Discover sine functions that fit the simulated datasets.
15. With the *Rocket* file in your system dynamics tool, for a particular situation, generate data for position (m) and velocity (m/sec) of a rocket versus time in seconds (see Module 3.4, "Up, Up, and Away—Rocket Motion," and Figure 3.4.2). Use a small time step, but take data over larger intervals. Discover functions that fit the simulated datasets. For velocity, employ two functions, one quadratic and one linear, pieced together. For position, fit a polynomial to the velocity data.
16. The Average Daily Temperature Archive (University of Dayton 2012) has datasets for the average daily temperatures over a period of several years for a number of cities in the United States. For one city, find a sine function that fits the data.

Answers to Quick Review Questions

1. The following are linear combinations of u and v:

A. $5u - 18v = (5)u + (-18)v$
B. $-18v + 5u = (-18)v + (5)u$
C. $7u = (7)u + (0)v$
D. $u/5 + v/3 = (1/5)u + (1/3)v$

2–4. The text's website has an *8_3QRQ.pdf* file, which contains these system-dependent Quick Review Questions and answers, available for download for various computational tools.

References

Giordano, Frank R., Maurice D. Weir, and William P. Fox. 2003. *A First Course in Mathematical Modeling*. 3rd ed. Pacific Grove, CA: Brooks/Cole–Thompson Learning.

Information Technology Laboratory. "NIST Statistical Reference Datasets." National Institute of Standards and Technology. http://www.itl.nist.gov/div898/strd/ (accessed December 27, 2012)

NIST Dataset Archives. "StRD Dataset BoxBOD." National Institute of Standards and Technology. http://www.itl.nist.gov/div898/strd/nls/data/boxbod.shtml. Originally from G. P. Box, W. G. Hunter, and J. S. Hunter. *Statistics for Experimenters*. New York: Wiley, 1978, pp. 483–487. (accessed December 27, 2012)

———. "StRD Dataset DanWood." National Institute of Standards and Technology. http://www.itl.nist.gov/div898/strd/nls/data/daniel_wood.shtml. Originally from C. Daniel, and F. S. Wood. *Fitting Equations to Data*. 2nd ed. New York: Wiley, 1980, pp. 428–431.

———. "StRD Dataset Filip." National Institute of Standards and Technology. http://www.itl.nist.gov/div898/strd/lls/data/Filip.shtml. Originally from A. Filippelli. National Institute of Standards and Technology.

———. "StRD Dataset Gauss1." National Institute of Standards and Technology. http://www.itl.nist.gov/div898/strd/nls/data/gauss1.shtml. Originally from B. Rust, National Institute of Standards and Technology, 1996.

———. "StRD Dataset Lanczos1." National Institute of Standards and Technology. http://www.itl.nist.gov/div898/strd/nls/data/lanczos1.shtml. Originally from C. Lancoz. *Applied Analysis*. Englewood Cliffs, NJ: Prentice Hall, 1956, pp. 272–280.

———. "StRD Dataset Lanczos3." National Institute of Standards and Technology. http://www.itl.nist.gov/div898/strd/nls/data/lanczos3.shtml. Originally from C. Lancoz. *Applied Analysis*. Englewood Cliffs, NJ: Prentice Hall, 1956, pp. 272–280.

———. "StRD Dataset MGH10." National Institute of Standards and Technology. http://www.itl.nist.gov/div898/strd/nls/data/mgh10.shtml. Originally from R. R. Meyer, *Theoretical and Computation Aspects of Nonlinear Regression. In Nonlin-*

ear Programming. J. B. Rosen, O. L. Mangasarian, and K. Ritter (eds.). New York: Academic Press, 1970, pp. 465–486.

———. "StRD Dataset MGH17." National Institute of Standards and Technology. http://www.itl.nist.gov/div898/strd/nls/data/mgh17.shtml. Originally from M. R. Osborne, *Some Aspects of Nonlinear Least Squares Calculations. In Numerical Methods for Nonlinear Optimization*. F. A. Lootsma (ed.). New York: Academic Press, 1972, pp.171–189.

———. "StRD Dataset Misra1a." National Institute of Standards and Technology. http://www.itl.nist.gov/div898/strd/nls/data/misra1a.shtml. Originally from D. Misra, "Dental Research Monomolecular Adsorption Study." National Institute of Standards and Technology, 1978.

———. "StRD Dataset NoInt1." National Institute of Standards and Technology. http://www.itl.nist.gov/div898/strd/lls/data/NoInt1.shtml. Originally from K. Eberhardt, National Institute of Standards and Technology.

———. "StRD Dataset Norris." National Institute of Standards and Technology. http://www.itl.nist.gov/div898/strd/lls/data/Norris.shtml. Originally from J. Norris, "Calibration of Ozone Monitors." National Institute of Standards and Technology.

———. "StRD Dataset Pontius." National Institute of Standards and Technology. http://www.itl.nist.gov/div898/strd/lls/data/Pontius.shtml. Originally from P. Pontius, "Load Cell Calibration." National Institute of Standards and Technology.

———. "StRD Dataset Wampler1." National Institute of Standards and Technology. http://www.itl.nist.gov/div898/strd/lls/data/Wampler1.shtml. Originally from R. H. Wampler, "A Report of the Accuracy of Some Widely Used Least Squares Computer Programs." *Journal of the American Statistical Association*, 65(1970): 549–565.

University of Dayton. 2012 Average Daily Temperature Archive. http://academic.udayton.edu/kissock/http/Weather/citylistUS.htm (accessed January 12, 2013)

9

SIMULATING WITH RANDOMNESS

MODULE 9.1

Computational Toolbox—Tools of the Trade: Tutorial 4

Prerequisite: Module 8.1, "Computational Toolbox—Tools of the Trade: Tutorial 3."

Download

From the textbook's website, download Tutorial 4 in the format of your computational tool or in PDF format. We recommend that you work through the tutorial and answer all Quick Review Questions using the corresponding software.

Introduction

This fourth computational toolbox tutorial, which is available from the textbook's website in your system of choice, prepares you to use the system to complete projects for this and subsequent chapters. The tutorial introduces the following functions and concepts:

- Random numbers
- Modulus function
- *If* statement
- Counting
- Loading a file
- The mean and standard deviation functions
- Histograms

The module gives computational examples and Quick Review Questions for you to complete and execute in the desired software system.

MODULE 9.2

Simulations

Download

The text's website has an *Area* file, which contains the model of this module, available for download for various system dynamics tools.

Introduction

Modeling is the application of methods to analyze complex, real-world problems in order to make predictions about what might happen with various actions. When it is too difficult, time-consuming, costly, or dangerous to perform experiments, the modeler might resort to **computer simulation**, or having a computer program imitate reality, in order to study situations and make decisions. Simulating a process, he or she can consider various scenarios and test the effect of each.

For example, a scientist might simulate the effects of ozone depletion on global warming. Scientists at Los Alamos National Laboratory used simulations to predict the behavior of nuclear reactions before physically testing a nuclear bomb during World War II (LANL 2012). Lawrence Livermore National Laboratory scientists have used molecular dynamics simulations to study the total energy and other quantities associated with molecules as they interact with one another. At the same laboratory, they have studied the greenhouse effect, making predictions based on levels of various pollutants (LLNL 2012). Before the Gulf War, military experts simulated a number of scenarios to test preparedness. The National Oceanographic and Atmospheric Administration performs simulations to predict the path and intensities of hurricanes (NOAA 2012). The Boeing Company designed the Boeing 777 airplane completely using computer-aided design and tested the designs using computer simulations before construction began. Also, flight simulators allow pilots to practice emergency situations under safe conditions (Boeing 2012).

We use simulations if one or more of the following statements is true:

- It is not feasible to do the actual experiment, as in the study of the greenhouse effect.
- The cost in money, time, or danger of the actual experiment is prohibitive, as with the study of nuclear reactions.
- The system does not exist yet, as in the development of an airplane.
- We want to test various alternatives, as with hurricane predictions

Disadvantages of Computer Simulations

Despite the many applications and advantages, the following are some disadvantages of computer simulations:

- The simulation may be expensive in time or money to develop.
- Because it is impossible to test every alternative, we can provide good solutions but not the best solution.
- The results may be difficult to verify because often we do not have real-world data.
- We cannot be sure we understand what the simulation actually does.
- When a simulation is probabilistic, involving an element of chance, we should be careful of our conclusions.

Element of Chance

At the core of most simulations is random number generation. The computer generates a sequence of numbers, called **random numbers,** or **pseudorandom numbers**. An algorithm actually produces the numbers, so they are not really random, but they appear to be random. Because of the element of chance, we often call a simulation a **Monte Carlo simulation**, named after the gambling capital. A Monte Carlo simulation is a probabilistic model involving an element of chance. Hence, such a simulation is not deterministic but is probabilistic or stochastic, and the results of each execution can vary from those of other runs.

Definition A **Monte Carlo simulation** is a probabilistic model involving an element of chance.

To illustrate the difference between a Monte Carlo simulation model and a purely mathematical model, consider a problem of finding the area between the curve $f(x)$ and the x-axis on the interval from $x = 0$ to $x = 2$, as in Figure 9.2.1. The area is certainly deterministic; exactly one answer exists for the area. Moreover, if the function f has an antiderivative from 0 to 2, then we can determine the area by integrating, $\int_0^2 f(x)dx$.

Alternatively, although Monte Carlo simulation is probabilistic, the technique can model deterministic behavior, such as area under a curve. One method used to

Historical Note: Genesis of Monte Carlo Simulations

Computer random number generators, which are essential for Monte Carlo simulations, have been available since some of the earliest days in the development of computers. John von Neumann, who introduced the idea of storing programs as well as data in the memory of the computer, also helped to develop the first algorithm for generating pseudorandom numbers with the computer.

Born in Budapest, Hungary, in 1903, von Neumann received his Ph.D. in mathematics at the age of 22. He contributed significantly to a variety of areas: the mathematical foundation of quantum theory, logic, the theory of games, economics, nuclear weapons, and meteorology, as well as theory and applications in early computer science. Many stories tell of his phenomenal memory, reasoning ability, and computational speed. He could memorize a column of the telephone book at a glance, and he had mastered calculus by age 8. Halmos wrote, in *Legend of John von Neumann*,

> When his electronic computer was ready for its first preliminary test, someone suggested a relatively simple problem involving powers of 2. (It was something of this kind: what is the smallest power of 2 with the property that its decimal digit fourth from the right is 7? This is a completely trivial problem for a present-day computer: it takes only a fraction of a second of machine time.) The machine and Johnny started at the same time, and Johnny finished first. (Halmos 1973).

During World War II, physicists on the Manhattan Project developed the concept of Monte Carlo simulation. Scientists knew the behavior of one neutron, but they did not have a formula for how a system of neutrons would behave. Although they needed to understand such behavior to construct dampers and shields for the atomic bomb, experimentation was too time consuming and dangerous. John von Neumann and Stanislaus Ulam developed the technique of Monte Carlo simulation to solve the problem.

estimate the area is to enclose the region in a rectangle of known dimensions, as in Figure 9.2.2. We have picked a rectangle of an arbitrary height, such as 1.5, higher than f in the interval. Then, we hypothetically throw darts at the rectangle, counting the total number of darts thrown and the number of darts that hit below the graph. To simulate a dart throw, we generate a random floating-point number, *randomX*, between 0.0 and 2.0 for the x-coordinate and a random floating-point number, *randomY*, between 0 and 1.5 for the y-coordinate. If the simulated dart hits on the curve, then its y-coordinate, *randomY*, would be $f(randomX)$; while $randomY < f(randomX)$ if and only if the dart strikes the board below the curve. To estimate the desired area, we take the proportion of dart hits below the curve times the total area of the rectangle, which we calculate as follows:

$$\text{area} \approx (\text{area of enclosing rectangle}) \left(\frac{\text{number of darts below}}{\text{number of darts}} \right)$$

We can easily compute the area of the rectangle as width times height; in the case of Figure 9.2.2, the area is (2)(1.5) = 3.0. If we throw 1000 darts and 778 of them hit

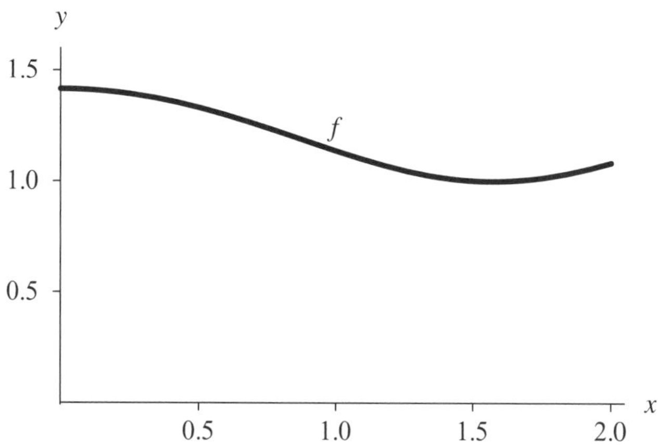

Figure 9.2.1 Graph of a function $f(x)$ on the interval between $x = 0$ and $x = 2$

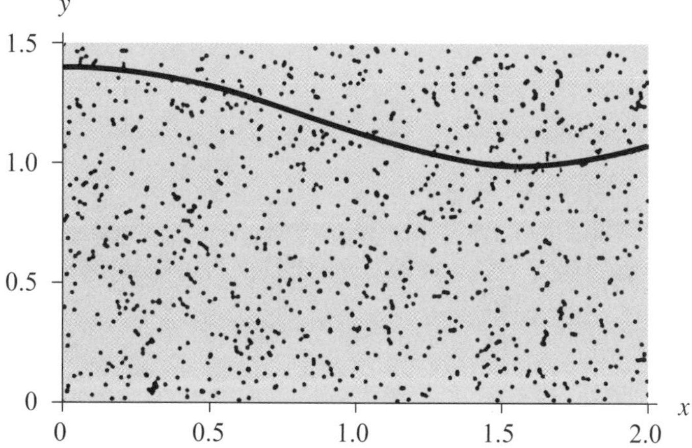

Figure 9.2.2 "Dartboard" enclosing area in Figure 9.2.1

below the graph, then 778/1000 = 0.778 = 77.8% of the total lands below f. This fraction is an estimate of the portion of the rectangle that is below f; about 77.8% of the rectangle's area rests between f and the x-axis. Thus, we estimate this smaller area by taking 77.8% of the total area of the rectangle,

$$\text{area} \approx 3.0 * 778/1000 = 2.334$$

Quick Review Question 1

Consider the linear function $g(x) = 3x + 1$ from $x = 1$ to 5.

a. Give the height of the smallest rectangle enclosing the desired area.

b. Suppose a random hit of a simulated dart is at location (3, 11). Does the dart hit above, below, or on the curve?

c. Using geometry, compute the area under g.
d. Compute the area of the smallest enclosing rectangle.
e. Suppose a simulation "throws" a million "darts" at the rectangle. Although any value is possible, from the following choices, indicate the most likely percentage of darts landing below the graph: 33%, 48%, 59%, 62%, 71%.
f. If 600,000 of a million darts land below the graph, estimate the area.

Notice that with a mathematical model, we do not necessarily need a computer. However, with a computer simulation, we do. The latter has much more of an experimental flavor. Each time we run this Monte Carlo simulation, we are very likely to get a different result, and one that is not the exact value. However, over many simulations using the same parameters, in this and many problems, the results tend to an equilibrium solution, which here is the area under the curve.

Of course, if we can compute the exact area analytically, we should. However, it is impossible to find an analytic form for the integral of many functions, such as $f(x) = \sqrt{\cos^2(x) + 1}$, which Figure 9.2.1 pictures. In such situations, we must employ other techniques, such as a Monte Carlo simulation, to estimate the area. Moreover, as we see in this and the following chapters, computer simulation is a useful technique for examining innumerable problems that do not have exact answers.

Measure of Quality

Theoretically, for the area problem, we can obtain a better estimate by throwing a larger number of darts. In this case, we define the number of darts in one simulation to be larger or run the simulation many times, taking the mean (average) of the area estimates from all the runs. The latter technique has the advantage of enabling us to use the **standard deviation** (σ) of the estimates from the different executions as a measure of the quality of the overall estimate. About 68.3% of the estimates are within $\pm\sigma$ of the mean. Thus, for a mean of 6.2838 and a standard deviation of 0.442276, 68.3% of the area estimates are between 6.2838 – 0.442276 = 5.79602 and 6.2838 + 0.442276 = 6.72608. A small standard deviation relative to the mean indicates a certain consistency for most of the simulations and gives us more confidence in the mean as an estimate of the area.

Quick Review Question 2

a. Suppose the mean of a number of simulations to compute the area under a curve is 40.10 and the standard deviation is 0.20. Is the estimate 39.94 within 1 standard deviation of the mean?
b. Is it better for the standard deviation to be smaller or larger?

Simulation Development

As with system dynamics modeling, our first step should *not* be to start typing on the computer. In the long run, we will save time and improve the quality of our work if we start by analyzing the problem and formulating the simulation model on paper—considering data, making simplifying assumptions, determining the variables and units, establishing relationships among variables and submodels, determining equations, and breaking the problem into small tasks, which will become functions.

Upon implementation, we should be conscious of making our code readable and understandable. Meaningful names for variables and constants help to make the program more **self-documenting**, or self-explanatory. By convention and for self-documentation, constant names, such as *NUM_SIMULATIONS*, are usually in all capitals. As we are writing a function, we should always include comments in the body and an opening comment describing what the function does, input parameters, and returned values. It is amazing how easy it is to forget what a function or other segment of code does, even when we wrote it only a few hours earlier.

Each function should have no more than 30, and often many fewer, lines of code. It is much easier to think about a small function than become mired in the details of a much longer one or of an entire program. Particularly if we are using virtually the same segment of code in several places, we should consider developing a function to handle the task. Thus, if we think of a better algorithm or need to correct the function's code, we need only adjust the instructions in one place.

Moreover, it is much easier to find a bug in one function than in an entire program. After writing a function, we should test the implementation immediately and thoroughly before developing another function. Besides testing for typical situations, we should ensure that the function behaves properly at the boundaries, or extremes, of possible parameter values.

With most computational tools, we can easily test individual functions and perform rapid prototyping of a problem's solution. With **prototyping**, we implement a preliminary version of the solution or part of the solution, which we later can refine to be more complete. Of course, as well as testing the individual functions, we should test that all the functions work properly together in the entire program.

With a prototype or full implementation, if at all possible, we should use real data to verify our results. Does our simulation match reality? We often find that we must

Historical Note: Origin of the Terms *Bug* and *Debugging*

The term bug originates with the first electromechanical, general-purpose computer, the Mark I, completed during World War II. The lack of air-conditioning meant that the windows were left open on hot summer days. On one such day, the computer malfunctioned, and after hours of testing, the laboratory workers walked into the machine (really—it was that big!) and found the problem. A moth lodged on the contacts was preventing the flow of electricity. The insect was carefully pasted it into the logbook with the note, "First actual case of bug being found." Grace Murray Hopper, who worked on the Mark I and developed the first program (a compiler) to translate from a computer language to a machine's language, is credited with popularizing the terms bug and debugging. The famous bug is now on display at the Smithsonian Institution's National Museum of American History (Kidwell 1998).

return to earlier steps in the modeling process, perhaps reconsidering simplifying assumptions and refining our model to improve the results.

When we have a good working model, we can exercise the simulation with different sets of parameters and use the results to explain why the real system we are simulating behaves as it does, to predict what would happen under various circumstances, and to make decisions about how to control or modify that system.

Multiplicative Linear Congruential Method (Optional)

Many people have contributed to the theory of random numbers, which are so useful in computer simulations. In 1949, D. J. Lehmer presented one of the best techniques for generating uniformly distributed pseudorandom numbers, the **linear congruential method**.

One simple linear congruential random number generator that generates values between 0 and 10, inclusive, is as follows:

$$r_0 = 10$$
$$r_n = (7r_{n-1} + 1) \bmod 11, \text{ for } n > 0$$

The initial value in the sequence of random numbers, $r_0 = 10$, is the **seed**. The **mod** function returns the remainder. For example, 71 mod 11 is 5, the remainder in the division of 11 into 71. Thus, substituting $r_0 = 10$ on the right-hand side of the second line of the definition, the **generating function**, we calculate $r_1 = (7 \cdot 10 + 1) \bmod 11 = 5$. After we calculate one "random number," to evaluate the next, we substitute that value into the expression on the right-hand side. Consequently, the next random number is $r_2 = (7 \cdot \mathbf{5} + 1) \bmod 11 = 36 \bmod 11 = 3$.

Quick Review Question 3

Using this random number generator and $r_2 = 3$, calculate the next random number, r_3, in the sequence.

Continuing in this fashion, we obtain ten pseudorandom numbers 5, 3, 0, 1, 8, 2, 4, 7, 6, 10 before the sequence starts repeating. A maximum of 11 nonnegative integers is generated for computation with mod 11.

Should we desire floating-point numbers between 0 and 1, we divide each number in the sequence by the **modulus**, 11, to obtain the following sequence:

$$\frac{5}{11}, \frac{3}{11}, \frac{0}{11}, \frac{1}{11}, \frac{8}{11}, \frac{2}{11}, \frac{4}{11}, \frac{7}{11}, \frac{6}{11}, \frac{10}{11}$$

or

$$0.454545, 0.272727, 0.0, 0.0909091, 0.727273, 0.181818,$$
$$0.363636, 0.636364, 0.545455, 0.909091$$

For this computation, the smallest possible pseudorandom floating-point number is 0.0 and the largest is $(modulus - 1)/modulus = 10/11$. Thus, floating-point numbers that we generate by dividing by the modulus are in the interval [0.0, 1.0), or the interval between the two values that includes 0.0 but not 1.0.

The general form for the **linear congruential method** to generate pseudorandom integers from 0 up to, but not including, *modulus* is as follows:

$$r_0 = seed$$
$$r_n = (multiplier\ r_{n-1} + increment) \bmod modulus, \text{ for } n > 0$$

where *seed*, *modulus*, and *multiplier* are positive integers and *increment* is a nonnegative integer

Not all choices of *multiplier* and *modulus* are good. For example, consider a similar function with a multiplier of 5, $r_0 = 10$ and $r_n = (5r_{n-1} + 1) \bmod 11$ for $n > 0$. This function produces only five numbers—7, 3, 5, 4, 10—before returning to 7. The random number generator should give as long a sequence as possible. Another desirable characteristic of such functions is that the sequence appears random. For example, using the function $r_0 = 10$ and $r_n = (2\ r_{n-1}) \bmod 11$, we obtain the sequence 9, 7, 3, 6, **1, 2, 4, 8**, 5, 10. With a subsequence containing powers of 2, the sequence does not appear random.

Much research has been done to discover choices for *multiplier* and *modulus* that give the largest possible sequence that appears random. For built-in random number generators, *modulus* is often the largest integer a computer can store, such as $2^{31} - 1 = 2{,}147{,}483{,}647$ on some machines. For this modulus, a multiplier of 16,807 and an increment of 0 produce a sequence of $2^{31} - 2$ elements.

Different Ranges of Random Numbers

Many computational tools have generators that can produce uniformly distributed integer or real random numbers in various ranges. Other software systems have limited options, such as a generator for only nonnegative random integers or no generator at all. The previous section described the linear congruential method for generating a random integer from 0 up to the modulus, where by up to we mean not including the modulus. We also saw how to obtain a floating-point counterpart with value from 0.0 up to 1.0 by dividing by the modulus. In this section, we discuss how to obtain uniformly distributed integer or real random numbers in any range. Module 9.3 considers how to generate random numbers from other distributions.

For this discussion, suppose that *rand* is a uniformly distributed random floating-point number from 0.0 up to 1.0. Suppose, however, that we need a random floating-point number from 0.0 up to 5.0. Because the length of this interval is 5.0, we multiply *rand* by this value, 5.0, to stretch the interval of numbers. Mathematically, we have the following:

$$0.0 \le rand < 1.0$$

Thus, multiplying by 5.0 throughout, we obtain the correct interval, as shown:

$$0.0 \le 5.0\ rand < 5.0$$

If the lower bound of the range is different from 0, we add that bound. For example, if we need a random floating-point number from 2.0 up to 7.0, we multiply by the length of the interval, 7.0 – 2.0 = 5.0, to expand the range. Then, we add the lower bound, 2.0, to shift, or translate, the result so that the following inequalities hold:

$$2.0 \le (7.0 - 2.0)\ rand + 2.0 < 7.0$$

or

$$2.0 \le 5.0\ rand + 2.0 < 7.0$$

Specifying Random Floating-Point Numbers in Other Ranges

If *rand* is a random floating-point number such that $0.0 \le rand < 1.0$, then $(max - min)rand + min$ is a random floating-point number from *min* up to *max* that satisfies the following inequality:

$$min \le \mathbf{(max - min)\ rand + min} < max$$

Quick Review Question 4

Suppose *rand* is a random floating-point number from 0.0 up to 1.0.

a. Give an expression to obtain a random floating-point number from 14.5 up to 24.5.
b. Give the range of random numbers for the expression $73.9rand + 21.2$.

Frequently, we need a more-restricted range of random integers than from 0 up to *modulus*. For example, a simulation might require random integer temperatures between 0 and 99, inclusive. One method of restricting the range is to multiply a floating-point random number between 0.0 and 1.0 by 100 (the number of integers from 0 through 99, or 99 + 1) and then return the **integer part** (the number before the decimal point). For example, suppose *rand* is 0.692871. Multiplying by 100, we obtain $100 \cdot 0.692871 = 69.2871$. Truncating, we obtain an integer (69) between 0 and 99.

Sometimes we want the range of random integers to have a lower bound other than 0, for example, from 100 to 500, inclusive. Because we include 100 and 500 as options, the number of integers from 100 to 500 is one more than the difference in these values, (500 – 100 + 1) = 401. As with the last example, we multiply this value by *rand* to expand the range. Then, we add the lower bound, 100, to the product to translate the range to start at 100 as follows:

$$100.0 \le 401rand + 100 < 501.0$$

Finally, we take the integer part of the result, which we write here as applying a function INT.

$$100 \le \text{INT}(401\textit{rand} + 100) < 501$$

or

$$100 \le \text{INT}(401\textit{rand} + 100) \le \mathbf{500}$$

Because the floating-point numbers ($401\textit{rand} + 100$) are less than 501.0, after truncation, the largest possible integer part is 500.

Specifying Random Integers in Other Ranges

If *rand* is a random floating-point number such that $0.0 \le \textit{rand} < 1.0$, then $\text{INT}((\textit{max} - \textit{min} + 1)\textit{rand} + \textit{min})$ is a random integer from *min* to *max*, inclusive, that satisfies the following inequality:

$$\textit{min} \le \textbf{INT}(\,(\textit{\textbf{max}} - \textit{\textbf{min}} + \mathbf{1})\cdot\textit{\textbf{rand}} + \textit{\textbf{min}}) \le \textit{max}$$

where INT is a function that returns the integer part of a number.

Quick Review Question 5

Suppose *rand* is a random floating-point number from 0.0 up to 1.0. Assume that INT is a function that returns the integer part of a floating-point number.

a. Give an expression to obtain a random integer from 28 to 41, inclusive.
b. Give the range of random numbers for the expression INT(73 *rand* + 21).

Exercises

Evaluate Exercises 1–3.

1. 349 mod 7 **2.** 4621 mod 100 **3.** 11,382 mod 542

4. Consider the following linear congruential random number generator:

$$r_0 = 8697$$
$$r_n = (229r_{n-1}) \bmod 349, \text{ for } n > 0$$

a. Compute the next three random numbers.
b. From the sequence of integers in Part a, compute an appropriate sequence of floating-point numbers between 0 and 1.
c. Give the maximum number of random numbers this function can generate.

5. Repeat Exercise 4 for the following random number generator:

$$r_0 = 1021$$
$$r_n = (467r_{n-1}) \bmod 1024, \text{ for } n > 0$$

6. Repeat Exercise 4 for the following random number generator:

$$r_0 = 8367$$
$$r_n = (229r_{n-1} + 1) \text{ mod } 10{,}000, \text{ for } n > 0$$

7. The following guidelines for choice of modulus, multiplier, and increment have been developed through computer testing of random number generators:

- The modulus is a positive integer.
- The multiplier and increment are nonnegative integers less than the modulus.
- If working on a decimal machine, choose the modulus to be a large power of 10 for easy computation of the mod function. Most computers, however, are binary machines, so that the modulus should be a large power of 2, such as 2^{32}. Division by 2^{32} moves the binary point 32 places to the left in a binary number.
- On a binary computer, choose *multiplier* such that *multiplier* mod 8 = 5 and 0.01*modulus* < *multiplier* < 0.99*modulus*.
- No integer great than 1 should divide both the increment and the modulus.

Give three choices for a multiplier that meets these suggested criteria with a modulus of 2^{20}.

8. **a.** If the modulus is 2^{32} and the increment is a nonnegative integer less than 10, list the choices for the increment based on the guideline in Exercise 7 that no integer greater than 1 should divide both the modulus and the increment.

b. List choices for a nonnegative increment less than 10 if the modulus is 10^{19}.

For the following exercises, assume that rand is a random floating-point number from 0.0 up to 1.0 and that INT is a function that returns the integer part of a number. For each exercise, write an expression to return a random number in the given interval.

9. random floating-point number from 0.0 up to, but not including, 20.0
10. random floating-point number from 6.0 up to, but not including, 26.0
11. random floating-point number from 35.8 up to, but not including, 73.4
12. random floating-point number from –8.0 up to, but not including, 4.0
13. random integer between 0 and 20, inclusive—thus, in {0, 1, 2, . . ., 20}
14. random integer between 6 and 26, inclusive
15. random integer between 35 and 73, inclusive
16. random integer between –8 and 4, inclusive

Projects

For Projects 1–7, do the following:

a. *In the case of 2D problems, plot the function, f.*

b. *Using the Monte Carlo technique, discussed in the section "Element of Chance," with your computational tool, define a function with a parameter for the number of darts that returns an estimate of the indicated value. Use practices as discussed in the section "Simulation Development."*

c. *Define a function that calls the function from Part b 1000 times and returns the mean and standard deviation of the results.*

d. *Using your computational tool, calculate the answer with integration.*

1. the area between the curve for $f(x) = \sqrt{\cos^2(x) + 1}$ and the x-axis from $x = 0$ to $x = 2$
2. the area between the curve for $f(x) = x^2$ and the x-axis from $x = 2$ to $x = 3$
3. the area between the curve for $f(x) = e^{x^2}$ and the x-axis from 0 to 1
4. An estimate of π. The area of a circle is πr^2, where r is the radius. The equation of a circle of radius r with center at the origin is $x^2 + y^2 = r^2$. Use a circle of radius 1. Consider the quarter of the circle in the first quadrant with $0 \le x \le 1$ and $0 \le y \le 1$, and multiply your result by 4.
5. An estimate of the volume of a sphere of radius 1 whose equation is $x^2 + y^2 + z^2 \le 1$. Consider the portion of the sphere with $x \ge 0$, $y \ge 0$, and $z \ge 0$; and multiply your result by 8. In the case of three dimensions, a point has three coordinates (x, y, z). Notice that Monte Carlo integration is useful for dimensions beyond two.
6. An estimate of $\int_2^3 \sin(x^2)dx$. Note that the function is not entirely above or entirely below the x-axis, so we must adjust the algorithm in the text to estimate the integral. Recall that where a function is negative (below the x-axis), its integral is the negative of the area between the curve and the x-axis.
7. Repeat Project 6 for $\int_0^{2\pi} (1 - x^{\sin(x)})dx$.
8. Create a file containing definitions of your own random number seeding and generating functions, as the following parts describe. Do the requested parts with an appropriate computational tool without using its built-in random number generator. Starting with Part b, define several versions of your own random number generator, if possible, each with the name *myRandom*. In a call to *myRandom*, the parameters determine which form the computational tool uses. If your computational tool does not allow the use of the same function name for different versions of the function, use different function names. Each definition's format should mimic that of your tool's corresponding built-in function. The definitions in Parts c–h should call the function of Part b. The function *myRandom* should assign a new pseudorandom number to *myRandValue*. Employ the old value of *myRandValue* to generate the new value. Test each function thoroughly by generating and checking a number of values.

   a. Give the function *mySeedRandom* two definitions that seed your random number generator by assigning a value to a variable, *myRandValue*. Because the value of *myRandValue* must persist and be available to both definitions of *mySeedRandom*, depending on your computational tool, you may need to declare *myRandValue* as a global variable. First, define *mySeedRandom* that assigns an argument to *myRandValue*. Thus, *mySeedRandom* with an argument of 12345 assigns 12345 to *myRandValue*. Second, define *mySeedRandom* with no argument so that *myRandValue* becomes an integer associated with the date and time of day. For testing, clear out any value of *myRandValue*, and test that *mySeedRandom* with various arguments assigns the correct value to *myRandValue*.

b. Define a function *myRandom*, as in Exercise 4a, to return a random integer between 0 and 348, inclusive. If *myRandValue* does not have a value because we did not call *mySeedRandom*, use an initial value of 1 for *myRandValue*.

For each of the following parts, define a function, if possible, called myRandom, to return the designated kind of random number. In each definition, directly or indirectly (by calling a function that calls this function) invoke the function in Part b.

c. a random floating-point number from 0.0 up to 1.0, expressed as a decimal number, not a fraction
d. a random floating-point number from 0.0 up to *max*
e. a random floating-point number from *min* up to *max*
f. 0 or 1 at random
g. a random integer between 0 and *max*, inclusive
h. a random integer between *min* and *max*, inclusive

9. Do Project 8 using the generator in Exercise 5.
10. Do Project 8 using the generator in Exercise 6.
11. Do Project 8 for a modulus of $2^{31} - 1$, a multiplier of 16,807, and an increment of 0.

Answers to Quick Review Questions

1. a. 16 = 3(5) + 1
b. above because 3(3) + 1 = 10 < 11
c. 40 = (16 + 4)(4)/2 is the area of the trapezoid
d. 64 = (5 – 1)(16)
e. 62% because 40/64 = 62.5%
f. 38.4 square units = (64)(600,000/1,000,000) = (64)(0.6)

2. a. Yes, numbers between 40.10 – 0.20 and 40.10 + 0.20—that is, between 39.90 and 40.30—are within one standard deviation of 40.10.
b. smaller

3. 0 because r_2 = (7 * 3 + 1) mod 11 = 22 mod 11 = 0

4. a. 10.0*rand* + 14.5
b. 21.2 up to 95.1 = 21.2 + 73.9

5. a. INT(14*rand* + 28) = INT((41 – 28 + 1)*rand* + 28)
b. integers from 21 to 93, inclusive, or {21, 22, 23, . . ., 93}

References

Boeing. 2012. "The Boeing Company." http://www.boeing.com/ (accessed December 27, 2012)

Halmos, P. R. 1973. "Legend of John von Neumann." *American Mathematical Monthly*, 80: 382–394.

Kidwell, Peggy Aldrich. 1998. "Stalking the Elusive Computer Bug." *IEEE Annals of the History of Computing*. 20(4), October–December.

LANL (Los Alamos National Laboratory). 2012. "Los Alamos National Laboratory Homepage." The University of California and the U.S. Department of Energy. http://www.lanl.gov/ (accessed December 27, 2012)

LLNL (Lawrence Livermore National Laboratory). 2012. "Lawrence Livermore National Laboratory Homepage." The University of California and the U.S. Department of Energy. http://www.llnl.gov/ (accessed December 27, 2012)

NOAA (National Oceanographic and Atmospheric Administration). 2012. "NOAA Homepage." U.S. Department of Commerce. http://www.noaa.gov/ (accessed December 27, 2012)

Park, Stephen K., and Keith W. Miller. 1988. "Random Number Generators: Good Ones Are Hard to Find." *Communications of the ACM*, 31(10): 1192–1201.

MODULE 9.3

Random Numbers from Various Distributions

Downloads

For several computational tools, the text's website has a *Distributions* file, which contains descriptions of built-in random number generators for various probability distributions, and a *9_3QRQ.pdf* file, which contains system-dependent text along with Quick Review Questions and answers, available for download.

Introduction

Monte Carlo simulations are important tools in scientific work and yield solutions to problems unobtainable by other means. Moreover, where alternative solutions are possible, such simulations often provide greater precision for the same computer cost.

A Monte Carlo simulation requires the use of unbiased random numbers. The **distribution** of these numbers is a description of the portion of times each possible outcome or each possible range of outcomes occurs on the average over a great many trials. However, the distribution that a simulation requires depends on the problem. In this module, we discuss the algorithms for generating random numbers from several types of distributions.

> **Definition** A **distribution** of numbers is a description of the portion of times each possible outcome or each possible range of outcomes occurs on the average.

Statistical Distributions

In Module 9.2, "Simulations," we considered the linear congruential method to generate pseudorandom numbers with a uniform distribution. Suppose a specified range

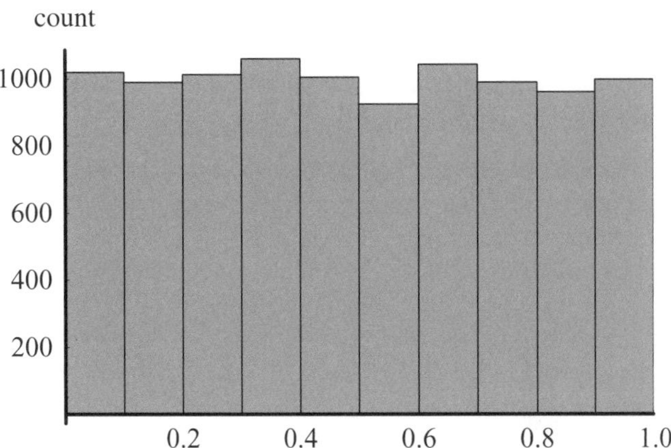

Figure 9.3.1 Histogram of 10,000 random floating-point numbers, uniformly distributed from 0.0 up to 1.0

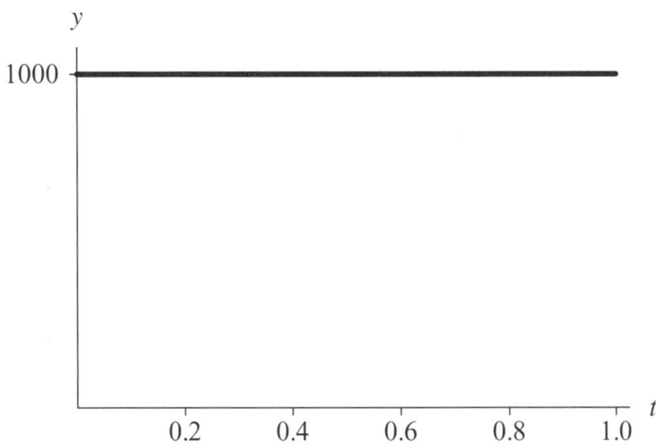

Figure 9.3.2 Horizontal line at height 1000 approximately goes across the top of the histogram in Figure 9.3.1

is partitioned into intervals of the same length. With a **uniform distribution**, the generator is just as likely to return a value in any of the intervals. Equivalently, in a list of many such random numbers, on the average each interval contains the same number of generated values. For example, Figure 9.3.1 presents a histogram with 10 intervals of length 0.1 of a table of 10,000 random floating-point numbers, uniformly distributed from 0.0 up to 1.0. As expected, approximately one-tenth of the 10,000, or 1000, numbers appears in each subdivision. Thus, the curve across the tops of the bars is virtually a horizontal line of height 1000 (Figure 9.3.2). As we will see, methods for generating random numbers in other distributions depend on our ability to produce random numbers with a uniform distribution.

Quick Review Question 1

Suppose we have a uniformly distributed random number generator that returns a floating-point value from 0.0 to 1.0.

a. Suppose we break the interval from 0.0 to 1.0 into 5 subintervals. If a list contains 100 random numbers, give the number of values we expect in each subinterval on average.
b. In a histogram of this data, give the height of each bar on average.
c. In general, for a list of n random numbers between 0.0 and 1.0 and for i subintervals, give the number of values we expect in each subinterval on average.
d. In a histogram of this data, give the height of each bar on average.

A distribution can be **discrete** or **continuous**. To illustrate the difference between the terms discrete and continuous, a digital clock shows time in a discrete manner, from one minute to the next, while a clock with two hands indicates time in a continuous, unbroken way. Similarly, as you pass the time-and-temperature sign in front of a bank, one moment it might register 28 °C, the next it might jump to 29 °. As a continuous counterpart, a thermometer outside a house might have a column of liquid, smoothly rising and falling to indicate the temperature. In a simulation of pollution, we might generate a random integer to indicate the number of dust particles in a cubic meter of air. The distribution of such values is discrete. In the same simulation, for the velocities of the particles, we might generate random floating-point values that have a continuous distribution. However, as noted in Module 5.2, "Errors," the expression of numbers in a computer is discrete. Thus, at times we employ discrete numbers to represent continuous events.

Definitions A **discrete distribution** is a distribution with discrete values. A **continuous distribution** is a distribution with continuous values.

For a discrete distribution, a **probability function** (or **density function**, or **probability density function**) returns the probability of occurrence of a particular argument value. For example, $P(1382)$ might be the probability that the random number generator returns 1382, indicating 1382 dust particles. However, if a distribution is continuous, the probability of occurrence of any particular value is zero. Thus, for a continuous distribution, a probability function (or density function, or probability density function) indicates the probability that a given outcome falls inside a specific range of values. The integral of the probability function from the lower to the upper bound of the range, which is the area under that portion of the curve, gives the probability that the outcome is in that range. For example, the probability that the random velocity in the x-direction of a dust particle is between 3.0 and 4.0 mm/s is the integral of the probability density function from 3.0 to 4.0. Figure 9.3.3 presents a horizontal line of height 1 that is the graph of the probability density function ($P(x) = 1$) for uniformly generated random numbers with values from 0.0 up to 1.0. The probability that a uniform random floating-point number between 0.0 and 1.0 falls between 0.6 and 0.8 is the integral of the function $f(x) = 1$ from 0.6 to 0.8. Thus, the

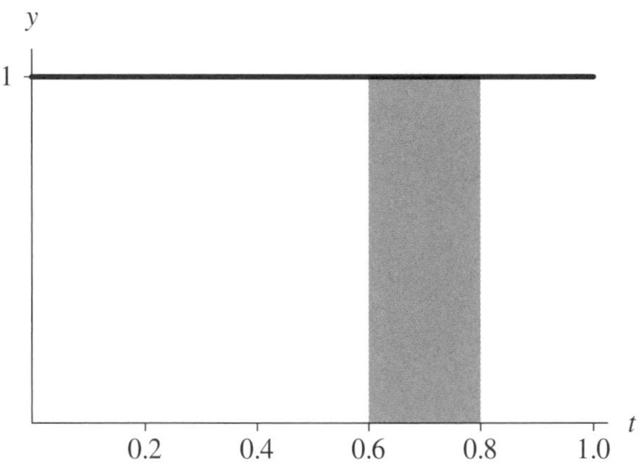

Figure 9.3.3 Probability density function for the distribution with histogram in Figure 9.3.1

probability is the area of the shaded region between 0.6 and 0.8, which is (0.8 – 0.6)(1.0) = 0.2. Such a random number is between 0.6 and 0.8 for 0.2 = 20% of the time.

Definitions For a discrete distribution, a **probability function** (or **density function**, or **probability density function**) returns the probability of occurrence of a particular argument. For a continuous distribution, a **probability function** (or **density function**, or **probability density function**) indicates the probability that a given outcome falls inside a specific range of values.

Quick Review Question 2

a. In the generation of random numbers to represent throws of a fair die (values 1, 2, 3, 4, 5, or 6), is the distribution discrete or continuous?
b. For the random numbers of Part a, give the value of the probability density function with an argument of 2.
c. For the probability density function with graph in Figure 9.3.3, give the probability that the random number falls between 0.2 and 0.7.
d. Can a probability density function ever have a negative value?
e. Give the value of the definite integral (i.e., area under the curve) of a probability density function over its entire range of values.

Discrete Distributions

If an equal likelihood of each of several discrete events exists, in a simulation we can generate a random integer to indicate the choice. For example, in a simulation of a pollen grain moving in a fluid, suppose at the next time step the grain is just as likely to move in any direction—north, east, south, west, up, or down—in a three-dimensional (3D) grid. A probability of 1/6 exists for the grain to move in any of the six directions. With these equal probabilities, we can generate a uniformly distributed integer between 1 and 6 to indicate the direction of movement.

To Generate Random Numbers in Discrete Distribution with Equal Probabilities for Each of *n* Events

Generate a uniform random integer from a sequence of *n* integers, where each integer corresponds to an event.

Quick Review Question 3

From the text's website, download your computational tool's *9_3QRQ.pdf* file for this system-dependent question to give a command for generating an appropriate random integer indicating a pollen grain's movement.

Frequently, however, the discrete choices do not carry equal probabilities, as in the following example. For example, in an initial 3D grid, suppose only 15% of the grid sites, or cells, contain pollen grains. Thus, a *probPollen* = 15% = 0.15 chance exists for a cell to contain a grain. If the location is to contain a pollen grain, we make the cell's value equal to *POLLEN* = 1; otherwise, the cell's value becomes *EMPTY* = 0. To initialize a grid for a simulation, we must designate for each cell if the location contains pollen or not. For each cell, we need to generate a uniformly distributed random floating-point number from 0.0 up to 1.0. On the average, 15% of the time this random number is less than 0.15, while 85% of the time the number is greater than or equal to 0.15 (Figure 9.3.4). Thus, to initialize the cell, if the random number is less than 0.15, we make the cell's value *POLLEN*; otherwise, we assign *EMPTY* to the cell's value. Thus, using the probabilities and cell values above, we employ the following logic to initialize each cell in the grid:

if a random number is less than *probPollen* (i.e., pollen grain at site)
    set the cell's value to *POLLEN*
else (i.e., no pollen grain at site)
    set the cell's value to *EMPTY*

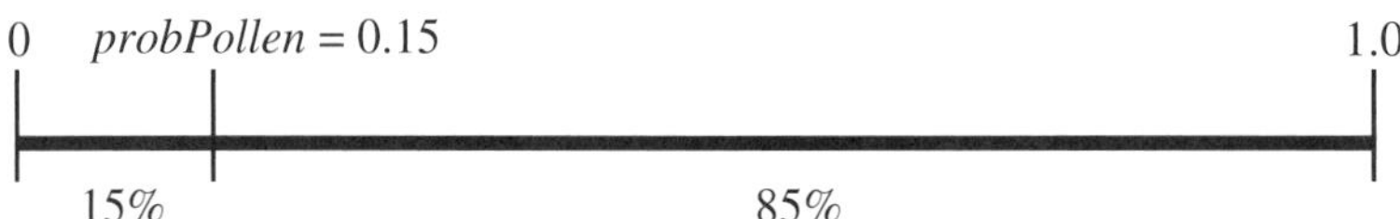

Figure 9.3.4 15% of floating-point values between 0 and 1 are less than *probPollen* = 0.15

Figure 9.3.5 15% of values less than 0.15; 20% between 0.15 and 0.35; 65% between 0.35 and 1.0

Quick Review Question 4

From the text's website, download your computational tool's *9_3QRQ.pdf* file for this system-dependent question to give a statement to implement initializing a cell, possibly with pollen.

In many situations, more than two choices exist. For example, suppose in a simulation involving animal behavior, a lab rat presses a food lever (*FOOD* = 1) 15% of the time, presses a water lever (*WATER* = 2) 20% of the time, and does neither (*NEITHER* = 3) the remainder of the time. For the simulation, we consider the range split into three parts, as in Figure 9.3.5, and again generate a uniformly distributed random floating-point number from 0.0 to 1.0. If the number is less than 0.15, which occurs 15% of the time, we assign *FOOD* = 1 to the rat's action. For 20% of the time, the uniformly distributed random number is greater than or equal to 0.15 and less than 0.35. With a random number in this range, we make the rat's action be *WATER* = 2. A random number is greater than or equal to 0.35 with a probability of 65%. In such a case, we assign *NEITHER* = 3 to the rat's action. Thus, with *rand* being a uniformly distributed random floating-point number from 0.0 to 1.0, we employ the following logic for determination of the rat's action:

if a random number, *rand*, is < 0.15
    the rat presses the food lever
else if *rand* < 0.35 (i.e., 0.15 ≤ *rand* < 0.35)
    the rat presses the water lever
else (i.e., 0.35 ≤ *rand*)
    the rat does neither

To Generate Random Numbers in Discrete Distribution with Probabilities $p_1, p_2, \ldots, p_n$ for Events $e_1, e_2, \ldots, e_n$, Respectively, Where $p_1 + p_2 + \cdots + p_n = 1$

Generate *rand*, a uniform random floating-point number in [0, 1).
If $rand < p_1$, then return e_1
else if $rand < p_1 + p_2$, then return e_2
. . .
else if $rand < p_1 + p_2 + \ldots + p_{n-1}$, then return e_{n-1}
else return e_n

Quick Review Question 5

Consider the following English description of a segment that returns the direction (N, E, S, or W) a simulated animal moves:

```
if a random number, rand, is < 0.12
    return N
else if rand < 0.26
    return E
else if rand < 0.69
    return S
else
    return W
```

Give the probability that the animal moves in each of the following directions:

a. N **b.** E **c.** S **d.** W

Normal Distributions

A **normal,** or **Gaussian, distribution**, which statistics frequently employs, has a probability density function $\frac{1}{\sqrt{2\pi\sigma}}e^{-(x-\mu)^2/(2\sigma)}$, where μ is the mean and σ is the standard deviation (Figure 9.3.6). Figure 9.3.7 displays a histogram of a set of 1000 random numbers in the Gaussian distribution with mean 0 and standard deviation 1. Without getting into a formal definition of standard deviation, 68.3% of the values in a normal distribution are within $\pm\sigma$ of the mean, μ; 95.5% are within $\pm 2\sigma$ of μ; and 99.7% are within $\pm 3\sigma$ of μ.

Many systems include a way to generate random numbers in a normal distribution with a given mean and standard deviation. For those that do not, we can employ the **Box-Muller-Gauss method**. The method first generates a uniformly distributed random number, a, between 0 and 2π. Then, the technique computes b, the product

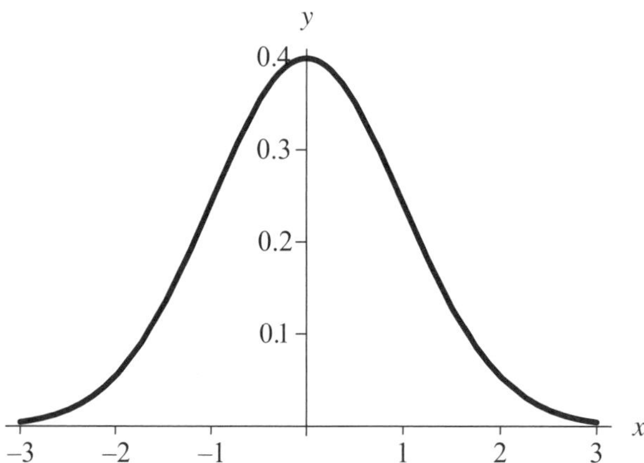

Figure 9.3.6 Probability density function for normal distribution with mean 0 and standard deviation 1

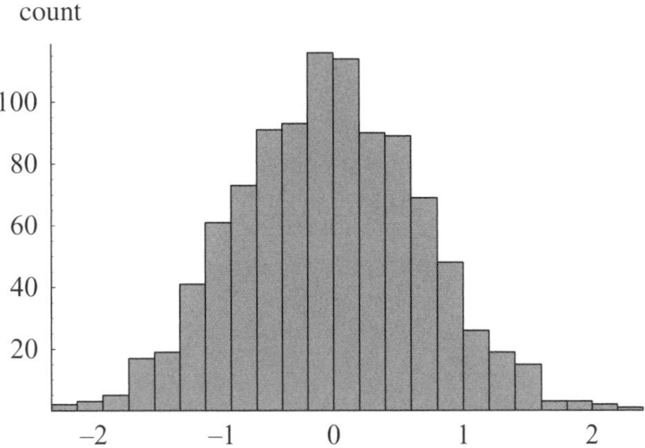

Figure 9.3.7 Histogram of a normal distribution with mean 0 and standard deviation 1

of the standard deviation (σ) and the square root of the negative natural logarithm of a uniformly distributed random number between 0.0 and 1.0, or $b = \sigma\sqrt{-2\ln(rand)}$, where *rand* is a uniformly distributed random number between 0.0 and 1.0. The two values $b \cdot \sin(a) + \mu$ and $b \cdot \cos(a) + \mu$ are normally distributed with mean μ and standard deviation σ.

Box-Muller-Gauss Method for Normal Distribution with Mean μ and Standard Deviation σ

compute $\boldsymbol{b}$ **sin**($\boldsymbol{a}$) + $\boldsymbol{\mu}$ and $\boldsymbol{b}$ **cos**($\boldsymbol{a}$) + $\boldsymbol{\mu}$ where
a = a uniform random number in $[0, 2\pi)$
rand = a uniform random number in $[0, 1)$
$b = \sigma\sqrt{-2\ln(rand)}$

Quick Review Question 6

Suppose for a simulation involving test scores, we need random numbers in a normal distribution with mean 70 and standard deviation 8. Suppose 5.32 and 0.754 are uniformly distributed random numbers between 0 and 2π and between 0.0 and 1.0, respectively. Using these values, evaluate the following, rounding to two decimal places:

a. a
b. b
c. The normally distributed number employing sine
d. The normally distributed number employing cosine

Quick Review Question 7

From the text's website, download your computational tool's *9_3QRQ.pdf* file for system-dependent text and a question to assign to n a random number in a normal distribution with mean 70 and standard deviation 8.

Exponential Distributions

A model for unconstrained growth or decay employs an exponential function e^{rt}, where t is time and r is the growth rate or $-r$ the decay rate, respectively. Functions of the form $\boldsymbol{f(t) = |r|e^{rt}}$ **with** $\boldsymbol{r < 0}$ **and** $\boldsymbol{t > 0}$ or $\boldsymbol{f(t) = |r|e^{rt}}$ **with** $\boldsymbol{r > 0}$ **and** $\boldsymbol{t < 0}$ are probability density functions in which the area under each curve is 1. Figure 9.3.8 contains the graph of a function in this category, $f(t) = 2e^{-2t}$. To obtain a number in such a distribution, the **exponential method** divides the natural logarithm of a uniformly distributed random number from 0.0 to 1.0 by the rate constant (r), that is, $\ln(rand)/r$, where *rand* is random between 0 and 1. For example, to generate numbers in the distribution $f(t) = 2e^{-2t}$, we calculate $\ln(rand)/(-2)$. Figure 9.3.9 displays a histogram of 1000 such exponentially distributed random numbers.

We employ the same algorithm to generate random numbers from minus infinity to 0 for probability density function $f(t) = |r|e^{rt} = re^{rt}$ with $r > 0$. Figure 9.3.10 shows the graph of one such function, $f(t) = 2e^{2t}$; and Figure 9.3.11 displays a histogram of 1000 pseudorandom numbers that the algorithm $\ln(rand)/2$ generates.

Exponential Method for Probability Density Function $\boldsymbol{|r|e^{rt}}$ **with** $\boldsymbol{r < 0}$ **and** $\boldsymbol{t > 0}$ or $\boldsymbol{f(t) = |r|e^{rt}}$ **with** $\boldsymbol{r > 0}$ **and** $\boldsymbol{t < 0}$

compute $\boldsymbol{\ln(rand)/r}$,
where *rand* = a uniform random number in [0.0, 1.0)

Quick Review Question 8

Suppose we are performing a simulation involving a radioactive substance with initial mass 0.1 mg and decay rate 0.1.

a. Give the probability density function.
b. Using 0.754 as a uniformly distributed random number between 0.0 and 1.0, determine a random number to 3 decimal places in this exponential distribution.

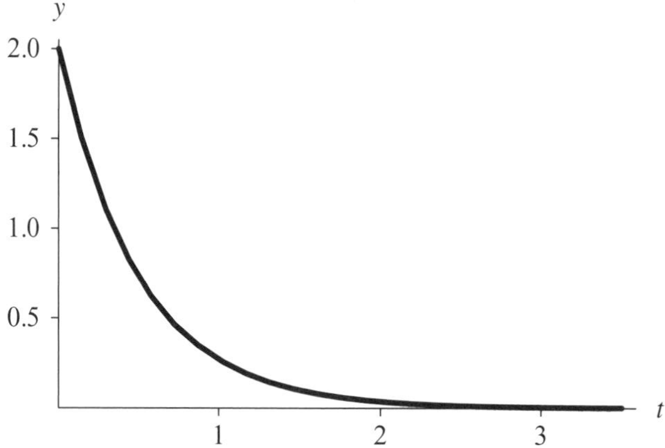

Figure 9.3.8 Probability density function $f(t) = 2e^{-2t}$ for $t > 0$

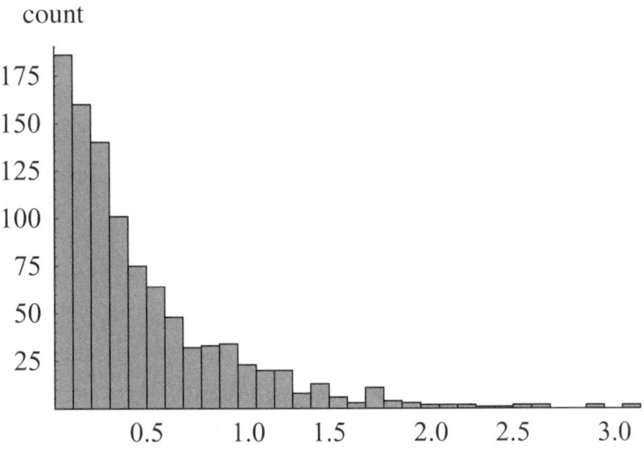

Figure 9.3.9 Histogram of 1000 random numbers ln(*rand*)/(–2), where *rand* is a uniformly generated random number in [0.0, 1.0)

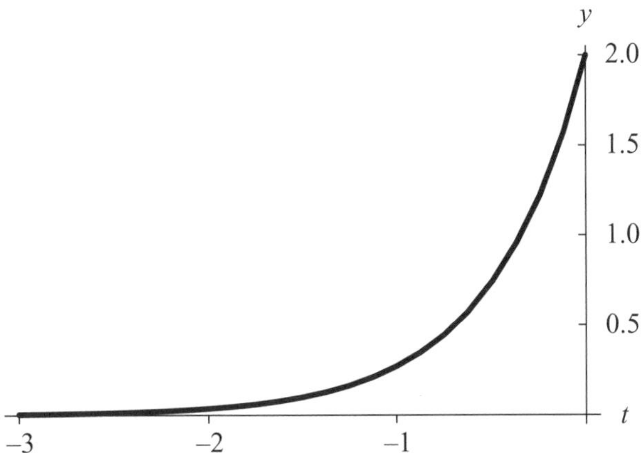

Figure 9.3.10 Probability density function $f(t) = 2e^{2t}$ for $t < 0$

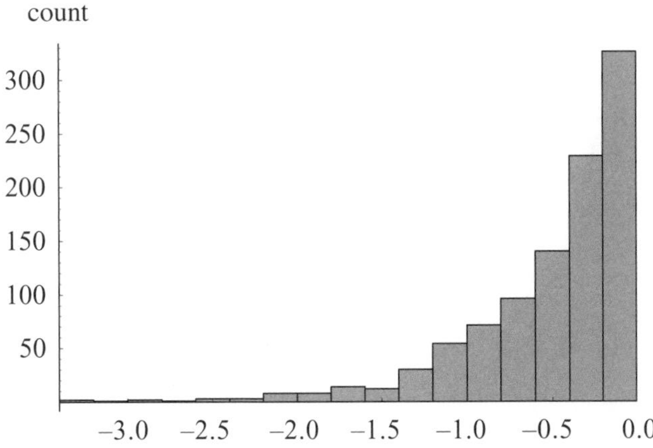

Figure 9.3.11 Histogram of 1000 numbers ln(*rand*)/2, where *rand* is a uniformly generated random number in [0.0, 1.0)

Quick Review Question 9

From the text's website, download your computational tool's *9_3QRQ.pdf* file for system-dependent text and a question about an exponential probability density function.

Rejection Method

The exercises and projects explore several methods for generating random numbers in other specific distributions. When these techniques do not apply, however, we can employ the **rejection method**. First, we obtain a uniformly distributed random number, *randInterval*, in the requested interval. If the probability density function at *randInterval* is greater than a uniform random number from 0.0 to an upper bound for the function, we return *randInterval*. Otherwise, we repeat the process.

Rejection Method for Random Numbers in Interval $[a, b)$ for Distribution $f(x)$

if *f(randInterval)* > *randUpperBound*, then return *randInterval*, where
randInterval is a uniform random number in interval $[a, b)$
randUpperBound is a uniform random number in [0, upper bound for f)
else repeat the process

Quick Review Question 10

Consider the probability density function $f(x) = 2\pi \sin(4\pi x)$ from 0.0 to 0.25 (Figure 9.3.12). For each of the following pairs of uniform random numbers in the indicated

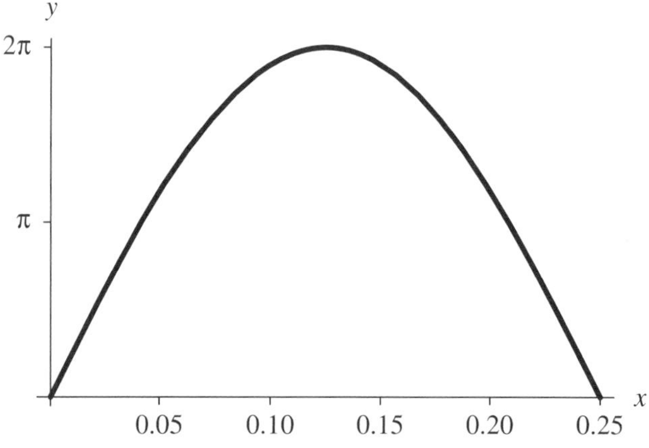

Figure 9.3.12 $f(x) = 2\pi \sin(4\pi x)$ from 0.0 to 0.25

intervals, determine the random number returned in the distribution $2\pi \sin(4\pi x)$, if any:

Part	*Interval from* 0.0 *to* 0.25	*Interval from* 0.0 *to* $2\pi\ sin(\pi/2) = 2\pi$
a	0.221	0.85
b	0.049	5.59
c	0.130	2.69

Exercises

Do Exercises 1–8 with an appropriate computational tool.

1. For the logic toward the end of the section "Discrete Distributions," used to determine the action of a rat, write a segment to return *FOOD*, *WATER*, or *NEITHER*, depending on the value of the random number.
2. This question refers to the Box-Muller-Gauss method, which generates random numbers from a normal distribution with mean = 9 and standard deviation = 2.
   a. Assign to a a uniformly distributed random number between 0 and 2π.
   b. Assign to b the product of the standard deviation ($stdDev = 2$) and the square root of the negative natural logarithm of a uniformly distributed random number between 0.0 and 1.0.
   c. Return a list of pairs of numbers with ($b \sin(a) + mean$) as the first coordinate and ($b \cos(a) + mean$) with $mean = 9$ as the second that the Box-Muller-Gauss method produces. Be sure to use the same a and b for both members of a pair.
   d. Assign to *tblGauss* a list of 1000 random numbers in the normal distribution with mean = 9 and standard deviation = 2. One way of accomplishing this task is to generate a table/array of 500 ordered pairs similar to Part c and to flatten the table/array to a corresponding list of 1000 numbers.
   e. Display a histogram of these values.
   f. If available, use a built-in method to generate the table in Part d.
   g. Display a histogram of these values.
3. **a.** Write a statement to assign to *tblExp* a table/array of 1000 random numbers in the exponential distribution $7e^{-7t}$ using the exponential method. Display a histogram of *tblExp*.
   b. If available, use a built-in method to generate the table in Part a. Display a histogram of these values.
4. Using the exponential method, give an expression to generate numbers with the probability density function $2e^{2(t-3)}$ for $t < 3$. Display a histogram of these values.
5. The expression $\ln(rand)/(-9) + 4$, where *rand* is a random number from 0 to 1, generates pseudorandom numbers for what exponential probability density function and interval?
6. This question develops code for the rejection method with the probability density function $f(x) = 2\pi \sin(4\pi x)$.

a. Define the function $f(x) = 2\pi \sin(4\pi x)$. See Figure 9.3.12.
b. Plot $f(x)$ from 0.0 to 0.25.
c. Assign to variable *rand* a uniform random number from 0.0 to 0.25, the interval of interest in Figure 9.3.12.
d. Define the function *rej* with no arguments to return a random number using the rejection method. If $f(rand)$ is greater than a uniform random number from 0 to $2\pi \sin(4\pi/8) = 2\pi \sin(\pi/2) = 2\pi$, which is the maximum value of $f(x)$, return *rand*. If the condition is false, we must reject *rand* and search for another candidate. To do so, we call the function *rej* again. Be sure *rand* gets a new value with each function call. (The process of a function, such as *rej*, calling itself is **recursion**.)
e. Write a statement to generate a list of 1000 random numbers from 0.0 to 0.25 with the probability density function $f(x) = 2\pi \sin(4\pi x)$.
f. Display a histogram of these values.

7. The **maximum method** is for distributions of the form nx^{n-1}, with x being from 0.0 to 1.0 and n being a positive integer less than 17. The method calls for taking the maximum of n uniformly distributed random numbers.
a. Define a function $f(x) = 3x^2$, and plot f from 0.0 to 1.0.
b. Write a statement to return a list of three uniform random numbers in [0.0, 1.0).
c. Write a statement to return a random number in the distribution $3x^2$ using the maximum method.
d. Write a statement to assign to *tblMax* a list of 1000 numbers in the distribution $3x^2$.
e. Display a histogram of the table in Part d.
f. Define a function *randMax* with parameter n to return a random number in the distribution nx^{n-1}, with x being a number from 0.0 to 1.0 and n being a positive integer less than 17. Use the maximum method.
g. Repeat Parts d and e using *randMax* from Part f.

8. The **root method** is for distributions of the form nx^{n-1}, with x being from 0.0 to 1.0 and n being a nonnegative number not in $\{1, 2, 3, \ldots, 16\}$. The method calls for taking the nth root of a uniformly distributed random number between 0.0 and 1.0.
a. Define a function *randRoot* with parameter n to return a random number using the root method.
b. Define a function $f(x) = 3x^2$, and plot f from 0.0 to 1.0.
c. Using *randRoot*, write a statement to assign to *tblRoot* a list of 1000 numbers in the distribution $0.5x^{-0.5}$, where x is between 0 and 1.
d. Display a histogram *tblRoot* from Part c.

Projects

1. Using a computational tool, define your own package of random number generators with continuous distributions using the following methods: Box-Muller-Gauss method, exponential method, rejection method, maximum method (Exercise 7), and root method (Exercise 8). Do not use built-in functions other than a uniform random number generator. Test the package thoroughly.

2. Using a computational tool, define your own package of random number generators with *myRandom* definitions as in Project 8 of Module 9.2, "Simulations." Using these, do Project 1, except do not use any built-in functions.

Answers to Quick Review Questions

1. **a.** 20
b. 20
c. n/i
d. n/i

2. **a.** discrete
b. 1/6 = 0.1667
c. 0.5 = (0.7 – 0.2)(1)
d. no
e. 1

3–4. From the text's website, download your computational tool's *9_3QRQ.pdf* file for an answer to this system-dependent question.

5. **a.** 12%
b. 14% = 0.26 – 0.12
c. 43% = 0.69 – 0.26
d. 31% = 0.1 – 0.69

6. **a.** $a = 5.32$
b. $b = 8\sqrt{-2\ln(0.754)} = 6.01$
c. $b\sin(a) + \mu = 6.01\sin(5.32) + 70 = 65.1$
d. $b\cos(a) + \mu = 6.01\cos(5.32) + 70 = 73.4$

7. From the text's website, download your computational tool's *9_3QRQ.pdf* file for an answer to this system-dependent question.

8. **a.** $f(t) = 0.1e^{-0.1t}$
b. $\ln(0.754)/(-0.1) = 2.824$

9. From the text's website, download your computational tool's *9_3QRQ.pdf* file for an answer to this system-dependent question.

10. **a.** 0.221 because $f(0.221) = 2\pi\sin(4\pi(0.221)) = 2.239 > 0.85$
b. nothing because $f(0.049) = 2\pi\sin(4\pi(0.049)) = 3.629 < 5.59$
c. 0.130 because $f(0.130) = 2\pi\sin(4\pi(0.130)) = 6.271 > 2.69$

References

Einwohner, Theodore H., and Angela B. Shiflet. 1987. "RANDOM_NUMBER, A Syntax-Directed Package to Produce Random Numbers in User-Specified, Univariate Distributions: User's Guide." Unclassified Internal Document for Lawrence Livermore National Laboratory.

Weisstein, Eric. 2012. *Wolfram MathWorld.* Wolfram Research, Inc. http://mathworld.wolfram.com/

MODULE 9.4

Computational Toolbox—Tools of the Trade: Tutorial 5

Prerequisite: Module 9.1. "Computational Toolbox—Tools of the Trade: Tutorial 4."

Download

From the textbook's website, download Tutorial 5 in the format of your computational tool or in PDF format. We recommend that you work through the tutorial and answer all Quick Review Questions using the corresponding software.

Introduction

This fifth computational toolbox tutorial, which is available from the textbook's website in your system of choice, prepares you to use the system to complete projects for this and subsequent chapters. The tutorial introduces the following functions and concepts:

- Taking part of a list
- Maximum and minimum functions
- Animation
- Logical operators
- Array/list membership
- *While* loop

The module gives computational examples and Quick Review Questions for you to complete and execute in the desired software system.

MODULE 9.5

Random Walk

Downloads

For several computational tools, the text's website has a *RandomWalk* file, which contains code for the module's algorithms, available for download. The site also has a data file *AverageDistances.dat* for Project 3.

Introduction

One technique of Monte Carlo simulations that has many applications in the sciences is the random walk. **Random walk** refers to the apparently random movement of an entity. In a time-driven simulation, we depict the entity in a **cell** on a rectangular **grid**. At any time step, the entity can move, perhaps under certain constraints, at random to a neighboring cell.

> **Definition** **Random walk** refers to the apparently random movement of an entity.

A certain type of computer simulation involving grids is a cellular automaton. **Cellular automata** are dynamic computational models that are discrete in space, state, and time. We picture space as a one-, two-, or three-dimensional **grid**, or array, or lattice. A **site**, or **cell**, of the grid has a state, and the number of states is finite. **Rules**, or **transition rules**, specifying local relationships and indicating how cells are to change state, regulate the behavior of the system. An advantage of such grid-based models is that we can visualize the progress of events through informative animations. For example, we can view a simulation of the movement of ants toward a food source, the spread of fire, or the motion of gas molecules in a container. In this module, the next chapter, and various modules in Chapter 14, we consider many scientific applications involving cellular automata.

Definitions A **cellular automaton** (plural, **automata**) is a type of computer simulation that is a dynamic computational model and is discrete in space, state, and time. Space is a **grid**, or a one-, two-, or three-dimensional lattice, or array, of **sites**, or **cells**. A cell of the lattice has a state, and the number of states is finite. **Rules**, or **transition rules**, specifying local relationships and indicating how cells are to change state, regulate the behavior of the system.

A random walk cellular automaton can model **Brownian motion**, which is the behavior of a molecule suspended in a liquid. The phenomenon bears the name of the English botanist Robert Brown. In 1827, he observed the rapid, random motion of pollen particles in a liquid could not occur because of life within the pollen, as some conjectured. A generation later, the physicists Maxwell, Clausius, and Einstein explained the phenomenon as invisible liquid particles striking the visible particles, causing small movements. Because diffusion of many things, such as pollutants in the atmosphere and calcium in living bone tissue, exhibit Brownian motion, simulations using random walks can also model these processes (Encyclopedia Britannica 1997; Exploratorium 1995).

In genetics, random walks have been used to simulate mutation of genes. As another example, scientists use the method **polymerase chain reaction** (**PCR**) to make many copies of particular pieces of DNA. A strand of DNA contains sequences of four bases, A, T, C, and G. Using the random walk technique in simulations, computational scientists can determine good proportions of these bases in solution to speed replication of the DNA.

Algorithm for Random Walk

At each time step of a particular random walk simulation, suppose an entity moves in a random, diagonal direction—NE, NW, SE, or SW. To go in such a direction, the entity walks east or west one unit and north or south one unit, covering a diagonal distance of $\sqrt{2}$ units.

We develop a function, ***randomWalkPoints***, with parameter, n, for the number of steps. The function generates such a walk and returns a list or array of the coordinates of the steps. In the function body, variables x and y store the horizontal and vertical coordinates, respectively, of the current location, and variable *lst* holds a list of locations in the path of the entity. Because the walker starts at the origin, we initialize *lst* to be a list containing the point (0, 0). With parameter n being the number of steps to be taken, a loop to produce the path executes n times. Within the loop, we generate one random integer of 0 or 1 to determine if the entity turns to the east or west by incrementing or decrementing x by 1, respectively. Then, another such "flip of the coin" dictates north with an increment of y or south with a decrement. We then append the new point (x, y) onto the developing *lst*. After the loop at the end of the function, we return this list of points.

Following is pseudocode, or a structured English outline of the design, for the function *randomWalkPoints* with **left-facing arrows** (←) indicating assignment. The parameter, *n*, appears in parentheses after the function name and a description of the action of the function follows. **Preconditions**, or the conditions that must be true for the function to behave properly, appear after "Pre." Preconditions should include any assumptions and information, such as parameters and their descriptions, that the function needs to meet its objectives. **Postconditions**, which follow "Post," describe the state of the system when the function finishes executing, any error conditions, and the information the function returns or otherwise communicates.

***randomWalkPoints(n)*:**

Function to produce a random walk, where at each time step the entity goes diagonally in a NE, NW, SE, or SW direction, and to return a list of the points in the walk

***Pre*:** *n* is the number of steps in the walk.
***Post*:** A list of the points in the walk has been returned.
Algorithm:

```
x ← 0 and y ← 0
lst ← a list containing the origin
do the following n times:
    rand ← a random 0 or 1
    if rand is 0
        increment x by 1
    else
        decrement x by 1
    rand ← a random 0 or 1
    if rand is 0
        increment y by 1
    else
        decrement y by 1
    append point (x, y) onto end of lst
return lst
```

After calling *randomWalkPoints* to generate the list containing the points of a path, we can create and display a graphics representing the random walk. For example, we might show all the random walk locations as colored dots, the path as line segments, and the first and last points as black dots. One execution of this code displays a graphic similar to Figure 9.5.1. Because the walk is random, each run of the function *randomWalkPoints* will very probably produce a different walk.

Quick Review Question 1

The following questions refer to *randomWalkPoints:*

a. After execution of the loop, how many elements does *lst* have?
b. Is it possible for the points (3, 5) and (3, 6) to be adjacent to each other in *lst*?

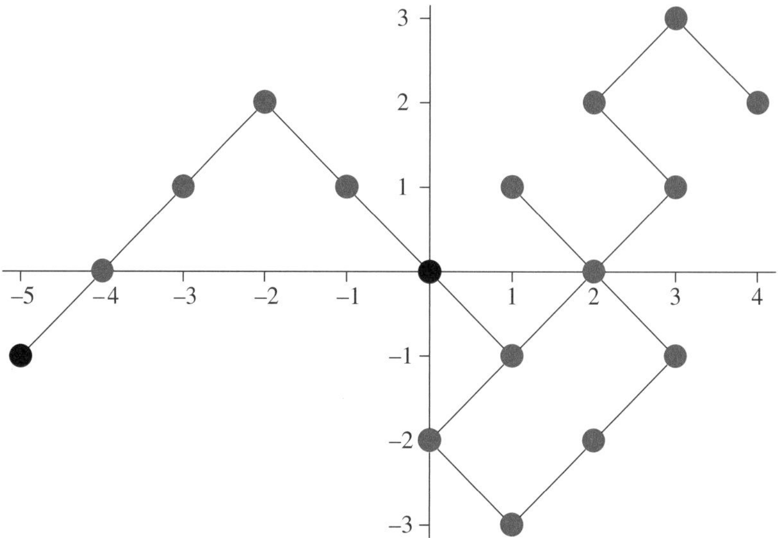

Figure 9.5.1 One possible display from execution of *randomWalkPoints*

Animate Path

Visualization of the path as it develops can aid in understanding the movement of the entity. Figure 9.5.2 presents several frames in such an animation.

To generate an animation, we develop a function, ***animateWalk***, which has as a parameter a list, *lst*, of $n + 1$ points in a random walk. For each i going from 1 through $n + 1$, we create a graphics of the first i points of the walk, which are in a sublist of the first i points of *lst*. Thus, we generate a sequence of $n + 1$ displays that we can animate with an appropriate computational tool. For the animation to be consistent, we specify that each graphics have the same axes, between the minimum and maximum of all x-coordinates on the x-axis and the minimum and maximum y-coordinates on the y-axis. The complete design of *animateWalk* follows.

animateWalk(lst)

Function to generate an animation of a random walk

***Pre*:** *lst* is the list of the points in the walk.
***Post*:** An animation of the walk has been generated.
Algorithm:

xMin ← minimum of x-coordinates in *lst*
xMax ← maximum of x-coordinates in *lst*
yMin ← minimum of y-coordinates in *lst*
yMax ← maximum of y-coordinates in *lst*
for i going from 1 through $n + 1$ do the following:
display a graphics of the first i points of *lst* with the display going from *xMin* to *xMax* in the x-direction and from *yMin* to *yMax* in the y-direction

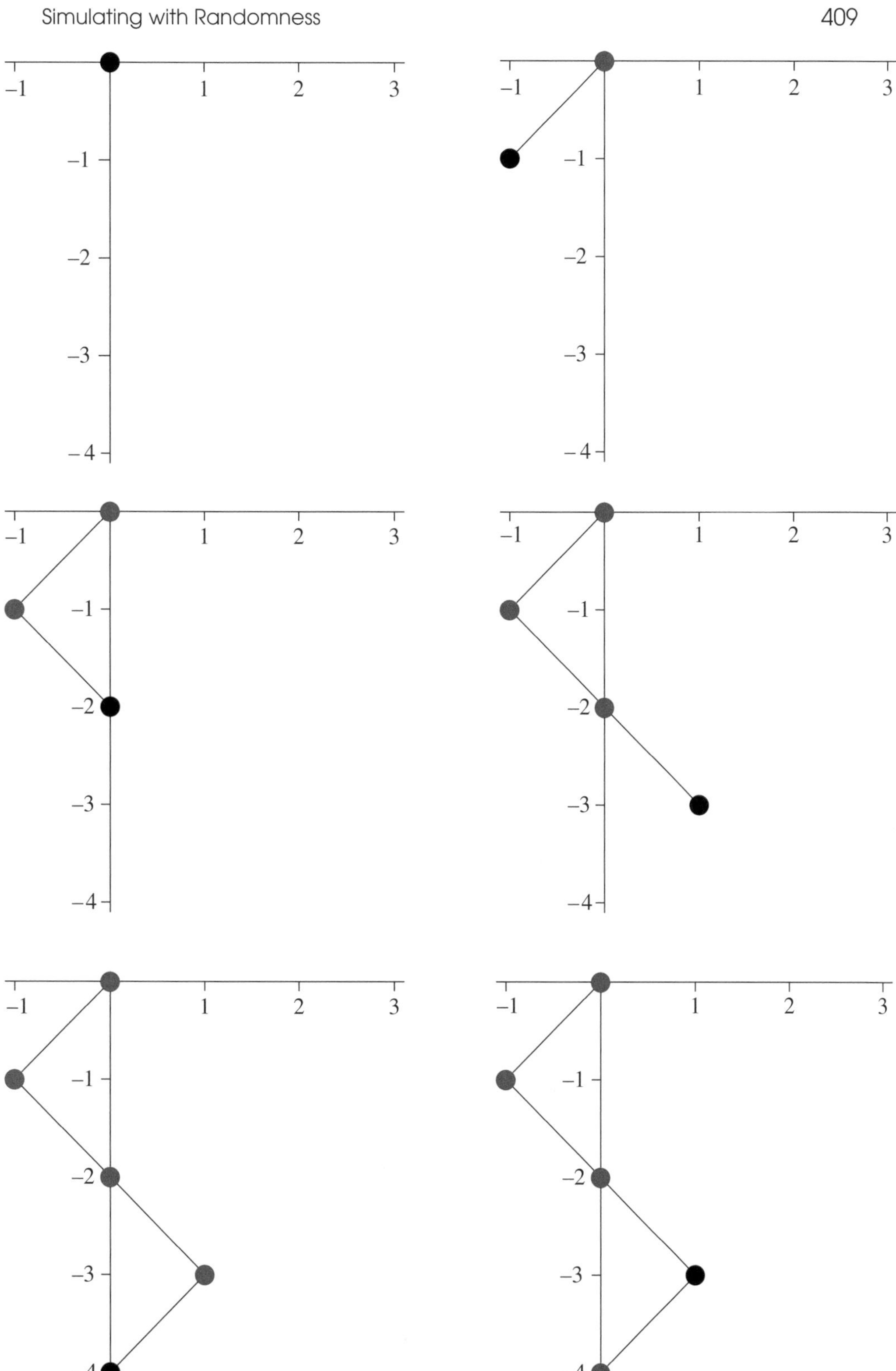

Figure 9.5.2 Several frames in an animation of the developing path from one random walk

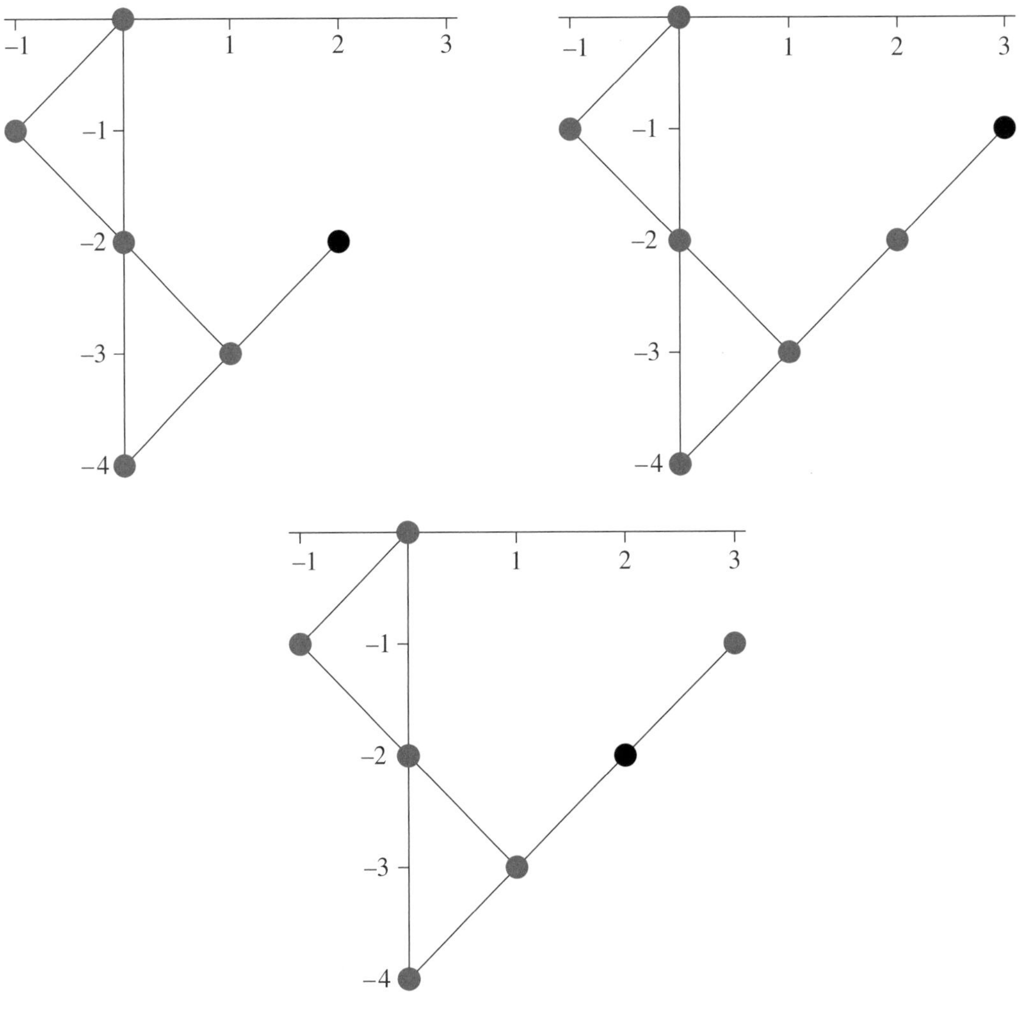

Figure 9.5.2 *(continued)*

Average Distance Covered

For the random walk in Figure 9.5.1, 5.09902 units is the distance between the final point and the initial one, which are the two black dots. However, because the walks are random, great variation can exist in both the paths and in the final distances from the starting point. Thus, to obtain an estimate of a typical distance between the starting and ending points of a random walk of n steps, we should run the simulation many times and take the average of all the distances. In such a case, we are not interested in viewing a random walk, so we first define another function, ***randomWalkDistance***, that is similar to *randomWalkPoints*, but which returns the desired distance instead of the list of points in a walk. Thus, in the loop that processes each step, we keep only the current point, (x, y), and, after the loop, we return the distance from the last point value of (x, y) and the origin, $\sqrt{x^2 + y^2}$. The next Quick Review Question designs this function.

Quick Review Question 2

Similar to *randomWalkPoints*, give a design for *randomWalkDistance*, a function with parameter, *n*, that returns the distance between the first and last point of a random walk of *n* steps.

For a function ***meanRandomWalkDistance***, which returns the average distance traveled over *numTests* number of random walks of *n* steps each, we place a call to *randomWalkPoints*(*n*) in a loop that iterates *numTests* number of times. A variable, *sumDist*, accumulates the distances covered by the random walks. Before the loop, *sumDist* is initialized to zero; after the loop, this sum is divided by *numTests* to return the average distance. One run of *meanRandomWalkDistance*(25, 100) might return an average distance of 5.75278 units for 100 simulations of random walks of 25 steps. The design of the function follows.

meanRandomWalkDistance(n, numTests)

Function to run a random walk simulation *numTests* number of times and to return the average distance between the first and last points

Pre: *n* is the number of steps in a walk.
numTests is the number of times to run the simulation.
Post: The average distance between the first and last points has been returned.
Algorithm:

let *sumDist*, the ongoing sum of distances, be 0
do the following *numTests* times:
  add *randomWalkDistance*(*n*) to *sumDist*
return *sumDist* / *numTests* as a floating point number

Quick Review Question 3

If we incorrectly move the initialization of *sumDist* inside the outer loop of *meanRandomWalkDistance*, select the final value of *sumDist:*

A. No change from current result.
B. *sumDist* would be 0.
C. *sumDist* would hold only the distance for the final path.
D. *sumDist* would be undefined.

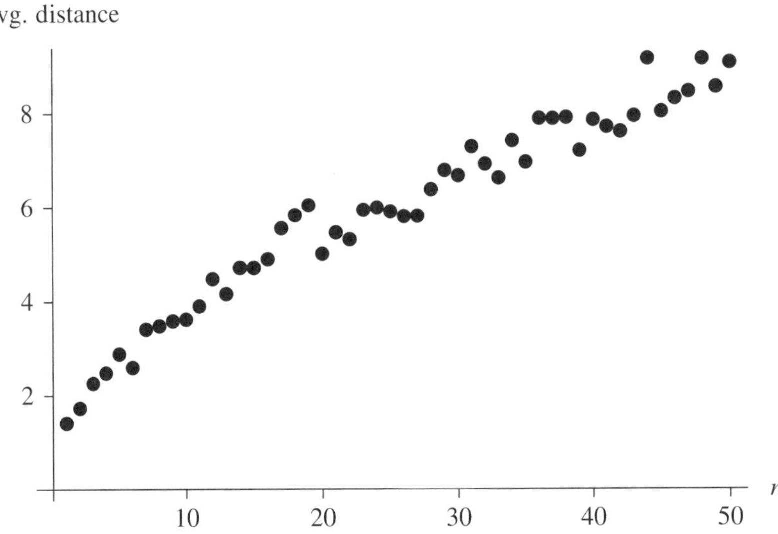

Figure 9.5.3 A plot of average distances traveled versus number of steps in a random walk

Relationship between Number of Steps and Distance Covered

To discern a relationship between the number of steps, n, and average distance covered in a random walk, we execute *meanRandomWalkDistance*(n, 100) for values of n from 1 to 50 and store each average distance in a list or array, *listDist*. Then, we employ the techniques of Module 8.3, "Empirical Models," to determine the relationship. Figure 9.5.3 shows a plot of the average distances traveled versus the number of steps. Projects 3 and 4 determine a formula for this relationship.

Exercises

On the text's website, RandomWalk files for several computational tools contain the code for the functions of this module. Complete the following exercises using your computational tool.

1. If possible in your computational tool, revise the code of *randomWalkPoints* to replace the loop with a call to a function to formulate *lst*.

2. Revise the code of *randomWalkPoints* or Exercise 1 to have the entity go with equal probability in a N, S, E, or W direction. *Hint:* Choose the direction based the value of a random integer, 0, 1, 2, or 3.

3. a. Revise the code of *randomWalkPoints* to have the entity go in an easterly direction (incrementing x) with probability of 30% and in a westerly direction (decrementing x) with probability of 70%.

b. Revise the code of Part a, to have the entity go in a northerly direction (incrementing y) with probability of 45% and in a southerly direction (decrementing y) with probability of 55%.

c. Give the probability for the entity going in each direction, NE, NW, SE, and SW.

4. Revise the code of *randomWalkPoints* or Exercise 1 to have the entity go in a N, S, E, or W direction with probabilities of 20%, 30%, 45%, or 5%, respectively.

Projects

On the text's website, RandomWalk files for several computational tools contain the code for the module's algorithms. Complete the following projects using your software system.

For additional projects, see Module 14.1, "Polymers—Strings of Pearls," and Module 14.2, "Solidification—Let's Make It Crystal Clear!"

1. Exercise 1
2. Exercise 2
3. Download the data file *AverageDistances.dat* of average distances covered for step sizes from 1 to 50 from the text's website. Using the techniques of Module 8.3, "Empirical Models," determine a relationship between the number of steps, n, and average distance covered in a random walk.
4. Develop code as discussed in the section on "Relationship between Number of Steps and Distance Covered" to obtain a list of average distances covered for random walks of step sizes from 1 to 50. Then, using the data the program generates, do the analysis of Project 3.
5. Develop code as discussed in the section on "Relationship between Number of Steps and Distance Covered" to obtain a list of average distances covered for random walks of step sizes from 1 to 50, where the entity travels E, W, N, or S with each step. Then, using the data the program generates, do the analysis of Project 3.
6. Develop code for Exercise 3 and run the simulation for 50 time steps. Include this code in a loop that runs the simulation 1000 or more times. Have the segment return the portion of time the entity ends on the 50th step in each of the four quadrants, NE, NW, SE, and SW. Do the figures seem to agree with your answer to Exercise 3c?
7. Develop code for Exercise 4 and run the simulation for 50 time steps. Include this code in a loop that runs the simulation 1000 or more times. Have the segment return the portion of time the entity ends on the 50th step in the N, S, E, or W direction from the starting location, the origin. On a particular run of the simulation, the 50th step could fall into one category, such as due north of the origin, or in two categories, such as N and E of the origin. Discuss the results in relationship to the probabilities of Exercise 4.
8. A hiker without a compass trying to find the way in the dark can step in any of eight directions (N, NE, E, SE, S, SW, W, NW) with each step. Studies

show that people tend to veer to the right under such circumstances. Initially, the hiker is facing north. Suppose at each step probabilities of going in the indicated directions are as follows: N, 19%; NE, 24%; E, 17%; SE, 10%; S, 2%; SW, 3%; W, 10%; NW, 15%. Develop a simulation to trace a path of a hiker, and run the simulation a number of times. Describe the results. (Note that other than at the initial step, this simulation simplifies the problem by ignoring the direction in which the hiker faces.)

9. Perform a simulation of Brownian motion of a pollen grain suspended in a liquid by generating a 3D random walk. Using documentation for your computational tool, investigate how to plot 3D graphics points and lines and create a 3D graphic of the walk.

Answers to Quick Review Question

1. a. $n + 1$ elements, (0, 0) and the n appended points
b. No, both coordinates are changed in the body of the loop.

2.

randomWalkDistance(n):

Function to produce a random walk, where at each time step the entity goes diagonally, and to return the distance between the first and last points

Pre: n the number of steps in the walk.
Post: The distance between the first and last points of a random walk of n steps was returned.
Algorithm:

$x \leftarrow 0$ and $y \leftarrow 0$
do the following n times:
  rand ← a random 0 or 1
  if *rand* is 0, increment x by 1; else decrement x by 1
  rand ← a random 0 or 1
  if *rand* is 0, increment y by 1; else decrement y by 1
return $\sqrt{x^2 + y^2}$

3. C. *sumDist* would hold only the distance for the final path.

References

Encyclopedia Britannica. 1997. "Brownian Motion." *Britannica Guide to the Nobel Prizes. Britannica Online*. http://www.britannica.com/nobel/micro/88_96.html

Exploratorium. 1995. "Brownian Motion." *Exploratorium Exhibit and Phenomena Cross-Reference*. http://www.exploratorium.edu/xref/phenomena/brownian_motion.html

10

CELLULAR AUTOMATON DIFFUSION SIMULATIONS

MODULE 10.1

Computational Toolbox—Tools of the Trade: Tutorial 6

Prerequisite: Module 9.4, "Computational Toolbox—Tools of the Trade: Tutorial 5."

Download

From the textbook's website, download Tutorial 6 in the format of your computational tool or in PDF format. We recommend that you work through the tutorial and answer all Quick Review Questions using the corresponding software.

Introduction

This sixth computational toolbox tutorial, which is available from the textbook's website in your system of choice, prepares you to use the system to complete projects for this and subsequent chapters. The tutorial introduces the following functions and concepts:

- Joining lists/arrays
- Finding the size of a list/array
- Visualizing a rectangular grid
- Matching patterns
- Position of a pattern in a list/array

The module gives computational examples and Quick Review Questions for you to complete and execute in the desired software system.

MODULE 10.2

Diffusion: Overcoming Differences

Downloads

For several computational tools, the text's website has a *Diffusion* file containing the simulation this module develops and a *10_2QRQ.pdf* file containing system-dependent Quick Review Questions and answers available for download.

Introduction

Heat energy is transferred by **thermal conduction** within or between objects where a temperature gradient exists. Particles or groups of particles with a higher temperature (more kinetic energy) transfer some of their energy to those at a lower temperature (less kinetic energy) upon collision. Thus, we have a **diffusion** of energy.

This diffusion of thermal energy presented a real problem for astronauts returning from a mission. As they brought their craft into the earth's atmosphere, the vehicle was traveling at about 40,000 km/h, generating tremendous friction. The temperatures on the exterior heat shield equaled 2760 °C, which is just over half the temperature of the sun's surface. Fortunately, the shield was effective enough to allow the cabin temperature to remain at 21°C (NASA Spinoff 1988, 2011).

During the late 1960s through the early 1970s, as part of the Apollo Mission, the United States sent manned spacecraft to the Moon. The heat shields for these vessels effectively fended off the diffusion of all that heat energy, generated upon atmospheric reentry, into the spacecraft. Each heat shield was coated with an ablative material—a substance that was allowed to char, dissipating energy and forming a protective coating, which did not allow the heat into the spacecraft itself (NASA Spinoff 1988, 2011).

A private company designed the heat shield for NASA, and the two entities collaborated in subsequent years to develop a number of fire-retardant paints and foams for military and civilian use. One of these, called Chartek, and derivative products are widely used by the oil and gas industries. Further product development led to Interchar, a fire-retardant commonly used to coat steel for construction. With a very thin layer (1–8 mm), Interchar does not hinder architectural design. Steel does not burn, but very high temperatures can weaken the metal. So, by delaying the transfer of heat energy to the steel, firefighters may be able to put out a fire before irreparable damage is done; and importantly, the coating delays loss of structural integrity for the evacuation of personnel (NASA Spinoff 1988, 2011).

Problem

In this module, we want to model the heat diffusion through a thin metal bar that has a constant application of heat and cold at designated locations on the bar (Cunningham 2007). We also want to develop an animated scientific visualization to depict the diffusion process.

Initializing the System

To simplify the situation, we apply heat and cold through the thickness of the bar and assume that each internal point on a line perpendicular to the top surface of the bar has the same temperature. If a point on the top surface has temperature 25 °C, then every point directly below that location is at 25 °C. Moreover, we assume that the bar is in a still room and that the immediate surroundings are at the same temperatures as the bar. Temperature diffuses within the bar, but external conditions do not affect the temperatures. Thus, we model the bar in two dimensions, length and width.

In many simulations, we model such a dynamic area with an $m \times n$ grid, or lattice, or a 2D rectangular array, or matrix, of numbers (Figure 10.2.1). Each cell in the lattice contains a value representing a characteristic of a corresponding location. For example, in a cellular automaton simulation of the diffusion of heat through a metal bar, a cell can contain that small square's average temperature in degrees Celsius. In a simulation involving a landscape, a cell might contain a moisture, nutrient, or veg-

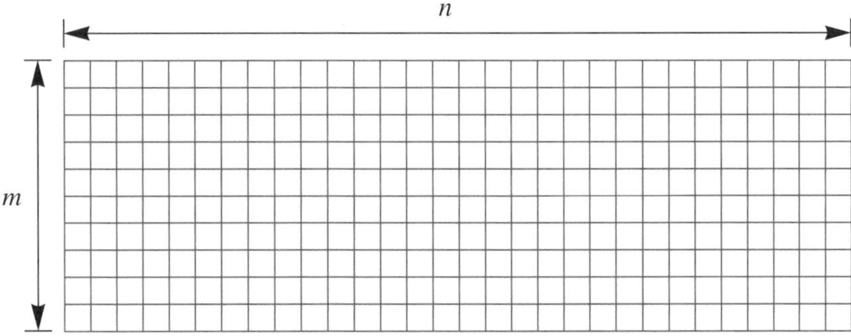

Figure 10.2.1 Cells to model area

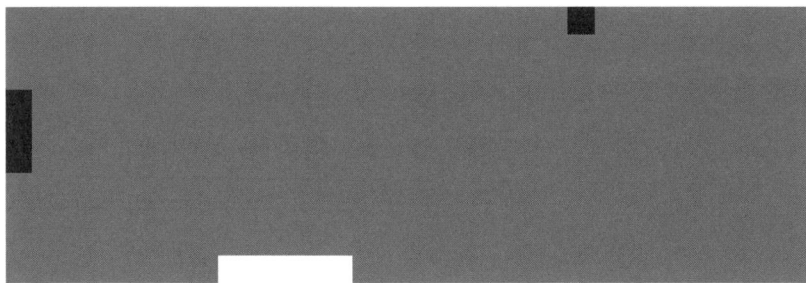

Figure 10.2.2 Initialized metal bar with black representing hot, white cold, and gray an intermediate temperature

etation level from 0.0 to 1.0. We can use a similar gradient to indicate the amount of pollution spreading through a lake.

In the case of heat diffusing through a thin metal bar, we might initialize each cell to be some ambient temperature, say *AMBIENT* = 25 °C, except for hot and cold spots, which might have the values *HOT* = 50 °C and *COLD* = 0 °C, respectively. The following algorithm ***initBar*** initializes the grid for such a bar with two hot spots—a larger one in the middle of the first column and a smaller one three-fourths the way on the first row—and one fairly large cold spot one-third of the way along the last row (Cunningham 2007). Because the hot and cold spots are always present, we define a function, ***applyHotCold***, which we can call elsewhere, to assign the values *HOT* and *COLD* to appropriate cells in a bar. Using black to represent *HOT*, white for *COLD*, and a proportional shade of gray for temperatures between these values, Figure 10.2.2 illustrates the top 2D surface of such an initialized bar with a 10 × 30 grid.

initBar(m, n, hotSites, coldSites)

Function to return an *m* × *n* grid of temperatures: Cells with coordinates in *hotSites* have the value *HOT*; cells with coordinates in *coldSites* have the value *COLD*; and all other cells have the value *AMBIENT*

Pre: *m* and *n* are positive integers.
hotSites and *coldSites* are lists of coordinates for hot and cold sites, respectively.
AMBIENT, *HOT*, and *COLD* are global constants, and *COLD* ≤ *AMBIENT* ≤ *HOT*.

Post: An *m* × *n* grid of values as described before has been returned.

Algorithm:

```
ambientBar ← m by n matrix of AMBIENT values
return applyHotCold(ambientBar, hotSites, coldSites)
```

applyHotCold(bar, hotSites, coldSites)

Function to accept a grid of temperatures and to return a grid with heat and cold applied at *hotSites* and *coldSites*, respectively

Pre: *bar* is a grid of values.
hotSites and *coldSites* are lists of coordinates inside the grid for hot and cold sites, respectively.
AMBIENT, *HOT*, and *COLD* are global constants, and *COLD* ≤ *AMBIENT* ≤ *HOT*.
Post: A grid of values as described above has been returned.
Algorithm:
newBar ← *bar*
assign *HOT* to every *newBar* cell with coordinates in *hotSites*
assign *COLD* to every *newBar* cell with coordinates in *coldSites*
return *newBar*

Quick Review Question 1

From the text's website, download your computational tool's *10_2QRQ.pdf* file for this system-dependent question on initializing the grid.

Heat Diffusion

At each simulation iteration, we apply a function, ***diffusion***, to each cell site to determine its temperature at the next time step. The cell's value at the next instant depends on the cell's current value (*site*) and the values of its four or eight nearest **neighbors**, as in Figure 10.2.3. The four neighbors along with the site itself in Figure 10.2.3a comprise the **von Neumann neighborhood** of a site, while the nine nodes in Figure 10.2.3b form the **Moore neighborhood** of a site. For diffusion of heat through a metal bar, we employ Moore neighborhoods.

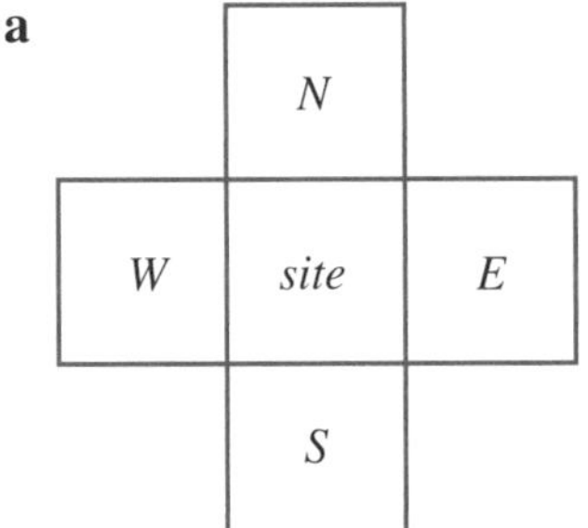

Figure 10.2.3 Cells that determine a site's next value

Definitions In a two-dimensional grid, the **von Neumann neighborhood** of a site is the set of cells directly to the north, east, south, and west of the site and the site itself. As well as these five cells, the **Moore neighborhood** of a site includes the corner cells to the northeast, southeast, southwest, and northwest of the site. The four or eight neighborhood cells not including the site are the site's **neighbors**.

We base our model of diffusion on **Newton's law of heating and cooling**, which states that the rate of change of the temperature with respect to time of an object is proportional to the difference between the temperature of the object and the temperature of its surroundings. Similarly, we can say that the change in a cell's temperature, $\Delta site$, from time t to time $t + \Delta t$ is a **diffusion rate parameter** (r) times the sum of each difference in the temperature of a neighbor ($neighbor_i$) and the cell's temperature ($site$), as follows:

$$\Delta site = r\sum_{i=1}^{8} (neighbor_i - site), \text{ where } 0 < r < 1/8 = 0.125$$

Thus, the site's temperature at time $t + \Delta t$ is the following:

$$site + \Delta site = site + r\sum_{i=1}^{8} (neighbor_i - site)$$

where $0 < r < 0.125$ and the sum is over the eight neighbors. With subtraction of $r \cdot site$ occurring 8 times, the formula simplifies to the following weighted sum of temperatures of the cell and its neighbors:

$$site + \Delta site = (1 - 8r)site + r\sum_{i=1}^{8} neighbor_i, \text{ where } 0 < r < 0.125$$

Similar diffusion formulas, which we explore in the projects, can have smaller coefficients for the corners than for the north, east, south, and west neighbors. However, the sum of the coefficients, which are fractions or percentages, for each of the nine cells in the neighborhood should be 1.0, or 100%.

Quick Review Question 2

Suppose the diffusion rate parameter is 0.1 and the temperatures in the cells are as in Figure 10.2.4. Calculate the temperature in the center cell at the next time step.

With diffusion rate (*diffusionRate*) and temperatures of a cell (*site*) and its eight neighbors (*N*, *NE*, *E*, *SE*, *S*, *SW*, *W*, *NW*) as parameters, the function *diffusion* computes and returns the new temperature for the cell.

2	3	4
0	5	6
1	3	7

Figure 10.2.4 Temperatures in a section of the grid for Quick Review Question 1

diffusion(diffusionRate, site, N, NE, E, SE, S, SW, W, NW)

Function to return the new temperature of a cell

Algorithm:

return (1 - 8*diffusionRate*)*site*
+ *diffusionRate*·(*N* + *NE* + *E* + *SE* + *S* + *SW* + *W* + *NW*)

Boundary Conditions

We must be able to apply the function *diffusion* to every grid point, such as in Figure 10.2.1, including those on the boundaries of the first and last rows and the first and last columns. However, the *diffusion* function has parameters for the grid point (*site*) and its neighbors (*N, NE, E, SE, S, SW, W, NW*). Thus, to apply *diffusion* we extend the boundaries by one cell in each direction, creating what we call **ghost cells**. Several choices exist for values in those cells:

- Give every extended boundary cell a constant value, such as 25. Thus, the boundary insulates. Figure 10.2.5 outlines an original square grid, which has white cells, with thick black lines, while the constant extension is in color. We call the situation where the boundary has a constant value an **absorbing boundary condition**. In the case of the diffusion of heat through a metal bar, the boundary is similar to the bar being placed in a well ventilated room at 25 °C.
- Give every extended boundary cell the value of its immediate neighbor. Thus, the values on the original first row occur again on the new first row of ghost cells. Similar situations occur on the last row and the first and last columns (Figure 10.2.6). Such immediate repetitions are called **reflecting,** or **reflective, boundary conditions**. In the case of the spread of temperature, the boundary tends to propagate the current local situation: The air in the room is still, and the air temperature around the bar tends to mimic the temperature of the bar.
- Wrap around the north-south values and the east-west values in a fashion similar to a donut, or torus. Extend the north boundary with a ghost row that is a copy of the original south boundary row, and extend the south boundary with a copy of the original north boundary row. Similarly, expand the column boundaries on the east and west sides. Thus, for a cell on the north boundary, its neighbor to the north is the corresponding cell to the south (Figure 10.2.7). Such conditions are called **periodic boundary conditions**. In the case of a simulation of heat diffusion, the area is a closed, continuous environment with the situation at one boundary effecting its opposite boundary cells.

In the application of heat diffusion, because we assume that the immediate surroundings are at the same temperatures as on the surface of the bar, we choose to employ reflecting boundary conditions to minimize the impact of the surroundings. In the beginning, we attach new first and last rows, as in Figure 10.2.8, by **concatenating**, or attaching, the original grid's first row, the original grid, and the last row to create a new lattice, *latNS*.

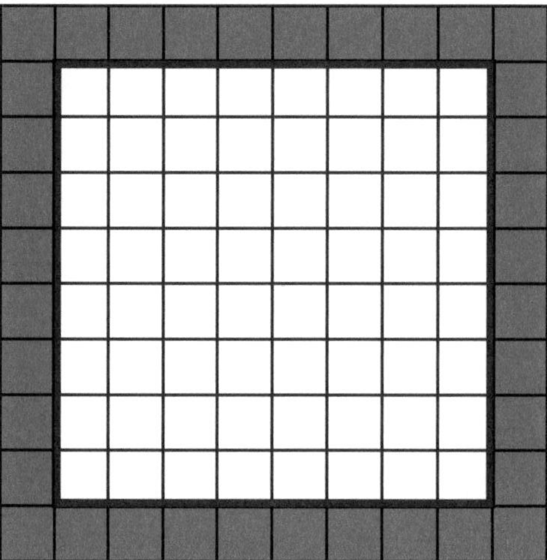

Figure 10.2.5 Absorbing boundary conditions: Grid with extended boundaries and each ghost having a constant value

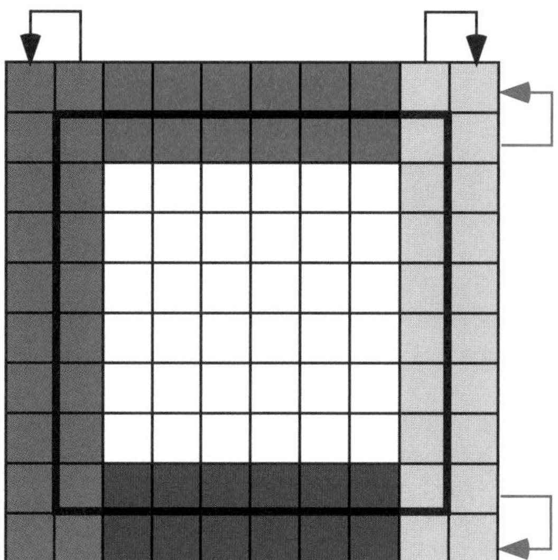

Figure 10.2.6 Reflecting boundary conditions: Grid with extended boundaries and each ghost cell having the value of its immediate neighbor in the original grid

Quick Review Question 3

Answer the following questions about Figure 10.2.4 as an extremely small entire thermal grid.

a. Give the size of the grid extended to accommodate boundary conditions.
b. Give the values in the first row of the extended matrix, assuming fixed boundary conditions with fixed value 0.
c. Give the values in the first row of the extended matrix, assuming reflecting boundary conditions, where we copy rows first.

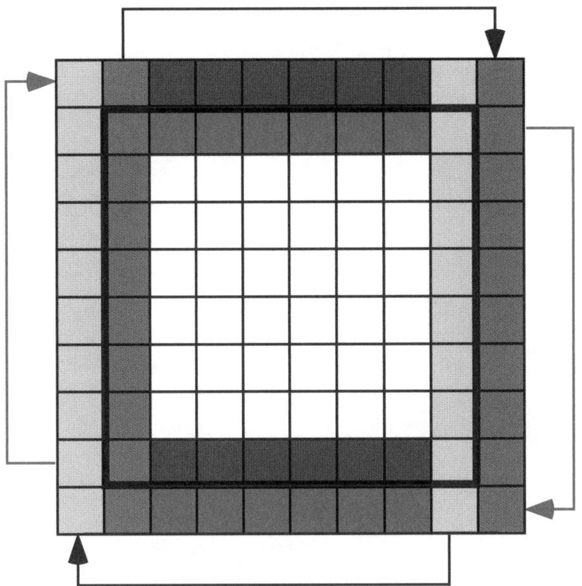

Figure 10.2.7 Extended grid with periodic boundary conditions

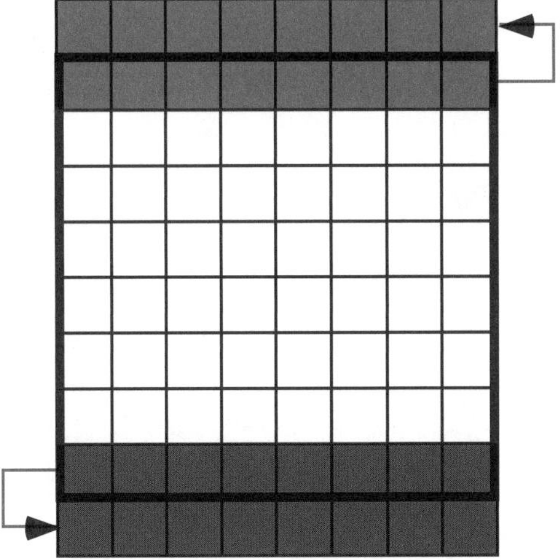

Figure 10.2.8 Grid extended by having a new first row that is a copy of the first row on the original grid and having a new last row that is a copy of the last row on the original grid

d. Give the values in the first row of the extended matrix, assuming periodic boundary conditions, where we copy rows first.

Quick Review Question 4

From the text's website, download your computational tool's *10_2QRQ.pdf* file for this system-dependent question that extends a grid as in Figure 10.2.8 by attaching a copy of the first row to the beginning and a copy of the last row to the end of the original grid to form a new grid, *latNS*.

To extend the grid with reflecting boundary conditions in the east and west directions, we concatenate the first column of *latNS* from Quick Review Question 4, *latNS*, and the last column of *latNS*. For some computational tools, it is easier to first transpose the lattice *latNS*, perform the same manipulation with the rows as in Quick Review Question 4, and then transpose the resulting lattice.

To consolidate these tasks, we define a function, ***reflectingLat***, using reflecting boundary conditions to extend by one cell in each direction the lattice. Pseudocode for the function follows.

reflectingLat(lat)

Function to accept a grid and to return a grid extended one cell in each direction with reflecting boundary conditions

Pre: *lat* is a grid.
Post: A grid extended one cell in each direction with reflecting boundary conditions was returned.
Algorithm:
latNS ← concatenation of first row of *lat*, *lat*, and last row of *lat*
return concatenation of first column of *latNS*, *latNS*, and last column of *latNS*

Quick Review Question 5

From the text's website, download your computational tool's *10_2QRQ.pdf* file for this system-dependent question that extends a lattice, as in Figure 10.2.9.

Applying a Function to Each Grid Point

After extending the grid by one cell in each direction using reflecting boundary conditions, we apply the function *diffusion* to each internal cell and then discard the boundary cells. We define a function, ***applyDiffusionExtended***, that takes an extended lattice, *latExt*, and returns the internal lattice with *diffusion* applied to each site. Figure 10.2.10 depicts an extended grid with the internal grid, which is a copy of the original lattice, in color. The number of rows of *latExt* is $m + 2$, while the number of columns is $n + 2$. As Figure 10.2.10 depicts, the number of rows (m) and columns (n) of the returned lattice is two less than the number of rows and columns of *latExt*, respectively. We apply the function *diffusion*, which has parameters *diffusionRate*, *site*, *N*, *NE*, *E*, *SE*, *S*, *SW*, *W*, and *NW*, to each internal cell in lattice *latExt*. If array indices in a computational tool begin with 0, these internal cells are in rows 1 through m and columns 1 through n. For array indices that start with 1, the internal

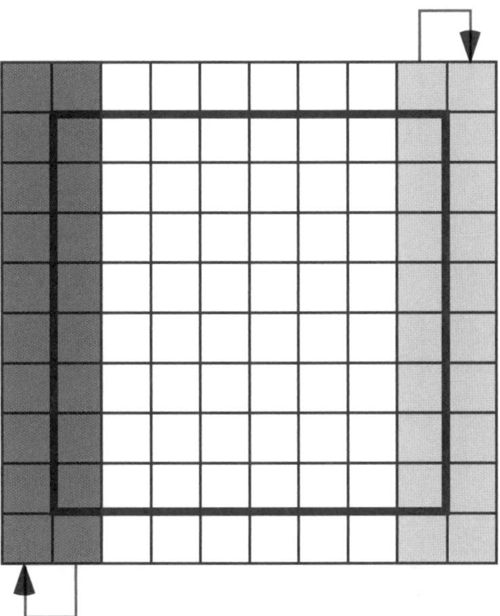

Figure 10.2.9 Grid from Figure 10.2.8 expanded by having a new first column that is a copy of the first column and a new last column that is a copy of the last column

cells are in rows 2 through $m + 1$ and columns 2 through $n + 1$. We added the boundary rows and columns to eliminate different cases for cells with or without one or more neighbors. Thus, for i going through the indices for the internal rows of the extended array and for j going through the internal column indices, *applyDiffusionExtended* obtains a value for each cell in a new $m \times n$ lattice by applying *diffusion* to each site with coordinates i and j. The site's neighbors with corresponding coordinates are as in Figure 10.2.11.

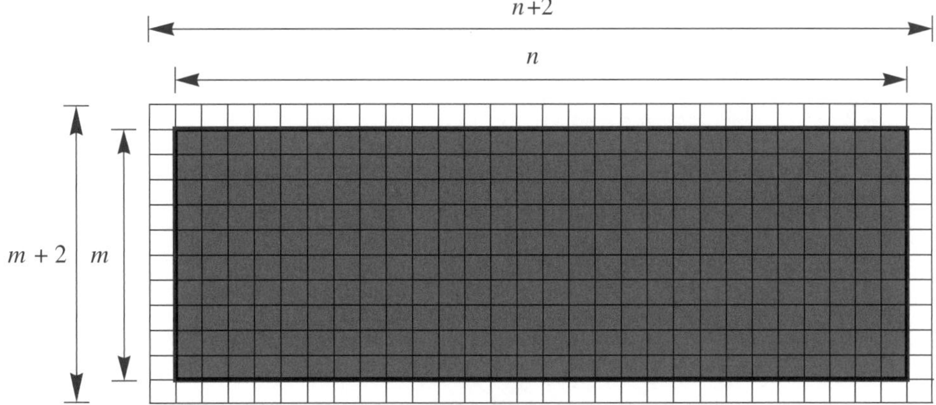

Figure 10.2.10 Internal grid in color that is a copy of the original grid (Figure 10.2.1) embedded in an extended grid

NW $(i - 1, j - 1)$	*N* $(i - 1, j)$	*NE* $(i - 1, j + 1)$
W $(i, j - 1)$	*site* (i, j)	*E* $(i, j + 1)$
SW $(i + 1, j - 1)$	*S* $(i + 1, j)$	*SE* $(i + 1, j + 1)$

Figure 10.2.11 Indices for a lattice site and its neighbors

Quick Review Question 6

Suppose *extMat* is an extended matrix of size 97 × 62.

a. Give the size of the matrix *applyDiffusionExtended* returns.
b. When $i = 33$ and $j = 25$, give the indices of the site's neighbor to the north.
c. For this site, give the indices of its neighbor to the southwest.

Quick Review Question 7

From the text's website, download your computational tool's *10_2QRQ.pdf* file for this system-dependent question that develops the function *applyDiffusionExtended.*

Simulation Program

To perform the simulation of diffusion of heat through a metal bar, we define a function, ***diffusionSim***, with parameters *m* and *n*, the number of grid rows and columns, respectively; *diffusionRate*, the rate of diffusion; and *t*, the number of time steps. The function *diffusionSim* returns a list of the initial lattice and the next *t* lattices in the simulation. Pseudocode for *diffusionSim* is presented on the following page.

Quick Review Question 8

From the text's website, download your computational tool's *10_2QRQ.pdf* file for this system-dependent question that implements the loop in the *diffusionSim* function.

diffusionSim(m, n, diffusionRate, t)

Function to return a list of grids in a simulation of the diffusion of heat through a metal bar

Pre: *m* and *n* are positive integers for the number of grid rows and columns, respectively.
diffusionRate is the rate of diffusion.
t is the number of time steps.
diffusion is a function to return a new temperature for a grid point.
Post: A list of the initial grid and the grid at each time step of the simulation was returned.
Algorithm:

bar ← *initBar*(*m*, *n*, *hotSites*, *coldSites*)
grids ← list containing *bar*
do the following *t* times:
  barExtended ← *reflectingLat*(*bar*)
  bar ← *applyDiffusionExtended*(*diffusionRate*, *barExtended*)
  bar ← *applyHotCold*(*bar*, *hotSites*, *coldSites*)
  grids ← the list with *bar* appended onto the end of *grids*
return *grids*

Display Simulation

Visualization helps us understand the meaning of the grids. For each lattice in the list returned by *diffusionSim*, we generate a graphic using grayscale or color. We define a function, ***animDiffusionGray***, with parameter *grids*, which is a list of lattices from the simulation, to produce a grayscale animation of the changing temperatures in the metal bar, with black representing the hottest locations and white the coldest. Starting with the initial bar from Figure 10.2.2 and a diffusion rate of 0.1 and displaying several frames of such an animation, Figure 10.2.12 shows that the bar quickly approaches equilibrium.

Quick Review Question 9

From the text's website, download your computational tool's *10_2QRQ.pdf* file for this system-dependent question that develops the function *animDiffusionGray*, which produces a grayscale graphic corresponding to each simulation lattice in a list (*grids*).

For a color display, we should employ a coloration that is evocative of the situation, such as red for hot and blue for cold. For display on a monitor, we usually employ the **red-green-blue (RGB) color model**. In the RGB color model, we specify the amounts between 0.0 and 1.0 of red, green, and blue light at each **pixel**, or picture element, or point in the graphics. For our heated bar, we employ only red and blue

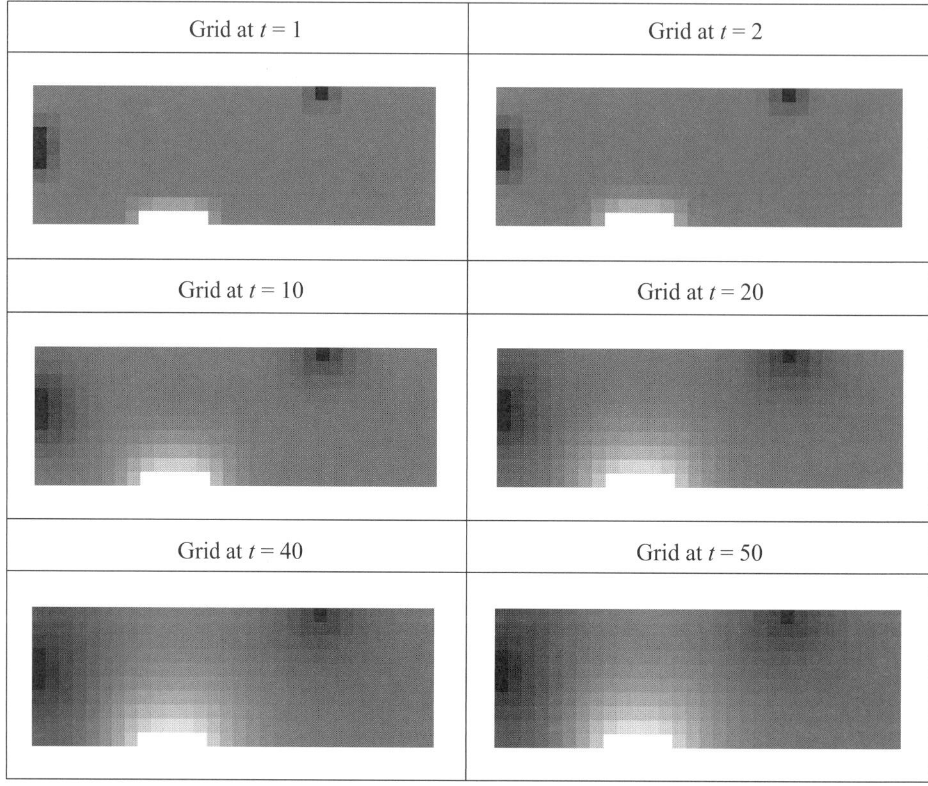

Figure 10.2.12 Several frames in an animation in grayscale of the spreading of heat through a metal bar

light, so the level of green light is 0.0. In going from the coldest to the hottest values, red increases from 0.0 to 1.0, while blue decreases from 1.0 to 0.0. To obtain a zero-to-one scale with the minimum temperature being *COLD* = 0, we divide a cell's temperature, *temp*, by the maximum temperature, *HOT*, so that the amount of red light is *temp*/*HOT*, expressed as a floating-point number. If *HOT* is 50.0 and *temp* is 0.0, so is *temp*/*HOT* = 0.0/50.0 = 0.0; while if *temp* is 50.0, then *temp*/*HOT* is 1.0. To have the amount of blue light decrease as the temperature decreases, we subtract the fraction from 1.0. If *temp* is 0.0, then 1.0 – *temp*/*HOT* is 1.0; and if *temp* is 50.0, then 1.0 – *temp*/*HOT* is 0.0. Using a temperature scale from 0 °C to 50 °C, Table 10.2.1 gives several RGB for this scaling.

Table 10.2.1
Several RGB Color Model Values of the Amounts of Red and Blue for Temperatures from 0 °C to 50 °C

Temperature (°C)	*0*	*10*	*25*	*40*	*50*
red fraction (temperature/50)	0.0	0.2	0.5	0.8	1.0
blue fraction (1.0 – temperature/50)	1.0	0.8	0.5	0.2	0.0

Quick Review Question 10

From the text's website, download your computational tool's *10_2QRQ.pdf* file for this system-dependent question that develops the function, ***animDiffusionColor***, which produces a color graphic corresponding to each simulation lattice in a list (*grids*).

Exercises

On the text's website, Diffusion *files for several computational tools contain the code for the simulation of the module. Complete the following exercises using your computational tool.*

1. Write a function to extend a grid using absorbing boundary conditions with the constant value on the boundary being 25.
2. Write a function to extend a grid using periodic boundary conditions.

Projects

On the text's website, Diffusion *files for several computational tools contain the code for the simulation of the module. Complete the following projects using your computational tool.*

For an additional project, see Project 4 from Module 13.4, "Probable Cause—Modeling with Markov Chains."

1. **a.** Determine how long it takes, t, for the bar modeled in this module to reach equilibrium, where from time t to time $t + 1$ the values in each cell vary by no more than plus or minus some small value, such as ±0.001.
   b. Repeat Part a, applying heat and cold for 10 time steps and then removing such heating and cooling.
2. Develop simulations and animations for the bar modeled in this module using several boundary conditions: three simulations of absorbing boundary conditions with constant values 0, 25, and 50 and periodic boundary conditions. Along with the reflecting boundary conditions, describe the results. Discuss the advantages and disadvantages of each approach and the situations, such as heat or pollution diffusion, for which each is most appropriate.
3. Instead of using the formula for diffusion in the section "Heat Diffusion," employ the filter in Figure 10.2.13. Thus, to obtain the value at a site for time $t + 1$, we add 25% of the site's temperature at time t, 12.5% of the north, east, south, and west cells at time t, and 6.25% of the corner cells to the northeast, southeast, southwest, and northwest. This sum is called a **weighted sum** with each nutrition value carrying a particular weight as indicated by the table. Revise the model using this configuration and compare the results with that of the module.
4. **a.** Model a bar at 100 °C that has a constant application of a 25 °C external source on its boundary. Generate plots of the temperatures at a corner

0.0625	0.125	0.0625
0.125	0.25	0.125
0.0625	0.125	0.0625

Figure 10.2.13 Filter for Project 3

and in the middle of the bar versus time. Describe the shapes of the graphs.

b. Repeat Part a with the bar being at –50 °C.

c. Discuss the results.

5. Consider a small, shallow body of water that initially has a constant amount of nutrient. A cypress toward one edge of the water consumes nutrients at a constant rate, so that at each time step the amount of nutrients in the corresponding cell decreases by a fixed amount. Suppose shore is on three sides and a larger body of water is on the fourth side. Nutrients from the larger body of water diffuse into the smaller area. Model and visualize the situation for the small body of water. Find a rate of diffusion and a rate of nutrient consumption so that the tree always has nourishment. Use the formula for diffusion in the section "Heat Diffusion" or the filter variation in Project 3.

6. Suppose an industry constantly spills pollutants into a containment pond, which initially has water. Using a diffusion rate of 0.1, how long will it take for the concentration of pollutants in the middle of the pond to reach 25%? Give your assumptions and discuss the results.

7. Model and visualize a situation in which diffusion tends to occur more in one direction than another, say more from the east than from the west. Thus, design a filter similar to that in Project 3 that favors directional diffusion. Such a configuration could be used in modeling diffusion on the surface of flowing water. Give your assumptions and discuss the results.

8. Suppose a dye is dissolved in water, which is poured on top of a gel. Model and visualize a cross section of the diffusion of the dye into the gel. Compare your results with the time-lapsed video at (Wikipedia Contributors, "Diffusion"). For your parameters, determine t to match the diffusion time in the video.

9. Often because of imperfections, variations in media, or other factors, diffusion does not proceed deterministically but varies slightly with an element of chance. Revise the function *diffusion*, which the section "Heat Diffusion" describes, to be stochastic. Instead of multiplying each $neighbor_i$ by r, the rate of diffusion, multiply each neighboring temperature by a different $(1 + rnd_i)r$, where rnd_i is a normally distributed random number with mean 0 and standard deviation 0.5. Adjust the coefficient of *site* so that the sum of all the coefficients is 1. Run the model 100 times for 20 time steps and determine the mean and range of temperatures for a designated cell towards the middle of the bar.

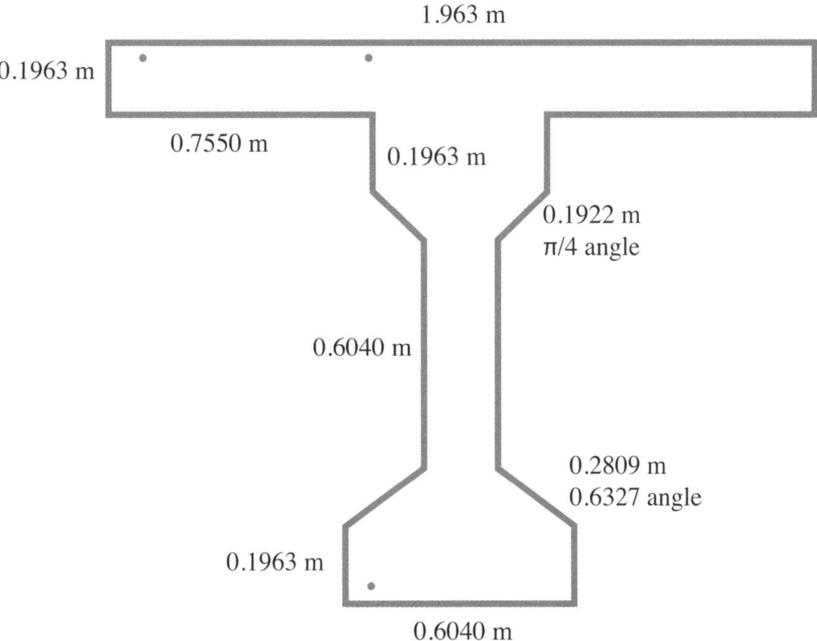

Figure 10.2.14 Cross section of a bridge support, based on Podroužek (2008)

10. Repeat Project 9 using the filter described in Project 3.

11. Application of deicing salts in the winter can degrade concrete reinforced structures, such as bridges, because of the ingress of harmful substances such as chloride ions. Engineers incase steel in concrete to protect against corrosion. However, when the concentration of chloride reaches a critical concentration, perhaps 0.4% Cl^- per unit of concrete content, the concrete no longer can protect the steel. Develop a cellular automaton simulation of the diffusion of chloride in a T-shaped cross section of a bridge support, as in Figure 10.2.14. Assume deicing salts can seep into the structure from all surfaces except the top. Referring to Project 9, employ stochastic diffusion with a basic diffusion rate of 0.125 and von Neumann neighborhoods. For 30 years of constant exposure, apply a 2% per unit chloride ion concentration from the salt to all external surfaces except the upper surface. Have the basic time step be 165 days. Small circles indicate where reinforcing bars intersect the T cross section. Averaging the results for many simulations, say 100 or 1000, determine the chloride ion concentration at the locations for these reinforcing bars after 30 years of continuous exposure (Podroužek 2008).

12. Model in 3D the diffusion of heat through a bar. Assume that the bar is sitting on a table in a room with good circulation. The part of the table on which the bar rests has approximately the same temperatures as the corresponding locations on the bottom of the bar, but the air around the bar remains almost constantly 25 °C.

Answers to Quick Review Question

From the text's website, download your computational tool's *10_2QRQ.pdf* file for answers to the system-dependent questions.

2. $3.6 = (1 - 8 \times 0.1)(5) + 0.1(2 + 3 + 4 + 0 + 6 + 1 + 3 + 7)$

3. a. 5×5

b. 0, 0, 0, 0, 0

c. 2, 2, 3, 4, 4

d. 7, 1, 3, 7, 1

5. a. 95×60

b. (32, 25)

c. (34, 24)

References

Cunningham, Steve. 2007. *Computer Graphics: Programming in OpenGL for Visual Communication*, Upper Saddle River, NJ: Prentice-Hall.

NASA Spinoff. 1988. "Spinoff from Mooncraft Technology." NASA.

NASA Spinoff. 2011. "Fire-Resistant Reinforcement Makes Steel Structures Sturdier." http://spinoff.nasa.gov/Spinoff2006/ps_3.html (accessed June 15 2012)

Podroužek, Jan, and Břetislav Teplý. 2008. "Modelling of Chloride Transport in Concrete by Cellular Automata." *Engineering Mechanics*, (15)3: 213–222.

Wikipedia Contributors, "Diffusion," *Wikipedia, The Free Encyclopedia*. http://en.wikipedia.org/wiki/Diffusion (accessed June 13, 2012)

Wikipedia Contributors, "Heat Transfer," *Wikipedia, The Free Encyclopedia*, http://en.wikipedia.org/wiki/Heat_transfer (accessed June 13, 2012)

MODULE 10.3

Spreading of Fire

Prerequisite: Module 10.2, "Diffusion: Overcoming Differences."

Downloads

For several computational tools, the text's website has available for download a *Fire* file containing the simulation this module develops and a *10_3QRQ.pdf* file containing system-dependent Quick Review Questions and answers.

Introduction

Human beings, with some justification, have considerable fear of fire. History is replete with disastrous losses of life and property from it. Nevertheless, fires in areas like the western United States are natural and, ecologists tell us, beneficial to the plant communities there. Periodic fires help to clear the forest floor of debris and promote the growth of sturdy, fire-resistant trees. Unfortunately, expanding human populations have intruded on previously uninhabited areas, establishing their own communities in "fire-prone" zones. Furthermore, human activities, such as fire suppression, livestock grazing, and logging, have increased the possibility of hotter and more destructive fires (NPS 2012).

During the fall of 2003, residents of Southern California faced a series of firestorms driven by powerful Santa Ana winds. After 3 days, the fires had destroyed more than 400,000 acres and 900 homes and had killed 15 people. Hundreds of firefighters battled a chain of fires that extended from Ventura County, north of Los Angeles, east into San Bernadino County and south to Tijuana, Mexico. A haze of toxins draped over the area like a pall (Wilson et al. 2003).

The Malibu region above Los Angeles is dominated by the Santa Monica Mountains and canyons that run from north to south. Much of the natural vegetation is dry **chaparral**, consisting of many small, oily, woody plants that are extremely flammable. This vegetation naturally would burn every 15 to 45 years, clearing out old and dead plant materials and returning nutrients to the soil. With the prevailing dry conditions, an illegal campfire can set off a ferocious blaze that may stop only after traveling many miles to the Pacific Ocean (SBCCBS 2010; Los Angeles Times 2010).

Fighting fires in Southern California or anywhere else is a very risky job, where loss of life is a real possibility. Proper training is essential. In the United States the National Fire Academy, established in 1974, presents courses and programs that are intended "to enhance the ability of fire and emergency services and allied professionals to deal more effectively with fire and related emergencies." The Academy has partnered with private contractors and the U.S. Forest Service to develop a 3D land fire-fighting training simulator. This simulator exposes trainees to a convincing fire-propagation model, where instructors can vary fuel types, environmental conditions, and topography. Responding to these variables, trainees may call for appropriate resources and construct fire lines. Instructors may continue to alter the parameters, changing fire behavior. Students can review the results of their decisions, where they can learn from their mistakes in the safety of a computer laboratory (DAS 2012).

This module develops a two-dimensional computer simulation for the spread of fire. The techniques can be extended to numerous other scientific examples involving contagion, such as the propagation of infectious diseases and distribution of pollution.

Problem

Our problem is to simulate the spread of fire from an initial landscape of empty ground, nonburning trees, and trees that are on fire. Moreover, the area can suffer from lightning strikes, which may or may not start additional fires.

Initializing the System

For our cellular automaton simulation of the spread of fire, a cell of an $n \times n$ grid can contain a value of 0, 1, or 2 indicating an empty cell, a cell with a nonburning tree, or a cell with a burning tree, respectively. Table 10.3.1 lists these values and meanings, along with associated constants, ***EMPTY***, ***TREE***, and ***BURNING***, which have values of **0**, **1**, and **2**, respectively. We initialize these constants at the beginning and employ the descriptive names throughout the program. Thus, the code is easier to understand and to change.

Table 10.3.1
Cell Values with Associated Constants and their Meanings

Value	*Constant*	*Meaning*
0	*EMPTY*	The cell is empty ground containing no tree.
1	*TREE*	The cell contains a tree that is not burning.
2	*BURNING*	The cell contains a tree that is burning.

To initialize this discrete stochastic system, we employ the following two probabilities:

probTree: The probability that a tree (burning or not burning) initially occupies a site. Thus, *probTree* is the initial tree density measured as a percentage.

probBurning: If a site has a tree, the probability that the tree is initially burning or that the grid site is *BURNING*. Thus, *probBurning* is the fraction of the trees that are burning when the simulation begins.

Using the preceding probabilities and cell values, we employ the following logic in a function, ***initForest***, to return an initialized grid for the forest. In the pseudocode, two slashes, //, indicate that the rest of the line is a comment.

initForest(n, probTree, probBurning)

Function to return an $n \times n$ grid of values—*EMPTY* (no tree), *TREE* (non-burning tree), or *BURNING* (burning tree)—where *probTree* is the probability of a tree and *probBurning* is the probability that the tree is burning

Pre: n is the size (number of rows or columns) of the square grid and is positive.
probTree is the probability that a site is initially occupied by tree.
probBurning is the probability that a tree is burning initially.
Post: A grid as described earlier was returned.
Algorithm:

```
for every cell in an n × n grid, forest, do the following:
   if a random number is less than probTree               // tree at site
      if another random number is less than probBuring  // tree is burning
         assign BURNING to the cell
      else                                               // tree is not burning
         assign TREE to the cell
   else  // no tree at site
      assign EMPTY to the cell
return forest
```

Quick Review Question 1

From the text's website, download your computational tool's *10_3QRQ.pdf* file for this system-dependent question that implements *initForest*.

Updating Rules

At every simulation iteration, we apply a function ***spread*** to each cell site to determine its value—*EMPTY*, *TREE*, or *BURNING*—at the next time step. The cell's value at the next instant depends on the values of the cells in its von Neumann neighborhood, as in Figure 10.2.3a—the cell's current value (***site***) and the values of its neighbors to the north (***N***), east (***E***), south (***S***), and west (***W***). For this simulation, the state of a diagonal cell to the northeast, southeast, southwest, or northwest does not have an impact on a site's value at the next iteration. Thus, we include five parameters—*site*, *N*, *E*, *S*, and *W*—for *spread*. (Shortly, we will see that *spread* should have two additional parameters.) In a call to this function, each neighborhood argument is one of three values: *EMPTY*, indicating an empty cell with no tree, *TREE* for a nonburning tree, or *BURNING* for a burning tree in that location.

Updating rules apply to different situations: If a site is empty (cell value *EMPTY*), it remains empty at the next time step. If a tree grows at a site (cell value *TREE*), at the next instant the tree may or may not catch fire (value *BURNING* or *TREE*, respectively) due to fire at a neighboring site or to a lightning strike. A burning tree (cell value *BURNING*) always burns down, leaving an empty site (value *EMPTY*) for the next time step. We consider each situation separately.

Quick Review Question 2

From the text's website, download your computational tool's *10_3QRQ.pdf* file for this system-dependent question that develops *spread*'s rule for the situation where a site does not contain a tree at this or any time step.

When a tree is burning, the first argument, which is the site's value, is *BURNING*. Regardless of its neighbors' situations, the tree burns down, so that at the next iteration of the simulation the site's value becomes *EMPTY*. Thus, the relevant rule for the *spread* function has a first argument of *BURNING*; each of the other four arguments are immaterial; and the function returns value of *EMPTY*.

Quick Review Question 3

From the text's website, download your computational tool's *10_3QRQ.pdf* file for this system-dependent question that develops *spread*'s rule for the situation where a site contains a burning tree.

To develop this dynamic, discrete stochastic system, we employ the following additional probabilities, which we include as parameters for *spread*:

probImmune: The probability of immunity from catching fire. Thus, if a site contains a tree (site value of *TREE*) and fire threatens the tree, *probImmune* is the probability that the tree will not catch fire at the next time step.

probLightning: The probability of lightning hitting a site.

When a tree is at a location (site value of *TREE*), at the next iteration the tree might be burning due to one of two causes, a burning tree at a neighboring site or a lightning strike at the site itself. Even if one of these situations occurs, the tree at the site might not catch fire. Separate rules apply to the two causes for fire.

For the first situation involving a neighboring burning tree, we employ the following logic:

```
if (site is TREE) and (N, E, S, or W is BURNING)
   if a random number between 0.0 and 1.0 is less than probImmune
     return TREE
   else
     return BURNING
```

Thus, even if a tree has the potential to burn because of a neighboring burning tree, it may not. Because of conditions such as wet weather, such a tree has a probability of *probImmune* of not burning.

Quick Review Question 4

From the text's website, download your computational tool's *10_3QRQ.pdf* file for this system-dependent question that develops *spread*'s rule for the situation where a site contains a nonburning tree that may catch fire because a neighboring site contains a burning tree.

A tree might also catch fire because of a lightning strike. The probability that the tree is struck by lightning is *probLightning*. However, with a probability of *probImmune*, the tree will not burn even if hit by lightning. In contrast, the probability that the tree is not immune to fire is (1 – *probImmune*). For example, if the probability of immunity (*probImmune*) is 0.4 = 40%, then a (1 – 0.4) = 0.6 = 60% chance exists for the tree not to be immune from burning. For the tree to catch fire due to lightning, it must be hit and not be immune. Thus, lightning causes a tree to catch fire with the probability that is the product *probLightning* * (1 – *probImmune*). For example, if a 0.2 = 20% chance exists for a lightning strike at the site of a tree, the tree burns with a probability of (0.2)(0.6) = 0.12 = 12%. Two things must happen: Lightning must strike, and the tree must not be immune from burning.

Quick Review Question 5

From the text's website, download your computational tool's *10_3QRQ.pdf* file for this system-dependent question that completes *spread*'s rule for the situation where a site contains a nonburning tree that may be hit by lightning and burn.

Periodic Boundary Conditions

For this simulation, we apply the function *spread* to every grid point, using periodic boundary conditions. Thus, to apply *spread* we extend the boundaries by one cell, as in Figure 10.2.7. The next two quick review questions extend the grid first to the north and south and then to the east and west.

Quick Review Question 6

From the text's website, download your computational tool's *10_3QRQ.pdf* file for this system-dependent question that extends a grid, as in Figure 10.3.1, by attaching

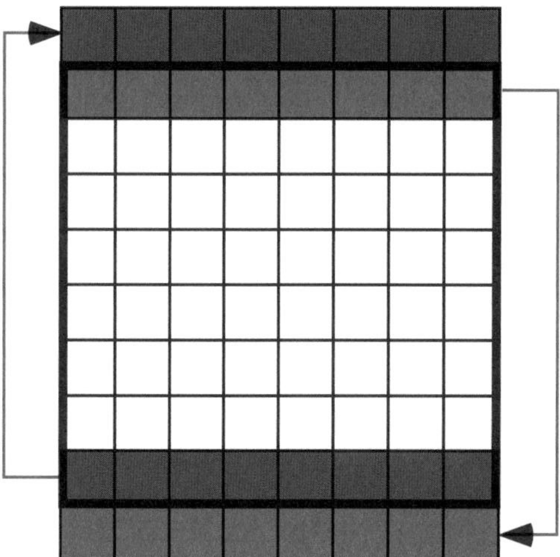

Figure 10.3.1 Grid (in bold square) extended by having a new first row that is a copy of the last row on the original grid and having a new last row that is a copy of the first row on the original grid

the last row to the beginning and the first row to the end of the original grid to form a new grid, *matNS*.

Quick Review Question 7

From the text's website, download your computational tool's *10_3QRQ.pdf* file for this system-dependent question that extends a lattice as in Figure 10.3.2.

To consolidate these tasks, we define a function, ***periodicLat***, using periodic boundary conditions to extend the square lattice by one cell in each direction. Pseudocode for the function follows.

periodicLat(lat)

Function to accept a grid and to return a grid extended one cell in each direction with periodic boundary conditions

Pre: *lat* is a grid.
Post: A grid extended one cell in each direction with periodic boundary conditions was returned.
Algorithm:
latNS ← concatenation of last row of *lat*, *lat*, and first row of *lat*
return concatenation of last column of *latNS*, *latNS*, and first column of *latNS*

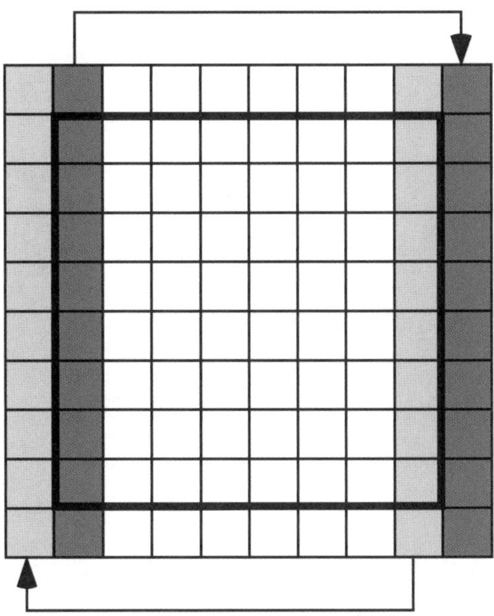

Figure 10.3.2 Grid from Figure 10.3.1 expanded by having a new first column that is a copy of the last column and a new last column that is a copy of the first column

Applying a Function to Each Grid Point

After extending the grid by one cell in each direction using periodic boundary conditions, we apply the function *spread* to each internal cell and then remove the boundary cells. Similar to *applyDiffuseExtended* in Module 10.2, a function ***applyExtended*** takes an extended square lattice (*latExt*) and two probabilities (*probLightning* and *probImmune*) that *spread* requires and returns the internal lattice with *spread* applied to each site.

Quick Review Question 8

From the text's website, download your computational tool's *10_3QRQ.pdf* file for this system-dependent question that develops the function *applyExtended.*

Simulation Program

To drive the simulation of spreading fire, we define a function *fire* with parameters n, the grid size, or number of grid rows or columns; *probTree*; *probBurning*; *probLightning*, the probability of lightning hitting a site; *probImmune*, the probability of immunity from catching fire; and t, the number of time steps. As with *diffusionSim* from Module 10.2, the function *fire* returns a list of the initial lattice and the next *t* lattices in the simulation. The functions *spread* and *fire* need the probabilities of lightning and immunity. Pseudocode for *fire* is as follows.

fire(n, probTree, probBurning, probLightning, probImmune, t)

Function to return a list of grids in a simulation of the spread of fire in a forest, where a cell value of *EMPTY* indicates the cell is empty; *TREE*, the cell contains a nonburning tree; and *BURNING*, a burning tree

Pre:
n is the size (number of rows or columns) of the square grid and is positive.
probTree is the probability that a site is initially occupied by tree.
probBurning is the probability that a tree is burning initially.
probLightning is the probability of lightning hitting a site.
probImmune is the probability of a tree being immune from catching fire.
t is the number of time steps
spread is the function for the updating rules at each grid point.
Post:
A list of the initial grid and the grid at each time step was returned.
Algorithm:
forest ← *initForest*(*n*, *probTree*, *probBurning*)
grids ← list containing *forest*
do the following *t* times:
  forestExtended ← *periodicLat*(*forest*)
  forest ← *applyExtended*(*forestExtended*, *probLightning*, *probImmune*)
  grids ← the list with *forest* appended onto the end of *grids*
return *grids*

Quick Review Question 9

From the text's website, download your computational tool's *10_3QRQ.pdf* file for this system-dependent question that implements the loop in the *fire* function.

Display Simulation

For each lattice in the list returned by *fire*, we generate a graphic for a rectangular grid, with yellow representing an empty site; green, a tree; and burnt orange, a burning tree. The function ***showGraphs*** with parameter *graphList* containing the list of lattices from the simulation produces these figures. We animate the sequence of graphics to view the changing forest scene.

Quick Review Question 10

From the text's website, download your computational tool's *10_3QRQ.pdf* file for this system-dependent question that develops the function *showGraphs*, which produces an animation with a graphic corresponding to each simulation lattice in a list (*graphList*).

Figure 10.3.3 displays several frames of a fire sequence in which empty cells are white; burning cells are in color; and cells with nonburning trees are gray. Clearly, different initial random number generator seeds result in different sequences. This simulation employs the parameters $n = 50$, $probTree = 0.8$, $probBurning = 0.0005$, $probLightning = 0.00001$, $probImmune = 0.25$, and $t = 50$. The initial graphic displays one fire toward the bottom of the grid. At time step $t = 2$, a lightning strike starts a fire at an isolated location toward the top of the grid. Subsequent frames show both fires spreading to neighboring cells. Grids for times starting at $t = 14$ reveal the influence of periodic boundary conditions as the fire at the bottom spreads to the top of the grid, and vice versa.

Exercises

On the text's website, Fire *files for several computational tools contain the code for the simulation of the module. Complete the following exercises using your computational tool.*

For Exercises 1–3, write update rules for spread*, where neighbor refers to a location in the von Neumann neighborhood other than the site itself. Revise grid values as necessary.*

1. A tree takes two time steps to burn completely.
2. A tree catches on fire from neighboring trees with a probability proportional to the number of neighbors on fire.
3. A tree grows instantaneously in a previously empty cell with a probability of *probGrow*.
4. Describe changes to the code to include diagonal elements as neighbors as well.
5. Write the code to assign the values to the northeast, southeast, southwest, and northwest to variables *NE*, *SE*, *SW*, and *NW*, respectively, of a site in the lattice *latExt*.
6. Suppose a lattice *g* has values for a forest grid, where a cell can be empty (value *EMPTY* = 0), a tree with the value (1 through 4) indicating the level of maturity from young to old, or a burning tree with the value indicating the intensity of the fire (5 for less intense or 6 for intense). Write code to show a graphic representing *g*, with yellow for an empty cell, a different level of green from pale to full green representing the age of a tree, light red for a less-intense fire and full red for an intense fire. Use constants, such as *EMPTY*, for the cell values.

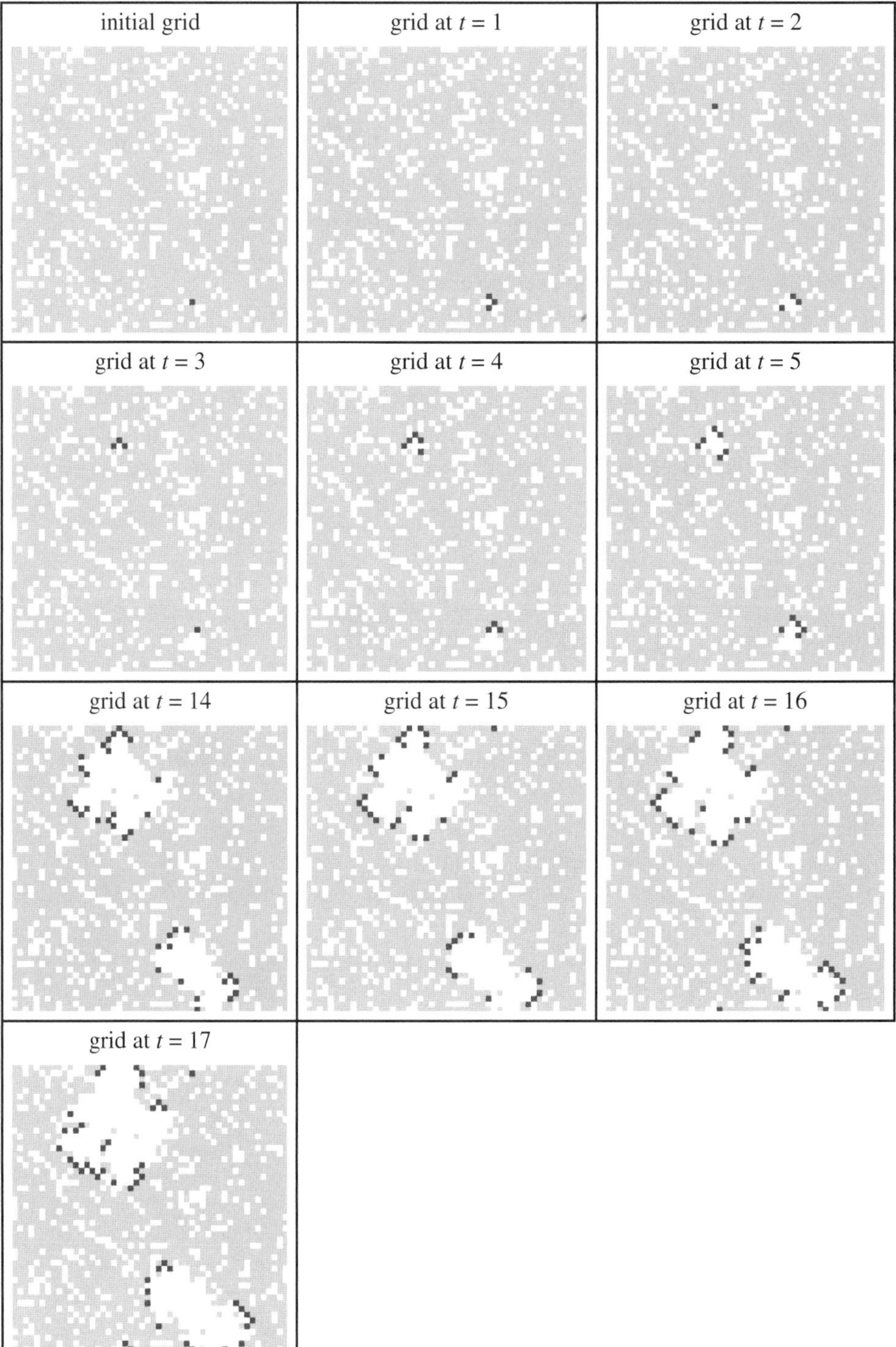

Figure 10.3.3 Several frames in an animation of the spreading of fire

Projects

On the text's website, Fire *files for several computational tools contain the code for the simulation of the module. Complete the following projects using your computational tool.*

For additional projects, see Module 14.3, "Foraging—Finding a Way to Eat"; Module 14.4, "Pit Vipers—Hot Bodies, Dead Meat"; Module 14.5, "Mushroom Fairy Rings—Growing in Circles"; Module 14.6, "Spread of Disease—Sharing Bad News"; Module 14.7, "HIV—The Enemy Within"; Module 14.8, "Predator-Prey—'Catch Me If You Can'"; Module 14.9, "Clouds—Bringing It All Together"; Module 14.11, "Spaced Out—Native Plants Lose to Exotic Invasives"; and Module 14.12, "Re-Solving the Problems with Cellular Automaton Simulations."

1. Run the simulation for fire several times for each of the following situations and discuss the results.
   a. *probBurning* is almost 0; *changeLightning* = *changeImmune* = 0
   b. *probBurning* is 0; *changeImmune* is 0
   c. *probBurning* is 0; *changeLightning* is 0
   d. Devise another situation to consider.

In each of Projects 2–8, revise the fire simulation to incorporate the change indicated in the exercise or boundary condition. Discuss the results.

2. Exercise 1 **3.** Exercise 2 **4.** Exercise 3
5. Exercise 4 **6.** Exercise 6
7. Absorbing boundary conditions **8.** Reflecting boundary conditions
9. Develop a fire simulation in which every cell in a 17 × 17 grid has a tree and only the middle cell's tree is on fire initially. Do not consider the possibility of lightning or tree growth. The simulation should have a parameter for *burnProbability*, which is the probability that a tree adjacent to a burning tree catches fire. The function should return the percent of the forest burned. The program should run eight experiments with *burnProbability* = 10%, 20%, 30%, . . ., and 90% and should conduct each experiment 10 times. Also, have the code determine the average percent burned for each probability. Plot the data and fit a curve to the data. Discuss the results (Shodor Educational Foundation, "Fire").
10. a. Develop a fire simulation that considers wind direction and speed. Have an accompanying animation. Do not consider the possibility of lightning. The simulation should have parameters for the probability (*probTree*) of a grid site being occupied by a tree initially, the probability of immunity from catching fire, the fire direction (value *N*, *E*, *S*, or *W*), wind level (value *NONE* = 0, *LOW* = 1, or *HIGH* = 2), coordinates of a cell that is on fire, and the number of cells along one side of the square forest. The function should return the percent of the forest burned (Shodor Educational Foundation, "Better Fire").
   b. With a wind level of *LOW* (1) and a fixed *probTree*, vary wind direction and through animations observe the affects on the forest burn. Discuss the results.

c. Develop a program to run three experiments with wind levels of *NONE* = 0, *LOW* = 1, and *HIGH* = 2. Have fixed wind direction and *probTree*. The program should conduct each experiment 10 times. Also, have the code determine the average percent burned for each level. Discuss the results.

d. Develop a program to run eight experiments with no wind and *probTree* = 10%, 20%, 30%, . . ., 90%. The program should conduct each experiment 10 times. Also, have the code determine the average percent burned for each probability. Plot the data and fit a curve to the data. Discuss the results.

11. Develop a fire simulation in which a tree once ignited or hit by lightning in one time step takes five additional time steps to burn. The fire can spread from the burning tree to a neighboring tree with different probabilities only on the second, third, and fourth time steps after catching fire. Assume a tree's fire is hottest the third time step after ignition.

12. Develop a fire simulation with accompanying animation in which a section of the forest is damper and, hence, harder to burn. Discuss the results.

Answers to Quick Review Question

From the text's website, download your computational tool's *10_3QRQ.pdf* file for answers to these system-dependent questions.

References

DAS (Dynamic Animation Systems, Inc.) 2012. U.S. Forest Service. http://www.d-a-s.com/node/36 (accessed December 26, 2012)

Dossel, B., and F. Schwabl. 1994. "Formation of Space-Time Structure in a Forest-Fire Model." *Physica Abstracts*, 204: 212–229.

Gaylord, Richard J., and Kazume Nishidate. 1996. "Contagion in Excitable Media." *Modeling Nature: Cellular Automata Simulations with Mathematica*. New York; TELOS/Springer-Verlag, pp. 155–171.

Los Angeles Times. 2010. "Two Men Who Started Malibu Corral Canyon Fire Sentenced to a Year in Jail." *LA Now*. September 9. http://latimesblogs.latimes.com/lanow/2010/09/two-men-who-started-malibu-corral-canyon-fire-sentenced-to-a-year-in-jail.html (accessed December 22, 2012)

NPS (National Park Service). 2012. "Fire Ecology." Yellowstone National Park. http://www.nps.gov/yell/parkmgmt/fireecology.htm (accessed December 22, 2012)

SBCCBS (Santa Barbara City College Biological Sciences). 2010. "Fire in the Chaparral." Biology 100 Concepts of Biology: Introduction to the Chaparral. http://www.biosbcc.net/b100plant/htm/fire.htm (accessed December 22, 2012)

Shodor Education Foundation. "Fire!!" The Shodor Education Foundation, Inc., 1997–2003. http://www.shodor.org/interactivate/activities/Fire/ (accessed December 22, 2012)

———. “A Better Fire!!” The Shodor Education Foundation, Inc., 1997–2003. http://www.shodor.org/interactivate/activities/ABetterFire/ (accessed December 22, 2012)

Wilson, Tracy, Stuart Pfeifer, and Mitchell Landsberg. 2003. “California Fires Threaten 30,000 More Homes.” *Pittsburg Post-Gazette*, October 28. http://www.post-gazette.com/stories/news/us/california-fires-threaten-30000-more-homes-520702/ (accessed December 22, 2012)

MODULE 10.4

Movement of Ants—Taking the Right Steps

Prerequisite: Module 10.2, "Diffusion—Overcoming Differences"

Downloads

For several computational tools, the text's website has an *Ants* file, which contains the simulation this module develops, and a *10_4QRQ.pdf* file, which contains system-dependent Quick Review Questions and answers, available for download.

Introduction

> *Everyone says stay away from ants. They have no lessons for us; they are crazy little instruments, inhuman, incapable of controlling themselves, lacking manners, lacking souls. When they are massed together, all touching, exchanging bits of information held in their jaws like memoranda, they become a single animal. Look out for that. It is a debasement, a loss of individuality, a violation of human nature, an unnatural act.*
>
> —Thomas (1979)

Ants are extremely successful constituents of the earth's fauna, but they seem so different from human beings and are generally regarded as pests. So, what can human beings learn from such lowly creatures?

Ants have occupied a variety of ecological niches for millions of years. They are the epitome of social insects, living in colonies of varying size. These colonies are generally made up of one or more queens, many workers, and various immature stages (egg, larvae, pupae). During most of the year, all the adults are female, and all but the queen are sterile. Seasonally, a few winged males and females (fertile) are produced, but normally most of the adults are sister workers.

The queen's responsibilities are fairly uncomplicated: she mates and lays eggs. Workers have a variety of chores: tending to young, nest construction, foraging, and protecting the nest. Their entire life is dedicated to sustaining the colony.

A nest of ants typically begins with only one individual, the queen. New, mature queens fly from the nest and search for mates from groups of males that have been produced during the same time. In selected meeting places, the queen mates with one or a few males, storing the sperm in special sacs until needed. Then she flies off to find suitable nest sites. Few of these queens successfully establish a new colony, and the males die right after their big moment.

Besides keeping herself alive, a queen must find a suitable site for the new colony, excavate the site, lay the eggs, and care for the developing young. She may also have to forage for food. A queen lives off of stored food reserves and some of her laid eggs until her young grow up. Once the first workers are produced, they take over all the queen's chores except laying eggs. The queen can now concentrate on her major role, although she also has some control over the sex ratios and new-queen production in the colony. The workers take care of everything else.

Gradually the colony grows as more and more young mature into workers. In many species, worker ants themselves become specialized for all the roles necessary to sustain the queen and the colony. Some remain in the nest, caring for the queen or the young. Others guard the nest, and still others forage for food.

There is quite a bit of variability in feeding strategies and food sources used by different species of ants, and many employ more than one type of feeding behavior. Ants may prey on small insects or eat dead insects. Others rely on seeds or raid other ant nests. One of the most interesting strategies is used by the leafcutter ants, which farm nutritious fungus.

Analysis of Problem

Most species of ants communicate their movements when carrying food by leaving trails with a chemical **pheromone**. Also, an ant can reinforce a trail by secreting additional pheromone. Thus, by following a scent, other ants can locate a food source. As expected, the pheromone dissipates and diffuses with time. In this module, we simulate the movements of such ants in the presence of a chemical trail, which spreads and evaporates over time. We do not include an ant carrying food, although the projects consider such an extension.

For the simulation, we use a model that incorporates aspects of cellular automaton simulations from Module 10.3, "Spreading of Fire," and Module 10.2, "Diffusion—Overcoming Differences." We hope to observe over time that the simulated ants tend to follow a chemical trail. Thus, the simulation should help us reflect on how behavior on the local level can lead to global behavior, which we can observe in some ants. Through the interactions of many separate individuals, a group of ants as a whole can exhibit **self-organizing** behavior that makes the group appear to have a single consciousness.

Formulating a Model: Gather Data

For the model we develop in this module, we employ empirical observations of ant species that leave pheromone trails. With each step, such an ant tends to turn to and move in the direction of the greatest amount of chemical. As time passes, the chemical diffuses away from an initial deposit; and with no ant in a location, the amount of pheromone diminishes there. For a professional model, we should obtain more exact data, such as the average amount of pheromone an ant deposits and the rates at which the chemical diffuses and decreases.

Formulating a Model: Make Simplifying Assumptions

In formulating a model, suppose that the ants are contained in a square area enclosed by glass. Moreover, we assume that an ant does not turn around completely in one time step, returning immediately to the location from which it just came, but otherwise tends to move toward an unoccupied neighboring location with the greatest amount of chemical. If no such move is available, we assume the ant waits in its current location. Thus, we employ an **avoidance-or-wait strategy** to prevent collision. With movement from a site that has a certain threshold of chemical, the ant deposits additional pheromone for reinforcement. However, the chemical diffuses and dissipates with time. For this problem, we start with a straight trail of increasing amounts of pheromone, perhaps laid by ants heading for food. We do not consider food or a nest, although various projects do.

Formulating a Model: Determine Variables

In Module 10.3, "Spreading of Fire," each cell of a grid contains an integer indicating the state of the cell—empty, tree, or burning tree; and in Module 10.2, "Diffusion—Overcoming Differences," we employ a grid of diffusing temperatures. In the current model, we have one grid to hold ant information, similar to the former, and another to store pheromone amounts, comparable to the later. To simulate a closed container, we assume absorbing boundary conditions. In the ant grid, each element of the first and last rows and columns has a constant value, ***BORDER*** = 6; and an empty cell has the value ***EMPTY*** = 0. A cell with an ant contains a constant—***NORTH*** (1), ***EAST*** (2), ***SOUTH*** (3), ***WEST*** (4), or ***STAY*** (5)—indicating direc-

Table 10.4.1
Cell Values with Associated Constants and Their Meanings

Value	*Constant*	*Cell Meaning*
0	*EMPTY*	Empty ground containing no ant
1	*NORTH*	Ant about to move to or just moved from the north
2	*EAST*	Ant about to move to or just moved from the east
3	*SOUTH*	Ant about to move to or just moved from the south
4	*WEST*	Ant about to move to or just moved from the west
5	*STAY*	Ant about to stay in or did not move from the current site
6	*BORDER*	Border

tional information. Before movement, such a constant indicates the von Neumann neighborhood cell where the ant intends to go; and after movement, the constant points back to the direction from which the ant just came. *STAY* denotes that the ant is staying in its current location for a time step. Table 10.4.1 enumerates the ant constants, their meanings, and suggested values.

For the initialization of this grid, we have a function, ***initAntGrid***, with parameters for the size, ***n***, of the internal part of the grid and the probability, ***probAnt***, that an ant initially occupies a cell. Thus, the function returns an $(n + 2) \times (n + 2)$ matrix of integers from Table 10.4.1. With probability *probAnt* a site contains an ant; should an ant be at a location, we assign a random integer—1, 2, 3, or 4—representing a direction—*NORTH*, *EAST*, *SOUTH*, or *WEST*, respectively.

Quick Review Question 1

From the text's website, download your computational tool's *10_4QRQ.pdf* file for this system-dependent question defines *initAntGrid*.

In the pheromone grid, a floating-point number represents the amount of chemical at a site. Because an ant is to move to a neighboring available cell with the maximum amount of chemical and because we are employing absorbing boundary conditions, in the pheromone grid, we have a border of slightly negative values, such as –0.01. Thus, an ant will never be tempted to step outside the grid. Moreover, these border values tend to diffuse inward, encouraging the ants to stay away from borders, which represent the walls. A function, ***initPherGrid***, initializes most of the interior cells as 0. However, for the pheromone trail, in the middle of the grid, we have a horizontal row of increasing pheromone values. With ***MAXPHER*** (say, 50.0) being the maximum initial chemical value, i starting at 1 and being a function of an internal column number, and n being the size (number of rows and number of columns) of the internal part of the grid (omitting the border), the amount of chemical in the trail is $MAXPHER \cdot i/n$. Thus, initially, the amount of pheromone gradually increases from left to right in the trail's row. If *MAXPHER* is 50.0 and n is 10, then in internal column 1 of the trail, the amount is $50.0 \cdot 1/10 = 5.0$; in column 5, the value is $50.0 \cdot 5/10 = 25.0$; and in column 10, we have the maximum pheromone amount of $50.0 \cdot 10/10 = 50.0$.

Quick Review Question 2

From the text's website, download your computational tool's *10_4QRQ.pdf* file for this system-dependent question that defines *initPherGrid*.

Formulating a Model: Establish Relationships and Submodels

Ant movement for one time step consists of two actions, sensing and walking. First, the ant tests the empty neighboring sites and turns to the one with the greatest amount of pheromone or decides to stay in its current location if no such site is available. Then, if possible to do so without colliding with another ant, the ant moves to the preferred location. After the reaction (sensing and walking) of the ants, a diffusion of the pheromone occurs. Thus, we have a **reaction-diffusion**-type simulation. As with diffusion in Module 10.2, "Diffusion: Overcoming Differences," we employ Moore neighborhoods with eight neighbors of a site and define a function, ***diffusion***, with parameters for a diffusion rate constant (*diffusionRate*) and pheromone values for the site and its neighbors. Because absorbing boundary conditions employ constant boundary values, all matrices are of the same size, $(n + 2) \times (n + 2)$. Thus, a function, ***applyDiffusionExtended***, applies *diffusion* to each internal cell and returns an $(n + 2) \times (n + 2)$ pheromone grid, keeping the border intact. The next two sections develop the *sense* and *walk* functions.

Quick Review Question 3

From the text's website, download your computational tool's *10_4QRQ.pdf* file for this system-dependent question that defines *applyDiffusionExtended.*

Formulating a Model: Determine Functions—Sensing

As with the fire simulation, for sensing we consider the neighbors to be the cells to the north, east, south, and west, that is, those neighbors in the von Neumann neighborhood. The rules for the function ***sense***, which points the ant towards its new location, are as follows:

1. An empty cell does not point toward any direction.
2. An ant does not turn to a cell from which the creature just came.
3. An ant does not turn to a location that is a border site.
4. An ant does not turn to a location that currently contains an ant.
5. Otherwise, an ant turns in the direction of the neighboring available (not the previous, an occupied, or a border cell) with the greatest amount of chemical. In the case of more than one neighbor having the maximum amount, the ant turns at random towards one of these cells.
6. If no neighboring cell is available, the ant will not move.

In the list, *lst*, of neighboring pheromone values of an ant that just moved, we assign an artificially small value, say –2, to the one corresponding to the direction from which it moved. Similarly, if another ant is in a neighboring site, we change *lst*'s corresponding value to –2. Such changes help to enforce Rules 2 and 4. To model Rule 5, we first form list, *posList*, of indices for maximum *lst* values. We randomly pick an index, *rndPos*, that has a maximum pheromone value in *lst*. For example, suppose *lst* contains adjusted pheromone values 9, –2, 9, and 8. With the maximum being 9 and assuming indexing begins with 1, *rndPos* could be 1 or 3 because the indices of 9 in *lst* are 1 and 3, which correspond to the directions north and south, respectively. The algorithm for *sense* follows.

sense(*site*, *na*, *ea*, *sa*, *wa*, *np*, *ep*, *sp*, *wp*)

Function to return the direction in which an ant is to turn (*NORTH* (1), *EAST* (2), *SOUTH* (3), or *WEST* (4)) or *STAY* (5) should the ant be planning to remain in its current location

Pre: *site*, *na*, *ea*, *sa*, and *wa* are the ant grid values for the current site and its neighbors to the north, east, south, and west, respectively. If a cell contains an ant, then its value represents the direction from which the ant came in the last time step. *site* is not *EMPTY* or *BORDER*.
np, *ep*, *sp*, *wp* are the pheromone grid values for the current site's neighbors to the north, east, south, and west, respectively.

Post: The function has returned *STAY* or the direction to which the ant turns.

Algorithm:

```
lst ← list with np, ep, sp, and wp
if site is not STAY, lst(site) ← –2       // Rule 2
if a neighboring cell contains an ant     // Rule 4
  assign –2 to the corresponding lst element
mx ← maximum value in lst                 // Rule 3 (pheromone < 0 on border)
if mx < 0                                 // Rule 6
  return STAY
else                                      // Rule 5
  posList ← list of positions in lst containing mx
  lng ← length of posList
  rndPos ← random integer between 1 and lng, inclusive
  return posList(rndPos)
```

Quick Review Question 4

From the text's website, download your computational tool's *10_4QRQ.pdf* file for this system-dependent question to define *sense*.

Similar to the models for diffusion and spreading of fire in earlier modules, we have a function, in this case ***applySenseExtended***, to process every cell of the internal grid. Unlike the application functions in those earlier modules but like *applyDiffusionExtended*, *applySenseExtended* returns $(n + 2) \times (n + 2)$ ant grid with the borders unchanged. In the function's definition, we first copy a parameter *antGrid* to a *newAntGrid* that the function returns after possible changes. Should an *antGrid* cell contain *EMPTY*, no further processing needs to be done on that location (Rule 1). Otherwise, *applySenseExtended* applies *sense* to that site, sending *sense* the ant grid value for the site and the ant and pheromone grid values for its four von Neumann neighbors.

Quick Review Question 5

From the text's website, download your computational tool's *10_4QRQ.pdf* file for this system-dependent question to define *applySenseExtended*.

Formulating a Model: Determine Functions—Walking

After applying the function *sense* to each cell of the grid, we call a function, ***walk***, which computes updated ant and pheromone grids for the next time step. The following additional rules relate to walking:

7. For a cell that remains empty, the amount of chemical decrements by a constant amount, ***EVAPORATE***, but does not fall below 0. Thus, the new amount is the maximum of 0 and the current amount minus *EVAPORATE*.
8. An ant facing in a certain direction will move into that neighboring cell as long as no other ant has already moved there.
9. Otherwise, the ant will stay in its current cell.
10. If an ant leaves a cell that has pheromone above a certain threshold, ***THRESHOLD***, the amount of chemical increments by a set amount, ***DEPOSIT***, to reinforce the trail.
11. If an ant stays in a cell, the amount of chemical remains the same.
12. After moving to a new location, the ant faces towards the cell from which the animal just came.

The design of the *walk* function follows, with details for one sense direction (*NORTH*). Behavior of the ant when facing another direction is comparable.

walk(antGrid, pherGrid)

Function to return a new ant and pheromone grids after each ant has moved or decided to remain in its current location

Pre: *antGrid* is an ant grid after application of *applySenseExtended* in a time step.
pherGrid is the corresponding pheromone grid.

Post: New ant and pheromone grids have been returned after application of the walk rules.

Algorithm:

```
n ← number of rows/columns in ant/pheromone grid minus 2
newAntGrid ← antGrid
newPherGrid ← pherGrid
for i going through each internal row index, do the following:
   for j going through each internal column index, do the following:
      if antGrid(i, j) is EMPTY                                  // Rule 7
         newPherGrid(i, j) ← maximum of 0 and
           (newPherGrid(i, j) – EVAPORATE)
      // Corresponding segments to the following occur for each direction:
      if antGrid(i, j) is NORTH
         if newAntGrid(i – 1, j) is EMPTY
           if newPherGrid(i, j) > THRESHOLD                      // Rule 10
              newPherGrid(i, j) ← newPherGrid(i, j) + DEPOSIT
           newAntGrid(i, j) ← EMPTY                              // Rule 8
```

```
            newAntGrid(i - 1, j) ← SOUTH                // Rule 12
        else
            newAntGrid(i, j) ← STAY                     // Rules 9 and 11
    // Corresponding segments for directions EAST, SOUTH, WEST go
    // here
return newAntGrid and newPherGrid
```

Quick Review Question 6

From the text's website, download your computational tool's *10_4QRQ.pdf* file for this system-dependent question to define *walk*.

Solving the Model—A Simulation

The simulation function, ***ants***, initializes the ant and pheromone grids and stores each in lists of grids, *antGrids* and *pherGrids*. After initialization, for each of the *t* time steps, reaction and diffusion occur. All the ants sense pheromone and walk toward the scent; and then, the pheromone diffuses. At each iteration, *antGrids* and *pherGrids* store the new grids. As the following algorithm reveals, the function finally returns these lists of grids.

ants(n, probAnt, diffusionRate, t)

Function to return a list of ant and pheromone grids in a simulation of ant movement, where ant cell values are as in Table 10.4.1 and pheromone cell values represent the levels of pheromone

Pre: *n* is the size (number of rows/columns) of the internal ant and pheromone grids.
probAnt is the probability that an ant initially occupies a cell.
diffusionRate is the diffusion rate.
t is the number of time steps.

Post: A list of the initial and subsequent ant grids at each time step of the simulation and a list of the initial and subsequent pheromone grids were returned.

Algorithm:

```
antGrid ← initAntGrid(n, probAnt)
pherGrid ← initPherGrid(n)
antGrids ← a list containing antGrid
pherGrids ← a list containing pherGrid
do the following t times:
    antGrid ← applySenseExtended(antGrid, pherGrid)
    antGrid and pherGrid ← walk(antGrid, pherGrid)
```

```
    pherGrid ← applyDiffusionExtended(pherGrid, diffusionRate)
    antGrids ← antGrids with antGrid appended
    pherGrids ← pherGrids with pherGrid appended
  return antGrids and pherGrids
```

Quick Review Question 7

From the text's website, download your computational tool's *10_4QRQ.pdf* file for this system-dependent question to define *ants*.

Verifying and Interpreting the Model's Solution—Visualizing the Simulation

We have a number of choices of how to communicate the information in an ant simulation for verification and interpretation of the model's solution. With constants $MAXPHER = 50.0$, $EVAPORATE = 1$, $DEPOSIT = 2$, and $THRESHOLD = 0$ and parameters $n = 17$, $probAnt = 0.1$, $diffusionRate = 0.01$, and $t = 11$ for the call to *ants*, Figure 10.4.1 presents a sequence of frames, with color representing ants and the level of gray indicating the strength of the chemical at a site with no ant. As the sequence shows, most ants have moved closer to the initial pheromone trail, while ants in contact with the a chemical trail have traveled along the path to levels of greater chemical strength. Initially, none of the 28 ants were on the trail. However, by time step 11, 8 of the 28 ants, or 29%, are on the trail, and 15 (54%) are within one unit of the path. Moreover, darkening near the path indicates the impact of pheromone reinforcement and diffusion. The simulation represents how this social insect can communicate chemically with its sisters for the common good.

For indicating the appropriate level of gray, whose values range from 0.0 to 1.0, we calculate the maximum amount of pheromone, *maxp*, throughout the list of pheromone grids, *pherGrids*. For each cell without an ant, we divide each pheromone value by *maxp* to obtain a normalized value from 0.0 to 1.0. The larger the pheromone amount, the closer this quotient is to 1.0. However, because a grayscale value of 0.0 represents black and 1.0 corresponds to white, we subtract this quotient from 1.0, 1.0 – pheromone/*maxp*, to obtain the appropriate grayscale number. Thus, the minimum amount of chemical, 0, yields RGB components of 1.0 – 0.0 = 1.0, while the maximum amount of chemical has grayscale value of 0.0. For example, if *maxp* is 50.0, the grayscale value is 1.0 – (50/50) = 0.0. A scientific visualization should impart information clearly while not misleading the viewer or suggesting more than is available.

Quick Review Question 8

From the text's website, download your computational tool's *10_4QRQ.pdf* file for this system-dependent question that develops a visualization for the simulation.

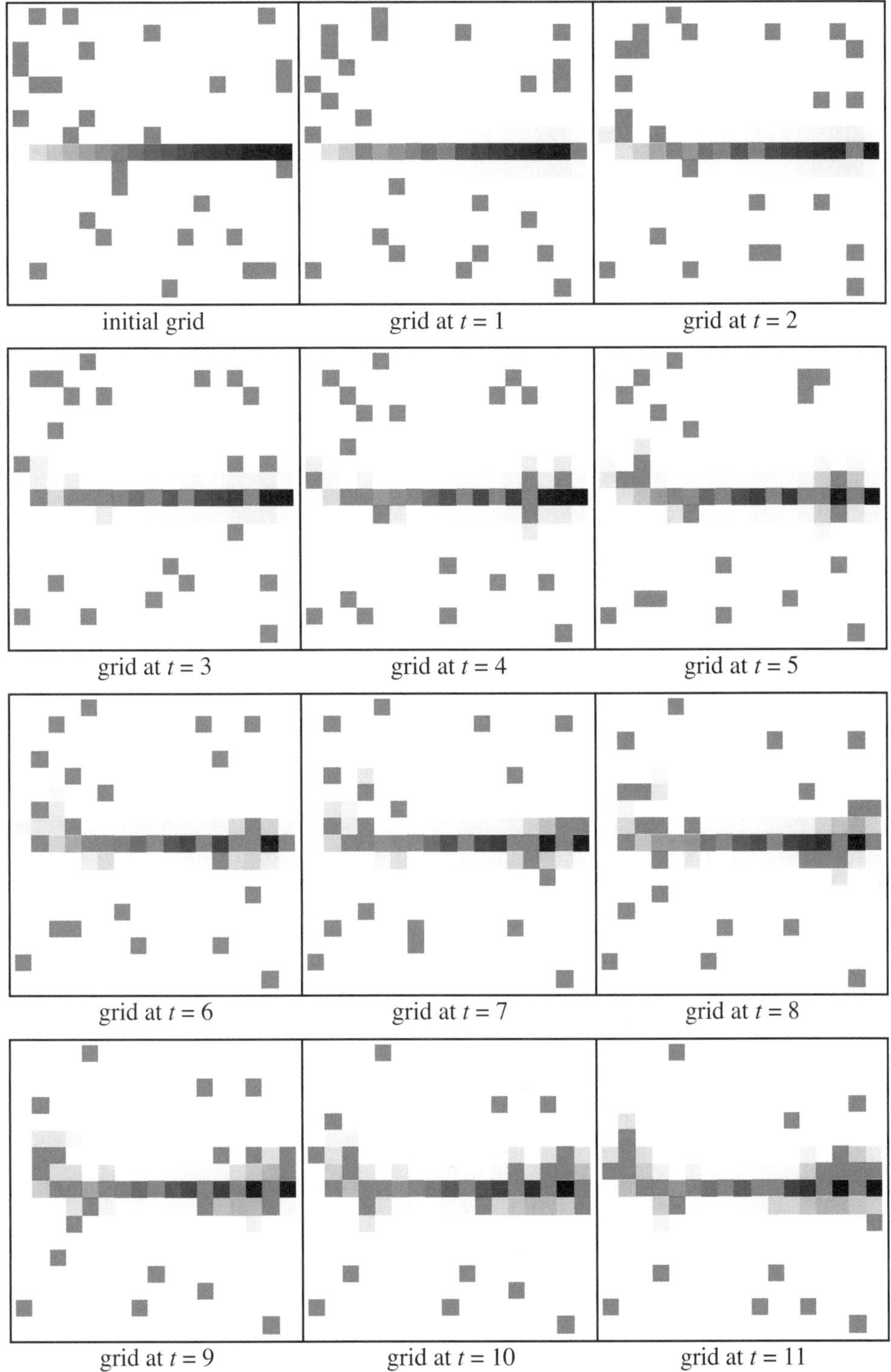

Figure 10.4.1 Several frames in an animation of ant simulation

Exercises

On the text's website, Ants *files for several computational tools contain the code for the simulation of the module. Complete the exercises below using your computational tool.*

1. Suppose the size of an internal grid is $n = 100$ and *MAXPHER* is 20. Using the initialization of the pheromone path in "Formulating a Model: Determine Variables," do the following.

a. Give the number of cells that will be initialized with pheromone in the path.

Give the pheromone amount in each column of the internal grid:

b. 2 **c.** 10 **d.** 50 **e.** 80 **f.** 100

2. In the simulation of this module, an ant cannot return immediately to a cell from which it just came. Without this rule, describe the movement of an ant in an area where no other ants are near and, initially, the ant is far from chemical deposits.

Projects

On the text's website, Ants *files for several computational tools contain the code for the simulation of the module. Complete the following projects using your computational tool.*

For additional projects, see Module 14.3, "Foraging—Finding a Way to Eat"; Module 14.4, "Pit Vipers—Hot Bodies, Dead Meat"; Module 14.5, "Mushroom Fairy Rings—Growing in Circles"; Module 14.6, "Spread of Disease—Sharing Bad News"; Module 14.7, "HIV—The Enemy Within"; Module 14.8, "Predator-Prey—'Catch Me If You Can'"; Module 14.9, "Clouds—Bringing It All Together"; Module 14.10, "Fish Schooling—Hanging Together, Not Separately"; Module 14.11, "Spaced Out—Native Plants Lose to Exotic Invasives"; and Module 14.12, "Re-Solving the Problems with Cellular Automaton Simulations."

1. For the ant simulation in the file *Ants* of this module, investigate the ant behavior in the following situations, keeping all parameters fixed, perhaps as in the section "Verifying and Interpreting the Model's Solution—Visualizing the Simulation," except as noted. Run each simulation at least 10 times, calculate the mean number of ants that are within two units of the chemical for each time step over a period of time, describe the results, and discuss the implications.

a. Use the original *Ants* file.

b. Have varying numbers, m, of areas of chemical concentrations with $m = 1$, 2, 3, 4, and 5.

c. Vary *probAnt* from 0.06 to 0.14 in increments of 0.02.

d. Vary n from 10 to 50 in increments of 10.

e. Vary *diffusionRate* from 0.01 to 0.10 in increments of 0.01.

f. Vary *MAXPHER* from 10 to 80 in increments of 10.

g. Vary *EVAPORATE* from 0.5 to 3.0 in increments of 0.5.

h. Vary *DEPOSIT* from 0.5 to 3.0 in increments of 0.5.
i. Vary *THRESHOLD* from 0 to 20 in increments of 2.

2. Consider an *Ants* model with a **decrease strategy for collision**. That is, two ants could go to the same cell, but after movement, that cell records only one ant, so that we have one less ant. Revise the simulation rules and implement the strategy. Running the simulation at least 10 times, plot the mean number of ants over time. Discuss the results.
3. Develop a simulation in which a single ant leaves the nest searching for a food source that is unknown to the ant and that is due north of the nest. Initially, the grid does not contain pheromones. With food, she returns directly to the nest in a straight-line fashion, leaving a chemical trail. As soon as one ant returns to the nest, another ant leaves, following the pheromone trail in the search for food. An ant following a pheromone trail emits a smaller chemical signal than one carrying food. Perform the process for a sequence of 10 ants, saving the points of each ant's path to the food. Plot each ant's path to the food. Discuss the results.
4. Develop a simulation with a nest, a food source, and ants that should not collide on a 20×20 grid. A sequence of 10 numbered ants leave the nest in search of food. Once an ant finds food, she returns to the nest carrying a morsel and depositing pheromones, with greater amounts closer to the food source. An ant seeking food, usually travels in the direction of maximum pheromone, but occasionally moves in a random direction. Have a large reinforcement threshold (e.g., 0.8 for pheromone values in the range 0 to 1), a small diffusion rate constant (e.g., 0.005), and a small evaporation constant (e.g., 0.001 or less). Besides ants and pheromone grids, have a nest grid, where the strength of a nest signal is related to the distance from the nest and is greatest close to the nest. Run the simulation 10 times, and plot the mean length of time for each ant to find the food. Do ants that leave the nest later find the food faster?
5. Develop a simulation with a nest, two food sources, and ants at random initial positions. Once an ant finds food, the amount of food at that location decreases by one unit and she returns to the nest with morsel. Besides ants and pheromone grids, have a nest grid, where the strength of a nest signal is proportional to the distance from the nest. Do ants exhaust one food source before focusing on the other?
6. An army ant raid can be 20 m wide and 200 m long and involve hundreds of thousands of ants. The raid is self-organizing, evolving from interactions on the local level into a global pattern. The pattern appears treelike with the forward part of the raid being branchlike. Develop a simulation with no food present that has the following rules, which are based on those of (Franks 2001):

    - Every ant deposits pheromone unless the cell is saturated, containing the maximum amount of chemical.
    - In new territory, where pheromones are not present, an ant goes randomly to the northeast or to the northwest.
    - When pheromone exists, with a certain probability an ant is more likely to follow the peromone trail.

- More than one ant can be in a cell, up to some maximum number of ants.
- Each time step, a constant number of ants leaves the nest, which is one cell.

7. Augment Project 6 to include the following rule: ants move faster in the presence of more pheromone. For example, you could consider that based on the amount of chemical, an ant makes a move per every one, two, or three time steps. Discuss the effects of varying the speed of the ants.

8. Augment Project 6 to include food and the following rule: once an ant finds food, she returns to the nest using the same rules as those of Project 6, except she goes to the southeast or southwest (Franks 2001). Discuss the difference in the self-organizing pattern between this simulation and that of Project 6.

For Projects 9–11, repeat the indicated project with the direction being relative to an ant's heading, front right and front left, instead of northeast and northwest, respectively. Have the nest be in one corner of the grid.

9. Project 6 **10.** Project 7 **11.** Project 8

12. Usually, trail following is not completely accurate. Introduce an additional stochastic element in the choice of direction in any of the earlier projects. For example, you might have an ant picking a random direction 25% of the time and face an available neighbor with the most chemical 75% of the time. Discuss the advantages and disadvantages of this lack of precision.

13. Adjust the grid on any of the earlier projects to contain obstacles.

14. Develop a cellular automaton simulation to illustate the exploitive competition of Argentine ants versus native ants, as described in Project 2 of Module 4.1, "Competition." Illustrate the competitive factor of discovery time. See Project 12 for an idea on simulating discovery time (Holway 1999).

15. Develop a cellular automaton simulation to illustrate the exploitive competition of Argentine ants versus native ants, as described in Project 2 of Module 4.1, "Competition." Illustrate the competitive factor of recruitment rate. See Project 7 for an idea on simulating rate of recruitment.

16. Develop a cellular automaton simulation to illustate the interference competition of Argentine ants versus native ants, as described in Project 2 of Module 4.1, "Competition."

Answers to Quick Review Questions

From the text's website, download your computational tool's *10_4QRQ.pdf* file for answers to these system-dependent questions.

References

Franks, Nigel R. 2001. "Evolution of Mass Transit Systems in Ants: A Tale of Two Societies." *Insect Movement: Mechanisms and Consequences Proceedings of the 20th Symposium of the Royal Entomological Society*. Wallingford, Oxford: CAB International, pp. 281–298.

Gaylord, Richard J., and Kazume Nishidate. 1996. "Chemotaxis." *Modeling Nature: Cellular Automata Simulations with Mathematica*. New York: TELOS/Springer-Verlag, chap. 12, pp. 121–130.

Hölldobler, B., and E. O. Wilson. 1990. *The Ants*. Cambridge, MA: Harvard University Press.

Holway, David A. 1999. "Competitive Mechanisms Underlying the Displacement of Native Ants by the Invasive Argentine Ant." *Ecology*, 80(1): 238–251.

Martinoli, Alcherio, Rodney Goodman, and Owen Holland. "Exploration, Exploitation, and Navigation in Ants." EE141: Swarm Intelligence, California Institute of Technology. http://www.coro.caltech.edu/Courses/EE141/Lecture/W3/ AM_EE141_W3ExplNav.pdf

Thomas, Lewis. 1979. *The Medusa and the Snail, More Notes of a Biology Watcher*. New York: The Viking Press.

Weimar, Jörg. 2003. "PredatorAgainstPrey." Source code, Technical University of Braunschweig. http://www-public.tu-bs.de:8080/~y0021323/ca/ PredatorAgainst Prey.cdl

MODULE 10.5

Biofilms—United They Stand, Divided They Colonize

Prerequisite: Module 10.2, "Diffusion: Overcoming Differences"

Downloads

For several computational tools, the text's website has a *Biofilm* file containing the simulation this module develops and a *10_5QRQ.pdf* file containing system-dependent Quick Review Questions and answers available for download.

Introduction

What do stones in streams, teeth, water and sewer pipes, and the breathing passages of cystic fibrosis patients have in common? At first, these may seem rather unrelated, but they are all linked by at least one commonality—all are covered or lined with biofilms. These assemblages may not be very familiar to you, but they prove absolutely critical to your life.

Scientists have been aware of biofilms for some time. Anton van Leeuwenhoek, who invented and handcrafted microscopes during the late seventeenth and early eighteenth centuries, saw remnants of a biofilm when he observed scrapings he made from his teeth (Donlan and Costerton 2002). We just haven't appreciated their importance until recently.

What is a biofilm, exactly? Simply, **biofilms** are communities of very small organisms that adhere to a surface (**substratum**) in an aqueous environment (Donlan and Costerton 2002). These organisms are usually bacteria, but algae or fungi may also form biofilms. Sometimes, these groups may even be mixed. The organisms are not only attached to a substratum, but they are linked with each other within a matrix of biopolymers (polysaccharides, proteins, lipids, and nucleic acids). Scientists now believe that the vast majority of microbes are not solitary, **planktonic**, as was once assumed. Most microbial life seems to be part of one of these communities, and the planktonic forms may be simply ways to colonize other surfaces. We also know now that the free-floating members of each species are phenotypically quite different from their socially connected counterparts (Boles et al. 2004).

When van Leeuwenhoek looked at the "animalicules" from his dental scrapings, what he was actually seeing is what we now call "plaque." Dental plaque is a fitting example of a biofilm (Overman 2000). Forming on the surfaces of the teeth and soft tissues of the oral cavity, this biofilm is linked to dental caries (cavities), periodontal diseases, and even cardiovascular disease (Genco et al. 2002). Examination of plaque reveals a complex architecture with a heterogeneous array and dispersal of cells within a matrix associated with fluid-filled spaces. The cells are mostly bacteria belonging to as many as 500 distinct species, but there are usually white blood cells and some epithelial cells as well. Bacteria associated with plaque are busy producing metabolites and various toxins. Some of these metabolites are organic acids, like lactic acid, which initiate the formation of caries by demineralization of the enamel. Furthermore, the production of enzymes, toxins, and other metabolites can cause a deeper deterioration of the support structures of the teeth. Severe periodontal disease is the leading cause of tooth loss in adults.

Biofilms are ubiquitous, and dental plaque is only one example. Given their prevalence, there must be some advantages for microbes to band together into such communities. In fact, there is quite a list of advantages. For a human pathogen like *Pseudomonas aeruginosa*, a common cause of respiratory diseases, living in biofilms greatly increases the success of infection (Boles et al. 2004). Biofilms afford greatly increased protection from antibiotics, the host's immune system, and physical injury. Scientists have discovered that biofilm organisms show great genetic diversity, which also stabilizes the community and promotes survival in the hosts (Boles et al. 2004). This is in conformance with the well-known ecological principle of the "insurance hypothesis"—diverse subpopulations increase the chances of survival of the community over a wider range of environmental conditions.

With the close association of biofilm constituents, there is the additional opportunity to share metabolites. Furthermore, such organisms are better able to communicate, coordinate behavior, and transfer genetic information (Harrison et al. 2005).

According to Costerton, 65% to 80% of all bacterial diseases in human beings are from chronic biofilm infections (Costerton et al. 1999; Costerton 2004). For years, most physicians and scientists conceived of bacterial diseases derived from the single-celled, planktonic form of the microbe. Medical treatment was gauged to combat such forms, not biofilms. So, it is little wonder that we are increasingly finding difficulties in combating pathogens.

For the most part, biofilms seem to be a threat to our species, although we are beginning to utilize them in positive ways for bioremediation and wastewater treatment for "green" buildings. Because these communities have such important impacts on our lives, it behooves us to better understand the structure and function of biofilms (Stewart 2003).

The Problem

In this module, we simulate the formation of the structure of a biofilm without regard to its function. We develop a 2D version, which we can extend to 3D. As the simulation time proceeds in a sequence of discrete steps, we consider the following phases at each time step:

1. Diffusion of nutrients
2. Growth and death of microbes
3. Consumption of nutrients by microbes

Thus, we have a **reaction-diffusion**-type simulation. We have a cycle of microbes reacting with the environment by consuming nutrients (as well as growing and dying) and nutrients diffusing through this environment. Projects consider additional phases, such as diffusion and release of microbial products, attachment to the biofilm of a microbe that is wandering in free space, and detachment of microbes from the biofilm. For simplicity, we consider the biofilm to be composed of only one type of bacterium, while projects at the end of this module consider more complex arrangements.

Nutrient Grid

To model the growth of a biofilm we employ a cellular automaton with two $m \times n$ grids, one for the biofilm and a corresponding one for nutrients. Assuming that we have a homogeneous nutrient that is completely mixed at a constant temperature, with a function, ***initNutrientGrid***, we initialize the nutrient grid to be an $m \times n$ matrix, with each element having a dimensionless constant value, ***MAXNUTRIENT***, with $0 < MAXNUTRIENT \leq 1$.

initNutrientGrid(m, n)

Function to return an $m \times n$ matrix, with each element having the value *MAXNUTRIENT*

Quick Review Question 1

From the text's website, download your computational tool's *10_5QRQ.pdf* file for this system-dependent question that defines *initNutrientGrid*.

Diffusion occurs on the nutrient grid at each time step. Although one of the projects considers another alternative, for now we assume that the nutrients diffuse at the same rate throughout the system. As in section "Heat Diffusion" from Module 10.2, we base our model of diffusion on Newton's Law of Heating and Cooling; so that for diffusion rate parameter (r), a cell's nutrient value at time $t + \Delta t$ is as follows:

$$site + \Delta site = (1 - 8r)site + r\sum_{i=1}^{8} neighbor_i, \text{ where } 0 < r < 0.125$$

That section also presents the algorithm for a function, ***diffusion***, which we employ in this module for the diffusion of nutrients. This function, which returns the new nutrient value for the site, has parameters for the diffusion rate (*diffusionRate*) and the nutrient values of a cell (*site*) and its eight neighbors (*N*, *NE*, *E*, *SE*, *S*, *SW*, *W*, *NW*).

Nutrient Boundary Conditions

In the biofilms model, we employ a combination of boundary conditions. Suppose that the surface to which the biofilm adheres, or substratum, is on the left and an infinite supply of nutrients occurs on the right. For this infinite supply, the expanded nutrient grid has an eastmost (right) column, with each cell having constant nutrient value. In this same grid with no nutrients present on the surface, we have a westmost column of all zeros. We use periodic boundary conditions in the north and south directions so that part of the nutrient in the north diffuses to the south, and vice versa.

Quick Review Question 2

From the text's website, download your computational tool's *10_5QRQ.pdf* file for this system-dependent question that defines a function ***extendNutrientGrid*** that takes an $m \times n$ nutrient grid and returns an extended $(m + 2) \times (n + 2)$ nutrient grid.

Biofilm Initialization

For modeling the bacteria in biofilms, we employ an identically shaped grid to that of the nutrient grid, and cells in the same position in the two grids represent the same location. For example, the cell in row 3 and column 7 of the bacteria grid indicates the bacterial state (empty, bacterium, dead bacterium), while the corresponding cell in the expanded nutrition grid indicates the nutrient amount there.

A cell of the bacteria grid can be in one of three states: contain a live bacterium, have a dead bacterium, or be empty and available for growth of a new bacterium, while a fourth border state is available in the extended bacteria grid. Table 10.5.1 lists the four bacteria cell state values—0, 1, 2, and 3—and meanings, along with associated constant names—***EMPTY***, ***BACTERIUM***, ***DEAD***, and ***BORDER***, respectively. By using such descriptive names, our program is easier to understand and to modify.

Table 10.5.1
Cell Values with Associated Constants and Their Meanings

Value	*Constant*	*Meaning*
0	*EMPTY*	The cell does not contain a live or dead bacterium or border.
1	*BACTERIUM*	The cell contains a live bacterium.
2	*DEAD*	The cell contains a dead bacterium.
3	*BORDER*	The cell is on the border and not under active consideration.

For initialization of a simulation's bacteria grid in a function, ***initBacteriaGrid***, we designate if each cell contains a bacterium or not. We can form this initial configuration in a variety of ways to study various situations. For this simulation, the initial bacteria grid is an $m \times n$ matrix with bacteria (value *BACTERIUM*) occurring at random in the first column and all other cells being empty (value *EMPTY*).

The initialization algorithm employs random numbers and probability in determining the first column's values. Suppose on the average only 15% of these cells contain bacteria. Thus, a *probInitBacteria* = 0.15, or 15%, chance exists for a cell to contain a bacterium. For each cell, we generate a uniformly distributed random floating-point number from 0.0 up to 1.0. If the random number is less than *probInitBacteria*, we make the cell's value *BACTERIUM*; otherwise, we assign *EMPTY* to the cell's value. Thus, using the preceding probability and cell values, we employ the following logic to initialize each cell in the grid.

initBacteriaGrid(m, n, probInitBacteria)

Function to return an initial $m \times n$ bacteria grid of all *EMPTY* values except for a first column, where the probability of a bacterium in a cell is *probInitBacteria*

Algorithm:

emptyMat ← $m \times (n - 1)$ matrix with each cell being *EMPTY*
onSurface ← $m \times 1$ matrix (column vector) with each element calculated as follows:
  if a random floating-point number is less than *probInitBacteria*
    set the cell's value to *BACTERIUM*
  else
    set the cell's value to *EMPTY*
return a matrix with *onSurface* as first column and *emptyMat* as rest of matrix

Quick Review Question 3

From the text's website, download your computational tool's *10_5QRQ.pdf* file for this system-dependent question that defines *initBacteriaGrid*.

Biofilm Boundary Conditions

As with the nutrient grid, we have periodic boundary conditions in the north-south direction. In an extended bacteria grid, which function ***extendBacteriaGrid*** returns, the far-west (left) direction has an edge of border cells indicating the substratum, and the far-east (right) direction also has an edge of border cells that do not accommodate growth from the interior. Thus, as designed next, we use periodic boundary conditions in the north-south directions, a first column, with each cell having the value *BORDER*, indicating a surface to the west and a last column, with each cell having the value *BORDER*, so that bacteria do not grow to the east.

extendBacteriaGrid(mat)

Function to take an $m \times n$ matrix parameter and return an $(m + 2) \times (n + 2)$ matrix with periodic boundary conditions in the north-south directions and with fixed boundary conditions in the east-west directions using constant value *BORDER*

Algorithm:

matNS ← concatenation of last row of *mat*, *mat*, and first row of *mat*
return concatenation of column of *BORDER*s, *matNS*, and column of *BORDER*s

Quick Review Question 4

From the text's website, download your computational tool's *10_5QRQ.pdf* file for this system-dependent question that defines *extendBacteriaGrid*.

Biofilm Growth

If a location with a bacterium has no nutrients, the bacterium dies of starvation. Cells with dead bacteria remain in that state from one time step to the next. In the projects, we consider other possibilities, such as decay of a dead bacterium to nutrients.

With a certain probability a live bacterium divides at random into a neighboring empty cell. Researchers have considered several calculations of the probability of such growth, usually related to the amount of available nutrients. For our model, we assume than this probability is proportional to the nutrients in the bacterium's cell and have the proportionality constant, *p*, as a parameter to the simulation.

For a *BACTERIUM* cell that is to divide, we must pick an empty neighbor to accept the daughter bacterium. While projects consider other alternatives, in this simulation if no empty neighbor exists, division does not occur. However, when possible, we select one of the empty neighbors at random. The function ***pickNeighbor*** has parameters of a cell's row (*i*) and column (*j*) in an extended matrix, the number of rows (*m*) of the corresponding unextended matrix, and the values of the (*i*, *j*) cell's four nearest neighbors (*N*, *E*, *S*, *W*). The function returns indices in the corresponding unextended bacteria grid. Thus, we first define *newi* and *newj* to be the indices in the unextended grid corresponding to the indices, *i* and *j*, in the extended grid. That is, *newi* and *newj* are one less than *i* and *j*, respectively. If no neighbor (*N*, *E*, *S*, *W*) is empty, we return the pair (*newi*, *newj*) so that division does not occur. Otherwise, we return the row and column in the unextended grid of the selected empty neighbor. We must be careful to consider north-south periodic boundary conditions. Thus, an empty cell north of a first-row bacteria grid cell is really on the grid's last row. For

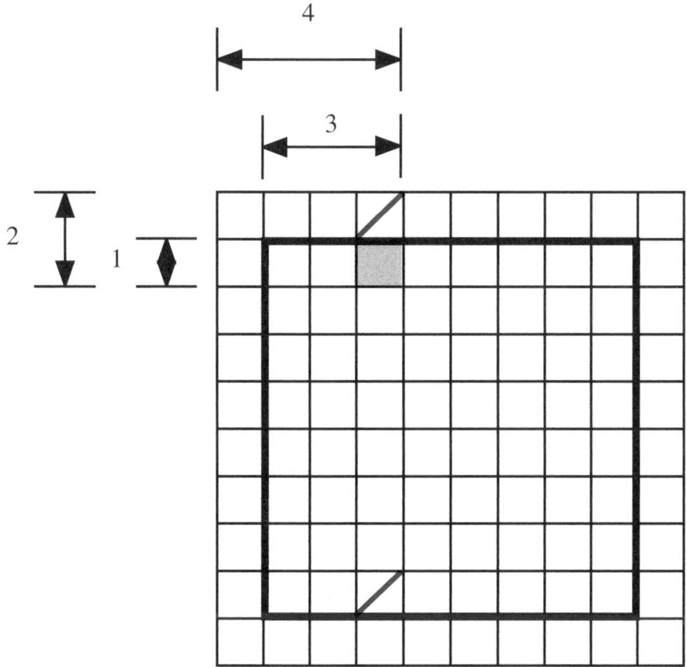

Figure 10.5.1 Extended and unextended bacteria grid with empty cell to the north of the site next to the boundary

example, if the minimum index is 1, suppose $i = 2$, $j = 4$, $m = 8$, and $N = EMPTY$, as in Figure 10.5.1. With coordinates (2, 4) in the extended bacteria grid, the corresponding coordinates in the unextended grid are (*newi*, *newj*) = (1, 3). Suppose the cell to the north is picked to accept the daughter bacterium in division. Wrapping around with periodic boundary conditions, this cell is really far south at location (8, 3) in the unextended grid. Similarly, we must consider periodic boundary conditions when a cell in the last row of the unextended grid has a selected empty neighbor to the south. Because the extended grid has a first and a last column of all *BORDER* values, we cannot pick a cell off the unextended bacteria grid to the west or east, which simplifies the code.

Quick Review Question 5

From the text's website, download your computational tool's *10_5QRQ.pdf* file for this system-dependent question that relates to *pickNeighbor*.

Quick Review Question 6

From the text's website, download your computational tool's *10_5QRQ.pdf* file for this system-dependent question that defines *pickNeighbor*.

pickNeighbor(i, j, m, N, E, S, W)

Function to return the row and column in the unextended bacteria grid of a randomly selected empty neighbor of a given cell. If an empty neighbor does not exist, the function returns the indices of the site in the unextended bacteria grid corresponding to the given cell.

Pre: *i*, *j* are indices of a site in extended bacteria grid.
m is the number of rows of an un-extended bacteria grid.
N, *E*, *S*, *W* are values of the nearest four neighbors of the site in extended bacteria grid.

Post: The function has returned the indices in the unextended bacteria grid of an empty neighbor or, if no such neighbor exists, the indices of the given site in the unextended bacteria grid

Algorithm:

```
lst ← list of N, E, S, W
pos ← list of positions (1 through 4) where EMPTY occurs in lst
newi ← i – 1                                  //indices in unextended grid
newj ← j – 1
if pos has no elements
   return (newi, newj)
else
   rand ← random integer representing an index of pos
   if pos(rand) is 1
      if newi is greater than the minimum index
         return (newi – 1, newj)              //north
      else
         return (maximum index, newj)         //wrap around because of
                                              //periodic boundary conditions
   else if pos(rand) is 2
      return (newi, newj + 1)                 //east
   else if pos(rand) is 3
      if newi less than the maximum index
         return (newi + 1, newj)              //south
      else
         return (minimum index, newj)         //wrap around because of
                                              //periodic boundary conditions
   else
         return (newi, newj - 1)              //west
```

Preliminary to the main iteration of the growth algorithm in function ***grow***, we make a copy (*bacGrid*) of the bacteria grid for updating, determine its number of rows (*m*) and columns (*n*), and expand the bacteria and nutrition grids to account for boundary conditions. Then, looking for bacteria, we iterate through every internal position of the extended bacteria grid (*extBacGrid*) by having an index (*i*) going through the internal row indices and an index (*j*) going through the internal column indices. (If indices begin with 0, *i* goes from row 1 through row *m*, and *j* goes from row 1 through row n. If indices begin with 1, the indices range from 2 through $(m + 1)$ and 2 through $(n + 1)$, respectively.) If a bacterium has no nutrition (nutrition value of 0), we change the corresponding element of *bacGrid* to be *DEAD*. Because *extBacGrid* is an expanded matrix of size $(m + 2) \times (n + 2)$, while *bacGrid* has size $m \times n$, an element of *extBacGrid* with indices i and j corresponds to an element of *bacGrid* with indices $(i - 1)$ and $(j - 1)$, respectively. For a position with a bacterium that is to live, we calculate the probability (p times its nutrition value) that the bacterium will divide. Thus, with this probability, we call *pickNeighbor* to obtain indices (*newi* and *newj*) of a daughter bacterium and then change the corresponding element of *bacGrid* from *EMPTY* to *BACTERIUM*. Thus, at the next time step, cell (*newi*, *newj*) in the unextended grid is to have a bacterium. However, in a call to *grow*, two growing bacteria may choose that same cell for expansion. If this is the case, as a simplifying assumption, we place only one daughter bacterium in that location. The algorithm for *grow* follows and assumes that the smallest matrix index is 1:

grow(bacteriaGrid, nutritionGrid, p)

Function to take a bacteria grid, a nutrition grid, and a partial probability p and return a bacteria grid for the next time step as follows: If a site with a bacterium has no nutrient, the bacterium dies; otherwise, if possible, with probability p times the cell's nutrient value, the bacterium divides and its daughter bacterium inhabits a randomly selected empty neighboring site.

Pre: *bacteriaGrid* and *nutritionGrid* are bacteria and nutrition grids, respectively.
The probability that a bacterium will divide is p times the cell's nutrition value.
Matrix indices begin with 1.

Post: The bacteria grid for the next time step was returned.

Algorithm:

```
bacGrid ← bacteriaGrid
m ← number of rows in nutritionGrid
n ← number of columns in nutritionGrid
extBacGrid ← extendBacteriaGrid(bacteriaGrid)
extNutGrid ← extendNutrientGrid(nutritionGrid)
for i going from 2 through m + 1, do the following: // indices starting with 1
  for j going from 2 through n + 1, do the following:
    if extBacGrid(i, j) is BACTERIUM,
       if extNutGrid(i, j) <= 0
          bacGrid(i – 1, j – 1) ← DEAD
       else if a random number is less than p*extNutGrid(i, j)
          (newi, newj) ← pickNeighbor(i, j, m, extBacGrid(i - 1, j),
                                            extBacGrid(i, j + 1),
                                            extBacGrid(i + 1, j),
                                            extBacGrid(i, j - 1))
          bacGrid(newi, newj) ← BACTERIUM
return bacGrid
```

Quick Review Question 7

From the text's website, download your computational tool's *10_5QRQ.pdf* file for this system-dependent question that defines *grow*.

Consumption of Nutrients

At each time step, we call a function, ***consumption***, in which each bacterium consumes a constant amount (***CONSUMED***) of nutrient. The amount of nutrient in a cell cannot fall below 0.

Quick Review Question 8

From the text's website, download your computational tool's *10_5QRQ.pdf* file for this system-dependent question that relates to consumption.

For this simulation, nutrition is consumed only in the cells containing bacteria. In each such cell, a bacterium eats a constant amount (*CONSUMED*) of nutrients, so that the new value for the cell's nutrient is the old value minus *CONSUMED*. However, a bacterium cannot consume more than is there; so that if the difference is negative, we use 0.0 instead. We can employ an *if* statement or can take the maximum of 0.0 and the old nutrient value minus *CONSUMED* to ensure that each result is nonnegative.

consumption(bacteriaGrid, nutritionGrid)

Function to return a new nutrition grid after bacteria have consumed nutrients in one time step

Algorithm:

```
m ← number of rows of nutritionGrid
n ← number of columns of nutritionGrid
nutGrid ← nutritionGrid
for i going from 1 through m // assuming matrix indices begin with 1
   for j going from 1 through n
      if bac(i, j) is BACTERIUM
         nutGrid(i, j) ← maximum of 0.0 and (nutGrid(i, j) – CONSUMED)
return nutGrid
```

Quick Review Question 9

Write pseudocode using an *if* statement instead of "maximum" in the nested loops to obtain a new value for *nutGrid(i, j)*.

For calculation of new values along the edges, we must extend the boundaries of a nutrient grid. As indicated in the "Nutrient Boundary Conditions" section, we have periodic boundary conditions in the north-south directions, constant 0 in the west direction containing the substratum, and constant *MAXNUTRIENT* in the east direction with its endless nutrient supply. The function ***extendNutrientGrid*** takes an $m \times n$ matrix, *mat*, and returns such an extended $(m + 2) \times (n + 2)$ matrix.

extendNutrientGrid(mat)

Function to take an $m \times n$ matrix parameter and return an $(m + 2) \times (n + 2)$ matrix with periodic boundary conditions in the north-south directions, a first column of zeros, and a last column with constant value *MAXNUTRIENT*

Algorithm:

matNS ← concatenation of last row of *mat*, *mat*, and first row of *mat*
return concatenation of column of zeros, *matNS*, and column of *MAXNUTRIENT*s

Quick Review Question 10

From the text's website, download your computational tool's *10_5QRQ.pdf* file for this system-dependent question that defines *extendNutrientGrid*.

After extending the grid by one cell in each direction using these boundary conditions, we apply the function *diffusion* to each internal cell and then discard the boundary cells. To do so, we employ a function ***applyDiffusionExtended*** similar to the function in section "Applying a Function to Each Grid Point" of Module 10.2 ("Diffusion: Overcoming Differences") that takes an extended square lattice (*matExt*) and a diffusion rate (*diffusionRate*) and returns the internal lattice with *diffusion* applied to each site.

Simulation Program

To perform the simulation of a biofilm's structural formation, we define a function ***biofilm*** with parameters *m* and *n*, the number of grid rows and columns, respectively; *probInitBacteria*, the probability of a bacterium in an initial bacteria grid's first column element; *diffusionRate*, the rate of diffusion of nutrients in the nutrient grid; *p*, the constant ($0 < p \leq 1$) used in the calculation of the probability that a bacterium divides; and *t*, the number of time steps. The function *biofilm* returns two lists, a list of the initial bacteria grid and the next *t* bacteria grids in the simulation and a corresponding list of nutrient grids. Pseudocode for *biofilm* is as follows.

biofilm (m, n, probInitBacteria, diffusionRate, p, t)

Function to return a list of bacteria grids and a list of nutrition grids in a simulation of the formation of the structure of a biofilm with one type of bacterium. In a bacteria grid, a cell value of *EMPTY* indicates the cell is empty;

BACERIUM, the cell contains a live bacterium; and *DEAD*, a dead bacterium. In a nutrition grid, cell values range from 0 (no nutrient) to 1.

Pre: *m* and *n* are the number of rows and columns, respectively, of the bacteria and nutrient grids.
probInitBacteria is the probability of a bacterium in an element of the initial bacteria grid's first column.
diffusionRate is the rate of diffusion of nutrients in the nutrient grid.
The probability that a bacterium will divide is *p* times the cell's nutrition value.
t is the number of time steps.
Matrix indices begin with 1.

Post: Two lists were returned: a list of the initial bacteria grid and the grid at each time step of the simulation and a corresponding list of nutrient grids.

Algorithm:

bacteriaGrid ← *initBacteriaGrid(m, n, probInitBacteria)*
initNutrientGrid ← *initNutrientGrid(m, n)*
bacGrids ← list containing *bacteriaGrid*
nutGrids ← list containing *nutrientGrid*
do the following *t* times:
  extNutrientGrid ← *extendNutrientGrid(nutrientGrid)*
  nutrientGrid ← *applyDiffusionExtended(extNutrientGrid, diffusionRate)*
  bacteriaGrid ← *grow(bacteriaGrid, nutritionGrid, p)*
  nutrientGrid ← *consumption(bacteriaGrid, nutritionGrid)*
  bacGrids ← the list with *bacteriaGrid* appended onto the end of *bacGrids*
  nutGrids ← the list with *nutrientGrid* appended onto the end of *nutGrids*
return *bacGrids* and *nutGrids*

Quick Review Question 11

From the text's website, download your computational tool's *10_5QRQ.pdf* file for this system-dependent question that defines *biofilm*.

Display Simulation

Visualization helps us understand the meaning of the grids. For each bacteria grid in the first list returned by *biofilm*, we generate a graphic for a rectangular grid with white representing an empty site; color, a bacterium; and dark gray, a dead bacterium. The function ***showBacteriaGraphs*** with parameter *graphList* containing the list of lattices from the simulation produces these figures. We animate the sequence of graphics to view the changing biofilm scene.

Quick Review Question 12

From the text's website, download your computational tool's *10_5QRQ.pdf* file for this system-dependent question that defines *showBacteriaGraphs*.

Because nutrient values are on a continuum from 0 to 1, we employ a grayscale for the animation of nutrient diffusion. On such a scale, 0 is black and 1 is white. So that the higher nutrient values appear darker, we subtract each nutrient value from 1 to obtain its degree of gray. For example, a low nutrient value of 0.2 converts to a grayscale value of $1 - 0.2 = 0.8$, which displays as light gray. In contrast, a high nutrient value of 0.8 has a grayscale value of 0.2 and appears as dark gray in the animation.

Quick Review Question 13

From the text's website, download your computational tool's *10_5QRQ.pdf* file for this system-dependent question that defines *showNutrientGraphs*.

Example Problem

Figure 10.5.2 displays several frames of a biofilm simulation with bacteria grids on one row and the corresponding nutrient grids on the next. Clearly, different initial seeds result in different sequences. This simulation employs the parameters $m = 50$, $n = 20$, $probInitBacteria = 0.5$, $diffusionRate = 0.1$, $p = 0.3$, and $t = 130$. In the simulation, the biofilm does not grow from $t = 0$ to $t = 1$, but subsequent frames show the biofilm spreading to neighboring cells. The nutrient grids illustrate the bacteria's gradual consumption of food as well as the diffusion of nutrients. Grids for times starting at $t = 55$ reveal the influence of north-south periodic boundary conditions as the biofilm at the top spreads to the bottom of the grid. The frame at $t = 100$ shows how some bacteria have consumed all their resources and died (in dark gray). As time advances, parts of the biofilm coalesce, and bacteria fill holes in the biofilm (see frame at $t = 130$). We must, of course, be careful not to allow the simulation to run so long that the biofilm reaches the east edge and starts filling in that end.

Assessment of the Model

As IWA (2006) points out, "Most biofilm models today capture only a small fraction of the total complexity of a biofilm system, but they are highly useful." We have chosen in the *biofilm* simulation only to model structural formation, not function.

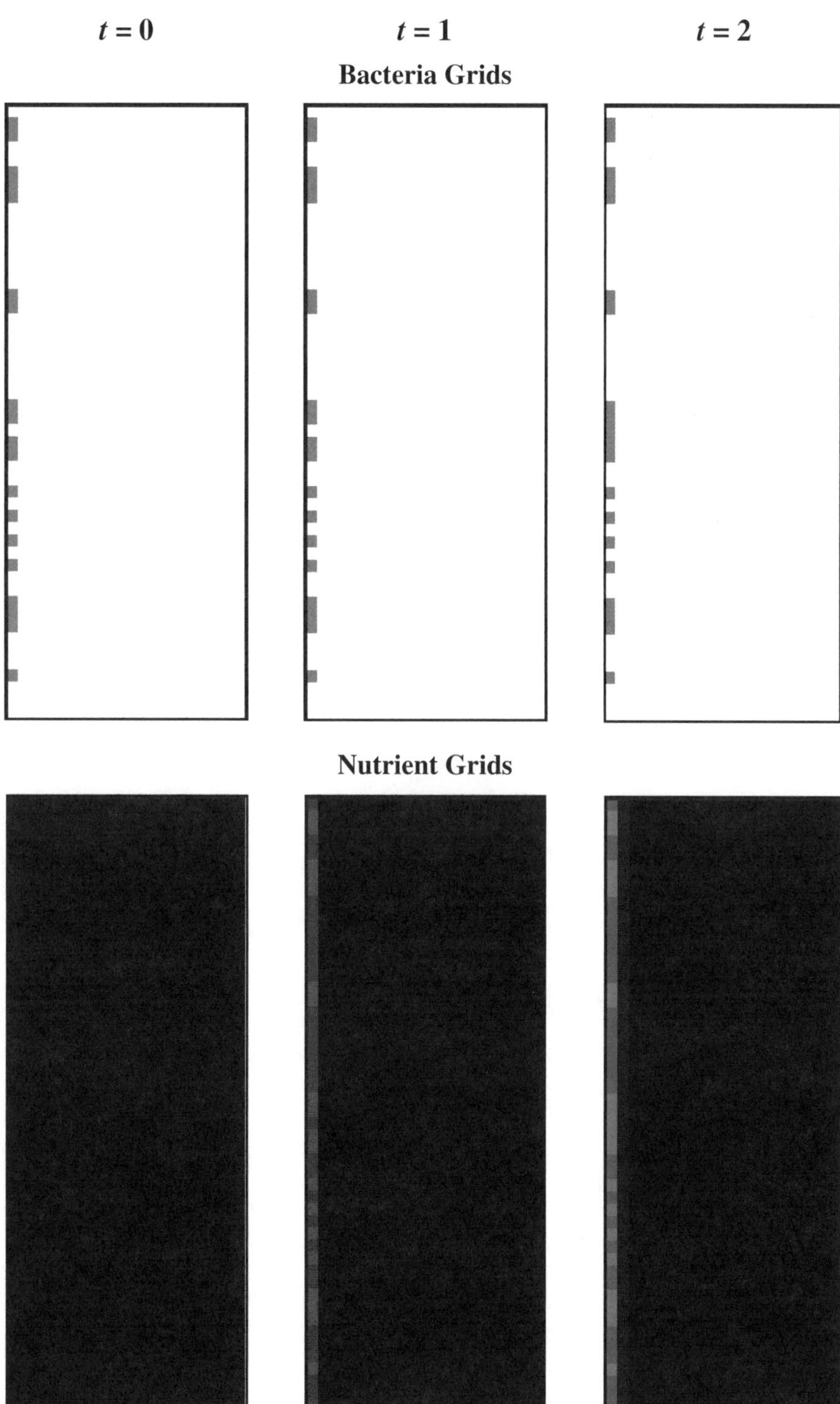

Figure 10.5.2 Several frames in an animation of the spreading of a biofilm

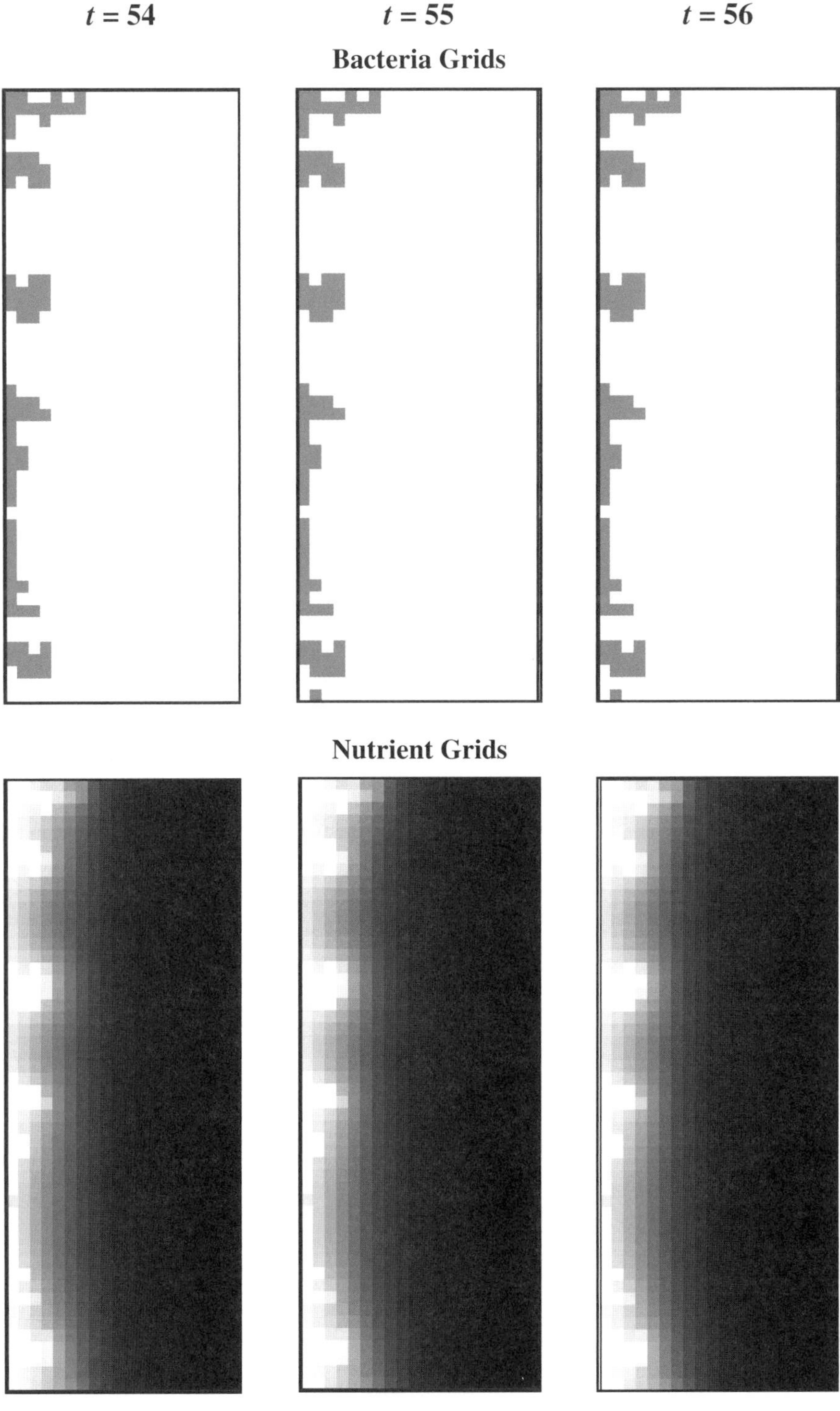

Figure 10.5.2 *(continued)*

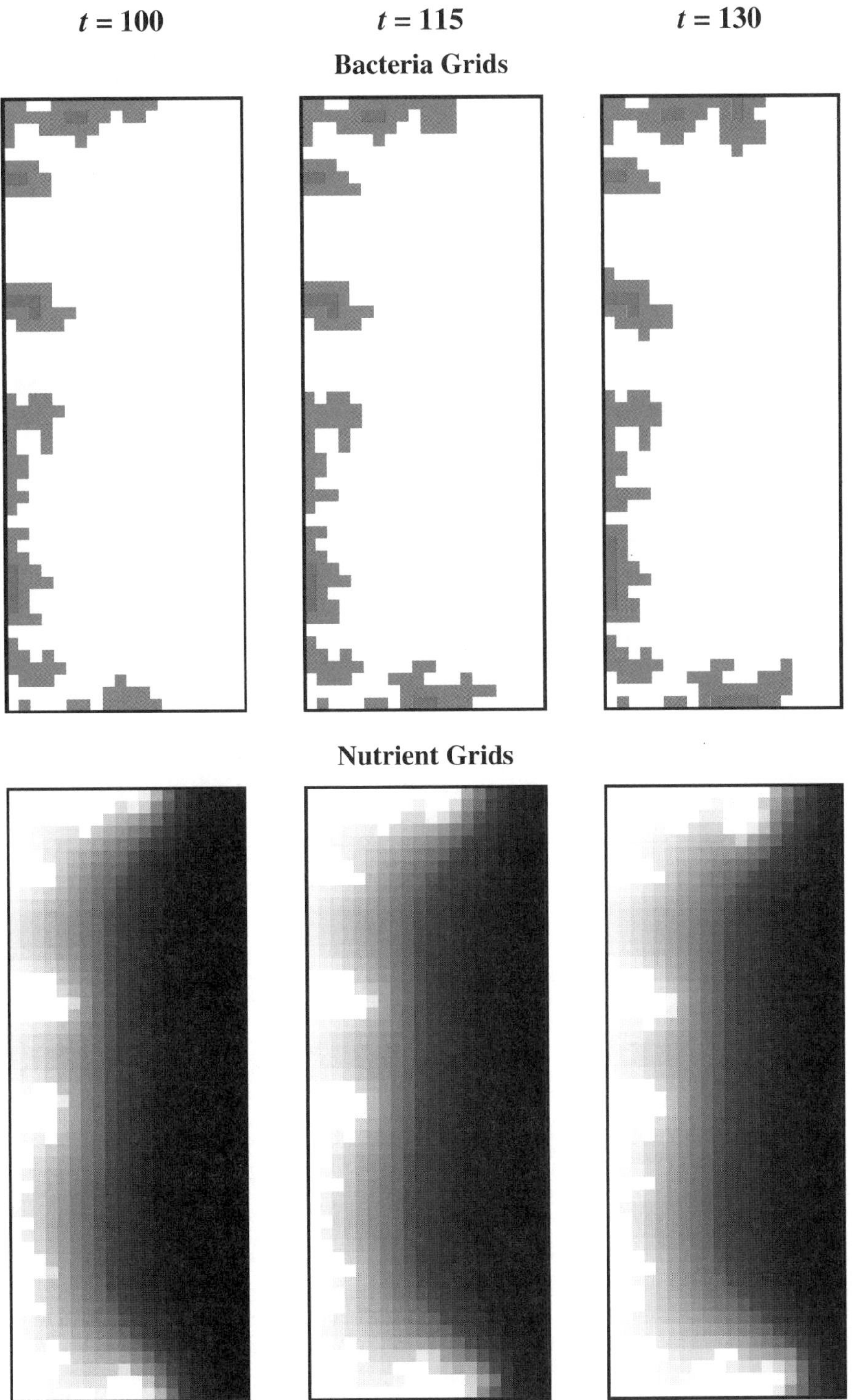

Figure 10.5.2 (*continued*)

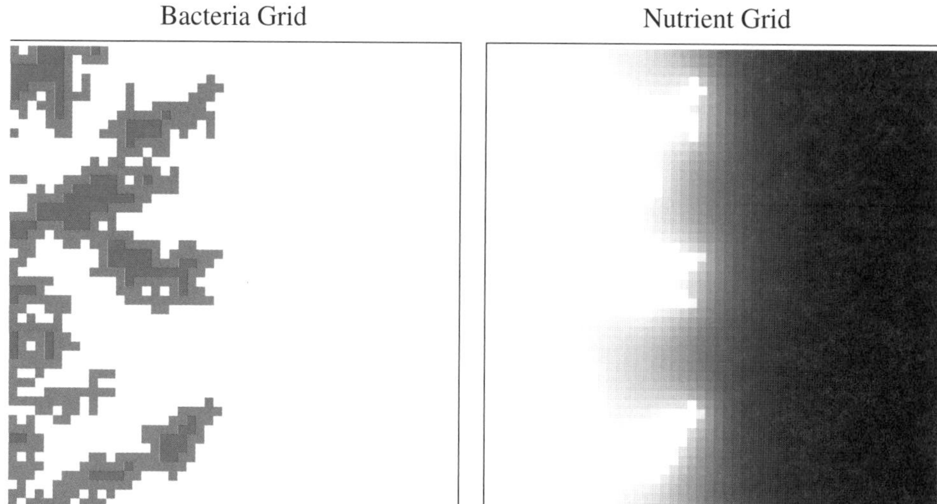

Figure 10.5.3 Mushroom shapes and pores in simulated biofilm along with corresponding nutrient grid

Simulation results agree with various features of biological biofilms. For example, as Figure 10.5.2 shows, with time, the overall thickness increases, and inert (dead) areas are greater near the substratum (Laspidou and Rittmann 2004a, b).

As Schaudinn and his coauthors (2007) so eloquently state, "magnified views reveal microcolonies in an English garden to topiary delights, taking shapes that resemble mushrooms, towers, and arboreal structures. . . ." Researchers have also observed other interesting features, such as pores, in biofilms (Harrison et al. 2005). Frames of simulation results in Figures 10.5.2 and 10.5.3 display these phenomena. The nutrient grid and both simulation rules based on reality provide explanations for some of these shapes. Bacteria have consumed much of the nutrients towards the substratum in the west; nutrients continually come from the east; and bacteria divide at higher rates in nutrient-rich environments.

However, allowing our simulation to run for many time steps can reveal some anomalies not generally present, such as very long dendritic structures. Refining the model to account for erosion of surface bacteria should ameliorate this situation (see Projects).

Also, the current model does not show water channels present in so many biofilms and does not indicate the biofilm's density, which is significantly greater near the substratum. Moreover, in our simplifying assumptions, we ignored important aspects, such as hydrodynamics, extracellular polymeric substances (EPS), chemical oxygen demand, and heterogeneity. Accounting for such features can enlighten our understanding of biofilms but can result in significantly more complex models that require much greater computing resources. Other types of models, such as continuum or discrete particle-based models, are advantageous for showing such features and for taking into account biofilm function.

Computing Power

Biofilms are highly complex with numerous features. We have chosen to consider form, not function, in 2D and have employed a number of simplifying assumptions. The simulations were run on a personal computer. However, with larger grids, conversion to 3D, and refinement to more complex models, simulations can stretch computing resources significantly. For example (IWA 2006), in referring to biofilm models involving hydrodynamics states, "Although such 2d models are now accessible for ordinary personal computers of nowadays (even for time-dependent problems a few minutes may be sufficient), the 3d problems of such type are at limit and better require parallel computing power." Thus, it is advantageous for the modeler to be able to use high-performance computing, which Chapter 12 introduces, when needed.

Projects

On the text's website, Biofilm files for several computational tools contain the code for the simulation of the module. Complete the following projects using your computational tool.

For additional projects, see Module 14.3 "Foraging—Finding a Way to Eat"; Module 14.4, "Pit Vipers—Hot Bodies, Dead Meat"; Module 14.5, "Mushroom Fairy Rings—Growing in Circles"; Module 14.6, "Spread of Disease—Sharing Bad News"; Module 14.7, "HIV—The Enemy Within"; Module 14.8, "Predator-Prey—'Catch Me If You Can'"; Module 14.9, "Clouds—Bringing It All Together"; Module 14.11, "Spaced Out: Native Plants Lose to Exotic Invasives"; and Module 14.12, "Re-Solving the Problems with Cellular Automaton Simulations."

1. Adjust the biofilm simulation to show attachment to the biofilm of bacteria floating in the nutrients. To do so, have *biofilm* execute a fourth phase (after consumption) at each time step. We can use the technique of **diffusion-limited aggregation** (**DLA**) for the attachment. One at a time, "bacteria" are released from random positions on the east boundary (or at least at a random position east of the biofilm) to go on random walks. For each time step of such a walk, a bacterium moves at random to a neighboring position. If the walker comes in contact with another particle (i.e., a neighbor to its north, east, south, or west), with a designated sticking probability, the walker adheres to the particle, resulting in a larger biofilm. If the walker travels too close to the east boundary of the grid, the simulation deletes that walker and releases another random walker. To speed attachment, we can have such a bacterium move eastward with a smaller probability than it moves in the other directions. Notice that such free-floating bacteria only adhere to the surface of the biofilm. For simplicity, define another state and constant for an unattached bacterium, and do not allow a floating bacterium to divide. However, such a bacterium does consume nutrients. Discuss the impact on the structure by allowing attachment. Discuss the effect of consumption of nutrients by a wandering bacterium.

2. One problem with the module's simulation is the formation of long dendritic structures, which do not occur so frequently in biofilms (see Figure 10.5.2, $t = 125$). We have not considered the loss of pieces from the biofilm due to erosion, abrasion, grazing, or sloughing (Picioreanu et al. 1996). Adjust the biofilm simulation so that it shows the erosion of surface bacteria. Similar to Project 1, release an inert particle one at a time from a random location east of the biofilm and have the particle go on a random walk. If the particle touches a bacterium (i.e., a neighbor to its north, east, south, or west), remove the bacterium, making its cell empty. Have *biofilm* execute one step of this random walk (after consumption) at each time step. Notice that erosion only occurs on the surface of the biofilm. Discuss the impact of allowing erosion on the structure. Which bacteria are more likely to erode?
3. Develop an alternative to the biofilm detachment model of Project 2 by eliminating any bacterium above a designated height from the surface. An example of such situation is a constant-depth film fermentor, a device that periodically removes the surface growth to maintain a biofilm with a constant geometry. Researchers use this system to grow and study oral biofilms (dental plaque) in the laboratory (Picioreanu et al. 2004; UCL Eastman Dental Clinic 2008). Run the simulation long enough to observe the pores gradually filling and formation of a compact biofilm.
4. Expand the biofilms simulation to have both an attachment phase (see Project 1) and an erosion phase (see Project 2 or 3) at each time step. Discuss the results.
5. Revise the *diffusion* algorithm so that as indicated in Project 3 of Module 10.2, "Diffusion: Overcoming Differences," diffusion of nutrients into a site is less likely to occur from its corner neighbors. Discuss any differences in diffusion and biofilm growth between this model and that of the module.
6. One classical growth model has a bacterium that is to divide die if its cell has no empty neighbors. Develop this model and compare the results to *biofilm* of this module.
7. One method to calculate the chance that a bacterium will divide is to multiply a positive constant $p \leq 1$ by the cell's fraction of the total nutrition available to bacteria, or p times the nutritional value of the bacterium's cell divided by the sum of the nutritional values of all cells with bacteria:

$$p\left(\frac{cell\text{'}s\ nutrition}{\sum_i nutrition_i}\right) = (cell\text{'}s\ nutrition)\left(\frac{p}{\sum_i nutrition_i}\right)$$

where the sum is over all cells with live bacteria and $0 < p \leq 1$. For example, suppose $p = 0.4$ and three cells have bacteria with corresponding nutrition values of 0.7, 0.5, and 0.8. The probability of the first bacterium dividing is

$$0.4\left(\frac{0.7}{0.7+0.5+08}\right) = 0.14 = 14\%$$

Because the last term is the same for all cells, define a function, *probGrow*, with parameters *bacteriaGrid*, *nutritionGrid*, and p to return $p/\sum_i nutrition_i$. However, if the grid does not contain any bacteria, we should be careful not to divide by zero and return zero instead. Have the function *grow* call *probGrow* and use the value as described to calculate the probability of growth. Compare the results with that of the version in the module.

8. The model in this module has a bacterium consuming nutrition only from its own site. However, it is reasonable to consider that the bacterium might consume some nutrition nearby. Adjust *consumption* so that a bacterium consumes a proportion of the nutrition from its own site and smaller proportions of nutrition from its four nearest neighbor sites. Adjust *grow* so that a bacterium dies if its available nutrition falls below a given threshold.

9. Create a variation of Project 1 where initially no bacteria are attached to the surface. Free-floating bacteria can attach to the surface or the biofilm. Run the simulation several times and discuss the variety of initial patterns of colonization.

10. Revise the biofilm simulation so that a dead bacterium decays with time, forming additional nutrients. To do so, we can have degrees of dead, such as *DEAD1*, *DEAD2*, and *DEAD3*, for a bacterium that decays in three time steps.

11. Revise the biofilm simulation so that if an empty neighbor does not exist for a dividing bacterium, a random walk of a given maximum number of steps occurs to search for an empty location. In a walk, a north, east, south, or west direction is repeatedly selected at random until success or the maximum number of steps is achieved. Such a random walk in search of an available location is comparable to a daughter bacterium adhering to the mother bacterium and pushing other bacteria out of the way. Discuss the impact of this revision on the biofilm structure.

12. Division requires energy. Thus, revise the simulation so that a dividing cell consumes nutrition from its own and, to a lesser extent, its neighboring cells. Discuss the impact of this revision on the biofilm structure.

13. Develop a simulation where the biofilm has two types of bacteria, Types 1 and 2, which are competing for resources. Have the bacteria grow at different rates. That is, have a Type 1 bacterium divide with a certain probability and a Type 2 bacterium divide with another probability. Also, have Type 1 bacteria consume nutrients at a different rate than Type 2 bacteria. Examine different initial situations, such as having the number of Type 1 bacteria being greater than the number of Type 2 bacteria, or vice versa, or having diverse initial configurations. Discuss the results concerning competition and the developing structure. The visualization should display the two bacteria with different colors.

14. Model a biofilm composed of two organisms, Type 1 that grows fast in an oxygen-rich environment, which occurs at the surface, and Type 2 that thrives in a low-oxygen setting deeper within the biofilm (Picioreanu et al. 2004). Have detachment (see Projects 2 and 3) as a phase of the simulation. Discuss the results.

15. According to Stewart (2003), "a biofilm that is 10 cells thick will exhibit a diffusion time 100 times longer than that of a lone cell." Adjust the diffusion algorithm to model slower diffusion deeper within the biofilm. Compare the results of this model to *biofilm* of this module.

16. Revise Project 1 or 3 to account for a phenomenon observed in some biofilms of the necessity of a critical neighborhood density for growth. To do so, we could adjust the rule for growth so that division cannot occur unless a bacterium has at least one neighbor (Picioreanu et al. 1996). Discuss the results.

17. Model the formation of filamentous bacteria, which grow in long threadlike strands, by having preferred growth in the direction away from the surface of the biofilm. Filamentous bacteria occur in wastewater treatment activated sludge flocs, which are large aggregates of adherent bacteria. These flocs can be filtered out for drinking water purification and sewage treatment (Picioreanu et al. 2004).

18. The module's model does not account for the bacterial products, such as chemical signals, metabolites, and antibiotic chemicals. In the same phase as consumption, model such product release.

19. Using a computational tool, develop a 3D version of the *biofilm* model.

20. Using a computational tool, develop a 3D version of any of the projects.

Answers to Quick Review Questions

From the text's website, download your computational tool's *10_5QRQ.pdf* file for answers to the system-dependent questions.

9. *nutGrid*(*i*, *j*) ← (*nutGrid*(*i*, *j*) – *CONSUMED*)
if *nutGrid*(*i*, *j*) < 0.0
    nutGrid(*i*, *j*) ← 0.0

References

Boles, B. R., M. Thoendel, and P. K. Singh. 2004. "Self-Generated Diversity Produces 'Insurance Effects' in Biofilm Communities." Proceedings of the National Academy of Science 101(47): 16630–16635.

Chicurel, M. 2000. "Slimebusters." *Nature*, 408: 284–286.

Costerton, B. 2004. "Microbial Ecology Comes of Age and Joins the General Ecology Community." *Proceedings of the National Academy of Science*, 101(49): 16983–16984.

Costerton, J. W., P. S. Stewart, and E. P. Greenberg. 1999. "Bacterial Biofilms: A Common Cause of Persistent Infections." *Science,* 284: 1318–1322.

Cunningham, Alfred B., John E. Lennox, and Rockford J. Ross, eds. 2011. *Biofilms: The Hypertextbook 4.3.* http://www.hypertextbookshop.com/biofilmbook/v004/r003/ (accessed January 8, 2013)

Donlan, R. M., and J. W. Costerton. 2002. "Biofilms: Survival Mechanisms of Clinically Relevant Microorganisms." *Clin. Microbiol. Rev.*, 15(2): 167–193.

Fux, C. A., J. W. Costerton, P. S. Stewart, and P. Stoodley. 2005. "Survival Strategies of Infectious Biofilms." *Trends Microbiol.*, 13(1): 34–40.

Genco, R., S. Offenbacher, and J. Beck. 2002. Periodontal Disease and Cardiovascular Disease: Epidemiology and Possible Mechanisms. *J. Am. Dent. Assoc.*, 133: 14S–22S.

Harrison, J. J., R. J. Turner, L.L.R. Marques, and H. Ceri. 2005. "Biofilms." *American Scientist*, 93: 508–515.

IWA (International Water Association). 2006. "Mathematical Modeling of Biofilms." IWA Task Group on Biofilm Modeling. IWA Publishing. 208 pp. 2006_Book_IWA-STR18_Wanner-et-al.pdf

Laspidou, Chrysi S., and Bruce E. Rittmann. 2004a. "Evaluating Trends in Biofilm Density using the UMCCA Model." *Water Research*, 38: 3362–3372.

———. 2004b. "Modeling the Development of Biofilm Density Including Active Bacteria, Inert Biomass, and Extracellular Polymeric Substances." *Water Research*, 38: 3349–3361.

Nadell, C. D., J. B Xavier, S. A. Levin, and K. R. Foster. 2008. "The Evolution of Quorum Sensing in Bacterial Biofilms." *PLoS Biology*, 6(1):171–179.

Overman, Pamela R. 2000. "Biofilms: A New View of Plaque." *Journal of Contemporary Dental Practice*, 1(3): 1–7.

Picioreanu, Cristian, Jan-Ulrich Kreft, and Mark C. M. van Loosdrecht. 2004. "Particle-Based Multidimensional Multispecies Biofilm Model." *Applied and Environmental Microbiology*, 70(5): 3024–3040.

Picioreanu, Cristian, Jan-Ulrich Kreft, M.C.M. van Loosdrecht, and J. J. Heijnen. 1996. "Cellular Automata Models for Biofilm Growth." Presented at the "Bioprocess Engineering Course," June14–18, Stockholm.

Sauer, K., A. H. Rickard, and D. G. Davies. 2007. "Biofilms and Biocomplexity." *Microbe*, 2(7): 347–353.

Schaudinn, Christoph, Paul Stoodley, Aleksandra Kainovic, Teresa O Keeffe, Bill Costerton, Douglas Robinson, Marc Baum, Garth Ehrlich, and Paul Webster. 2007. "Bacterial Biofilms, Other Structures Seen as Mainstream Concepts," *Microbe*, 2(5): pp. 231–237.

Stewart, Philip S. 2003. "Guest Commentaries, Diffusion in Biofilms" *J. Bacteriol.* 185(5): 1485–1491.

UCL Eastman Dental Institute, Microbial Diseases. 2008. "In Vitro Models, Biofilms and Ecology: Constant Depth Film Fermentor." http://www.eastman.ucl.ac.uk/research/MD/biofilms_ecology_models/index.html

Watnick, P., and Kolter, R. 2000. "Biofilm, City of Microbes." *J. Bacteriol.*, 182(10): 2675–2679.

11

AGENT-BASED MODELS

MODULE 11.1

Agent-Based Tool—Tutorial 1

Download

From the textbook's website, download Tutorial 1 for your agent-based tool. We recommend that you work through the tutorial and answer all Quick Review Questions using the corresponding software.

Introduction

This first agent-based tool tutorial, which is available from the textbook's website in your system of choice, prepares you to use the tool to complete many projects in Modules 11.2, 11.4, and 14.13. The tutorial introduces the following functions and concepts:

- Getting started
- Agents and their states
- Creating and destroying agents
- Behaviors
- Movement and animation
- Neighbors and interactions
- Variables and assignments
- Built-in functions
- Probability and random numbers
- Program testing
- User-defined functions
- Graphs
- Documentation
- Stopping simulation

The module gives examples and Quick Review Questions for you to complete and execute in the desired software system.

MODULE 11.2

Agents of Interaction: Steering a Dangerous Course

Downloads

The text's website has a file, *CattleAndDisease*, available for download for various agent-based tools. The file contains the following: *CattleAndDiseaseV1* and *CattleAndDiseaseV2* files for simulations this module develops; *CattleAndDiseaseV2Data.xls*, which contains data from Version 2 of the program; and an *11_2QRQ.pdf* file, which contains system-dependent Quick Review Questions and answers.

Introduction

Doug Taylor and his son were driving past their front pasture admiring a bumper crop of new calves when Doug noticed something odd about a young bull calf that their best breeder had birthed. The calf was now 3 weeks old, and Doug was surprised to see the animal squinting in the sun. In fact, the farmer also noted that the animal didn't seem to be thriving or wanting to leave the shade of a tree. He stopped his truck, went to take a better look, and didn't like what he saw. Scrutinizing the calf, he saw that its left eye was watery, with a small ulcer in the center. Doug knew he needed to act fast because he recognized that the calf had pinkeye. They quickly loaded the calf into the back of the truck to take it to an isolation pen.

Pinkeye (*infectious bovine keratoconjunctivitis*) is a highly infectious disease that afflicts cattle and is the most prevalent illness among breeding beef cattle females. It is also especially common in bull calves. Several strains of the bacterium *Moraxella bovis* cause most cases of the disease. Because early on in the infection the animal is shedding large numbers of bacteria, the disease can be spread fairly easily from animal to animal from direct contact, through indirect contact with contaminated surfaces, or via an insect vector, commonly, face flies. The infectious agent binds to the conjunctiva and cornea, where its various toxins damage the eye. If not treated, blindness may ensue in less than 72 h. Because topical antibiotics require repeated application, farmers instead routinely inject antibiotics and steroids under the eyelid or make intramuscular injections of antibiotics, such as oxytetracycline. These treatments are quite effective if administered soon enough. Besides blindness, infections usually result in less food intake and, therefore, less weight gain. Regardless of size, a blind animal must be sold at a discount. There are various estimates, but through loss of productivity and reduced animal value, the economic impacts of pinkeye in the United States may exceed $150 million (USDA 2011).

To control pinkeye, experts suggest three basic approaches, and perhaps best is a program that involves all three. The first approach is to reduce irritants that might make the eyes more vulnerable to infection. Irritants include the seed and pollen of grass and weeds in grazing areas. Trimming these areas to prevent the flowering/seeding of these plants can help reduce the irritation. Other irritating factors are UV radiation and dust, which are difficult to moderate. The second approach is to reduce exposure and to immunize. (Doug isolated the sick calf to reduce exposure to other animals.) There are vaccines for various strains of this bacterium; but these strains often become resistant, and the vaccines may not be effective. The third approach, and probably the most effective control measure, is to reduce the number of vectors (flies), which is usually a challenge. Flies pick up bacteria when they feed on secretions from the eyes of infected calves. Then, the flies transfer the bacteria to susceptible individuals, rapidly spreading the disease. Farmers can apply various insecticides, but the flies also can become resistant to them. So, often a farmer must try several insecticides (Irsik 2012; Kirkpatrick 2012; Powell 2004).

Problem

Calves are born in the spring and weigh between 60 and 100 lb. In the beef cattle industry in America, a calf roams freely on a farm or ranch for 6 to 9 months until reaching a weight of 600 lb. Then calves from many sources are brought to sale barns to be sold as stockers. As stockers, steers and heifers gain weight to 900 lb (Liu et al. 2012). Feedlots buy these cattle at sale barns and fattened them in pens for 4 to 6 months until the animals reach weights between 1200 and 1400 lb and then go to market (Cattlemen's Beef Board 2009).

Cattle from many sources are in a common pen, and contagious diseases, such as foot-and-mouth disease (FMD), can reach epidemic proportions. Modeling can help in devising strategies to prevent epidemics and to avoid costly loss of animals. To accommodate various diseases, we consider a hypothetical disease, spread by physical contact, in which an animal is infectious for a certain number of days, and once recovery occurs, the animal cannot become sick again. We follow the cattle born in one spring until they reach market and determine situations that lead to epidemics.

Agent-Based Modeling

One technique of modeling the movement of individual beef cattle and the spread of disease among the animals is a cellular automaton simulation, such as what we employed with the movement of ants in Module 10.4, "Movement of Ants—Taking the

Right Steps." A related alternative is a grid-based, **agent-based (individual-based) simulation**.

For a *cellular automaton simulation*, the state of a grid cell might indicate the number of cattle at that location as well as attributes, such as weight(s), associated with the animal(s). Transition rules that specify the relationship of a cell with its neighbors determine the state of the cell at the next time step. For each time step, a cellular automaton simulation sweeps through every cell of the grid, updating its state.

With an *agent-based simulation*, each animal is modeled as an autonomous, decision-making **agent** that has a **state**, which is represented by a set of **state variables,** or attribute values, and **behaviors**, which control its actions. A **method** or **procedure**, which is associated with a class, or breed or group, of agents, is a function that captures some or all of an agent's behavior. A simulation frequently includes several **global simulation variables**, which all agents can access. Agents often operate in an **environment** that arranges cells in a rectangular **grid**. (In Module 13.5, "The Next Flu Pandemic: Old Enemy—New Identity," we consider an individually based model that is not grid based.) The environment, its neighboring agents, and the states and behavior of an agent determine the agent's new state. For each time step, instead of iterating through each grid cell, an agent-based simulation proceeds through each agent, revising its state.

With both cellular automaton and agent-based simulations, individual actions and local interactions can help us to access their effects on the whole system, and we can easily visualize any emerging patterns. Both simulation techniques can be effective in modeling dynamic, spatially complex situations. These models can help us understand systems, evaluate various scenarios, and make informed decisions about actions to take.

Quick Review Question 1

Indicate to which each of the following applies, cellular automaton (CA) simulations, agent-based (AB) simulations, or both:

a. Autonomous, decision-making entity has a state and behaviors.
b. Grid cell has state and transition rules specify next state.
c. Relationship with neighbors determines next state.
d. Can use grid.
e. For each time step, iteration is over each grid cell.
f. For each time step, iteration is over each autonomous, decision-making entity.
g. Local interactions can cause global change.

Formulating the Simulation Model

We want to simulate the effects of disease on bringing cattle to market. A bovine might be infected at any time. An infection lasts 40 days; and when an infection ends, the bovine is immune to subsequent infections. For the simplified model in this module, we do not consider immunity, isolation, or reduction of weight due to illness, but various projects do.

All cattle in the simulation have the same life cycle. Each appears first as a calf weighing between 60 and 100 lb and lives in a pasture until reaching a weight of 600 lb. The animal is then taken to a sale barn and sold to a stocker for further fattening in pastures. Upon reaching a weight of 900 lb, the beef cow is returned to the sale barn and sold to a feedlot. The bovine remains in a feedlot pen until achieving a weight of 1300 lb. The animal is then transferred to an abattoir for processing.

We make the following simplifying assumptions so that the simulation is easier to implement, but later we can refine the model to remove some of the assumptions:

- Birthing cows and new births are not considered. We begin with all newborn calves in a pasture.
- Feed is always available so a bovine does not have to move far to graze.
- Except when traveling to another location, cattle move at random in the pasture and the stocker and do not tend to congregate.
- Instead of being trucked, a bovine moves on its own along a one-way road to the sale barn.
- For each trip to the sale barn, a beef cow is in the facility for at most for 2 days.
- The disease is spread only through direct contact, not vectors.
- A bovine that recovers from the disease cannot become susceptible again.
- No immunity from the disease occurs.
- No cattle die before reaching the abattoir.
- Infected and noninfected cattle gain weight at the same rate.

The simulation has the following input parameters:

- The probability that a particular section of pasture on a farm contains a bovine
- The probability that a bovine will become infected when close to another infected bovine
- The time step

The simulation provides the following results:

- The total number of cattle
- The number of susceptible, infected, and recovered cattle at each time step
- The cumulative total of infected cattle at each time step

Overall Design of the Simulation

Our agent-based simulation needs to represent cattle as well as their world. For simplification, we choose to model an area that has six farms, one sale barn, one stocker, one feedlot, and one abattoir. We have agents to represent cattle (bovines). Each type of environmental area—farm, one-way road that goes to the east, one-way road that goes to the west, sale barn, stocker, feedlot, and abattoir—is composed of a grid of cells/patches/tiles, with each square **cell/patch/tile** being an agent. For example, one farm consists in a grid of farm tile agents that have certain behaviors and comprise a pasture on which the cattle feed.

The simulation proceeds by computing what happens over a series of discrete time steps of length 1⁄4 day. During each time step, a bovine on a farm, stocker, or feedlot gains weight; upon reaching a weight threshold, the animal moves to another designated location; a susceptible animal may become sick due to contact with an infected beef cow; and if infected, the animal becomes a little closer to recovery. The simulation continues until the user stops execution.

Model Environment

There are seven types of environmental agents, each with a different depiction, on which the cattle move: farm pasture (***Farm***), one-way road to the east (***RoadEast***), one-way road to the west (***RoadWest***), sale barn (***SaleBarn***), stocker (***Stocker***), feedlot (***Feedlot***), and abattoir (***Abattoir***). As the following algorithm indicates, the initializations of *Farm* agents place susceptible cattle agents, *Susceptible* (discussed in the next section), at random on the farm patches. For initialization purposes, we define a global simulation variable, ***INIT_CATTLE_ PROBABILITY***, which is the approximate initial fraction of susceptible cattle and, in a sense, is a measure of the initial density of cattle on the farms. Thus, a *Farm* agent, which is one grid patch of a farm, has approximately an *INIT_CATTLE_ PROBABILITY* chance of having a *Susceptible* agent on top of it. (By convention, constants are in all uppercase letters. Moreover, we begin the names of agents with uppercase letters, while the names of methods/procedures and variables start with lowercase letters.)

Farm Initialization

Procedure to initialize a *Farm* agent possibly to have a *Susceptible* agent on top

Pre: *INIT_CATTLE_PROBABILITY* is a global simulation variable with value between 0.0 and 1.0.

Algorithm:

with a chance of *INIT_CATTLE_PROBABILITY*
create a new *Susceptible* agent on top of the *Farm* agent

Quick Review Question 2

From the text's website, download your agent-based tool's *11_2QRQ.pdf* file for this system-dependent question to write the initialization of a *Farm* agent.

Figure 11.2.1 presents one possible environment with six farms, each 95 tiles long and approximately 16 tiles wide, for a total of 7505 *Farm* agents. The white background appears in certain areas, such as between the farms, which are in black. For the scenario shown in the figure, each *Farm* agent initially had *INIT_CATTLE_*

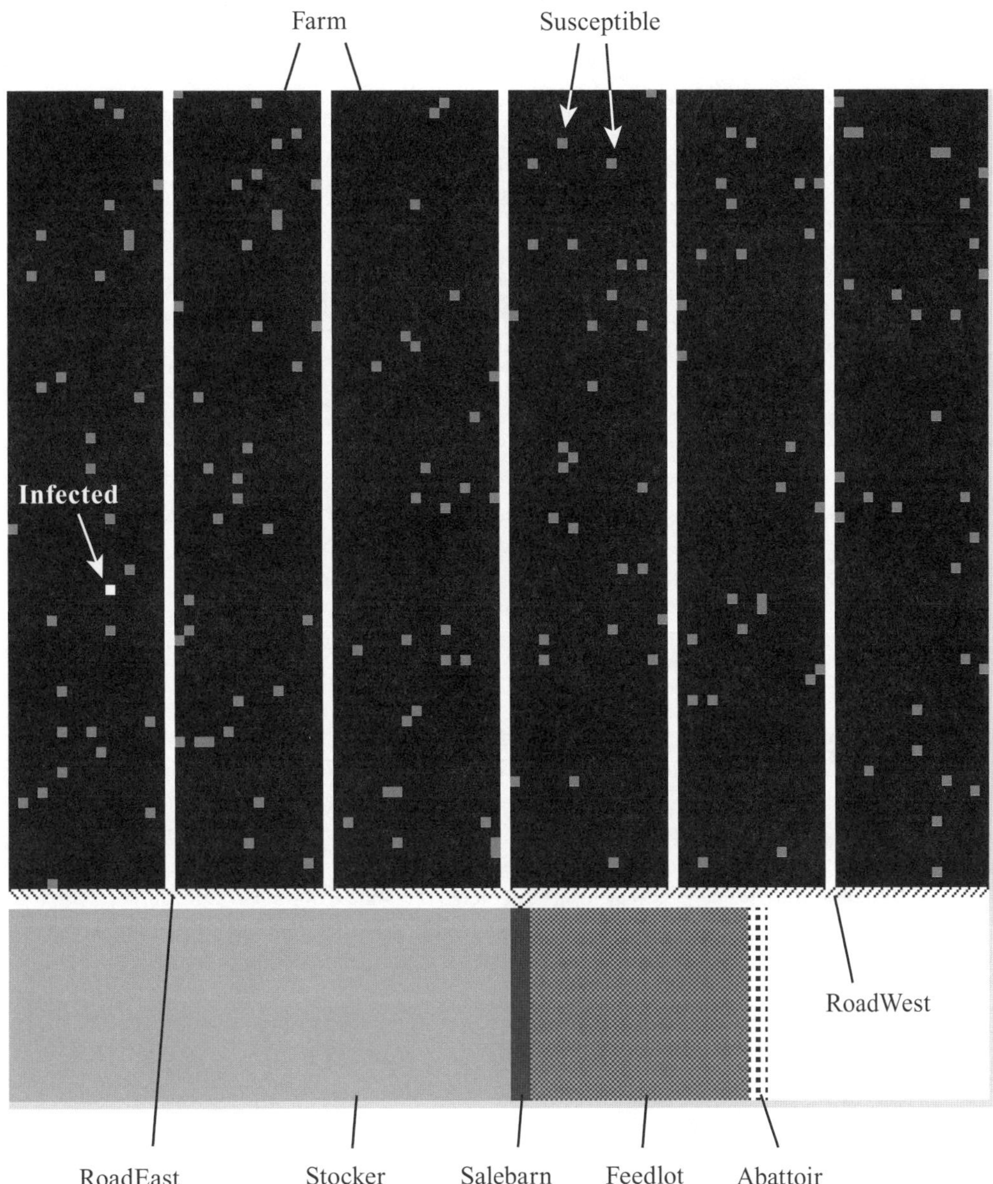

Figure 11.2.1 Environment

PROBABILITY = 0.02 = 2% chance of creating a new *Susceptible* agent on top of a tile. Although cattle often inhabit only one small portion of a much larger pasture at any one time, the configuration does not capture the tendency for these animals to congregate. To this landscape, we added one infected bovine agent in the bottom half of the first farm from the left. After initialization, the scenario has 154 susceptible cattle and one infected bovine.

Agents and Their States

A *Susceptible* cattle agent, depicted in bright color, is one heifer or one steer or a group of cattle susceptible to disease. If such an agent gets the disease and is infectious, it becomes an *Infected* cattle agent, depicted in white. Upon recovery, this agent becomes a *Recovered* cattle agent, shown with a duller color. Upon going to the abattoir, the dead animal becomes a *Processed* cattle agent, having a dark color.

Because weight is the primary determining factor in the location of the cattle (farm, road east or west, sale barn, stocker, feedlot, abattoir), the various cattle agents have a *weight* attribute. For simplification, we follow the herds from just after the birthing season in the spring. Thus, at the beginning of the simulation, we initialize all instances of *Susceptible* to be a uniformly distributed random weight between 60 and 100 lb. To simplify the model, we do not consider birthing cows, which typically remain on the farms.

All cattle agents have a ***weight*** attribute, and various states require additional attributes. For example, attributes ***time1InSale*** and ***time2InSale*** come into existence when the agent enters the sale barn for the first or second time, respectively, and serve to regulate the length of time the animal remains in the sale barn.

As part of its state, an *Infected* agent has the attribute ***daysSick***. For most scenarios, we start the simulation with one *Infected* agent, which has an initial random weight from 60 to 100 lb and an initial *daysSick* value of 0 days to indicate the calf has just become sick.

We define global simulation properties to count the number of each type of live cattle agent (***numSusceptible***, ***numInfected***, and ***numRecovered*** for the number of susceptible, infectious, and recovered cattle, respectively), the cumulative total of infected cattle (***cummulativeInfected***), and the total number of cattle agents (***numCattle***). All these are initialized to zero before appearance of any cattle. With a **left-pointing arrow** (←) indicating assignment, the following shows the initialization algorithm for a newly created cattle agent. This initialization calls a method, ***countSIR***, which asks each entity to increment the appropriate counters by one.

Cattle Initialization

Procedure to initialize new calf with a random weight between 60 and 100 lb, to establish the days sick to be 0 for an infected calf, and to establish various category counters

Algorithm:

if the entity is an *Infected* agent
    weight ← a random number between 60.0 and 100.0
    daysSick ← 0
    call *countSIR*
else if the entity is any other cattle agent
    weight ← a random number between 60.0 and 100.0
    call *countSIR*

countSIR

Procedure to update *numSusceptible*, *numInfected*, *numRecovered*, *cummulativeInfected*, and *numCattle* after addition of a new bovine

Pre: Global simulation variables *numSusceptible*, *numInfected*, *numRecovered*, *cummulativeInfected*, and *numCattle* are the number of susceptible (*Susceptible*), infected (*Infected*), recovered (*Recovered*), cumulative infected cattle, and cattle agents, respectively, before initialization of this entity.

Algorithm:

```
if entity is a Susceptible agent,
    increment numSusceptible and numCattle by 1
else if entity is Infected agent,
    increment numInfected, cummulativeInfected, and numCattle by 1
else if entity is a Recovered agent,
    increment numRecovered and numCattle by 1
```

Quick Review Question 3

From the text's website, download your agent-based tool's *11_2QRQ.pdf* file for this system-dependent question to write the initialization of a *Susceptible* agent and the associated *countSIR* method.

Agent Behaviors

Locations and weights of the cattle agents are the primary determinants of the simulation. At each time step, which has length ***dt*** = 0.25 day, each cattle agent executes a cattle scheduler algorithm, which defines the overall behavior of *Susceptible*, *Infected*, *Recovered*, and *Processed* agents. Because all other agents (*Farm*, *RoadEast*, *RoadWest*, *SaleBarn*, *Stocker*, *FeedLot*, and *Abattoir*) are environmental agents that do not have any methods (other than *Farm*'s initialization procedure), the following cattle scheduler algorithm is the main driver of the simulation. This scheduler calls various methods, whose subsequent algorithms provide greater detail.

Cattle Scheduler to Be Executed Each Time Step

Pre: The entity is a *Susceptible*, *Infected*, *Recovered*, or *Processed* agent.

Algorithm:

```
if the agent is on a farm and weighs less than 600 lb
    call sir
    call inFarm
```

```
else if agent is on a farm
    call sir
    call farm2Sale
else if weight < 900 and agent is in a sale barn
    call sir
    call inSalebarn1
else if agent is in a stocker
    call sir
    call inStocker
else if weight ≥ 900 and agent is in a sale barn
    call sir
    call inSalebarn2
else if agent is in a feedlot
    call sir
    call inFeedlot
else if agent is in an abattoir
    call sirAbattoir
    move agent to east
    change agent to Processed
```

Quick Review Question 4

From the text's website, download your agent-based tool's *11_2QRQ.pdf* file for this system-dependent question to write a cattle scheduler method.

At every step in the run of a simulation, the disease processing ***sir*** method is called for animals that are not being taken to market. (The method's name stands for SIR: Susceptible, Infected, Recovered.) In their agent-based simulation, Liu et al. (2012) employed two main attributes related to the nonfatal disease: **rate of infection, β,** and **rate of recovery**, μ. Assuming that a diseased animal is sick for ***INFECTIOUS_PERIOD*** = 40 days, $\mu = 1/(40 \text{ days}) = 0.025/\text{day}$; approximately 0.025 = 2.5% of the infected cattle recover each day. For our simulation, an *Infected* agent who has been sick for the indicated duration of the disease, say 40 days, becomes a *Recovered* agent. Otherwise, *daysSick* for the ill animal is incremented by the length of one time step, assumed to be $dt = 0.25$ days. Thus, a sick bovine requires 160 time steps to recover.

At any one time step, the probability that a *Susceptible* agent will catch the disease from an adjacent *Infected* individual is the rate of infection (per day) times the length of a time step ($dt = 0.25$ days), $\beta \cdot dt$. To illuminate the outcomes from various diseases and situations, we will execute the simulation for several of values of β. As the following algorithm prescribes, a susceptible cattle agent who comes in physical contact with (i.e., is next to) an infectious cattle agent can become sick with probability $\beta \cdot dt$, perhaps $\beta \cdot dt = 0.125$ for one scenario. We make this probability a global simulation property, such as ***INFECTION_PROBABILITY*** = 0.125, known to all agents.

As the following algorithm indicates, besides determining the progression of the disease, we must also increment/decrement simulation properties appropriately. An animal having been sick for *INFECTIOUS_PERIOD* = 40 days recovers, so we decrement by one the current number of infected cattle (*numInfected*) and increment by one the current number of recovered cattle (*numRecovered*). Should a susceptible animal become infected, we decrease the count of current *Susceptible* agents (*numSusceptible*) and increase the count of those currently infected (*numInfected*) and the cumulative total of infected cattle.

sir

Procedure to advance an infected bovine's illness, possibly to recovery, and determine if a susceptible bovine agent becomes sick

Pre: The entity is a cattle agent.
INFECTIOUS_PERIOD and *INFECTION_PROBABILITY* are global constants and *numSusceptible*, *numInfected*, *numRecovered*, and *cummulativeInfected* are global variables.
If the bovine is an *Infected* agent, *daysSick* is the number of days sick.

Post: The state of the bovine and the variables related to infection have been updated.

Algorithm:

if entity is an *Infected* agent and *daysSick* > *INFECTIOUS_PERIOD*
    change agent to be a *Recovered* agent
    decrement *numInfected* by 1
    increment *numRecovered* by 1
else if entity is an *Infected* agent
    add 0.25 to *daysSick*
else if entity is a *Susceptible* agent and is next to an *Infected* agent, with *INFECTION_PROBABILITY* chance
    change agent to be an *Infected* agent
    set 0 to *daysSick*
    decrement *numSusceptible* by 1
    increment *numInfected* and *cummulativeInfected* by 1

Quick Review Question 5

From the text's website, download your agent-based tool's *11_2QRQ.pdf* file for this system-dependent question to write the *sir* method.

When a beef cow goes to the abattoir, we should no longer count the animal in one of the SIR categories. Thus, ***sirAbattoir*** adjusts the appropriate counter, *numSusceptible*, *numInfected*, or *numRecovered*.

sirAbattoir

Procedure to adjust appropriate system variables when a beef cow is slaughtered

Pre: The entity is a cattle agent on top of an *Abattoir* agent.
Post: The appropriate counter (*numSusceptible*, *numInfected*, or *numRecovered*) has been decremented by 1.
Algorithm:
if *Susceptible* agent, decrement *numSusceptible* by 1
else if *Infected* agent, decrement *numInfected* by 1
else if *Recovered* agent, decrement *numRecovered* by 1

Quick Review Question 6

From the text's website, download your agent-based tool's *11_2QRQ.pdf* file for this system-dependent question to write the *sirAbattoir* method.

The remaining algorithms deal mainly with cattle movement. On a farm, we assume that cattle graze freely, so that each cattle agent performs a random walk around its farm. On the average, a 60- to 100-lb newborn calf takes 6 to 9 months (180–270 days) to gain about 540 to 500 lb, necessary for sale. Thus, over a 200-day period, a calf must gain approximately 2.5 lb/day, or 0.625 lb/*dt*. As the following algorithm for ***inFarm*** indicates, to account for variability in weight gain, we increment *weight* by a random number between 0.50 and 0.75 lb/*dt* while the animal is on the farm.

inFarm

Cattle agent's behavior on a *Farm* patch

Pre: Cattle agent is on a farm and weighs less than 600 lb.
Algorithm:
move agent at random to adjacent *Farm* patch
add a random number between 0.50 and 0.75 to *weight*

Quick Review Question 7

From the text's website, download your agent-based tool's *11_2QRQ.pdf* file for this system-dependent question to write the *inFarm* method.

Although farmers transport their 600-lb calves to sale in trucks, for simplicity, we have the cattle move independently to the sale barn using the environment as a guide. (A project explores using truck agents to transport the animals.) Cattle-movement rules from one major location to the next are designed using cues, such as one-way west and east roads, for the specific configuration of Figure 11.2.1. In particular, as the following algorithm for the *farm2Sale* indicates, a calf of at least 600 lb travels to the south on the farm until encountering a *RoadWest* or a *RoadEast* agent, which indicates the subsequent movement direction. Traveling along the indicated route, the bovine proceeds into the sale barn upon seeing that area to its south.

Unfortunately, some agent-based tools allow an agent to move only to an adjacent cell in one time step. However, this disadvantage has limited consequences for our simulation since we do not have the animal gaining weight or increasing *daysSick* during movement from the farm to the sale barn. Truck transport can expose susceptible cattle to infected animals with which they are not usually in contact. Thus, we allow sick bovines to infect susceptible animals that are on their way to the sale barn.

Most cattle are brought to the sale barn the day before the sale, but some arrive 2 or more days before and a few arrive the day of the sale (Davie 1997). For our simulation, we assume that each cattle agent spends at least 3 and no more than 8 quarter-days (time steps) in the sale barn before moving to the stocker, which is to the west of the sale barn in Figure 11.2.1. To help manage the time in the sale barn, we initialize a state variable, *time1InSale*, to be a random integer between 1 and 5, inclusive, upon entering the sale barn. While in the barn, we increment *time1InSale* with each time step; and when *time1InSale* becomes greater than 8, we start moving the animal toward the stocker.

farm2Sale

Cattle agent's behavior in moving from *Farm* patches to *SaleBarn* patches

Pre: Cattle agent is on a *Farm*, *RoadEast*, or *RoadWest* patch and weighs at least 600 lb.
Post: Cattle agent is on a *SaleBarn* patch.
Algorithm:

```
if Farm is to the south, move agent to south
else if RoadWest is to the south, move agent to southwest
else if RoadWest is to the west, move agent to west
else if RoadEast is to the south, move agent to southeast
else if RoadEast is to the east, move agent to east
else if sale barn is to the south
    move agent to south
    time1InSale ← a random integer between 1 and 5, inclusive
```

Quick Review Question 8

From the text's website, download your agent-based tool's *11_2QRQ.pdf* file for this system-dependent question to write the *farm2Sale* method.

For method ***inSalebarn1*** with *time1InSale* incrementing at each iteration, a heifer or steer remains in the sale barn for up to 8 time steps, or 2 days, until *time1InSale* becomes greater than 8. The method ***inSalebarn2*** defines the behavior of the agent in the sale barn for the second time, after the steer or heifer has gained weight to 900 lb in the stocker. This behavior parallels that of *inSalebarn1* by incrementing another variable, *time2InSale*, instead of *time1InSale*, and moving the animal to the east, instead of west, when that variable reaches 8.

inSalebarn1

Cattle agent's behavior when in sale barn for the first time

Pre: Cattle agent is in sale barn for the first time.
Post: Cattle agent is in the stocker.
Algorithm:

```
if time1InSale > 8, move agent to west
else
    increment time1InSale by 1
    call moveInSalebarn
```

Quick Review Question 9

From the text's website, download your agent-based tool's *11_2QRQ.pdf* file for this system-dependent question to write the *inSaleBarn1* method.

To simulate close quarters, agents are moved at random and allowed to stack on top of each other in the sale barn, as the following algorithm indicates:

moveInSalebarn

Procedure for a cattle agent's random movement in a sale barn

Pre: Cattle agent is in sale barn.
Algorithm:

```
if agent is next to SaleBarn, Susceptible, Infected, or Recovered,
    move agent at random on top of that item, respectively
```

Quick Review Question 10

From the text's website, download your agent-based tool's *11_2QRQ.pdf* file for this system-dependent question to write the *moveInSalebarn* method.

While grazing at the stocker, the steer or heifer gains approximately 300 lb over a 4- to 6-month (120- to 180-day) period. Thus, we have the animal gaining a random weight between 0.4 and 0.6 lb per time step (0.25 day).

Similar to its behavior on the farm, the animal moves at random while in the stocker. Upon reaching 900 lb, the agent moves again toward the sale barn, which is to the east in Figure 11.2.1. As with farm transport, we have the animal moving on its own instead of by truck.

Similar to the transition from the farm to the sale barn, as a cattle agent enters the sale barn from the stocker, a state variable, *time2InSale*, with an initial value of a random number between 1 and 5, inclusive, comes into existence. We employ this variable to regulate the length of time in the sale barn.

inStocker

Procedure for a cattle agent's behavior in stocker

Pre: Cattle agent is in stocker.
Post: Cattle agent is in sale barn for the second time.
Algorithm:

```
if weight ≥ 900 and stocker is to east
    move agent to east
if weight ≥ 900 and sale barn is to east
    move agent to east
    set time2InSale to a random integer between 1 and 5, inclusive
else
    move agent to east on stocker
    add a random number between 0.4 and 0.6 to weight
```

Quick Review Question 11

From the text's website, download your agent-based tool's *11_2QRQ.pdf* file for this system-dependent question to write the *inStocker* method.

Feedlots usually place animals in pens. Thus, we do not move the animals at random in the feedlot but line them up as far to the east as possible. Staying at the feedlot for from 4 to 6 months (120 to 180 days), the animal fattens to 1200 to1400 lb, an additional 300 to 500 lb. Thus, we have the animal gaining between 0.5 and 1.0 lb/*dt*. Upon reaching a weight of 1300 lb, the animal is moved to the abattoir.

inFeedlot

Procedure for a cattle agent's behavior in feedlot

Pre: Cattle agent is in a feedlot.
Post: Cattle agent is in an abattoir.
Algorithm:

```
if weight ≥ 1300
    move to east
else if see feedlot to east, north, or south
    move in that direction, respectively
    add a random number between 0.5 and 1.0 to weight
else
    add a random number between 0.5 and 1.0 to weight
```

Quick Review Question 12

From the text's website, download your agent-based tool's *11_2QRQ.pdf* file for this system-dependent question to write the *inFeedlot* method.

Example Problem

With Figure 11.2.1 having the initial configuration, Figure 11.2.2 presents a typical sequence of frames from one simulation in which the disease becomes epi-

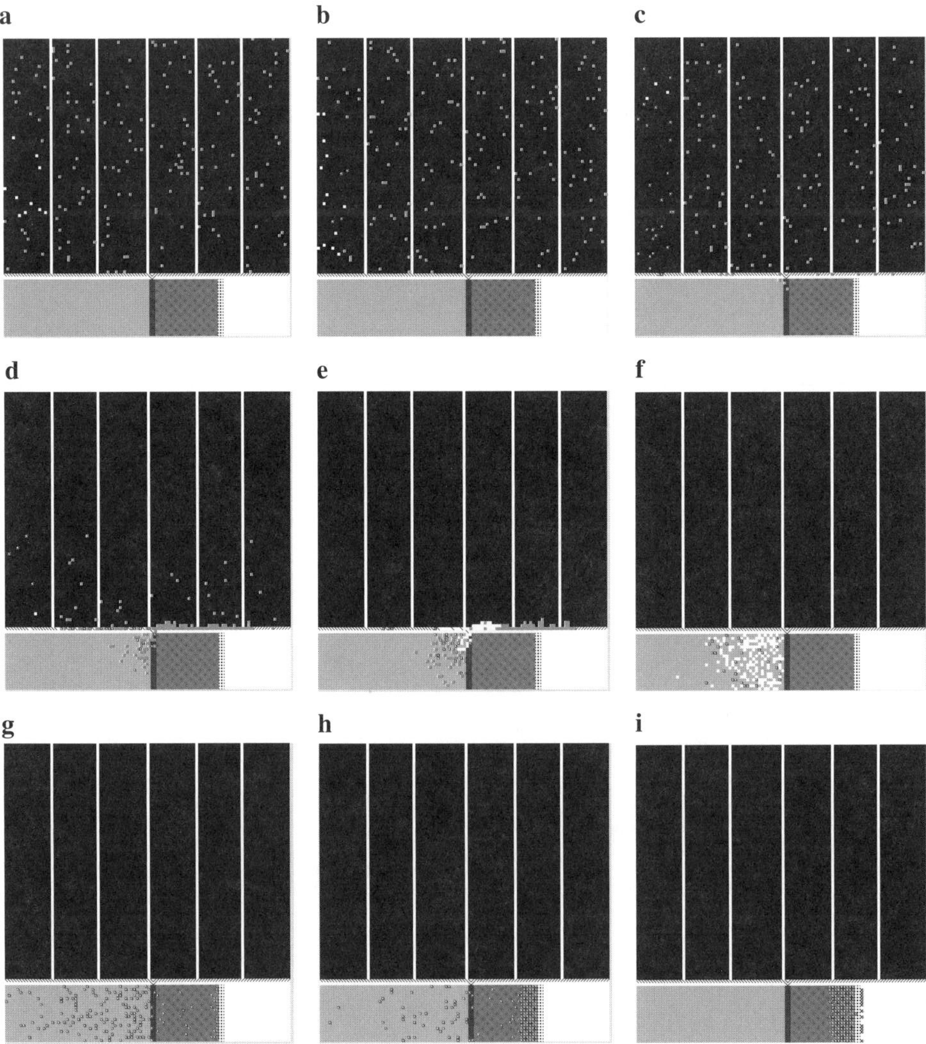

Figure 11.2.2 Result of one simulation resulting in an epidemic

demic in the cattle population. With the smaller, duller-colored dots representing recovered beef cows, Frames a–c show the random movement of cattle and the spread of the disease in one farm. For this simulation, the disease does not die out before at least one infectious animal (white) begins its movement to the sale barn. Packed in close proximity to each other, many susceptible cattle (bright color) become sick during transport and sale (Frame e). Thus, in the stocker (Frame f), all cattle are infected or recovered. In Frames g and h, recovered cattle of the appropriate weight are sold in the sale barn and moved into close quarters in the feedlot. Finally, the bottom right of Frame i shows bovines weighing at least 1300 lb being processed for market.

Repeated Simulations

With a variety of possible values for *INFECTION_PROBABILITY* and *INFECTIOUS_PERIOD* and with weight, movement, and disease spread occurring with elements of chance, one simulation does not capture all possibilities. Thus, for each of several combinations of parameter values, we should run the simulation a number of times, averaging the results. The following section, "Model Refinement," discusses how to automate random creations of *Susceptible* agents and an *Infected* agent on an existing farm configuration so that redrawing is unnecessary.

As with Liu et al. (2012), we performed the simulations 40 times each for a disease that lasts for 40 days (rate of recovery, $\mu = 1/(40 \text{ days}) = 0.025/\text{day}$) and for probabilities of transmission by contact from 0.625 to 0.125/*dt* (rates of infection, β, from 2.5 to 0.5/day, respectively). The **epidemic ratio, β/μ**, is of particular importance. If the rate of infection in the numerator is larger or the rate of recovery in the denominator is smaller, the epidemic ratio is greater and the disease is more virulent. For a fixed recovery rate of 0.025/day and a time step of 0.25 day, we run the simulation for the reciprocal of the epidemic ratio, μ/β, having values ranging from 0.01 to 0.05 (β/μ from 100 to 20, β from 2.5 to 0.5/day, probability of transmission by contact 0.625 to 0.125/*dt*, respectively). With no immunity, three situations arise: The disease affects only a few animals on one farm; the disease reaches epidemic proportions on one farm but does not spread to animals at other locations; or the disease becomes epidemic, eventually infecting all the cattle. Figure 11.2.3 gives histograms of the frequencies of outbreaks for a variety of μ/β values, where the environment has six farms with an average of 24.9 beef cattle each.

Figure 11.2.4 plots the average outbreak sizes (*cummulativeInfected*) versus μ/β in situations where the disease did not spread beyond one farm. Clearly, smaller values of μ/β with its larger rates of infection (β) resulted in more serious outbreaks.

For $\mu/\beta = 0.1$, Figure 11.2.5 plots the number of infected cattle versus time for outbreaks on one farm and for outbreaks that escaped to the whole population. The mixing of cattle in trading situations drives the spread of the disease. For spread of the disease beyond one farm, the disease must persist until trading of animals, so the average length of the disease is an important factor. For these simulations, we have maintained a recovery rate of 1/(40 days) while considering other rates and stochastic situations.

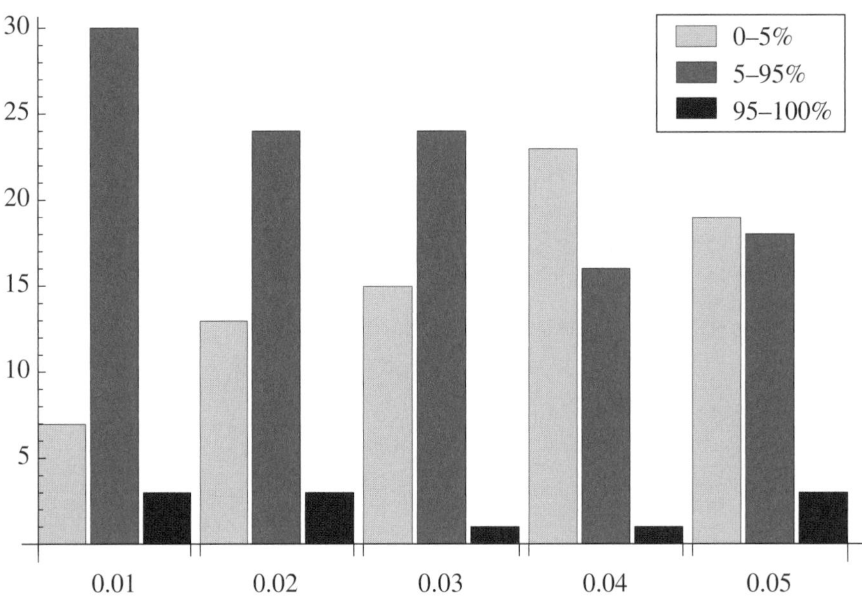

Figure 11.2.3 Frequencies of outbreaks for $\mu/\beta = 0.1, 0.2, 0.3, 0.4, 0.5$ (rate of recovery, $\mu = 0.025$/day; rate of infection, β)

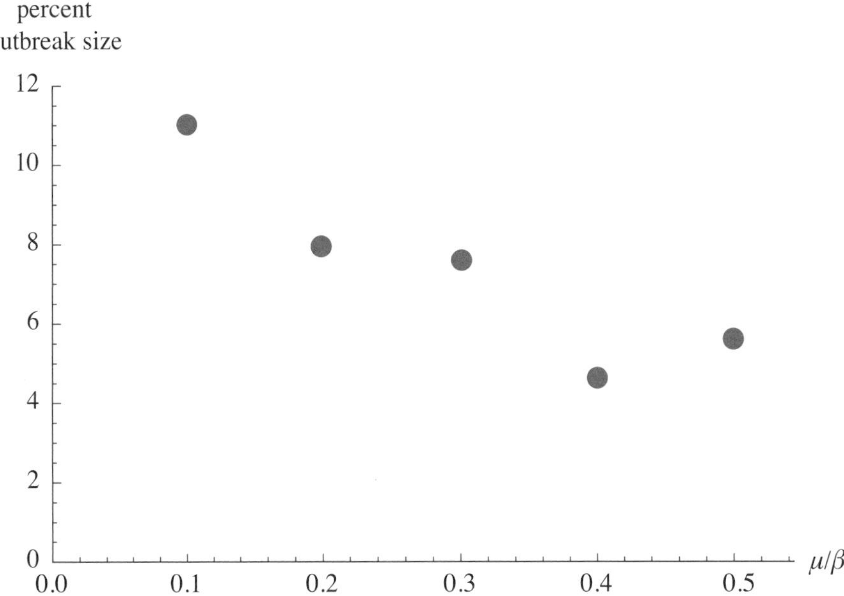

Figure 11.2.4 Average percent outbreak sizes versus μ/β, where the disease does not spread beyond one farm (rate of recovery, $\mu = 0.025$/day; rate of infection, β)

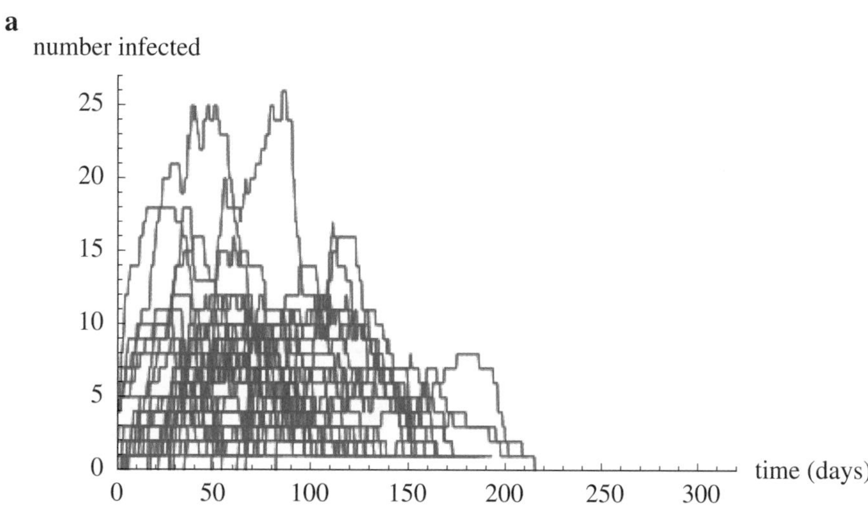

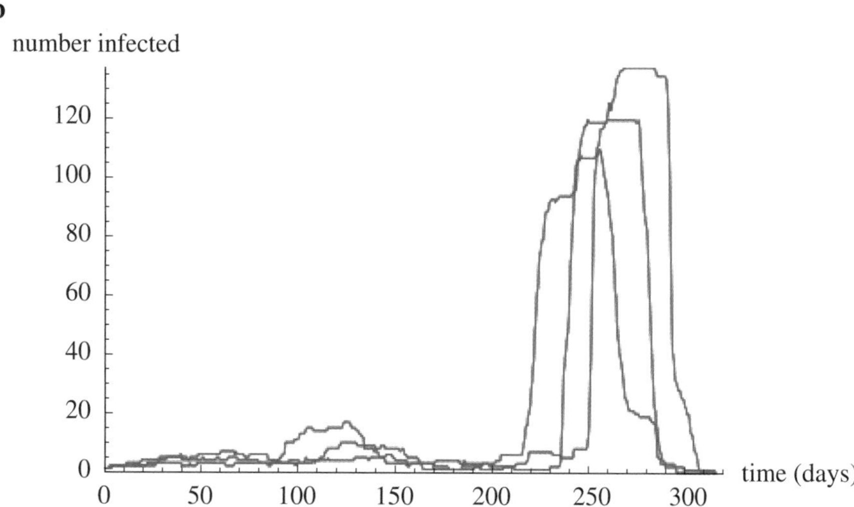

Figure 11.2.5 Number of infected cattle versus time for $\mu/\beta = 0.1$ (rate of recovery, μ = 0.025/day; rate of infection, β) (a) outbreaks on one farm (b) epdimics

Model Refinement

The creation of random initial configurations of cattle for numerous simulations can be inefficient to redraw. Thus, in this section we revise our model so that for a fixed environment, such as Figure 11.2.6, the simulation creates a random configuration of cattle and at random changes one of these to be an infected bovine.

To manage the simulation, we create a new driver to schedule activities for each time step. To implement this driver, some agent-based tools would employ a built-in procedure that makes the simulation go, while others would create a new agent, say,

Figure 11.2.6 Starting environment for version 2 of the simulation

SimulationDriver. For the latter, we might place exactly one *SimulationDriver* agent in a blank space, such as in the bottom-right corner of the environment of Figure 11.2.6. Using three phases, this driver, which executes each time step, initializes the susceptible cattle, changes one of these to an infected bovine, and instructs the cattle to follow their behavior rules.

For this version, initialization of the cattle occurs during the running of the simulation and not in creation of the environment. Thus, we change "Farm Initialization" (see the "Model Environment" section) and "Cattle Scheduler" (see the "Agent Behaviors" section) to be the methods ***randomCattle*** and ***cattleBehave***, respectively.

In Phase 0, the Simulation Driver asks every farm agent to execute its *randomCattle* method, which places *Susceptible* agents at random on the initially empty *Farm* patches. Once the cattle are placed in a pasture, one of the *Susceptible* agents needs to be changed to an *Infected* agent. Thus, after changing to Phase 1 and until one of the *Susceptible* agents becomes an *Infected* agent, the Simulation Driver instructs each cattle agent to execute ***initInfected***. In *initInfected*, if all the cattle are susceptible (i.e., *numSusceptible* equals *numCattle*) with probability 1/*numCattle*, the method changes the *Susceptible* agent being processed to an *Infected* agent and updates *numSusceptible*, *numInfected*, and *cummulativeInfected* appropriately. For example, if the farms initially have *numCattle* = 200 susceptible cattle, then the probability that a particular bovine is the initial infected animal is 1/200 = 0.005 = 0.5%.

After initialization of the cattle, the program moves to Phase 2 and remains in this phase for each subsequent step of the simulation. Because we are mainly interested in observing the progression of the disease and obtaining the cumulative total of infected cattle over repeated executions, we stop the simulation when no animals are sick, that is, when *numInfected* is 0. Otherwise, we instruct each cattle agent to execute its *cattleBehave* method, which specifies the animal's behavior during that time step. Pseudocode for the Simulation Driver, which executes each time step, and *initInfected* follows:

Simulation Driver to Be Executed Each Time Step

Driver for version 2 of the simulation

Algorithm:

```
if phase is 0
    request each farm agent to execute randomCattle
    phase ← 1
else if phase is 1 and numInfected is 0
    request each cattle agent to execute initInfected
else if phase is 1
    phase ← 2
else if phase is 2 and numInfected is 0
    stop simulation
else if phase is 2
    request each cattle agent to execute cattleBehave
```

initInfected

Method to change *Susceptible* agent to *Infected* agent with probability 1/*numCattle*

Pre: The agent is a *Susceptible* or an *Infected* agent.
All cattle agents are susceptible to disease, or one bovine is infected and all others are susceptible.

Algorithm:

```
if entity is a Susceptible agent and numSusceptible equals numCattle, with
  a 1/numCattle chance
    change the entity to be an Infected agent
    decrement numSusceptible by 1
    increment numInfected and cummulativeInfected by 1
```

Quick Review Question 13

From the text's website, download your agent-based tool's *11_2QRQ.pdf* file for this system-dependent question to write the Simulation Driver scheduler method.

Quick Review Question 14

From the text's website, download your agent-based tool's *11_2QRQ.pdf* file for this system-dependent question to write the *initInfected* method.

For analyzing the results of our multiple simulations, we should generate histograms and graphs, such as Figures 11.2.3–11.2.5. If an agent-based tool does not have the capability of producing such figures, after each simulation, we could record appropriate data, such as *numCattle* and *cummulativeInfected*, in a spreadsheet or other computational tool for later processing into histograms, such as Figure 11.2.3, and graphs, such as Figure 11.2.4. One of the projects explores creation of a simulation driver to manage multiple simulations and averaging.

A graph of the number of each category of cattle, particularly the number of infected cattle (*numInfected*), versus time for one simulation can also be instructive. If an agent-based tool does not have the ability to plot the results of multiple simulations on one graph, we can usually export the data to spreadsheets for later generation of more complex graphs, such as Figure 11.2.5.

The projects consider other refinements of the model, such as spreading of the disease through vectors, variation in weight gain, and quarantine.

Exercise

1. Adjust the second version of the simulation *CattleAndDiseaseV2* so that an infected bovine is sick for a random period of time, from 5 to 60 days.

Projects

On the text's website, files CattleAndDiseaseV1 *and* CattleAndDiseaseV2 *for several agent-based tools contain code for the simulations of the module. Complete the following projects using your agent-based tool.*

For additional projects, see Module 14.13, "Re-Solving the Problems with Agent-Based Simulations."

1. Holding the *INFECTION_PROBABILITY* constant at one of the values in the module, vary the *INFECTIOUS_PERIOD* so μ/β values vary from 0.01 to 0.05 in simulation *CattleAndDiseaseV2*. From the text's website, download the data (*CattleAndDiseaseV2Data.xls*) for *INFECTIOUS_PERIOD* = 40 days. Generate figures such as in Figures 11.2.3–11.2.5. Discuss the results.
2. Modify *CattleAndDiseaseV2* so that with a certain probability a farmer isolates an infected beef cow. Run the simulation a number of times and discuss the impact of isolations on the results.

3. Modify *CattleAndDiseaseV2* so that infected calves gain less weight. After recovery, an animal resumes normal weight gain. Compare your results with those of *CattleAndDiseaseV2*.
4. Do the modifications from Projects 2 and 3.
5. Employing the discussion in the section "Introduction," model the progression of pinkeye through a cattle population.
6. Modify *CattleAndDiseaseV2* to automate the process of calculating the mean percent of the final cumulative number of infected cattle for *NUM_SIMS* number of simulations with a fixed set of parameters. Omit graphing.
7. Adjust the first simulation, *CattleAndDiseaseV1*, to have one or more trucks transport the cattle to the sale barn.
8. Pig fever is a serious problem in the swine population of sub-Saharan Africa and Russia. For example, a 2011 outbreak in the Russian Federation resulted in the death of 300,000 of the country's 19 million pigs. This viral disease can be spread in a variety of ways, including infected human food scraps, direct contact, virus particles left on transport vehicles, infected wild boar. No cure or vaccine exists, so mass culls, careful hygiene, and quarantine are primary defenses. Although frequently fatal, the pigs that do survive are immune (Callaway 2012). Model an outbreak of pig fever along with efforts to hinder the spread of the disease.
9. The Classical Swine Fever Virus (CSFV), after 16 years of effort, was eliminated from U.S. pig farms in 1978. However, CSFV is a recurring problem for pig farmers in parts of Europe. The virus is easily transmitted to susceptible animals, usually through direct contact with infected animals. Moreover, the virus also can be spread from contaminated transport vehicles, food, or pens. There are also vectors—such as birds, flies, and human beings—that have been known to transfer the virus. The seriousness of the disease varies, depending on the strain. In the severe, acute form, incubation ranges from 3 to 6 days, and death occurs 10 to 20 days postinfection (APHIS 2008).

   Minimal control measures required by the European Union include enforcing restriction zones and transport regulations, accompanied by culling of affected herds. Animals can be individually vaccinated, but one test vaccine does not provide any protection for the first 7 days. After 7 days, immunity gradually increases over time (Backer et al. 2008).

   Model an outbreak of a severe, acute form of CSFV along with efforts, including vaccination, to hinder the spread of the disease. Indicate the modes of transmission and interventions you are considering. Run the model for a variety of parameters indicating measures that are most successful. Discuss the results.

Answers to Quick Review Questions

From the text's website, download your agent-based tool's *11_2QRQ.pdf* file for answers to the system-dependent questions.

1. **a.** AB **b.** CA **c.** both **d.** both
   e. CA **f.** AB **g.** both

References

APHIS (United States Department of Agriculture Animal and Plant Health Inspection Service, Veterinary Services). 2008. "Classical Swine Fever." Fact Sheet (August).

Backer, Jantien A., Thomas J. Hagenaars, Herman J. W. van Roermund, and Mart C. M. de Jong. 2008. "Modelling the Effectiveness and Risks of Vaccination Strategies to Control Classical Swine Fever Epidemics." *J. R. Soc. Interface*, published online December 3, 2008. http://www.ncbi.nlm.nih.gov/pmc/articles/PMC2838352/ (accessed 11/10/12)

Callaway, Ewen. 2012. "Pig Fever Sweeps across Russia," *Nature* 488 (August 30).

Cattlemen's Beef Board and National Cattlemen's Beef Association. 2009. "Modern Beef Production Fact Sheet."

Davie, Roland. 1997. "Learning Your Way Around Your Local Auction Barn." http://www.luckysnlranch.com/articles/38.html (accessed 07/31/12)

Irsik, Max. 2012. "Pinkeye in Beef Cattle." University of Florida Institute of Food and Agricultural Sciences Extension. Publication #VM141. http://edis.ifas.ufl.edu/vm141 (accessed 08/12/12)

Kirkpatrick, John G., and David Lalman. 2012. "Pinkeye in Cattle Infectious Bovine Keratoconjunctivitis (IBK)." Oklahoma Cooperative Extension Service VTMD-9128.

Liu, Hong, Phillip Schumm, Anton Lyubinin, and Caterina Scoglio. 2012. "Epirur_Cattle: A Spatially Explicit Agent-Based Simulator of Beef Cattle Movements." ICCS 2012, Procedia Computer Science 9 (2012) 857 – 865

Powell, Jeremy. 2004. "Pinkeye (Livestock Health Series)." University of Arkansas Division of Agriculture. FSA3087-PD-1-10RV.

USDA (United States Department of Agriculture). 2011. "Peering at Genes To Detect Origin of Cattle Diseases." http://www.ars.usda.gov/is/AR/archive/sep11/cattle0911.htm (accessed 08/12/12)

MODULE 11.3

Agent-Based Tool: Tutorial 2

Prerequisite: Module 11.1, "Agent-Based Tool: Tutorial 1."

Download

From the textbook's website, download Tutorial 2 for your agent-based tool. We recommend that you work through the tutorial and answer all Quick Review Questions using the corresponding software.

Introduction

This second agent-based tool tutorial, which is available from the textbook's website in your system of choice, prepares you to use the tool to complete projects in Modules 11.4 and 14.13. The tutorial introduces the following functions and concepts:

- Grid inspection and communication
- Color
- Minimum and maximum

The module gives examples and Quick Review Questions for you to complete and execute in the desired software system.

MODULE 11.4

Introducing the Cane Toad—Able Invader

Download

The text's website has the file *CainToads*, which contains a model for this module, and a *11_4QRQ.pdf* file, which contains system-dependent Quick Review Questions and answers, available for download for various agent-based tools.

Introduction

At a time when biologists are concerned about a worldwide decline in amphibian populations, it is ironic that an introduced species of toad is rapidly increasing in abundance in Australia. In the first global assessment studies, scientists found 43% of known amphibian species are in decline (Stuart et al. 2004), while Australian populations of this toad have increased from an original population of 3000 to millions (up to 2000 per hectare; Freeland 1986). And, they are expanding their range at up to 50 km per year into habitats previously thought too restrictive for their survival (Phillips et al. 2006). These toads are voracious predators and nimble competitors. Their large populations have spread widely through several Australian states, threatening native species and disrupting the existing biological communities.

The toad, commonly called the **marine**, or **cane**, **toad** (***Bufo* (*Chaunus*) *marinus***), was introduced into various countries, but in Australia it has become a major concern. Why would anyone introduce such an animal to a country that already had such a rich, unique fauna? To answer that question, we must examine Puerto Rico during the early part of the twentieth century. At that time, sugar cane growers there were desperately seeking something to control beetle grubs (larvae) that were destroying the roots of their crops. In response, the U.S. Department of Agriculture imported cane toads from Barbados. Within 10 years, the beetle grub numbers were reduced to the level of a mere nuisance. This relatively rare example of a positive outcome from introducing species to new geography encouraged other cane-growing regions to mimic this "successful" strategy. Cane toads were introduced to Australia in 1935 to fight the damage done to cane crops by gray-backed beetles (*Dermolepida albohirtum*) and Frenchi beetles (*Lepidiota frenchi*) (Freeland and Martin 1985; Alford et al. 1995). However, the toads proved to be ill chosen as a cure:

- The adult beetles attack the top of the cane instead of the roots, and toads do not fly.
- Beetle grubs are active during the day, while toads are active at night.
- The toads do not like the hot cane fields, where there is a high danger of **desiccation**, or drying.
- The amphibian had too many other tasty prey alternatives.

Sadly, cane toad numbers continue to increase, and models predict that they will eventually occupy twice the 1 million km^2 of Australia they presently do (Urban et al. 2007).

In Australia, female cane toads are prodigious producers of eggs (8000–35,000 eggs/clutch; Hero and Stoneham 2009), laying eggs once or twice each year, beginning during their second wet season (Cohen and Alford 1993). These eggs are produced in long gelatinous strings attached to shallow vegetation. After approximately 48 h, tadpoles emerge and initially feed on algae (Hinkley 1962). After 37 to 40 days, metamorphosis into toadlets normally occurs. This time is variable, however, dependent on various climatic factors, competition, and predation (from previous tadpole cohorts). Growth rates are strongly density dependent, with higher growth rates and maturation at lower densities (Alford et al. 1995). Because they are small and poorly developed, young toadlets must initially stay near water to prevent desiccation (Cohen and Alford 1993). As they age and mature, young toads move farther from the water, but the first dry season takes its toll with only 10% to 47% surviving (Alford et al.). Freeland and Martin (1985) found that young toads are the primary colonizers, with dispersion occurring at the edges of the toads' distribution.

Juvenile and mature toads are active at night, feeding on insects attracted to lighted areas (Wright and Wright 1949). They can often be found in gardens, around houses, and in other disturbed areas (Krakauer 1968). Cane toads are relatively aggressive and somewhat undiscerning predators. Although favoring certain beetles, they will also occasionally dine on ants, crabs, spiders and other arthropods (Krakauer; Easteal 1982). Various researchers have observed cane toads feeding on snakes, birds, small mammals, and other amphibians (Rabor 1952; Krakauer; Oliver 1949; Bartlett and Bartlett 1999). Researchers are constantly investigating the actual impacts the toads may have on competitor, prey, and predator populations.

Because these animals seem to fancy relatively open, disturbed areas associated with human activity, residents in many areas of eastern and northern Australia are quite likely to encounter them. Imagine yourself and your family living in a quiet suburb near Cairns, Queensland. You might have a pet dog or cat you feed every evening on the back patio. If your pet does not eat all the food, an opportunistic cane toad might finish off the remains. If your dog happens upon and mouths a toad, it would be in for a nasty surprise; and you might find your pet drooling profusely or foaming at the mouth, staggering, twitching, seizing, and/or vomiting. Your dog would be exposed to one of the toad's major defenses against predation—toxic secretions.

The members of the genus *Bufo*, including the cane toad, are somewhat notorious for their capacity and propensity to produce bufadienolides and other toxic chemicals that they secrete in quantity from prominent, specialized parotoid glands on their heads. Cane toads release these secretions defensively when trifled with by curious or potentially predaceous animals. The bufadienolides are cardiotoxic steroids that act much like digitalis, interfering with membrane sodium-potassium pumps, thereby increasing the contraction of the heart (Halliday et al. 2009). The mouth lining rapidly absorbs the toxins, so it is little wonder your dog soon regrets the folly of its appetite or curiosity. For heart patients, digitalis may restore normal heart muscle function, but in large-enough quantities, cane toad toxin may lead to cardiac failure. On the other hand, *Bufo* toxin (*Chansu*) has been used in China since the seventh century (Xiao 2002) and is still used in Chinese traditional medicine as an analgesic and to treat heart failure (Ma et al. 2007). Additionally, toad toxin preparations are used for certain types of cancer (Meng et al. 2009). It is little wonder that enterprising Australians are exporting *Bufo marinus* to the Chinese (BBC News 2010).

The Problem

Invasive species do not often expand along a continuous front, but satellite groups establish themselves at **invasion hubs**, such as advantageous habitat areas. In the vast parts of arid Australia during the dry season, **artificial water points** (**AWPs**), such as troughs or dams for livestock, can serve as invasion hubs for cane toads. Thus, restricting the toads' access to AWPs might help prevent the spread of this invasive species. To study this hypothesis, scientists erected toad-proof fences around AWPs in an experimental zone, removing toads that were already in the AWPs and excluding others from these water sources. They also performed simulations to model the potential dispersal ability of toads under various climate conditions with and without AWPs (Florance et al. 2011). In this module, we are interested in developing a similar, but simplified, model to study the effect of fencing AWPs on adult cane toad invasion. Ignoring climate, topography, periods of sleep, moist areas (except around AWPs), and the cane toad life cycle for the basic model, we wish to examine the impact of fenced and unfenced AWPs on the migration of cane toads through an area.

Grid-Based Individual-Based Model

For our simulation, a **grid-based individual-based model** seems appropriate. An **individual-based model** (**agent-based model,** or **entity-based model**) follows individuals of a population in an **environment**, considering the global consequences of their local interactions. Individuals are described in terms of their **behaviors** (**rules,** or **transition rules**) and their **state** (or set of characteristic parameters). With a **grid-based model**, the environment consists of one or more **grids**, or rectangular arrangements of **cells** (or **sites**), and an individual moves discretely from one cell to another instead of continuously to any point (Reynolds 1999).

Grid-based individual-based simulations are related to **cellular automaton simulations**, which are also dynamic computational models, or models that change with time and that are discrete in space, state, and time. Cellular automaton grids represent environments, and rules regulate the behavior of the system by specifying local relationships and indicating how cells change states. Recall from our discussion in Module 11.2 ("Agents of Interaction: Steering a Dangerous Course"), the prime distinction between a grid-based individual-based model and a cellular automaton model is that a simulation for the former loops through all *individuals* one at a time, while a simulation for the latter loops through all *grid elements* one at a time.

Model of Environment

For our grid-based individual-based simulation of the movement of cane toads through an area with AWPs, the environment consists of an $m \times n$ grid of cells, which are desert agents. Each such agent stores wetness (***moisture***) and nutrient (***food***) values between 0.0 and 1.0, representing characteristics of that location.

We can assume a fixed or random amount of nutrition at each desert agent. For example, we may initialize all food values to be *FOOD_CELL* = 0.05.

Throughout the grid, we have a low, constant value, such as 0.0 for moisture, except at or near AWPs. AWPs have values of 1.0 (*AMT_AWP*). Cells immediately adjoining such water areas, whether fenced or unfenced, have lower positive moisture values, such as 0.4 (*AMT_AWP_ADJACENT*), while those two locations away have even lower positive values, such as 0.2 (*AMT_AWP_OVER2*). Figure 11.4.1 displays part of a grid's *moisture* values and a visualization of the corresponding area around an AWP.

To prevent toads from going off the grid in certain directions or traveling to fenced AWPs, we give moisture and food values of –1 to cells on those directional boundaries and in prohibited areas, and we define toad behaviors so that the animal never moves to a cell that has negative values. Thus, if we want to allow toads to migrate out of the environment only to the west, we assign –1 to all *moisture* and *food* values of cells in the first and last rows (north and south borders, respectively) and last column (east border). By making such a restriction, we require toads starting on the east border to migrate the entire width of the desert area, mimicking much of the migration that is occurring in Australia. Consequently, survival calculations

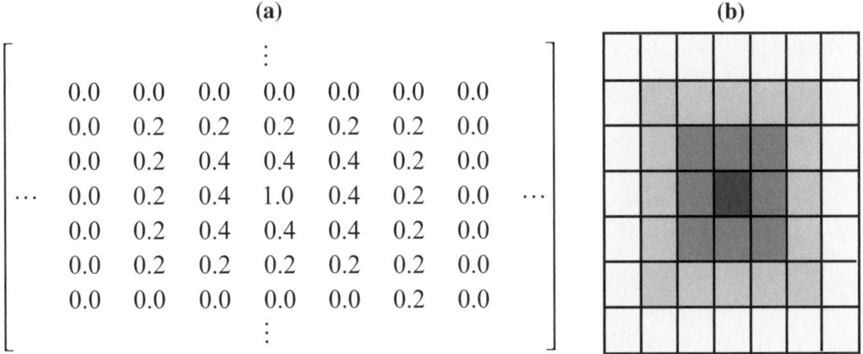

Figure 11.4.1 (a) Part of a grid's *moisture* values and (b) visualization of area around an AWP

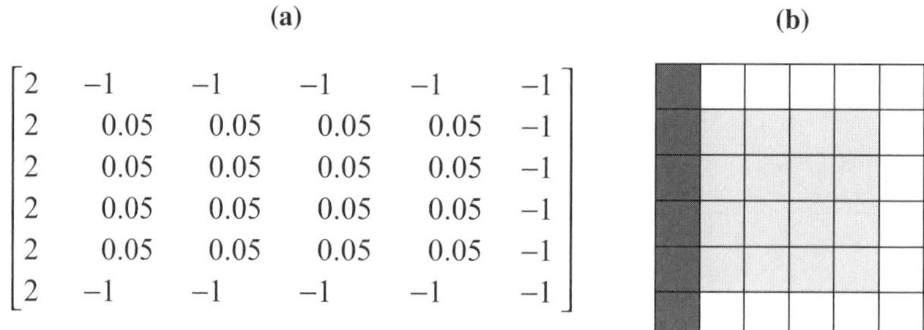

$$\begin{bmatrix} 2 & -1 & -1 & -1 & -1 & -1 \\ 2 & 0.05 & 0.05 & 0.05 & 0.05 & -1 \\ 2 & 0.05 & 0.05 & 0.05 & 0.05 & -1 \\ 2 & 0.05 & 0.05 & 0.05 & 0.05 & -1 \\ 2 & 0.05 & 0.05 & 0.05 & 0.05 & -1 \\ 2 & -1 & -1 & -1 & -1 & -1 \end{bmatrix}$$

Figure 11.4.2 Example of (a) a grid's initial *food* values with (b) visualization

would tend to be more accurate. By parallel reasoning, to attract toads in cells adjacent to the west border, we assign a value greater than 1, say 2, to both moisture and food in the first column. Figure 11.4.2 presents a 6 × 6 grid's *food* values and a corresponding visualization with darker shades of gray representing larger values.

Agents

Figure 11.4.3 depicts a desert area (***Desert*** agents) in light gray with 14 unfenced AWPs (***Awp***) in black and 4 fenced AWPs (***FencedAwp***), each with a small black dot surrounded by white. The AWPs and fenced AWPs are surrounded by ***AwpAdjacent*** grid agents (grey), and in turn these are surrounded by ***AwpOver2*** agents (hatched). Thirty-five (35) ***Toad*** agents in color are on ***StartBorder*** patches on the far-right (east) column. The top and bottom rows are ***Border*** cells, while the far-left (west) column has ***FinishBorder*** agents. A ***SimulationDriver***, which is useful for some agent-based tools, appears in the top left corner of the environment.

Toad's State

For each toad, we store certain state variables, or characteristic parameters, or attributes, such as the following, which represent the toad's state at that instant:

energy—value from 0.0 to 1.0 indicating toad's amount of energy from low to high
water—value from 0.0 to 1.0 indicating toad's amount of water from low to high
availableFood—*food* value of the cell on which the toad is located
availableMoisture—*moisture* value of the cell on which the toad is located

To facilitate detection that a cell is an undesirable location because of occupation by another toad, each toad also has associated variables ***food*** and ***moisture***, with values of –1 that do not change throughout the simulation. For example, suppose the cell to the north of toad *A* has a toad, *B*. Because the *food* and *water* values to *A*'s north, which are *B*'s values, are both –1, indicating an inhospitable location, *A* will not hop to that cell until the location is vacant.

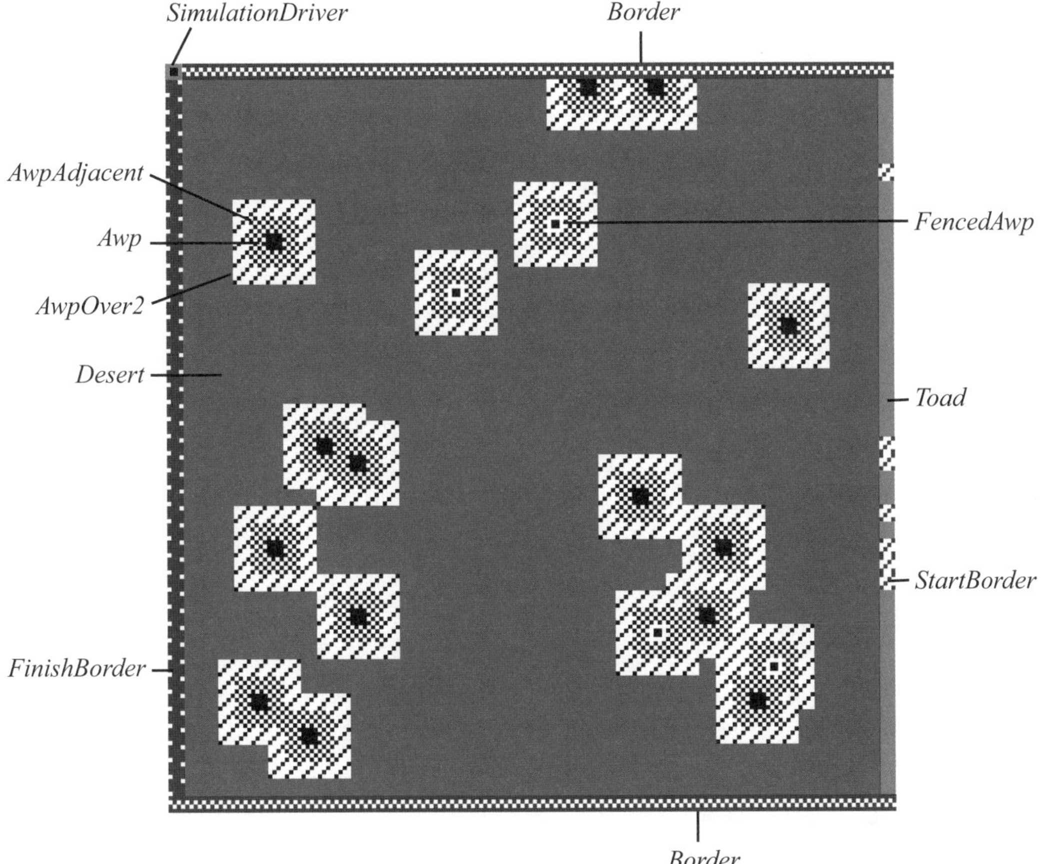

Figure 11.4.3 Visualization with AWPs (fenced and unfenced), borders toads, and simulation driver

Toad Behavior

A toad's behavior is regulated by its own state, particularly its amount of water and energy, and the moisture and food conditions around it. For a current location (cell) of a toad on the grid, our simulation might use the site's **von Neumann neighborhood**, which includes the four nearest neighbors to the north, east, south, and west (see Figure 10.2.3a), or might employ the **Moore neighborhood** of all eight neighbors (see Figure 10.2.3b).

At times, the model of a toad's behavior involves a random walk, which refers to the animal's apparent random movement (Module 9.5, "Random Walk"). With this simulation technique, perhaps under certain constraints, at any time step an agent can move at random to a neighboring cell. Relevant **toad behavior rules**, which include aspects of a random walk in the movement rules, are as follows:

1. If the toad's energy value is such that it would like to eat (*energy* value below *WOULD_LIKE_EAT*) and there is food at the site, it eats.

2. A toad gains energy (***amtEat***) eating. *amtEat* is at most *AMT_EAT* but no more than the amount of food in the toad's *food* site; moreover, the toad's energy value cannot exceed 1.0.
3. A toad gains some water (*FRACTION_WATER* of *amtEat*) eating. However, the toad's water value cannot exceed 1.0.
4. If the toad is in water and would like to drink, it does.
5. A toad gains water (*AMT_DRINK*) drinking. However, the toad's water value cannot exceed 1.0.
6. If the toad is thirsty (*water* value below *WOULD_LIKE_DRINK*) but is not in water, it hops to the nearest neighboring site with no other toad and the most moisture.
7. Else if the toad is hungry (*energy* value below *WOULD_LIKE_EAT*), it hops to the nearest neighboring site with no other toad and the most food.
8. Else with probability *MAY_HOP*, the toad hops to a neighboring site.
9. A toad uses energy (*ENERGY_HOPPING*) hopping.
10. While a toad uses less energy (50% of *ENERGY_HOPPING*) sitting.
11. A toad uses water (*WATER_HOPPING*) hopping into a dry area.
12. While a toad uses less water (50% of *WATER_HOPPING*) sitting in a dry area.
13. A toad that crosses the west border is considered migrated and is removed from the simulation.
14. If a toad's water value falls below *DESICCATE* or food value falls below *STARVE*, the toad dies.

Constants and Global Simulation Variables

The simulation employs a number of constants shown in Table 11.4.1. Some constants, such as *DESICCATE* and *FRACTION_WATER*, are from the literature, although death from desiccation may occur at slightly different percentages and the fraction of prey that is water varies, depending on the source. Many of the values from Table 11.4.1 can be adjusted to explore alternative conditions.

Besides constants, the simulation has several global simulation variables, which Table 11.4.2 lists. Each of these is initialized to be zero.

Initial Environment

To facilitate creation of random environments and random toads, the simulation begins with an area that contains a 42 × 42 desert, borders, and for some agent-based tool, a simulation driver, but no AWPs or toads (see Figure 11.4.5). Thus, to represent a square field that is 210 m long and 210 m wide, each cell corresponds to a $(210 \times 210)/(42 \times 42) = 25\text{-m}^2$ area. Each *Desert* agent has a state variable of *food* with an initial value of *FOOD_CELL*. *StartBorder* and *Border* agents have inhospitable *food* and *moisture* values of –1, while *FinishBorder* agents have attractive such values of 2.

Table 11.4.1
Cane Toad Simulation Constants

Constant	*Value*	*Meaning*
AMT_AWP	1	*moisture* value for water, such as an AWP
AMT_AWP_ADJACENT	0.4	*moisture* value of neighboring cell to water
AMT_AWP_OVER2	0.2	*moisture* value of cell 2 cells away from water
AMT_DRINK	0.05	maximum amount toad drinks in 1 time step
AMT_EAT	0.01	maximum amount toad eats in 1 time step
AMT_MIN_INIT	0.88	minimum initial toad *energy* and *water* values
DESICCATE	0.6	level at which desiccation occurs
ENERGY_HOPPING	0.002	maximum energy used by toad in a hop
FOOD_CELL	0.05	food value for initializing constant *food* grid
FRACTION_WATER	0.6	fraction of prey that is water
INIT_PERCENT_TOADS	80(%)	percent chance a *StartBorder* agent forms a toad
INIT_RANGE	0.12	range of initial toad *energy* and *water* values
MAY_HOP	0.5	probability of hopping if not thirsty or hungry
PERCENT_AWPS	1.0(%)	percent chance a desert cell has an AWP
PERCENT_AWPS_FENCED	0-100(%)	percent chance an AWP is fenced
STARVE	0.6	level at which starvation occurs
WATER_HOPPING	0.002	maximum water used by toad in a hop
WOULD_LIKE_DRINK	0.9	water level at which toad would like to drink
WOULD_LIKE_EAT	0.9	food level at which toad would like to eat

Table 11.4.2
Global Simulation Variables

Variable	*Meaning*
numAlive	number of live toads in grid area
numCroaked	number of dead toads
numMigrated	number of toads that have migrated off the grid
phase	simulation phase

Figure 11.4.4 Visualization of initial environment

Simulation Driver

A simulation driver or a built-in procedure that makes the simulation go repeatedly guides the simulation through three phases: (1) toad consumption, (2) toad movement, and (3) removal, or cleanup. We say that one sequence involving Phases 1, 2, and 3 constitutes a **consumption-movement-removal cycle**, or a **cycle**. The phase is stored in a global simulation variable, ***phase***, that is 0 initially, to indicate the simulation (toads and environment) needs to be initialized. The simulation progresses through the phases, as follows:

Simulation Driver to Be Executed Each Time Step

Algorithm:

if *phase* is 0, perform initialization: Place toads at random on the *StartBorder* cells. Second, the desert gains random AWPs and fenced AWPs. Then, set *phase* to 1. (*Note:* The simulation never returns to Phase 0.)
else if *phase* is 1, perform consumption: All toads eat and drink; *phase* updates to 2.
else if *phase* is 2, perform movement: All toads move. After each toad has moved or decided to remain in its current location, *phase* changes to 3.
else if *phase* is 3, complete the cycle: All dead and migrated toads are removed from the simulation. The simulation may terminate because no toads remain. Otherwise, cycling back, *phase* changes to 1.

Phase 0: Initialization

In Phase 0, the simulation driver asks each *StartBorder* agent to execute its ***createToads*** method to place toads on the east border. After formation of the toads, the simulation driver directs the *Desert* agents to complete the landscape with unfenced and fenced AWPs, and then the driver changes *phase* to be 1.

Phase 0 of Simulation Driver

Initialization phase

Pre: A grid exists with *StartBorder* cells on the east, *FinishBorder* cells on the west, *Border* cells to the north and south, and *Desert* cells in the middle.
Global simulation constants (Table 11.4.1) have been initialized.
phase is 0.

Post: A random number of toads are on the east border.
The desert landscape has been initialized with fenced and unfenced AWPs.
phase is 1.

> ***Algorithm:***
> request each *StartBorder* agent to execute *createToads*
> request each desert agent to execute *placeAwps*, *placeFencedAwps*, *initAwp*, and *initAwp2*
> *phase* ← 1

This version of the model requests that each *StartBorder* patch generate a toad on top of it with *INIT_PERCENT_TOADS* percent chance. Thus, if global variable *INIT_PERCENT_TOADS* is 80, there is a 0.80 probability that a *StartBorder* agent will create a new *Toad* agent; and, after initialization, toads will cover approximately 80% of the starting border, here on the east.

Quick Review Question 1

From the text's website, download your agent-based tool's *11_4QRQ.pdf* file for this system-dependent question to write the method *createToads*.

Toad initialization establishes the toad's *energy* and *water* to be random floating-point numbers in a certain range and its available food and moisture to be –1. Moreover, each new toad increments the global *numAlive* by 1, so that this simulation variable maintains a running total of the number of live toads.

> **Toad Initialization**
>
> Procedure to initialize a new toad with random energy and water
>
> ***Pre:*** *AMT_MIN_INIT* and *INIT_RANGE* are global variables indicating the minimum energy/water amounts and length of interval of values, respectively.
> ***Algorithm:***
> increment *numAlive* by 1
> *energy* ← random floating-point number between *AMT_MIN_INIT* and *AMT_MIN_INIT* + *INIT_RANGE*
> *water* ← random floating-point number between *AMT_MIN_INIT* and *AMT_MIN_INIT* + *INIT_RANGE*
> *availableFood*, *availableMoisture*, *food*, and *water* ← –1

Quick Review Question 2

From the text's website, download your agent-based tool's *11_4QRQ.pdf* file for this system-dependent question to write the *Toad* initialization method.

After toads are generated on the east border, desert features—AWPs and fenced AWPs—are created with various *Desert* procedures. The method ***placeAwps*** changes approximately *PERCENT_AWPS* percent of the *Desert* agents to *Awp* agents; and then, ***placeFencedAwps*** "fences" about *PERCENT_AWPS_FENCED* percent of these AWPs. To initialize the areas around AWPs to have the gradations moisture, which attract thirsty toads, we have methods ***initAwp1*** and ***initAwp2***. First, *initAwp1* changes any *Desert* agent next to a fenced or unfenced AWP to be an *AwpAdjacent* agent. A subsequent call to *initAwp2* converts *Desert* agents adjacent to *AwpAdjacent* agents to be *AwpOver2* agents. Thus, an inner ring of *AwpAdjacent*s and an outer ring of *AwpOver2*s surround each AWP. Because only *Desert* agents change, the simulation does allow AWPs to be next to or one cell away from one another.

Quick Review Question 3

Suppose *PERCENT_AWPS* is 0.3 (i.e., representing 0.003%), *PERCENT_AWPS_FENCED* is 25, and the grid is 100 × 40 cells. On the average, after the initialization phase, how many of the following would we expect on the grid?

a. *Awp* agents immediately after the call to *placeAwps*
b. *FencedAwp* and *Awp* agents immediately after the call to *placeFencedAwps*
c. *AwpAdjacent* agents if there are 5 *Awp* and 2 *FencedAwp* agents, none of which are within 3 cells of a border or each other

Quick Review Question 4

From the text's website, download your agent-based tool's *11_4QRQ.pdf* file for this system-dependent question for completion of the landscape.

Phase 1: Consumption

Phase 1 is the consumption stage of a cycle of the simulation. The simulation driver directs the toads to consume food and water as needed and instructs the grid to update its food values after such consumption. For simplicity, we assume that the amount of water does not change at any location on the grid. Upon completion of this phase, which occurs in one time step, the simulation advances to Phase 2. The behaviors of the simulation driver, toad, and desert agents in Phase 1 follow.

Phase 1 of Simulation Driver

Consumption phase of the simulation driver

Pre: The desert landscape has been initialized with AWPs and toads, and *phase* is 1.
Post: Toads have had the opportunity to eat and drink, and *phase* is 2.
Algorithm:

```
request each Toad agent to execute toadMayEat and toadMayDrink
request each desert agent to execute updateFood
phase ← 2
```

Method ***toadMayEat***, which follows, implements Rule 1 of toad behavior, which states a toad that would like to eat, does. For a hungry toad, we call method ***eat***, which updates the toad's state, including toad attribute *amtEat*. Otherwise, we set *amtEat* to 0.

toadMayEat

Toad behavior regarding eating

Pre: The agent is a toad, and the *phase* is 1.
Post: The toad may have eaten. Its *amtEat* state variable has an updated value.
Algorithm:

```
if the toad is in the desert and its energy is less than WOULD_LIKE_EAT
    request this toad to execute eat
else
    set the toad's amtEat to 0
```

AMT_EAT is the most that an animal consumes in one time step. However, the toad cannot consume more food than is available in its current location. Moreover, its energy level (*energy*) should not exceed 1, so the additional food cannot be more than 1 – *energy*. Thus, the toad eats *amtEat*, which is the minimum of *AMT_EAT*, the *food* available in the toad's desert cell (*availableFood*), and 1 – *energy* (Rule 2). This value is added to the energy value of the toad and eventually subtracted from the value of *food* in the corresponding desert cell. Food, such as a beetle, contains water, too; and we assume toad prey averages *FRACTION_WATER* = 0.60 = 60% water. Thus, after eating, we add *FRACTION_WATER* * *amtEat* to a toad's water quantity. However, we must again be careful that the sum does not exceed 1.0, so we change *water* by the minimum of *water* + *FRACTION_WATER* * *amtEat* and 1.0 (Rule 3). The algorithm for *eat* follows:

eat

Function to update a toad's energy and water after it eats

Pre: The agent is a toad, and *AMT_EAT* and *FRACTION_WATER* are global variables.
Post: The toad's energy and water levels have been adjusted after eating.
Algorithm:

```
amtEat ← minimum of AMT_EAT, availableFood, and 1 - energy
energy ← energy + amtEat
water ← minimum of (water + FRACTION_WATER * amtEat) and 1.0
```

A corresponding *Desert* method, ***updateFood***, reduces the amount of food in a grid cell under a toad to reflect the amount the toad has eaten from that location.

Quick Review Question 5

Suppose *AMT_EAT* = 0.01 and *FRACTION_WATER* = 0.6. Assume a toad is on top of a desert cell. Give the values of a toad's *energy* and *water* and a desert cell's *food* after execution of *eat* and *updateFood* for each of the following situations:

a. *energy* = 0.9, *water* = 0.8, and *availableFood* = 0.03
b. *energy* = 0.9, *water* = 0.8, and *availableFood* = 0.005
c. *energy* = 0.999, *water* = 0.8, and *availableFood* = 0.03
d. *energy* = 0.9, *water* = 0.999, and *availableFood* = 0.03

Method ***toadMayDrink***, which implements Rule 4 concerning a thirsty toad in water, is similar to *toadMayEat*. Because we assume the amount a toad drinks from a water source is negligible relative to the water source, the algorithm for ***drink*** is simpler than that for *eat*. Thus, in one time step, a drinking toad adds no more than *AMT_DRINK* to its internal water amount, being careful that the total does not exceed 1.0 (Rule 5). Recall our simplifying assumption that water, if present in a cell, is continuously available in that cell.

Quick Review Question 6

From the text's website, download your agent-based tool's *11_4QRQ.pdf* file for this system-dependent question on consumption methods.

Phase 2: Movement

After each toad has had the opportunity to consume a certain amount of the available food and water on one step, the simulation proceeds to Phase 2, where the toads may move. The following algorithm indicates the simulation driver's actions during Phase 2:

Phase 2 of Simulation Driver

Movement phase of the simulation driver

Pre: *phase* is 2.
Post: All toads have had the opportunity to move, and *phase* is 3.
Algorithm:
  request each toad agent to execute *toadMove*
  phase ← 3

The main driver for toad movement is the method ***toadMove***. We assume that a toad's highest priority is to satisfy its thirst and secondarily its hunger. In the case of thirst, the toad moves toward the nearest unoccupied neighbor with the most moisture (Rule 6). Otherwise, if hungry, the amphibian surveys its surroundings and heads to an empty neighborhood location with the most food (Rule 7). If neither thirsty nor hungry, the animal might hop to a neighboring cell or remain in its current location (Rule 8). The algorithm for *toadMove* and the methods this movement procedure invokes follow.

toadMove

Possibly have the toad move

Pre: The agent is a toad, and *phase* is 2.
Post: The toad has moved or decided to remain in its current location.
Algorithm:
 if *water* < *WOULD_LIKE_DRINK*
  call the toad's *thirsty* method
 else if *energy* < *WOULD_LIKE_EAT*
  call the toad's *lookForFood* method
 else if a random number is less than *MAY_HOP*
  call the toad's *hopForFun* method
 else call the toad's *stayHere* method

Quick Review Question 7

From the text's website, download your agent-based tool's *11_4QRQ.pdf* file for this system-dependent question on *toadMove*.

If after possibly eating and/or drinking, the toad is still thirsty with water value less than *WOULD_LIKE_DRINK*, we call the function ***thirsty***, whose algorithm follows, to process the animal's next behavior. If in water, the cane toad does not change locations, executing method ***stayHere***; while if above another desert agent, *thirsty* calls ***lookForMoisture*** to move the toad to a neighboring vacant cell with the most moisture. In the case where more than one neighbor has the maximum moisture, we choose at random one of those with the maximum value. When the toad is on a *StartBorder* agent, whose *moisture* and *food* values are –1, the animal moves to the left with method ***moveW*** as long as no other toad already occupies that location. If such movement is not possible, *thirsty* calls ***stayHere***.

thirsty

Function to change the position of a very thirsty toad

Pre: The agent is a toad.
Algorithm:
 if the toad is in an AWP
  stayHere
 else if the toad is above another desert agent
  lookForMoisture
 else if toad is above a *StartBorder* agent and a desert agent without a toad is to west
  moveW
 else if the toad is above a *StartBorder* agent
  stayHere

Both *moveW* and *stayHere* adjust the toad's *availableFood* and *availableMoisture*. Because the toad exerts energy and possibly uses water in hopping to an adjacent cell, *moveW* calls ***useWaterEnergyHopping***. This procedure decreases the toad's *energy* by amount *ENERGY_HOPPING* (Rule 9); and, if the amphibian is not in water, the method decrements the toad's *water* by amount *WATER_HOPPING* (Rule 11). Because a toad uses 50% less water and energy by sitting, *stayHere* calls ***useWaterEnergySitting*** (Rules 10 and 12). In the case of diminishing water or energy, the toad will die before the danger of *water* or *energy* becoming negative.

moveW

Procedure to move toad west and update its state variables

Pre: The agent is a toad and cell to the west is unoccupied.
Post: The toad was moved west, and its state variables have been updated.
Algorithm:
move toad to west one cell
availableFood ← cell's *food* value
availableMoisture ← cell's *moisture* value
call toad's *useWaterEnergyHopping* method

Quick Review Question 8

From the text's website, download your agent-based tool's *11_4QRQ.pdf* file for this system-dependent question related to movement for moisture.

Even if not needing to drink, a toad may be hungry, in which case, we call *lookForFood* (Rule 7). Similar to *lookForMoisture*, a hungry toad on the start border attempts to move to the left and one elsewhere moves to a random vacant neighbor that has the most food.

Quick Review Question 9

From the text's website, download your agent-based tool's *11_4QRQ.pdf* file for this system-dependent question related to movement for food.

If not thirsty or hungry, this amphibian may still move, which ***hopForFun*** processes (Rule 5). In *hopForFun*, a toad on the starting boundary moves directly west, provided that cell is unoccupied. If elsewhere, at random, the amphibian moves on top of a neighboring vacant *Desert* agent, calling ***hopHere*** to establish its new *availableFood* and *availableMoisture* values and calling *useWaterEnergyHopping*.

Quick Review Question 10

From the text's website, download your agent-based tool's *11_4QRQ.pdf* file for this system-dependent question related to hopping for fun.

hopForFun

Function to update a toad's location to hop in a random "legal" direction if possible

Pre: The agent is a toad.
Post: The toad may have moved at random on top of a vacant *Desert* neighbor.
Algorithm:
if the toad is above a *StartBorder* agent and can move to the left
moveW
else if the toad is above a *StartBorder* agent or has no empty *Desert* neighbors
stayHere
else
move on top of a random empty *Desert* neighbor
hopHere

Phase 3: Complete Cycle

After directing toads to eat, drink, and move in Phases 1 and 2, the simulation driver eliminates migrated, desiccated, and starved toads from the simulation in Phase 3 (Rules 13 and 14). (That is, violators will be "toad"!) Then, the driver checks if the simulation should be terminated because no more toads are alive on the grid. If toads remain, the simulation driver changes *phase* to 1, so that on the next time step the process can loop back to Phase 1 and a new cycle of eating, drinking, moving, and updating counts, such as the number of live toads, can begin. The algorithm for Phase 3 follows:

Phase 3 of Simulation Driver

Removal phase of the simulation driver

Pre: *phase* is 3.
Post: Migrated, desiccated, and starved toads are eliminated. If the simulation continues, *phase* is 1.
Algorithm:
request each *Toad* agent to execute its *changeCounts* method
checkTerminate
phase ← 1

The simulation driver has each toad execute its ***changeCounts*** method to eliminate migrated, desiccated, and starved toads and adjust global simulation variables *numAlive*, *numCroaked*, and *numMigrated*, as necessary. By calling ***checkTerminate***, the driver stops the simulation if *numAlive* is zero.

changeCounts

Method to eliminate a toad that should be dead or migrated

Pre: The agent is a *Toad* agent.
DESICCATE and *STARVE* are global constants.
numAlive are the number of live toads.
numCroaked are the number of toads that have died.
numMigrated are the number of toads that have migrated.

Post: If the toad has desiccated or starved, the agent has been erased, *numAlive* has been decremented by 1, and *numCroaked* has been incremented by 1.
If the toad has migrated off the grid to the west, the agent has been erased, *numAlive* has been decremented by 1, and *numMigrated* has been incremented by 1.

Algorithm:

if *water* < *DESICCATE* or *energy* < *STARVE*
    erase toad agent
    numAlive ← *numAlive* – 1
    numCroaked ← *numCroaked* + 1
else if the toad is above a *FinishBorder* agent
    erase toad agent
    numAlive ← *numAlive* – 1
    numMigrated ← *numMigrated* + 1

Quick Review Question 11

From the text's website, download your agent-based tool's *11_4QRQ.pdf* file for this system-dependent question to write *changeCounts* and *checkTerminate*.

Visualization of Example Problem

With Figure 11.4.3 containing the initial configuration of 4 fenced AWPs and 14 unfenced AWPs, Figure 11.4.5 contains several frames of an animation of a simulation with constant definitions as in Table 11.4.1. To simulate the activity of toads over a 12-h night using 1200 cycles, each cycle is of length 36 s:

$$\frac{12\text{ h}}{1200\text{ cycles}} = \frac{12\text{ h}}{1200\text{ cycles}} \times \frac{60\text{ min}}{\text{h}} \times \frac{60\text{ s}}{\text{min}} = \frac{36\text{ s}}{\text{cycles}}$$

We assume activity occurs during evening hours, so the illustrated simulation spans more than one night (1977 cycles, or 1.64 nights). Initially, all toads are in the rightmost column. Frames a–c of Figure 11.4.5 show the toads as they start on their journey from the east border. Frames d–f demonstrate thirsty toads taking turns in an AWP toward the northeast, but other toads being "frustrated" in their efforts to obtain water from a fenced AWP in the southeast. In the latter case, the toads "sense" that water is close but cannot get to the AWP. Lingering around a fenced AWP, they tend to eat all the resources in the area without getting the necessary amount of hydration. In general, diminished food supplies, indicated by lighter shades of gray in frames such as g–i, help to drive the toads westward. By Frames j and k, most of the toads have migrated west or died from desiccation or starvation, while Frame l contains a depiction of the final food grid. For this particular execution of the model, 9 of the 35 original toads migrated and 26 died.

Multiple Simulations

The stochastic nature of this model is such that we should not take any one simulation as indicative of what will happen in general. Thus, we could also develop a ***MultipleSimulationDriver*** agent that instructs *SimulationDriver* to carry out the simulation a designated number of times, *numSimulations*. At the end of each simulation, a method, ***calculateNumAlive***, accumulates the number of dead toads (*numCroaked*) and the number of toads that have migrated (*numMigrated*) in ongoing totals (***totalDead*** and ***totalMigrated***, respectively). After execution of all the simulations, *MultipleSimulationDriver* calculates the average of each total, or the totals divided by *numSimulations*.

Averages for such simulations indicate that appropriate fencing of AWPs should help to curb cane toad migration. We can start with the same basic configuration as in Figure 11.4.3 for each simulation. One execution with *numSimulations* = 10 yielded a mean of 26.8 of 35 (76.6%) toads dying and 8.2 (23.4%) successfully migrating. With no fencing, more migrated (mean 13.8 of the 35 toads, or 39.4%); while, if all 18 AWPs were fenced in another experiment, on the average only 1.7 (4.9%) migrated and 33.3 (95.1%) died. The projects explore calculating such averages for various configurations.

Assessment of Model

The results do show the impact of fencing all or some AWPs in general agreement with the experimental results from Florance et al. (2011). Moreover, the toads do migrate from the direction of release on the east to the west. A search for food and water with diminishing food availability to the east drives the invasion front westward.

However, it is probably unrealistic to consider our simulated food resources being depleted quite so completely without additional prey moving into the area. For instance, if we allow the simulation of the enclosed area with 20 toads to run long enough, all toads die as they eat all the food.

The simulation of this module makes many other simplifying assumptions. We have not considered the impact of climate, particularly wet and dry seasons and temperature. Moreover, we have not adequately addressed the adult cane toad's inclination to sleep in moist places during the day and to travel, often to water, during the night.

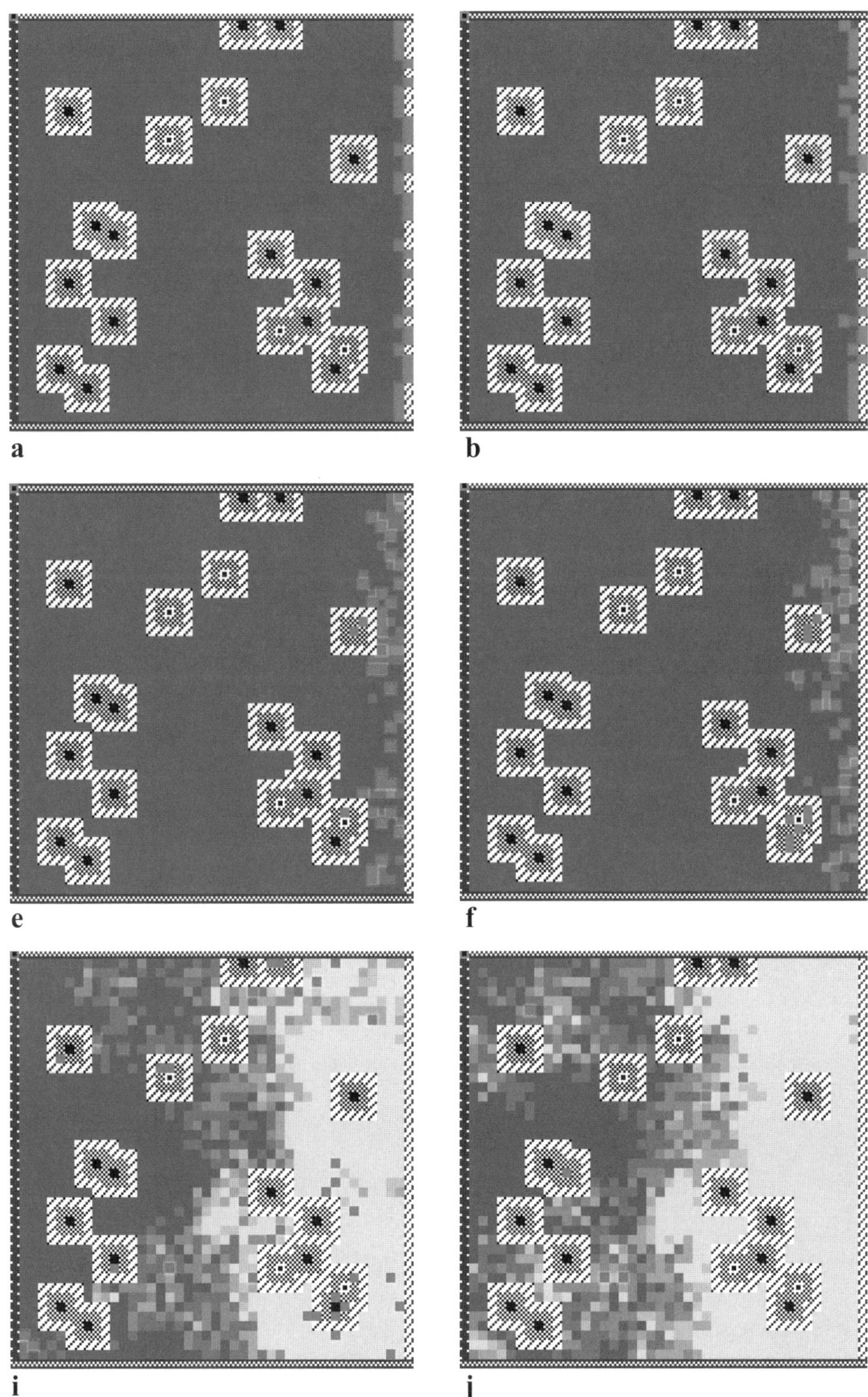

Figure 11.4.5 Several frames of the animation of one simulation

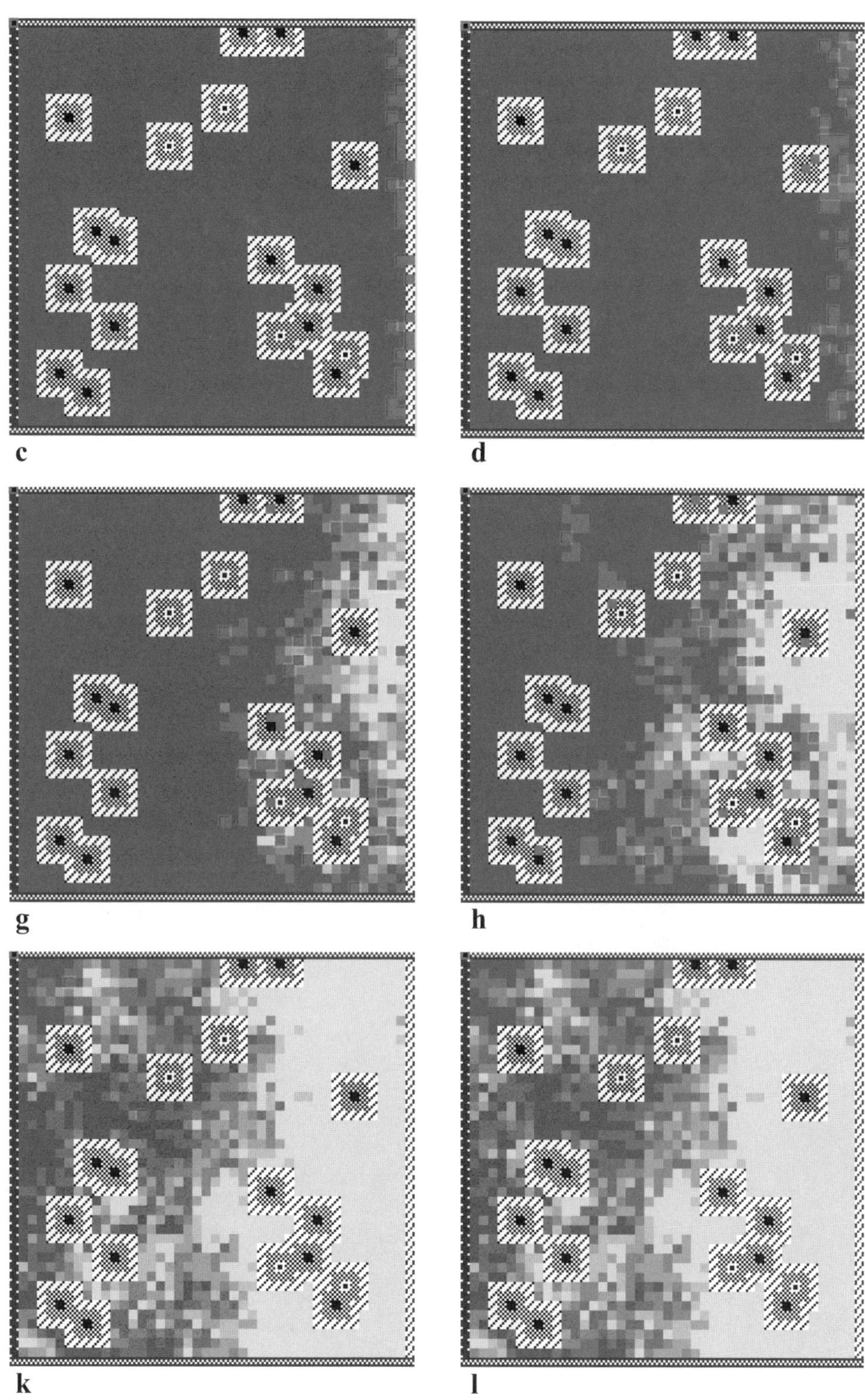

Figure 11.4.5 (*continued*)

Additionally, we have used an arbitrary landscape for the simulation. A high-quality simulation employed to make predictions upon which to take actions must incorporate a map of the distribution of permanent waters and AWPs, as in Florance et al. (2011). With their experiments and their model that uses such a map as well as climate data, but no propagation of toads, Florance et al. predict "that systematically excluding toads from AWP would reduce the area of arid Australia across which toads are predicted to disperse and colonize under average climatic conditions by 38% from 2,242,000 to 1,385,000 km^2."

Exercise

1. Develop an algorithm for an alternative *StartBorder* method, *createMultipleToads*, which generates a *Toad* agent with probability 0.3 as long as the number of toads is less than 50. Multiple toads can be in a cell.

Projects

For additional projects, see Module 14.13, "Re-Solving the Problems with Agent-Based Simulations."

For Projects 1–12, adjust the simulation of this module as indicated. Perform multiple simulations as described in the section "Multiple Simulations" to determine the mean number of toads that are alive, dead, and migrated. Discuss the results, including how closely your results match the experimental data when provided.

1. Run the simulation 100 times with constant definitions, as in Table 11.4.1, each time with a different random configuration, and average the results.
2. Run the simulation 100 times with constant definitions, as in Table 11.4.1, using three fixed configurations: unfenced, half fenced, and all AWPs.
3. Simulate releasing 21 toads in a 20 × 20 enclosed area that does not allow migration and that has only one AWP in the middle. Average the results over 100 simulations, each of which runs for 7200 cycles, representing six 12-h nights. Repeat the experiment with the one AWP being fenced. In a field experiment, with an unrestricted AWP and a control group of 20 toads, 19 toads survived a 72-h period. Predation, which our simulation does not consider, accounted for the 1 toad's death. Then, in another field experiment, 20 of 21 toads died overnight in an enclosed area with a fenced AWP (Florance et al. 2011). Can you adjust constants to obtain a closer match to the experimental data?
4. Incorporate a rule to help prevent toads from remaining for an extended period of time around a fenced AWP. For example, you might write a rule where occasionally a thirsty toad hops in a random direction or in the direction of minimum moisture. Compare your results to the existing model.
5. Adjust the model to allow for rapid food regeneration.
6. Adjust the model so that toads are released at random times and not all at once.

7. Initialize each *Desert* agent to have a random amount of food within a designated range. Using a diffusion algorithm, as described in the section "Heat Diffusion Model" of Module 10.2, "Diffusion: Overcoming Differences," diffuse the food at the end of each cycle. Does the change have an impact on depletion of food around fenced AWPs?
8. Have the grid include a larger body of permanent water, such as a river. Describe the impact on survivability of the toads.
9. Explore the difference between dry and rainy seasons on the effectiveness of fencing AWPs. For rainy seasons, the moisture grid has more small puddles and larger ponds that tend to evaporate between rainstorms.
10. Refine the model to include moist areas, where toads can take refuge from the heat. Assume in such an area a toad's water amount does not change. A thirsty toad in a moist area does not automatically look for water, but the longer the toad is in that location, the more likely it is to move to the nearest neighboring site with the most moisture. One technique to implement this rule is to have a toad state variable, *numCycles*, to count the number of cycles a toad is in a moist area and have the probability of embarking to search for water be the reciprocal of *numCycles*.
11. Adjust Rule 6 so that a thirsty toad does not immediately return to its previous location, and compare the results of your refined model to those of the text.
12. Refine the model to include the changes from Projects 10 and 11.

For Projects 13–17, use the information in the project description and the "Introduction" section, as necessary.

13. Scientists continue to search for effective measures to restrain cane toad populations. Dr. Rick Shine, a biologist at the University of Sydney, and his colleagues are experimenting with various control measures. One of Professor Shine's honor students, Georgia Ward-Fear, has come upon a remarkable possibility. Toadlets of this species, unlike those of other anuran species, are active by day. To avoid desiccation, they confine themselves to the areas around water, which a species of meat ant (*Iridomyrmex reburrus*) favors for foraging. Most other species of young frogs hop away to avoid ants—but not cane toadlets. Their ancestors never had to deal with such large, predatory ants, so the escape behavior has not evolved. Thus, the toadlets often provide a nutritious morsel for the ants, and these predators successfully reduce the young toad population (up to 90%; Ward-Fear et al. 2009).

    Develop a simulation that contains a grid with water, land, toadlets, and meat ants. The toadlets can stay where they are or move in random directions, but they stay close to or in water. Meat ants remain on land. Assume that when a meat ant is adjacent to a toadlet on land, with a certain probability the ant "eats" the toadlet, that is, the toadlet disappears from the simulation. Perform an animation of the simulation, and plot the number of toadlets versus time.
14. Sometime between September and March, Australian cane toads work their way toward aquatic sites to breed. Because the toads are so widely distributed, from New South Wales to the Northern Territory, the timing depends on the particular climatic zone and habitat conditions. The water can be tem-

porary or permanent, brackish or fresh, but they prefer relatively clear water with rather neutral pH and sufficient emergent, submergent, and/or floating plants for cover to lay their eggs. As noted in the introduction, clutch size varies widely but is correlated positively with the size of the female.

If a female cane toad lays 10,000 eggs, about 7000 will survive to produce tadpoles. Tadpoles live normally for 10 to 100 days (average = 38), and the wide range results from various environmental factors (e.g., temperature), food levels, and density. For instance, high levels of intraspecific competition for food can delay development, or impede growth rate.

Tadpole survival is strongly density dependent (Hearnden 1991) but is also influenced by predation from older cohorts of *B. marinus* tadpoles, climatic conditions, and food levels. Field data suggests that the maximum survival for tadpoles (σ_{tmax}) to be about 0.35 tadpoles/L. Using a coefficient of intraspecific competition ($d = 0.5771$ tadpoles/L), we can fit the following function to field data to reveal the relationship between tadpole survival (σ_t) and initial density (T).

$$\sigma_t(T) = \sigma_{tmax}/(1 + d \cdot T) \qquad \text{(Lampo and De Leo 1998)}$$

Predation by older tadpoles can reduce survival from 88% to 1.7% (Hero and Stoneham 2009). Surviving tadpoles become **metamorphs** (**toadlets**), which must make the transition to a terrestrial lifestyle.

a. Develop an individual-based model with a food grid to simulate development of tadpoles in a pond. Initialize tadpoles in the pond to be of a variety of reasonable ages and locations. Running the simulation a number of times, determine the mean number of tadpoles surviving to become toadlets for various densities. Attempt to adjust parameters to match $\sigma_t(T)$. Indicate simplifying assumptions you make, and discuss your results.

b. Using the values from Part a, have tadpoles emerge at random locations and times around the edge of the pond. Running the simulation a number of times, determine the mean number of tadpoles surviving to become toadlets.

15. Tadpoles may survive at temperatures between 17 °C and 42 °C, with maximum survival at 29 °C. Refine Project 14b to take into account the impact of pond temperature on tadpole survival. For simplicity, assume the water temperature is the same throughout the pond but is 1 °C to 3 °C lower at night. Run the simulation for low, high, and optimum temperatures.

16. Because cane toad toadlets (see Project 14) are initially quite small (9–13 mm) and lack the extreme toxicity of other life stages, they are quite vulnerable to predation. Metamorphs grow very rapidly at first (0.647 mm/da; Zug and Zug 1979), but the rate of growth is density dependent. The earliest metamorphs are generally found within 1 m of the water (Cohen and Alford 1993). Survival is influenced by desiccation and predation, varying from 1.2% to 17.6% (Lampo and De Leo 1998). Susceptibility to desiccation is reduced with increased numbers of retreat sites available to the toadlets.

Develop an individual-based simulation of toadlets near a pond that includes a moisture grid and predators and that does not allow the toadlets migrate off the grid. Attempt to adjust parameters so that survival is as indicated. Running the simulation a number of times, determine the percent that survive and the mean toadlet size at the end of 1 year. Indicate your assumptions and discuss your results.

17. Surviving toadlets are considered **juveniles** at 1 year (Lampo and De Leo 1998) and become breeding adults at 2 years. Adult survival depends on a number of environmental factors, especially desiccation. Toads obtain much of their water from their prey (~69%) and lose water via evaporation, respiration, and excretion (Kearney et al. 2008). Although these animals can sustain substantial water loss, if they lose 40% of their body mass or more, they are much more likely to die of dehydration (Florance et al. 2011). Adult survival rates vary between 30 and 70% (Lampo and De Leo). Juveniles are assumed to have only 10% of the adult survival rates.

Using this information, develop a grid-based individual-based simulation involving juvenile and adult cane toads. Initialize the grid with juveniles and adults in random locations and juveniles of random ages. Have new juveniles entering the simulation at random times from around a pond. Have new adult toads entering the simulation at random times from grid boundaries. Allow toads to migrate out of the area in any direction. Because young toads are primary colonizers, young adults should be more likely to move than older toads and young toadlets. Running the simulation a number of times, determine the mean number of juveniles and adults that survive, die, and migrate.

For each of the following projects, using the information, develop a simulation with an animation. Also, perform multiple simulations to determine the mean number of toads that are alive, dead, and migrated. Discuss the results.

18. Temperature can have a big impact on migration of *Bufo marinus* as the animal favors warmer weather but tends to desiccate faster under such conditions. The threshold temperatures for population growth are estimated as 14 °C and 40 °C, while the optimal temperature range for population growth is estimated as 31 °C to 35 °C (Sutherst et al. 1996). Incorporate a temperature gradient grid into your simulation, where temperatures are cooler to the south and gradually warm for cells further north, as generally happens in Australia. Thus, *Desert* agents to the north have warmer temperature-state variable values than those further south. Write and incorporate rules using this grid. Have the toads released either gradually or all at once from the middle part of the south border and allow them to migrate off the grid anywhere. Save the temperatures where toads die and migrate, and display two histograms of the numbers of dead and migrated toads versus temperature.

19. Through experimentation and curve fitting, Kearney et al. (2008) developed the following model for the hopping speed, S (km/h), of the cane toad as a function of its core body temperature, T_b (°C), from 15 °C to 35 °C:

$$S = -25.48396 + 4.51222T_b - 0.29052T_b^2 + 0.0082619T_b^3 - 0.000086431T_b^4$$

Core body temperature is directly proportional to air temperature, and the two temperatures are almost equal on rainy nights. Moreover, the scientists estimated the proportion of time a toad moves as having a median of 3.84% and interquartile (middle) range of 1.4% to 6.8%. They assumed activity is limited only by temperature and not rainfall. Their analysis predicted foraging rates generally less than 1 g/h throughout the present range. Assuming that a diet of crickets has 69% water content, they also predicted that toads would need to drink more than 1 L/yr in cooler areas and 10 L/yr in arid northern regions. Incorporating these findings from Kearney et al., develop a grid-based individual-based model of the spread of toads.

20. Extend the previous project to account for cane toad metabolic rates. With experimentation and curve fitting, Kearney et al. (2008) derived a formula for resting metabolic rate, *M* in watts (W), or joules (J) per second, as a function of core body temperature, T_b in degrees Celsius, and *mass* in grams, as follows:

$$M = 0.0056 - 10.0^{(0.038\ T_b - 1.771)} mass^{0.82}$$

Field metabolic rates for active toads was assumed to be 2.5 that for resting toads in the laboratory. Moreover, Kearney et al. determined that a diet of crickets has an energy density of about 6.3 kJ/g wet mass and that cane toads can assimilate about 85% of this amount, or 5.355 kJ/g = 5355 J/g. For instance, a 120-g toad with body temperature 25 °C has resting metabolic rate $M = 0.0429$ W = 0.0429 J/s. Dividing by the assimilated energy density of crickets, we find that this resting cane toad requires about 8×10^{-6} g/s. Assume cane toad body masses between 50 g and 500 g. Letnic et al. (2008) indicate that most of the cane toads in colonizing-front populations in the Northern Territory are adults. Moreover, the scientists estimated mean masses of 170 g for males and 290 g for females, with some as large as 2 kg.

21. For free-ranging cane toads, Halsey and White (2010) obtained estimates of energetics, such as the rate of the change the volume of oxygen in the blood (rate of energy expenditure or metabolic rate), *dV/dt* (mL/h), calibrated to overall dynamic body acceleration (*ODBA*) in grams, a metric for body motion. Using data and curve fitting, they developed the following equation for *dV/dt* as a function of *ODBA* and *mass* in grams at 25 °C body temperature:

$$dV/dt = 555.9\ ODBA + 0.372\ mass - 19.98$$

6.4 mL O_2 h^{-1} represents about 3200 J/da = 0.037 J/s. Assume such a relationship is proportional. Moreover, Kearney et al. (2008) determined that a diet of crickets has an energy density of about 6.3 kJ/g wet mass and that cane toads can assimilate about 85% of this amount, or 5.355 kJ/g = 5355 J/g. Dividing by the assimilated energy density of crickets, we find that a cane toad with *dV/dt* of 6.4 mL/h requires about 6.9×10^{-6} g/s.

Studying eight cane toads deployed in the field, they determined the following values ± **standard error of the mean** (**SEM**):

- Mean body mass = 136 ± 13 g; minimum mass = 97 g; maximum = 204 g
- Mean *ODBA* over recording time = 0.0384 ± 0.0044 g; minimum = 0.0232 g; maximum *ODBA* = 0.054 g
- Maximum *ODBA* over 5 min = 0.086 ± 0.016 g

They also determined the following proportions (percentages) of recording time:

- Proportion spent resting = 84.0%
- Proportion spent in low-activity behavior = 13.87 ± 2.3%
- Proportion spent hopping = 2.10 ± 0.7%

Toads typically hop for less than 5% of the time, moving on the average 4% of the time at a rate of 18 m/h.

Incorporate energetics into a model of free-ranging cane toads.

22. Extend the previous project to account for temperature, as in Project 15. The equation estimated the rate of the change the volume of oxygen in the blood, for cane toad with body temperature 25 °C (Halsey and White 2010). To obtain the rate at other body temperatures, we can employ a Q_{10} correction, as follows:

$$Q_{10} = (r_2/r_1)^{10/(t_2 - t_1)}$$

where r_1 and r_2 are metabolic rates and t_1 and t_2 are corresponding temperatures. Thus, using the formula for *dV/dt* from the previous project to estimate $r_1 = dV/dt$ at $t_1 = 25$ °C and knowing the toad's body temperature t_2 and Q_{10}, we can calculate its r_2, or metabolic rate, at t_2. A Q_{10} of 2 results in a doubling of the metabolic rate with each increase in temperature of 10 °C. For the eight toads in the study, Q_{10} values ranged from 2.0257 to 7.5960 and averaged 3.4426 (standard deviation = 1.8316). With body temperatures ranging from 13.4 to 19.7, the mean ± SEM for the eight toads was 17.1 ± 0.9 °C.

Answers to Quick Review Questions

From the text's website, download your agent-based tool's *11_4QRQ.pdf* file for answers to these system-dependent questions.

3. a. 12 = (0.003)(100)(40)

b. 3 *Awp*'s and 9 *FencedAwp*'s: using 12 from Part a, 3 = (0.25)(12); 9 = 12 – 3

c. 56 = (8)(5 + 2) because each *Awp* and *FencedAwp* agent is surrounded by 8 *AwpAdjacent* agents.

5. a. *energy* = 0.91, *water* = 0.806, and *food* = 0.02 because *amtEat* = 0.01, so *energy* = 0.9 + 0.01, *water* = 0.8 + 0.6*0.01, and *food* = 0.03 – 0.01

b. *energy* = 0.905, *water* = 0.803, and *food* = 0.0 because *amtEat* = *availableFood* = 0.005, so *energy* = 0.9 + 0.005, *water* = 0.8 + 0.6*0.005, and *food* = 0.005 – 0.005

c. *energy* = 1.0, *water* = 0.8006, and *food* = 0.029 because *amtEat* = 1 – *energy* = 0.001, so *energy* = 0.9 + 0.001, *water* = 0.8 + 0.6*0.001, and *food* = 0.03 – 0.001

d. *energy* = 0.91, *water* = 1.0, and *food* = 0.02 because *amtEat* = 0.01, so *energy* = 0.9 + 0.01, *water* = the minimum of 0.999 + 0.6*0.01 = 1.005 and 1.0, and *food* = 0.03 - 0.01

References

Alexander, T.R. 1964. "Observations on the Feeding Behavior of *Bufo marinus* (Linne)." *Herpetologica*, 20: 255–259.

Alford, R. A., M. Lampo, and P. Bayliss. 1995. "The Comparative Ecology of *Bufo marinus* in Australia and South America." CSIRO *Bufo* Project: An Overview of Research Outcomes, Unpublished report, CSIRO, Australia.

Bartlett, R. D., and P. P. Bartlett. 1999. *A Field Guide to Florida Reptiles and Amphibians*. Houston: Gulf Publishing Company.

BBC News. 2010. http://news.bbc.co.uk/2/hi/8480041.stm (accessed December 26, 2012)

Cohen, M. P., and R. A. Alford. 1993. "Growth, Survival and Activity Patterns of Recently Metamorphosed *Bufo marinus*." *Wildlife Research*, 20: 1–13.

Easteal, S. 1982. "The Genetics of Introduced Populations of the Marine Toad, *Bufo marinus (Linnaeus), (Amphibia: Anura);* A Natural Experiment in Evolution." Ph.D. diss. Griffith University, School of Australian Environmental Studies, Brisbane, Queensland, Australia.

———. 1985. "The Ecological Genetics of Introduced Populations of the Giant Toad, *Bufo marinus*." III. Geographical Patterns of Variation. *Evolution*, 39: 1065–1075.

———. 1986. *Bufo marinus*. Catalogue of American Amphibians and Reptiles. Society for the Study of Amphibians and Reptiles, St. Louis, MO, pp. 395.1–395.4

Florance, Daniel, Jonathan K. Webb, Tim Dempster, Michael R. Kearney, Alex Worthing, and Mike Letnic. 2011. "Excluding Access to Invasion Hubs Can Contain the Spread of an Invasive Vertebrate" *Proc. R. Soc. B.* Published online February 23, 2011.

Freeland, W. J. and K. C. Martin. 1985. "The Rate of Range Expansion by *Bufo marinus* in Northern Australia, 1980–84." *Australian Wildlife Research* 12: 555–559.

Freeland, W. J. 1986. "Populations of Cane Toad, *Bufo marinus*, in Relation to Time since Colonization." *Australian Wildlife Research*, 13: 321–329

Gao, Huimin, Martin Zehl, Alexander Leitner, Xiyan Wu, Zhimin Wang, and Brigitte Kopp. 2010. "Comparison of Toad Venoms from Different 'Bufo' Species by HPLC and LC-DAD-MS/MS.". *J Ethnopharmacology*, 131, 368–376.

Halliday, Damien C. T., Daryl Venables, David Moore, Thayalini Shanmuganathan, Jackie Pallister, Anthony J. Robinson, and Alex Hyatt. 2009. "Cane Toad Toxicity: An Assessment of Extracts from Early Developmental Stages and Adult Tissues using MDCK Cell Culture." *Toxicon*, 53, 385–391.

Halsey, Lewis G., and Craig R. White. 2010. "Measuring Energetics and Behaviour Using Accelerometry in Cane Toads *Bufo marinus*," *PLoS ONE*, 5(4): e10170. doi:10.1371/ journal.pone.0010170

Hayes, R. A., Piggott, A. M., Dalle, K., and Capon, R. J. 2009. "Microbial Biotransformation as a Source of Chemical Diversity in Cane Toad Steroid Toxins." *Bioorganic and Medicinal Chemistry Letters*, 19: 1790–1792.

Hearnden M. N. 1991. Reproductive and Larval Ecology of *Bufo marinus (Anura: Bufonidae)*. Ph.D. diss. Townsville (Australia): James Cook University of North Queensland.

Hero, J. M., and M. Stoneham. 2009. "*Bufo marinus*." AmphibiaWeb. http://amphibiaweb.org/cgi/amphib_query?where-genus=Bufoandwhere-species=marinus (accessed August 21, 2011)

Hinckley, A. D. 1962. "Diet of the Giant Toad, *Bufo marinus* (L.), in Fiji." *Herpetologica*, 18: 253–259.

Kearney, Michael, Ben L. Phillips, Christopher R. Tracy, Keith A. Christian, Gregory Betts, and Warren P. Porter. 2008. "Modelling Species Distributions without Using Species Distributions: The Cane Toad in Australia under Current and Future Climates" *Ecography*, 31: 423–434.

Kearney, Michael, and Warren P. Porter. 2004. "Mapping the Fundamental Niche: Physiology, Climate, and the Distribution of a Nocturnal Lizard. " *Ecology* 85(11): 3119–3131.

Krakauer, T. 1968. "The Ecology of the Neotropical Toad, *Bufo marinus*, in South Florida." *Herpetologica*, 24: 214–221.

———. 1970. "Tolerance Limits of the Toad, *Bufo marinus*, in South Florida." *Comparative Biochemistry and Physiology*, 33: 15–26.

Lampo, Margarita, and Giulio A. De Leo. 1998. "The Invasion Ecology of the Toad *Bufo marinus*: from South America to Australia. " *Ecological Applications* 8(2): 388–396.

Letnic, Mike, Jonathan K. Webb, and Richard Shine. 2008. "Invasive Cane Toads (Bufo marinus) Cause Mass Mortality of Freshwater Crocodiles (Crocodylus johnstoni) in Tropical Australia." *Biological Conservation*, 141: 1773–1782.

Lever, C. 2001. *The Cane Toad: The History and Ecology of a Successful Colonist.* York, UK: Westbury Academic and Scientific.

Ma, H. Y., J. P. Kou, J. R.Wang and B. Y. Yu. 2007. "Evaluation of the Anti-inflammatory and Analgesic Activities of Liu-Shen-Wan and Its Individual Fractions." *Journal of Ethnopharmacology*, 112: 108–114.

Meng, Zhiqiang, Peiying Yang, Yehua Shen, Wenying Bei, Ying Zhang, Yongqian Ge, Robert A. Newman et al. 2009. "Pilot Study of Huachansu in Patients with Hepatocellular Carcinoma, Nonsmall-Cell Lung Cancer, or Pancreatic Cancer. *Cancer* 115(22): 5309–5318.

Markula, Anna, Steve Csurhes, and Martin Hannan-Jones. 2010. Cane Toad (*Bufo marinus*). Pest Animal Risk Assessment. Biosecurity Queensland Department of Employment, Economic Development and Innovation. GPO Box 46, Brisbane 4001. http://www.daff.qld.gov.au/documents/Biosecurity_EnvironmentalPests/IPA-Cane-Toad-Risk-Assessment.pdf (accessed December 26, 2012)

Oliver, J. A. 1949. "The Peripatetic Toad." *Natural History*, 58: 30–33.

Phillips, B. L., G. P. Brown, M. Greenlees, J. K. Webb, and R. Shine. 2007. "Rapid Expansion of the Cane Toad (*Bufo marinus*) Invasion Front in Tropical Australia." *Austral Ecology*, 32: 169–176.

Phillips, B. L., G. P. Brown, J. K. Webb, and R. Shine. 2006. "Invasion and the Evolution of Speed in Toads." *Nature*, 439: 803.

Rabor, D. S. 1952. "Preliminary Notes on the Giant Toad *Bufo marinus* (Linn.) in the Philippine Islands." *Copeia*. 1952: 281–282.

Reynolds, Craig. 1999. "Individual-Based Models, an Annotated List of Links." http://www.red3d.com/cwr/ibm.html (accessed July 18, 2011)

Rossi, J. V. 1983. "The Use of Olfactory Cues by *Bufo marinus*." *Journal of Herpetology*. 17: 72–73.

Schwarzkopf, Lin, and Ross A. Alford. 2002. "Nomadic Movement in Tropical Toads." *OIKOS*. 96: 492–506.

———. 2005. "Movement and Dispersal in Established and Invading Toad Populations." Cane Toad Forum held in Kununurra. http://www.canetoads.com.au/forumprocedespp4.htm (accessed August 9, 2011)

Shiflet, Angela B., and George W. Shiflet. 2009. "Biofilms: United They Stand, Divided They Colonize." UPEP Curriculum Module. http://shodor.org/petascale/materials/UPModules/biofilms/ (accessed December 26, 2012)

———. 2010. "Time after Time: Age- and Stage-Structured Models." UPEP Curriculum Module. http://shodor.org/petascale/materials/UPModules/ageStructuredModels/ (accessed January 24, 2012)

Stuart, Simon N., Janice S. Chanson, Neil A. Cox, Bruce E. Young, Ana S. L. Rodrigues, Debra L. Fischman, and Robert W. Waller. 2004. "Status and Trends of Amphibian Declines and Extinctions Worldwide." *Science*, 306(5702): 1783–1786.

Sutherst, Robert W., Robert B. Floyd, and Gunter F. Maywald. 1996. "The Potential Geographical Distribution of the Cane Toad, *Bufo marinus* L. in Australia" *Conservation Biology*, 10(1, February): 294–299.

Urban, Mark C., Ben L. Phillips, David K. Skelly and Richard Shine. 2007. "The Cane Toad's (*Bufo marinus*) Increasing Ability to Invade Australia Is Revealed by a Dynamically Updated Range Model." *Proc. R. Soc B.*, 274(1616): 1413–1419.

Ward-Fear, G., G. P. Brown, M. Greenlees, and R. Shine. 2009. "Maladaptive Traits in Invasive Species: in Australia, Cane Toads Are More Vulnerable to Predatory Ants Than Are Native Frogs." *Functional Ecology*, 23: 559–568

Wright, A. H., and A. A. Wright. 1949. *Handbook of Frogs and Toads of the United States and Canada*, 3rd ed. Ithaca, NY: Comstock Publishing Associates..

Xiao, P. G., 2002. *Modern Chinese Materia Medica*, vol. IV. Beijing, China: Chemical Industry Press, p. 253.

Zug, George R., and Patricia B. Zug. 1979. *The Marine Toad, Bufo marinus: A Natural History Resume of Native Populations*. Washington: Smithsonian Institution Press, 1979.

12

HIGH-PERFORMANCE COMPUTING

MODULE 12.1

Concurrent Processing

Introduction

> *"Because of our ability to collect and analyze vast quantities of data, scientists now have the potential to solve some of the world's biggest problems. . .By utilizing 21st-century computing power, human expertise, and a systematic approach to storing and mining information, scientists are beginning to achieve real breakthroughs."*
>
> — Tony Hey, "The Next Scientific Revolution"

Before humankind lies a vast, fascinating, and crucial collection of knowledge—knowledge that will change our lives. With precision instrumentation, modern laboratory techniques, and ever-increasing computational abilities, we will be able to investigate and understand physical, chemical, and biological systems from the most fundamental elements of the universe to the largest and most complex systems. The possibilities for major research breakthroughs, significant technological innovations, medical and health advances, better economic competitiveness, and the like, are unfathomable.

Enhanced computer technology and power are crucial to progress on this new frontier. In 2002, the Japanese government began to simulate the earth's climate and geological activity using what was at that time the world's fastest supercomputer—the Earth Simulator. This remarkable machine, occupying a building that would hold four tennis courts, at that time could perform 35.86 Tflops (almost 36 trillion floating-point calculations per second). This achievement was the first time the fastest supercomputer had been built outside the United States (Habata et al. 2003). Responding to this challenge and realizing the associated opportunities, then–U.S. Secretary of Energy Spencer Abraham announced a new project in 2004 called the "Leadership Class Computing Facility for Science" to build the fastest supercomputer in the world (ORNL 2004). This emphasis on supercomputer development led to the Titan supercomputer, now housed at Oak Ridge National Laboratory (ORNL).

The Titan has a theoretical peak performance of more than 20 petaflops (20,000 trillion floating-point calculations per second; OLCF 2011).

It is difficult to find a large employer who does not utilize the capabilities of such **high-performance computing** (**HPC**). As the following examples illustrate, industrial companies, biomedical research institutions and corporations, government agencies and laboratories, pharmaceutical firms, filmmakers, and so on, all benefit or will benefit from the use of HPC.

If you had been a "safety tester" for General Motors back in the 1930s, your working clothes would have been made of leather, and your job would have been to steer a moving test car toward an obstacle, like a big tree, and to jump off before the car hit. Safety testing at GM has come a long way. Now, GM employs IBM Blue-Gene high-performance computing systems to simulate various forms of car wrecks, which reduce significantly the number of physical test crashes (piloted remotely) and saves the company millions of dollars. Testers who work for GM today do not wear protective clothing and are more likely to be computer or computational scientists. Successful application of HPC to automobile safety has now been extended to other areas of the automobile production, including improved fuel efficiency and product marketing. In this way, HPC contributes significantly to the production of safer, more efficient automobiles and makes GM more competitive with other automakers (King 2010).

As another industrial example, Boeing is one of the world's two major commercial aircraft makers, and its success or failure has significant impact on the U.S. economy. In 2011, the company exported manufactured goods valued at $34 billion (Commerce.gov 2012). In 2009, the newest of Boeing's commercial airliners, the 787 Dreamliner, took its first flight (Boeing 2012). The design of this aircraft with its numerous technical innovations utilized more than 800,000 processor hours on Cray supercomputers. Using computer-aided design (CAD) and computer-aided engineering software, the company was able to validate their designs before building a physical prototype. For example, engineers, using this virtual prototyping, needed to test only 11 wing designs for the 787, as opposed to 77 used for the earlier Boeing 767. Best of all, the finished product is lighter, is more fuel-efficient, and produces lower emissions (Cray 2012).

HPC has also dramatically improved the entertainment industry. Movie director James Cameron conceived of the phenomenally successful movie *Avatar* more than 15 years before its release. At that time, his vision could not have been realized, given the visual effects tools available. Eventually, the technical capabilities became reality, and he began the screenplay in 2006. The film was released in December 2009 and has earned more than $2.75 billion (gross) worldwide (Box Office Mojo 2012). The film is noted for its groundbreaking and stunning visual effects, for which it won an Oscar in 2010. Weta Digital, already noted for its involvement in producing *The Lord of the Rings* trilogy, provided the digital production and 3D animation. Central to their effort was a considerable amount of high performance computing. Rendering the data into images is computationally demanding, and Weta employed a large server farm made up of 4000 quad-core Hewlitt-Packard blades with 35,000 processor cores, 104 terabytes (space for 104×10^{12} characters) of RAM (random-access memory), and three petabytes (space for 3×10^{15} characters) of network storage. During the film development, Weta ran about 10,000 jobs (1.3 –1.4 million individual tasks) per day. So, in just about 10 years, HPC had changed significantly, making possible what previously was impossible (Ericson 2010; Swan 2010; Wikipedia 2012).

Numerous medical applications arise from the advances we have made in the field of genomics, and supercomputers are crucial players in those advances. For instance, we have long known that individuals have differing susceptibilities to disease and responses to therapies. Physicians could treat or even prevent the development of some human diseases if they knew each person's genetic make-up—personal genomics. We have the technology to determine individual gene sequences, but the cost has previously been too great. Being able to obtain an individual's sequence at a reasonable cost ($\leq$\$1000) would facilitate a whole new approach to health care. Physicist Aleksei Aksimentiev at the University of Illinois–Urbana-Champaign has used more than 10 million processor-hours on the Jaguar supercomputer at ORNL to develop a sequencing system involving a nanopore (protein pore one-billionth of a meter wide). As DNA moves through the pore, a detector deciphers the nucleotide sequence of the DNA. This system drastically reduces the time and costs for sequencing (OLCF 2011).

Cancer comes in many forms and affects many diverse body tissues. Cancers also respond differently to various forms of chemotherapy. If we understand the genetics and metabolic pathway of each type of cancer, we can devise the most effective therapies, and we could likely even prevent its development. NantHealth, a collaborative effort among various insurers, research institutes, and businesses, has devised a high-speed fiber network that will provide partnering oncologists with important, detailed information about patient cancers in a very short time. Instead of treating a cancer, say, of the breast, in an organ-specific way, the physician can treat the disease based on knowledge of the patient's genetics and the cancer's developmental pathway, which may be identical to a cancer of another organ type. The collaborating scientists of NantHealth collected data on thousands of exomes (expressed sequences of DNA) from tumors. These data were collected from more than 3000 cancer patients and were stored in 96.5 gigabytes (space for 96.5 billion characters) that could be processed by a supercomputer in less than 3 days. 1.8 million cancer cases were projected for 2012 in the United States. These computational tools will enable analysis of 5000 of such cases every day. As Dr. Chan Soon-Shiong of the Soon-Shiong Institute for Advanced Health says, "Doctors will finally be able to provide higher-quality treatment in a dramatically more efficient, effective, and affordable manner" (Business Wire 2012).

This chapter is not meant to be an in-depth study of high performance computing but is intended to give an idea of some of the applications, architecture, concepts, challenges, and algorithms.

Analogy

A **processor**, or a **processor core** or a **central processing unit** (**CPU**), performs the arithmetic and logic in a computer and is its brain. **Concurrent processing** involves having associated, multiple CPUs working concurrently, or simultaneously, on the same or different problems. To achieve the type of high performance for the problems discussed in the introduction, concurrent processing is essential. For an examination of some of the options and problems involved, we consider an analogous situation.

Definitions A **processor**, or **processor core** or **central processing unit (CPU)**, of a computer performs the arithmetic and logic of a computer. **Concurrent processing** involves having associated, multiple CPUs working concurrently, or simultaneously, on the same or different problems.

Suppose a scout leader is taking a group of 10 scouts in a van on a camping trip. Before the fun begins, the troop must shop for provisions—about a hundred different items at a grocery store. What are some of the options for shopping? Before reading further, list some your ideas for the task.

1. One option is for the leader to leave all the scouts in the van and to do the shopping alone. This option is analogous to a single processor working on a single program. Meanwhile, 10 processors (i.e., scouts), who could be helping, are doing nothing to speed the process.
2. Another alternative is for the leader to tear the grocery list into 10 parts and have each scout gather the items on his or her list and meet at a cash register, where the leader is to pay. What difficulties might arise?
    - Initially, each scout must wait for a partial list; finally, he or she must wait for all the other scouts to finish shopping and for the leader to pay. We have a bottleneck because on part of the overall task, only one processor (i.e., the leader) is working.
    - Perhaps only three shopping carts are at the front of the store, so that adequate resources are not immediately available.
    - Moreover, suppose a scout cannot find peanut butter. With this scenario, everyone must wait while the child wanders through the store without assistance from anyone else. A synchronization problem exists. The shopping lists probably could have been divided differently to shorten the wait.
3. A better choice might be that as soon as a scout finishes gathering his or her groceries, he or she helps someone else. However, a speedy scout must know where to go, and both scouts must agree how to divide the work.
4. Another way to help might be to separate scouts into pairs, each consisting of a seeker and a shopper, who have cell phones. The seeker finds an item on the list and tells the location to the shopper. While the shopper is gathering the item into the basket, the seeker searches for the next item. This pipeline system still has situations in which a processor (i.e., scout) must wait. The shopper must wait for the seeker to find the first item, and the seeker must wait for the shopper to gather the last item. At intermediate stages, the seeker might have difficulty finding an item, causing the shopper to be idle; or the shopper might take a while loading cans of Spam™ into the basket, while the seeker has already found the Vienna sausage, the next item on the list.
5. To avoid the bottleneck at checkout, the leader might give each child money as well as part of the list. However, the leader must have an excess of resources (i.e., money) to distribute to the group.
6. The leader can also do some preprocessing on the grocery list. For example, shopping would be faster if each scout had a list of items on a single aisle.

This task, too, could be accelerated with the help of some of the scouts. This scenario would work best if the leader and some of the scouts were familiar with the locations of items in the store.

7. Another alternative consists in the leader texting a partial grocery list to each scout. Scouts are responsible for buying their parts of the groceries and meeting at the van at a certain time. Upon receiving a receipt, the leader reimburses a scout. The scouts, who might be located at great distances from each other, do not have to shop at the same grocery store. Difficulties can still arise, such as a scout not receiving the message or a scout getting sick and not being able to shop. It would be advisable for the leader to make sure all scouts read their texts and shop. If a scout is sick, the leader can redistribute the workload.

Consider other alternatives along with their advantages and disadvantages.

Types of Processing

Three types of processing exist: sequential, parallel, and distributed. **Sequential processing** involves a single processor working on one program. Such processing is analogous to the leader being the only shopper.

Parallel processing consists in a collection of connected processors in close physical proximity working concurrently. Several examples of a vanload of scouts with the leader shopping together at one grocery store provide analogues to parallel processing.

Distributed processing involves several (possibly many) processors, perhaps at great distances from each other, communicating via a network and working concurrently. The example of the leader texting the partial lists to the scouts for them to shop at a variety of stores is analogous to distributed processing.

Definitions **Sequential processing** involves a single processor working on one program. **Parallel processing** consists in a collection of connected processors in close physical proximity, or **tightly coupled**, working concurrently. **Distributed processing** involves several processors, perhaps at great distances from each other, communicating via a network (hence, **loosely coupled**) and working concurrently.

Quick Review Question 1

Indicate the type(s) of processing—sequential, parallel, distributed, or none—for each of the following:

a. Can involve execution of more than one program at a time
b. Can involve execution of one program
c. Can have processors in different countries

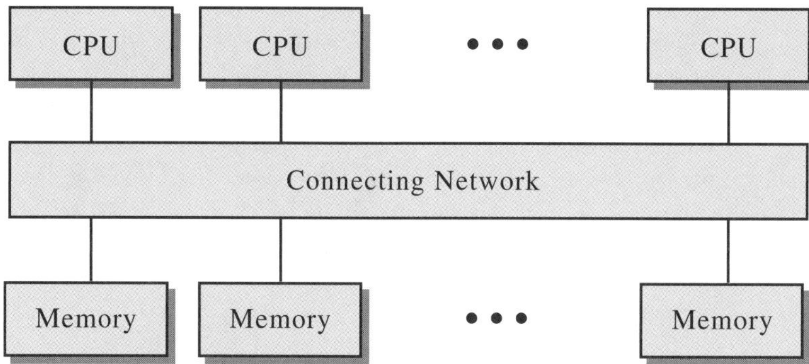

Figure 12.1.1 Shared-memory MIMD architecture

Communication

Concurrent computers usually have an **MIMD** (multiple instruction streams, multiple data streams) architecture in which their processors can execute programs or subprograms at the same time. Communication among processors is accomplished through shared memory or message passing.

Figure 12.1.1 presents a diagram of a traditional **shared-memory MIMD architecture**. Although a memory module might be associated with an individual processor, all processors can access all memory modules.

One difficulty with this architecture is maintaining consistency of the data. For example, suppose processor A reads a value, say, 2, for a shared variable x; and while A is performing computations with the value, processor B writes a different value, say 3, to x. The values are not consistent. Shared-memory systems must provide mechanisms for the programs to ensure consistency of shared data.

A shared-memory system is specifically designed as a parallel computer. However, another kind of architecture can be constructed with a network of workstations that communicate with each other through message passing. Figure 12.1.2 gives a diagram of such a **distributed-memory MIMD architecture**, in which each processor, or **node**, has its own associated memory that is inaccessible to other processors. **Computer clusters**, which have this type of architecture, can range from small systems with a few processors to supercomputers with thousands of processors.

Communication in such an architecture is accomplished through **message passing**. With message passing, programmers must explicitly divide a program into pieces, called **processes**, for concurrent execution. However, in the case of a message-passing system, computer A cannot access directly a variable, say, x, stored in computer B's memory. Instead, A sends a message to B requesting the value of x; if acceptable, B sends a message to A with the value of the variable.

To handle these operations, a message-passing system does require a programmer to write special message-passing calls. However, a shared-memory system must have its own mechanism to ensure the consistency of shared data, and execution of this mechanism can add significantly to execution time of a program. Additionally, a message-passing system with its network of complete individual computers has the

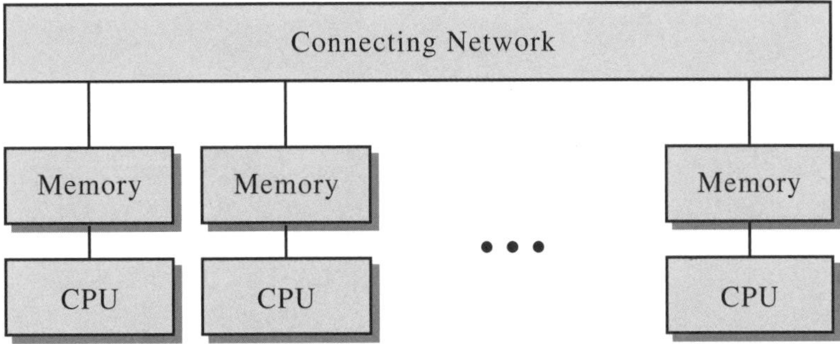

Figure 12.1.2 Distributed-memory MIMD architecture

advantage of **scalability**—with the usually easy addition of more processors to the system, a program's execution speed increases. Also, with improvements in technology, a faster, commercially available workstation can easily be swapped for a slower one in a message-passing system.

Advances in computer architectures, including hybrid systems, are occurring all the time. However, discussion of these is beyond the scope of this text.

Definitions In an **MIMD** (multiple instruction streams, multiple data streams) architecture, processors can execute programs or subprograms concurrently. In a **shared-memory MIMD architecture**, processors communicate through a shared memory. In a **distributed-memory MIMD architecture**, each processor has its own associated memory not directly accessible by other processors, but processors communicate through **message passing**. A **process** is a task or a piece of a program that is executing separately. **Scalability** is the capability of a computer system with expanded hardware resources to exhibit better performance.

Quick Review Question 2

Indicate which type of MIMD computer system—shared-memory, message-passing, both, or neither—exhibits the characteristic for each of the following:

a. The system is more scalable than others.
b. The system must provide a way to ensure consistency of data.
c. Processors of the system can work on a problem concurrently.
d. System can be upgraded more easily.
e. Usually, a manufacturer develops the system as a parallel computer.
f. A programmer splits a program into parts for execution on different processors.
g. A programmer writes a call to request that a processor send data from its memory to another processor for its memory.
h. Processors can execute several independent programs at the same time.
i. A processor can write directly to the memory of another processor.

Metrics

We can employ several metrics, or measures, to indicate the improvement achieved using various configurations of a parallel computer instead of a sequential machine.

For execution on a concurrent machine, a program is divided into separate processes, or tasks, to be executed in parallel on various processors. The **granularity** of parallelism refers to the number of components. We say that a machine has **fine granularity** if it contains many processors, such as a system with thousands of simple processors, each executing relatively few instructions. A machine with **coarse granularity** contains a small number of processors, such as a system with a dozen very fast and complex processors, each executing many instructions simultaneously. A measure of the granularity is the ratio of computation to communication:

$$\textbf{ratio of computation to communication} = \frac{\textbf{computation time}}{\textbf{communication time}}$$

This ratio is large in the case of coarse granularity and small for fine granularity. Fine granularity has the advantage that many processors can execute the program simultaneously, but the larger number of processes has the disadvantage of requiring greater communication time. Coarse granularity reduces communication—an advantage—but reduces concurrency—a disadvantage. Thus, the programmer seeks a balance between the extremes of granularity by achieving a larger ratio of computation to communication along with suitable parallelism.

Definition A granularity metric is as follows:

$$\textbf{ratio of computation to communication} = \frac{\textbf{computation time}}{\textbf{communication time}}$$

Quick Review Question 3

Suppose communication consumes 10% of the execution time for a concurrent program. Determine the ratio of computation to communication.

A commonly used metric for a parallel computer's performance is the speedup factor. For a system with n processors, the **speedup factor** $S(n)$ is as follows:

$$S(n) = \frac{\textbf{execution time on sequential computer}}{\textbf{execution time on system with } n \textbf{ processors}}$$

or

$$S(n) = \frac{\textbf{number of computational steps on sequential computer}}{\textbf{number of computational steps in parallel with } n \textbf{ processors}}$$

Often algorithms to accomplish some computation are different on a sequential computer and a parallel or distributed system, and we employ the times for the best algorithms available on each system in measuring speedup. For example, suppose the best sequential algorithm for a particular task takes 100 ms, while the corresponding work with a 200-processor system requires 2.5 ms. In this case, the speedup is $S(200) = (100\text{ ms})/(2.5\text{ ms}) = 40$.

Definition For a system with n processors, the **speedup factor $S(n)$** is as follows:

$$S(n) = \frac{\textbf{execution time on sequential computer}}{\textbf{execution time on system with } n \textbf{ processors}}$$

or

$$S(n) = \frac{\textbf{number of computational steps on sequential computer}}{\textbf{number of computational steps in parallel with } n \textbf{ processors}}$$

Quick Review Question 4

Suppose a sequential algorithm takes 24 ms, while the speedup on a parallel computer with 8 processors is $S(8) = 4$. Determine the execution time on the parallel computer.

Usually, the maximum speedup possible with n processors is $S(n) = n$, which we call **linear speedup** because the graph is a straight line.[1] With linear speedup, the time required for execution with n processors is $1/n$ of time for execution on a sequential computer. For example, suppose the time for a sequential algorithm is 1 ms. With linear speedup and two processors, the execution time is 1/2 ms = 0.5 ms, so that $S(2) = 1/0.5 = 2$. For three processors, the execution time is 1/3 ms; with four processors, 1/4 ms; and so on.

Quick Review Question 5

Suppose maximum speedup is achieved with an eight-processor system for an algorithm that executes in 24 ms on a sequential computer. Determine the execution time on the parallel computer.

Linear speedup is rarely achieved because of several overhead factors, including the following:

1. Communication time between processors
2. Times when some of the processors are idle
3. Additional computations necessary in the parallel version and unnecessary in the sequential version

For algorithms in the next module, we consider such overheads and speedup factors.

1 Occasionally, a speedup better than $S(n) = n$ can be achieved through comparison with an inferior sequential algorithm or through a special multiprocessor architectural feature, such as a very large amount of memory.

Exercises

1. Complete the matching related to the shopping scouts example in the "Analogy" section. An answer may be used more than once.

a Grocery list	**A** Data item
b Item on grocery list	**B** Distributed processing
c Leader goes through checkout alone	**C** Memory
d Peanut butter	**D** Message passing
e Scout	**E** MISD
f Scout leader	**F** Parallel processing
g Scouts shopping in same store	**G** Processor
h Seeker talks to shopper	**H** Sequential processing

2. Give the advantages and disadvantages of a shared-memory system and of a message-passing system.
3. The best sequential sorting algorithm that compares elements requires $n \log(n)$ computational steps. Suppose a sorting algorithm on a parallel system requires $4n$ computational steps. Determine the speedup factor (Wilkinson and Allen 1999).
4. For each of the overhead factors inhibiting linear speedup, give an analogous example using the shopping scouts.
5. Draw the graph for linear speedup.

Project

1. Write a paper with references on a scientific application that is advanced by high-performance computing.

Answers to Quick Review Questions

1. **a.** parallel processing and distributed processing
   b. sequential processing, parallel processing, and distributed processing
   c. distributed processing
2. **a.** message passing
   b. both (Although the problem is more obvious with shared-memory MIMD architectures, the problem exists with both architectures.)
   c. both
   d. message passing
   e. shared memory
   f. message passing
   g. message passing
   h. both
   i. shared memory
3. ratio of computation to communication = 0.9/0.1 = 9

4. 6 ms, because $S(8) = 4 = (24 \text{ ms})/t$; the parallel algorithm on an eight-processor system is four times faster than the sequential one. Thus, $t = (24 \text{ ms})/4 = 6$ ms.
5. 3 ms, because for linear speedup, $S(8) = 8 = 24$ ms/(execution time on parallel computer); thus, execution time on parallel computer = (24 ms)/8 = 3 ms

References

Boeing. 2012. 787 Dreamliner. http://www.boeing.com/commercial/787family/background.html (accessed December 28, 2012)

Box Office Mojo. 2012. "Worldwide Grosses." http://boxofficemojo.com/alltime/world/ (accessed December 28, 2012)

Business Wire. 2012. "Launch of the Nation's Fastest Genomic Supercomputing Platform Reduces Cancer Genome Analysis from Months to Seconds—One Patient Every 47 Seconds." October 3. http://www.businesswire.com/news/home/20121003005513/en/Launch-Nation's-Fastest-Genomic-Supercomputing-Platform-Reduces (accessed December 28, 2012)

Commerce.gov. 2012. "President Obama Announces New Steps to Promote Manufacturing, Increase U.S. Exports." U.S. Department of Commerce. February 23. http://www.commerce.gov/blog/2012/02/23/president-obama-announces-new-steps-promote-manufacturing-increase-us-exports (accessed December 28, 2012)

Cray. 2012. "Cray Solutions for Manufacturing."http://www.cray.com/IndustrySolutions/Manufacturing.aspx (accessed December 28, 2012)

Ericson, Jim. 2010. "Overheard: Avatar's Data Center." Information Management, January 2. http://www.information-management.com/issues/20_1/avatars-data-center-10016926-1.html (accessed December 28, 2012)

Habata, Shinichi, Mitsuo Yokokawa and Shigemune Kitawaki. 2003. "The Earth Simulator." *NEC Res. & Develop.*, 44(1): 21–26. http://www.eecg.toronto.edu/~amza/ece1747h/papers/earth-sim-nec.pdf (accessed December 28, 2012)

Hey, Tony. 2010. "The Next Scientific Revolution." *Harvard Business Review*, 88(11): 56–150.

King, Rachael. 2010. "At GM, High-Performance Computing Curbs Test Costs." *Bloomberg Business Week. Technology*, October 5. http://www.businessweek.com/stories/2010-10-05/at-gm-high-performance-computing-curbs-test-costsbusinessweek-business-news-stock-market-and-financial-advice (accessed December 28, 2012)

OLCF (Oak Ridge Leadership Computing Facility). 2011. "Whole-Genome Sequencing Simulated on Supercomputers." *Science*, February 14. http://www.olcf.ornl.gov/2011/02/14/whole-genome-sequencing-simulated-on-supercomputers/ (accessed December 28, 2012)

ORNL (Oak Ridge National Laboratory). 2004. "Department of Energy Awards $25 Million to Oak Ridge National Lab to Lead Effort in Building World's Largest Computer." News Release, May 12. http://www.ornl.gov/info/press_releases/get_press_release.cfm?ReleaseNumber=mr20040512-00 (accessed December 28, 2012)

Swan, Georgina. 2010. "How do they do IT? Avatar's Special Effects." *Computerworld*, January 13. http://www.computerworld.com.au/article/332337/how_do_they_do_it_avatar_special_effects/ (accessed December 28, 2012)

Wikipedia. 2012. *Avatar* (2009 film). http://en.wikipedia.org/wiki/Avatar_(2009_film) (accessed December 28, 2012)

Wilkinson, Barry, and Michael Allen. 1999. *Parallel Programming*. Upper Saddle River, NJ: Prentice-Hall, p. 431.

MODULE 12.2

Parallel Algorithms

Introduction

High-performance computing (HPC) does not just involve computers but includes the algorithms—many for computational models and simulations—that unleash their capabilities. The following applications illustrate some of the triumphs and potential of HPC algorithms.

Scientists at the Oak Ridge National Laboratory are using a powerful supercomputer, named "Gaea," the mother of earth in Greek mythology, to better understand the earth's climate. With Gaea, scientists can devise, test, and enhance climate models; and better models will improve our understanding of climate and its changeableness and variation. Such improvements will be useful for making better projections to inform the public and to help leaders in sectors, such as government, industry, agriculture, and transportation, in planning, making decisions, and creating policy (NOAA 2012).

Although climate changes are of some concern to most people, almost all people are concerned with weather, especially when natural disasters like hurricanes and tornadoes threaten lives and property. In late October 2012, a tropical storm developed from a weather disturbance in the Caribbean, south of the island of Hispañola. Named Sandy, it headed northward past Jamaica, Cuba, the Bahamas, and Mid-Atlantic States of the United States, developing into a Category 1 and, then, a Category 2 hurricane (NWS 2012). Although not the most powerful storm, it turned into one of the most devastating hurricanes the northeastern United States had ever experienced. Several factors served to make the storm more damaging: very warm water, a high-pressure system that nudged it westward, and a cold front from the west, with which it merged. Coincidentally, there was a full moon, which augmented the effects of high tide (Jacobson 2012).

Various sensors and observers—balloons, satellites, airplanes, oceanic buoys, and weather stations—monitor storms like Sandy and generate enormous amounts of data, creating a computationally challenging situation that we call **big data**. In the case of Sandy, scientists at the National Hurricane Center (NHC) and at weather-forecasting/monitoring facilities in other parts of the world fed the data into supercomputer models that help predict the storm's track and power. Such models can show great variation in their predictions, and those for this storm were no exception. Some models predicted that the storm would head out to sea, while others predicted that Sandy would do exactly what it did. The fastest machine provided the most accurate model. Even though there was loss of life and incredible damage, without the warnings from the NHC, made possible by the computer models, the devastation would have been much worse (Jacobson 2012).

As another example, Sequoia, a supercomputer housed at Lawrence Livermore National Laboratory, has been used to create the fastest and most detailed simulation of the electrophysiological activity of the human heart. This simulation, which has the spatial resolution of one heart cell, should give scientists great insights into various types of heart disease and the efficacy of various treatments. For instance, with such models, scientists can perform virtual experiments to test the effects of various drugs used to treat cardiac arrhythmia. Associated with thousands of deaths in the United States each year, abnormalities in the electrical activity of the heart induce arrhythmia. Scientists at Johns Hopkins University claim that physicians can use computational models to guide preventive treatment, which can reduce the number of cardiac deaths. These types of computationally based tools will enable doctors to render treatment, personalized for each patient, for heart disease as well as for other diseases, such as cancer (Giordani 2012; Winslow et al. 2012).

Computer scientist, Chandrajit Bajaj, at the University of Texas Advanced Computing Center, has led a team that employed biophysical algorithms and parallel-processing supercomputers to create 3D models of cellular binding sites for viruses, targets for drug therapy. These computing tools reduce the time required to select drugs that are most likely to target effectively the disease-prone binding sites from months to days. As Dr. Bajaj says, "Computers are a good way to accelerate the process of drug design. . . . It takes 10 years to proof out a drug, and a billion dollars or more. Hence computational drug discovery is not only timesaving, but economics tells you this is the way we should be going" (TACC 2012; McBride 2012). Pharmaceutical firms of all sizes employ enormous amounts of computing power for drug discovery, which saves both time and money.

In this module, we examine some of the algorithms for solving classical problems with scientific applications in parallel. In doing so, we investigate speedups and some of the challenges of parallel programming. While not considering actual code, we can gain an appreciation for some of the aspects of designing parallel algorithms.

Embarrassingly Parallel Algorithm: Adding Two Vectors

Some algorithms are so easy to partition onto noncommunicating processes that we call them embarrassingly so. An **embarrassingly parallel algorithm** can divide computation into many completely independent parts that have virtually no communication.

Definition An **embarrassingly parallel algorithm** can divide computation into many completely independent parts that have virtually no communication.

Addition of two vectors is an example of such an algorithm. On the Cartesian plane, to add two vectors, represented by ordered pairs such as (1, 3) and (2, 5), we add **componentwise**. That is, to obtain the result's first component, or coordinate, we add the first components of the ordered pairs; the sum of the second components yields the second coordinate of the result. Thus, the sum of (1, 3) and (2, 5) is as follows:

$$(1, 3) + (2, 5) = (1 + 2, 3 + 5) = (3, 8)$$

An ordered pair is a special case of a **vector**, which is an ordered n-tuple of numbers, $\mathbf{v} = (v_1, v_2, \ldots, v_n)$. (Notice that we boldface the name of a vector, such as **v**, but not the name of a scalar, such as v_1.) To obtain the sum of two vectors $\mathbf{v} = (v_1, v_2, \ldots, v_n)$ and $\mathbf{u} = (u_1, u_2, \ldots, u_n)$, with n elements each, we also compute the sum componentwise, as follows:

$$\begin{aligned}\mathbf{v} + \mathbf{u} &= (v_1, v_2, \ldots, v_n) + (u_1, u_2, \ldots, u_n)\\ &= (v_1 + u_1, v_2 + u_2, \ldots, v_n + u_n)\end{aligned}$$

With n processes, numbered 1 to n, each process can perform the sum of two corresponding coordinates without communication with other processes. Thus, the algorithm for Process i is as follows:

Algorithm for Process *i* in Calculation of Vector Sum w = v + u

$$w_i = v_i + u_i$$

Definitions A **vector** is an ordered n-tuple of numbers, $\mathbf{v} = (v_1, v_2, \ldots, v_n)$. A **componentwise** vector operation is performed component by component, or coordinate by coordinate. For n-tuples $\mathbf{v} = (v_1, v_2, \ldots, v_n)$ and $\mathbf{u} = (u_1, u_2, \ldots, u_n)$, their sum is $\mathbf{v} + \mathbf{u} = (v_1 + u_1, v_2 + u_2, \ldots, v_n + u_n)$.

In this case, assuming an "ideal" concurrent system in which communication is not a consideration, the speedup is linear with $S(n) = n$. Of course, if n processes are not available, some processes must calculate more than one coordinate of the result, and speedup is less.

Quick Review Question 1

Suppose we wish to perform the sum of two vectors of 24 elements each. Compute the most number of coordinate sums per process and the speedup, expressed as a rounded integer, if the following number of processes is available:

a. 24 **b.** 6 **c.** 5 **d.** 4 **e.** 40

Data Partitioning: Adding Numbers

Many applications exist that must compute the sum of a sequence of numbers, x_0, x_1, . . ., x_{n-1}. A sequential algorithm mirrors how we usually add a column of numbers using a calculator. Initially, we enter the first number (x_0) into the calculator. Correspondingly, with the sequential algorithm, we have a variable, say, *sum*, that accumulates the ongoing sum and has an initial value of x_0. On the calculator, we repeatedly press the + key and enter the next number from the sequence. With the algorithm, we also add one element at a time to the old value of *sum*, obtaining a new value for *sum*. On the calculator, we complete the process by pressing the = key; and in the algorithm we return *sum*. The sequential algorithm is as follows.

Sequential Algorithm to Calculate the Sum of a Sequence of Numbers, $x_0, x_1, \ldots, x_{n-1}$

```
sum ← x0
for i going from 1 through n – 1
    sum ← sum + xi
return sum
```

One parallel technique of adding a set of numbers uses **partitioning**. With (often) one process per processor, a specially designated process, called the **root,** or **root process**, splits the list of numbers into nonoverlapping subsets and sends the subsets to different processes, keeping one subset for itself. Each process computes the sum of its subset using a sequential algorithm, such as the preceding one. The processes send their partial sums to the root, which adds these values to obtain the overall sum.

Parallel Data-Partitioning Addition Algorithm with Message-Passing Root's Algorithm

Partition set of *n* numbers and send *n/p* numbers to each of *p* - 1 other processes[1]
Compute sum of remaining *n/p* numbers
Receive *p* partial sums from processes
Compute sum of these *p*-values

[1] Some processes might get a slightly larger list if *p* does not divide into *n* evenly.

Other Process's Algorithm

Receive set of numbers from root
Compute sum of these numbers
Send sum to root

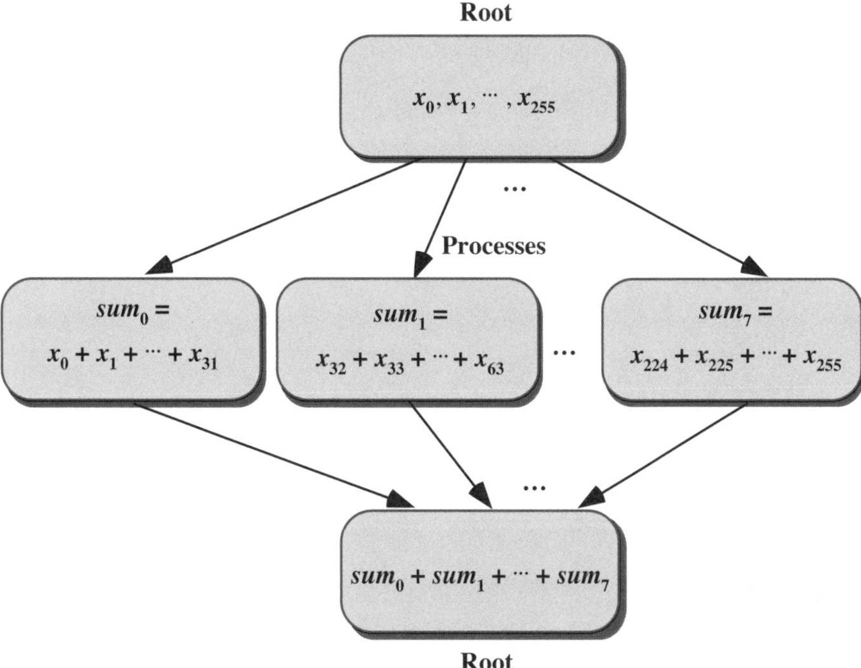

Figure 12.2.1 Sum of 256 numbers using partitioning with root and 8 processes

For example, if there are 256 numbers in the sequence and 8 processors, each with its own process, each process computes the sum of 256/8 = 32 numbers, and the root calculates the final sum involving the 8 partial sums from the processes. This technique is analogous to the scout leader splitting the grocery list, giving each scout a sublist, and keeping one sublist. The scouts and leader work individually to gather their parts of the groceries, which they bring to the leader to purchase. Figure 12.2.1 presents a diagram of the partitioning process, and the algorithms designate the duties of all the processes.

Definitions A parallel **data-partition algorithm** uses one process, the **root**, to partition the data into subsets and to send the subsets to processes, often keeping one subset for itself. Each process performs the appropriate computations with its subset and sends the result to the root for final processing.

Table 12.2.1
Time for Parallel Addition by Partitioning Algorithm

Time to send 256 numbers	Time to add 32 numbers	Time to send 8 numbers	Time to add 8 numbers

Let us perform a rough analysis of the time involved in this parallel addition by partitioning algorithm assuming one process per processor. As Table 12.2.1 illustrates, the total time has four phases: initial communication of n data items from root to p processes, calculation of the sum of n/p numbers by each process, communication of p partial sums from the processes to the root, and sum of p numbers by the root. Ignoring communication time, the speedup factor, $S(p)$, for p processes is roughly as follows:

$$\text{speedup without communication} = \frac{n}{\frac{n}{p} + p}$$

As an exercise shows, this speedup tends to p for large n. In our example, the speedup is as follows:

$$\frac{256}{\frac{256}{8} + 8} = \frac{256}{40} = 6.4$$

We could achieve additional speedup by having the root perform additions as partial sums arrive from the processes.

In the preceding computation of speedup, we are ignoring the time for communication. For a worst-case analysis of communication time, assume the processors do not share memory; each processor has exactly one process; communication is sequential; messages cannot overlap; a message can contain at most one number; and the root distributes all of the numbers and does not keep a subset. In this case, we must move n numbers one at a time before the parallel computation and p numbers afterwards. This communication time might consume as much time as adding the numbers sequentially in one process. Moreover, all process processes are idle while the root performs the final addition of p partial sums. If at all possible, we seek to avoid such idle times by so many processes. The divide-and-conquer approach, which the next section discusses, provides an alternative that is useful for many applications.

Quick Review Question 2

Suppose we need to compute the sum of $1024 = 2^{10}$ numbers, and all communication is sequential. Determine how many values are transferred to and from the root with partitioning for each of the following number of processes:

a. 2 **b.** 8 **c.** 256

Quick Review Question 3

For the situations in Quick Review Question 2, determine how many addition operations occur at the same time. Consider the parallel computations by the processes and the computations by the root.

Divide and Conquer: Adding Numbers

Divide-and-conquer algorithms are widely used in computer science, particularly in parallel processing. With such an algorithm, the problem is divided into subproblems of the same form. We continue dividing the problems into smaller and smaller problems. Then, we solve the small problems and reassemble the solutions.

Figures 12.2.2 and 12.2.3 diagram a divide-and-conquer solution of adding 256 numbers on 8 processes, $p_0, p_1, \ldots, p_7$, with 1 process per processor. Root process, p_0, which initially has all the numbers, transmits half of the numbers to process p_4, so that each is in charge of 128 numbers. Concurrently, p_0 and p_4 send half their numbers (64 numbers each) to p_2 and p_6, respectively. Then, these four processes (p_0, p_2, p_4, p_6) pass half the values (32 numbers each) to the remaining processes (p_1, p_3, p_5, p_7). In all, this tree of divisions to 8 processes in Figure 12.2.2 has $\log_2 8 = 3$ levels of divisions. In Figure 12.2.2, each arrow indicates a message containing half a process's values, and the relative thickness of the arrow represents the amount of data. The next-lower level of the figure shows that a process that sent a message with data becomes responsible for the remaining half of its data.

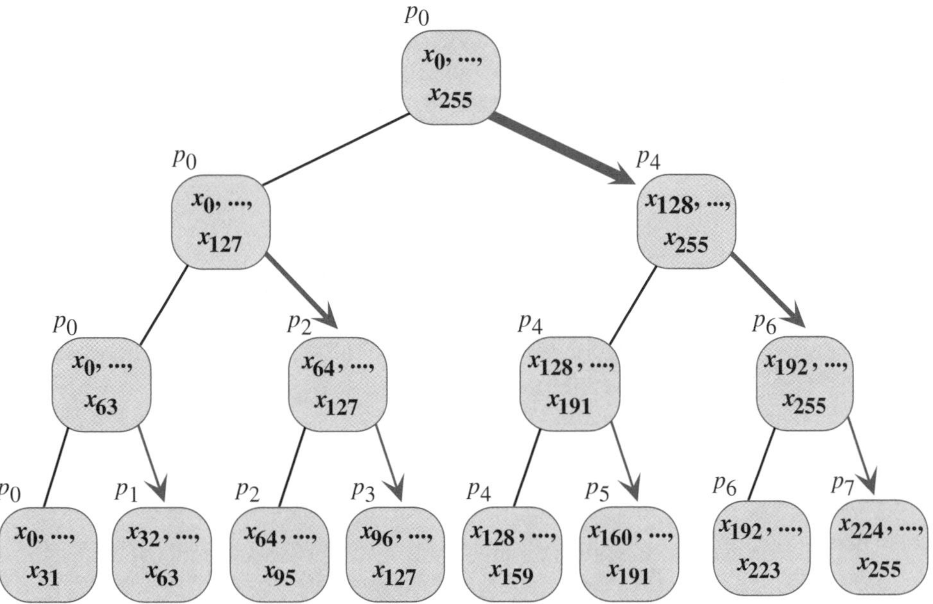

Figure 12.2.2 "Divide" phase of divide-and-conquer algorithm for sum of 256 numbers with 8 processes. Each arrow represents a message containing half the values from the sender.

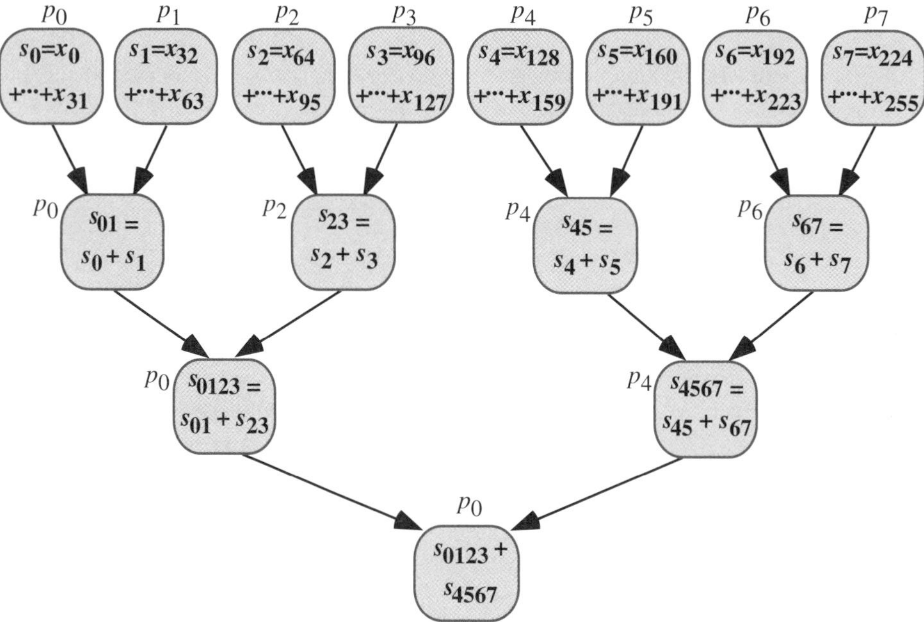

Figure 12.2.3 "Conquer" phase of divide-and-conquer algorithm for sum of 256 numbers with 8 processes. Each arrow represents a single sum value.

After the "divide" phase comes the "conquer" phase (Figure 12.2.3). Each process calculates the sum of its 32 numbers. The odd-numbered processes, p_1, p_3, p_5, p_7, send their results to the even-numbered processes, p_0, p_2, p_4, p_6, respectively. Each of the even-numbered processes adds its answer to the result that an odd-numbered process communicated. Retracing the path when the processes were dividing the data set, p_2 and p_6 send their answers to p_0 and p_4, respectively. Each of processes p_0 and p_4 adds its two numbers, and finally, p_0 computes the sum of its and p_4's results. In Figure 12.2.3, each arrow represents a single sum value.

Definition A **divide-and-conquer algorithm** divides a problem into subproblems of the same form and then divides these into subproblems of the same form, and so on. The small problems are solved, and the final solution is assembled.

To analyze the time involved, let us initially ignore communication and assume one process per processor. At the first step of the "conquer" phase, each process is adding $n/p = 256/8 = 32$ numbers, or performing $(n/p) - 1 = 31$ addition operations. However, at each level of the tree thereafter, we have only simultaneous sums of pairs of numbers. Thus, after time for the initial additions by the 8 processes, we need only the time to compute 3 more sums. The number of these sums (3) is the same as the number of levels of divisions. In general, a system with p number of processes has $\log_2 p$ of these division levels. Thus, in all we have $(n/p) - 1 + \log_2 p$

sums; without communication, the speedup factor $S(p)$ for p processes is roughly as follows:

$$\text{speedup without communication} = \frac{n}{\frac{n}{p} - 1 + \log_2 p}$$

This speedup tends to p for large n.

For 256 numbers and 8 processes, during the "divide" phase, communication time includes time to move half the numbers initially, and because of concurrency, one-fourth and one-eighth of the numbers, respectively, at the next two levels. Thus, in all, the process must have time to communicate (256)(7)/8 = 224 numbers, as the following illustrates:

$$\frac{256}{2} + \frac{256}{4} + \frac{256}{8} = \frac{(256)(7)}{8}$$

In general, for p being a power of 2, the divide phase communicates the following number of values:

$$\frac{n}{2} + \frac{n}{4} + \frac{n}{8} + \ldots + \frac{n}{p} = \frac{n(p-1)}{p}$$

For the processes sending their results, communication time is small and approximately proportional to the number of levels, $\log_2 p$. Thus, ignoring startup times for processes, the total communication time is approximately proportional to the following expression:

$$\frac{n(p-1)}{p} + \log_2 p$$

This value is smaller than the communication time for the partitioning algorithm, which is approximately proportional to $n + p$. Moreover, the time in which processes are idle is smaller.

Quick Review Question 4

Suppose we need to compute the sum of $1024 = 2^{10}$ numbers, and all communication is sequential. Determine how many values are transferred concurrently in the divide-and-conquer algorithm of this section for each of the following number of processes:

a. 2 **b.** 8 **c.** 256

Quick Review Question 5

For the situations in Quick Review Question 4, determine how many addition operations occur at the same time.

Quick Review Question 6

Compare the results of Quick Review Questions 2 and 3 with Quick Review Questions 4 and 5, respectively. For the given situations, determine the better summation algorithm, partitioning or divide and conquer.

a. For a small number of processes, which algorithm uses less communication?
b. For a large number of processes, which algorithm performs fewer concurrent additions?

Parallel Random Number Generator

Another example of a nearly embarrassingly parallel algorithm is the Monte Carlo estimation of area under a curve (Module 9.2). For n darts, n processors, one process per processor, and shared memory, each process can compute the coordinates of one "dart" hit and increment a shared counter if the "dart" hits below the curve. In general, for n darts and p processes, each process can compute the number of hits for n/p darts. One process performs the final step of estimating the area using this count.

However, a problem exists. Each process must generate random numbers, but generation of the same pseudorandom number sequence by two processes would skew the result.

We illustrate a solution to the problem using the following simple sequential random number generator, which produces the sequence 1, 7, 5, 2, 3, 10, 4, 6, 9, 8 before cycling back to 1:

$$r_0 = 1$$
$$r_n = (7\ r_{n-1}) \bmod 11, \text{ for } n > 0$$

Suppose for simplicity that two processes are available ($n = 2$), and each needs pseudorandom numbers. Using the number of processes as the exponent of the coefficient, 7, of r_{n-1}, we have the following:

$$7^2 \bmod 11 = 5$$

Instead of using 7 as the coefficient in the generating function, we employ 5 as follows:

$$r_n = (5r_{n-1}) \bmod 11$$

For one computer, we use the seed $r_0 = 1$, while for the other computer we employ $r_0 = 7$, the second number in the original sequence. Thus, one process generates the sequence 1, 5, 3, 4, 9 with the following random number generator:

$$r_0 = 1$$
$$r_n = (5\ r_{n-1}) \bmod 11$$

The second process generates alternate random numbers from the original sequence 7, 2, 10, 6, 8 by using the following random number generator:

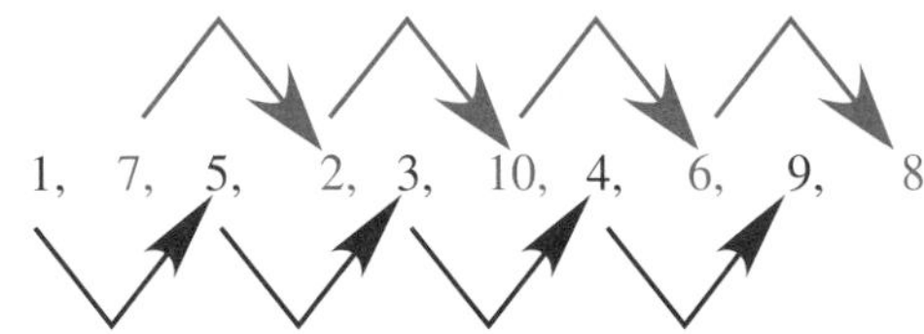

Figure 12.2.4 Parallel random number sequences for two processes

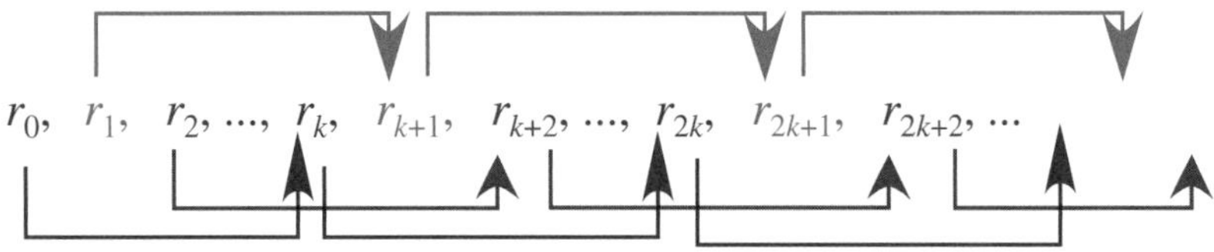

Figure 12.2.5 Parallel random number sequences for ***k*** processes

$$r_0 = 7$$
$$r_n = (5\ r_{n-1}) \bmod 11$$

Figure 12.2.4 shows the sequence developed with the generating function $r_n = (7\ r_{n-1}) \bmod 11$ and the sequences for the two processes with the generating function $r_n = (5\ r_{n-1}) \bmod 11$.

In general, suppose we have a pseudorandom number generator of the following form:

$$r_0 = 1$$
$$r_n = (a\ r_{n-1}) \bmod m$$

For k processes, we compute the new coefficient as $A = a^k \bmod m$. With the processes numbered 0, 1, 2, . . ., $k - 1$, the seed for process i is $a^i \bmod m$. In the preceding example, for process 0, the seed is $7^0 = 1$; and for process 1, the seed is $7^1 = 1$. The algorithm is given next, and Figure 12.2.5 shows the sequences for this general case.

Parallel Random Number Generator Algorithm for Process *i* of *k* Processes from Sequential Generator $r_0 = 1$, $r_n = (a\ r_{n-1}) \bmod m$

$r_0 = a^i \bmod m$
$r_n = (A\ r_{n-1}) \bmod m$, where $A = a^k \bmod m$

Quick Review Question 7

Consider the following very small, sequential random number generator:

$$r_0 = 1$$
$$r_n = (7\ r_{n-1}) \bmod 11, \text{ for } n > 0$$

Suppose we wish to have the corresponding parallel version for five processes. Determine the following:

a. The generating function
b. Seeds for the five processes
c. The sequence for the process with seed 1

Sequential Algorithm for the *N*-Body Problem

The ***N*-body problem**, which concerns simulations of the interactions and movements of a number of objects, or bodies, in space, has many applications, including fluid dynamics, evolution of the galaxy, and molecular dynamics. A sequential algorithm for such a simulation has the following general design.

General Sequential Algorithm for Solving the *N*-Body Problem

```
initialize positions and velocities of objects
for time going from start to finish by step size Δt
    calculate forces
    move bodies
```

Gravity causes acceleration and movement. The **magnitude of the gravitational force between two bodies** with masses m_1 and m_2 at a distance r apart is

$$F = \frac{Gm_1m_2}{r^2}$$

where $G = 6.67 \times 10^{-11} \mathrm{m}^3\mathrm{kg}^{-1}\mathrm{s}^{-2}$ is **Newton's gravitational constant**. For example, suppose body 1 with mass $m_1 = 2$ kg is located at position (1, 3, 0) and body 2 with mass $m_2 = 4$ kg is at (2, 0, 3), where distances are in meters. We calculate the distance between the objects in a similar fashion to the way in which we compute the distance between points on the plane; we take the square root of the sum of

Definitions The **distance between two points**, (x_1, y_1, z_1) and (x_2, y_2, z_2), is

$$r = \sqrt{(x_2 - x_1)^2 + (y_2 - y_1)^2 + (z_2 - z_1)^2}$$

The **magnitude of the gravitational force between two bodies** with masses m_1 and m_2 at a distance r apart is

$$F = \frac{Gm_1m_2}{r^2}$$

where $G = 6.67 \times 10^{-11}\, \mathrm{m}^3\,\mathrm{kg}^{-1}\,\mathrm{s}^{-2}$ is **Newton's gravitational constant**.

the squares of the differences in corresponding coordinates, as in the following example:

$$r = \sqrt{(2-1)^2 + (0-3)^2 + (3-0)^2} = \sqrt{19}\,\text{m}$$

Thus, the following expression gives the magnitude of the gravitational force attracting the two bodies:

$$F = \frac{(6.67\times10^{-11}\,\text{m}^3\text{kg}^{-1}\text{s}^{-2})(2\,\text{kg})(4\,\text{kg})}{19\,\text{m}^2} = 2.8 \times 10^{-11}\ \text{N}$$

Quick Review Question 8

Consider body 1 of mass 35×10^9 kg at location (4000, 0, 5000) and body 2 of mass 14×10^9 kg at location (2000, 3000, –1000), with distances in meters. Compute the following:

a. The distance between the two bodies

b. The magnitude of gravitational force between two bodies

The **direction of force** on body 1 at (x_1, y_1, z_1) by body 2 at (x_2, y_2, z_2) is the unit vector, or vector of length 1, from body 1 to body 2, namely,

$$\mathbf{d} = \left(\frac{x_2 - x_1}{r}, \frac{y_2 - y_1}{r}, \frac{z_2 - z_1}{r}\right)$$

where r is the distance between the bodies. For example, with bodies at positions (1, 3, 0) and (2, 0, 3), the direction of force on body 1 by body 2 is

$$\mathbf{d} = \left(\frac{2-1}{\sqrt{19}}, \frac{0-3}{\sqrt{19}}, \frac{3-0}{\sqrt{19}}\right) = \left(\frac{1}{\sqrt{19}}, \frac{-3}{\sqrt{19}}, \frac{3}{\sqrt{19}}\right)$$

Definition The **unit direction vector from (x_1, y_1, z_1) to (x_2, y_2, z_2)**, or **direction of the force** on body 1 by body 2 at those points, respectively, is as follows:

$$\mathbf{d} = \left(\frac{x_2 - x_1}{r}, \frac{y_2 - y_1}{r}, \frac{z_2 - z_1}{r}\right),$$

where $r = \sqrt{(x_2 - x_1)^2 + (y_2 - y_1)^2 + (z_2 - z_1)^2}$ is the distance between the points.

Quick Review Question 9

For the bodies from Quick Review Question 8, determine the direction of force on body 1 by body 2.

With magnitude and direction of the force on body 1 by body 2, we can compute the force vector as the magnitude of the force times the direction of force. Thus, with F being the magnitude of the gravitational force, the force vector $\mathbf{F}$ is as follows:

$$\mathbf{F} = (F_x, F_y, F_z) = F\left(\frac{x_2 - x_1}{r}, \frac{y_2 - y_1}{r}, \frac{z_2 - z_1}{r}\right)$$
$$= \left(\frac{F(x_2 - x_1)}{r}, \frac{F(y_2 - y_1)}{r}, \frac{F(z_2 - z_1)}{r}\right)$$

This vector indicates that the force that body 2 exerts on body 1 in the x-direction is $F_x = \frac{F(x_2 - x_1)}{r}$, in the y-direction is $F_y = \frac{F(y_2 - y_1)}{r}$, and in the z-direction is $F_z = \frac{F(z_2 - z_1)}{r}$. Thus, for the preceding example with magnitude of the gravitational force between them being $F = 2.8 \times 10^{-11}$ N and the direction vector being $\mathbf{d} = \left(\frac{-1}{\sqrt{19}}, \frac{3}{\sqrt{19}}, \frac{-3}{\sqrt{19}}\right)$, the force exerted by body 2 on body 1 is the vector

$$\mathbf{F} = \left(\frac{-2.8\times10^{-11}}{\sqrt{19}}, \frac{8.4\times10^{-11}}{\sqrt{19}}, \frac{-8.4\times10^{-11}}{\sqrt{19}}\right) \approx (0.64 \times 10^{-11}, 1.9 \times 10^{-11}, -1.9 \times 10^{-11})$$

Quick Review Question 10

For the bodies from Quick Review Questions 8 and 9, determine the force vector indicating the force that body 2 exerts on body 1.

The total force on a body is the vector sum of all forces on the body. The sequential algorithm for computing the total force on each of the N bodies, body 1, body 2, . . ., body N, is given next. For each body i except the last, we compute the force vector from it to each body j, where $j > i$, and vice versa, and add these force vectors to the appropriate ongoing sums. Thus, when the algorithm is complete, for each i, $\mathbf{F}_i$ is the total accumulated sum of all force vectors on body i.

Sequential Algorithm to Determine the Total Force on Each of the *N* Bodies

assume $\mathbf{F}_i$ is 0 N for i from 1 to N
for i going from 1 to $N - 1$
  for j going from $i + 1$ to N
    calculate distance ($r_{i,j}$) between body i and body j
    calculate direction vector ($\mathbf{d}_{i,j}$) from body i to body j
    calculate magnitude ($F_{i,j}$) of force between them
    add force body j exerts on body i ($F_{i,j}\,\mathbf{d}_{i,j}$) to $\mathbf{F}_i$
    add force body i exerts on body j ($-F_{i,j}\,\mathbf{d}_{i,j}$) to $\mathbf{F}_j$

Quick Review Question 11

In the sequential algorithm to determine the total force on each of the N bodies, suppose the number of bodies is 15. Give the following:

a. The range of numbers for the bodies
b. The sequence of values for j when i is 5

The next major part of the simulation is to move all bodies. To do so, we must first calculate their velocities. Recall that by Newton's second law of motion (Module 3.1, "Modeling Falling and Skydiving," the section "Physics Background"), a force **F** on a body of mass m creates an acceleration **a** on that body, and

$$\mathbf{F} = m\mathbf{a}$$

or

$$\mathbf{a} = \frac{\mathbf{F}}{m}$$

However, acceleration is the instantaneous rate of change of velocity with respect to time, or the derivative of velocity,

$$\mathbf{a} = d\mathbf{v}/dt$$

Thus, for small change in time Δt, we have the following approximation of the small change in the velocity vector, $\Delta\mathbf{v}$:

$$\Delta\mathbf{v} \approx \mathbf{a}\Delta t = \frac{\mathbf{F}\,\Delta t}{m}$$

The rate of change of velocity is approximately **a** for one time unit, while the change in velocity is about $\mathbf{a}\Delta t$ for Δt. We estimate the velocity vector ($\mathbf{v}_{t+\Delta t}$) at time step $t + \Delta t$ as the sum of the velocity vector at the previous time step ($\mathbf{v}_t$) and the change in velocity ($\Delta\mathbf{v}$), as follows:

$$\mathbf{v}_{t+\Delta t} \approx \mathbf{v}_t + \Delta\mathbf{v} \approx \mathbf{v}_t + \frac{\mathbf{F}\,\Delta t}{m}$$

We use this velocity vector to calculate the new position of a body. Recall that velocity is the instantaneous rate of change of position with respect to time, or the derivative of position with respect to time,

$$\mathbf{v} = d\mathbf{s}/dt$$

Thus, for a small change in time Δt, the small change in position is as follows:

$$\Delta\mathbf{s} \approx \mathbf{v}\Delta t$$

In one time unit, the change in position is **v**, while in Δt time units the change in position is $\mathbf{v}\Delta t$. As with the new velocity, the new position ($\mathbf{s}_{t+\Delta t}$) for the body at time

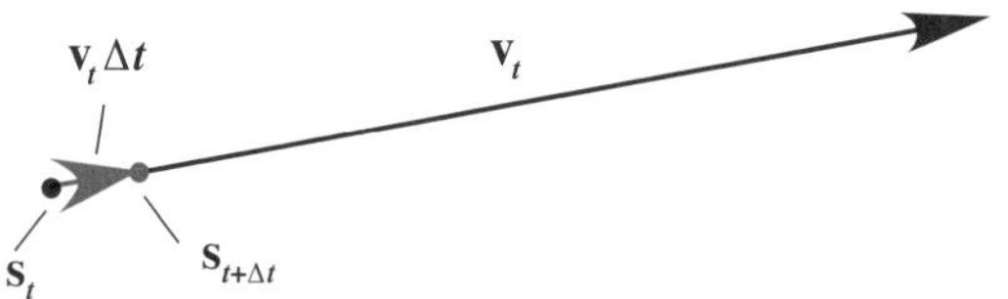

Figure 12.2.6 New position from old

$t + \Delta t$ is approximately the sum of the position at the previous time step (s_t) and the change in position ($\Delta\mathbf{s}$), as follows:

$$\mathbf{s}_{t+\Delta t} \approx \mathbf{s}_t + \Delta\mathbf{s}$$
$$\approx \mathbf{s}_t + \mathbf{v}_t\Delta t$$

That is, as Figure 12.2.6 illustrates, we estimate the new position as the old position plus the product of the old velocity and the change in time. We put these various aspects together in the sequential algorithm for computing the new positions of all the bodies.

Sequential Algorithm to Move *N* Bodies

for i going from 1 to N
  calculate change in velocity vector, $\Delta\mathbf{v}$, as $\mathbf{F}_i(\Delta t/m)$
  calculate change in position vector, $\Delta\mathbf{s}$, as $\mathbf{v}\Delta t$
  add $\Delta\mathbf{v}$ to $\mathbf{v}$
  add $\Delta\mathbf{s}$ to $\mathbf{s}$
  assign 0 to $\mathbf{F}_i$

Quick Review Question 12

Consider body 1 of mass 35×10^9 kg at location $\mathbf{s} = (4000, 0, 5000)$ with velocity $\mathbf{v} = (500, 300, -100)$. Suppose $\Delta t = 0.1$ s and $\mathbf{F}_1 = (3500 \times 10^9, -1400 \times 10^9, -7000 \times 10^9)$. Evaluate the following:

a. $\Delta\mathbf{v}$
b. $\Delta\mathbf{s}$
c. The new $\mathbf{v}$
d. The new $\mathbf{s}$

Barnes-Hut Algorithm for the *N*-Body Problem

Suppose N processes are available so that each process can be responsible for exactly one body. After computation of the force vector for a body on a time step, the movement of a body is completely independent of the other bodies. Thus, this phase of the simulation can be embarrassingly parallel.

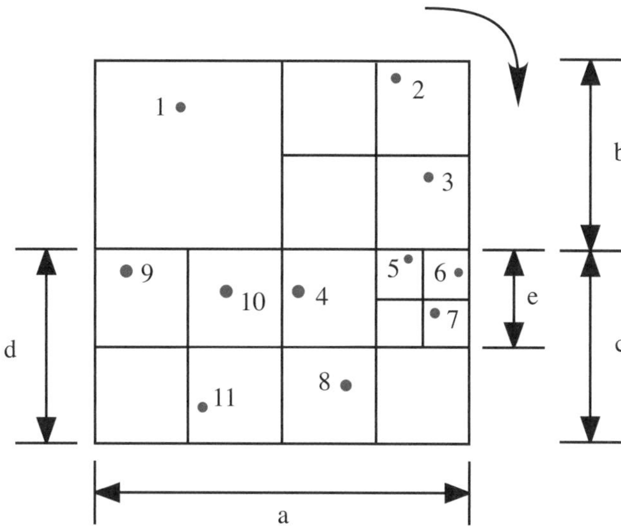

Figure 12.2.7 Partitioning of square into subsquares using the Barnes-Hut algorithm

Overall this simulation consumes a great deal of time, particularly if the number of bodies N is large and the time step Δt is small. For each time step, the computation of the total force uses a nested loop whose body executes approximately N^2 times. Concurrency can help in the movement phase. However, if we have one process responsible for one body's computations, for the force phase, communication times would be approximately proportional to N^2 because of the information required about interactions with all other bodies. A simplification that can reduce communication and speed the process is **clustering**, in which we approximate several bodies as a **cluster** that we consider to be one body.

The **Barnes-Hut algorithm**, another divide-and-conquer parallel algorithm, performs a simulation of the N-body problem employing clustering. For each time step, the algorithm divides space, which we can consider as a cube, into eight subcubes. Any subcube that does not contain a body is eliminated from further consideration. The algorithm continues the partitioning process on any subcube that contains more than one body. Eventually, we have a collection of cubes of varying sizes that each contains one or no body. Figure 12.2.7 presents an example of the 2D counterpart to this process, in which squares are divided into subsquares.

Quick Review Question 13

Partition the square in Figure 12.2.8 into subsquares using the Barnes-Hut algorithm.

While performing the partitioning process, the algorithm generates an **octtree,** with each node, or tree vertex, corresponding to a cube and with branches going to at most 8 nodes, representing subcubes of that cube. Figure 12.2.9 illustrates the 2D counterpart, a **quadtree**, to accompany the partitioned square from Figure 12.2.7. To develop the quadtree, for each subdivided square, we start with the top left square and travel in a clockwise fashion, generating a child node for each square.

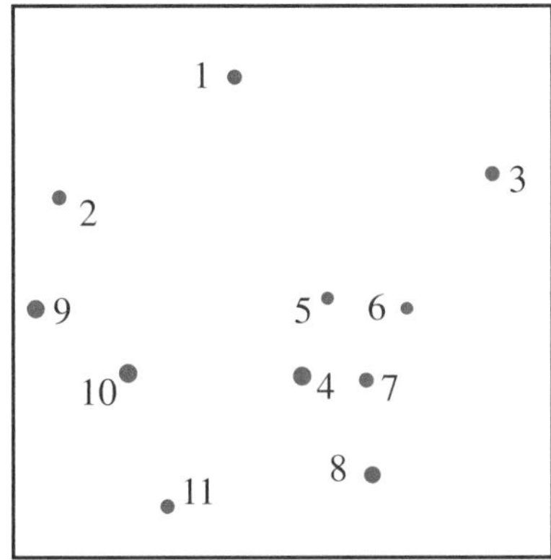

Figure 12.2.8 Square of bodies for Quick Review Question 13

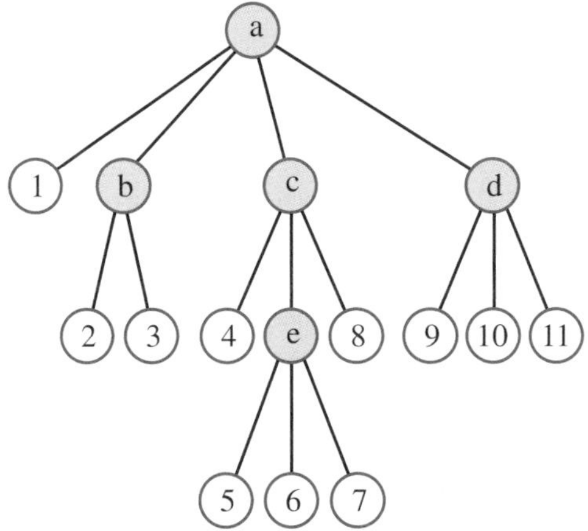

Figure 12.2.9 Quadtree to accompany Figure 12.2.7, in which a filled circle with a letter represents a square and an unfilled circle with a number represents a body

Quick Review Question 14

Generate a quadtree for the partition of Figure 12.2.8.

In each node, we store the total mass and center of mass for all the bodies in that subdivision (cube for 3D or square for 2D). As we are computing the total force on a body, say, body 1 in Figure 12.2.9, we **traverse**, or travel through, the tree, starting at the top node, called the **root**, and accumulate the force exerted by the other bodies on body 1. In determining interactions, if a node representing several bodies is sufficiently far, we do not consider the bodies individually but as a clustered body. For example, in Figure 12.2.9, the cluster that node e represents might be at a distance greater than some predetermined distance from body 1. Thus, we do not consider body 1's interactions with bodies 5, 6, and 7 individually but perform the computations with the information in node e as if it were one object interacting with body 1.

Thus, a small-enough threshold distance can significantly decrease the amount of communication between processes during the force computation phase.

As with the sequential algorithm, after computing the forces, we move the bodies. Because this phase does not require communication with other processes, this part of the simulation can run very quickly in parallel. Periodically, we reformulate the octtree. The following presents the general Barnes-Hut algorithm.

Barnes-Hut Algorithm for Solving the *N*-Body Problem in Parallel

```
generate octtree as follows:
   repeatedly subdivide cube containing more than one body into 8 subcubes
   generate tree with nodes representing nonempty subcubes
   store in each node the total mass and center of mass of its bodies
for time going from start to finish by step size of Δt
   calculate forces as follows:
       for each body, traverse tree starting at root
           perform computations with node when center of mass is "far"
               or when node has no children
   move bodies
   update center of mass
   when "appropriate," rebuild octtree
```

To calculate the total force on a body requires time on the order of log(N) for a fairly bushy tree. Because we perform this process for each of the N bodies, the total time is approximately proportional to N log(N), a significant improvement over the sequential force computation algorithm, which takes about N^2 steps. The periodic reformulation of an octtree also takes time on the order of N log(N). Moreover, the movement phase can easily be preformed in parallel without communication. Thus, overall the Barnes-Hut Algorithm is usually an improvement over the sequential version of simulation of the N-body problem. Difficulties do exist, however, in attempts to parallelize. For example, the distribution of the bodies is usually nonuniform, which leads to an unbalanced tree and, consequently, longer traversal times.

Quick Review Question 15

Consider the bodies in Figure 12.2.7 with quadtree in Figure 12.2.9.

a. In the sequential algorithm to compute the total force vector for each body, determine the number of times the body inside the nested loop is executed.

b. Suppose in execution of the Barnes-Hut Algorithm the threshold is such that a body interacts only with nodes (b, c, d, e) and other bodies in its same subsquare from the first partition. The two exceptions to this rule are that the distances between body 4 and node e and between body 8 and node e are considered to be beyond the threshold distance. For example, body 1 interacts with nodes b, c, and d; Body 2 interacts with bodies 1 and 3 and nodes c and d; body 4 interacts with bodies 1 and 8 and nodes b, e, and d; and body 5 interacts with bodies 1, 4, 6, 7, and 8 and nodes b and d. Determine the total number of interactions for the force computation.

Exercises

1. Give an embarrassingly parallel algorithm to compute a scalar times a vector. For example, 3(4, 2, –1) = (3 · 4, 3 · 2, 3 · (–1)) = (12, 6, –3).
2. Suppose a computer graphics screen has resolution 1024 × 768, that is, 1024 **pixels** (dots on screen) wide and 768 pixels high. Each pixel is has levels of red, green, and blue, where each value is between 0.0 and 1.0. Suppose 16 processes are available for parallel computation. Give an algorithm for a nearly embarrassingly parallel algorithm to add 0.05 to the red component up to a maximum of 1.0 for every pixel.
3. As the section "Data Partitioning: Adding Numbers" indicates, ignoring communication, the speedup factor, $S(p)$, for p processes is roughly $\dfrac{n}{\dfrac{n}{p}+p}$. Show that for large n, this speedup tends to p. One way to do so is first to simplify the expression by obtaining a common denominator for n/p and p and by inverting the resulting fraction and multiplying. Use the fact that for very large n and relatively small p, $n + p^2 \approx n$.
4. Similar to Exercise 3, show that the speedup without communication for the divide-and-conquer addition algorithm, $\dfrac{n}{\dfrac{n}{p}-1+\log_2 p}$, of the section "Divide and Conquer: Adding Numbers" tends to p for large n.
5. **a.** Give a parallel partitioning algorithm to compute the maximum of n numbers with p processes.
   b. Analyze the communication cost and the speedup.
6. Repeat Exercise 5 for a divide-and-conquer algorithm.
7. Develop a divide-and-conquer algorithm to find the number of occurrences of a particular element in an array, or vector.
8. **a.** Develop a divide-and-conquer algorithm to perform a parallel merge sort of an array. *Hint:* After division, each process sorts its part of the array using an efficient algorithm. Then, the subarrays are merged into larger sorted subarrays.
   b. Analyze the communication and computation times if the number of processes is equal to the number of array elements, n.
   c. Repeat Part b if the number of processes is less than the number of array elements. Assume that the computation time for the sequential sorting algorithm employed is proportional to $m \log(m)$, where m is the number of elements being sorted.
9. Consider the following random number generator:

$$r_0 = 1$$
$$r_n = (59\, r_{n-1}) \bmod 349, \text{ for } n > 0$$

a. Suppose four processes working concurrently need to generate random numbers. Give the corresponding generating function.
b. Give the seeds for the four processes.
c. Give the most number of pseudorandom values the original function generates.
d. Give the most number of pseudorandom values the parallel function generates.

10. Repeat Exercise 9 for the following random number generator and 3 processes:

$$r_0 = 1$$
$$r_n = (523\ r_{n-1}) \bmod 1021, \text{ for } n > 0$$

11. Suppose a pseudorandom number generator of the following form generates the most number of values possible:

$$r_0 = 1$$
$$r_n = (a\ r_{n-1}) \bmod m$$

a. Give the number of pseudorandom numbers this function generates.
b. Suppose $m - 1 = pq$, where p and q are positive integers. Give the number of pseudorandom numbers the following function generates:

$$r_0 = 1$$
$$r_n = (a^p\ r_{n-1}) \bmod m$$

12. Consider the following generating function for a pseudorandom number generator:

$$r_n = (a\ r_{n-1} + c) \bmod m$$

For k processes, the parallel version is as follows:

$$r_n = (A\ r_{n-1} + C) \bmod m$$

where $A = a^k \bmod m$ and $C = c(1 + a + a^2 + \cdots + a^{k-1}) \bmod m$. Notice that the value of the coefficient is the same as with the version in the text, where $c = 0$ (Wilkinson and Allen 1999).

a. Determine the parallel version for two processes and the generating function $r_n = (7\ r_{n-1} + 4) \bmod 11$, for $n > 0$.
b. Repeat Part a for five processes.

13. Repeat Exercise 12a for three processes and the generating function $r_n = (229\ r_{n-1} + 1) \bmod 10{,}000$, for $n > 0$.
14. Suppose in a Barnes-Hut algorithm, the measure of "far" is relative. Instead of visiting its children, we use the information from a node if the following is true for some number, *threshold:*

(width of subsquare for node)/(distance to body) < *threshold*

Suppose in Figure 12.2.7 that the smallest square, such as the one containing body 6, is 1 unit wide; so that the square with body 4 has width 2 units; and the square with body 1 is 4 units wide. Suppose *threshold* is 0.5. Estimate (width of subsquare for node)/(distance to body) and if the node information would be used in the force computation for the following situations:

a. Body 4 and node e
b. Body 1 and node e
c. Body 1 and node d

Projects

For an additional project, see Project 8 from Module 14.9, "Clouds—Bringing It All Together." See the text's website for additional high-performance computing material.

Complete the following projects using your computational tool.

1. Develop a sequential program to simulate the Parallel Algorithm for Addition Using Data Partitioning. Create a function for the root's algorithm with parameters for the number of processes (p) and an array of numbers. Create another function for the section "Other Process's Algorithm," which has parameters for the process number (or **rank**) and an array of numbers. Assume the number of summands is evenly divisible by p. Display communications, such as "Process 3 sending partial sum 537 to Root."

*In each of Projects 2-4, develop a sequential program to simulate the indicated parallel divide-and-conquer algorithm. Create a function with the following parameters: the number of processes (*p*), the number of participating processes at this level (*participants*; see Figure 12.2.2), the process number (or *rank*), the array, left index (or one less) and right index (or one less) of the subarray under consideration (see Figure 12.2.2).*

Assume the number of array elements is a power of 2 that is evenly divisible by p. *Display communications, such as "Process 0 receiving work, # participants = 1, left = 0, right = 255"; "Process 0 sending work to process 4, requesting sum from 128 to 255"; and later, "Process 4 sending sum (sum = 131) of its part." For the function's algorithm, if all processes are participating (*participants = p*), perform the operation sequentially on its subarray. Otherwise, perform the required operation (addition or maximum) on two calls to the function with appropriate arguments. Thus, the function is* ***recursive****, or is a function that calls itself.*

2. Addition of a list of numbers
3. Finding the maximum of a list of numbers
4. Determining the number of occurrences of a value in a list of numbers
5. For the situation in Exercise 9, develop the sequential random number generator and the random number generators for each process. Display the complete sets of random numbers generated.
6. Develop a simulation in 2D or 3D of the N-body problem using a sequential algorithm. Generate an animation of the simulation.
7. Suppose a pipeline of p processes operates on a stream of integers, 2, 3, 4, . . ., passed from one process to the next. Each process remembers the first number, N, it receives and passes to the next process all remaining numbers

in the sequence that are not multiples of N. When the last process receives a number, the algorithm stops. (This algorithm is a parallel version of the sequential sieve of Eratosthenes algorithm.)

a. Determine the task of this algorithm.
b. Develop a program for a sequential version of this algorithm.
c. Develop a program to simulate the pipeline version.
d. Write an analysis of the amounts of computation and communication for the sequential and pipeline versions and of the speedup.

Answers to Quick Review Questions

1. **a.** 1, 24
b. 4, 6
c. 5, 5
e. 6, 4
e. 1, 24

2. **a.** 1026 = 1024 + 2
b. 1032 = 1024 + 8
c. 1280 = 1024 + 256
Notice that as the number of processes increases, the communication cost increases.

3. **a.** 512 = 511 + 1
b. 134 = 127 + 7
c. 258 = 3 + 255

4. **a.** 513 = 512 + 1
b. 899 = 512 + 256 + 128 + 3 = 1024(7)/8 + $\log_2 8$
c. 1028 = 512 + 256 + 128 + 64 + 32 + 16 + 8 + 4 + 8 = 1024(255)/256 + $\log_2 256$

5. **a.** 512 = 511 + 1 = (1024/2) - 1 + $\log_2 2$
b. 130 = 127 + 3 = (1024/8) - 1 + $\log_2 8$
c. 11 = 3 + 8 = (1024/256) - 1 + $\log_2 256$
Notice the improvement using 256 processes for the divide-and-conquer algorithm over the partitioning algorithm.

6. **a.** divide-and-conquer
b. divide-and-conquer

7. **a.** $r_n = (10\ r_{n-1}) \bmod 11$, for $n > 0$ because $7^5 \bmod 11 = 10$
b. 1, 7, 5, 2, 3
c. 1, 10

8. **a.** 7000 m because $r = \sqrt{(2000-4000)^2 + (3000-0)^2 + (-1000-5000)^2}$

$$= \sqrt{4\times10^6 + 9\times10^6 + 36\times10^6} = 7000$$

b. 667.0 N = $F = \dfrac{(6.67\times10^{-11})(35\times10^9)(14\times10^9)}{49\times10^6} = 6.67 \times 10^2$

9. (–2/7, 3/7, –6/7) = ((2000 – 4000)/7000, (3000 – 0)/7000, (–1000 – 5000)/7000)

10. Approximately (–190.6, 285.9, –571.7) = 667.0(–2/7, 3/7, –6/7)

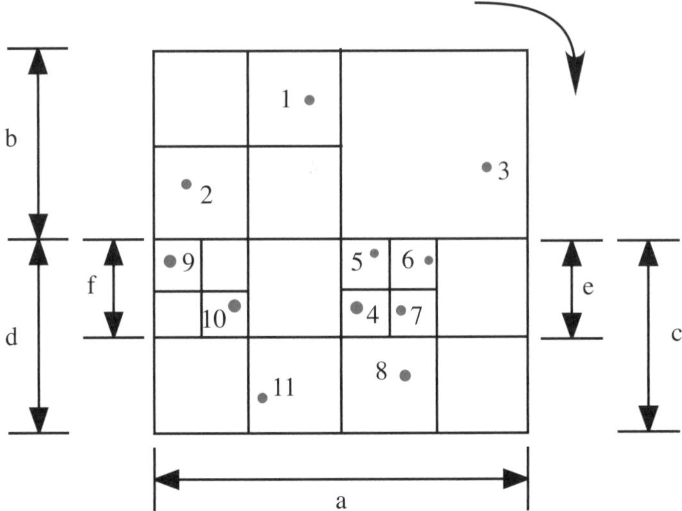

Figure 12.2.10 Partition for Quick Review Question 13

11. **a.** body 1 through body 15
b. integers 6 through 15

12. **a.** (10, –4, –20) because $\Delta\mathbf{v} = \mathbf{F}_1(\Delta t/m) = ((3500 \times 10^9)(0.1)/(35 \times 10^9), (-1400 \times 10^9)(0.1)/(35 \times 10^9), (-7000 \times 10^9)(0.1)/(35 \times 10^9))$
b. (50, 30, –10) because $\Delta\mathbf{s}$ = (old $\mathbf{v}$) Δt = (500, 300, –100) 0.1
c. (510, 296, –120) because new $\mathbf{v}$ = old $\mathbf{v} + \Delta\mathbf{v}$ = (500, 300, –100) + (10, –4, –20)
d. (4050, 30, 4990) because new $\mathbf{s}$ = old $\mathbf{s} + \Delta\mathbf{s}$ = (4000, 0, 5000) + (50, 30, –10)

13. See Figure 12.2.10 for the answer.

14. See Figure 12.2.11 for the answer.

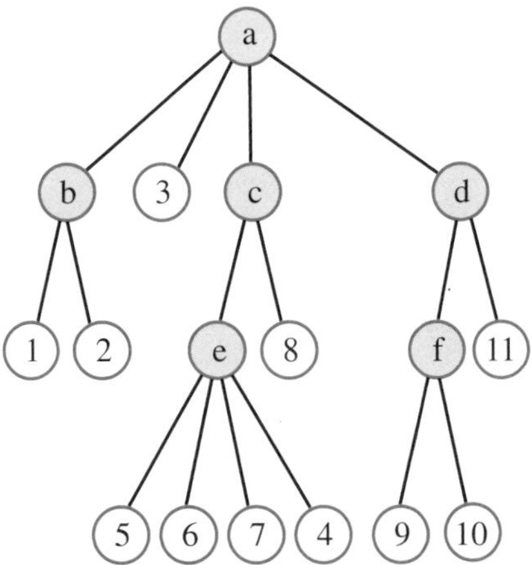

Figure 12.2.11 Quadtree for Quick Review Question 14

Table 12.2.2
Table of Interactions for Figure 12.2.9 and Quick Review Question 15

Body	*Interaction with Body*	*Interaction with Node*	*Number of Interactions*
1		b, c, d	3
2	1, 3	c, d	4
3	1, 2	c, d	4
4	1, 8	b, d, e	5
5	1, 4, 6, 7, 8	b, d	7
6	1, 4, 5, 7, 8	b, d	7
7	1, 4, 5, 6, 8	b, d	7
8	1, 4	b, d, e	5
9	1, 10, 11	b, c	5
10	1, 9, 11	b, c	5

15. a. 55 = 10 + 9 + 8 + 7 + 6 + 5 + 4 + 3 + 2 + 1 = 10(11)/2 because for $i = 1$, j goes from 2 through 11; for $i = 2$, j goes from 3 through 11; and so forth, until for $i = 10$, j goes from 11 through 11.

b. 52 because of the evaluation in Table 12.2.2

References

Aik, Selim G. 1989. *The Design and Analysis of Parallel Algorithms*. Upper Saddle River, NJ: Prentice,Hall, p. 401.

Andrews, Gregory R. 2000. *Foundations of Multithreaded, Parallel, and Distributed Programming*. Reading, MA: Addison-Wesley–Longman, Inc., p. 664.

Giordani, Adrian. 2012. "World's Second Most Powerful Supercomputer Goes Straight for the Heart." Feature, November 14. http://www.isgtw.org/feature/world's-second-most-powerful-supercomputer-goes-straight-heart (accessed December 28, 2012)

Jacobson, Rebecca. 2012. "Satellites, Supercomputers and the Challenge of Forecasting Storms." *Science*, Thursday, November 8. http://www.pbs.org/newshour/rundown/2012/11/the-challenge-of-forecasting-hurricane-sandy.html (accessed December 28, 2012)

McBride, Ryan. 2012. "Univ. of Texas Group Speeds Up Drug Discovery with Supercomputers." *Discovery IT*, May 30. http://www.fiercebiotechit.com/story/univ-texas-group-speeds-drug-discovery-supercomputers/2012-05-30#ixzz2GSGv5XYl (accessed December 27, 2012)

Miller, Russ, and Laurence Boxer. 2000. *Algorithms Sequential & Parallel, A Unified Approach*. Upper Saddle River, NJ: Prentice Hall, p. 330.

NOAA (National Oceanic and Atmospheric Administration). 2012. "Superfast 'Gaea' Supercomputer Helps Scientists Model the Earth's Climate." *News*. April 3. http://researchmatters.noaa.gov/news/Pages/gaea1.aspx (accessed December 28, 2012)

NWS (National Weather Service Weather Forecast Office). 2012. "Hurricane Sandy." http://www.srh.noaa.gov/mfl/?n=sandy (accessed December 29, 2012)

TACC (Texas Advanced Computer Center). 2012. "Molecular Matchmaking for Drug Discovery: From Image Processing, to 3D Modeling, to Search Algorithms, Computer and Computational Science Help Improve the Drug Discovery Pipeline." *News*, May 16. http://www.tacc.utexas.edu/news/feature-stories/2012/molecular-matchmaking (accessed December 28, 2012)

Wilkinson, Barry and Michael Allen. 2004. *Parallel Programming: Techniques and Applications Using Networked Workstations and Parallel Computers* (2nd ed.). Upper Saddle River, NJ: Prentice Hall.

Winslow, R. L., N. Trayanova, D. Geman, M. I. Miller. 2012. "Computational Medicine: Translating Models to Clinical Care." *Sci. Transl. Med.* **4**, 158rv11.

13

MATRIX MODELS

MODULE 13.1

Computational Toolbox—Tools of the Trade: Tutorial 7

Prerequisite: Module 10.1, "Computational Toolbox—Tools of the Trade: Tutorial 6." Alternative Tutorial 7 has no prerequisite for those who wish to cover Chapter 13 before (or instead of) cellular automaton simulations.

Download

From the textbook's website, download Tutorial 7 or Alternative Tutorial 7 in the format of your computational tool or in PDF format. We recommend that you work through the tutorial and answer all Quick Review Questions using the corresponding software.

Introduction

This seventh computational toolbox tutorial, which is available from the textbook's website in your system of choice, prepares you to use the system to complete projects for this chapter. The tutorial introduces the following functions and concepts:

- Vectors operations, such as addition, scalar multiplication, and dot product
- Matrix operations, such as addition, scalar multiplication, matrix multiplication, and power
- Eigenvalues and eigenvectors for Modules 13.3 and 13.4
- Sorting of ordered pairs for Module 13.5
- Timing for Module 13.5

The module gives computational examples and Quick Review Questions for you to complete and execute in the desired software system.

MODULE 13.2

Matrices for Population Studies—Linked for Life

Prerequisite: Module 13.1, "Computational Toolbox—Tools of the Trade: Tutorial 7," or "Alternative Tutorial 7" (up to section on "Eigenvalues and Eigenvectors"). Additional high-performance computing materials related to this module are available on the text's website.

Downloads

The text's website has the file *PopsAndMatOps*, which contains examples from this module, available for download for various computational tools and project data files *cell_trajectory_file.txt*, *cell_types_file.txt*, and *cell_vel_file.txt*.

Population Matrices and High-Performance Computing

Blue crabs (*Callinectes sapidus*) are very important to life along the Gulf Coast of the United States. Essential to the complex, estuarine food webs, these animals also represent the second-largest commercial fishery in the area, and thereby provide livelihoods for many—and they are delicious! Because the ecological and economic impact of population fluctuations of this species is immense, our understanding of the dynamics of crab populations is crucial. Human intrusion (oil spills, pollution, overfishing, habitat degradation, etc.) and natural disasters (hurricanes, etc.) in addition to natural oscillations (climate cycles, dispersal, etc.) all impact populations. Some environmentalists may decry the emphasis on the crabs' economic importance, but the reality is that proper management is necessary for healthy, sustained populations. We need to understand better the factors critical to the quality and quantity of native populations.

Actually, we do know quite a bit about blue crabs. They are found in the western Atlantic Ocean, from Nova Scotia to Argentina, in the Gulf of Mexico and in the

Caribbean. Also, blue crabs have been introduced into the North Sea, southwest of France, parts of the Mediterranean, and Japan. Found to depths of 90 m, they eat about anything—plants, benthic invertebrates, small fish, detritus, carrion—and do a great deal of cannibalism (FAO 2012; Zinski 2006).

Females mate only one time, right after what is called a "pubertal" molt. As she readies for this molt, she calls male crabs with chemical signals. Drawn to her allure (and attractants), as with many animals, the males may squabble over mating rights. The winner often cradles the female until her key molt. Once she is "softened up" for mating, sperm is transferred to storage sacs (seminal receptacles), from which the female will fertilize her eggs. When her shell hardens, the female migrates to estuaries, where she buries herself in mud to overwinter. With the arrival of spring, the female fertilizes and transfers her eggs to form a mass (sponge) attached to her body, which often contains about 2 million individuals (some up to 8 million; Zinski 2006).

Hatching into larvae (**zoeae**) in about 2 weeks, they are carried out into the open ocean, where some feed, grow, and molt several times over a month or more before they are transformed into **megalops**. Over 1 to 3 weeks, these swimming larvae are transported closer to shore. On shore, they molt into juvenile crabs and then head up the estuaries (primary habitat along the Gulf Coast), where they reside, grow, and undergo numerous molts. Maturity normally is achieved by the following summer. Adult males tend to stay in the upper estuaries (lower salinity), whereas adult females, after mating, remain in the lower reaches (higher salinity). Of course, most of the millions of fertilized eggs/larvae never reach adulthood, because they become food for other organisms—including their own kind (Zinski 2006).

Although we know a great deal, we still do not know enough to understand the population dynamics of this or any other species that is passively dispersed over large areas. The countless larval stages are of small sizes and at the mercy of predators, currents, and winds. There is considerable drift of immatures from their birth estuary among other estuaries connecting the adults of different sites. How is it possible to understand the population dynamics of this species when we so obviously do not understand dispersal?

Gulf coast populations are considered **metapopulations**, which means that they are spatially fragmented. The extent of **connectivity**, or exchange of individuals among these populations, is extremely important for population stability and recolonization following local extirpation events. Larval dispersal is very much influenced by mortality, duration of planktonic stages, and behavior in the water column and upon settling. To assess connectivity, scientists must quantify the controlling influences for transport, stocks, and maintenance.

Given the scale and complexity of this problem, scientists are turning to computer modeling and simulation to work out spatially explicit models for blue crab populations. These multifaceted, ecological models are now possible because finely tuned hydrodynamic models of coastal areas are available. Before they can develop any useful population model, scientists must determine the influence of larval dispersal, settlement, and survival rates on fluctuations in blue crab numbers and also a connectivity matrix for the estuaries, which is a rectangular array of numbers indicating contacts various estuary populations have with one another.

Using the Northern Gulf of Mexico Nowcast-Forecast System of the U.S. Navy, biologists at Tulane University are able to use a particle-tracking model (PTM) to simulate larval dispersal. The Navy system incorporates tides, freshwater runoff, winds, sea height, sea temperatures, and 3D current velocities. So, using this type of input and PTM, they can follow the trajectory of individual particles (larvae) with time. The resulting dispersal model can then be incorporated into the larger population model.

Over a 3-year period, scientists have collected more than a terabyte (space for 10^{12} characters) of data for their study. On a 2010 sequential computer, they estimate that the simulation time for one larva is 5 min and for 2000 larvae is 1 week. Thus, averaging the results of 300 simulations involving 2000 larvae each would take about 5.7 years! With such massive amounts of data and such intensive computations, researchers must use high-performance computing with multiple computer processors to store the data and large matrices and to perform the needed simulations in a reasonable amount of time (Taylor 2010).

In this module, we examine populations that change with time. To make long-term predictions about these populations, we store their data in structures called **vectors** and **matrices** and perform calculations on these structures.

Vectors

A **data structure** is a formal skeleton that can hold data and on which we can perform specific operations. One data structure that most computational tools and computer languages have is a **vector**, or a one-dimensional array. Vectors allow us to collect a great deal of similar data together under one name instead of thinking of perhaps hundreds of individual variable names. For example, Table 13.2.1 indicates simulated changes in populations of competing whitetip reef sharks (WTS) and blacktip sharks (BTS) in an area (from the model in Module 4.1). We can summarize these values in two vectors, $\mathbf{w} = (20.00, 6.57, 4.69, 3.08, 0.99, 0.02)$, or $\begin{bmatrix} 20.00 \\ 6.57 \\ 4.69 \\ 3.08 \\ 0.99 \\ 0.02 \end{bmatrix}$, and $\mathbf{b} = (15.00, 5.37, 4.84, 6.00, 10.83, 27.43)$, or $\begin{bmatrix} 15.00 \\ 5.27 \\ 4.84 \\ 6.00 \\ 10.83 \\ 27.43 \end{bmatrix}$, for WTS and BTS, respectively. We use boldface, such as **w**, or a line over the letter, such as $\overline{w}$, to indicate a vector. Each of the vectors **w** and **b** has six numbers, or **elements,** or **members**, so the **size** of each is 6. A subscript, or **index** (plural **indices**), indicates the particular item of the vector, and indices begin with 1 or 0. For a starting index of 1, $w_1 = 20.00$ is the number of whitetip sharks at month 0. By month 5, the simulated population dwindles to 0.02, which is w_6. Another advantage of vectors is the ability

Table 13.2.1
Simulated Populations

Time (months)	*WTS*	*BTS*
0	20.00	15.00
1	6.57	5.37
2	4.69	4.84
3	3.08	6.00
4	0.99	10.83
5	0.02	27.43

to use a variable like i as a subscript instead of a constant like 6. In a computational tool, we can employ this index to change values in a loop, allowing us to perform the same operations on all array elements. In mathematics, we can employ a variable index to express a general case.

Definitions A **vector v** is an ordered n-tuple, written as a row or a column,

$$\mathbf{v} = (v_1, v_2, \ldots, v_n) = \begin{bmatrix} v_1 \\ v_2 \\ \vdots \\ v_n \end{bmatrix}$$

where $v_1, v_2, \ldots, v_n$ are numbers, called **elements**, or **members**. The **size** of a vector is the number of elements, here n. A subscript is an **index** (plural **indices**), and in vector notation, indices begin with 1 or 0.

Quick Review Question 1

For $\mathbf{b} = (15.00, 5.37, 4.84, 6.00, 10.83, 27.43)$, where indices begin with 1, give the value of b_4.

A vector **equal** to **w** has size 6 and its numbers are identical to and in the same order as those of **w**. Thus, two vectors are equal if and only if they are of the same size and corresponding elements are equal.

Definition Vectors $(x_1, x_2, \ldots, x_n)$ and $(y_1, y_2, \ldots, y_n)$ are **equal** if and only if $x_i = y_i$ for $i = 1, 2, \ldots, n$.

Vector Addition

Suppose the only sharks in the given area are BTS and WTS. To obtain the total number of sharks each month (vector **s**), we add corresponding values of the vectors element by element, as follows:

$$\mathbf{s} = \mathbf{w} + \mathbf{b} = \begin{bmatrix} 20.00 \\ 6.57 \\ 4.69 \\ 3.08 \\ 0.99 \\ 0.02 \end{bmatrix} + \begin{bmatrix} 15.00 \\ 5.27 \\ 4.84 \\ 6.00 \\ 10.83 \\ 27.43 \end{bmatrix} = \begin{bmatrix} 20.00 + 15.00 \\ 6.57 + 5.27 \\ 4.69 + 4.84 \\ 3.08 + 6.00 \\ 0.99 + 10.83 \\ 0.02 + 27.43 \end{bmatrix} = \begin{bmatrix} 35.00 \\ 11.94 \\ 9.53 \\ 9.08 \\ 11.82 \\ 27.45 \end{bmatrix}$$

For instance, initially, the number of sharks is $s_1 = b_1 + w_1 = 20.00 + 15.00 = 35.00$, or the sum of the number of BTS and the number of WTS at the start of the simulation. The two vectors must be of the same size for their sum to make sense.

Definition Let $\mathbf{x} = (x_1, x_2, \ldots, x_n)$ and $\mathbf{y} = (y_1, y_2, \ldots, y_n)$ be vectors of n elements each. The **sum** of **x** and **y** is the vector

$$\mathbf{x} + \mathbf{y} = (x_1 + y_1, x_2 + y_2, \ldots, x_n + y_n).$$

Quick Review Question 2

Suppose two scientists, Drs. Chang and Morris, are leading research teams studying red-footed boobies on two islands in Galapagos. Each group counts the number of eggs, hatchlings, juveniles, and nesting pairs over a 1-week period. Suppose the values for Dr. Chang's team are 35, 16, 240, and 351 and for Dr. Morris' team are 18, 10, 103, and 153, respectively.

a. Express the values for Dr. Chang's team in a vector, **c**, and for Dr. Morris' team in a vector, **m**.
b. Compute **t** = **c** + **m**.
c. What does **t** represent?

Multiplication by a Scalar

Suppose that each member of the vectors **w** and **b** is in hundreds of sharks. In this case, $w_1 = 20.00$ indicates that the initial number of whitetip sharks is $100 \cdot 20.00 = 2000$ WTS. To carry the process through every month, we use **scalar multiplication**, as follows:

$$100(20.00, 6.57, 4.69, 3.08, 0.99, 0.02) = (2000, 657, 469, 308, 99, 2)$$

We multiply the **scalar**, or number, 100 by each element.

Definitions A **scalar** is a real number. Let $\mathbf{x} = (x_1, x_2, \ldots, x_n)$ be a vector. The **scalar product** of a scalar a and the vector $\mathbf{x}$ is the vector

$$a\mathbf{x} = a(x_1, x_2, \ldots, x_n) = (ax_1, ax_2, \ldots, ax_n).$$

Quick Review Question 3

The scientists in Quick Review Question 2 estimate that the actual numbers of boobies in each category to be 1.1 as many as they observed. The vector of data for Dr. Chang's team is $\mathbf{c} = (35, 16, 240, 351)$.

a. Using 1.1 and the variable name **c**, give the expression for the vector of estimated values.
b. Give the vector with values rounded to the nearest integers.

Dot Product

As part of the effort to keep them from extinction, scientists around the world have studied the magnificent green sea turtle and used mathematics and computer science to make predictions about their populations. We deal with a different type of multiplication in estimating the number of eggs laid by Hawaiian green sea turtles in one year. We can consider their life cycle to be in five stages, with egg layers in two stages, novice breeders of age 25 years, and mature breeders from ages 26 through 50 years. On the average, a novice breeder lays 280 eggs in a year, and a mature breeder lays 70 eggs per year. We can combine these data in a vector $\mathbf{e} = (280, 70)$. Suppose also that there are 291 novice and 9483 mature breeders, which we store in the vector $\mathbf{b} = (291, 9483)$. To approximate the total green sea turtle egg production in a year, we multiply together corresponding terms and add the results, as follows:

$$\begin{aligned} \mathbf{e} \cdot \mathbf{b} &= (280, 70) \cdot (291, 9483) \\ &= 280 \cdot 291 + 70 \cdot 9483 \\ &= 81{,}480 + 663{,}810 \\ &= 745{,}290 \text{ eggs} \end{aligned}$$

This type of multiplication, the **dot product**, involves two vectors of the same size and results in a *number*, *not* another vector (Green Sea Turtle).

Definition Let $\mathbf{x} = (x_1, x_2, \ldots, x_n)$ and $\mathbf{y} = (y_1, y_2, \ldots, y_n)$ be vectors of n elements each. The **dot product** (or **scalar product**, or **inner product**) of **x** and **y** is

$$\mathbf{x} \cdot \mathbf{y} = x_1 \cdot y_1 + x_2 \cdot y_2 + \cdots + x_n \cdot y_n.$$

Often in writing the dot product, the first term is written as a row vector, such as (280, 70), while the second is written as a column vector, such as $\begin{bmatrix} 291 \\ 9483 \end{bmatrix}$. Multiplication of elements follows the arrows, elements from left to right corresponding to elements from top down:

$$\mathbf{e} \cdot \mathbf{b} = \overrightarrow{(280, 70)} \cdot \begin{bmatrix} 291 \\ 9483 \end{bmatrix} \downarrow = 280 \cdot 291 + 70 \cdot 9483 = 745{,}290 \text{ eggs}$$

This notation will be useful for computations we will be doing shortly.

Quick Review Question 4

The first stage in the life of the Hawaiian green sea turtle, consisting of eggs and hatchlings, occurs during the first year. Stage 2, juveniles, extends from year 1 to year 16. Suppose 23% of the hatchlings survive and move to stage 2, while 67.9% of those in Stage 2 remain in that stage each year. In one year, suppose Stage 1 has 808,988 individuals, and Stage 2 has 715,774 (Green Sea Turtle).

a. Give a vector, **p**, with real-number elements representing the percentages.
b. Give a vector, **s**, storing the individuals in Stages 1 and 2.
c. Using variables **p** and **s**, not the data, give the vector operation to determine the number of individuals that will be in Stage 2 the following year.
d. Calculate this value.

Matrices

In the section "Vectors," we considered the data structure of a one-dimensional array, or vector. One example involved vector **w**, which stored under that one name the simulated number of whitetip sharks from 0 through 5 months. Quite often, however, more features need to be stored and manipulated. In such cases 2D arrays may be helpful. For example, we can store the data from Table 13.2.1 for the number of whitetip reef sharks (WTS) and blacktip sharks (BTS) in the following 2D array:

$$S = \begin{bmatrix} 20.00 & 15.00 \\ 6.57 & 5.27 \\ 4.69 & 4.84 \\ 3.08 & 6.00 \\ 0.99 & 10.83 \\ 0.02 & 27.43 \end{bmatrix}$$

The name used in mathematics and in many computational tools for a 2D array is matrix. A **matrix** (plural, **matrices**) is a rectangular array of numbers, and we can think of a matrix as a table of numbers.

The symbol for an individual matrix element has two subscripts indicating its row and column. Assuming that the row and column indices for S shown previously begin with 1, the value 3.08, which is the number of WTS at month 3, is element s_{41}, the value in the fourth row and first column. Usually, we represent a matrix with an uppercase letter and an element with the corresponding lowercase letter and indices. Thus, we can abbreviate the array as $S = [s_{ij}]$. The **size** of a matrix is the number of rows by the number of columns. Thus, S is a 6×2, or a 6-by-2, matrix.

Definitions A **matrix** (plural, **matrices**), $S = [s_{ij}]$, is a rectangular array of numbers. Element s_{ij} is in row i and column j. A matrix with m rows and n columns has **size** $\boldsymbol{m \times n}$, or $\boldsymbol{m}$ **by** $\boldsymbol{n}$.

Quick Review Question 5

For matrix S, assume the row and column indices begin with 1. Give each of the following.

a. The value of s_{12}
b. The notation for the element with value 6.00

Scalar Multiplication and Matrix Sums

As with vectors, two matrices are equal if they have the same size and corresponding elements are identical. To compute the sum of two matrices that have the same size, we add corresponding elements. For the product of a scalar times a matrix, we multiply each element by the scalar.

Definitions Let $A = [a_{ij}]$ and $B = [b_{ij}]$ be two $m \times n$ matrices. A and B are **equal** if and only if $a_{ij} = b_{ij}$ for all i and j. The **product of scalar** $\boldsymbol{c}$ **and matrix** $\boldsymbol{A}$ is $m \times n$ matrix

$$cA = [ca_{ij}]$$

That is, each element of A is multiplied by c. The **matrix sum** of A and B is an $m \times n$ matrix

$$A + B = [a_{ij} + b_{ij}]$$

That is, corresponding elements are added.

Quick Review Question 6

For the following, calculate the indicated matrices:

$$A = \begin{bmatrix} 1 & 3 & 9 \\ 0 & 5 & 6 \end{bmatrix}, B = \begin{bmatrix} -1 & 2 & 0 \\ 1 & 4 & 3 \end{bmatrix}, \text{ and } C = \begin{bmatrix} 7 & 4 \\ -2 & 8 \end{bmatrix}$$

a. $3B$
b. $A + B$
c. $C + 2A + B$

Matrix Multiplication

The ability to multiply matrices allows us to model many problems, including those involving changes in populations. The foundation for the operation of matrix multiplication is the concept of dot product of vectors. Suppose we wish to estimate the total shark mass at each month of the simulation involving whitetip reef sharks (WTS) and blacktip sharks (BTS). An estimate of the average mass of a whitetip is 20 kg, while that of a blacktip is 18 kg. Suppose we started the simulation with 20 WTS and 15 BTS. Thus, the initial total shark mass is the following dot product:

$$(20, 15) \cdot (20, 18) = 20 \cdot 20 + 15 \cdot 18 = 670 \text{ kg}$$

As pointed out before, we can write the second vector as a column vector. Then, we are actually multiplying a 1 × 2 matrix by a 2 × 1 matrix to find a 1 × 1 matrix, as follows:

$$\begin{bmatrix} 20 & 15 \end{bmatrix} \begin{bmatrix} 20 \\ 18 \end{bmatrix} = [670]$$

To compute the shark mass totals at each month, we multiply the shark matrix S by the mass vector $\mathbf{g} = (20, 18)$, written as a column. We take the dot product of each row of S with $\mathbf{g}$ to compute a 6 × 1 matrix of monthly masses (kg) rounded to the nearest integer, as follows:

$$\begin{array}{c} \\ \text{month 0} \\ \text{month 1} \\ \text{month 2} \\ \text{month 3} \\ \text{month 4} \\ \text{month 5} \end{array}
\begin{array}{c} \begin{array}{cc} \text{WTS} & \text{BTS} \end{array} \\ \begin{bmatrix} 20.00 & 15.00 \\ 6.57 & 5.27 \\ 4.69 & 4.84 \\ 3.08 & 6.00 \\ 0.99 & 10.83 \\ 0.02 & 27.43 \end{bmatrix} \end{array}
\begin{array}{c} \text{mass} \\ \begin{bmatrix} 20 \\ 18 \end{bmatrix} \\ \\ \\ \\ \\ \end{array}
\begin{array}{c} \\ \text{WTS} \\ \text{BTS} \\ \\ \\ \\ \\ \end{array}
=
\begin{bmatrix} 20.00 \cdot 20 + 15.00 \cdot 18 \\ 6.57 \cdot 20 + 5.27 \cdot 18 \\ 4.69 \cdot 20 + 4.84 \cdot 18 \\ 3.08 \cdot 20 + 6.00 \cdot 18 \\ 0.99 \cdot 20 + 10.83 \cdot 18 \\ 0.02 \cdot 20 + 27.43 \cdot 18 \end{bmatrix}
=
\begin{array}{c} \text{mass} \\ \begin{bmatrix} 670 \\ 228 \\ 181 \\ 170 \\ 215 \\ 494 \end{bmatrix} \end{array}
\begin{array}{c} \\ \text{month 0} \\ \text{month 1} \\ \text{month 2} \\ \text{month 3} \\ \text{month 4} \\ \text{month 5} \end{array}$$

As we move from left to right on a row of the first matrix, we go down on the second, multiplying corresponding elements and then adding the results. Consequently, the number of columns in the first matrix must equal the number of rows in the column vector. The resulting vector has the same number of rows as the first matrix and the same number of columns as the second. In this example, S has size 6 × 2 while $\mathbf{g}$ has size 2 × 1, and the result is a 6 × 1 matrix.

Quick Review Question 7

a. For the vector $\mathbf{v} = (5, 0, -1)$, written as a column vector, and the matrix $A = \begin{bmatrix} 1 & 3 & 9 \\ 0 & 5 & 6 \end{bmatrix}$, calculate $A\mathbf{v}$.

b. For a 5×8 matrix B, give the size of a vector $\mathbf{w}$ for which we can calculate $B\mathbf{w}$.

c. Give the resulting size of $B\mathbf{w}$.

Suppose scientists observed that 25% of the WTS have wounds, while none of the BTS do. Such wounds could contribute to an animal's decreased hunting ability. In calculating the total number of wounded sharks, we need to consider only the WTS. Again, the computation can be accomplished with a dot product. At the start of the simulation, we have the following computation:

$$\begin{bmatrix} 20.00 & 15.00 \end{bmatrix} \begin{bmatrix} 0.25 \\ 0.00 \end{bmatrix} = [5.00]$$

Zero in the second row eliminated the effect of the number of BTS.

Additionally, suppose scientists noted that 30% of the WTS and 20% of the BTS have lesions. The total number of sharks with lesions at a particular month is the dot product of a vector of the numbers of sharks and a column vector with these percentages. For example, as the following computation shows, 9 sharks have lesions in month 0:

$$\begin{bmatrix} 20.00 & 15.00 \end{bmatrix} \begin{bmatrix} 0.30 \\ 0.20 \end{bmatrix} = [9.00]$$

Certainly, we can take the shark-numbers matrix, S, and multiply by any 2×1 attribute vector, such as $\begin{bmatrix} 20 \\ 18 \end{bmatrix}$, $\begin{bmatrix} 0.25 \\ 0.00 \end{bmatrix}$, or $\begin{bmatrix} 0.30 \\ 0.20 \end{bmatrix}$. However, we can perform all these calculations together. We form a 2×3 attribute matrix, $A = \begin{bmatrix} 20 & 0.25 & 0.30 \\ 18 & 0.00 & 0.20 \end{bmatrix}$, and we dot each row of the first matrix (S) by each column of a second matrix (attribute matrix, A) to get a resulting totals matrix (T) containing the totals for shark mass, wounded sharks, and sharks with lesions by month. There are six months, thus six rows, one for each month, in the resulting totals matrix. Because there are three attributes to total, the totals matrix has three columns, or three totals, for each month. Six rows in the first matrix along with three columns in the second yield a 6×3 totals matrix, as follows:

$$SA = \begin{array}{r} \\ \text{month 0} \\ \text{month 1} \\ \text{month 2} \\ \text{month 3} \\ \text{month 4} \\ \text{month 5} \end{array} \begin{array}{c} \begin{array}{cc} \text{WTS} & \text{BTS} \end{array} \\ \begin{bmatrix} 20.00 & 15.00 \\ 6.57 & 5.27 \\ 4.69 & 4.84 \\ 3.08 & 6.00 \\ 0.99 & 10.83 \\ 0.02 & 27.43 \end{bmatrix} \end{array} \begin{array}{c} \begin{array}{ccc} \text{mass} & \text{\% wounded} & \text{\% lesions} \end{array} \\ \begin{bmatrix} 20 & 0.25 & 0.30 \\ 18 & 0.00 & 0.20 \end{bmatrix} \begin{array}{l} \text{WTS} \\ \text{BTS} \end{array} \end{array}$$

$$
= T = \begin{array}{l} \\ \text{month 0} \\ \text{month 1} \\ \text{month 2} \\ \text{month 3} \\ \text{month 4} \\ \text{month 5} \end{array}
\begin{array}{c} \begin{array}{ccc} \text{mass (kg)} & \text{\# wounded} & \text{\#lesioned} \end{array} \\ \begin{bmatrix} 670 & 5.00 & 9.00 \\ 228 & 1.64 & 3.03 \\ 181 & 1.17 & 2.38 \\ 170 & 0.77 & 2.12 \\ 215 & 0.25 & 2.46 \\ 494 & 0.005 & 5.49 \end{bmatrix} \end{array}
$$

Usually we write the matrix product without row and column headings, as follows:

$$
SA = \begin{bmatrix} 20.00 & 15.00 \\ 6.57 & 5.27 \\ 4.69 & 4.84 \\ 3.08 & 6.00 \\ 0.99 & 10.83 \\ 0.02 & 27.43 \end{bmatrix} \begin{bmatrix} 20 & 0.25 & 0.30 \\ 18 & 0.00 & 0.20 \end{bmatrix} = \begin{bmatrix} 679 & 5.00 & 9.00 \\ 228 & 1.64 & 3.03 \\ 181 & 1.17 & 2.38 \\ 170 & 0.77 & 2.12 \\ 215 & 0.25 & 2.46 \\ 494 & 0.005 & 5.49 \end{bmatrix} = T
$$

In the totals matrix, T, the third-row, first-column element ($t_{31} = 181$) indicates that at month 2 of the simulation, the total mass of WTS and BTS in the area is 181 kg. The rounded second-row, second-column element ($t_{22} = 1.64$, rounded to 2) indicates that in month 1, two of the sharks have wounds. The rounded sixth-row, third-column element ($t_{63} = 5.49$, rounded to 5) shows that five of the sharks have lesions in month 5. For the dot products to be possible, the number of columns in the first matrix (here S) and the number of rows in the second (here A) have to be identical; here both are 2.

Definition Let $A = [a_{ij}]_{m \times q}$ be an $m \times q$ matrix and $B = [b_{ij}]_{q \times n}$ a $q \times n$ matrix. The **matrix product** of A and B is an $m \times n$ matrix AB, or $A \cdot B = C = [c_{ij}]_{m \times n}$, where c_{ij} is the dot product of the ith row of A and the jth column of B.

Quick Review Question 8

Consider the following matrices:

$$
A = \begin{bmatrix} 8 & 5 & 3 & -4 \\ -5 & 1 & 0 & 1 \end{bmatrix}, \quad B = \begin{bmatrix} 2 & -6 \\ 7 & 1 \\ 4 & 3 \\ -9 & -2 \end{bmatrix}, \quad C = \begin{bmatrix} 6 & 5 \\ 1 & -3 \\ 2 & -8 \end{bmatrix}, \quad I_2 = \begin{bmatrix} 1 & 0 \\ 0 & 1 \end{bmatrix}, \text{ and}
$$

$$
I_3 = \begin{bmatrix} 1 & 0 & 0 \\ 0 & 1 & 0 \\ 0 & 0 & 1 \end{bmatrix}
$$

Evaluate each of the following:

a. AB
b. BA
c. CI_2
d. I_3C

Square Matrices

Each of I_2 and I_3 in Parts c and d of the last Quick Review Question is a **square matrix**, having the same number of rows as columns. Moreover, each has 1s along its **diagonal**, the line of elements from the top left to the bottom right. A number of examples in biology involve square matrices. For example, the hypothetical data in Table 13.2.2 presents the distribution of ABO blood types for mothers and newborns (multiple births omitted) in a county over a year. The corresponding matrix is as follows:

$$M = \begin{bmatrix} \mathbf{1068} & 53 & 68 & 516 \\ 37 & \mathbf{273} & 88 & 601 \\ 60 & 58 & \mathbf{21} & 0 \\ 491 & 189 & 0 & \mathbf{2059} \end{bmatrix}$$

According to the datum in the second row, first column, only 37 mothers with type B blood gave birth to a child with type A blood in that county during the year of the study. The diagonal values, which are in boldface, indicate the numbers of mother-newborn pairs that share the same blood type. For instance, 1068 type A mothers gave birth to type A children in the county that year.

Definitions An $n \times n$ matrix is called a **square matrix**. In an $n \times n$ square matrix M, the **diagonal** is the set of elements $\{m_{11}, m_{22}, \ldots, m_{nn}\}$.

As another example involving a square matrix, Table 13.2.3 presents similarity measures (specifically, Euclidean distances) of the 18S rRNA sequences of pairs of animals, where smaller numbers indicate closer relationships. Thus, with a Euclid-

Table 13.2.2
Hypothetical Data for the Distribution of ABO Blood Types of Mothers and Newborns (Multiple Births Omitted) in a County Over a Year. (Similar to Table 1 in Bottini et al. 2001)

Mother\Newborn	*A*	*B*	*AB*	*O*
A	1068	53	68	516
B	37	273	88	601
AB	60	58	21	0
O	491	189	0	2059

Table 13.2.3
Similarity Measures (Specifically, Euclidean Distances) of the 18S rRNA Sequences of Pairs of Animals (Lockhart et al. 1994, Table 3)

	Frog	*Bird*	*Human*	*Rabbit*
Frog	0	0.316	0.350	0.336
Bird	0.316	0	0.130	0.102
Human	0.350	0.130	0	0.028
Rabbit	0.336	0.102	0.028	0

ean distance of 0.028, the rRNA sequences of a human and a rabbit are more closely related than that of a human and a frog (distance 0.350). Because the distance from animal A's rRNA sequence to animal B's sequence is the same as the distance from B to A, the table and the corresponding matrix, which follows, are **symmetric** around the diagonal:

$$M = \begin{bmatrix} 0 & 0.316 & 0.350 & 0.336 \\ 0.316 & 0 & 0.130 & 0.102 \\ 0.350 & 0.130 & 0 & \mathbf{0.028} \\ 0.336 & 0.102 & \mathbf{0.028} & 0 \end{bmatrix}$$

Thus, as the boldface emphasizes, the distance in row 3, column 4, namely, 0.028, is the same as the number in row 4, column 3. In general, elements $m_{ij} = m_{ji}$. For symmetry, the values on the diagonal do not have to be zero as they are in this example.

Definition An $n \times n$ square matrix M is **symmetric** if $m_{ij} = m_{ji}$ for all i and j.

Matrices and Systems of Equations

The section "Dot Product" indicates that a Hawaiian green sea turtle novice breeder lays an average of 280 eggs per year, while a mature breeder only lays 70. We considered a specific number of turtles in each category, 291 and 9483, respectively, and calculated the total yearly egg production as the following dot product:

$$\mathbf{e} \cdot \mathbf{b} = (280, 70) \cdot (291, 9483)$$

Instead of specifying the number of turtles in each category, let n be the number of novice breeders and m the number of mature breeders with $\mathbf{b} = (n, m)$. In general, the average annual egg production, a, is computed as follows:

$$\begin{aligned} \mathbf{e} \cdot \mathbf{b} &= (280, 70) \cdot (n, m) \\ &= 280n + 70m = a \end{aligned}$$

Thus, the dot product translates into one side of a linear equation.

The following are examples of linear equations:

$$6x = 1$$
$$5x + 7y = 3$$
$$-2x + \pi y + \sqrt{3}\,z = 9$$
$$1/2x_1 + 33.2x_2 + 15x_3 + 13x_4 = 33/4$$

The equations derive their name from the fact that when they have only one, two, or three variables, as in the first three examples, their graphs are straight lines. The general **linear equation** is

$$a_1x_1 + a_2x_2 + \cdots + a_nx_n = c$$

where a_i and c are numbers for $i = 1, 2, \ldots, n$.

While we can employ a dot product in representing one linear equation, we can use matrix multiplication for a system of linear equations. Returning to the example involving whitetip sharks (WTS) and blacktip sharks (BTS) from the section "Matrix Multiplication," suppose the number of each kind of shark by month is as in Table 13.2.1. That section represented the data in the following matrix, with the number of WTS in the first column and the number of BTS in the second:

$$S = \begin{bmatrix} 20.00 & 15.00 \\ 6.57 & 5.27 \\ 4.69 & 4.84 \\ 3.08 & 6.00 \\ 0.99 & 10.83 \\ 0.02 & 27.43 \end{bmatrix}$$

Let x be the percentage of whitetip sharks with lesions, y be the percentage of blacktip sharks with lesions, and $\mathbf{h} = (x, y)$. Suppose the total number of sharks with lesions from month 0 through 5 is 9.00, 3.04, 2.38, 2.12, 2.46, and 5.49, respectively, with vector representation $\mathbf{v} = (9.00, 3.04, 2.38, 2.12, 2.46, 5.49)$. Thus, we have the following linear equation for the total number of sharks with lesions in month 0:

$$20.00x + 15.00y = 9.00$$

which we can write as the following dot product:

$$(20.00, 15.00) \cdot (x, y) = 9.00$$

or

$$[20.00, 15.00] \cdot \begin{bmatrix} x \\ y \end{bmatrix} = [9.00]$$

Quick Review Question 9

Use the preceding shark data for month 1.

a. Write the linear equation.

b. Write the corresponding equation using a dot product.

Instead of writing each equation separately, we can employ matrix multiplication to system of six equations, as follows:

$$\begin{bmatrix} 20.00 & 15.00 \\ 6.57 & 5.27 \\ 4.69 & 4.84 \\ 3.08 & 6.00 \\ 0.99 & 10.83 \\ 0.02 & 27.43 \end{bmatrix} \begin{bmatrix} x \\ y \end{bmatrix} = \begin{bmatrix} 9.00 \\ 3.04 \\ 2.38 \\ 2.12 \\ 2.46 \\ 5.49 \end{bmatrix}$$

or

$$S\mathbf{h} = \mathbf{v}$$

As we see in other modules, besides providing a useful abbreviation for systems of equations, matrices can simplify the process of finding solutions.

Quick Review Question 10

Express the following system of equations using a matrix-vector notation:

$$\begin{cases} 2x - y = 7 \\ 6x \quad\;\; = 5 \end{cases}$$

Exercises

Given the scalars a = *7 and* b = *3 and the vectors* **u** = *(3, –4, 8, 0),* **v** = *(–9, 4, 21, 2),* **y** = *(8, 8, 1, –2), and* **x** = *(7, 17, 6), where possible, compute the values of Exercises 1–20. Check your work with a computational tool.*

1. $a\mathbf{u}$ **2.** $b\mathbf{v}$ **3.** $a\mathbf{u} + b\mathbf{v}$ **4.** $\mathbf{u} + \mathbf{v}$
5. $v + u$ **6.** $(\mathbf{u} + \mathbf{v}) + \mathbf{y}$ **7.** $\mathbf{u} + (\mathbf{v} + \mathbf{y})$ **8.** $\mathbf{u} + \mathbf{x}$
9. $(a + b)\mathbf{y}$ **10.** $a\mathbf{y} + b\mathbf{y}$ **11.** $0\mathbf{x}$ **12.** $\mathbf{u} \cdot \mathbf{v}$
13. $\mathbf{y} \cdot (2\mathbf{v})$ **14.** $2(\mathbf{y} \cdot \mathbf{v})$ **15.** $(2\mathbf{y}) \cdot \mathbf{v}$ **16.** $\mathbf{x} \cdot \mathbf{y}$
17. $\mathbf{v} - \mathbf{y}$ **18.** $a(\mathbf{u} + \mathbf{y})$ **19.** $a\mathbf{u} + a\mathbf{y}$ **20.** $(0, 0, 0) \cdot \mathbf{x}$

Compute, if possible, the dot products in Exercises 21–23. Check your work with a computational tool.

21. $(5,7)\cdot\begin{bmatrix} -1 \\ 4 \end{bmatrix}$ **22.** $(6,2,3)\cdot\begin{bmatrix} 1 \\ 1 \end{bmatrix}$ **23.** $(7,-7,1)\cdot\begin{bmatrix} 3 \\ 3 \\ 1 \end{bmatrix}$

24. Suppose the following items are for sale one week by a scientific supply house at the indicated prices: a particular bacterial culture, \$17; case of pipettes, \$310; case of Petri dishes, \$190; case of beakers, \$40.
a. Write these prices in a vector **v**.
b. Suppose there is a 25%-off sale. What scalar is multiplied by **v** to give the sale prices?

c. Perform this scalar multiplication.

d. Suppose 83 bacterial cultures, 18 cases of pipettes, 145 cases of Petri dishes, and 108 cases of beakers are sold during the sale. Write a dot product of vectors to calculate the amount of money from the sale, and evaluate this dot product.

e. Suppose the next week the store sells 20 bacterial cultures, 3 cases of pipettes, 76 cases of Petri dishes, and 37 cases of beakers. Write the vector sum to indicate the number of each item sold during the 2-week period, and evaluate this addition.

Determine the values of the unknowns to make the vectors equal in Exercises 25–27.

25. $(3, 5, 7) = (a, b, 7)$ **26.** $(-6, 2, 1) = (-6, 2, 1, a)$ **27.** $2(6, 1, a) = b(3, c, 4)$

28. Consider the matrix $A = [a_{ij}] = \begin{bmatrix} 6 & 3 & -2 \\ 0 & -8 & 4 \end{bmatrix}$

a. What is A's size? b. Find a_{21}, a_{12}, a_{31}, and a_{13}.

Using $A = \begin{bmatrix} 6 & 3 & -2 \\ 0 & -8 & 4 \end{bmatrix}$, $B = \begin{bmatrix} -1 & 2 & 3 \\ -7 & 2 & 1 \end{bmatrix}$, *and* $C = \begin{bmatrix} 0 & -4 & 1 \\ 3 & 1 & -8 \end{bmatrix}$, *calculate the matrices in Exercises 29–40. Check your work with a computational tool.*

29. $3A$ **30.** $3B$ **31.** $3A + 3B$ **32.** $A + B$

33. $3(A + B)$ (Compare to Exercise 31.) **34.** $B + A$ (Compare to Exercise 32.)

35. $-A$ **36.** $B + C$ **37.** $(A + B) + C$ (Use Exercise 32.)

38. $A + (B + C)$ (Use Exercise 36; compare to Exercise 37.)

39. $2(3A)$ (Use Exercise 29.) **40.** $6A$ (Compare to Exercise 39.)

If possible, compute the matrices in Exercises 41–43 using $A = [3 \quad 5]$ *and*

$B = \begin{bmatrix} 1 & 1 \\ 3 & 4 \\ -1 & 4 \end{bmatrix}$. *Check your work with a computational tool.*

41. $2A$ **42.** $A + A$ **43.** $A + B$

44. How many elements are in matrix A if it is of each given size?

a. 20×5 **b.** $m \times n$ **c.** 5×5 **d.** $n \times n$

45. Consider the square matrix

$$A = \begin{bmatrix} 1 & 5 & 3 \\ 5 & 4 & 6 \\ 3 & 6 & 8 \end{bmatrix} = [a_{ij}]_{3\times 3}.$$

a. Find a_{21} and a_{12}. **b.** Why is A symmetric?

c. Give the diagonal elements of A.

d. Suppose B is a symmetric matrix. Fill in the blanks.

$$B = \begin{bmatrix} -7 & - & - & - \\ 2 & 3 & - & - \\ -1 & 4 & -4 & - \\ 6 & 5 & 0 & -3 \end{bmatrix}$$

46. A lab is using spectrophotometer to indicate the number of bacteria in a broth. From a reading, they determine absorbance, a value between 0.0 and 2.0. As the number of bacteria increases, so does the absorbance. Each team takes measurements at 10-min intervals. Suppose following measurements are made for *E. coli* at 15 °C from 70 min to 130 min: 0.041, 0.055, 0.064, 0.062, 0.089, 0.097, 0.103. The following measurements are for *E. coli* at 21 °C: 0.055, 0.070, 0.077, 0.095, 0.105, 0.115, 0.124. Place the values in a matrix, and indicate the meanings of the rows and columns (Johnson and Case 2009).

47. Suppose a certain animal has a maximum life span of 3 years. This example predicts populations in each age category: year 1 (0–1 year), year 2 (1–2 years), and year 3 (2–3 years). We consider only females. A year 1 female animal has no offspring; a year 2 female has 3 daughters on the average; and a year 3 female has a mean of 2 daughters. A year 1 animal has a 0.3 probability of living to year 2. A year 2 animal has a 0.4 probability of living to year 3. Suppose at one instance, the number of year 1, 2, and 3 females are 2030, 652, and 287, respectively.
   a. Write a row vector of three elements giving the mean number of female offspring in each age category.
   b. Write a row vector triple giving the probabilities that in the next year a year 1 animal lives to years 2, 3, and 4.
   c. Write a row vector triple giving the probabilities that in the next year a year 2 animal lives to years 2, 3, and 4.
   d. Place the row vectors from Parts a–c in a matrix, L.
   e. Write a column vector, **c**, of the female counts in each year.
   f. Using Parts d and e, estimate the female numbers in each age category a year later.
   g. Using Parts d and f, estimate the female numbers in each age category 2 years after the initial counts.
   h. Using Part d, calculate L^2, or LL.
   i. Using Parts h and e, calculate $L^2\mathbf{c}$.
   j. How do your answers from Parts g and i compare?

48. Consider the matrix

$$T = \begin{bmatrix} 0 & 50 & 20 \\ 100 & 150 & 120 \\ 90 & 170 & 200 \end{bmatrix}.$$

For any 3 × 3 matrix M with elements from the set of nonnegative integers, apply the function f to each element, defined as

$$f(m_{ij}) = \begin{cases} 0, & \text{if } m_{ij} < \text{corresponding threshold value, } t_{ij} \\ 1, & \text{if } m_{ij} \geq \text{corresponding threshold vlaue, } t_{ij}. \end{cases}$$

T is called a **threshold matrix**, and each t_{ij} is a threshold value. An element of the matrix M is mapped to 1 if and only if it is at least as big as the corresponding threshold value. Fill in the blanks for the image of the elements of the following matrix M:

$$M = \begin{bmatrix} 110 & 112 & 100 \\ 100 & 70 & 75 \\ 90 & 80 & 90 \end{bmatrix} \rightarrow \begin{bmatrix} - & 1 & - \\ - & 0 & - \\ - & - & - \end{bmatrix}.$$

49. A dither matrix can be used to enhance a digital image, such as a medical image from a CT (computerized tomography) scan of the body. A computer can analyze the degree of grayness of each dot, or **pixel**, of one such black-and-white image and assign it a value for intensity, say from 0 (white) to 255 (black). One method of enhancing the picture is **dithering**. Each digitized pixel is compared with an individual threshold value to determine if a dot will or will not be placed at that point on the reconstructed picture. There are no gray dots in the reconstructed picture; a black dot is either present or not present at each position, depending on the presence of a 1 or 0 in the corresponding position of the final matrix. To accomplish this procedure a threshold matrix, called a **dither matrix**, is needed. Much experimentation has been done in dithering to find the best threshold matrix to help produce a clear, apparently continuous image using black dots on a white background. The construction of one dither matrix is presented in this problem. Let

$$D_2 = \begin{bmatrix} 0 & 2 \\ 3 & 1 \end{bmatrix} \quad \text{and} \quad V_2 = \begin{bmatrix} 1 & 1 \\ 1 & 1 \end{bmatrix}.$$

To develop the dither matrix, find the following matrices:

a. $4D_2$ **b.** $4D_2 + 2V_2$ **c.** $4D_2 + 3V_2$ **d.** $4D_2 + V_2$

e. Construct the 4×4 matrices D_4 and V_4 with the 2×2 matrices from the previous parts placed in the indicated positions.

$$D_4 = \begin{bmatrix} 4D_2 & 4D_2 + 2V_2 \\ 4D_2 + 3V_2 & 4D_2 + V_2 \end{bmatrix}, \quad V_4 = \begin{bmatrix} V_2 & V_2 \\ V_2 & V_2 \end{bmatrix}$$

f. Calculate the dither matrix $16D_4 + 8V_4$, which is the threshold matrix that will be used in reconstructing a picture below.

g. Consider the 4×4 matrix M containing pixel intensities transmitted from space.

$$M = \begin{bmatrix} 100 & 145 & 100 & 178 \\ 111 & 60 & 250 & 102 \\ 40 & 200 & 20 & 73 \\ 254 & 198 & 223 & 204 \end{bmatrix}.$$

With function f defined as in Exercise 48, find the image of M after applying f to every point.

h. Draw the picture in a 4×4 array. *Note:* If the picture were larger, we could use the same dither matrix by applying that threshold matrix in a checkerboard fashion over the entire picture.

Let $A=\begin{bmatrix} 1 & 2 \\ 3 & 4 \end{bmatrix}$, $B=\begin{bmatrix} 6 & 2 & 0 \\ 0 & -1 & 4 \end{bmatrix}$, $C=\begin{bmatrix} -2 & 1 & -5 \\ 7 & 1 & 0 \end{bmatrix}$. *Where possible, perform the indicated operation or answer the question in each of the Exercises 50–63. Check your work with a computational tool.*

50. AB **51.** AC **52.** $AB + AC$ (Use Exercises 50 and 51.)
53. $B + C$ **54.** $A(B + C)$ (Use Exercise 53. Compare to Exercise 52.)
55. BA **56.** BC **57.** $3(AC)$ (Use Exercise 51.)
58. $3A$ **59.** $(3A)C$ (Use Exercise 58. Compare to Exercise 57.)
60. $A^2 = A \cdot A$ **61.** B^2 **62.** $\mathbf{0}_{2\times2} \cdot A$, where $\mathbf{0}_{2\times2}$ is a 2×2 matrix of all zeros.
63. $B \cdot \mathbf{0}_{3\times3}$, where $\mathbf{0}_{3\times3}$ is a 3×3 matrix of all zeros.

For Exercises 64–67, perform the indicated matrix multiplication. Check your work with a computational tool.

64. $\begin{bmatrix} 0.1 & 0.2 & 0.9 \\ 1.3 & 0.5 & 0.7 \end{bmatrix}\begin{bmatrix} 2.2 & 3.9 \\ 0.6 & 0.4 \\ 1.1 & 2.8 \end{bmatrix}$ **65.** $\begin{bmatrix} 1 & -3 \\ 8 & 5 \\ 0 & 7 \\ 2 & -2 \end{bmatrix}\begin{bmatrix} -6 & 5 & 5 \\ -5 & 1 & 0 \end{bmatrix}$

66. $\begin{bmatrix} 9 & 0 \\ 3 & 2 \end{bmatrix}\begin{bmatrix} -6 & 10 & 3 & 8 \\ -3 & 11 & 20 & 7 \end{bmatrix}$ **67.** $\begin{bmatrix} 0.4 & 3.2 & 4.9 & 1.1 \\ 8.4 & 2.6 & 3.6 & 8.8 \end{bmatrix}\begin{bmatrix} 2 & 1 & 0 \\ 3 & 0 & -9 \\ -7 & 5 & 5 \\ 3 & 9 & 8 \end{bmatrix}$

Using the matrix $A = \begin{bmatrix} 1 & 2 \\ 3 & 4 \end{bmatrix}$, *where possible, perform the indicated operation or answer the question in each of the Exercises 68–74.*

68. Find the matrix H such that $HA = \begin{bmatrix} 5\cdot1 & 5\cdot2 \\ 7\cdot3 & 7\cdot4 \end{bmatrix}$.

69. Find the matrix J such that $AJ = \begin{bmatrix} 1\cdot4 & 2\cdot9 \\ 3\cdot4 & 4\cdot9 \end{bmatrix}$.

70. $[6 \quad 1]A$ **71.** $A\begin{bmatrix} 6 \\ 1 \end{bmatrix}$ **72.** $A\begin{bmatrix} x \\ y \end{bmatrix}$ **73.** $\begin{bmatrix} x \\ y \end{bmatrix}A$ **74.** $A[6 \quad 1]$

75. Write the following system of equations as $AX = B$ using a matrix and vectors:

$$4x_1 + 5x_2 = -3$$
$$7x_1 + 9x_2 = 4$$

Projects

To complete the following projects, use a computational tool.

1. The Network Dynamics and Science Simulation Laboratory (NDSSL) at Virginia Technical University generated from real data a synthetic dataset for the activities of the population of Portland, Ore. Various NDSSL datasets are available at http://ndssl.vbi.vt.edu/opendata/download.php (NDSSL 2009a, b, c, d). Scientists use such data "for simulating the spread of epidemics at the level of individuals in a large urban region, taking into account realistic contact patterns and disease transmission characteristics" (VBI 2008). Files in the *Details* column describe the datasets; files in the *Samples* column display small example files; and files in the *Download* column are compressed large datasets. Omitting the header line in the file, cut and paste a Data Set Release 1 or 2 sample activities file into a text file. For this dataset develop code to accomplish the following tasks, which can be used for epidemiological studies:

   a. Form the vector, *personIdLst*, of person IDs.
   b. Form the vector, *locationIDLst*, of location IDs.
   c. Generate connection matrix, *connMat*, with people indices representing row labels and location indices representing column labels. The *ij* element of *connMat* is 1 if the *i*th person visits the *j*th location; otherwise, the element is 0.
   d. Write a function to return the number of locations a person visits.
   e. Write a function to return the number of people that visit a location.
   f. Generate a people-to-people connection matrix, *connPeopleMat*, with people indices representing row labels and column labels. The *ij* element of *connPeopleMat* is 1 if the *i*th and the *j*th people visit the same location in a day but not necessarily at the same time; otherwise, the element is 0.
   g. Using *connPeopleMat* from Part f, write a function to return the degree of a person ID, that is, to return the number of people that go to locations visited by the individual.
   h. Calculate the square of the matrix *connPeopleMat* from Part f. Develop a function that returns true if two people, A and B, have direct or indirect contact, that is if A and B were at the same location in a day or if there is a person C such that A and C were at the same location and C and B were at the same location in a day. As before, ignore times people visit locations. Explain why the square of *connPeopleMat* and the sum of *connPeopleMat* and its square are useful for this task.

2. Using the NDSSL site listed in Project 1, download and uncompress a Data Set Release 1 or 2 activities file. Generate 1000 random unique person IDs. From the dataset, create a data file with the activities lines for only these individuals. Repeat Project 1 with this new dataset.
3. Using the NDSSL site listed in Project 1, download and uncompress a Data Set Release 1 or 2 activities file. Repeat Project 1 using high-performance computing.

*Projects 4–7 use data from simulations with Cancer Chaste. "Chaste (****C****ancer, **H**eart **a**nd **S**oft **T**issue **E**nvironment) is a general purpose simulation package aimed at multi-scale, computationally demanding problems arising in biology and physiol-*

ogy. Current functionality includes tissue and cell level electrophysiology, discrete tissue modelling, and soft tissue modelling. The package is being developed by a team mainly based in the Computational Biology Group at Oxford University Computing Laboratory, and development draws on expertise from software engineering, high-performance computing, mathematical modeling and scientific computing. While Chaste is a generic extensible library, software development to date has focused on two distinct areas: continuum modelling of cardiac electrophysiology (Cardiac Chaste); and discrete modeling of cell populations (Systems Biology Chaste), with specific application to tissue homeostasis and carcinogenesis (Cancer Chaste)" (Chaste 2012). The initial focus of Cancer Chaste was on colorectal cancer, which it is believed originates in tiny ***crypts of Lieberkühn*** *that descend from the colon's epithelium into the underlying connective tissue (Cancer Chaste 2012).*

At Oxford, using Cancer Chaste, Ornella Cominetti and Angela Shiflet, in consultation with George Shiflet, developed simulations to see the impact of ***differential cell adhesion****, or variations in the level of adhesion between cells of various types, in the crypt. The categories of cells are* ***stem*** *(generation 0);* ***transit*** *categories* ***TA1*** *(generation 1),* ***TA2*** *(generation 2),* ***TA3*** *(generation 3), and* ***TA4*** *(generation 4); and* ***differentiated*** *(generation 5). Stem cells are anchored at the bottom of the crypt. Except for differentiated cells, cells of all other categories can divide. Using Cancer Chaste, the researchers' work attempts to reproduce the work of Wong et al. (2010) using a cellular Potts model. Files* cell_trajectory_file.txt, cell_types_file.txt, and cell_vel_file.txt *generated by some of the Oxford simulations are available for download from the text's website (Cominetti et al. 2010).*

4. (See the italicized description immediately before Project 4.) The file *cell_trajectory_file.txt*, which is available for download from the text's website, has simulated data about the location of a cell in the crypt each simulation hour until the cell leaves the crypt. Each line of the file contains the simulation time, generation, and x- and y-coordinates of the cell's location. Plot the trajectory of the cell using a different color for each generation. Have a legend indicating the generation. See Figure 4b of Wong et al. (2010) for a similar figure. Discuss the results.

5. (See the italicized description immediately before Project 4.) The file *cell_types_file.txt*, which is available for download from the text's website, has simulated data for 20 runs (experiments) of the simulation about the total number of each cell type every half hour for times 70 to 170 simulated hours. Each line of the file contains the simulation time and the number of cells in each category (stem, TA1, TA2, TA3, TA4, differentiated). Generate a stacked bar chart of the average number of cells in each category by time. See Figure 7 of Wong et al. (2010) for a similar figure. Are there any anomalies in the figure? Discuss the results.

6. (See the italicized description immediately before Project 4.) The file *cell_vel_file.txt*, which is available for download from the text's website, has simulated data about the velocities of cells in the crypt one simulation hour before the end of the simulation and at the end of the simulation for 20 runs (experiments) of the simulation. Each line of the file contains the simulation time, cell number, generation number, and x- and y-coordinates of the cell's location. Averaging over the 20 datasets, generate a plot of the mean migration velocities (change in y-coordinate over 1 h) of cells at different heights (y-coordinates) in the crypt. See Figure 5b, graph with triangles, of Wong et al. (2010) for a similar figure. Discuss the results.

7. (See the italicized description immediately before Project 4.) Scientists have found that cultured epithelial cells move collectively in sheets. For a simulation of cells in the crypt, we can use **spatial correlation of velocity**, $C(r)$, as a metric of the amount of coordinated movement of the cells. With r being the distance between two cell centroids, or centers of mass for the cells, for all pairs of cells at distance r from each other, we add the cosines of the angles between their velocity vectors and divide by the number of such pairs. For cells i and j with average velocities over 1 h (change in position from one hour to the next), $\mathbf{v}_i$ and $\mathbf{v}_j$, the cosine of the angle between $\mathbf{v}_i$ and $\mathbf{v}_j$ is $\frac{(\mathbf{v}_i \cdot \mathbf{v}_j)}{|\mathbf{v}_i||\mathbf{v}_j|}$, where $|\mathbf{v}_i|$ is the length of vector $\mathbf{v}_i$. Because the cos(0) = 1, its maximum, the fraction is largest when the angle is zero and the two velocity vectors point in the same direction, or the two cells are headed in the same direction. Thus, a large value for the spatial correlation of velocity, $C(r) = \frac{1}{N_r} \sum_{i,j}^{r=|r_i - r_j|} \frac{(\mathbf{v}_i \cdot \mathbf{v}_j)}{|\mathbf{v}_i||\mathbf{v}_j|}$, where N_r is the number of cell pairs with distance r, indicates a high correlation of velocities of pairs of cells at distance r from each other (Haga et al. 2005).

The file *cell_vel_file.txt*, which is available for download from the text's website, has simulated data about the velocities of cells in the crypt one simulation hour before the end of the simulation and at the end of the simulation for 20 runs (experiments) of the simulation. Each line of the file contains the simulation time, cell number, generation number, and x- and y-coordinates of the cell's location. Using only data for differentiated cells, produce a plot of the mean $C(r)$ values along with **standard error bars**, or symmetric error bars that are two standard deviation units in length, and use intervals of length 1/6 for r. We consider only differentiated cells for two reasons. Because differentiated cells do not divide, cell division does not affect their velocities as much as it does cells of other types. Moreover, differentiated cells compose the largest category of cells. See Figure 6b, graph with circles, of Wong et al. (2010) for a similar figure.

Answers to Quick Review Questions

1. 6.00

2. **a.** $\mathbf{c}$ = (35, 16, 240, 351), $\mathbf{m}$ = (18, 10, 103, 153)
b. $\mathbf{t}$ = (53, 26, 343, 504)
c. data totals by category

3. **a.** 1.1$\mathbf{c}$
b. (39, 18, 264, 386)

4. **a.** (0.23, 0.679)
b. (808988, 715774)

c. $\mathbf{p} \cdot \mathbf{s}$

d. $672078 = 0.23 \cdot 808988 + 0.679 \cdot 715774$

5. a. 15.00

b. s_{42}

6. a. $\begin{bmatrix} -3 & 6 & 0 \\ 3 & 12 & 9 \end{bmatrix}$

b. $\begin{bmatrix} 0 & 5 & 9 \\ 1 & 9 & 9 \end{bmatrix}$

c. cannot be done because C has size 2×2, not 2×3, the size of $2A + B$

7. a. $A\mathbf{v} = \begin{bmatrix} 1 & 3 & 9 \\ 0 & 5 & 6 \end{bmatrix} \begin{bmatrix} 5 \\ 0 \\ -1 \end{bmatrix} = \begin{bmatrix} -4 \\ -6 \end{bmatrix}$ because

$(1)(5) + (3)(0) + (9)(-1) = -4$

$(0)(5) + (5)(0) + (6)(-1) = -6$

b. 8×1

c. 5×1

8. a. $A \cdot B = \begin{bmatrix} 8 & 5 & 3 & -4 \\ -5 & 1 & 0 & 1 \end{bmatrix} \begin{bmatrix} 2 & -6 \\ 7 & 1 \\ 4 & 3 \\ -9 & -2 \end{bmatrix} = \begin{bmatrix} 99 & -26 \\ -12 & 29 \end{bmatrix}$ because

$(8)(2) + (5)(7) + (3)(4) + (-4)(-9) = 99$

$(8)(-6) + (5)(1) + (3)(3) + (-4)(-2) = -26$

$(-5)(2) + (1)(7) + (0)(4) + (1)(-9) = -12$

$(-5)(-6) + (1)(1) + (0)(3) + (1)(-2) = 29$

b. $B \cdot A = \begin{bmatrix} 2 & -6 \\ 7 & 1 \\ 4 & 3 \\ -9 & -2 \end{bmatrix} \begin{bmatrix} 8 & 5 & 3 & -4 \\ -5 & 1 & 0 & 1 \end{bmatrix} = \begin{bmatrix} 46 & 4 & 6 & -14 \\ 51 & 36 & 21 & -27 \\ 17 & 23 & 12 & -13 \\ -62 & -47 & -27 & 34 \end{bmatrix}$

c. C

d. C

9. a. $6.57x + 5.27y = 3.04$

b. $(6.57, 5.27) \cdot (x, y) = 3.04$

10. $\begin{bmatrix} 2 & -1 \\ 6 & 0 \end{bmatrix} \begin{bmatrix} x \\ y \end{bmatrix} = \begin{bmatrix} 7 \\ 5 \end{bmatrix}$

References

Bottini, Nunzio, Gian Franco Meloni, Andrea Finocchi, Giuseppina Ruggiu, Ada Amante, Tullio Meloni, and Egidio Bottini. 2001. "Maternal-Fetal Interaction in the ABO System: A Comparative Analysis of Healthy Mothers and Couples with

Recurrent Spontaneous Abortion Suggests a Protective Effect of B Incompatibility." *Human Biology*, 73(2): 167–174.

Cancer Chaste. 2012. "Cell-Based Chaste: A Multiscale Computational Framework for Modeling Cell Populations." http://www.cs.ox.ac.uk/chaste/cell_based_index.html (accessed December 30, 2012)

Chaste (Cancer, Heart and Soft Tissue Environment). 2012. http://www.cs.ox.ac.uk/chaste/about.html (accessed December 30, 2012)

Cominetti, Ornella, Angela Shiflet, and George Shiflet. 2010. Files *cell_trajectory_file*, *cell_types_file*, and *cell_vel_file* available on this website.

FAO (Food and Agriculture Organization of the United Nations). 2012. "*Callinectes sapidus* Species Fact Sheets," Fisheries and Aquaculture Department. http://www.fao.org/fishery/species/2632/en (accessed December 30, 2012)

Green Sea Turtle. Demographics of the Hawaiian Green Sea Turtle: Modeling Population Dynamics Using a Linear Deterministic Matrix Model: The Leslie Matrix. http://isolatium.uhh.hawaii.edu/linear/ch6/green.htm (accessed December 30, 2012)

Haga, H., Irahara, C., Kobayashi, R., Nakagaki, T. and Kawabata, K. 2005. "Collective Movement of Epithelial Cells on a Collagen Gel Substrate," *Biophys. J.*, 88: 2250– 2256.

Johnson, Ted R., and Christine L. Case. 2009. *Laboratory Experiments in Microbiology*. San Francisco, CA: Benjamin Cummings.

Lockhart, Peter J., Michael A. Steel, Michael D. Hendy, and David Penny. 1994. "Recovering Evolutionary Trees under a More Realistic Model of Sequence Evolution." *Mol. Biol. Evol.,* 11(4): 605–6 12.

LANL (Los Alamos National Laboratory). " EpiSimS: Epidemic Simulation System" http://www.lanl.gov/programs/nisac/episims.shtml (accessed January 26, 2013)

NDSSL (Network Dynamics and Simulation Science Laboratory, Virginia Polytechnic Institute and State University). 2009a. "NDSSL Proto-Entities" http://ndssl.vbi.vt.edu/opendata/ (accessed December 30, 2012)

———. 2009b. Synthetic Data Products for Societal Infrastructures and Proto-Populations: Data Set 1.0. ndssl.vbi.vt.edu/Publications/ndssl-tr-06-006.pdf (accessed December 30, 2012)

———. 2009c. Synthetic Data Products for Societal Infrastructures and Proto-Populations: Data Set 2.0. ndssl.vbi.vt.edu/Publications/ndssl-tr-07-003.pdf (accessed December 30, 2012)

———. 2009. Synthetic Data Products for Societal Infrastructures and Proto-Populations: Data Set 3.0. ndssl.vbi.vt.edu/Publications/ndssl-tr-07-010.pdf (accessed December 30, 2012)

Taylor, Caz, and Erin Grey. 2010. "Population Dynamics of Gulf Blue Crabs" Tulane University. http://leag.tulane.edu/PDFs/Grey-LEAG-4.28.10.pdf (accessed December 30, 2012)

VBI (Virginia Bioinformatics Institute at Virginia Tech). 2008. "EpiSims." http://ndssl.vbi.vt.edu/episims.php (accessed December 30, 2012)

Wong, Shek Yoon, K.-H. Chiam, Chwee Teck Lim, and Paul Matsudaira. 2010. "Computational Model of Cell Positioning: Directed and Collective Migration in the Intestinal Crypt Epithelium." *J. R. Soc. Interface*, 7: S351–S363. First published online March 31, 2010; *doi: 10.1098/rsif.2010.0018.focus)

Zinski, Steven C. 2006. "Blue Crab Life Cycle." http://www.bluecrab.info/lifecycle.html (accessed December 30, 2012)

MODULE 13.3

Time after Time—Age- and Stage-Structured Models

Prerequisites: Module 13.1, "Computational Toolbox—Tools of the Trade: Tutorial 7" or "Alternative Tutorial 7" (through section on "Eigenvalues and Eigenvectors") and Module 13.2, "Matrices for Population Studies—Linked for Life." Additional high-performance computing materials related to this module are available on the text's website.

Downloads

The text's website has the file *AgeStructured*, containing the models in this module, available for download for various computational tools.

Introduction

> The worst thing that can happen—will happen—is not energy depletion, economic collapse, limited nuclear war, or conquest by a totalitarian government. As terrible as these catastrophes will be for us, they can be repaired within a few generations. The one process ongoing. . .that will take millions of years to correct, is the loss of genetic and species diversity by the destruction of natural habitats. This is the folly our descendants are least likely to forgive us.
>
> –E. O. Wilson (Bean 2005)

If you were sitting on a beach on one of the 12 islets of French Frigate Shoals in northwestern Hawaii admiring the April moon, you might be surprised to see a rather large body crawling deliberately up the sand. It is likely a female green sea turtle (*Chelonia mydas*) on her way to deposit her eggs. Although these turtles may normally feed around other Hawaiian islands, they usually return to the beach where they hatched (natal beach) to nest. Ninety percent of green turtle nests in Hawaii are found on these islets.

Nesting is an arduous process for this animal, and she may make the journey more than once this season. Though she may have several more clutches to lay, she digs a hole with her front flippers to a depth of about 2 ft, deposits her 100± eggs, covers the eggs, and returns to the water. She might return 2 weeks or so later to build another nest and deposit more eggs. Fortunately for her, she does this only every 3 years or so.

Undisturbed eggs deposited this night incubate below the surface for about 2 months. After escaping from their leathery cases, 1-oz hatchlings work together to emerge from their sandy womb. All this occurs at night, when temperatures are lower and the turtles are less conspicuous. Once out of the nest, they sprint toward the bouncing glints of light on the ocean surface. Many do not make it, intercepted by birds, crabs, or other predators, which have learned that these hatching events provide tasty meals. Even if they make it to the water, no matter how fiercely they swim, carnivorous fish may eat them. Then, as adults, turtles have two main predators—sharks and human beings, the latter being more of a threat.

Those that survive the beach dash and shallow waters swim out to sea, where they feed on various floating plants and animals. As they become adults they utilize large, shallow sea grass beds for much of their diet. Such a diet results in the development of body fat that is green, which gives this animal its name. Long lived, this animal may not become sexually mature for 20 years or more. Few from that original clutch of eggs, however, will make it to return to this beach for breeding and nesting.

Marine predators are not the only obstacles to survival and breeding success. Turtles and their eggs are still consumed in many places in the world. Coastal development and subsequent habitat destruction also devastate breeding and nesting. For example, in St. Croix various species of sea turtles nest primarily in the Jack, East End and Isaac Bays, and Buck Island, where there is no development and the beaches are relatively undisturbed. Information from satellites has proved invaluable in collecting a wide variety of environmental data that help in protecting important, unique habitats, understanding environmental changes, and ensuring the survival of endangered species. NASA with the CNES (Centre National d'Etudes Spatiales, the French space agency) and NOAA (the National Oceanic and Atmospheric Administration) established Argos, a satellite-based system that helps to collect, process, and disseminate environmental data for various platforms (ARGOS 2013).

Pollution of various sorts may not only cause turtle mortality directly but also induce an ever-increasing incidence of fibropapilloma. This disease results in the development of large tumors that interfere with normal life activities of the animals, resulting in death.

On December 28, 1973, the Endangered Species Act became law in the United States. This act provides programs that promote the conservation of threatened and endangered plant and animal species and the habitats where they are found. **Endangered** organisms are species that are in danger of extinction throughout all or over a sizeable portion of their range. **Threatened** species are those likely to become endangered in the foreseeable future. Currently, there are almost 2000 threatened and endangered species worldwide found on the list maintained and published by the

Fish and Wildlife Service of the U.S. Department of Interior. Of the approximately 1200 animals on the list are six of the seven species of sea turtles. Green sea turtles were added to the list in 1978.

Many studies have been attempted to ascertain the status of green sea turtle populations worldwide. Various interventions, primarily aimed at protecting the nests and hatchlings, have been attempted. However, there is much we do not know about the biology and demography of these animals that need to be understood to make appropriate conservation efforts. Sea turtle life cycles are long and complex; because growth stops at sexual maturity, it has been difficult to determine the age of turtles. Also, it has been virtually impossible to mark hatchlings so that we can identify them as adults. Detailed information regarding the population demography of turtles is vital if we are to establish the status of wild populations and to implement effective management procedures. Decisions and conservation efforts we make today may be crucial to preventing their extinction. But, how can we make effective decisions if we do not understand how various management alternatives affect turtle populations?

One approach to studying sea turtle populations is the use of mathematical models, specifically Leslie and Lefkovitch matrix population projections. The Leslie matrix projection, developed by P. H. Leslie in 1945, uses mortality and fecundity rates to develop population distributions. These distributions are founded on initial population distribution of age groups. Because the age of adult turtles is difficult to determine, some researchers have used a Lefkovitch matrix, which divides the populations into stage classes. Some of the life stages are easily recognizable (eggs, hatchlings, nesting adults), but the juvenile stages are long lasting, and age is difficult to determine. So, size (length of carapace or shell) is used to define stages.

Resulting population projections have indicated that we may need to increase protective measures to juveniles and adults if we really want to increase the numbers of sea turtles. Crowder et al. (1994) published a stage-based population model for the loggerhead turtle (*Caretta caretta*) that projected the effects of the use of turtle-exclusion devices (TEDs) in trawl fisheries. These devices allow young turtles to escape the trawls that trap shrimp, and the model predicted that the required use of TEDs for offshore trawling would allow a gradual increase in Loggerheads by an order of magnitude in about 70 years. Such regulations may save thousands of turtles each year and help to save sea turtle species from extinction (Bjorndal et al. 2000; Crouse et al. 1987; Crowder et al. 1994; Earthtrust 2009; Forbes 1992; Zug 2002).

The Problem

We can classify many animals by discrete ages to determine reproduction and mortality. For example, suppose a certain bird has a maximum life span of 3 years. During the first year, the animal does not breed. On the average, a typical female of this hypothetical species lays 10 eggs during the second year but only 8 during the third. Suppose 15% of the young birds live to the second year of life, while 50% of the birds from age 1 to 2 years live to their third year of life, age 2 to 3 years. Usually, we consider only the females in the population; and in this example, we assume that half the offspring are female.

For such a situation, we are interested in the answers to several questions:

- Can we determine the projected population growth rate?
- In the case of declining populations, what is the predicted time of extinction?
- As time progresses, does the population reach a stable distribution?
- If so, what is the proportion of each age group in such a stable age distribution?
- How sensitive is the long-term population growth rate or predicted time of extinction to small changes in parameters?

Age-Structured Model

Figure 13.3.1 presents a state diagram for the problem with the states denoting ages (year 1, 2, or 3) of the bird. The information indicates that an **age-structured model** might be appropriate. In age-structured models we ignore the impact of other factors, such as population density and environmental conditions. We can use such models to answer questions about the rate of growth of the population and the proportion of each age group in a stable age distribution.

For the example in the previous section, three clear age classes emerge, one for each year. Thus, in formulating this deterministic model, we employ the following variables: x_i = number of females of such a bird at the beginning of the breeding season in year i (age $i - 1$ to i) of life, where $i = 1, 2,$ or 3. Thus, x_1 is the number of eggs and young birds in their first year of life.

Time, t, of the study is measured in years immediately before breeding season, and we use the notation $x_i(t)$ to indicate the number of year i females at time t. For example, $x_2(5)$ represents the number of females during their second year, ages 1 to 2 yr old, at the start of breeding season 5. Some of these survive to time $t + 1 = 6$ yr and progress to the next class, those females in their third year of life. At that time (at time 6 yr of the study), the notation for number of year 3 females is $x_3(6)$.

To establish equations, we use these data to project the number of female birds in each category for the following year. The number of eggs/chicks depends on the number of adult females, x_2 and x_3. Because on the average a year 2 (ages 1 to 2 years old) mother has 5 female offspring and a year 3 (ages 2 to 3 years old) mother has 4 female offspring, the number of year 1 (ages 0 to 1 year old) female eggs/chicks at time $t + 1$ is as follows:

$$5x_2(t) + 4x_3(t) = x_1(t + 1) \qquad (1)$$

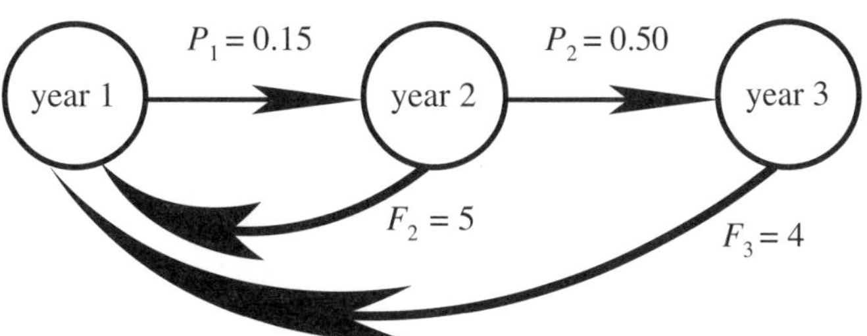

Figure 13.3.1 State diagram for problem

However, at time $t + 1$, the number of year 2 (ages 1 to 2 years old) females, $x_2(t + 1)$, depends only on the number of year 1 (ages 0 to 1 year old) females this year, $x_1(t)$, that live. The latter survives with a probability of $P_1 = 15\% = 0.15$, so that we estimate next year's number of year 2 females to be as follows:

$$0.15x_1(t) = x_2(t + 1) \tag{2}$$

Similarly, to estimate the number of year 3 (ages 2 to 3 years old) females next year, we need to know only the number of year 2 (ages 1 to 2 years old) females, $x_2(t)$, and their survival rate (here, $P_2 = 50\% = 0.50$). Thus, the number of year 2 females next year will be approximately the following:

$$0.50x_2(t) = x_3(t + 1) \tag{3}$$

Placing Equations 1, 2, and 3 together, we have the following system:

$$\begin{cases} \quad 5x_2(t) + 4x_3(t) = x_1(t+1) \\ 0.15x_1(t) \qquad\qquad\qquad\quad = x_2(t+1) \\ \qquad\quad 0.50x_2(t) \qquad\quad\; = x_3(t+1) \end{cases}$$

This system of equations translates into the following matrix-vector form:

$$\begin{bmatrix} 0 & 5 & 4 \\ 0.15 & 0 & 0 \\ 0 & 0.50 & 0 \end{bmatrix} \begin{bmatrix} x_1(t) \\ x_2(t) \\ x_3(t) \end{bmatrix} = \begin{bmatrix} x_1(t+1) \\ x_2(t+1) \\ x_3(t+1) \end{bmatrix}$$

or

$$L\mathbf{x}(t) = \mathbf{x}(t + 1), \text{ where}$$

$$L = \begin{bmatrix} 0 & 5 & 4 \\ 0.15 & 0 & 0 \\ 0 & 0.50 & 0 \end{bmatrix}, \mathbf{x}(t) = \begin{bmatrix} x_1(t) \\ x_2(t) \\ x_3(t) \end{bmatrix}, \text{ and } \mathbf{x}(t + 1)$$

$$= \begin{bmatrix} x_1(t+1) \\ x_2(t+1) \\ x_3(t+1) \end{bmatrix}.$$

Suppose an initial population distribution has 3000 female eggs/chicks, 440 year 2, and 350 year 3 female birds, so that $\mathbf{x}(0) = \begin{bmatrix} 3000 \\ 440 \\ 350 \end{bmatrix}$ is the **initial age-distribution vector**. The next year, because of births, aging, and deaths, the number of females in each age class changes. The following vector gives the calculation for the estimated population at time $t = 1$ year:

$$\mathbf{x(1)} = L\mathbf{x(0)} = \begin{bmatrix} 0 & 5 & 4 \\ 0.15 & 0 & 0 \\ 0 & 0.50 & 0 \end{bmatrix} \begin{bmatrix} 3000 \\ 440 \\ 350 \end{bmatrix} = \begin{bmatrix} 3600 \\ 450 \\ 220 \end{bmatrix}$$

Thus, at $t = 1$ year, we project a population of more eggs/chicks but fewer year 3 female adults than initially present.

Quick Review Question 1

Suppose an insect has maximum life expectancy of 2 months. On the average, this animal has 10 offspring in the first month and 300 in the second. The survival rate from the first to the second month of life is only 1%. Assume half the offspring are female. Suppose initially a region has 2 females in their first month of life and 1 in her second.

- **a.** Define the variables of the model.
- **b.** Construct a system of equations for the model.
- **c.** Give the matrix representation for the model.
- **d.** Using matrix multiplication, determine the number of females for each age at time $t = 1$ month expressed to two decimal places.
- **e.** Determine the number of females for each age at time $t = 2$ months.

Leslie Matrices

L is an example of a **Leslie matrix**, which is a particular type of **projection matrix,** or **transition matrix**. Such a square matrix has a row for each of a finite number (n) of equal-length age classes. Suppose F_i is the average **reproduction,** or **fecundity, rate** of class i; and P_i is the **survival rate** of those from class i to class $(i + 1)$. With $x_i(t)$ being the number of females in class i at time t, $x_1(t)$ is the number of females born between time $t - 1$ and time t and living at time t. The model has the following system of equations:

$$\begin{cases} F_1x_1(t) + F_2x_2(t) + \cdots + F_{n-1}x_{n-1}(t) + F_nx_n(t) = x_1(t+1) \\ P_1x_1(t) = x_2(t+1) \\ P_2x_2(t) = x_3(t+1) \\ \vdots \\ P_{n-1}x_{n-1}(t) = x_n(t+1) \end{cases} \quad (4)$$

where

F_i is the average reproduction rate (fecundity rate) of class i,
P_i is the survival rate of from class i to class $(i + 1)$, and
$x_i(t)$ is the number of females in class i at time t.

Therefore, the corresponding $n \times n$ Leslie matrix is as follows:

$$L = \begin{bmatrix} F_1 & F_2 & F_3 & \cdots & F_{n-1} & F_n \\ P_1 & 0 & 0 & \cdots & 0 & 0 \\ 0 & P_2 & 0 & \cdots & 0 & 0 \\ \vdots & \vdots & \vdots & & \vdots & \vdots \\ 0 & 0 & 0 & \cdots & P_{n-1} & 0 \end{bmatrix}$$

F_i and P_i are nonnegative numbers, which appear along the first row and the **subdiagonal**, respectively; all other entries are zero.

Definition In an $n \times n$ square matrix B, the **subdiagonal** is the set of elements

$$\{b_{21}, b_{32}, \ldots, b_{n(n-1)}\}.$$

With **x**(*t*) being the population female distribution vector at time t, $(x_1(t), x_2(t), \ldots, x_n(t))$, and **x**($t$ + **1**) being the female distribution vector at time $t + 1$, $(x_1(t + 1), x_2(t + 1), \ldots, x_n(t + 1))$, both expressed as column vectors, we have the following matrix equivalent of the system of Equations 4: $L\mathbf{x}(t) = \mathbf{x}(t + 1)$

Definition A **Leslie matrix** is a matrix of the following form, where all entries F_i and P_i are nonnegative:

$$\begin{bmatrix} F_1 & F_2 & F_3 & \cdots & F_{n-1} & F_n \\ P_1 & 0 & 0 & \cdots & 0 & 0 \\ 0 & P_2 & 0 & \cdots & 0 & 0 \\ \vdots & \vdots & \vdots & & \vdots & \vdots \\ 0 & 0 & 0 & \cdots & P_{n-1} & 0 \end{bmatrix}$$

Quick Review Question 2

Give the Leslie matrix for a system with four classes, where the (female) reproduction rates are 0.2, 1.2, 1.4, and 0.7 for classes 1 to 4, respectively, and the survival rates are 0.3, 0.8, and 0.5 for classes 1 to 3, respectively.

Age Distribution over Time

Let us now consider the population distribution as time progresses. In the section "An Age-Structured Model," we considered the initial female age distribution of a hypothetical bird species to be $\begin{bmatrix} 3000 \\ 440 \\ 350 \end{bmatrix}$ and calculated the distribution at time $t = 1$ to be $\mathbf{x(1)} = L\mathbf{x(0)} = \begin{bmatrix} 3600 \\ 450 \\ 220 \end{bmatrix}$. Repeating the process, we have the following results at time $t = 2$ years:

$$\mathbf{x(2)} = L\mathbf{x(1)} = \begin{bmatrix} 0 & 5 & 4 \\ 0.15 & 0 & 0 \\ 0 & 0.50 & 0 \end{bmatrix} \begin{bmatrix} 3600 \\ 450 \\ 220 \end{bmatrix} = \begin{bmatrix} 3130 \\ 540 \\ 225 \end{bmatrix}$$

Summing the elements of the result gives us a total female population at that time of 3895. The percentage of females in each category is as follows:

$$\begin{bmatrix} 3130/3895 \\ 540/3895 \\ 225/3895 \end{bmatrix} = \begin{bmatrix} 0.803594 \\ 0.138639 \\ 0.0577664 \end{bmatrix} = \begin{bmatrix} 80.36\% \\ 13.86\% \\ 5.78\% \end{bmatrix}$$

We note that the calculation $\mathbf{x(2)} = Lx\mathbf{(1)} = L(L\mathbf{x(0)}) = L^2\mathbf{x(0)}$. Similarly, $\mathbf{x(3)} = L\mathbf{x(2)} = L(L^2\mathbf{x(0)}) = L^3\mathbf{x(0)}$. In general, $\mathbf{x}(t) = L^t\mathbf{x(0)}$.

For several values of t, Table 13.3.1 indicates the population change in the three classes by presenting the distributions, $\mathbf{x}(t) = L^t\mathbf{x(0)}$, and the percentage of female animals in each class. As time goes on, although the numbers of birds in each class changes, the vector of percentages of animals in each category converges to $\mathbf{v} = \begin{bmatrix} 0.8206 \\ 0.1205 \\ 0.0590 \end{bmatrix} = \begin{bmatrix} 82.06\% \\ 12.05\% \\ 5.90\% \end{bmatrix}$. From time $t = 20$ years on, the percentages expressed to two decimal places do not change from one year to the next. Over time, the percentage of eggs/chicks stabilizes to 82.06% of the total population, while year 2 birds comprise 12.05% and year 3 birds are 5.90% of the population. This convergence to fixed percentages is characteristic of such age-structured models. Because we are assuming the number of females (or males) to be a fixed proportion (half) the population, the convergence of category percentages for females (or males) is the same as the convergence of category percentages for the entire population (females and males).

Projected Population-Growth Rate

Interestingly, if we divide corresponding elements of the population distribution at time $t + 1$, $\mathbf{x}(t + 1)$, by the members of the distribution at time t, $\mathbf{x}(t)$, we have convergence of the quotients to the same number. Table 13.3.2 shows several of these quotients, which converge in this example to 1.0216, which we call λ. Thus, eventually each age group changes by a factor of $\lambda = 1.0216$ (102.16%) from one year to

Table 13.3.1
Population Distributions and Class Percentages of the Total Population

Time, t	*Distribution* $x(t) = L^n x(0)$	*Class Percentages*
0	$\begin{bmatrix} 3000 \\ 440 \\ 350 \end{bmatrix}$	$\begin{bmatrix} 79.16\% \\ 11.61\% \\ 9.23\% \end{bmatrix}$
1	$\begin{bmatrix} 3600 \\ 450 \\ 220 \end{bmatrix}$	$\begin{bmatrix} 84.31\% \\ 10.54\% \\ 5.15\% \end{bmatrix}$
2	$\begin{bmatrix} 3130 \\ 540 \\ 225 \end{bmatrix}$	$\begin{bmatrix} 80.36\% \\ 13.86\% \\ 5.78\% \end{bmatrix}$
3	$\begin{bmatrix} 3600 \\ 469.5 \\ 270 \end{bmatrix}$	$\begin{bmatrix} 82.96\% \\ 10.82\% \\ 6.22\% \end{bmatrix}$
$\vdots$	$\vdots$	$\vdots$
9	$\begin{bmatrix} 3913.31 \\ 574.45 \\ 281.813 \end{bmatrix}$	$\begin{bmatrix} 82.04\% \\ 12.04\% \\ 5.91\% \end{bmatrix}$
10	$\begin{bmatrix} 3999.5 \\ 586.997 \\ 287.225 \end{bmatrix}$	$\begin{bmatrix} 82.06\% \\ 12.04\% \\ 5.89\% \end{bmatrix}$
$\vdots$	$\vdots$	$\vdots$
20	$\begin{bmatrix} 4950.87 \\ 726.933 \\ 355.783 \end{bmatrix}$	$\begin{bmatrix} 82.06\% \\ 12.05\% \\ 5.90\% \end{bmatrix}$
21	$\begin{bmatrix} 5057.8 \\ 742.631 \\ 363.467 \end{bmatrix}$	$\begin{bmatrix} 82.06\% \\ 12.05\% \\ 5.90\% \end{bmatrix}$
$\vdots$	$\vdots$	$\vdots$
100	$\begin{bmatrix} 27353.5 \\ 4016.29 \\ 1965.7 \end{bmatrix}$	$\begin{bmatrix} 82.06\% \\ 12.05\% \\ 5.90\% \end{bmatrix}$
101	$\begin{bmatrix} 27944.3 \\ 4103.03 \\ 2008.15 \end{bmatrix}$	$\begin{bmatrix} 82.06\% \\ 12.05\% \\ 5.90\% \end{bmatrix}$

Table 13.3.2
$\boldsymbol{x}(t+1)/\boldsymbol{x}(t)$ for Table 13.3.1

Time, t	$x(t+1)/x(t)$
0	$\begin{bmatrix} 3600/3000 \\ 450/440 \\ 220/350 \end{bmatrix} = \begin{bmatrix} 1.2 \\ 1.02273 \\ 0.628571 \end{bmatrix}$
1	$\begin{bmatrix} 3130/3600 \\ 540/450 \\ 225/220 \end{bmatrix} = \begin{bmatrix} 0.869444 \\ 1.2 \\ 1.02273 \end{bmatrix}$
2	$\begin{bmatrix} 3600/3130 \\ 469.5/540 \\ 270/225 \end{bmatrix} = \begin{bmatrix} 1.15016 \\ 0.869444 \\ 1.2 \end{bmatrix}$
⋮	⋮
9	$\begin{bmatrix} 3999.5/3913.31 \\ 586.997/574.45 \\ 287.225/281.813 \end{bmatrix} = \begin{bmatrix} 1.02202 \\ 1.02184 \\ 1.01921 \end{bmatrix}$
⋮	⋮
20	$\begin{bmatrix} 5057.8/4950.87 \\ 742.631/726.933 \\ 363.467/355.783 \end{bmatrix} = \begin{bmatrix} 1.0216 \\ 1.02159 \\ 1.0216 \end{bmatrix}$
⋮	⋮
100	$\begin{bmatrix} 27944.3/27353.5 \\ 4103.03/4016.29 \\ 2008.15/1965.7 \end{bmatrix} = \begin{bmatrix} 1.0216 \\ 1.0216 \\ 1.0216 \end{bmatrix}$

the next. For instance, in going from time $t = 100$ years to $t + 1 = 101$ years, Table 13.3.1 shows that the number of year 1 females increases 2.16%, from 27,353.5 to 1.0216(27,353.5) = 27,944.3. Similarly, the number of year 2 females changes from 4,016.29 to 1.0216(4,016.29) = 4,103.03, and the year 3 females also goes up by the same factor, from 1,965.7 to 1.0216(1,965.7) = 2,008.15. Thus, with each age group ultimately changing by a factor of 1.0216 = 102.16% annually, eventually we estimate the population will increase by 2.16% per year. Thus, for an initial total population of P_0, the estimated population at time t is $P = P_0(1.0216)^t$.

With an annual increase in population of 2.16% per year and, correspondingly, $\lambda = 1.0216 > 1$, we expect that this bird population will increase with time. Had the population been projected to decline each year with $0 < \lambda < 1$, we would expect the birds eventually to become extinct. A value of $\lambda = 1$ would signal a stable population in which, on the average, an adult female produces one female offspring that will live to adulthood. Thus, λ is an important concept related to the stability of a population.

Interestingly, multiplying the constant λ by the vector of percentages to which the category distributions converge, **v**, has the same effect as multiplying the Leslie matrix L by **v**, or $L\mathbf{v} = \lambda\mathbf{v}$, as the following calculations indicate:

$$L\mathbf{v} = \begin{bmatrix} 0 & 5 & 4 \\ 0.15 & 0 & 0 \\ 0 & 0.50 & 0 \end{bmatrix} \begin{bmatrix} 0.8206 \\ 0.1205 \\ 0.0590 \end{bmatrix} = \begin{bmatrix} 0.84 \\ 0.12 \\ 0.06 \end{bmatrix}$$

$$\lambda\mathbf{v} = 1.0216 \begin{bmatrix} 0.8206 \\ 0.1205 \\ 0.0590 \end{bmatrix} = \begin{bmatrix} 0.84 \\ 0.12 \\ 0.06 \end{bmatrix}$$

Multiplying both sides of the equation by a constant, c, maintains the equality, $cL\mathbf{v} = c\lambda\mathbf{v}$ or $L(c\mathbf{v}) = \lambda(c\mathbf{v})$. The formula holds for any constant, c, and, consequently, for any population distribution where the percentages of the total for the three classes are 82.06%, 12.05%, and 5.90%, respectively. Thus, multiplication of the population distribution vector by the constant 1.0216 is identical to the product of the Leslie matrix by the distribution vector. λ is an **eigenvalue** for the matrix L, and **v** is a corresponding **eigenvector** for L.

Quick Review Question 3

Consider the Leslie matrix $L = \begin{bmatrix} 5 & 150 \\ 0.01 & 0 \end{bmatrix}$ from Quick Review Question 1c with the initial population distribution vector $\mathbf{x(0)} = \begin{bmatrix} 2 \\ 1 \end{bmatrix}$.

a. Using a computational tool, for each age class give the value to which its percentage of the total population converges as time progresses. Express your answer to six significant figures.

b. Using a computational tool, give the number, λ, to which the quotient of each class population at time t over the class population at time $t - 1$ converges as time progresses. Express your answer to six significant figures.

c. Using the values from Parts a and b, give a vector **v** that satisfies $L\mathbf{v} = \lambda\mathbf{v}$.

d. Give another vector **v** that satisfies the equation from Part c, where a million times more insects are in their first month of life.

Using mathematics, which we do not show here, or a computational tool, we can obtain three eigenvalues and three corresponding eigenvectors for the 3×3 matrix L (Table 13.3.3). Two of the three eigenvalues are imaginary numbers, and the corresponding two eigenvectors contain imaginary numbers. The eigenvalue with the largest magnitude (for real numbers, the largest absolute value), 1.0216, is the **dominant eigenvalue** and is the projected annual growth rate associated with the Leslie matrix. Leslie matrices always have such a unique positive eigenvalue. The sum of

Table 13.3.3
Eigenvalues and Vectors for **L**

Eigenvalue	*Eigenvector*
1.0216	(–0.9869, –0.144906, –0.0709212)
–0.510798 + 0.180952 *i*	(0.935827, –0.244171 – 0.0864985 *i*, 0.185709 + 0.150458 *i*)
0.510798 – 0.180952 *i*	(0.935827, –0.244171 + 0.0864985 *i*, 0.185709 – 0.150458 *i*)

the components of the corresponding eigenvector, (–0.9869, –0.144906, –0.0709212), is –1.20273. Dividing this sum into each element, we obtain another eigenvector, (0.8206, 0.1205, 0.0590), which is the preceding vector of projected proportions for the three classes.

Definition For square matrix M, the constant λ is an **eigenvalue** and **v** is an **eigenvector** if multiplication of the constant by the vector accomplishes the same results as multiplying the matrix by the vector; that is, the following equality holds:

$$M\mathbf{v} = \lambda\mathbf{v}$$

The **dominant eigenvalue** for a matrix is the largest eigenvalue for that matrix.

Stage-Structured Model

An age-structured model, where we distinguish life stages by age, is a special case of a **stage-structured model**, where we divide the life of an organism into stages. Frequently, it is convenient or necessary to consider the life of a species in stages instead of equally spaced time intervals, such as years. Perhaps the animal, such as a loggerhead sea turtle, typically lives for a number of years, and we cannot accurately determine the age of an adult. Conceivably the stages differ greatly in lengths of time. Also, rates may be strongly associated with developmental stages or animal size.

Morris, Shertzer, and Rice generated a stage-structured model of the Indo-Pacific lionfish *Pterois volitans* to explore control of this invasive and destructive species to reef habitats (Morris et al. 2011). Such consideration is very important because in a Caribbean region study, Albins and Hixon found lionfish reduced recruitment of native fishes (addition of new native fishes) by an average of 79% over a 5-week period (Albins and Hixon 2008). A lionfish goes through three life stages: larva (L, about 1 month), juvenile (J, about 1 year), and adult (A). With 1 month being the basic time step, the probability that a larva survives and grows to the next stage is $G_L = 0.00003$, while the probability that a juvenile survives and remains a juvenile in a 1-month period is $P_J = 0.777$. In 1 month, $G_J = 0.071$ of the juveniles mature to the adult stage, while $P_A = 0.949$ of the adults survive in a month. Only adults give birth, and the number of female larvae she produces per month is $R_A = 35{,}315$. Figure 13.3.2 presents a state diagram for these circumstances.

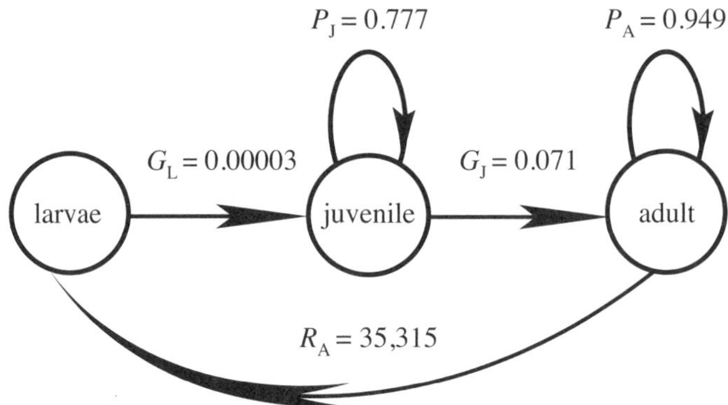

Figure 13.3.2 State diagram for lionfish (Morris et al. 2011)

Thus, if $x_L(t)$, $x_J(t)$, and $x_A(t)$ represent the number of female larvae, juveniles, and adults at time t, respectively, we have the following system of equations for the distribution at time $t + 1$:

$$\begin{cases} 35315x_A(t) = x_L(t+1) \\ 0.00003x_L(t) + 0.777x_J(t) = x_J(t+1) \\ 0.071x_J(t) + 0.949x_A(t) = x_A(t+1) \end{cases}$$

Thus, we have the following transition matrix, called a **Lefkovitch matrix**:

$$\begin{bmatrix} 0 & 0 & 35315 \\ 0.00003 & 0.777 & 0 \\ 0 & 0.071 & 0.949 \end{bmatrix}$$

Using these values, the lionfish monthly growth rate (λ) is about 1.13. Because adult lionfish reproduce monthly over the entire year, adult survivorship has a great impact on the population's growth rate. With all else being the same, not until the probability of an adult surviving in a 1-month period is reduced approximately 30%, from $P_A = 0.949$ to $P_A = 0.66$ or less, could a negative population growth be achieved. Harvesting 30% of the adult lionfish each month is quite a challenge. However, simultaneous reductions of 17% for the probabilities of juvenile and adult survivorship could also produce a declining population. Thus, "results indicate that an eradication program targeting juveniles and adults jointly would be far more effective than one targeting either life stage in isolation" (Morris et al. 2011).

Algorithms

For an age-structured or a stage-structured problem, we form the appropriate matrix, L, and the vector representing initial female population distribution, $\mathbf{x}$, and determine the distribution at time t by calculating $L^t\mathbf{x}$.

For the projected population growth rate, we calculate the eigenvalue, λ, which is available through a command with many computational tools. When it is not available, to estimate the projected population growth rate to within m decimal places, we keep calculating the ratio of age distributions, $\mathbf{x}(t+1)/\mathbf{x}(t)$, until two subsequent ratios differ by no more than 10^{-m}. For the preceding example with birds, to estimate population growth to within four decimal places, we consider any one of the components, say the first, of $\mathbf{x}(t+1)/\mathbf{x}(t) = L^{t+1}\mathbf{x} / L^t\mathbf{x}$. After repeated calculation, we discover with $\mathbf{x}(15)/\mathbf{x}(14) = L^{15}\mathbf{x}/L^{14}\mathbf{x} = (1.02153, 1.02173, 1.02136)$ and $\mathbf{x}(16)/\mathbf{x}(15) = L^{16}\mathbf{x}/L^{15}\mathbf{x} = (1.02162, 1.02153, 1.02173)$ that the first components are sufficiently close to each other:

$$| 1.02153 - 1.02162 | = 0.00009 < 10^{-4} = 0.0001$$

These first elements, 1.02153 and 1.02162, differ by no more than 10^{-4}, so our projected population growth rate is 1.0216. Similarly, we can determine the category percentages of the total to within m decimal places by finding when each of the corresponding elements of $\mathbf{x}(t)$/(total population) and $\mathbf{x}(t+1)$/(total population) differ by no more than 10^{-m}.

Sensitivity Analysis for the Age-Structured Example

We can use **sensitivity analysis** to examine how sensitive values, such as long-term population growth rate (dominant eigenvalue λ) or predicted time of extinction, are to small changes in parameters, such as survivability and fecundity. Suppose in the preceding bird example, we wish to examine the sensitivity of the long-term population growth rate to small changes in survivability of year 1 and year 2 birds, P_1 and P_2, respectively. Adjusting P_1 and P_2 individually by ±10% and ±20%, Table 13.3.4 shows the corresponding new values of λ and the change in projected population growth rate, $\lambda_{\text{new}} - \lambda$, as calculated using a computational tool. **Relative sensitivity** or **sensitivity of λ with respect to P_i** measures the numeric impact on λ of a change in P_i, or the instantaneous rate of change of λ with respect to P_i (the partial derivative of λ with respect to P_i, $\partial\lambda/\partial P_i$). Thus, to approximate this relative sensitivity, we divide the change in projected population growth rate by the corresponding small change in P_i:

$$\textbf{sensitivity of } \boldsymbol{\lambda} \textbf{ to } \boldsymbol{P_i} = \frac{\partial\lambda}{\partial P_i} \approx \frac{\lambda_{\text{new}} - \lambda}{P_{i,\text{new}} - P_i},$$

where $P_{i,\text{new}}$ is the new value of P_i and λ_{new} is the resulting new value of λ. For example, $P_1 = 0.15$, and $P_1 + (10\% \text{ of } P_1) = P_1 + 0.10P_1 = 1.10P_1 = 0.15 + 0.015 = 0.165$. With $P_1 = 0.15$, the original dominant eigenvalue λ is 1.0216. Replacing the chance of a year 1 bird surviving with $P_{1,\text{new}} = 1.10P_1 = 0.165$, the new dominant eigenvalue λ_{new} is 1.06526, and $\lambda_{\text{new}} - \lambda = 1.06526 - 1.0216 = 0.04366$. Thus, the relative sensitivity of P_1 using +10% is approximately $(\lambda_{\text{new}} - \lambda)/(P_{1,\text{new}} - P_1) = (\lambda_{\text{new}} - \lambda)/(0.10P_1) = 0.04366/0.015 = 2.9067$. Similarly for −10% of P_1, the sensitivity is 3.0677. However, the relative sensitivity of λ to small changes in P_2 (+10% and −10%) is much smaller (0.2480 and 0.2562, respectively). From these calculations in

Table 13.3.4
Sensitivity of λ (Originally 1.0216) to Changes in Survivability

Survivability Parameter	*Percent Change*	$P_{i,\text{new}}$	λ_{new}	$\lambda_{\text{new}} - \lambda$	$P_{i,\text{new}} - P_i$	$\frac{\lambda_{\text{new}} - \lambda}{P_{i,\text{new}} - P_i}$
$P_1 = 0.15$	+10%	0.165	1.0653	0.0437	0.015	2.9111
$P_1 = 0.15$	+20%	0.180	1.1069	0.0853	0.030	2.8435
$P_1 = 0.15$	–10%	0.135	0.9756	–0.0460	–0.015	3.0677
$P_1 = 0.15$	–20%	0.120	0.9268	–0.0948	–0.030	3.1599
$P_2 = 0.50$	+10%	0.550	1.0340	0.0124	0.050	0.2480
$P_2 = 0.50$	+20%	0.600	1.0460	0.0244	0.100	0.2443
$P_2 = 0.50$	–10%	0.450	1.0088	–0.0128	–0.050	0.2562
$P_2 = 0.50$	–20%	0.400	0.9955	–0.0261	–0.100	0.2607

Table 13.3.4, we see that λ is most sensitive to changes in survivability of year 1 birds, P_1. This analysis indicates that conservationists might concentrate their efforts on helping eggs and nestlings survive.

Definition The **relative sensitivity,** or **sensitivity,** of λ to parameter P in a transition matrix is the partial derivative of the dominant eigenvalue of the matrix (λ) with respect to P, $\partial\lambda/\partial P$, or the instantaneous rate of change of λ with respect to P. Thus, this relative sensitivity of λ with respect to P is approximately the change in λ divided by the corresponding small change in P:

$$\textbf{sensitivity of } \boldsymbol{\lambda} \textbf{ to } \boldsymbol{P} = \frac{\partial \lambda}{\partial P} \approx \frac{\lambda_{\text{new}} - \lambda}{P_{\text{new}} - P},$$

where P_{new} is the new value of P close to P and λ_{new} is the resulting new value of λ.

Sensitivity Analysis for the Stage-Structured Example

Using sensitivity analysis, Morris et al. (2011) determined that lionfish population growth λ is very sensitive to lower-level mortality parameters of larval, juvenile, and adult mortality and is "most sensitive to the lower-level parameter of larval mortality." However, the larvae have venomous spines, probably making them less appealing prey than many of the native reef fish. A project explores a lionfish sensitivity analysis and the model of Morris et al. model more closely.

Applicability of Leslie and Lefkovitch Matrices

Leslie or Lefkovitch matrices are appropriate to use when we can classify individuals in a species by age or stage, respectively. The dynamics of the populations are based only on the females, and an adequate number of males for fertilization is assumed. The models in this module accommodate only population growths that do not depend on the densities of the populations so that the fecundity and survival rates remain constant. However, we can extend the models to incorporate density dependence by dampening values in the matrix. Unfortunately, estimations of fecundity and survival rates can be difficult. If appropriate, however, an age- or a stage-structured model can allow us to use matrix operations to determine the projected population growth rate and the stable-age distribution (Horne 2008).

Need for High-Performance Computing

Typically, a Leslie matrix is small enough so that high-performance computing (HPC) is unnecessary to model the long-term situation for one type of animal. However, one species might be a small part of a much bigger network of other species of animals and plants and their environment. Execution of models of such larger problems involves extensive computation that can benefit from HPC.

For example, PALFISH is a parallel, age-structured population model for freshwater small planktivorous fish and large piscivorous fish, structured by size, in south Florida. The model contains 111,000 landscape cells, with each cell corresponding to a 500-m by 500-m area and containing an array of floating-point numbers representing individual fish density of various age classes. Researchers reported a significant improvement in runtime of PALFISH over the corresponding sequential version of the program. The mean simulation time of the sequential model was about 35 h, while the parallel version with 14 processors and dynamic load balancing was less than 3 h (Wang et al. 2006).

Another use of HPC can be in **parameter sweeping**, or executing a model for each element in a set (often a large set) of parameters or of collections of parameters. The results can help the modeler obtain a better overall picture of the model's behavior, determine the relationships among the variables, find variables to which the model is most sensitive, find ranges where small variations in parameters cause large output changes, locate particular parameter values that satisfy certain criteria, and ascertain variables that might be eliminated to reduce model complexity (Luke et al. 2007).

Definition **Parameter sweeping** is the execution of a model for each element in a set of parameters or of collections of parameters to observe the resulting change in model behavior.

For example, suppose in our simple example of the bird, which has a maximum life span of 3 years, we are interested in determining the impact on the projected growth rate (positive eigenvalue) of changing the probabilities of the animal living from year 1 to 2 and from year 2 to 3. Such a problem is embarrassingly parallel on a high-performance system; we can divide computation into many completely independent experiments with virtually no communication. Thus, we could have multiple nodes on a cluster running the same program with different probability pairs and with their own output files. After completion, we can compare the results, perhaps using these to predict the impact of various interventions to improve the one, the other, or both probabilities. For more computationally intensive programs that require significant runtime, HPC can be particularly useful for such parameter sweeping.

For example, researchers are modeling biological metabolism at a kinetic level for a green alga, *Chlamydomonas reinhardtii* (Chang et al. 2008). However, limited knowledge exists of parameters for enzymes with known kinetic responses. Consequently, the researchers have developed the High-Performance Systems Biology Toolkit, HiPer SBTK, to perform sensitivity analysis and fitting of differential equations to the data. One problem involves 64 parameters and approximately 450,000 calculations for a full sensitivity matrix. Chang et al. (2008) wrote, "In moving from desktop-scale simulations of a small set of biochemical reactions to genome-scale simulations in the high-performance computing (HPC) arena, a paradigm shift must occur in the way we think of biological models. A complete representation of metabolism for a single organism implies model networks with thousands of nodes and edges." Because parallelism of the calculations is extremely well balanced, where each process has approximately the same amount of work as any other process, the scientists are optimistic that the code will scale to thousands of processors. Thus, ultimately, they plan to develop an *in silico* cell model of metabolism that contains all reliable experimental data for *C. reinhardtii* with problem sizes perhaps thousands of times larger than the current problem.

Exercises

1. **a.** Suppose a Leslie matrix associated with an age-structured population model has an eigenvalue of 0.984. Is the equilibrium population growing or shrinking?
   b. By how much?
   c. Suppose a corresponding eigenvector is (–2.35, –1.04, –0.87, –0.69). For each age class, give the estimated percentage of the total population to which the class converges as time progresses.
2. Suppose certain animal has a maximum life span of 3 years. A year 1 (0–1-year) female has no offspring; a year 2 (1–2-years) female has 3 daughters on the average; and a year 3 (2–3-years) female has a mean of 2 daughters. Thirty percent of year 1 animals live to year 2, and 40% of year 2 animals live to year 3. Suppose the numbers of year 1, 2, and 3 females are 2030, 652, and 287, respectively.
   a. Determine the corresponding Leslie matrix, L.
   b. Give the initial female population distribution vector $\mathbf{x(0)}$.
   c. Calculate the population distribution at time $t = 1$, vector $\mathbf{x(1)}$.
   d. Calculate the class percentages of the total population at time $t = 1$.
   e. Give the vector for class percentages of the total population, $\mathbf{v}$, expressed to two decimal places, to which $\mathbf{x}(\boldsymbol{t})/T(t)$ converges, where $T(t)$ is the total population at time t.

f. Find the number λ, expressed to two decimal places, to which $\mathbf{x}(t + 1)/\mathbf{x}(t)$ converges.

g. Using answers from Parts a, d, and f, verify that $L\mathbf{v} = \lambda\mathbf{v}$.

3. Consider the following Leslie matrix representing a population, where the basic unit of time is 1 year:

$$\begin{bmatrix} 0 & 0.2 & 1.3 & 3.5 \\ 0.1 & 0 & 0 & 0 \\ 0 & 0.2 & 0 & 0 \\ 0 & 0 & 0.4 & 0 \end{bmatrix}$$

a. Give the animal's maximum life span, and describe the meaning of each positive number in the matrix.

b. Determine the dominant eigenvalue and the annual growth rate. Do you expect the animal's numbers to grow or decline?

c. Draw a state diagram for the animal.

d. Using –10% of the parameter, determine the sensitivity of λ to the second row, first column parameter (0.1).

e. Determine the sensitivity of λ to the third row, second column parameter.

f. Determine the sensitivity of λ to the fourth row, third column parameter.

g. Based on your answers to Parts d–f, where should conservation efforts focus?

4. Crouse et al. (1987) considered the following seven stages in the life of loggerhead sea turtles (*Caretta caretta*): (1) eggs and hatchlings (< 1 year); (2) small juveniles (1–7 years); (3) large juveniles (8–15 years); (4) subadults (16–21 years); (5) novice breeders (22 years); (6) first-year emigrants (23 years); (7) mature breeders (> 23 years). Only the last three stages reproduce with female fecundities of 127, 4, and 80 per year, respectively. Table 13.3.5 gives the probabilities per year of a stage *From* turtle surviving and remaining at or advancing to stage *To*.

Table 13.3.5
The Probabilities per Year of a Stage *From* Loggerhead Sea Turtle Surviving to Stage *To*

From *Stage*	To *Stage*	*Probability per Year*
1	2	0.6747
2	2	0.7370
2	3	0.0486
3	3	0.6611
3	4	0.0147
4	4	0.6907
4	5	0.0518
5	6	0.8091
6	7	0.8091
7	7	0.8089

a. Give the Lefkovitch matrix for this model.
b. Determine the dominant eigenvalue and the annual growth rate. Do you expect the animal's numbers to grow or decline?
c. Give the annual mortality rate for stage 1 animals.
d. Give the annual mortality rate for stage 2 animals.
e. Give the annual mortality rate for stage 3 animals.
f. Give the annual mortality rate for stage 4 animals.
g. Give the annual mortality rate for stage 5 animals.
h. Give the annual mortality rate for stage 6 animals.
i. Give the annual mortality rate for stage 7 animals.
j. Draw a state diagram for the animal.
k. Determine the sensitivity of λ to each parameter indicated in Table 13.3.5.

5. Consider the following Lefkovitch matrix representing a population, where the basic unit of time is 1 year:

$$\begin{bmatrix} 0 & 0 & 3.4 & 7.5 \\ 0.1 & 0.2 & 0 & 0 \\ 0 & 0.3 & 0 & 0 \\ 0 & 0 & 0.4 & 0.5 \end{bmatrix}$$

a. Describe the meaning of each positive number in the matrix.
b. If possible, give the animal's maximum life span. Give the length of time for any stage that you can determine.
c. Determine the dominant eigenvalue and the annual growth rate. Do you expect the animal's numbers to grow or decline?
d. Draw a state diagram for the animal.
e. Determine the sensitivity of λ to the second-row, first-column parameter (0.1).
f. Determine the sensitivity of λ to the second-row, second-column parameter.
g. Determine the sensitivity of λ to the third-row, second-column parameter.
h. Determine the sensitivity of λ to the fourth-row, third-column parameter.
i. Determine the sensitivity of λ to the fourth-row, fourth-column parameter.
j. Based on your answers to Parts e–j, where should conservation efforts focus?

Projects

1. In the 1960s and 1970s, scientists did an experimental reduction in population density of Uinta ground squirrels in three types of habitats in Utah: lawn, nonlawn, and edge (Oli et al. 2001). For 4 years, they collected life table data; then for 2 years, they reduced the population by about 60%, keeping the same sex and age composition. Subsequently, they collected new life-table data. Data was collected postbreeding, after the birth pulse. Table 13.3.6 presents their data for the nonlawn habitat in three categories: Young (< 1 year), Yearling (1–2 years), and Adult (> 2 years).

Table 13.3.6
Pre- and Postreduction Survival and Fertility data for Nonlawn Uinta Ground Squirrels (Oli et al. 2001)

	Prereduction		*Postreduction*	
Category	*Survival*	*Fertility*	*Survival*	*Fertility*
Young	0.375	0.353	0.474	0.792
Yearling	0.419	0.741	0.481	0.981
Adult	0.500	0.885	0.588	1.200

a. Use a **partial life cycle model** to analyze the effect of the population reduction. Consider five age groups with year 1 being young, year 2 being yearling, and years 3, 4, and 5 being adults. Do not consider any of these animals after age 5 years. Determine the projected population growth rate (λ) pre- and postreduction.

b. By changing each survival rate in the prereduction data one at a time by $\pm 10\%$ and $\pm 20\%$, determine to which parameter λ is most sensitive. Discuss the results.

2. People in Europe and Asia enjoy eating skates, which are closely related to sharks. Consequently, the animal has declined since the 1970s. Frisk, Miller, and Fogarty did a study of little skates, winter skates, and barndoor skates to determine sustainable harvest levels and strategies (Frisk et al. 2002). For the little skate (*Leucoraja erinacea*), the scientists used an age-structured model incorporating 1-year age categories with 8-year longevity. Data from a previous study indicated age of 50% maturity to be 4 with annual female fecundity of 15 for mature females. They assumed this level to be constant for mature females. For age-specific survival (P_i), they adopted an exponential decay based on natural mortality (M_i) and fishing mortality (H_i): $P_i = e^{-(M_i + H_i)}$, $i = 1, 2, \ldots, 8$. The original analysis considered skates to be large enough for fishing by age 2, at which time the fishing mortality became 0.35. The probability of death by natural causes was assumed be 0.45 for these fish and 0.70 for year 1 skates.

a. Develop a Leslie matrix for the little skate and determine the long-term annual population growth rate, λ. The intrinsic rate of population increase, r, is the natural logarithm of λ, which the researchers calculated as 0.21 for little skates. Do you get the same value? What is the meaning of r? Interpret λ and r for the long-term forecast of little skates.

b. The researches also performed a stochastic analysis to test the sensitivity of their model to parameter estimation. For little skate, they drew adult fecundity, first-year survival, and adult survival from normal distributions with means and standard deviations as indicated in Table 13.3.7. Develop a stochastic version of the model. Run the simulation 1000 times for at least 200 years on each simulation. Determine average values for λ and r.

c. Researchers did a similar study for winter skate using an adult female fecundity mean of 17.5 and standard deviation of 5. Due to a lack of adequate information about adult mortality, they held M_1 and M constant (Frisk et al. 2002). Repeat Part a using this information.

Table 13.3.7
Means and Standard Deviations for Adult Fecundity, First-Year Survival, and Adult Survival of Little Skate Used for Stochastic Analysis (Frisk et al. 2002)

	Mean	*Standard Deviation*
little skate		
adult fecundity (female)	15	2.5
first-year survival (M_1)	1.21	0.4
adult-year survival (M)	0.45	0.05

3. Scientists conducted a 4-year study, "Population Viability Analysis for Red-Cockaded Woodpeckers in the Georgia Piedmont," to evaluate the risk of extinction for this endangered species and to recommend management to minimize this danger (Maguire et al. 1995). They considered five age groups: < 1 year (juvenile; class 1), 1 year (class 2), 2 years (class 3), 3 years (class 4), and > 3 years (class 4+). From observed data, they modeled the population and performed various simulations. For the survival rate of class i (P_i), they calculated the observed number of class i females surviving to class $i + 1$ divided by the number of females in class i. To consider the situation at postbreeding time (postbirth pulse sampling), they calculated fecundity for class i as the average number of female nestlings born to mothers of class $i + 1$ (m_{i+1}) multiplied by the proportion of females entering class i that will survive to class $i + 1$ (P_i). Because the study placed all female red-cockaded woodpeckers from age 4 years on into the same class, class 4+, they calculated the number in that group at time $t + 1$, $x_{4+}(t + 1)$, as $P_4x_4(t) + P_{4+}x_{4+}(t)$. Table 13.3.8 presents data for newly banded birds (NB) and for newly banded birds and nonbanded birds (NBU). The researchers found an initial distribution of (20, 10, 9, 9, 6) for the five classes. In their simulations, they considered extinction to be the time at which the total population was less than or equal to 1.

a. Develop a deterministic model for each set of birds, NB and NBU, using a stage-structured 5 × 5 Leslie matrix. What happens to the population over a period of time? Assuming extinction when a population is less than

Table 13.3.8
Data on Red-Cockaded Woodpeckers in the Georgia Piedmont, 1983–1988, for Newly Banded Birds (NB) and for Newly Banded Birds and Nonbanded Birds (NBU), Where P_i is the Proportion of Females Entering Class i That Will Survive to Class $i + 1$ and m_i Is the Average Number of Female Offspring per Female of Class i (Maguire et al. 1995)

	NB		**NBU**	
Class	P_i	m_i	P_i	m_i
1	0.380	0.000	0.401	0.000
2	0.653	0.133	0.734	0.126
3	0.850	1.082	0.961	1.023
4	0.400	1.194	0.456	1.129
4+	0.589	1.590	0.667	1.504

1, when does extinction occur? How might you explain the difference between the outcomes of NB and NBU populations?

b. By changing each survival rate, P_i, one at a time by ±10% and ±20%, determine which parameter poses the greatest sensitivity to extinction risk. Use NB data. By determining the parameter that has the greatest impact, ecologists can focus their efforts on improving that group's survival.

c. Researchers determined that in 1987–88, the area contained 41 potential nesting sites. Develop a **habitat saturation model** by limiting the number of breeding woodpeckers each year to 41. Give nesting preference to older birds. Thus, with n_{Bi} being the number of class i breeding females, n_i being the number of class i females, and min being the minimum function, we have $n_{B4+} = min(n_{4+}P_{4+}, 41)$. That is, the number of potential class 4+ breeders is $n_{4+}P_{4+}$, but at most 41 can breed. If $n_{4+}P_{4+}$ is greater than 41, no nesting sites remain for birds in other classes. If $n_{4+}P_{4+}$ is less than 41, the model allows $n_{B4} = min(n_4P_4, 41 - n_{B4+})$ woodpeckers in class 4 to breed in the remaining number of sites. Similarly, $n_{B3} = min(n_3P_3, 41 - n_{B4+} - n_{B4})$, and so on.

d. Table 13.3.9 gives the researchers' calculations for P_1 along with the corresponding probability for each of the 4 years of the study. Starting with an initial population distribution of (20, 10, 9, 9 6), which was their 1988 estimate, develop a stochastic version of the model for NB or NBU birds. To do so, at each year with the given probabilities, randomly select the juvenile (class 1) survival rate from the estimations in the table. That is, at each time step, generate a uniformly distributed random number, r, between 0 and 1. For the NB data, if r is less than the first probability, 0.295, use the first value, 0.3708, for P_1; else if r is less than 0.295 + 0.310 = 0.605, use P_1 = 0.4131; and so on. Run the simulation 1000 times for at least 200 years on each simulation. Determine the range of extinction, the average extinction, and the probability of extinction within 100 years. Discuss the results.

These simulations correspond to **environmental stochasticity**, or variation in parameters caused by random environmental changes. The researchers simplified the model to use variations in P_1 to reflect this environmental stochasticity. Why might they make such an assumption?

Table 13.3.9
Yearly Estimates of Juvenile Survival Rates (P_1) and Corresponding Probabilities for Red-Cockaded Woodpeckers in the Georgia Piedmont, 1983–1988, for Newly Banded Birds (NB) and for Newly Banded Birds and Nonbanded Birds (NBU) (Maguire 1995).

	NB		*NBU*	
Year	P_1	*Probability*	P_1	*Probability*
1984	0.3708	0.295	0.3793	0.285
1985	0.4131	0.310	0.4220	0.318
1986	0.2176	0.135	0.2353	0.095
1987	0.4354	0.260	0.4508	0.302

Table 13.3.10
Values from (Morris et al. 2011) Assuming 30 Days/Month

Symbol	*Meaning*	*Value*
M_E	Egg mortality	9.3/month
M_L	Larval mortality	10.5/month
M_J	Juvenile mortality	0.165/month
M_A	Adult mortality	0.052/month
D_E	Egg duration	0.1 month
D_L	Larval duration	1 month
ρ	Proportion female	46%
f	Fecundity	194,577 eggs/month/ female

e. Repeat Part c at each time step selecting randomly a juvenile survival rate in an appropriate range as discussed in Part d.

f. For NB or NBU, perform a sensitivity analysis to determine the parameters P_i to which the growth rate λ is most sensitive. Using these results, make recommendations about where to concentrate conservation efforts.

4. This project considers the lionfish example in the section "Stage-Structured Model." With data from the literature on average mortalities of eggs and lionfish in the various stages, durations of eggs and larvae, fecundity, and proportion female as in Table 13.3.10, Morris et al. (2011) calculated various probabilities using the following exponentially decreasing models:

$$G_L = e^{-M_L D_L},\ G_J = e^{-M_J}/12,\ P_J = 11e^{-M_J}/12,\ P_A = e^{-M_A},\ R_A = \rho f e^{-M_E D_E}$$

a. With these models and the values in Table 13.3.10, recalculate G_L, G_J, P_J, P_A, and R_A and use these calculations for a Lefkovitch matrix. Revise the last paragraph in the section "Stage-Structured Model" for this matrix. That is, update the growth rate λ, the percentage that reduces probability of an adult surviving to produce negative population growth, and the percentages of simultaneous reductions of probabilities of juvenile and adult survivorship to produce a declining population.

b. Perform a sensitivity analysis to determine the higher-level parameters—G_L, G_J, P_J, P_A, and R_A—to which the monthly growth rate λ is most sensitive. Using these results, make recommendations for controlling this menace.

c. Perform a sensitivity analysis to determine the lower-level parameters from Table 13.3.10 to which the monthly growth rate λ is most sensitive. Using these results, make recommendations for controlling this menace.

d. Adult mortality, M_A, is dependent upon fishing intensity. Draw 1000 values of adult mortality from a normal distribution with mean 0.052/month and standard deviation 5% of this mean, excluding numbers beyond two standard deviations from the mean. Generate 1000 Lefkovitch matrices and calculate the resulting growth rates λ for these matrices. (Morris et al. 2011) used similar calculations to "illustrate sensitivity to misspecification of parameter values." Discuss your results.

5. Typically, spawning (breeding) Pacific salmon travel up the same river where they were born, breed, lay their eggs, and then die. The eggs hatch; the young salmon develop for 1 or 2 years in the streams; the juvenile salmon travel downstream to the ocean; then, as smolts they enter the ocean, where they may remain for several years, continuing to grow.

Between 1961 and 1975, four dams were constructed on the lower Snake River. Unfortunately, the dams inhibited the usual migration of spring/summer chinook salmon, so officials made various dam passage improvements, including transportation of spawning salmon upstream and of juvenile salmon downstream. Kareiva, Marvier, and McClure studied the situation using age-structured models (Kareiva et al. 2000). They tested the effectiveness of various implemented management interventions and examined whether improving the survival of any of the life stages could stop population declines. The study assumed a 5-year life expectancy, equal proportion of male and female salmon, and breeding at year 3 or later. Table 13.3.11 contains the study's parameters. As we will see, parameters s_2 (probability of surviving from year 1 to year 2) and μ (probability of surviving upstream migration) are calculated from the parameters indented immediately below them.

Researchers are continuing to gather data, to further develop population models, and to publish their results (Zabel et al. 2006; Interior Columbia

Table 13.3.11
Parameters from Tables 1 and 2 (Kareiva et al. 2000)

Symbol	*Meaning*	*Value*
s_1	Probability of surviving from year 0 to year 1	0.022
s_2	Probability of surviving from year 1 to year 2	
z	Proportion of fish transported	0.729
s_z	Probability of surviving transportation	0.98
s_d	Probability of surviving in-river migration (no transportation)	0.202
s_e	Probability of surviving in estuary and during ocean entry	0.017
s_3	Probability of surviving from year 2 to year 3	0.8
s_4	Probability of surviving from year 3 to year 4	0.8
s_5	Probability of surviving from year 4 to year 5	0.8
b_3	Probability of a year 3 female to breed	0.013
b_4	Probability of a year 4 female to breed	0.159
b_5	Probability of a year 5 female to breed	1.0
μ	Probability of surviving upstream migration	
h_{ms}	Harvest rate in main stem of columbia river	0.020
s_{ms}	Probability of survival of unharvested spawner from bonneville dam to spawning basin	0.794
h_{sb}	Harvest rate in subbasin	0
s_{sb}	Probability of survival of unharvested adult in subbasin before spawning	0.9
m_3	Number of eggs per year 3 female spawner	3257
m_4	Number of eggs per year 4 female spawner	4095
m_5	Number of eggs per year 5 female spawner	5149

Technical Recovery Team 2007). Moreover, these results are playing a part in salmon recovery planning (NOAA 2011).

a. To survive to year 2, a year 1 fish must travel downstream past the dam and survive in one of the following two ways:
1. have transportation over the dam and survive;
2. migrate on its own past the dam and survive.

In either case, the fish must then survive journeys in the estuary and into the ocean. With z being the proportion of fish transported, give the formula for the proportion of fish that migrate in-river past the dam without transportation. Using z, s_z, s_d, and s_e from Table 13.3.11, develop a model for s_2, the probability of a salmon surviving from year 1 to year 2. Using the indicated parameter values from Table 13.3.11, evaluate s_2.

b. With h_{ms} being the harvest rate in main stem of Columbia River, give the formula for the proportion not harvested in the river. Similarly, with h_{sb} being the harvest rate in the subbasin, give the formula for the proportion not harvested in the subbasin. To survive upstream migration, a fish must survive in the river and the subbasin. Thus, it must survive the danger of harvest and travel in both locations. Using h_{ms}, s_{ms}, h_{sb}, and s_{sb} from Table 13.3.11, develop a model for μ, the probability of survival of an unharvested spawner from Bonneville Dam to spawning basin. Using the indicated parameter values from Table 13.3.11, evaluate μ.

c. With b_3 being the probability of a year 3 female to breed, give the formula for the proportion of year 3 females that do not breed. Using this formula and s_4, the probability of surviving from year 3 to year 4, determine the proportion of females that survive to year 4, that is, the probability that a female does not breed and survives to year 4. Using the indicated parameter values, evaluate the proportion of females that survive to year 4. Why do we not include the proportion that spawn? Using the indicated parameter values from Table 13.3.11, evaluate the proportion of females that survive to year 4.

d. Similarly to Part c, determine the proportion of females that survive to year 5.

e. Determine a formula for the fecundity of year 3 salmon; that is, the average number of female young from a year 3 mother. For your formula consider the probability that a year 3 salmon breeds, the probability that the salmon then survives the upstream journey, the average number of eggs for a 3-year-old, the proportion of those that are female offspring, and the probability that the egg hatches and the offspring survives the first year. Using the indicated parameter values from Table 13.3.11, calculate the fecundity of year 3 salmon.

f. Similarly to Part e, determine the fecundity of year 4 salmon.

g. Similarly to Part e, determine the fecundity of year 5 salmon.

h. Using the previous parts, determine the Leslie matrix. After calculating its dominant eigenvalue, discuss the long-term prospects for the chinook salmon on the Snake River if the situation does not change.

i. Kareiva et al. (2000) examined the impact on long-term population growth had authorities not taken the following actions: (i) "reductions of harvest rates, from approximately 50% in the 1960s to less than 10% in the 1990s"; (ii) "engineering improvements that increased juvenile downstream migration survival rates from approximately 10% just after the last turbines were installed to 40 to 60% in most recent years"; (iii) "the transportation of approximately 70% of juvenile fish from the uppermost dams to below Bonneville Dam, the lowest dam on the Columbia River." Based on calculations, they concluded, "If such improvements had not been made, the rates of decline would likely have been 50 to 60% annually. . .." Discuss which parameters would need to be adjusted for their calculations.

j. Many conservation efforts have been focused on transportation through the dams. If such efforts were completely successful (an impossible goal), determine the long-term growth. Would such actions be enough to reverse the population declines? Based on these results, should conservation efforts focus solely on transportation?

k. Justify the following statement (Kareiva et al. 2000): "management actions that reduce mortality during the first year by 6% or reduce ocean/estuarine mortality by 5% would be sufficient" to reverse the population declines.

l. Justify the statement (Kareiva et al. 2000) that "a 3% reduction in first-year mortality and a 1% reduction in estuarine mortality" would be sufficient to reverse the population declines.

m. Perform a sensitivity analysis to determine the s_i parameters (Table 13.3.11) to which the monthly growth rate λ is most sensitive. Using these results, make recommendations on where to focus conservation efforts.

6. Furbish's lousewort, *Pedicularis furbishiae*, is an endangered herbaceous plant that grows along a 140-mi stretch of the St. John River in northern Maine and New Brunswick, Canada. This perennial does well in areas having little cover from woody plants and little riverbank disturbance. Wetter conditions promote growth and colonization, but moist soil is more likely to slide into the river. River ice flows scrape the banks, advantageously removing woody vegetation but also disturbing *P. furbishiae*. If disturbances occur too frequently (more frequently than every 6 to 10 years), the lousewort does not have adequate time to reestablish itself. Thus, success of the plant appears to depend on a delicate balance of conditions.

To examine the long-term prospects of the species' survival, Eric Menges performed a 3-year (1983–1986), spring-to-spring study of *P. furbishiae*, recording plant and environmental data. Then, he used stage-based modeling with the following six stages: seedling; juvenile, which is below minimum flowering size; vegetative, which is not flowering but is above minimum flower size; small repro.—flowering plant with one **scape**, or leafless flower stalk; medium repro.—flowering plant with two to four scapes; and large repro.—flowering plants with more than four scapes. Table 13.3.12 gives probabilities of transitioning from one stage to another based on the data from 1984–1985. The plants reproduce only sexually, so fecundity as presented in Table 13.3.13 was determined using an estimate of the number of seedlings produced (Menges 1990).

a. Draw a state diagram for the model.

Table 13.3.12
Probabilities with Standard Errors of *P. furbishiae* Changing from One Stage to Another Based on Data from Spring 1984 to Spring 1985 (Menges 1990)

From	*To*	*Probability*
Seedling	Juvenile	0.39
Seedling	Vegetative	0.01
Juvenile	Juvenile	0.47
Juvenile	Vegetative	0.21
Juvenile	Small repro.	0.11
Juvenile	Medium repro.	0.00
Vegetative	Juvenile	0.14
Vegetative	Vegetative	0.24
Vegetative	Small repro.	0.45
Vegetative	Medium repro.	0.11
Small repro.	Juvenile	0.09
Small repro.	Vegetative	0.24
Small repro.	Small repro.	0.36
Small repro.	Medium repro.	0.21
Small repro.	Large repro.	0.01
Medium repro.	Juvenile	0.04
Medium repro.	Vegetative	0.16
Medium repro.	Small repro.	0.26
Medium repro.	Medium repro.	0.42
Medium repro.	Large repro.	0.10
Large repro.	Vegetative	0.01
Large repro.	Medium repro.	0.28
Large repro.	Large repro.	0.61

Table 13.3.13
Fecundities of *P. furbishiae* Based on Data from Spring 1984 to Spring 1985 (Menges 1990)

Stage	*Fecundity*
Small repro.	2.45
Medium repro.	7.48
Large repro.	29.93

b. Develop a Lefkovitch matrix model, *L84to85*, using data from Tables 13.3.12 and 13.3.13 and determine the finite rate of population change, λ. If the plant could maintain such annual population growth, would you anticipate the population of *P. furbishiae* to increase or decrease over time?

c. The growing season in 1984–85 was advantageous for *P. furbishiae*. However, disturbance from ice scour and riverbank slumping in 1983–84 were challenging and resulted in a transition matrix *L83to84* with dominant eigenvalue $\lambda = 0.77$. In 1985–86, the environmental conditions resulted in a transition matrix *L85to86* with $\lambda = 1.02$. Using these values and your result from Part b, discuss the wisdom of using data from one year to make long-term predictions.

d. Because environmental conditions can vary greatly from year to year, using one year's data can be misleading for making long-term predictions. To account for such environmental stochasticity, Menges (1990) performed 100 simulations following the population for 100 simulated years, where each year he used a Lefkovitch matrix selected at random from the observed matrices for 1983–84, 1984–85, and 1985–86. Perform these simulations; for each simulation multiply the matrices together and calculate the finite rate of population change for the resulting final matrix. Assuming an initial population distribution of (156, 158, 82, 55, 44, 5) for 500 individuals, calculate the final population distribution and total population for each simulation. Discuss your results.

Menges (1990) did not give all the data for 1983-84 and 1985-86. For crude estimates of the Lefkovitch matrices (say, *L83to84* and *L85to86*) for these years, multiply matrix *L84to85* from Part b by appropriate constants; in each case, multiply by the desired dominant eigenvalue (0.77 and 1.02, respectively) and divide by the dominant eigenvalue, $\lambda = 1.27$, for *L84to85*.

7. During the early part of the twentieth century, sugar cane growers in Puerto Rico were desperately seeking something to control beetle grubs (larvae) that were destroying the roots of their crops. In response, the U.S. Department of Agriculture imported some rather large toads, *Bufo marinus*, from Barbados. Within 10 years, the beetle grubs numbers were reduced to the level of a mere nuisance. This was a relatively rare example of a positive outcome from introducing species to new geography. The toad, commonly called the cane toad, was introduced to cane-growing areas in other countries, including Australia; but in Australia they have become a major pest. Dispersing widely through several Australian states, these voracious predators and nimble competitors are threatening native species and disrupting biological communities (Markula et al. 2010).

One method that has been successful in controlling certain insect pests, particularly for initial invasions into an area, is the release of sterile males. With the release of a large number of such males relative to the number of fertile males, the hope is that many nonproductive matings will occur, resulting in a population reduction. However, usually the female insect causes most of the damage, whereas the male cane toad is as destructive as the female. Moreover, typically insects have very short life spans, but a large influx of sterile male toads that live for several years can increase the population size significantly and cause extensive environmental damage.

Stage-based models in McCallum (2006) with data from Lampo and De Leo (1998) demonstrate the impracticality of using sterile males to control the cane toad population in Australia. Table 13.3.14 summarizes the model probability parameters for the following stages: egg, tadpole, juvenile, and adult. With data indicating a range of from 7500 to 20,000 eggs in a clutch, the models use a clutch size of 15,000 eggs, half of which are assumed to be female.

a. Draw a state diagram for the model.

b. Develop a Lefkovitch matrix model, L, using the mean probabilities from Table 13.3.14 with a fecundity of 7500 female eggs, and determine the

Table 13.3.14
Australian Cane Toad Data (Lampo and De Leo 1998)

From	*To*	*Mean Probability*	*Probability Range*
Egg	Tadpole	0.718	0.688–0.738
Tadpole	Juvenile	0.05	0.012–0.176
Juvenile	Adult	0.05	0.03–0.07
Adult	Adult	0.50	0.3–0.7

finite rate of population change, λ. If the animal could maintain such annual population growth, would you anticipate the cane toad population to increase or decrease over time?

c. Repeat Part b for the lower and upper extremes of the probability and fecundity ranges. Discuss the results.

d. Determine an eigenvector associated with the dominant eigenvalue, λ, for the matrix of Part b. Scale the vector so that the number of adult cane toad females is 100. Plot the number of adult females versus time over a 15-year period. Because of the exponential growth involved, create another plot of the common logarithm (logarithm to the base 10) of the number of adult females versus time. Plot the log of the number of adults versus time.

e. Suppose a control effort releases 5000 sterile males into the population each year. Develop a program to estimate the number of adult females per year for 15 years. Each simulation year, adjust the Lefkovitch matrix of Part b so that the female fecundity is the probability that a male is fertile multiplied by the mean female clutch size of 7500. Thus, each year, calculate the number of sterile males in the population; besides an additional release of 5000 sterile males, the data indicate that on the average an adult has a 0.50 chance of surviving from one year to the next. Also, calculate the total number of males (fertile and sterile) in the population each year. Assume that the number of fertile males equals the number of females in the population. Plot the common log of the number of adult females versus time on the same graph as the corresponding plot for Part d. Plot the common log of the number of adults versus time on the same graph as the corresponding plot for Part d. Does the model predict that such a control effort would be successful?

f. Repeat Part e for a sterile male release of 10,000 per year. Discuss the results and the practicality of such a control effort.

g. Repeat Part e where 10,000 sterile males are released each year until the number of females falls below 50, half the original number of females. Discuss the results.

h. Perform a sensitivity analysis to determine the parameters to which the annual growth rate λ is most sensitive. Using these results, make recommendations on where to focus conservation efforts.

8. For several constants in Table 11.4.2 of the agent-based Module 11.4, "Introducing the Cane Toad—Able Invader," perform a sensitivity analysis to determine how sensitive the death percentage is to each of the constants with value less than 1.

Answers to Quick Review Questions

1. **a.** $x_i(t)$ = number of this female insect in the ith month of life alive in the area at time t, where $i = 1$ or 2
   b. Assuming that an insect gives birth to half females, 5 and 150 in the first or second month of life, respectively, we have the following system of equations:

$$5x_1(t) + 150x_2(t) = x_1(t+1)$$

$$0.01x_1(t) \qquad\qquad = x_2(t+1)$$

   c. $$L = \begin{bmatrix} 5 & 150 \\ 0.01 & 0 \end{bmatrix}$$

   d. 160 month 1 female insects and 0.02 month 2 female insects because

$$\mathbf{x(1)} = L\mathbf{x(0)} = \begin{bmatrix} 5 & 150 \\ 0.01 & 0 \end{bmatrix}\begin{bmatrix} 2 \\ 1 \end{bmatrix} = \begin{bmatrix} 160.00 \\ 0.02 \end{bmatrix}$$

   e. 803 month 1 female insects and 1.6 month 2 female insects because

$$\mathbf{x(2)} = L\mathbf{x(1)} = \begin{bmatrix} 5 & 150 \\ 0.01 & 0 \end{bmatrix}\begin{bmatrix} 160.00 \\ 0.02 \end{bmatrix} = \begin{bmatrix} 803.00 \\ 1.6 \end{bmatrix}$$

2. $$\begin{bmatrix} 0.2 & 1.2 & 1.4 & 0.7 \\ 0.3 & 0 & 0 & 0 \\ 0 & 0.8 & 0 & 0 \\ 0 & 0 & 0.5 & 0 \end{bmatrix}$$

3. **a.** 99.811103%, 0.188897%
   b. 5.28388
   c. $$\begin{bmatrix} 0.99811103 \\ 0.00188897 \end{bmatrix}$$
   d. $$\begin{bmatrix} 998111.03 \\ 1888.97 \end{bmatrix}$$

References

Albins, M. A., and M. A. Hixon. 2008. "Invasive Indo-Pacific Lionfish (*Pterois volitans*) Reduce Recruitment of Atlantic Coral-Reef Fishes." *Marine Ecology Progress Series*, 367: 233–238.

"ARGOS, 2013. Worldwide Tracking and Environmental Monitoring by Satellite." http://www.argos-system.com/welcome_en.html?nocache=0.9022361093666404 (accessed July 26, 2013)

Bean, Michael J. 2005. "The Endangered Species Act: Success or Failure?" Incentive Paper No.2, May 2005. http://www.dodpif.org/kiwa/kw-articles/2005 Bean. ESA Success or Failure%20(Environmental Defense).pdf (accessed December 30, 2012)

Bjorndal, Karen A., Alan B. Bolten, and Milani Y. Chaloupka. 2000. "Green Turtle Somatic Growth Model: Evidence for Density Dependence. Ecological Applications." 10(1): 269–282

Chang, Christopher H., Peter Graf, David M Alber, Kwiseon Kim, Glenn Murray, Matthew Posewitz, and Michael Seibert. 2008. "Photons, Photosynthesis, and High-Performance Computing: Challenges, Progress, and Promise of Modeling Metabolism in Green Algae." *J. Phys.: Conf. Ser.* 125 012048

Crouse, D. T., L. B. Crowder, and H. Caswell. 1987. "A Stage-based Population Model for Loggerhead Sea Turtles and Implications for Conservation." *Ecology*, 68:1412–1423.

Crowder, L. B., D. T. Crouse, S. S. Heppel, and T. H. Martin. 1994. "Predicting the Impact of Turtle Excluder Devices on Loggerhead Sea Turtle Populations." *Ecological Applications* 4:437– 445.

Earthtrust. 2009. "Green Sea Turtles." http://earthtrust.org/wlcurric/turtles.html (accessed July 26, 2013)

Forbes, Gregory A. 1992. "A Most Amazing Animal, The Green Sea Turtle." *Tortuga Gazette* 28(6): 1–3, June 1992. http://www.tortoise.org/archives/green.html (accessed July 26, 2013)

Frisk, M. G., T. J. Miller, and M. J. Fogarty. 2002. "The Population Dynamics of Little Skate Leucoraja erinacea, Winter Skate Leucoraja ocellata, and Barndoor Skate *Dipturus laevis*: Predicting Exploitation Limits Using Matrix Analyses." *ICES Journal of Marine Science*, 59: 576–586.

Horne, J. S. 2008. "Lab 10:Leslie Matrices" in *Fish & Wildlife Population Ecology*. http://www.cnr.uidaho.edu/wlf448/Leslie1.htm (accessed December 30, 2012)

Interior Columbia Technical Recovery Team, and R. W. Zabel. 2007. "Assessing the Impact of Environmental Conditions and Hydropower on Population Productivity for Interior Columbia River Stream-type Chinook and Steelhead Populations." "Matrix Model" linked from http://www.nwfsc.noaa.gov/trt/col/trt_ic_viability_survival.cfm (accessed December 30, 2012)

Kareiva, P., M. Marvier, and M. McClure. 2000. "Recovery and Management Options for Spring/SummerChinook Salmon in the Columbia River Basin." *Science*, 290: 977–979.

Lampo, Margarita, and Giulio A. De Leo. 1998. "The Invasion Ecology of the Toad *Bufo marinus*: From South America to Australia." *Ecological Applications*, 8: 388–396.

Luke, Sean, Deeparka Sharma, and Gabriel Catalin Balan. 2007. "Finding Interesting Things: Population-based Adaptive Parameter Sweeping.". In *GECCO '07: Proceedings of the 9th Annual Conference on Genetic and Evolutionary Computation*, pp. 86–93. ACM.

Maguire, Lynn A., George F. Wilher, and Quan Don. 1995. "Population Viability Analysis for Red-Cockaded Woodpeckers in the Georgia Piedmont." *The Journal of Wildlife Management*, 59(3): 533–542.

Markula, Anna, Steve Csurhes, and Martin Hannan-Jones Pest. 2010. "Animal Risk Assessment: Cane Toad, *Bufo marinus*." The State of Queensland, Department of Employment, Economic Development and Innovation.

McCallum, Hamish. 2006. "Modelling Potential Control Strategies for Cane Toads." *Proceedings of the Invasive Animals CRC/ CSIRO/QLD NRM&W Cane Toad Workshop*, Brisbane, pp. 123–133.

Menges, Eric S. 1990. "Population Viability Analysis for an Endangered Plant." *Conservation Biology*, 4(1): 52–62.

MeSsAGE (The Monash eScience and Grid Engineering). 2012. https://messagelab.monash.edu.au/ (accessed December 30, 2012)

Morris, James A., Kyle W. Shertzer, and James A. Rice. 2011. "A Stage-Based Matrix Population Model of Invasive Lionfish with Implications for Control." *Biol Invasions*, 13:7–12.

Nimrod Toolkit. 2011. https://messagelab.monash.edu.au/Nimrod (accessed December 30, 2012)

NOAA (National Oceanic and Atmospheric Administration). 2011. "Salmon Recovery Planning." National Marine Fisheries Service, Northwest Regional Office. http://www.nwr.noaa.gov/Salmon-Recovery-Planning/ (accessed December 30, 2012)

Oli, Madan K., Norman A. Slade, and F. Stephen Dobson. 2001. "Effect of Density Reduction on Uinta Ground Squirrels: Analysis of Life Table Response Experiments" *Ecology*, 82(7): 1921–1929.

Wang, D., M. W. Berry, N. Buchanan, and L. J. Gross. 2006. A GIS-enabled Distributed Simulation Framework for High Performance Ecosystem Modeling. *Proceedings of ESRI International User Conference*, August 7–11, 2006.

Zabel, Richard W., Mark D. Sceuerell, Michelle M. McClure, and John G. Williams. 2006. "The Interplay between Climate Variability and Density Dependence in the Population Viability of Chinook Salmon" *Conserv Biol*, 20(1): 190–200.

Zug, George R., George H. Balazs, Jerry A. Wetherall, Denise M. Parker, and Shawn K. K. Murakawa. 2002. "Age and Growth of Hawaiian Green Seaturtles (*Chelonia mydas*): An Analysis Based on Skeletochronology." *Fish. Bull.* 100:117–127. http://fishbull.noaa.gov/1001/zug.pdf (accessed July 26, 2013)

MODULE 13.4

Probable Cause—Modeling with Markov Chains

Prerequisite: Module 13.1, "Computational Toolbox—Tools of the Trade: Tutorial 7" or "Alternative Tutorial 7" (through the section "Eigenvalues and Eigenvectors"); Module 13.2, "Matrices for Population Studies: Linked for Life"; and definitions of "eigenvalue" and "eigenvector" from Module 13.3, "Age- and Stage-Structured Models." Additional high-performance computing materials related to this module are available on the text's website.

Introduction

To the U.S. Navy and shipping companies around the world, barnacles can be a real drag, and they are out to get rid of the creatures. How can such a small animal be so despised by so many? Though seemingly insignificant, barnacles are one of the main causes of fouling of ship hulls. Growing on the submerged hull surfaces, they interfere with the smooth movement of ships through the water, causing ships to use more fuel, which adds up to tremendous costs. Millions of dollars have been expended to find ways to eliminate or at least greatly inhibit attachment. Ship owners have tried various types of paints, but many of them leach toxic compounds into the water. Recently, researchers have developed some nontoxic coatings, which help to change the mechanical properties of the hull surface, so that barnacle larvae and other fouling organisms are less likely to attach.

Incidentally, barnacles also help to foul intake pipes for coastal power stations. So, finding an effective, nontoxic method to prevent such fouling would be a significant benefit to human populations.

As adults, barnacles are mostly small, sessile animals—they remain attached to firm surfaces. They adapted to various naturally occurring surfaces before human beings began exploring and harvesting the seas. The 900 or so species can be found on whale skin, crab and mollusk shells, and rocky shores. You may have seen them as you explored a rocky beach or examined a seashell on a sandy beach. Barnacles are prominent members of a community of organisms that call the intertidal zone home.

Intertidal regions, which lie between high- and low-tide lines, represent a transition between the marine and terrestrial ecosystems. Although these regions include sand beaches, estuaries, and bays, barnacles particularly like rocky shores. In fact, rocky intertidal areas include very dense and diverse communities, highly adapted to the periodic exposure to drying, wave action, and extremes of temperature. The organisms of this habitat are often found in distinct, vertical zones, arranged according to degree of exposure—low, middle, high intertidal, and splash zones. The width of each zone is determined somewhat by the degree of protection from wave action—narrower in more protected areas.

So, barnacles are important members of these communities, which are rich in numbers of taxa. Like their neighbors in this zone, barnacles must live under some fairly extreme physical conditions (e.g., heavy wave action, desiccation, high temperature), while trying to supply themselves with sufficient food and available oxygen, overcoming competition and predation, and producing gametes for reproduction. They are unable to control the physical environment. Moreover, none of the biological challenges is easily met. Any additional physical or biological stress would put even organisms as hardy as barnacles in jeopardy.

What if environmental conditions changed so that a barnacle species disappears? Many scientists suggest that the world oceans are warming. What effects might ocean temperature change have on intertidal communities? Well, increasing temperature would add to the other extremes that these organisms already have to endure, and the animals might not be able to withstand higher temperatures. Temperature cues are also important for development and reproduction of many animals. From 1993 through 1996, researchers at Hopkins Marine Station in California surveyed transects in a rocky intertidal community that was first surveyed in the 1930s. They found a dramatic shift in species, where southern species (warm-adapted) increased significantly over northern species (cold-adapted) during a time period where ocean and summer air temperatures had both increased over the 60-year span of time. What this study suggests is that such a change can eliminate some species from the community—perhaps a barnacle species. So what? (California Coastal Commission 1987; Foster 2009; "Intertidal Stressors" 2007)

Barnacles are filter feeders that occur in large numbers in their communities. They form hiding places for small animals and serve as food for others. Their role, or niche, is interwoven into the community structure and function, and their loss might have serious ramifications. Each species is integrated so that it has multiple interactions with other community members and the environment. The extinction of a barnacle species would certainly affect other constituents of the ecosystem ("Barnacles" 2012; Cornell University 2003; Natural History Museum 2012; Stout 2009).

Understanding the effects of losses in diversity is and will continue to be critical to the implementation of judicious conservation policies, but that understanding is problematic in the multifarious, natural ecosystems. Mathematical models may offer us an effective approach to estimating the impact of species losses to a community. For this type of study, we can employ **Markov chain models** (**MCM**), which are based on the probability of passing from one state to another. Normally, the parameters of these models depend on the observed and experimental data available, but MCMs allow us to utilize parameters without extensive experimentation.

Problems from Psychology to Genetics

Besides predicting effects of species loss to a community, Markov chain models are useful in quite a variety of problems, from predicting the behavior of animals to locating genes in the DNA. In this module, we start with a problem from psychology in which we have observed the various activities of an animal and the likelihood of moving from one pursuit to another. Using this information, with MCMs we can estimate the average amount of time a typical animal spends performing each endeavor and, given the activities of a group of animals, predict their behavior in the near future. In Module 14.14, "Computational Code—Breaking: Deciphering Our Own Mysteries," we employ this same modeling technique to pursue vastly different problems in genetics.

Probability

Markov chain models involve matrices in which all the elements are probabilities, so we start with a brief introduction (some of which is a review) to probability theory. The **probability** of an event, or the occurrence of something, is a number between 0 and 1, inclusive, indicating the chance of the event happening. A probability of 0 means that the event can never occur, while 1 says that that the situation is always true. As an example, suppose a certain kind of seed has a 50–50 chance of germinating. Thus, the probability or chance of germinating is P(germinating) = ½ = 0.5 = 50%. For each seed, one of two events can occur, germination or no germination; and the results are equally likely to occur. We expect that if we observe many seeds, about half the seeds will germinate.

> **Definition** The **probability** of an event, E, written $\boldsymbol{P(E)}$, is the chance of its occurrence and is a number between 0 and 1, inclusive.

Quick Review Question 1

Suppose at a site on a strand of DNA, an equal likelihood exists for any of the four bases (A, C, T, G). Give the probability of the base T occurring at a particular site.

The sum of all the possible events for a situation, such as germinating and not germinating, sums to 1. If a seed has only a 30% chance of germinating, P(germinating) = 0.3, then it has a 70% chance of not germinating: P(not germinating) = 1 – P(germinating) = 1 – 0.3 = 0.7.

> **Rule** The probability of an event not occurring is 1 minus the probability of the event,
>
> $$P(\text{not } E) = 1 - P(E)$$

Quick Review Question 2

Suppose at a site on a strand of DNA, an equal likelihood exists for any of the four bases (A, C, T, G). Give the probability of T *not* being at a particular site.

Suppose an ant is equally likely to go in any one of eight directions, N, NE, E, SE, S, SW, W, or NW. For example, $P(\text{N}) = 1/8$ and $P(\text{S}) = 1/8$. The probability that the ant will move in the north or south direction is $P(\text{N or S}) = P(\text{N}) + P(\text{S}) = 1/8 + 1/8 = 1/4$. The ant cannot move in two directions at the same time, so moving to the north and moving to the south are **mutually exclusive**; the events cannot occur at the same time. If events E_1 and E_2 are mutually exclusive, then the probability of E_1 *or* E_2 is the sum of the probabilities of the individual events, that is, $P(E_1 \text{ or } E_2) = P(E_1) + P(E_2)$.

Rule If events E_1 and E_2 are **mutually exclusive** and, thus, cannot occur at the same time, then the probability of E_1 or E_2 is the sum of the probabilities of the individual events,

$$P(E_1 \text{ or } E_2) = P(E_1) + P(E_2)$$

Rule If $E_1, E_2, \ldots, E_n$ are **all possible mutually exclusive events** for a situation so that no two of the events can occur at the same time, then

$$P(E_1) + P(E_2) + \cdots + P(E_n) = 1$$

Quick Review Question 3

Suppose at a site on a strand of DNA, an equal likelihood exists for any base. Give the probability of a site containing A or T.

To calculate the probability that the ant will go in a northerly (N, NE, NW) or westerly (W, NW, SW) direction, we must subtract the probability of where the events overlap, going NW, as follows:

$$P(\text{northerly or westerly}) = P(\{\text{N, NE, NW}\}) + P(\{\text{W, NW, SW}\}) - P(\text{NW})$$
$$= 3/8 + 3/8 - 1/8 = 5/8$$

We subtract $P(\text{NW})$ to avoid counting that direction twice. The two events, heading in a northerly direction and heading in a westerly direction, are not mutually exclusive. If events are not mutually exclusive, then for the probability of one or the other we must subtract the probability of overlap from the sum of the probabilities.

Rule $P(E_1 \text{ or } E_2) = P(E_1) + P(E_2) - P(E_1 \text{ and } E_2)$

Quick Review Question 4

Suppose a certain medicine causes nausea in 1 out of every 10 patients. On the average, 4% of those taking the drug experience diarrhea. The probability that a patient who is using the drug experiences nausea and diarrhea is 0.01. Give the probability that a patient taking the drug has nausea or diarrhea.

Considering again the seeds that have a 30% chance of germinating, suppose we have two seeds, S_1 and S_2. Each has 0.3 probability of germinating, and the state of one seed has no bearing on the state of the other. We say these events are **independent**. Certainly, the probability of both seeds germinating is even less likely than any one germinating. In fact, the probability of S_1 germinating *and* S_2 germinating is the product of their individual probabilities:

$$\begin{aligned} P(S_1 \text{ germinating and } S_2 \text{ germinating}) &= P(S_1 \text{ germinating}) \cdot P(S_2 \text{ germinating}) \\ &= (0.3)(0.3) = 0.09 \end{aligned}$$

Only a 9% chance exists of both seeds germinating.

Definition Events are **independent** if the occurrence of one event has no impact on the occurrence of the other.

Rule For **independent events** E_1 and E_2, the probability of both events occurring is the product of their individual probabilities:

$$\boldsymbol{P(E_1 \text{ and } E_2) = P(E_1) \cdot P(E_2)}$$

Quick Review Question 5

Suppose at a site on a strand of DNA, an equal likelihood exists for any of the four bases. Give the probability of one site containing A and another unrelated site containing T.

Frequently, we wish to know the probability of one event, E_2, given the occurrence of another event, E_1. The notation for such a **conditional probability** is $\boldsymbol{P(E_2 | E_1)}$. For example, suppose a public health agency wages an aggressive campaign to stop the spread of a particular disease by trying to quarantine any individual who has come in contact with someone who has the disease. The probability that an exposed individual is quarantined can be written as a conditional probability, P(quarantined | exposed), the probability of quarantine given exposure. This quantity is equal to probability of the individual being quarantined and exposed divided by the probability of being exposed:

$$P(\text{quarantined} \mid \text{exposed}) = P(\text{quarantined and exposed})/P(\text{exposed})$$

For example, suppose in a group of 100 people, 10 have been exposed and 2 have been exposed and quarantined. Thus, picking an individual at random from the group of 100, we have a 10/100 = 10% = 0.10 chance of selecting an exposed person and a 2/100 = 2% = 0.02 chance of the person being quarantined and exposed. However, if our selection is only from the subset of 10 exposed people, then the probability of picking one of the 2 individuals who is also quarantined is 2/10 = 0.20 = 20%; the probability that an exposed individual is quarantined is 0.02/0.10 = 0.2 = 20%.

Rule Conditional probability of event E_2 given event E_1 is

$$P(E_2 \mid E_1) = P(E_2 \text{ and } E_1)/P(E_1)$$

Thus,

$$P(E_2 \text{ and } E_1) = P(E_2 \mid E_1)P(E_1)$$

Quick Review Question 6

Suppose the DNA for a certain animal contains the sequence, s_1, of 20 bases (A, C, T, G) that evolves to another sequence, s_2, as follows:

s_1	C	A	C	T	T	G	T	G	A	G	C	C	C	A	C	T	T	C	G	T
s_2	C	A	T	T	T	G	T	G	A	C	C	C	T	A	C	T	T	A	G	T

For Parts a–d, determine the probabilities.

a. That C occurs in s_1, written $P(E_1 = \text{C})$
b. That C occurs in s_2
c. That C occurs in s_1 and T occurs in the corresponding site in s_2, written $P(E_2 = \text{T and } E_1 = \text{C})$
d. That T occurs in the corresponding site in s_2, given that C occurs in s_1, written $P(E_2 = \text{T} \mid E_1 = \text{C})$
e. Calculate $P(E_2 = \text{T and } E_1 = \text{C}) \,/\, P(E_1 = \text{C})$, which is your answer from Part a divided into your answer from Part c.
f. How do your answers from Parts d and e compare?

Transition Matrix

We can employ a matrix of conditional probabilities to estimate the long-term behavior of an animal. For example, the red howler monkey's primary food is leaves. Because leaves are hard to digest, the monkey spends about half of its waking hours resting. Resting requires less energy than other activities and gives time for digestion. Suppose we consider a simplified system where the monkey is in only two states, eating (E) and resting/sleeping (R); and $S = \{\text{E, R}\}$ is the **state space**, or set of possible states.

Let us consider some hypothetical data. If one state (X_n) of the monkey is eating, then the probability that the state of the monkey 1 h later (X_{n+1}) is eating is 0.6. We express this information as a conditional probability, $P(X_{n+1} = \text{E} \mid X_n = \text{E}) = 0.6$. Because we assume the monkey is either eating or resting at any time, the probability that the monkey is resting 1 h after eating is $P(X_{n+1} = \text{R} \mid X_n = \text{E}) = 1 - 0.6 = 0.4$.

Quick Review Question 7

With a state of resting at time n, let us suppose that 1 h later the monkey is eating with a probability of 0.2.

a. Express this information in conditional probability notation.
b. Give the conditional probability notation and value for the monkey resting 1 h later.

We can express the data in the preceding paragraph and Quick Review Question 7 with the following matrix, T:

$$
\begin{array}{ccc}
 & X_{n+1} \backslash X_n & \text{E} \quad \text{R} \\
T = & \begin{array}{c}\text{E}\\ \text{R}\end{array} & \begin{bmatrix} 0.6 & 0.2 \\ 0.4 & 0.8 \end{bmatrix}
\end{array}
$$

The first column indicates the probabilities of the indicated values (E or R) of state X_{n+1} given that the monkey is initially eating, $X_n = \text{E}$. Note that the sum of this column's values is 1, because we are considering only one of two possible states for the monkey at any time. Similarly, the second column sums to 1 and presents the probabilities of the monkey eating or resting/sleeping, given that the animal was resting the previous hour. Figure 13.4.1 presents a **state diagram** of the system, with the nodes representing the states and probabilities of going from one state to another labeling the directed edges.

We call T a **transition matrix** (**Markov matrix**, **probability matrix**, or **stochastic matrix**). A **Markov chain** consists of a sequence of variables $X_1, X_2, X_3, \ldots$ in which the value of any variable, X_{n+1}, depends only on the value of its immediate predecessor, X_n. That is, $P(X_{n+1} = x \mid X_n = x_n, \ldots, X_2 = x_2, X_1 = x_1) = P(X_{n+1} = x \mid X_n = x_n)$.

Definitions A **transition matrix** (**Markov matrix**, **probability matrix**, or **stochastic matrix**) is a matrix in which all the entries are nonnegative and the sum of the elements in each column (or each row) is 1. A **Markov chain** consists of a sequence of variables $X_1, X_2, X_3, \ldots$ in which the value of any variable, X_{n+1}, depends only on the value of its immediate predecessor, X_n.

Suppose initially 90% of a group of howler monkeys are eating and 10% are resting, represented by the **probability vector** $\mathbf{v}_0 = \begin{bmatrix} 0.9 \\ 0.1 \end{bmatrix}$, where the components are nonnegative and sum to 1. We can predict the percentage of monkeys eating and resting 1 h later by evaluating $T\mathbf{v}_0$, as follows:

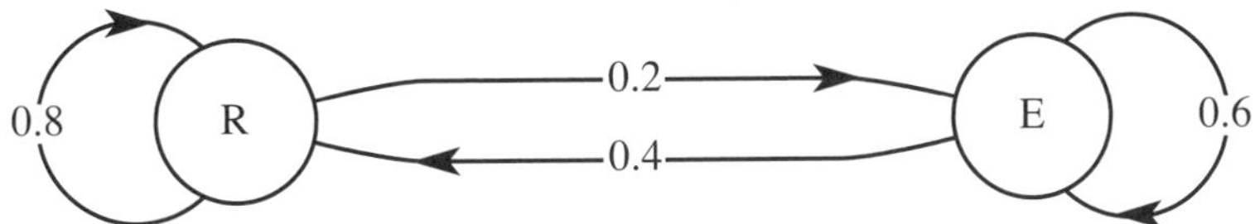

Figure 13.4.1 State diagram of the system

$$\mathbf{v}_1 = T\mathbf{v}_0 = \begin{bmatrix} 0.6 & 0.2 \\ 0.4 & 0.8 \end{bmatrix} \begin{bmatrix} 0.9 \\ 0.1 \end{bmatrix} = \begin{bmatrix} 0.56 \\ 0.44 \end{bmatrix}$$

The calculations predict that at the next hour 56% will be eating, while 44% will be resting.

> **Definition** A **probability vector** is a vector whose components are nonnegative and sum to 1.

Using T and $\mathbf{v}_1$, we can predict the situation at hour 2, as follows:

$$\mathbf{v}_2 = T\mathbf{v}_1 = \begin{bmatrix} 0.6 & 0.2 \\ 0.4 & 0.8 \end{bmatrix} \begin{bmatrix} 0.56 \\ 0.44 \end{bmatrix} = \begin{bmatrix} 0.424 \\ 0.576 \end{bmatrix}$$

Thus, we predict 42.4% of the monkeys will be eating and 57.6% resting at hour 2.

Note that by substitution of $\mathbf{v}_1 = T\mathbf{v}_0$ in $\mathbf{v}_2 = T\mathbf{v}_1$, we see that $\mathbf{v}_2 = T(\mathbf{v}_1) = T(T\mathbf{v}_0) = TT\mathbf{v}_0 = T^2\mathbf{v}_0$. Similarly, at the next hour, the vector is $\mathbf{v}_3 = T\mathbf{v}_2 = T(T^2\mathbf{v}_0) = T^3\mathbf{v}_0 = \begin{bmatrix} 0.3696 \\ 0.6304 \end{bmatrix}$. In general, $\mathbf{v}_n = T^n\mathbf{v}_0$. Table 13.4.1 presents several calculations for T^n and $\mathbf{v}_n$. Notice that as n gets larger and larger, written $n \to \infty$, T^n approaches, or **converges to,** $\begin{bmatrix} 1/3 & 1/3 \\ 2/3 & 2/3 \end{bmatrix}$ and $\mathbf{v}_n$ converges to $\mathbf{v} = \begin{bmatrix} 1/3 \\ 2/3 \end{bmatrix}$, an **equilibrium**, or **steady-state, vector** associated with T. Thus, $\mathbf{v}$ is a probability vector with $T\mathbf{v} = \mathbf{v}$, where each coordinate of $\mathbf{v}$ is the long-term probability that the system will be in the corresponding state. As time progresses, at any one time approximately one-third of the monkeys will be eating and two-thirds resting. Moreover, regardless of the starting vector giving the percentages in each category, with time the percentages will approach 33⅓% and 66⅔% for eating and resting, respectively. Even if all monkeys are eating initially, eventually about one-third will be eating at any one time. When all the entries of a transition matrix are positive, it can be shown that T^n will converge to a matrix M and $\mathbf{v}_n = T^n\mathbf{v}_0$ will converge to a steady-state vector. (We will cover a technique for calculating these limiting steady-state values shortly.)

> **Definition** An **equilibrium**, or **steady-state, vector**, $\mathbf{v}$, of the Markov chain associated with the transition matrix T is a probability vector, where $T\mathbf{v} = \mathbf{v}$.

Table 13.4.1
Markov Matrix, ***T***, and Probability Vector, **v**, to Several Powers

n	$v_n = T^n v_{n-1}$	T^n
0	$\begin{bmatrix} 0.9 \\ 0.1 \end{bmatrix}$	
1	$\begin{bmatrix} 0.56 \\ 0.44 \end{bmatrix}$	$\begin{bmatrix} 0.6 & 0.2 \\ 0.4 & 0.8 \end{bmatrix}$
2	$\begin{bmatrix} 0.424 \\ 0.576 \end{bmatrix}$	$\begin{bmatrix} 0.44 & 0.28 \\ 0.56 & 0.72 \end{bmatrix}$
3	$\begin{bmatrix} 0.3696 \\ 0.6304 \end{bmatrix}$	$\begin{bmatrix} 0.376 & 0.324 \\ 0.624 & 0.688 \end{bmatrix}$
4	$\begin{bmatrix} 0.34784 \\ 0.65216 \end{bmatrix}$	$\begin{bmatrix} 0.3504 & 0.3248 \\ 0.6496 & 0.6752 \end{bmatrix}$
10	$\begin{bmatrix} 0.333393 \\ 0.666607 \end{bmatrix}$	$\begin{bmatrix} 0.333403 & 0.333298 \\ 0.666597 & 0.666702 \end{bmatrix}$
100	$\begin{bmatrix} 0.333333 \\ 0.666667 \end{bmatrix}$	$\begin{bmatrix} 0.333333 & 0.333333 \\ 0.666667 & 0.666667 \end{bmatrix}$

Theorem 1 If all the entries of a Markov matrix are positive, then as n gets larger and larger, T^n converges to a matrix, M, and $\mathbf{v}_n = T^n\mathbf{v}_0$ converges to a vector, $\mathbf{v} = M\mathbf{v}_0$.

Quick Review Question 8

Suppose baboons are observed to be eating (E), grooming (G), or resting (R). A biologist records their activities every 15 min and estimates that if a baboon is eating in one period, in the next 15-min period the animal will be eating or resting with the probabilities 0.3 and 0.6, respectively. If grooming at one observation, in 15 min they are likely to be grooming with a 0.3 probability or eating with a 0.4 probability. If resting in one time period, at the next observation the probabilities a baboon will still be resting or will instead be eating are 0.8 and 0.2, respectively.

a. Using the order E, G, and R for rows and columns, develop a transition matrix, T, for this problem.

b. Suppose when the study began, 30% of the baboons were eating, 10% were grooming, and 60% were resting. Using the model from Part a, give estimates for the percentages of baboons in each state 15 min later.

c. Using a computational tool, estimate the matrix to which T^n converges as n gets larger and larger.

d. Using a computational tool, estimate the vector to which a probability vector for the system converges n goes to infinity.

Using a computational tool, we can calculate that the dominant eigenvalue of the Markov matrix, T, for the howler monkey example is $\lambda = 1$, and a corresponding eigenvector is $\mathbf{x} = (-0.447214, -0.894427)$, so that

$$\begin{bmatrix} 0.6 & 0.2 \\ 0.4 & 0.8 \end{bmatrix} \begin{bmatrix} -0.447214 \\ -0.894427 \end{bmatrix} = 1 \cdot \begin{bmatrix} -0.447214 \\ -0.894427 \end{bmatrix}$$
$$T\mathbf{x} = \lambda\mathbf{x}$$

The ratio of the first coordinate of $\mathbf{x}$ to the second coordinate is one-third to two-thirds. That is, if we add the coordinates of $\mathbf{x}$, $s = -0.447214 + (-0.894427) = -1.34164$, and divide the sum s into each coordinate of $\mathbf{x}$, we obtain $-0.447214/-1.34164 = 1/3 = 0.33\overline{3} = 33\tfrac{1}{3}\%$ and $-0.894427/(-1.34164) = 2/3 = 0.66\overline{6} = 66\tfrac{2}{3}\%$. These values are the exact proportions to which the components of $\mathbf{v}_n = T^n\mathbf{v}_{n\text{-}1}$ tend as n goes to infinity (see Table 13.4.1). The vector $\mathbf{x} = (1/3, 2/3) = (0.33\overline{3},\ 0.66\overline{6}) = (33\tfrac{1}{3}\%, 66\tfrac{2}{3}\%)$ is the equilibrium vector associated with the transition matrix T.

In general, $\lambda = 1$ is always an eigenvalue for the transition matrix of a Markov chain. Moreover, if each of the components of the corresponding eigenvector $\mathbf{x}$ is nonnegative and s is the sum of these components, then $(1/s)\mathbf{x}$ is the equilibrium vector for T, and this vector is a probability vector. If we start with a probability vector, $\mathbf{v}_0$, where each component gives the fraction in each corresponding state, such as eating (E) and resting/sleeping (R), then $T^n\mathbf{v}_0$ converges to $\mathbf{v} = (1/s)\mathbf{x}$ as n becomes larger and larger. Moreover, each coordinate of this equilibrium vector, $\mathbf{v}$, is the ultimate proportion of the corresponding state.

Theorem 2 Suppose T is a Markov chain transition matrix. Then, T has an eigenvalue $\lambda = 1$. Moreover, if each of the components of the corresponding eigenvector, $\mathbf{x}$, is nonnegative and s is the sum of these components, then $(1/s)\mathbf{x}$ is a steady-state vector for T.

Theorem 3 Suppose T is a Markov chain transition matrix. If T^n has all positive entries for some positive integer n, then T has a unique equilibrium vector $\mathbf{v}$. Moreover, if $\mathbf{y}$ is a probability vector, then $T^n\mathbf{y}$ converges to $\mathbf{v}$ as n becomes larger and larger (Agnew and Knapp 2002).

Quick Review Question 9

For the baboon example in Quick Review Question 8, using a computational tool, determine each value.

a. The dominant eigenvalue.

b. The principal eigenvector.

c. The steady-state vector associated with T.
d. The ultimate percentages, expressed in whole numbers, in each state.

Exercises

1. In this problem we consider the animal community on a vertical rock wall of a middle intertidal zone. Suppose we have data for large (> 2 cm) and small (≤ 2 cm) mussels *Mytilus californianus* (B and SMC, respectively), goose barnacles *Pollicipes polymerus* (PP), and other crustaceans (other). Suppose at a fixed point the transition probabilities from ecological state B to ecological states B, SMC, and PP are 0.84, 0.04, and 0.03, respectively; from SMC to B, SMC, and PP are 0.55, 0.26, and 0.03, respectively; from PP to B, SMC, and PP are 0.40, 0.06, and 0.35, respectively; and from other to B, SMC, and PP are 0.15, 0.07, and 0.02, respectively.
   a. Develop the 4×4 transition matrix for this model, where the sum of the elements in each column is 1.0.
   b. Determine the equilibrium vector.
   c. Interpret the results.
2. The **Jukes-Cantor model** for DNA sequence evolution uses a constant, α, for the probability of substitution of one base (A, C, G, or T) for a different base, such as G for T.
   a. Under this model, give the formula for the probability that a base at a particular position does not mutate from one evolutionary time step to the next.
   b. Give the general transition matrix for this model.
   c. Give the transition matrix in the situation where $\alpha = 0.25$, and determine the ultimate distribution of bases.
   d. Give the transition matrix in the situation where $\alpha = 0.3$, and determine the ultimate distribution of bases.
   e. Give the transition matrix in the situation where $\alpha = 0.1$, and determine the ultimate distribution of bases.
   f. Determine the ultimate distribution of bases for the general matrix of Part b.
   g. What conclusions do you draw from your calculations?
3. The **Kimura model** for DNA sequence evolution gives a higher probability for a transition (from A to G, from G to A, from T to C, or from C to T; probability α) than a transversion (from A to C, from C to A, from T to G, or from G to T; probability β) with $\alpha > \beta$ (Sinha 2007).
   a. Under this model, give the formula for the probability that a base at a particular position does not mutate from one evolutionary time step to the next.
   b. Give the general transition matrix for this model.
   c. Give the transition matrix in the situation where $\alpha = 0.25$ and $\beta = 0.10$, and determine the ultimate distribution of bases.
   d. Determine the ultimate distribution of bases for the general matrix of Part b.
   e. What conclusions do you draw from your calculations?

Projects

For additional projects, see Project 14 from Module 14.11, "Spaced Out: Native Plants Lose to Exotic Invasives," and Module 14.14, "Computational Code-Breaking—Deciphering Our Own Mysteries."

1. **Epithelial tissue**, composed of layers of cells, is a covering or lining. For example, the outer portion of the skin, linings of the gastrointestinal system and the lungs, and the outer surface of the cornea are all epithelial tissue. Usually when a cell divides, the daughter cells have one less side than the parent cell but neighboring cells gain sides. It has been observed that virtually no cells are triangular.

   Markov chains can be used to model cell shape, specifically the number of sides of their 2D polygonal structure, in dividing sheets of epithelial cells. A Markov chain model for the number of sides in dividing sheets of epithelial cells hypothesizes that the distribution of sides from a dividing cell to two daughter cells follows a binomial distribution with its coefficients from Pascal's triangle, as indicated in Table 13.4.2. The table gives a model of the relative odds of a cell of one shape becoming a cell of another shape after division of that cell and its neighbors. For example, the value in row 7, column 8 is 6; and the sum of the values in column 8 is 16. Thus, 6/16 is the probability that a cell with 8 sides will become a cell with 7 sides after its and its neighbors' divisions. The table incorporates the distribution of sides of a dividing cell to its daughter cells and the observed average gain of one side from the division of neighbors. (Gibson et al. 2006a, 2006b)

   a. Develop a Markov chain model for the number of sides in dividing sheets of epithelial cells where the state of a cell is its number of sides, $s > 3$. In developing the model, draw a state diagram, form a transition matrix, determine the stable equilibrium percentages for categories of the number of cell sides, and the average number of sides.

   b. Verify the model by comparing these percentages and this average with observations from time-lapse microscopy of three very different animals: *Drosophila* wing disk epithelium, the outer epidermis of the freshwater cnidarian *Hydra*, and the tadpole tail epidermis of the frog *Xenopus* (Table 13.4.3). Employ a histogram for your comparisons.

Table 13.4.2
A Model of the Relative Odds of a Cell of One Shape Becoming a Cell of Another Shape after Division of That Cell and Its Neighbors

After Division	*Before Division*						
	4	**5**	**6**	**7**	**8**	**9**	**10**
4							
5	1	1	1	1	1	1	
6		1	2	3	4	5	
7			1	3	6	10	
8				1	4	10	
9					1	5	
10						1	

Table 13.4.3
Observed Number of Cell Sides in *Drosophila* Wing Disk Epithelium, the Outer Epidermis of the Freshwater Cnidarian *Hydra*, and the Tadpole Tail Epidermis of the Frog *Xenopus* (Gibson et al. 2006a)

	Number of Cell Sides							
	3	*4*	*5*	*6*	*7*	*8*	*9*	*10*
Drosophila	0	64	606	993	437	69	3	0
Hydra	0	16	159	278	125	23	1	0
Xenopus	2	40	305	451	191	52	8	2

c. Based on your work, are the scientists who developed this model justified in concluding that "the distribution of polygonal cell types in epithelia is not a result of cell packing, but rather a direct mathematical consequence of cell proliferation"?

2. H. S. Horn used Markov chains to model succession in a forest, perhaps from a virgin forest or from a forest after a catastrophic event, such as fire. Using a tree-by-tree replacement process with synchronous replacement of all trees by a new generation, he assumed that "the probability that a given species will be replaced by another given species is proportional to the number of saplings of the latter in the understory of the former." Besides synchrony, he makes additional simplifying assumptions, such as sapling abundance predicts survival to reach the canopy and transition probabilities are constant. A study of Institute Woods in Princeton, New Jersey, yielded the data in Table 13.4.4.

a. Using Table 13.4.4's data, develop a Markov chain model of this forest's succession and determine the stable equilibrium percentages.

Table 13.4.4
Transition Matrix for Institute Woods in Princeton: Percent Saplings under Various Species of Trees; BTA, Bigtooth Aspen; GB, Gray Birch; SF, Sassafras; BG, Blackgum; SG, Sweetgum; WO, White Oak; OK, Red Oak, HI, Hickory; TU, Tuliptree: RM, Red Maple; BE, Beech (Horn 1975b, Table 1, p. 199)

		BTA	*GB*	*SF*	*BG*	*SG*	*WO*	*OK*	*HI*	*TU*	*RM*	*BE*
	BTA	3	—	3	1	—	—	—	—	—	—	—
	GB	5	—	1	1	—	—	—	—	—	—	—
	SF	9	47	10	3	16	6	2	1	2	13	—
	BG	6	12	3	20	0	7	11	3	4	10	2
Sapling	*SG*	6	8	6	9	31	4	7	1	4	9	1
species	*WO*	—	2	3	1	0	10	6	3	—	2	1
(%)	*OK*	2	8	10	7	7	7	8	13	11	8	1
	HI	4	0	12	6	7	3	8	4	7	19	1
	TU	2	3	—	10	5	14	8	9	9	3	8
	RM	60	17	37	25	27	32	33	49	29	13	6
	BE	3	3	15	17	7	17	17	17	34	23	80
Species Counts		104	837	68	80	662	71	266	223	81	489	405

Table 13.4.5
Longevity (Years) of Trees in Institute Woods (Horn 1975b, Table 2, p. 200)

BTA	*GB*	*SF*	*BG*	*SG*	*WO*	*OK*	*HI*	*TU*	*RM*	*BE*
80	50	100	150	200	300	200	250	200	150	300

b. Using the distribution of species in the last row of Table 13.4.5 as the initial distribution and the transition matrix from Part a, plot the estimated number of trees of each species for 20 generations.

c. Trees, however, do not have the same life expectancy, as Table 13.4.5 indicates. Thus, Horn weighted (i.e., multiplied) the stationary distribution by the longevities in the table, normalized the result (i.e., divided by the sum of the components), and obtained percentages (i.e., multiplied by 100). Perform these calculations on your stable equilibrium distribution to obtain an age-corrected distribution, which is the analog of a climax community.

d. Calculate the relative invasiveness of each species as the sum of the percent saplings under other trees divided by the maximum such sum. That is, to calculate this metric, for each row, calculate the row sum minus the diagonal element; find the maximum of these sums; and divide each row sum minus the diagonal element by this maximum. Discuss how the beech's ability to invade under other species is evident in the probabilities of Table 13.4.4.

e. Evaluate a metric for each species' resistance to invasion by other species as follows: Calculate the sum of percentages of other saplings under its canopy (column sum excluding diagonal element); determine the minimum such sum; and for each species, compute this minimum divided by the sum of percentages of other saplings under its canopy. Discuss how the beech's resistance to invasion by other species is evident in the probabilities of Table 13.4.4.

f. Calculate a metric for each species' self-replacement as the percentage of its own saplings under its canopy (diagonal element) divided by the maximum such percentage (maximum diagonal element). Discuss how the beech's copious self-replacement is evident in the probabilities of Table 13.4.4.

g. Compare your results to those of Horn's data for several subforests of varying ages in Institute Woods (see Table 13.4.6).

Table 13.4.6
"The empirical approach results from independent measurements of 639 trees in stands that have been fallow for at least the number of years indicated. The percentages are of total basal area, calculated from diameters measured at breast height." (Horn 1975b, Table 2, p. 200)

Years fallow	*BTA*	*GB*	*SF*	*BG*	*SG*	*WO*	*OK*	*HI*	*TU*	*RM*	*BE*
25	0	49	2	7	18	0	3	0	0	20	1
65	26	6	0	45	0	0	12	1	4	6	0
150	—	—	0	1	5	0	22	0	0	70	2
350	—	—	—	6	—	3	—	0	14	1	76

h. Discuss the climax abundance of each species in relationship to its possession of the characteristics of Parts d, e, and f.

3. Fecal shedding is the elimination of a pathogen through an animal's fecal matter. Because many diseases spread by fecal shedding, an understanding of the dynamics of contagiousness is important in disease prevention and control. Ivanek and others (2007) used Markov chain models to study, in dairy cattle, the dynamics of fecal shedding of the pathogen *Listeria monocytogenes* (LM), a bacterium that causes listeriosis, a disease of the central nervous system. Models with two states, shedding (of LM) and nonshedding, were developed for overall (all subtypes) *L. monocytogenes* shedding considering various combinations of time-dependent risk factors, or **covariates** that can change with time. These covariates include silage (feed) contaminated with LM and stress, such as from antiparasitic treatment.

Using data and statistics and considering the situations of presence or absence of contaminated silage and stress, the scientists estimated the probability of fecal shedding or nonshedding one day (time $t - 1$) leading to the presence or absence of LM in a cow's feces the next day (time t). Thus, they determined $2^3 = 8$ probabilities (Table 13.4.7). With 1 indicating presence and 0 absence of each of the three conditions (contaminated silage, stress, and fecal shedding) the day before, Table 13.4.7 gives the probabilities of fecal shedding of LM. For example, the first two rows under the headings consider the situation in which silage contamination and stress did not exist at time $t - 1$. In this case, the probability of changing from a nonshedding state at time $t - 1$ to a shedding state at time t is $p_{01} = P(\text{shedding at time } t \mid \text{nonshedding at time } t - 1) = 0.038$, while the probability of remaining in a shedding state is $p_{11} = P(\text{shedding at time } t \mid \text{shedding at time } t - 1) = 0.116$. Using these two probabilities, we can develop a 2×2 Markov matrix. In Table 13.4.7, each pair of rows below the headings results in a different model.

a. Develop four Markov chain models $\begin{bmatrix} p_{00} & p_{10} \\ p_{01} & p_{11} \end{bmatrix}$ for each covariant situation in Table 13.4.7. Starting with an initial distribution at time $t = 0$ of

Table 13.4.7

For all subtypes of *Listeria monocytogenes*, presence (1) or absence (0) of overall LM contamination of silage, stress, and LM fecal shedding at time ***t*** – 1 with the probability of LM fecal shedding the next day (time ***t***)

	At time t – 1			*At time t*
Subtypes	*Silage Contam.*	*Stress*	*Fecal Shedding*	*Probability of Fecal Shedding*
All	0	0	0	$p_{01} = 0.038$
	0	0	1	$p_{11} = 0.116$
	0	1	0	$p_{01} = 0.174$
	0	1	1	$p_{11} = 0.410$
	1	0	0	$p_{01} = 0.358$
	1	0	1	$p_{11} = 0.648$
	1	1	0	$p_{01} = 0.746$
	1	1	1	$p_{11} = 0.907$

100% nonshedding cows, for each situation, plot the percent of shedding cows from day 0 through day 10. Determine the long-term distributions, which give the equilibrium probabilities of being in nonshedding and shedding states, or the eventual proportion of time in each state. Discuss the results.

b. The time spent in a state of this model has a geometric distribution. If the initial day (day 0) is nonshedding, the probability of the next day (day 1) being nonshedding, or the proportion of time of a nonshedding day 1, is p_{00}; the probability of days 1 and 2 being nonshedding is $p_{00}p_{00} = (p_{00})^2$; the probability of days 1–3 being nonshedding is $(p_{00})^3$; and so on. Thus, the mean time spent in the nonshedding state over a period of $n - 1$ days is $1 + p_{00} + (p_{00})^2 + \cdots + (p_{00})^{n-1}$. This sum is a **finite geometric series**, which equals $\frac{1-p_{00}{}^n}{1-p_{00}}$ (discussed in Module 2.5, "Drug Dosage"). As n goes to infinity, $(p_{00})^n$ goes to 0 because $0 \le p_{00} < 1$. Thus, for a Markov chain model of Part a, we can estimate the mean time for a cow to spend in a nonshedding state as $\frac{1}{1-p_{00}}$. Similarly, we can estimate the time for a cow to spend in a shedding state as $1/(1 - p_{11})$. Make such estimates for each of the covariant situations in Table 13.4.7, and discuss the results.

c. The models of Part a are **homogenous Markov chain models**, which use the same transition matrix throughout. However, we can employ a **nonhomogenous Markov chain model**, where we vary the transition matrix depending on the presence or absence of the time-varying covariates (contaminated silage and stress). Thus, for a real or assumed pattern of time-varying covariates, by employing the appropriate transition matrices, we can examine the changing distributions. Develop a program to accept a sequence of time-varying covariates for a period of 20 days and to plot the percentage of shedding cows versus day. Discuss the results for several patterns.

4. (*Prerequisite*: From Module 10.2, "Diffusion: Overcoming Differences," sections "Heat Diffusion" and "Boundary Conditions") The **stepping-stone model** is useful in the study of genetics. For the model, we start with an $n \times n$ grid (matrix) with each cell (element) having one of k integer values. Repeatedly, we select a cell at random and choose one of its eight neighbors at random. We then change the value at the cell to be the value of the selected neighbor. Periodic boundary conditions are employed. A grid represents a state of the system. Thus, with each grid having n^2 cells and each cell having k possible values, the system has k^{n^2} possible states. For example, a small 10×10 grid with 100 cells and values of only 0 and 1 has $2^{100} = 1.2677 \times 10^{30}$ possible states. A transition matrix with this number of states would have an excessive number of elements: $10^{30} \times 10^{30} = 10^{60}$ elements. However, we can employ cellular automaton simulations to simulate the Markov chain (Grinstead and Snell 2003).

a. Develop the stepping-stone model using $n = 20$ and $k = 2$ (values *num1* = 1 and *num2* = 2). Employ a random initial configuration with a probability of p for one of the cell values, *num1*. Using visualizations of the grid with white representing one cell value and black representing the other, develop an animation. Run the animation a number of times with different values of p and observe regions of color and the ultimate "winner." Does the winner seem related to p? Discuss the results.

b. Repeat Part a without the animation but plotting the number of each color at each time step. Discuss the results.

c. Repeat Part a using $k > 2$.

d. Repeat Part b using $k > 2$.

Answers to Quick Review Questions

1. ¼ = 0.25 = 25%

2. 0.75 = 1 – 0.25 = 1 – P(T); alternatively, 0.75 = P(A) + P(C) + P(G).

3. ½ = 0.5 = 50%

4. 0.13 = 0.10 + 0.04 – 0.01

5. 1/16 = (¼)(¼)

6. **a.** 7/20 = 0.35

b. 5/20 = 0.25

c. 2/20 = 0.10

d. 2/7 = 0.286 because C occurs in s_1 7 times

e. 2/7 = (2/20)/(7/20)

f. They are equal.

7. **a.** $P(X_{n+1} = \text{E} \mid X_n = \text{R}) = 0.2$

b. $P(X_{n+1} = \text{R} \mid X_n = \text{R}) = 0.8 = 1 - 0.2$

8. **a.** $$\begin{bmatrix} 0.3 & 0.4 & 0.2 \\ 0.1 & 0.3 & 0.0 \\ 0.6 & 0.3 & 0.8 \end{bmatrix}$$

b. E: 25%, G: 6%, R: 69% because the product of T from Part a and (0.3, 0.1, 0.6), expressed as a column vector, is (0.25, 0.6, 0.69), expressed as a column vector.

c. $$\begin{bmatrix} 0.229508 & 0.229508 & 0.229508 \\ 0.0327869 & 0.0327869 & 0.0327869 \\ 0.737705 & 0.737705 & 0.737705 \end{bmatrix}$$

d. (0.229508, 0.0327869, 0.737705)

9. **a.** 1

b. Any nonzero multiple of (–0.296799, –0.0423999, –0.953998)

c. (0.229508, 0.0327869, 0.737705) obtained by multiplying the vector from Part b by 1 over the sum of its elements, $s = -1.2932$

d. E: 23%, G: 3%, R: 74% obtained by expressing as percentages the elements of the vector from Part c

References

Agnew, Jeanne, and Robert C. Knapp. 2002. *Linear Algebra with Applications*. Monterey, CA.: Brooks/Cole.

"Barnacles." 2012. *Science Encyclopedia*. http://science.jrank.org/pages/752/Barnacles.html (accessed December 30, 2012)

California Coastal Commission. 1987. "California's Rocky Intertidal Zones, " exerpts from the *California Coastal Resource Guide*. http://ceres.ca.gov/ceres/calweb/coastal/rocky.html (accessed December 30, 2012)

Cornell University. 2003. "Barnacles Will Cling No More with Self-Cleaning, Non-Toxic Coating for Ships Developed by Cornell researchers," *Cornell News*. http://www.news.cornell.edu/releases/March03/ACS.Ober.deb.html (accessed December 30, 2012)

Foster, Rick. 2009. "Intertidal Communities Overview." ADF&G. 2000. Alaska Department of Fish and Game, February 11. http://svp.soic.indiana.edu/svp/4970813/FID1/html/ecosys/estuarin/intertdl.htm (accessed December 30, 2012)

Gibson, Matthew C., Ankit B. Patel, Radhika Nagpal and Norbert Perrim. 2006a. "The Emergence of Geometric Order in Proliferating Metazoan Epithel." *Nature* 442: 1038–1041.

———. 2006b. "The Emergence of Geometric Order in Proliferating Metazoan Epithel," Supplementary Material. http://genepath.med.harvard.edu/~perrimon/papers/GibsonM_Supp_Nature.pdf (accessed December 30, 2012)

Grinstead, Charles, and Laurie Snell. 2003. *Introduction to Probability*, "Markov Chains." American Mathematical Society.

Gropl, Clemens, and Daniel Huson, 2005. "Hidden Markov Models." February17. http://www.inf.fu-berlin.de/inst/ag-bio/FILES/ROOT/Teaching/Lectures/WS0405/aldabi/script-hmm.pdf (accessed December 30, 2012)

Horn, H. S. 1975a. "Forest Succession," *Scientific American*, 232: 90–98.

———.1975b. "Markovian Properties of Forest Succession," in *Ecology and Evolution of Communities*, M. L. Cody and J. M. Diamond, eds. Cambridge, MA: Harvard University Press, pp. 196–211.

"Intertidal Stressors," 2007. *Ocean News*. http://oceanlink.island.net/ONews/ONews7/intertidal.html (accessed December 30, 2012)

Ivanek, Renata, Yrjo T. Grohn, Alphina Jui-Jung Ho, and Martin Wiedmann. 2007. "Markov Chain Approach to Analyze the Dynamics of Pathogen Fecal Shedding—Example of *Listeria monocytogenes* Shedding in a Herd of Dairy Cattle." *J. Theor. Biol.* 245: 44–58.

Natural History Museum. 2012. "The Secret Life of "Barnacles," http://www.nhm.ac.uk/nature-online/life/other-invertebrates/barnacles/index.html (accessed December 30, 2012)

Sinha, Saurabh. 2007. "Evolutionary Models," lecture notes in CS 498. www.cs.uiuc.edu/class/fa07/cs498ss/lectures/LectureEvoModel.ppt (accessed December 30, 2012)

Stout, Prentice. 2009. "Barnacle," Rhode Island Sea Grant Fact Sheet. http://seagrant.gso.uri.edu/factsheets/597barnacle.html (accessed December 30, 2012)

National Human Genome Research Institute. "Talking Glossary of Genetic Terms." http://www.genome.gov/glossary/index.cfm? (accessed December 30, 2012)

Tang, Haixu. 2007. "Probabilistic Sequence Modeling II: Markov Chains," Lecture Notes for Bioinformatics in Molecular Biology and Genetics: Practical Applications http://darwin.informatics.indiana.edu/col/courses/I529/Lecture/lec-4.ppt (accessed December 30, 2012)

MODULE 13.5

The Next Flu Pandemic—Old Enemy, New Identity

Prerequisites: Module 13.1, "Computational Toolbox—Tools of the Trade: Tutorial 7" or "Alternative Tutorial 7" (except section on "Eigenvalues and Eigenvectors") and Module 13.2, " Matrices for Population Studies—Linked for Life." Additional high-performance computing materials related to this module are available on the text's website.

Downloads

For several computational tools, the text's website has a *SocialNetworks* file, with implementations of the functions of this module, available for download.

Introduction

Charlie Bates is a college sophomore who wakes up this morning feeling really bad. He assumes that it is just a hangover. He had a pretty wild night of drinking at his fraternity's welcome-back party that traditionally begins the spring semester. His head is pounding, and he is exhausted.

Charlie feels that he can sleep this one off, so he decides to cut his 9 o'clock economics class. He resets his alarm and rolls over. Four hours later he is distressed to find that he has also slept through his 11:00 government class. Even more disturbing is that he feels even worse. He has never had a sore throat from a hangover, and he is feeling very achy. So, he gets up, dresses, and stumbles over to the campus infirmary. The nurse finds that he has a temperature of 102.5 °F. She thinks he has the flu.

Every year we hear the warnings from public health officials to get our flu shots. Some of us comply, but many of us do not. In fact, the CDC reported that less than 40% of the U.S. population was vaccinated during the 2008–2009 flu season (CDC 2009). Influenza can attack any age, race, or sex. It not only makes us feel miserable, it costs millions of days of lost productivity at school or at work. Although the highest rates of infection are in children, the most severe, even life-threatening effects are on those over the age of 65.

So, why don't we get the shot? Well, it may be a result of "flu myths." Some of these **misconceptions** are as follows (NFID 2008a, b):

- *Flu is not a serious disease.*
  Flu is not the common cold, which also can certainly make you miserable. CDC estimates that the flu is the cause of an average of 36,000 deaths and hundreds of thousands of hospitalizations per year in the United States (CDC 2009).
- *The vaccination is not necessary.*
  Because the influenza virus is so genetically pliable, it changes from year to year. A vaccination received one year will offer you little to no protection from the influenza virus the next year.
- *You can get the flu from the shot.*
  Not likely. The vaccine is made from inactivated or killed viruses. The worst side effect you may obtain from a flu shot is a sore arm. However, if you are allergic to eggs, you should *not* get a flu shot that is egg based.

Public health organizations worldwide are trying to find ways of effectively blunting the inevitable epidemics/pandemics of this disease. As part of their efforts, officials are using the results of computational science models, which employ computer science and mathematics along with the science, to make informed decisions on how to combat the menace.

The Problem

Computational scientists model the spread of disease in a number of ways. **System dynamics models** consider the changing sizes of complex interrelated systems, such as susceptibles, infecteds, and recovereds, as time progresses. **Cellular automaton simulations** model reality with 1D, 2D, or 3D grids that change with time. A grid site has a state, such as susceptible, infected, or recovered. Rules, such as an infected person recovers in 5 days, regulate the behavior of the system. The results of cellular automaton simulations are challenging to verify, and system dynamics models do not provide some of the specificity that would be helpful in making public health decisions in the face of an epidemic. As Bisset and Madhav (2009) write, "these modeling approaches were limited in their ability to capture the complexity of human interaction that underlies disease transmission." **Grid-based agent-based models**, which simulate and visualize the behaviors and interactions of individuals, can overcome some of these limitations; but the restricted movement of an individual from one cell to a neighboring cell at a discrete time step and the inability for such systems to handle large datasets hinder their utility. A related modeling technique, **individual-based** (or **network-based**) **epidemiology simulation**, which employs matrices, tracks the simulated behavior of individuals in a community. Thus, this method provides the desired specificity and is easier to verify, but its simulations incorporate massive amounts of data that require extensive effort to gather and need massive computing power to process.

To help meet this challenge, scientists at Los Alamos National Laboratory developed the Epidemiological Simulation System (EpiSims) to simulate the spread of epidemics in a large city at the individual level, considering realistic contacts among the people and characteristics of disease transmission (VBI 2008). Employing transportation information, census data, and activities surveys from a sample of about 2000 people, researchers have generated hypothetical data that model the movements and demographics of Portland, Ore. With EpiSims, scientists can study a number of important issues: the efficacy of prevention measures, such as vaccinating particular segments of the population; the value of early detection measures, such as placing fever sensors at high traffic buildings; the effectiveness of public health interventions, such closing schools; and fundamental questions, such as patterns of the spread of disease.

Individual-based epidemiology simulations can estimate some of the following metrics:

- A smallest set of locations (**minimum dominating set**) that a given proportion of the population visits

  Such information can be helpful in determining sites for fever sensors or in closing of particular public buildings during an epidemic.
- The distribution of the number of contacts people have with other people (**degree distribution)**

  Targeted vaccination of individuals who have many contacts can offer significantly better results than vaccinating people at random (Mason and Verwoerd 2007).
- The probability that two contacts of a randomly chosen person have contact with one another (**clustering coefficient**; Newman et al. 2002)

  A large probability indicates that a disease can spread rapidly through a community.
- The average smallest number of contacts for a disease to spread from one arbitrary individual to another (**mean shortest path length**)

  A small mean also indicates the probable rapid spread of a disease.

Graphs

Employing an area of mathematics called **graph theory**, individual-based epidemiology simulations use graphs that represent the contacts between people to predict the spread of disease and to analyze health-care interventions. By a **graph** we do not mean a graph of a function, such as $f(x) = x^2$, but rather a set of **nodes** with undirected or directed **edges** connecting some of the points, as illustrated in Figure 13.5.1. In that **contact network**, or **social network**, which is a type of graph, the nodes represent people or groups of people, such as members of a household that can become infected, and places, where the disease can spread from an infected person to a susceptible individual. Numbers indicate the households, while the places are school, hospital, work, shop, and cloister. Each edge represents an association that can lead to transmission of the disease. For example, one or more individuals in household 6 go to work, shop, and school, locations where they can contract or spread the disease.

An **undirected graph** $G = (V, E)$ consists of a set V of **vertices** (singular, **vertex**), or **nodes** or **points**, and a set E of **edges**, or **arcs**, connecting pairs of points. In the graph of Figure 13.5.1, $V = \{1, 2, 3, 4, 5, 6, 7,$ *School*, *Hospital*, *Work*, *Shop*,

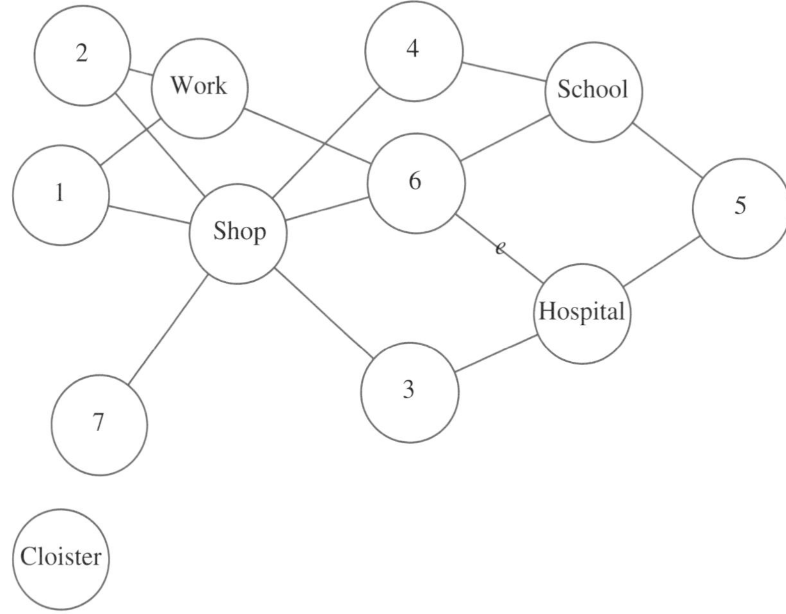

Figure 13.5.1 Contact network of households and places

Cloister}. We denote an undirected edge *e* between nodes *u* and *v* as (*u*, *v*) or as (*v*, *u*). Here, (*u*, *v*) is not an ordered pair. In the preceding figure, *e* = (6, *Hospital*) = (*Hospital*, 6) is an edge because there is contact between at least one member of household 6 and people who are at the *Hospital*. The **size** of the graph is its number of nodes, so the graph in Figure 13.5.1 has size 12.

Definitions An **undirected graph** *G* = (*V*, *E*) consists of a set *V* of **vertices** (singular, **vertex**), or **nodes** or **points**, and a set *E* of **edges**, or **arcs**, connecting pairs of points. An undirected edge between nodes *u* and *v* is denoted as the unordered pair (*u*, *v*) or (*v*, *u*). The number of nodes in a graph is the **size** of the graph.

Quick Review Question 1

Referring to Figure 13.5.2, give the following:

a. The set of vertices, *V*, for the graph
b. Two notations for the edge connecting 3 and 6
c. The graph's size

We need to know several terms to speak the language of graph theory. In Figure 13.5.1, points 6 and *Hospital* are **adjacent** because there is an edge, *e*, connecting them. We say that edge *e*, which can be written as (6, *Hospital*) or as (*Hospital*, 6), is **incident** to points 6 and *Hospital*. Vertex *Cloister* is **isolated**, having **degree** 0, or no incident lines, while point 5 has degree 2, or is the endpoint of two edges.

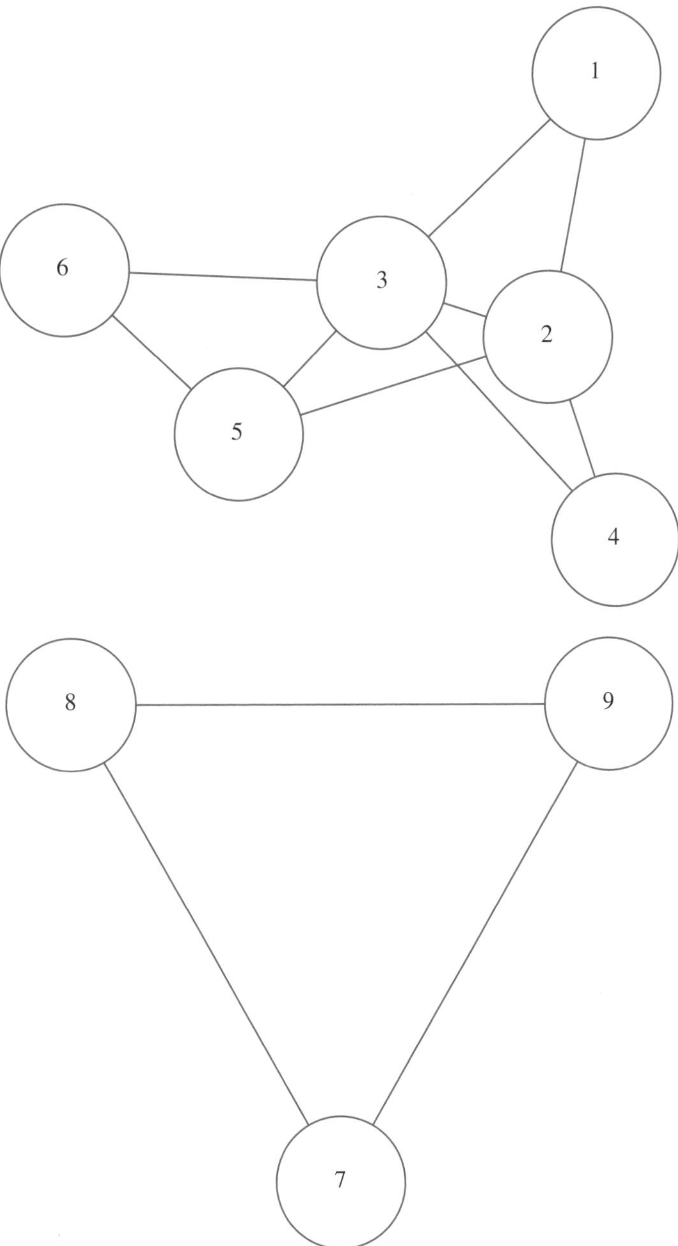

Figure 13.5.2 Graph for Quick Review Question 1

Definitions Two vertices u and v of a graph are **adjacent** if there exists an edge (u, v) connecting them. An edge e is **incident** to vertex v if v is an endpoint of e. The **degree** of a vertex v, **deg(v)**, is the number of times v is an endpoint of an edge. If $\deg(v) = 0$, then v is called an **isolated point**.

Quick Review Question 2

Referring to Figure 13.5.2, give each of the following:

a. The vertices adjacent to node 3
b. The edge(s) incident to node 1
c. The degree of node 3
d. The isolated point(s)

Much work on the structural properties of biological networks, such as social networks, has focused on the distribution of degrees. If n is the number of nodes in a network and n_k is the number of nodes of degree k, then the **degree distribution** is $\boldsymbol{P(k) = n_k/n}$, which is the proportion of nodes having degree k, for $k = 0, 1, 2, \ldots$. For example, in Figure 13.5.1, which is a graph of size $n = 12$, one node (*Cloister*) has degree 0 and one node (node 7) has degree 1, so $P(0) = P(1) = 1/12$. With five nodes (nodes 1, 2, 3, 4, and 5) having degree 2, five-twelfths of the nodes ($P(2) = 5/12$) are incident to two nodes.

Studies of this and many other biological networks, such as the central metabolic networks of 43 organisms and protein interaction networks for various organisms, have degree distributions that appear to follow power laws. A function f follows a **power law** if $f(x)$ is proportional to x^b for some constant b; that is, $f(x) \propto x^b$, or $f(x) = cx^b$, for some constants c and b. In the case of many biological networks, the degree distribution $P(k)$ is proportional to k^{-r} for some constant r. Specifically for metabolic networks, $P(k) \propto k^{-r}$ for $2 < r < 3$, or $P(k) = ck^{-r}$ for $2 < r < 3$ and some constant c. A degree distribution following this power law implies that nodes with small degree are extremely common, while nodes with large degree are quite rare. Figure 13.5.3 shows the graph of $P(k) = k^{-2.5}$ with the typical broad-tail, or long, stretched-out portion to the right, of such a power law form.

Networks that follow the power law $P(k) \propto k^{-r}$ with $r > 1$ are called **scale-free networks**. Interestingly, the Internet is a scale-free network. In a scale-free network, most nodes have relatively low degree, but a few nodes, called **hubs**, have high degrees. Removal of hub nodes can easily result in the network being disconnected. Thus, scale-free networks are particularly vulnerable to attack and failure at the hubs. Biologists have suggested that in a genetic or protein network, a hub node, which is a gene or protein that participates in a large number of interactions, may be more significant for the survival of an organism than nodes that have small degrees (Mason and Verwoerd 2007). In a social network that is scale-free, a hub location, which has numerous visitors each day, is a prime site for the spread of disease.

Definitions A function f follows a **power law** if $f(x)$ is proportional to x^b for some constant b; that is, $f(x) \propto x^b$, or $f(x) = cx^b$, for some constants c and b. If n is the number of nodes in a graph and n_k is the number of nodes of degree k, then the **degree distribution** is $\boldsymbol{P(k) = n_k/n}$, which is the proportion of nodes having degree k. Networks that follow the power law $P(k) \propto k^{-r}$ with $r > 1$ are called **scale-free networks**. **Hubs** are nodes with high degrees in scale-free networks.

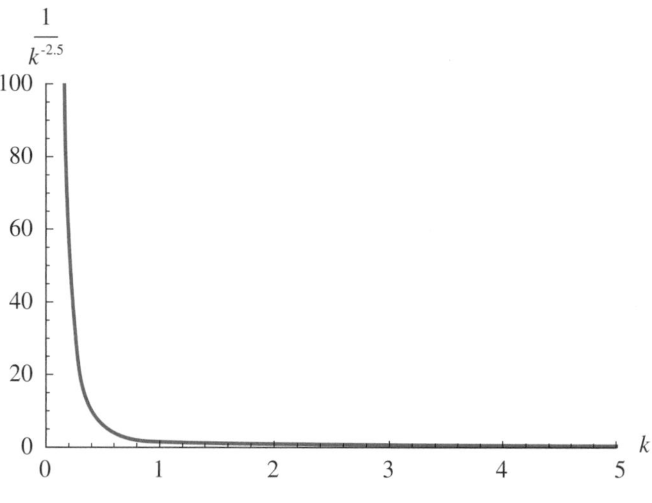

Figure 13.5.3 Graph of $P(k) = k^{-2.5}$

Paths

Paths through the contact network in Figure 13.5.1 can help illuminate the epidemiology of the disease. Suppose initially someone in household 1 has the disease. One way for someone from household 5 to contract the disease indirectly from 1 is by the path 1, *Shop*, 6, *School*, 5. Someone from household 1 goes shopping, infecting someone at the shop. Likewise, an individual from household 6 goes shopping and catches the disease. That person or someone in household 6 who becomes ill from contact with that individual goes to school, spreading the disease further. An individual from household 5 also attends the school and contracts the disease there. With four edges along this path, we say the **path length** is 4.

Definitions In a graph G, a **path** from vertex v_0 to v_n along edges e_0 to e_{n-1} is the sequence

$$v_0, e_0, v_1, e_1, \ldots, v_{n-1}, e_{n-1}, v_n$$

where $e_i = (v_i, v_{i+1})$ for $i = 0, 1, \ldots, n-1$. If no ambiguity exists, the path can be represented with just the vertices as the sequence $v_0, v_1, \ldots, v_n$ or just the edges as the sequence, $e_0, e_1, \ldots, e_{n-1}$. A path of n edges is said to be of **length** n.

Quick Review Question 3

Give two paths of length 3 from node 2 to node 6 in Figure 13.5.2.

In general, for biological networks, such as protein, gene, or metabolic networks, the average length of a path between nodes is small in comparison to the size of the graph. Thus, we say such networks exhibit the **small-world property**. Specifically, if such a graph has n nodes, by definition the average shortest-path length is on the order of magnitude of log n or smaller. For example, metabolic networks have between 200 and 500 metabolites (nodes), but the average path length is between 3 and 5. Genetic networks contain about 1000 genes (nodes) and 4000 interactions (edges), with an average path length of 3.3. Because average path length indicates how readily the network can communicate information, biological networks are efficient communicators. For instance, a metabolic network needs few interactions for one metabolite to influence the behavior of another metabolite (Mason and Verwoerd 2007). One of the projects considers an algorithm to calculate mean path length.

Definition A graph with n nodes exhibits the **small-world property** if the average shortest-path length is on the order of magnitude of log n or smaller.

Clustering

Figure 13.5.4 is a **subgraph** of Figure 13.5.1 because every node and every edge of Figure 13.5.4 is in Figure 13.5.1. The subgraph of Figure 13.5.1 that includes every node and edge except *Cloister* is **connected** because there is a way to get from any point to any other point in that subgraph by following edges. Thus, the disease has the potential of spreading to every node in the subgraph.

Suppose household 6 of Figure 13.5.4 represents a family of two parents and three children in which every member of the family has contact with every other member. Figure 13.5.5 illustrates the graph of this household, which is **complete**, having every point (vertex or node) connected to every other point directly by exactly one edge. An ill member of the household will expose everyone in the house.

In the graph of Figure 13.5.5, each of the five points has four incident edges connecting that node to the remaining nodes. In other words, each node has degree 4.

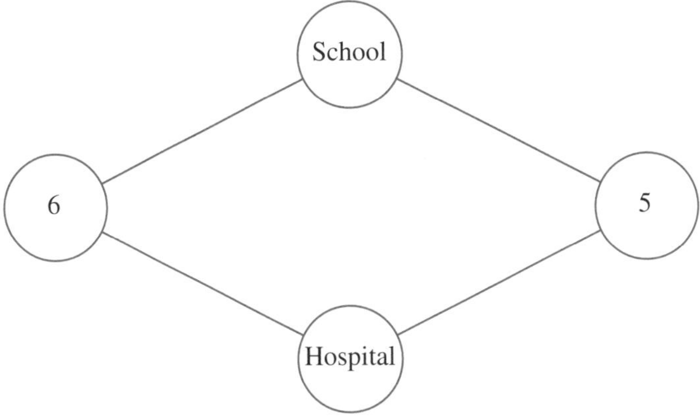

Figure 13.5.4 One subgraph of Figure 13.5.1

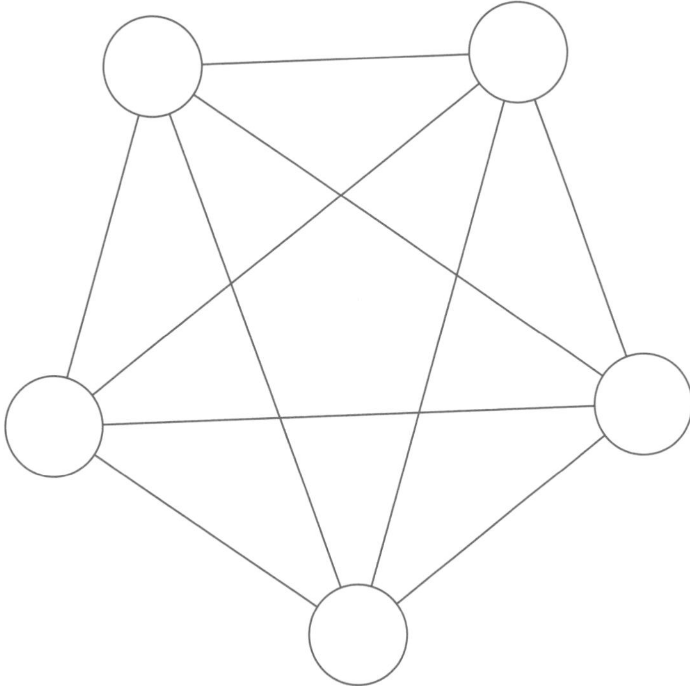

Figure 13.5.5 Complete graph of a household with five individuals

Definitions S is a **subgraph** of graph G if S is itself a graph and every node and edge of S is in G. A graph is **connected** if there exists a path from any vertex to any other vertex. A graph is **complete** if each point is adjacent to every other point with exactly one edge between each pair of nodes.

Therefore, the sum of the degrees of all the points is $(5)(4) = 20$. Because in summing the degrees we count each edge twice, once for each endpoint, the number of edges in this complete graph is half the sum, $(5)(4)/2 = 20/2 = 10$. In general, a complete graph with n points has $n(n - 1)/2$ edges.

Theorem A complete graph with n nodes has $n(n - 1)/2$ edges.

Quick Review Question 4

Referring to Figure 13.5.2, give each of the following:

a. The edge sets of all connected subgraphs with $V = \{7, 8, 9\}$
b. The complete subgraph $G = (V, E)$ of three points containing node 7
c. The edge(s) to add to have a complete subgraph with $V = \{2, 3, 4, 5\}$
d. The number of edges in the complete subgraph formed by this addition

The concept of completeness is central to the calculation of clustering coefficients, a measure of how rapidly a disease can spread. The **clustering coefficient** for a vertex, v, is the probability that two nodes adjacent to v are themselves adjacent. That is, the chance that two arbitrary edges incident to v are part of a triangle of edges in the graph is the clustering coefficient for v. Thus, if A is the set of nodes adjacent to v, then the clustering coefficient is the quotient of the number of edges in the subgraph with points from A and the number of edges in a complete graph with that number of points. If a node has degree zero or one, then its clustering coefficient is 0. The clustering coefficient of v indicates of how close v and its adjacent nodes are to being a complete graph. For example, node 2 in Figure 13.5.2 is adjacent to 4 nodes, nodes 1, 3, 4, and 5. Three edges appear in the subgraph with these adjacent nodes, $V = \{1, 3, 4, 5\}$. However, a complete graph with four nodes has $(4)(3)/2 = 6$ edges. Thus, the clustering coefficient for node 2 is $3/6 = 0.5$. A 50% probability exists that two neighbors of node 2 are themselves adjacent. One research project calculated the clustering coefficient of people using the Internet to be 0.1078; but with an edge connecting two actors if they were in a movie together, the clustering coefficient of movie actors is significantly higher, 0.79 (Eggemann and Noble 2008).

Definition Suppose A is the set of nodes adjacent to node v in graph G, and $n(A)$ is the number of points in A. The **clustering coefficient** for v, $\boldsymbol{C(v)}$, is the number of edges of G in the subgraph with points from A divided by the number of edges in a complete graph with $n(A)$ nodes:

$$C(v) = \frac{\text{number of edges of } G \text{ in subgraph with set of nodes } A}{\text{number of edges in complete graph with } n1(A) \text{ nodes}}$$

Quick Review Question 5

Give the clustering coefficient for each of the following:

a. Node 3 from Figure 13.5.2
b. Node 7 from Figure 13.5.2
c. Node 5 from Figure 13.5.1
d. Node *Cloister* from Figure 13.5.1

Typically, a small-world network not only has a small mean path length, but it also has a large mean clustering coefficient. With these characteristics, disease can spread rapidly in social networks.

Bipartite Graphs

Figure 13.5.6 shows a contact network of Wards A, B, and C in a psychiatric hospital with health-care workers, indicated by numbers. This **bipartite graph** has its vertices split into two sets, a set of wards and a set of workers, with edges only between

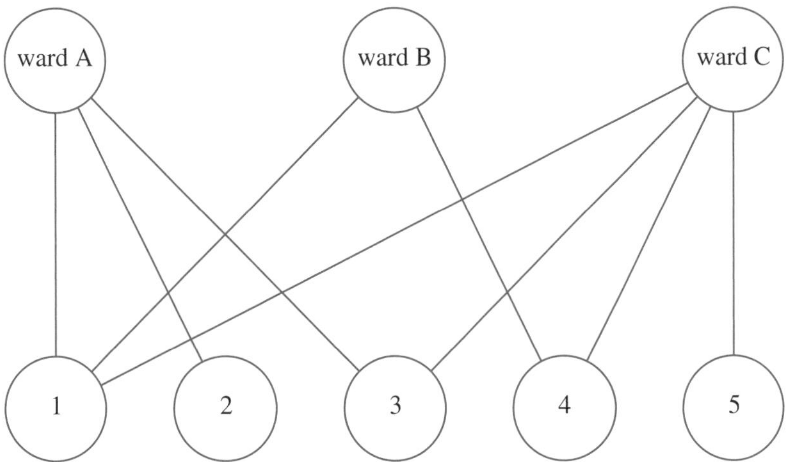

Figure 13.5.6 Bipartite contact network

vertices in different sets. As another example, a metabolic network is a directed bipartite graph. Its nodes are partitioned into the set of metabolites and the set of reactions catalyzed by the metabolism's enzymes.

Definition A **bipartite graph** is a graph with vertices partitioned into two sets, V_1 and V_2, where arcs are only between vertices in different sets.

Quick Review Question 6

Is Figure 13.5.1 a bipartite graph? If not, explain why not. If so, give two sets of vertices, V_1 and V_2, where arcs are only between vertices in different sets.

Matrix Representation of Graphs

For a graph to be manipulated in a computer, its structure must be stored in some convenient manner. Often, we use one- and two-dimensional arrays, or vectors and matrices, respectively, to represent graphs. Such arrays, depending on the representation, can specify such information as adjacent nodes, data stored at nodes or along edges, and existence of paths between nodes.

We will be using **adjacency matrices** and **connection matrices**, which can store graphs where we are not interested in the values at the nodes but only in the graphs themselves. An associated vector can store nodal values. Take, for example, the graph Figure 13.5.7.

In the **adjacency matrix**, A, with indexing beginning at 1, the element in row i and column j, a_{ij}, indicates the number of edges between node i and node j for i and

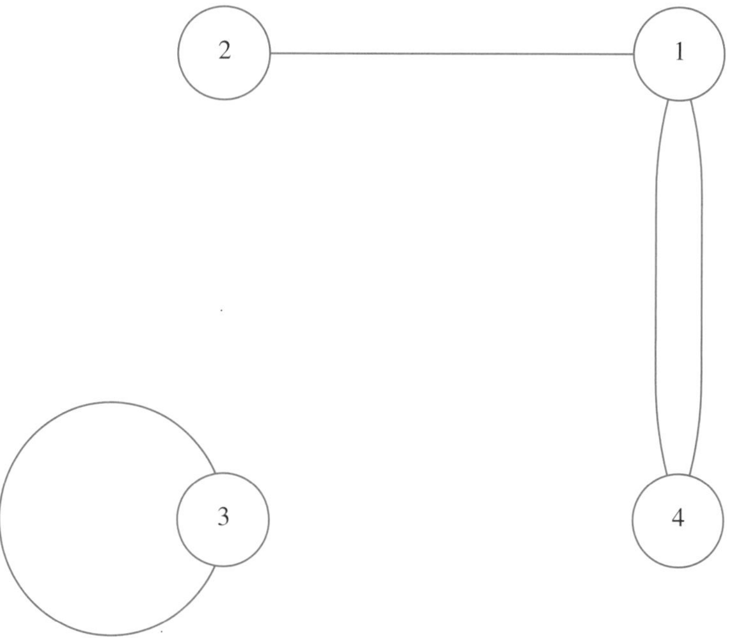

Figure 13.5.7 Example graph

$j = 1, 2, 3, 4$. For instance, because two edges connect points 1 and 4 in Figure 13.5.7, elements a_{14} and a_{41} are both 2. As the following matrix illustrates, the adjacency matrix for an undirected graph is symmetric about the diagonal:

$$\begin{array}{c|c} & \begin{array}{cccc} 1 & 2 & 3 & 4 \end{array} \\ \hline \begin{array}{c} 1 \\ 2 \\ 3 \\ 4 \end{array} & \begin{bmatrix} \underline{0} & 1 & 0 & 2 \\ 1 & \underline{0} & 0 & 0 \\ 0 & 0 & \underline{1} & 0 \\ 2 & 0 & 0 & \underline{0} \end{bmatrix} \end{array}$$

A **connection matrix**, C, indicates only existence of an edge from one point to another, not the number of such edges. As the following connection matrix for Figure 13.5.7 illustrates, if at least one edge exists between node i and node j, then the element in row i and column j, c_{ij}, is 1; while the value is 0 otherwise:

$$\begin{array}{c|c} & \begin{array}{cccc} 1 & 2 & 3 & 4 \end{array} \\ \hline \begin{array}{c} 1 \\ 2 \\ 3 \\ 4 \end{array} & \begin{bmatrix} 0 & 1 & 0 & 1 \\ 1 & 0 & 0 & 0 \\ 0 & 0 & 1 & 0 \\ 1 & 0 & 0 & 0 \end{bmatrix} \end{array}$$

Definitions An **adjacency matrix** for a graph with n nodes is an $n \times n$ matrix, where the element in row i and column j indicates the number of edges between node i and node j. A **connection matrix** for this graph is an $n \times n$ matrix, where the element in row i and column j is 1 if an edge exists between node i and node j and is 0 otherwise.

People-Location Graphs

Suppose we have a file of activities of people in an area, where each record includes, among other data, an identification number for a person (*personID*) and an identification number for a location (*locationID*) that person visited during a day. For example, suppose the file contains the following person-location pairs:

(7, 2938), (7, 27618), (7, 2938)
(8, 2938), (8, 6270), (8, 21032), (8, 2938), (8, 15370), (8, 2938)
(9, 10628), (9, 29740), (9, 10628)
(18, 2938), (18, 5212), (18, 2938), (18, 19815), (18, 2938)

Thus, the person with ID 7 started out at location with ID 2938, presumably home; traveled to location 27618, perhaps work; and then returned home. The person with ID 8, who lived in the same home (2938), probably did a couple of errands in the morning (to locations 6270 and 21032), returned home for lunch, and made another trip (to 15370) in the afternoon before returning home for dinner. The activities file is likely to contain much more information, such as the times at each location, but we simplify the modeling by considering only people and locations. The corresponding bipartite graph indicating connections is in Figure 13.5.8. In this section, we develop a way of storing this information.

After reading the file information into an array of records, *activities*, we can employ the functions ***genPersonIDLst*** and ***genLocIDLst*** to return the list of *personIDs* and *locationIDs*, respectively. For the preceding activities, *genPersonIDLst*(*activities*) returns the list {7, 8, 9, 18}, while *genLocIDLst*(*activities*) returns the list of locations visited, {2938, 27618, 6270, 21032, 15370, 10628, 29740, 5212, 19815}.

genPersonIDLst(activities)

Function to return the list of IDs for people in array, *activities*, of activity records

genLocIDLst(activities)

Function to return the list of IDs for locations in list, *activities*, of activity records

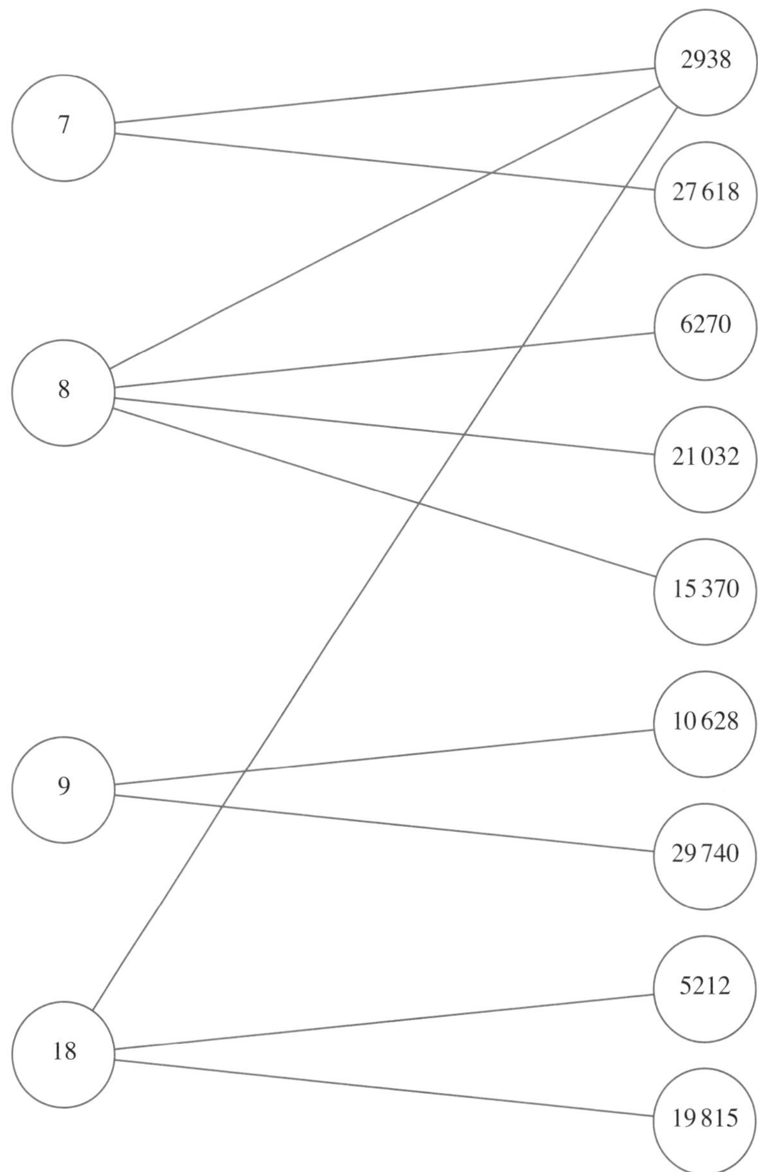

Figure 13.5.8 Bipartite graph of people and locations

A person's or a location's ID is different than its index in the list of people or locations, respectively. For example, assuming indices begin with 1 in *locationIDLst* given before, the index of location ID 2938 is 1, while location ID 21032 has index 4. To obtain an index, we define a function, ***index***, with parameters of an element and a list to return the index of the element in the list. Thus, for the preceding example, *index*(21032, *locationIDLst*) returns 4; and for *personIDLst* = {7, 8, 9, 18}, *index*(8, *personIDLst*) returns 2.

***index*(*el*, *lst*)**

Function to return the index of element *el* in list *lst*

We define function ***genPeopleLocConnMat*** to generate a type of connection matrix for the bipartite graph of people to locations they visit. Each person corresponds to a row and each location to a column. For example, the person with ID *personIDLst*(**4**) = 18 visited locations *locationIDLst*(**1**) = 2938, *locationIDLst*(**8**) = 5212, and *locationIDLst*(**9**) = 19815. Thus, row 4 of the bipartite graph's adjacency matrix has 1s in columns 1, 8, and 9 and 0s elsewhere. The entire matrix to represent the connections of this **people-location graph** appears in Figure 13.5.9.

Person ID \ Location ID	2938	27618	6270	21032	15370	10628	29740	5212	19815
7	1	1	0	0	0	0	0	0	0
8	1	0	1	1	1	0	0	0	0
9	0	0	0	0	0	1	1	0	0
18	1	0	0	0	0	0	0	1	1

Figure 13.5.9 Connection matrix for people-location graph in Figure 13.5.8, where each person corresponds to a row and each location to a column

***genPeopleLocConnMat*(*personIDLst*, *locationIDLst*, *activities*)**

Function to return connection matrix for a people-location graph

Pre:

personIDLst is a list of IDs of people.
locationIDLst is a list of IDs of locations.
activities is a list of activities of people with each record including a person's ID and the ID of a location he or she visited.

Post:

The function has returned a connection matrix with each row corresponding to a person and each column corresponding to a location.

Algorithm:

numPersons ← length of *personIDLst*
numLocations ← length of *locationIDLst*

```
numActivities ← length of activities
connMat ← numPersons-by-numLocations matrix of zeros
for i going from 1 through numActivities do the following:
    personID ← ID of person in activities(i)
    locationID ← ID of location in activities(i)
    row ← index(personID, personIDLst)
    column ← index(locationID, locationIDLst)
    connMat(row, column) ← 1
return connMat
```

Minimal Dominating Set

With a looming epidemic, public health might want to quickly place fever detectors or vaccination sites in a relatively small set of locations to which a high percentage of people travel. With realistic contact patterns from a community, determination of such a minimum dominating set is in the realm of graph theory and computer science.

To compute a **minimum dominating set**, or a smallest set of locations that a given proportion of the population visits, we can use the ***FastGreedy* Algorithm** to obtain an approximation. First, we arrange the locations in nondecreasing order of degree in the people-location matrix. For example, in Figure 13.5.8, because location 2938 has the largest degree, 3, that place is first in any arrangement of degrees in nondecreasing order. Because all other locations have degree 1, the remainder of the list can be in any order. With the *FastGreedy* Algorithm, we keep selecting locations from largest degree down until a given population percentage has visited the set of selected locations (Eubank et al. 2004). Thus, returning to Figure 13.5.8, if we want to "dominate" 75% or less of the people, which in this case is less than or equal to three people, we need to select only the first location, 2938, from our list. *FastGreedy* is called a **greedy algorithm** because at each iteration, the most advantageous near-term choice is picked.

Definitions A **minimum dominating set** for a people-locations graph is a smallest set of locations that a given proportion of the population visits.

For the design of the *FastGreedy* Algorithm, we break the technique into several smaller functions. First, we define a function, ***degLocation***, to calculate the degree of a location index in the people-location graph. For example, in the graph of Figure 13.5.8 with connection matrix, *connMat*, in Figure 13.5.9, location 2938 with index 1 has degree 3 because 3 people visit that location. Thus, *degLocation*(*connMat*, 1) returns 3.

degLocation(connMat, j)

Function to return the degree of the location with a given index in a people-location graph

Pre: *j* is the index of a location in connection matrix, *connMat*, for a people-location graph.
Post: The function has returned the degree of the location with index *j*.
Algorithm:
Return the number of ones in column *j* of *connMat*.

We must sort the location degrees while continuing to associate the location index with its degree. Thus, we form a list of ordered pairs of location indices and their degrees. A function, ***sortSecond***, sorts this list by its second coordinates. Depending on the sorting algorithm, the function might return the following list for the graph in Figure 13.5.8:

$$\{(1, \mathbf{3}), (9, \mathbf{1}), (8, \mathbf{1}), (7, \mathbf{1}), (6, \mathbf{1}), (5, \mathbf{1}), (4, \mathbf{1}), (3, \mathbf{1}), (2, \mathbf{1})\}$$

The order of the first coordinates (location indices) is unimportant, but the order of the second coordinates (location degrees) must be nonincreasing.

sortSecond(pairLst)

Function to return *pairLst*, a list of ordered pairs, sorted by the second coordinates

To determine coverage, we also have a function, ***adjacentPeopleLst***, to return a list of indices for people adjacent to a particular location with a given index. Thus, the function indicates the people who travel to a particular location. For example, with *connMat* being the connection matrix in Figure 13.5.9 for the graph in Figure 13.5.8, *adjacentPeopleLst*(*connMat*, 1) returns the list {1, 2, 4} because people with IDs 7, 8, and 18 (with indices 1, 2, 4, respectively, in *personIDLst*) visit location 2938 (with index 1 in *locationIDLst*).

adjacentPeopleLst(connMat, j)

Function to return a list of the indices of people who visit the location with a given index in a people-location graph

Pre:
j is the index of a location in connection matrix, *connMat*, for a people-location graph.
Post:
The function has returned a list of indices of people who visit the location with index *j* in a people-location graph with connection matrix *connMat*.
Algorithm:
Return a list of the row indices where ones occur in column *j* of *connMat*.

Using the *FastGreedy* Algorithm, we now define a function, ***minDominating***, to send back a list of locations that "dominates" a given percentage of people; that is, the function returns a list of locations to which that percentage of people visits.

minDominating(locationIDLst, connMat, percentPeople)

Function to return a dominating set for a proportion of the people in a people-location graph using the *FastGreedy* Algorithm

Pre:

locationIDLst is a list of locations in a people-location social network.
connMat is a connection matrix associate with the graph.
percentPeople is a proportion of the people.

Post:

A dominating set has been returned.

Algorithm:

```
if percentPeople is not between 0 and 1, percentPeople ← 1
people ← empty list
locations ← empty list
locDegPairLst ← list of location indices (j = 1, 2, 3, . . ., number of col-
  umns in connMat) and associated degrees obtained by calling
  degLocation(connMat, j) for each j
sortedLocDegPairLst ← sortSecond(locDegPairLst) // list of pairs
locDegPair ← 1
percentLength ← percentPeople * (number of rows in connMat)
while (length of people) < percentLength
    (locIndex, locDeg) ← sortedLocDegPairLst(locDegPair) // next pair
    loc ← locationIDLst(locIndex)
    locations ← locations with loc appended
    people ← union of people and adjacentPeopleLst(connMat, locIndex)
    locDegPair ← locDegPair + 1
return locations
```

For example, with *locationIDLst* = {2938, 27618, 6270, 21032, 15370, 10628, 29740, 5212, 19815}, *connMat* being as in Figure 13.5.9, and *percentPeople* = 0.75 = 75%, *minDominating*(*locationIDLst*, *connMat*, *percentPeople*) returns {2938}; 3 out of 4 people go to the location with ID 2938. With *percentPeople* = 1.0 = 100%, the function can return {2938, 5212, 19815, 29740}. One hundred percent of the people go to these 4 locations. Examining Figure 13.5.8, however, we note that smaller sets, such as {2938, 29740} and {2938, 10628}, dominate 100% of the people. The *FastGreedy* Algorithm, which is fast and greedy, provides a heuristic for determining a dominating set. The algorithm yields a good approximation but not necessarily the best answer, and algorithms exist that are even faster.

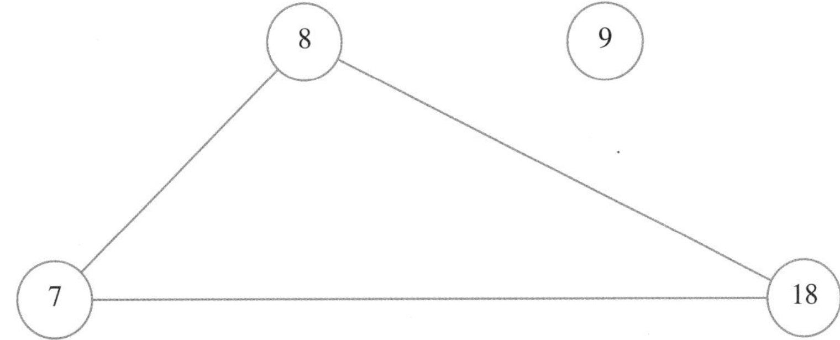

Figure 13.5.10 People-people graph derived from social network in Figure 13.5.8

Location ID

Person ID	7	8	9	18
7	0	1	0	1
8	1	0	0	1
9	0	0	0	0
18	1	1	0	0

Figure 13.5.11 Connection matrix of Figure 13.5.10

Degree Distribution

For the distribution of the number of contacts people have with other people (**degree distribution**), we first generate a connection matrix for a people-people graph associated with a social network, such as the people-location graph of Figure 13.5.8. For simplicity, we ignore timing and assume two people are adjacent if they visited the same location in a day. Thus, Figure 13.5.10 is the people-people graph corresponding to Figure 13.5.8, and the associated connection matrix is given in Figure 13.5.11.

We define a function, ***peopleToPeople***, to return the connection matrix (*connPeopleMat*) for a people-people graph. Going through every column of the people-location connection matrix, *connMat*, we go down each column looking for 1s. For every 1, we look through the rest of the column for 1s. The people corresponding to these indices are adjacent. For example, *connMat* of Figure 13.5.9 has 1 in element *connMat*(1, 1). Because *connMat*(1, 2) is also 1, we define the 1–2 and 2–1 elements of *connPeopleMat*, *connPeopleMat*(1, 2) and *connPeopleMat*(2, 1), to be 1. That is, people with indices 1 and 2 are adjacent. Similarly, in the developing symmetric matrix, we assign 1 to *connPeopleMat*(1, 4) and *connPeopleMat*(4, 1). Then, with 1 in the first column, second row of *connMat* and another 1 in the last row, we define *connPeopleMat*(2, 4) and *connPeopleMat*(4, 2) as 1. Because the matrix is symmetric, we have already indicated that people with indices 2 and 1 are adjacent.

peopleToPeople(connMat)

Function to return a connection matrix for the people-people graph associated with the people-location social network with a given connection matrix

Pre:
connMat is a connection matrix for a people-location graph.
Post:
The function has returned a connection matrix for the corresponding people-people graph.
Algorithm:

```
maxPersonIndex ← number of rows of connMat
connPeopleMat ← maxPersonIndex-by-maxPersonIndex matrix of 0s
for locIndex going from 1 through the number of columns of connMat:
    for i going from 1 through maxPersonIndex:
        if connMat(i, locIndex) is 1
            for j going from i + 1 through maxPersonIndex:
                if connMat(j, locIndex) is 1
                    connPeopleMat(i, j) ← 1
                    connPeopleMat(j, i) ← 1
return connPeopleMat
```

To obtain a degree distribution, we have a function, ***degPersonPPG***, to calculate the degree of each node in the person-person graph. For the connection matrix *connPeopleMat* in Figure 13.5.11 with graph in Figure 13.5.10, *degPersonPPG*(*connPeopleMat*, 1) = *degPersonPPG*(*connPeopleMat*, 2) = *degPersonPPG*(*connPeopleMat*, 4) = 2, while *degPersonPPG*(*connPeopleMat*, 3) = 0. That is, the people with IDs 7, 8, and 18 and indices 1, 2, and 4, respectively, have degree 2, or 2 adjacent nodes. ID 9 (with index 3) is isolated because that person does not go anywhere the other 3 people visit.

degPersonPPG(connPeopleMat, i)

Function to return the degree of the person with a given index in a people-people graph

Pre: *i* is the index of a location in connection matrix, *connPeopleMat*, for a people-people graph.
Post: The function has returned the degree of the person with index *i*.
Algorithm:
Return the number of ones in row *i* of *connPersonMat*.

Finally, we define a function, ***pLst***, that returns the a list of ordered pairs, (k, $P(k)$), of a degree, k, and the corresponding **degree distribution value**, $\boldsymbol{P(k)}$. For $k = 0, 1, 2, \ldots$, maximum(degree), $P(k) = n_k/n$, where n_k is the number of nodes of degree k and n is the total number of nodes in a people-people graph. Thus, for connection matrix of Figure 13.5.11, *pLst*(*connPeopleMat*) returns {(0, ¼), (1, 0), (2,

¾)}. For the $n = 4$ nodes, one-fourth of the nodes have degree 0; no nodes have degree 1; and three-fourths of the nodes have degree 2.

pLst(connPeopleMat)

Function to return a list of ordered pairs, $(k, P(k))$, for $k = 0, 1, 2, \ldots,$ maximum(degree), where $P(k)$ is the degree distribution value

Pre: *connPeopleMat* is the connection matrix for a people-people graph.
Post: The function has returned a list of ordered pairs $(k, P(k))$, $k = 0, 1, 2, \ldots,$ maximum(degree).
Algorithm:
numPeople ← number of rows of *connPeopleMat*
degreeLst ← list of *degPersonPPG*(*connPeopleMat*, *i*), $i = 1, 2, 3, \ldots,$ *numPeople*
return list of ordered pairs with first coordinate being *deg* and second coordinate being (number of occurrences of *deg* in *degreeLst*)/*numPeople*, *deg* = 0, 1, 2, . . ., (maximum value in *degreeLst*)

The plot of the list *pLst*(*connPeopleMat*), {(0, ¼), (1, 0), (2, ¾)}, for the graph of Figure 13.5.10 is certainly stilted because of the small number of nodes (Figure 13.5.12). Projects investigate some realistic social networks, revealing the more characteristic broad-tailed power law form for probability distributions of biological networks (Figure 13.5.3). Additionally, the mean degree of the graph in Figure 13.5.10 is $1.5 = (2 + 2 + 0 + 2)/4$. However, a typical scale-free network has a few nodes with significantly higher degrees than the graph's mean degree.

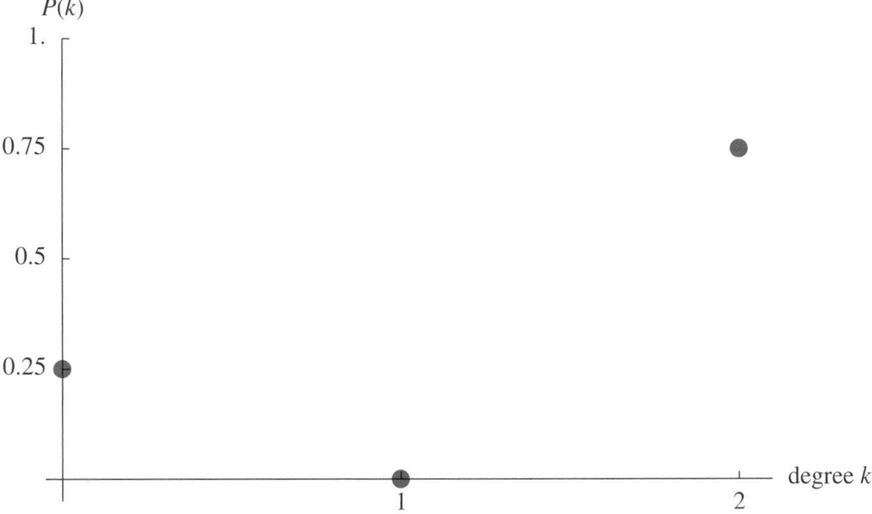

Figure 13.5.12 Plot of degree distribution for the graph of Figure 13.5.10 with connection matrix in Figure 13.5.11

Clustering Coefficient

The clustering coefficient of a person provides an indication of the rapidity with which a disease can spread among his or her associates, and the mean clustering coefficient is a metric on local connectivity of a population. To determine the clustering coefficient of a node with a given index in a connection matrix, we start by defining a function, ***adjacentPeople***, to return a list of indices of its adjacent nodes. Another function, ***numPeopleEdges***, returns the number of edges in a subgraph with a given collection of vertex indices. Thus, for this function, we count the number of ones in the connection matrix with row and column indices in this set. Because each edge is counted twice, the result is divided by 2.

adjacentPeople(connPeopleMat, i)

Function to return a list of indices of the people adjacent to the person with index *i* in the people-people graph with connection matrix *connPeopleMat*

numPeopleEdges(connPeopleMat, vertices)

Function to return the number of edges in a subgraph with a given collection of vertex indices

Pre: *connPeopleMat* is the connection matrix for a people-people graph.
vertices is a list of indices of nodes for a subgraph.
Post: The function has returned the number of edges in this subgraph.
Algorithm:
return half the number of ones in elements of *connPeopleMat* with row and column indices from *vertex*

With these two functions and *degPersonPPG*, we can define a function, ***clusteringCoeff***, to return the clustering coefficient of a vertex, given its index. Using *clusteringCoeff*, we determine that the mean clustering coefficient for our very small people-people graph in Figure 13.5.11 is the high value of 0.75.

clusteringCoeff(connPeopleMat, v)

Function to return the clustering coefficient of a node

Pre: *connPeopleMat* is the connection matrix for a people-people graph.
v is a index of a node in this graph.
Post: The function has returned the clustering coefficient for *v*.
Algorithm:

```
deg ← degPersonPPG(connPeopleMat, v)
if deg < 2
    return 0
else
```

```
adj ← adjacentPeople(connPeopleMat, v)
numerator ← numPeopleEdges(connPeopleMat, adj)
denominator ← deg*(deg - 1)/2.0
    // number of edges in complete graph with deg number of nodes
return numerator/denominator
```

Example Problems

Activities with four people and nine locations served as an example in the "Algorithms" part of this module. We discovered that the function ***minDominating***, which uses the heuristic *FastGreedy* Algorithm, can return a set of four locations that dominates all the people and a set with only one location to which 75% of the people visit. With a mean degree of 1.5 in the people-people graph, a plot of the degree distribution (Figure 13.5.12) has only three points. The mean clustering coefficient (0.75) is also exaggerated with such a small set.

A more realistic example derives from the activities in a synthetic data set (*activities-portland-1-v1.dat*) for the population of Portland, Ore., that the Network Dynamics and Science Simulation Laboratory (NDSSL) at Virginia Technical University generated from real data (NDSSL 2009a). The set contains 8,922,359 activities involving 1,615,860 people. So that we can perform calculations on a sequential machine, we select 1000 people at random, using all their activities.

One such set of people has 5511 activities involving 3458 different locations. Execution of *minDominating* to cover 100% of the people yields a set of 3455 locations, while 594 locations can cover 50% of the population.

A plot of the degree distribution of a people-people graph for the 1000 individuals (Figure 13.5.13) with a fitted function $f(k) = -0.0219242 + 0.259918k^{-1.2}$ reveals a more realistic situation than that of Figure 13.5.12.

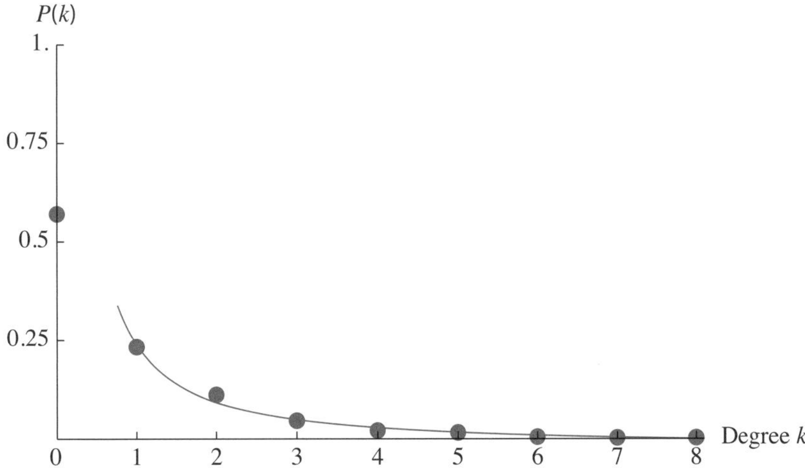

Figure 13.5.13 Degree distribution of a 1000 randomly selected people with fitted function $f(k) = -0.0219242 + 0.259918k^{-1.2}$

The mean clustering coefficient for this group of 1000 individuals is 0.118524 = 11.8524%, slightly more than that of people on the Internet (Eubank et al. 2004).

Assessment of Model

Clearly, the results involving 1000 randomly selected people and their activities from the NDSSL synthetic data set are more realistic than the illustrative example with 4 people. Although the *FastGreedy* Algorithm does not necessarily yield a minimum dominating set even for the larger data set above, Eubank et al. (2004) proved that the technique does provide a fast method for obtaining a good dominating set. Thus, public health officials can use the results to designate which locations should have fever-detection sensors or should close during epidemics.

The degree distribution of the larger data set with fitted function $f(k) = -0.0219242 + 0.259918k^{-1.2}$ (Figure 13.5.13) does approximate the power law, $P(k) = ck^{-r}$, common for scale-free networks. The shape reveals only a few nodes, in this case 8 out of 1000, or 0.8% of the population, have degree 6 or more. Using further demographic information, public health officials might target such "well-networked" people for immediate vaccination. Moreover, with simulations, they can investigate the impact of this and other vaccination policies on combating the spread of influenza.

The mean clustering coefficient of 0.118524 for our group of 1000 individuals is indicative of small-world networks, which exhibit higher cliquishness of an average neighborhood. By contrast, random graphs with about 300 to 5000 nodes have clustering coefficients of approximately 0.05 to 0.005, respectively, exhibiting almost no clustering (Watts and Strogatz 1998).

Computing Power

As Bisset and Marathe (2009) state, "These far more complex network-based models present a new set of computational challenges that require the use of high-performance computing." Several reasons exist for requiring this power:

- Social networks are very large, have unstructured shapes, and change frequently. Thus, computing systems require enormous storage to hold and significant power to manipulate the data.
- Because these models are stochastic, involving chance, scientists must execute a simulation with various intervention scenarios and compliance rates and must replicate each simulation run many times to obtain meaningful results on which to base policy recommendations.
- Incorporation of the diverse characteristics of people and their activities is important in studying disease propagation in space and time. By contrast, the study of physical systems usually does not require such diversity.

Computational scientists at such institutions as Los Alamos National Laboratory and Virginia Tech have developed high-performance individual-based simulation models to study the nature of epidemics and the impacts of policy decisions on controlling epidemics in urban environments (Bisset and Marathe 2009; Eubank et al. 2004). With petascale computing platforms, researchers anticipate developing simulations to model global pandemics (Bisset and Marathe).

Because of the direction of much research in the study of epidemics, it is advantageous for the modeler to be able to use high-performance computing. The text's website has additional high-performance computing materials related to this module.

Projects

For additional projects, see Module 14.15, "Social Networks: Value in Being Well Connected."

For each of the following projects, use one of the NDSSL data sets at http://ndssl.vbi.vt.edu/opendata/ (NDSSL 2009a) or another data set with activity data.

1. Develop the following functions:
   a. *genPeopleLocAdjMat*(*personIDLst*, *locationIDLst*, *activities*) to return an adjacency matrix for a people-location graph
   b. *degPerson* to return the degree of a person index in a people-location graph
   c. *SPD*(*connMat*, *i*) to return the number of pairs of vertices in a people-people graph with connection matrix, *connMat*, where the length of shortest path (or the **distance**) between the vertices is *i*
   d. *isolated* to return a list of isolated nodes in a graph
2. This project involves computation of the metric mean shortest-path length. A biological network typically exhibits the small-world property with a small average path length, making the system an efficient communicator of information or disease.

   The **distance** between two points in a graph is the length, or the number of edges, of the shortest path between those points. **Floyd's algorithm** is a method for finding the distance between each pair of vertices in a network. We start by replacing each off-diagonal zero in the square connection matrix with ∞, a value greater than any number. For example, for the graph in Figure 13.5.14, the initial matrix follows:

$$D = \begin{bmatrix} 0 & \infty & \infty & 1 \\ \infty & 0 & 1 & \infty \\ \infty & 1 & 0 & 1 \\ 1 & \infty & 1 & 0 \end{bmatrix}$$

Figure 13.5.14 Graph for Project 2

For $k = 1, 2, 3, 4$, we consider each element d_{ij}; if $d_{ik} + d_{kj}$ is smaller than d_{ij}, we replace d_{ij} with the sum. For example, when $k = 3$, we replace d_{24} (∞) by $d_{23} + d_{34} = 1 + 1 = 2$; the length of the shortest path from v_2 to v_4 is 2 and goes through v_3. Symmetrically, d_{42} is replaced by $d_{43} + d_{32} = 1 + 1 = 2$. At the end of that iteration, the distance matrix is as follows:

$$D = \begin{bmatrix} 0 & \infty & \infty & 1 \\ \infty & 0 & 1 & 2 \\ \infty & 1 & 0 & 1 \\ 1 & 2 & 1 & 0 \end{bmatrix}$$

When $k = 4$, d_{12} becomes $d_{14} + d_{42} = 1 + 2 = 3$; and $d_{14} + d_{43} = 1 + 1 = 2$ replaces d_{13}. The final distance matrix, which the algorithm returns, is as follows:

$$D = \begin{bmatrix} 0 & 3 & 2 & 1 \\ 3 & 0 & 1 & 2 \\ 2 & 1 & 0 & 1 \\ 1 & 2 & 1 & 0 \end{bmatrix}$$

Floyd's algorithm for finding the length of the shortest path between each pair of vertices in a network of n nodes:

```
D ← connection matrix for the network
for i going from 1 through n, do the following:
    for j going from 1 through n, do the following:
        if i ≠ j and d_ij is 0
            d_ij ← ∞
for k going from 1 through n, do the following:
    for i going from 1 through n, do the following:
        for j going from 1 through n, do the following:
            if d_ik + d_kj < d_ij then
                d_ij ← d_ik + d_kj
```

Use Floyd's algorithm to calculate the distance matrix for a social network, and generate a histogram of these distances. Compute the mean smallest number of contacts for a disease to spread from one arbitrary individual to another. Also calculate the **diameter**, which is the largest distance in the graph and is a measure of the extent of the graph.

Studies involving metabolic networks of between 200 and 500 nodes discovered mean smallest-path lengths between 3 and 5 (Mason and Verwoerd 2007). Genetic networks of about 1000 genes and 4000 interactions were discovered to have a mean-smallest path length of 3.3 (Mason and Verwoerd). Do your data set exhibit the small-world property with a small mean smallest-path length? Does your data set have a fairly large mean clustering coefficient?

3. Define a function $C(k)$ to return the mean clustering coefficient of nodes that have degree k. For metabolic networks of 43 organisms, $C(k)$ is approximately proportional to k^{-1}, so that as the degree increases, the clustering coefficient decreases. Thus, nodes with small degrees tend to be clustered densely, and nodes with large degrees are not (Mason and Verwoerd 2007). Does your data set exhibit a similar attribute?
4. Let $N(v)$ be the set of nodes adjacent to the person or location vertex, v, in a people-location network. For V being a subset of vertices, let $N(V) = \bigcup_{v \in V} N(v)$, the union of all $N(v)$ for v in V; that is, $N(V)$ is the collection of all nodes adjacent to at least one vertex in V. The **overlap ratio** for a subset of locations, S, is the number of elements in $N(S)$ divided by the sum of the degrees of nodes in S. This value is greater than 0 and less than or equal to 1. A smaller overlap ratio indicates a greater overlap, or a greater probability that two people visit the same location. Eubank et al. (2004) shows that the *FastGreedy* Algorithm yields better results for higher overlap ratios of the locations. Develop a function to compute the overlap ratio for a set of locations. Calculate the overlap ratios of the locations for the sample data set and a random subset of 1000 people and their activities from the *activities-portland-1-v1.dat* at NDSSL (2009a). Based on your results and the work of Eubank et al. (2004), on which data set should the *FastGreedy* algorithm give the best results?
5. The *FastGreedy* algorithm gives a good approximation for the dominating set problem. Develop an algorithm to return a minimum dominating set. Compare the speeds of the revised and the original algorithms on small graphs of increasing sizes (numbers of nodes). Plot these speeds verses size, and fit functions to the data.
6. When data sets are very large, we may wish to approximate metrics using random subsets of some proportion, p $(0 < p \leq 1)$, of the total size. Implement each of the techniques below for returning a random sublist of distinct integers from 1 through n, where the size of the sublist is approximately $\lceil pn \rceil$, the smallest integer greater than or equal to pn. For example, if n = 1,615,860, which is the size of *activities-portland-1-v1.dat* from NDSSL (2009A), and $p = 0.001$, then the size of the sublist is $\lceil pn \rceil$ = (0.001)(1,615,860) = 1615.86 = 1616. Note that *getSublist1*(n, p) returns exactly $\lceil pn \rceil$ elements, whereas *getSublist2*(n, p) returns approximately $\lceil pn \rceil$ elements.

Technique 1—*getSublist1*(n, p):

size ← *pn*
personIdLst ← empty list
while (number of elements in *personIdLst*) is less than *size* do the following:
    randInt ← random integer in {1, 2, 3, . . ., *n*}
    if *randInt* is not in *personIdLst*
        append *randInt* to *personIdLst*
return *personIdLst*

Technique 2—*getSublist2*(*n*, *p*):

personIdLst ← empty list
for *i* going from 1 through *n*, do the following:
  if (random floating point number between 0 and 1) is less than *p*
    append *i* to *personIdLst*
return *personIdLst*

b. Compute the amount of time to execute each function for $n = 1615860$ and $p = 0.001$.

c. For *i* going from 1 to 10, compute the time it takes to execute *getSublist1*(100000, 0.001**i*), and plot the time versus the number of elements in the sublist, 100000 * 0.001 * *i*. Fit a function to this graph. Repeat this task for *getSublist2*. Discuss the results, including the pros and cons of each technique. Make recommendations about when to use each method.

7. We can estimate the mean clustering coefficient by evaluating a metric for a random subset of $\log(|P|)$ number of people, where $|P|$ is the number of people in the data set (Eubank et al. 2004). For *i* varying from 1 to 50, select a random subset of size $\log(|P|) \cdot i$, compute the mean clustering coefficient, and use commands to determine the time the computer took to perform these two tasks. Plot the time versus subset size. Observe the mean clustering coefficients for your subsets. Discuss the results.

8. Develop a simulation of the spread of influenza using a people-people graph. For simplicity, assume people fall into one of three categories, susceptibles (*S*), infecteds (*I*), or recovereds (*R*); recovereds are immune from the disease. Besides a connection matrix, have a corresponding list of node values recording each person's state as an integer, such as 0 for susceptible, 1 for infected, and 2 for recovered. Start the simulation at time $t = 0$ with one infected person and all others being susceptible. At each time step, a healthy person can catch the flu from an infected adjacent node with a probability of *transmissionRate*. The logic for this segment is as follows:

   if (a random floating point number between 0 and 1) < *transmissionRate*
     that healthy person becomes sick

   A person is sick for only one time unit. Continue the simulation until the disease no longer exists in the population. Record the total number of people who caught the flu and the number of time steps for the simulation run. For a given value of *transmissionRate*, run the simulation at least 10 times and average the totals (Watts and Strogatz 1998).

9. Develop the simulation discussed in Project 8. Define a function to return the percent of the population who became sick. Have the program run nine experiments with *transmissionRate* = 10%, 20%, 30%, . . ., 100%, and conduct each experiment 10 times. Also, have the code determine the average percent of people who became ill for each probability. Plot the data. Discuss the results.

10. Repeat the development in Project 8. At each time step, count the number of people in each category, *S*, *I*, and *R*. Plot the number of people in *S*, *I*, and *R* versus time for one run.

11. Refine the model of Project 8 to account for deaths in some percentage of the infecteds and perform the task of Project 8, 9, or 10.

12. Repeat any of Projects 8–11 so that a person is sick for two time steps, making a longer infectious period. Thus, have two states for infected.

13. Develop Project 9 and the corresponding version where a person is sick for two time steps, making a longer infectious period. In the latter case, have two states for infected. Compare the results. Compute the clustering coefficient for your data set. Mason and Verwoerd (2007) states, "for networks with strong local connectivity the fittest strains are those that have high transmission rates and relatively short infectious periods." Do your results support this statement?

14. Develop one of Projects 8, 10, or 11 designating *percentVaccinated* percent of the people with the highest degree as being vaccinated against the flu. Have the program run five experiments with *percentVaccinated* = 0%, 10%, 20%, 30%, 40%, and conduct each experiment 10 times. Also, have the code determine the average percent of the population who became ill for each probability. Plot the data. Discuss the results.

15. This project tests the observation that the mean clustering coefficient (MCF) for networks exhibiting the small-world property, such as most social networks, is significantly higher than for random graphs (Mason and Verwoerd 2007). Using a realistic social network data set, calculate the MCF and the percentage (p) of 1s, indicating edges, in its people-people connection matrix. Develop a function with parameters for a size, n, and a probability, p, to return a random $n \times n$ connection matrix, where p is the probability of 1 in a position. Calculate the *MCF* of a generated random graph for the number of people and the percentage of ones in the realistic social network. Run this simulation a number of times, say 100 to 1000 times, to obtain an average value for MCF, and compare the results with that of the realistic social network.

16. Using the function from Project 15 and a fixed size, n, say 20, calculate the mean clustering coefficient (MCF) for networks with probabilities $p = 0.0$, 0.1, 0.2, . . ., 1.0. For each probability, run the simulation a hundred times to obtain an average for MCF. Plot MCF versus p. Discuss the results.

17. Repeat Project 15 for mean path length instead of MCF (see Project 2).

18. Repeat Project 15 for degree distribution instead of MCF. For the random graphs, average the number of nodes of degree k to obtain n_k.

19. Using distributions of movement patterns, Schwarzkopf and Alford (2002, 2005) developed individual-based, correlated random walk models to ascertain if adult cane toads move nomadically. At each time step, a simulated toad does the following:

- Decide if it is going to move or not.
- If it decides to move, decide if it returns to its most recent previous location.
- If it decides to move but not return to its most recent previous location, decide the angle to turn and the distance to move.

With this algorithm, develop a simulation following the movements of 100 toads and using a simulation time of 30 days with a time step of 1/2 day.

Consider each of three seasons (wet, late wet, dry) using movement parameters from those in Table 13.5.1, which are similar to values obtained from observing a group of tracked cane toads in the Heathlands region of Australia. Assume the wet season lasts 4 months, late wet lasts 2, and dry lasts 6 months. Ignore death, food, and moisture. For each season, using multiple simulations, compute the following path characteristics: mean total distance traveled, mean total displacement, and mean **straightness**, which is the total displacement divided by total distance traveled. Table 13.5.2 gives approximate mean straightness and mean total displacement per day for a group of traced cane toads in the region. Discuss your results, including when toads tend to move straighter and further. Hypothesize reasons for these results.

Table 13.5.1
In three seasons (wet, late wet, dry), probabilities that a toad moves; if moves, that it returns to most recent previous location; if moves but does not return, that it goes various distances at various angles; Values approximated for Heathlands from Schwarzkopf and Alford (2002)

Probability	*Wet season*	*Late wet season*	*Dry season*
moves	0.80	0.79	0.69
returns	0.03	0.10	0.10
distance 0 m	0.11	0.14	0.18
distance 5 m	0.25	0.34	0.16
distance 15 m	0.24	0.27	0.17
distance 35 m	0.16	0.15	0.18
distance 75 m	0.14	0.08	0.17
distance 155 m	0.05	0.02	0.12
distance 315 m	0.02	0.00	0.01
distance 635 m	0.02	0.00	0.01
distance 1275 m	0.01	0.00	0.00
angle –170°	0.12	0.22	0.28
angle –150°	0.03	0.08	0.05
angle –130°	0.03	0.07	0.03
angle –110°	0.03	0.02	0.03
angle –90°	0.03	0.03	0.03
angle –70°	0.03	0.03	0.03
angle –50°	0.02	0.02	0.03
angle –30°	0.07	0.02	0.03
angle –10°	0.06	0.02	0.04
angle 10°	0.05	0.04	0.02
angle 30°	0.06	0.02	0.03
angle 50°	0.06	0.03	0.02
angle 70°	0.06	0.02	0.04
angle 90°	0.03	0.04	0.02
angle 110°	0.05	0.04	0.04
angle 130°	0.07	0.04	0.03
angle 150°	0.05	0.08	0.05
angle 170°	0.15	0.18	0.20

Table 13.5.2
In three seasons (wet, late wet, dry), mean straightness and mean displacement (m) per day; values approximated for Heathlands from Schwarzkopf and Alford (2002)

Mean	*Wet season*	*Late wet season*	*Dry season*
Straightness	0.33	0.20	0.09
Displacement (m)	36	22	15

Answers to Quick Review Questions

1. a. $V = \{1, 2, 3, 4, 5, 6, 7, 8, 9\}$
b. (3, 6) = (6, 3)
c. 9
2. a. 1, 2, 4, 5, 6
b. (1, 2) = (2, 1) and (1, 3) = (3, 1)
c. 5
d. none
3. 2, 1, 3, 6; 2, 3, 5, 6; 2, 4, 5, 6; 2, 5, 3, 6
4. a. {(7, 8), (8, 9)}, {(7, 8), (7, 9)}, {(7, 9), (8, 9)}, {(7, 8), (7, 9), (8, 9)}
b. $V = \{7, 8, 9\}$, $E = \{(7, 8), (7, 9), (8, 9)\}$
c. (4, 5) = (5, 4)
d. 6 = (4)(3)/2
5. a. 0.4 = 4/((5)(4)/2)
b. 1.0 = 1/1
c. 0 = 0/1
d. 0
6. Yes, Figure 13.5.1 is a bipartite graph. For example, one possible partition is V_1 = {*School*, *Hospital*, *Work*, *Shop*, *Cloister*} and V_2 = {1, 2, 3, 4, 5, 6, 7}. However, the node *Cloister* can belong to either set.

References

Bisset, Keith, and Madhav Marathe. 2009. "A Cyber Environment to Support Pandemic Planning and Response." *SciDAC Review* 13: 36–47. http://www.scidacreview.org/0903/html/maranthe.html (accessed December 30, 2012)

CDC (Centers for Disease Control and Prevention). "Prevention and Control of Seasonal Influenza with Vaccines, Recommendations of the Advisory Committee on Immunization Practices (ACIP), 2009." July 24, 2009/58(Early Release);1–52. http://www.cdc.gov/mmwr/preview/mmwrhtml/rr58e0724a1.htm?s_cid=rr58e0724a1_e (accessed December 30, 2012)

Eggemann, Nicole, and Steven Noble. 2008. "The Clustering Coefficient of a Scale-Free Random Graph." www.newton.ac.uk/programmes/CSM/seminars/050114001.pdf (accessed December 30, 2012)

Eubank, S., V. S. Anil Kumar, M. Marathe, A. Srinivasan, and N. Wang. 2004. "Structural and Algorithmic Aspects of Large Social Networks." *Proc. 15th ACM-SIAM Symposium on Discrete Algorithms* (SODA): 711–720.

Mason, Oliver and Mark Verwoerd. 2007. "Graph Theory and Networks in Biology." *IET Systems Biology*, 1: 89–119. http://www.hamilton.ie/SystemsBiology/files/2006/graph_theory_and_networks_in_biology.pdf (accessed December 30, 2012)

NDSSL (Network Dynamics and Simulation Science Laboratory, Virginia Polytechnic Institute and State University). 2009a. "NDSSL Proto-Entities." http://ndssl.vbi.vt.edu/opendata/ (accessed December 30, 2012)

———. 2009b. Synthetic Data Products for Societal Infrastructures and Proto-Populations: Data Set 1.0. ndssl.vbi.vt.edu/Publications/ndssl-tr-06-006.pdf (accessed December 30, 2012)

———. 2009c. Synthetic Data Products for Societal Infrastructures and Proto-Populations: Data Set 2.0. ndssl.vbi.vt.edu/Publications/ndssl-tr-07-003.pdf (accessed December 30, 2012)

———. 2009d. Synthetic Data Products for Societal Infrastructures and Proto-Populations: Data Set 3.0. ndssl.vbi.vt.edu/Publications/ndssl-tr-07-010.pdf (accessed December 30, 2012)

Newman, M.E.J., D. J. Watts, and S. H. Strogatz. 2002. "Random Graph Models of Social Networks." *Proceedings of the National Academy of Science* 99 (Suppl 1): 2566–2572. http://www.pnas.org/content/99/suppl.1/2566.full (accessed December 30, 2012)

NFID (National Foundation for Infectious Diseases). 2008a. "Influenza Myths." http://www.nfid.org/INFLUENZA/consumers_myths.html (accessed December 30, 2012)

———. 2008b. "Mid-Season Flu Immunization Rates are Too Low." December 10. http://www.prweb.com/releases/2008/12/prweb1734534.htm (accessed December 30, 2012)

Schwarzkopf, Lin, and Ross A. Alford. 2002. "Nomadic Movement in Tropical Toads." *Oikos* 96(3): 492–506.

———. 2005. "Movement and Dispersal in Established and Invading Toad Paopulations." Cane Toad Forum, Kununurra. http://www.canetoads.com.au/forumprocedespp4.htm (accessed July 29, 2013)

Shiflet, Angela. 1984. *Discrete Mathematics for Computer Science*, St. Paul. MN: West.

VBI (Virginia Bioinformatics Institute at Virginia Tech). 2008. "EpiSims." http://ndssl.vbi.vt.edu/episims.php (accessed December 30, 2012)

Watts, D. J., and Steven Strogatz. 1998. "Collective dynamics of 'small-world' networks." *Nature* 393: 440–442.

14

ADDITIONAL CELLULAR AUTOMATA, AGENT-BASED AND MATRIX PROJECTS

Overview

In Chapters 8–13, we studied techniques, issues, and applications of computational science empirical models, random walk and cellular automaton simulations, agent-based simulations, high-performance computing, and matrix modeling. Projects usually built on or were closely related to the examples discussed and developed in the modules.

As with Chapter 7, Chapter 14 provides opportunities for students to enhance their computational science problem-solving abilities through completion of additional extensive projects. Although not containing examples, each module in the chapter does have sufficient scientific background for students to complete projects in the application area. At the beginning of a module, the prerequisite material is listed. Correspondingly, project sections in previous chapters suggest appropriate Chapter 14 modules for additional projects. Thus, students can work with projects in the current chapter at any time after covering the prerequisites.

As with earlier projects, those of Chapter 14 are appropriate for teamwork. Interdisciplinary teams perform most of the research and development in computational science. Thus, teamwork experiences developing models and simulations in a variety of application areas are important for students studying computational science.

Chapter 14's applications with projects are in a variety of scientific areas, including the following: polymers, solidification, foraging behavior, pit vipers and heat diffusion, growth of mushroom fairy rings, spread of disease, HIV and the immune system, predator-prey interactions, clouds, fish schooling, invasive plants, bioinformatics, social science, and so on.

MODULE 14.1

Polymers—Strings of Pearls

Prerequisites: Module 9.5, "Random Walk," and Module 8.3, "Empirical Models," for parts that involve fitting of functions.

Introduction

The Mesoamerican civilization of the Maya extended for about thirty-five hundred years, between 2000 BC and the AD 1500s. When the Spanish found them during the sixteenth century, the Mayans were playing a very interesting ballgame with political and religious significance that transcended the game as a sport. These games were played on large courts; the oldest known is one located in Chiapas, Mexico, which dates from about 1400 BC. The balls they used were large, between 12 and 18 in. in diameter, and were made of solid rubber, likely weighing 8 to 40 lb. Most Westerners think that rubber originated with the Charles Goodyear's vulcanization process, which cured natural rubber into a commercially useful product. As it turns out, people of Mesoamerica were processing rubber by 1600 BC, thirty-five hundred years before the first rubber patent in England. The Mayans used rubber not only for making balls, but also for paint, medicines, waterproofing, and so on. Interestingly, they took the latex from a local tree, *Castilla elastica*, and mixed it with the liquid from morning glory, *Ipomoea alba*. With stirring and the heat of the region, they produced rubber in a process very similar to Goodyear's (Armstrong; Hosler et al. 1999, Wikipedia 2012).

Rubber is a good example of what scientists term a "natural polymer." But, what is a polymer? The word is from the Greek *polumeres*, which means "having many parts" (poly, "many"; merous, "parts"; Marko). Thus, we use the term **polymer** to describe a class of chemical compounds composed of repeating chemical building blocks. These building blocks are identical or closely related chemical structures, each referred to as a **monomer**. Rubber is a polymer made up of repeating subunits (monomers) of *isoprene* (2-methyl-1,3-butadiene), as in Figures 14.1.1 and 14.1.2 (Michalovic 2000). In Figure 14.1.1, n is a chemist's shorthand to indicate that similar types of bonds link n of these subunits. So, rubber is *polyisoprene*. The monomer

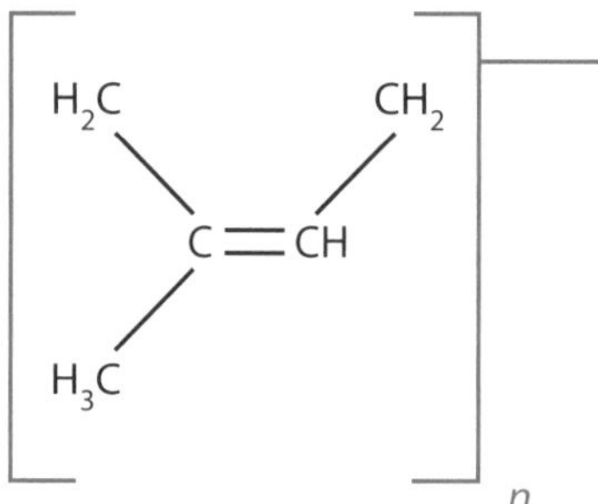

Figure 14.1.1 Rubber polymer with similar types of bonds linking *n isoprenes*

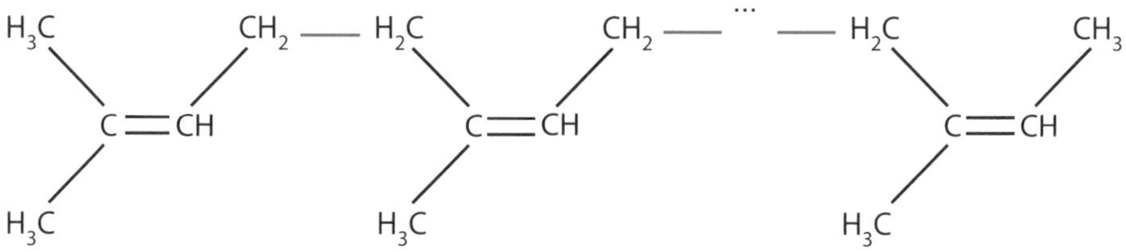

Figure 14.1.2 Rubber polymer

isoprene is also used to generate a number of biologically significant molecules, such as lanosterol (precursor to various animal sterols), vitamins A and E, and carotene pigments (Case Western Reserve University; Nelson and Cox 2012).

Definitions A **polymer** is a chemical compound composed of repeating building blocks, which are identical or closely related chemical structures, called **monomers**.

Rubber is not the only natural polymer. Natural polymers abound in nature and include cellulose, starch, chitin, nucleic acids (DNA, RNA), and proteins. Chains of monomers make up each of these polymers. Proteins, for instance, are strings of amino acids. Because each monomer in the protein chain can be any one of the 20 different naturally occurring amino acids, proteins are quite diverse in composition (Nelson and Cox 2012).

Scientists have learned to duplicate the process of polymer synthesis, **polymerization**, so that quite a variety of synthetic polymers, such as polyethylene and polystyrene, exist. We can now even synthesize rubber. These polymers are found in almost all manufactured products and their packaging (Case Western Reserve University).

Some polymers, like rubber, are **elastomers**. Elastomers are made up of molecules that are loosely cross-linked (1 in 100 molecules linked to other molecules). By contrast, **plastics** have 1 in 30 cross-linked molecules; and we can shape or mold these stiffer, nonelastic materials (Case Western Reserve University).

Some of the most interesting developments in polymer chemistry are medically applicable polymers. **Controlled drug delivery**, for instance, combines polymers with medicines to release medicines into the body in predetermined manners. We can manage release to be constant or cyclic over a prolonged time period, or we can cause the medicine to disperse in response to an environmental cue. These combinations improve effectiveness of chemical therapies, reduce the number of administrations, and ensure patient compliance. As an example, Nifedipine, classified as a calcium-channel blocker, may be prescribed for treatment of angina and hypertension and can be administered in an **extended-release tablet (ERT)**. This tablet looks like a normal tablet but consists of two layers, an inner, "active" drug layer surrounded by an external, inert polymer. This outer layer of polymer is osmotically sensitive. In the digestive tract, osmotic forces cause this outer layer to swell, pressing against the inner layer and forcing small amounts of the medicine out of a previously drilled hole in the tablet. Small, but constantly released, doses are absorbed into the circulatory system. This mechanism provides a controlled release of the medication over a 24-h period. Patients have to take the medicine only once a day, and levels of the drug in the blood remain high enough to be effective. Other drugs, like some for birth control, can be implanted for up to 5 years of delivery (Brannon-Peppas 1997; Tenanbaum 2003; Pfizer, Inc. 2011; Rxlist).

Scientists who study polymers and those who study biology and medicine are teaming up to develop a wide variety of medical applications. For instance, based on textile industry techniques, polymer scaffolds have been developed that have proven most useful for cell and tissue growth. Such scaffolds have been employed to generate artificial skin to treat burn patients. The "skin" helps to prevent infection and produces chemical factors that promote faster healing. In the future, scientists may use such scaffolds to grow nerve cells for spinal cord injuries or to grow insulin-secreting cells for diabetes patients (Case Western Reserve University).

Simulations

Scientists are very interested in predicting a protein's native structure from its amino acid sequence because the folding of a protein polymer with its resulting geometric shape determines its function. An understanding of the mechanism would be a major breakthrough in studying the basic science of the cell. Computational scientists are developing computer simulations of polymers to gain insight in solving the protein-folding problem and other polymer questions (Bastolla et al. 1997).

Random walks are commonly used to generate 2- and 3D models of polymers. However, two different monomers of a polymer cannot occupy the same space at the same time. Thus, such a simulation generally involves a **self-avoiding walk (SAW)**, which is a random walk that does not cross itself, that is, a walk that does not travel through the same cell twice. For example, Figure 9.5.1 of Module 9.5, "Random Walk," displays a walk that is not self-avoiding because at least once (and in this walk, more than once) the walker cycled back to an earlier position. In an algorithm to generate a self-avoiding walk, at each time step eliminating the direction from which the walker comes, the walker selects at random one of the remaining directions. The projects develop several such simulations (Gould and Tobochnik 1988).

Definition A **self-avoiding walk (SAW)** is a random walk that does not cross itself, that is, a walk that does not travel through the same cell more than once.

Projects

1. **a.** Develop a simulation and visualization to generate a 2D model of a polymer with a self-avoiding walk. Terminate the simulation when the random direction would cause the developing path to cross itself. The visualization should show the entire model of the polymer from the beginning of the random walk until the end.
   b. Run the simulation a number of times, recording the fraction, $f(n)$, of times the simulation generates a polymer of length (number of steps) at least n. Using empirical modeling, derive an equation for f as a function of n (Gould and Tobochnik 1988).
   c. The end-to-end distance is an important geometric property of a polymer that influences the texture and other physical properties of the polymer. Run the simulation a number of times, evaluating the **root-mean-square displacement R_n** as follows: For each walk of n steps, evaluate the square of the displacement (end-to-end distance) of the nth step (particle) from the initial position, the origin; then, over all such n-step walks, compute the average of the squares of the displacements; take the square root of this average to calculate R_n. If the nth step for trial i is to location $(x_{n,i}, y_{n,i})$, the square of the displacement from the origin is $(s_{n,i})^2 = (x_{n,i})^2 + (y_{n,i})^2$. The formula for the root-mean-square displacement is

$$R_n = \sqrt{\frac{\sum_{i=1}^{m} \left(s_{n,i}\right)^2}{m}}$$

   where $s_{n,i}$ is the displacement of the nth particle from the starting position during trial i of m successful trials. Using empirical modeling, derive an equation for R_n as a function of n (Gould and Tobochnik 1988).
2. Repeat Project 1 in three dimensions, where at each step the walker selects at random one of six directions.
3. Repeat Project 1 where bond angles are all 90°, so that the random walker turns to the right or left with each step (Gould and Tobochnik 1988).
4. Repeat Project 2 where bond angles are all 90°, so that the random walker turns to the right, left, up, or down with each step.
5. The technique in Project 1 is inefficient in calculating R_n (see Part c) for large n because of the number of aborted trials. An alternative method uses a **weight W_n** associated with each walk of n steps to skew the importance of underrepresented large chains. The weights in a SAW are as follows:

- $W_1 = 1$.
- If no step is possible at step n, $W_n = 0$. In this case, the walk terminates, and the program generates a new walk starting at the origin.
- If the walker can go in any of the three directions at step n, $W_n = W_{n-1}$.
- If the walker can go in exactly two directions at step n, $W_n = \frac{2}{3}W_{n-1}$.
- If the walker can go in exactly one direction at step n, $W_n = \frac{1}{3}W_{n-1}$.

We estimate the root-mean-square displacement as follows:

$$R_n = \sqrt{\frac{\sum_{i=1}^{m}\left(W_{n,i}\right)\left(s_{n,i}\right)^2}{\sum_{i=1}^{m} W_{n,i}}}$$

where $W_{n,i}$ is the value of W_n and $s_{n,i}$ is the displacement of the nth particle from the starting position during trial i of m successful trials. Using empirical modeling, derive an equation for R_n as a function of n (Gould and Tobochnik 1988).

6. Repeat Project 5, attempting to produce longer polymer models by aborting any one random walk when the weight becomes smaller than some threshold value, such as 0.15.
7. Using the technique of Project 1a, develop a collection of models for polymers of some length n. Write a program that attempts to construct models of polymers of length $2n$ by attaching the head of one "polymer" to the tail of another. The program should reject the walk if it crosses itself. This technique is at the basis of an accelerated method to calculate R_n (see Project 1c; Gould and Tobochnik 1988).

References

Armstrong, Joseph E. "Rubber Production: Tapping Rubber Trees, Latex Collection And Processing Of Raw Rubber." Illinois State University. http://bio.illinoisstate.edu/jearmst/syllabi/rubber/rubber.htm (accessed January 1, 2013)

Bastolla, Ugo, Helge Frauenkron, Erwin Gerstner, Peter Grassberger, and Walter Nadler. 1997. "Testing a New Monte Carlo Algorithm for Protein Folding." Proteins: Structure, Function, and Genetics, 32: 52–66. http://arxiv.org/pdf/cond-mat/9710030.pdf (accessed January 1, 2013)

Brannon-Peppas, Lisa. 1997. "Polymers in Controlled Drug Delivery." *Medical Plastics and Biomaterials Magazine*. http://www.mddionline.com/article/polymers-controlled-drug-delivery (accessed December 31, 2012)

Case Western Reserve University. *Virtual Textbook*. Cleveland, OH: Department of Physics, the Department of Macromolecular Science and Engineering, and the Center for Advanced Liquid Crystalline Optical Materials. http://plc.cwru.edu/tutorial/enhanced/files/textbook.htm (accessed December 31, 2012)

Dhandayuthapani, Brahatheeswaran, Yasuhiko Yoshida, Toru Maekawa, and D. Sakthi Kumar. 2011. "Polymeric Scaffolds in Tissue Engineering Application: A Review," *International Journal of Polymer Science*, 2011. Article ID 290602.

Gould, Harvey, and Jan Tobochnik. 1988. *An Introduction to Computer Simulation Methods, Applications to Physical Systems, Part 2*. Reading, MA: Addison-Wesley, p. 695.

Hosler, Dorothy, Sandra L. Burkett, and Michael J. Tarkanian. 1999. "Prehistoric Polymers: Rubber Processing in Ancient Mesoamerica." *Science,* 284: 1988–1991 [DOI: 10.1126/science.284.5422.1988] (in Reports).

Marko, John. "What Are Polymers And Why Are They Interesting?" Cornell University, Laboratory of Atomic and Solid State Physics. http://www.lassp.cornell.edu/marko/polymers.html (accessed January 1, 2013)

Michalovic, Mark. 2000. "The Story of Rubber—a Self-Guided Polymer Expedition." *Science Sidetrip: Meet Polyisoprene*. Polymer Science Learning Center and the Chemical Heritage Foundation. http://www.pslc.ws/macrog/exp/rubber/sepisode/meet.htm (accessed January 1, 2013)

Nelson, David L., and Michael M. Cox. 2012. Lehninger's Principles of Biochemistry. 6th ed. New York: W. H. Freeman, pp. 76–86, 859–875.

Pfizer, Inc. 2011. "Procardia XL (nifedipine)." New York: Pfizer Labs. http://labeling.pfizer.com/ShowLabeling.aspx?id=542 (accessed January 1, 2013)

Rxlist. "Procardia XL." http://www.rxlist.com/procardia-xl-drug.htm (accessed January 1, 2013)

Tenanbaum, David J. 2003. "Polymers and People." *Beyond Discovery: The Path from Research to Human Benefit*. Washington, DC: The National Academy of Sciences. http://www.beyonddiscovery.org/content/view.article.asp?a=203 (accessed December 31, 2012)

Wikipedia. 2012. "Mesoamerican ballgame." http://en.wikipedia.org/wiki/Mesoamerican_ballgame (accessed January 1, 2013)

MODULE 14.2

Solidification—Let's Make It Crystal Clear!

Prerequisite: Module 9.5, "Random Walk."

Introduction

What do snowflakes and steel have in common? At first glance, we probably would say, not much. However, if we could look closely enough, we would see that they both are crystalline, possessing amazing structural similarities. Each is made of treelike structures called **dendrites**, which are formed as substance cools during the process of **solidification**.

Snowflakes are composed of one or more **snow crystals**. Each crystal is built of water molecules arranged in a very specific, hexagonal **lattice**. These crystals form in the clouds by the condensation of water vapor into ice. At first, while very small, the crystals form as hexagonally shaped prisms, following the original, molecular symmetry. The edges of the facets of this prism grow out more rapidly than the facets themselves, leading to the formation of "limbs." These limbs may, and usually do, produce other branches, leading to the dendrite, or treelike, forms.

A number of factors determine the precise shape of the crystal, but temperature is the primary influence. As snowflakes blow and fall through the clouds, they encounter significant variations in temperature, humidity, and pressure. Each snowflake tends to have different environmental "experiences," which lead to the development of different shapes. Why snowflake shape is so temperature dependent is not completely understood (Libbrecht).

The solidification of snowflakes is fascinating, but the process of solidification has an impressive array of manufacturing applications. Despite the increased use of plastics, think of all the things we use everyday that are metal. Used to produce everything from soda cans to car engines, these metals and alloys are formed from liquids that have "frozen," or solidified. Solidification, therefore, is an important process for generating metal products as well as snowflakes.

Dendrites form within the molten metals/alloys as they solidify during the casting process. These dendrites vary greatly in shape, size, and orientation. Furthermore, the individual dendrites interconnect in various ways to generate a series of intricate **microstructures**. These individual and collective variations greatly influence the structural qualities (e.g., strength and flexibility) of the product (Glicksman et al. 1991). There are numerous horror stories of castings that have broken apart from internal defects that originated from thermal stresses occurring during solidification (Seetharamu et al. 2001). According to scientists, we would be able to understand (and, therefore, control) the properties of materials that solidify dendritically better if we could develop effective computational models of the behavior of individual dendrites.

Under the influence of Earth's gravity, liquid metal is subject to the influence of **convective currents** as it cools. These currents significantly alter the growth of the dendrites, which makes modeling of "normal" dendritic growth and the effects of convective currents on such growth virtually impossible. Confronting this difficulty, the National Aeronautics and Space Administration (Glicksman et al. 1991) has teamed with scientists at Rensselaer Polytechnic Institute in the Isothermal Dendritic Growth Experiment (IDGE). Experiments in this program, conducted in conditions of low gravity that Earth orbit offers, have already shed tremendous light on dendritic growth. For instance, scientists, using IDGE data, will be able to separate the effects of convection from other factors that impact solidification of metals and alloys. Such information will go far to improve computational models, which should guide us to improved industrial production of various metals/alloys.

Projects

1. a. We can use the technique of **diffusion-limited aggregation** (**DLA**) to build a dendritic structure. In one form of the algorithm, a **seed**, or initial location for the developing dendritic structure, is in the middle of an $m \times m$ **launching rectangle**. This launching rectangle is a region in the middle of an $n \times n$ grid, where $m < n$. For example, m might be 16 and n might be 40. One at a time, "particles" are released from random positions on the launching rectangle boundary to go on random walks. If the walker comes in contact with another particle (i.e., a neighbor to its north, east, south, or west), with a designated sticking probability, the walker adheres to the particle, resulting in a larger structure. If the walker travels too close to the boundary of the grid, the simulation deletes that walker and releases another random walker from the launching rectangle. Use the DLA algorithm to develop a simulation to generate dendritic structures, with the number of particles for the structure as a parameter (Panoff 2004).

b. Develop a visualization that shows the simulation one step at a time, including the random walks. Develop another animation that shows only the particles as a new particle attaches to the growing structure. An attractive enhancement is for the color of the particle to be a function of its distance from the seed. (Follow the link "Simple DLA Example" at the Shodor website for an example of such a simulation with animation (The Shodor Educational Foundation 2002).)

c. Run the simulation and visualization a number of times for several different sticking probabilities. Discuss the impact of the sticking probabilities on the resulting structures.

2. Repeat Project 1 considering the eight surrounding cells as a walker's nearest neighbors.
3. Repeat Project 1, Parts a and b, where the sticking probability is 0.33 for contact with one particle, 0.67 for simultaneous contact with two particles, and 1.0 for contact with three. Run the simulation a number of times and discuss the results (Panoff 2004).
4. a. Repeat Project 2, Parts a and b, where the sticking probability is based on the number of particles the walker contacts simultaneously. Run the simulation a number of times and discuss the results (Panoff 2004).
   b. Adjust the situation so that the sticking probability is 0.1 for contact with one particle, 0.5 for two particles, and 0.9 for three or more particles. Run the simulation and animation a number of times and discuss the results (Panoff 2004).
   c. Adjust the situation so that the sticking probability is 0.01 for contact with one or two particles, 0.03 for three particles, and 1.0 for more than three particles. Run the simulation a number of times and discuss the results (Panoff 2004).
5. Repeat Project 1, Parts a and b, where the sticking probability is greater for bonds continuing in a straight line. For example, a walker is more likely to adhere to a north neighbor if that particle is stuck to a particle to its north. Similar situations exist for the other directions. Run the simulation a number of times and discuss the results (Panoff 2004).
6. Repeat Project 5, considering the eight surrounding cells as a walker's nearest neighbors (Panoff 2004).
7. Changing conditions affect crystalline formation and cause a great variety in the shapes. During a simulation, we can vary the sticking probability to indicate such changing conditions. Do Project 2, starting with sticking probabilities as in Project 4, Part b. After forming an aggregate with a specified number (such as 100) of particles, use sticking probabilities, as in Project 4c, for a specified number (such as 100) of particles; then change to a different sticking probability configuration (Panoff 2004).
8. Repeat any of Projects 1–6, considering the impact of wind or gravity on dendritic growth by having the walker travel with a greater probability in a particular direction (Shodor 2002).
9. Repeat any of Projects 1-8, using a launching circle of radius m instead of a launching rectangle. (Follow the link "Diffusion Limited Aggregation Calculator" at the Shodor website for such a simulation example (The Shodor Educational Foundation 2002).)
10. Repeat any of Projects 1–8, using a launching circle instead of a launching rectangle, of radius $2r_{max}$, where r_{max} is the radius of the structure so far. Delete a walker if it travels too close to the boundary of the grid or beyond a distance of $3r_{max}$ from the seed. Such adjustments should speed the simulation (Gould and Tobochnik 1988).
11. Do Project 10, with the following additional adjustment to speed the simulation by having larger step sizes further away from the structure: If a walker is

at a distance $r > r_{max} + 4$ from the seed, where r_{max} is the radius of the structure so far, then have step sizes of length $r - r_{max} - 2$; otherwise, have step sizes of length 1 (Gould and Tobochnik 1988).

12. Repeat any of Projects 1 or 2, considering accumulation on a structure, such as the deposit of snow on a tree. Have the seed be a triangular tree-like structure or other type of structure on the bottom of the grid. Release random walkers from the north end of the grid with a greater likelihood of traveling south (Panoff 2004).

References

Glicksman, Martin E., M. B. Koss, R. C. Hahn, Ana Cris R. Veloso, A. Rojas, and E. Winsa. 1991. "Scientific Basis for the Isothermal Dendritic Growth Experiment: A USMP-2 Space Flight Experiment." In *Materials Science Forum*, 77: 51–60.

Gould, Harvey, and Jan Tobochnik. 1988. *An Introduction to Computer Simulation Methods, Applications to Physical Systems, Part 2*. Reading, MA: Addison-Wesley: 695.

Libbrecht, Kenneth G. "Snowflake Primer—The Basic Facts About Snowflakes and Snow Crystals." California Institute of Technology. http://www.its.caltech.edu/~atomic/snowcrystals/primer/primer.htm (accessed January 1, 2013)

Panoff, Robert. 2004. "Diffusion Limited Aggregation." Educational Materials for Undergraduate Compuational Science. Capital University. http://www.capital.edu/cs-computational-science/ (accessed January 1, 2013)

Seetharamu, K. N., R. Paragasam, Ghulam A. Quadir, Z. A. Zainal, P. Sthaya Prasad, and T. Sundararajan. 2001. "Finite Element Modeling of Solidification Phenomena." *Sadhana*, 26 (Parts 1 and 2): 103–120.

The Shodor Educational Foundation. 2002. "Software—Diffusion Limited Aggregation Calculator." Computational Science Education Reference Desk. http://www.shodor.org/refdesk/Resources/Models/DLA/ (accessed January 1, 2013)

MODULE 14.3

Foraging—Finding a Way to Eat

Prerequisite: One of Module 10.3, "Spreading of Fire"; Module 10.4, "Movement of Ants–Taking the Right Steps"; or Module 10.5, "Biofilms—United They Stand, Divided They Colonize."

Introduction

> If the brain was simple enough to be understood—
> we would be too simple to understand it!
>
> —Marvin Minsky, MIT

Some animals must navigate over long distances, such as in the migration of birds, butterflies, whales, or salmon. This **large-scale navigation** typically utilizes celestial and/or geomagnetic cues. In animals like migratory birds, circadian (endogenous, daily) rhythms not only help to initiate migration but may also influence an animal's spatial course of migration (Gwinner 1996).

Almost all animals must navigate over shorter distances, such as to forage or seek mates and/or nesting sites. Honeybees, for example, are **central place foragers**, often losing visual and auditory contact with the hive as they search for food. These animals must be able to return to the nest and to "remember" where they have been for return foraging. To do this, they rely on **path integration**, which means that they must take all the angles and distances they experience on their foraging trips and integrate them into "mean home vectors." Like the dead reckoning of human pilots, they perform continuous spatial updating to relate their current location to the hive. This process permits them to return home by more direct routes. Bees appear to use the sun's position and patterns of polarized light as components of a compass, but the calculations of distance and direction are internal (egocentric). Once the route is established, landmarks, which are forms of geocentric information, likely supplement the insect's internal calculator (Wehner et al. 1996; Wehner 1996).

Some birds, like the chickadee, store, or cache, excess food for retrieval in times of shortages. They are capable of remembering perhaps hundreds of locations over fairly large areas (up to 30 hectare (ha)) for at least a month. Evidence suggests that these birds might navigate to these storage sites using spatial relationships among objects (landmarks) short distances from the caches. Features of the cache sites themselves appear to be less significant. This series of markers form geocentric references that are stored in memory, likely in the hippocampus, as a neural representation of the bird's environment. Many scientists refer to this representation as a **cognitive map**. The idea of a cognitive map is a reasonably controversial proposal for those who study spatial navigation. It is intriguing, however, that the hippocampus regions of food-storing birds are larger than comparable bird species that do not store food (Doupe 1994; Sherry 1996; Sherry and Duff 1996). Evidence also exists that the regions of the human hippocampus enlarge in those who are highly dependent on navigational skills (Maguire et al. 2000).

Some extend the concept of a cognitive map to insects. Honeybees and ants certainly seem capable of remembering landmark locations, but whether they are capable of integrating these memories into a map seems improbable. It is more likely that they store a series of snapshots of the surroundings as they make their journeys, which they compare to the current landscape at a particular time (Wehner et al. 1996).

Simulations

The projects in this module develop cellular automaton methods for goal-directed spatial searches in which some form of adaptation occurs. Psychologists have used such simulations to elucidate qualitative properties of animal spatial orientation and learning, such as with a pigeon searching for food in an area where it found food previously, a rat learning its way through a changing maze, a badger returning to multiple food sites when food is no longer present, a dog finding a shortcut, and a shrew maneuvering around a detour. Such studies have revealed that a simulated animal can find its way, not through some "insight" of the problem as a whole, but through repeated local decisions based on its immediate surroundings, or values in neighboring cells (Reid and Staddon 1997, 1998). Besides applications to cognitive psychology, scientists are studying the use of such simulation algorithms in the guidance systems of autonomous agents (robots) searching for land mines (Staddon and Chelaru).

For the searching, the robot or animal possesses a cognitive map of the area. A grid represents the cognitive map, and each cell contains the following information:

- Whether the cell contains a barrier, an "animal," or neither
- An expectation value

Usually, only one cell in the grid has an animal, representing the one with the map.

A cognitive map reader is a searching algorithm that is similar to that of a dynamic diffusion model, such as of the diffusion of heat through a metal bar. The change in an empty cell's expectation value, ΔV_t, from time t to time $t + \Delta t$ is a **dif-**

fusion rate parameter r times the sum of the differences in the **expectation value** of the neighbor's ($V_{k,t}$) and the cell's expectation values (V_t), as follows:

$$\Delta V_t = r\sum_{k=1}^{8}\left(V_{k,t} - V_t\right), \text{ where } 0 < r < 1/8 = 0.125$$

As developed in Module 10.2, "Diffusion: Overcoming Differences," for a cell's temperature time $t + \Delta t$, the expectation value at time $t + \Delta t$ simplifies to the following:

$$V_{t+\Delta t} = (1 - 8r)V_t + r\sum_{k=1}^{8} V_{k,t}, \text{ where } 0 < r < 0.125 \quad (1)$$

If a cell contains an animal and a reward event occurs, such as finding food, then the expectation value becomes some reward value, say 1, usually for only one time step. In contrast, if a nonreward event, such as not finding food, occurs in an animal cell, then the expectation value becomes 0. In this situation, because of diffusion from neighbors, a cell's expectation value increases with time after the animal leaves that cell. The reward/nonreward system assures that the animal does not remain at a reward site. An animal consumes a reward, and the cell's expectation value becomes 1. However, an animal returning to the cell no longer finds the reward, so that the expectation value becomes 0.

In a deterministic version of the algorithm, called the **hill-climbing process**, at each time step, the animal moves to a neighboring cell that has the highest expectation value. In a stochastic version, the move to such a cell occurs with a certain probability.

The simulations involving cognitive maps to illustrate reinforced learning and adaptive behavior contain at least two phases. The first phase often has the animal exploring the grid until finding a reward as the expectation values diffuse. Then, the animal is removed from the grid. Perhaps the diffusion is allowed to continue without the presence of the animal. Finally, the animal is returned to the grid, which might have been altered, to search for its reward (Reid and Staddon 1997, 1998).

Projects

1. a. Develop a cellular automaton simulation and visualization of open-field foraging with an **area-restricted search**. In an experiment this work is to simulate, a psychologist places a foraging animal into an enclosure that contains a food stash. Once the food has been found, the animal is removed for a while. Then, the animal is returned to the area, which no longer contains any food. Experiments have shown that the animal goes directly to the former location of the food. Not finding the stash, the animal begins searching around the area where the food was. Its path appears erratic and looping and gradually moves further from the former stash site. Simulation results should mimic this behavior.

For the first phase of the simulation, initialize each cell of a 20 × 20 grid (cognitive map) to have some very low, uniform expectation value. Assume the food is in the middle of the grid and the diffusion rate parameter (r) is 0.05. Have the animal start at an edge or corner and search with a random walk for the food until found. As discussed in the section "Simulations," diffusion proceeds; each failure results in zeroing out the cell's expectation value; and success sets the value to 1. The second phase allows diffusion to continue in this grid without the animal, say, for 30 time steps. Third, place the animal in the same starting location as in the first phase, and allow the simulation to run for 60 time steps (Reid and Staddon 1997).

b. Run the simulation four times for diffusion rate parameters of 0.001, 0.01, 0.05, and 0.1, respectively. Discuss the search patterns, why they occur according to the algorithm, and how they parallel experimental results.

2. a. Develop a cellular automaton simulation and visualization of a desert ant (genus *Cataglyphis*) conducting an area-restricted search for its nest (see Project 1). The nest is in the ground with an opening less than 1 cm in diameter; and the ant finds the entrance, not with pheromones but with dead reckoning. In a psychological cognition experiment, an ant that is far from the nest is captured, placed in a different location, and released. By dead reckoning, the ant heads in the compass direction it should have taken from the location where it was captured. Not finding the nest in the expected location, the ant begins searching. The path is in erratic loops of increasing diameters, centered at the supposed nest. However, repeatedly the ant returns to this location before beginning another loop.

For the first phase of the simulation, initialize each cell of a 20 × 20 grid (cognitive map) to have some very low, uniform expectation value and use a diffusion rate parameter of 0.1. Assume the nest is in the middle of the grid. Have the ant start at a corner and search at random for the nest. As discussed in the section "Simulations," diffusion proceeds; each failure results in zeroing out the cell's expectation value; and success sets the value to 1. The second phase allows diffusion to continue in this grid without the ant, say, for 20 time steps. Third, place the ant at the former nest location and allow the simulation to run (Reid and Staddon 1997).

b. Graph the ant's distance from the origin in the third phase as a function of time for several hundred time steps. Discuss the meaning of the graph.

3. Develop a cellular automaton simulation and visualization of a European badger learning to search for multiple sources of food in specific locations. The original psychological experiment was on an open field with three sections. Peanuts were spread out at the same places on the left and right sections in the field for six nights. The badger learned to find the peanuts from these sections, spending very little time in the middle, empty section.

For the first phase of the simulation, initialize each cell of a grid (cognitive map), which is three times longer than wide, to have some very low, uniform expectation value and use a diffusion rate parameter of 0.001. Consider a total of 16 designated food locations in the left and right thirds of the grid. Have the badger start at an edge and search at random for the food. As discussed in the section "Simulations," diffusion proceeds; each failure results in zeroing out the cell's expectation value; and success sets the value to 1. The second phase allows diffusion to continue in this grid without the badger, say, for 300 time steps, simulating the 24-h period between searches. Third, place the badger again at the edge of the grid, and allow the simulation to run until the badger leaves the grid. Discuss the results (Reid and Staddon 1998).

4. a. Develop a cellular automaton simulation and visualization of a dog with an appropriate cognitive map exhibiting what appears to be spatial "insight" in finding a shortcut to food. The psychological experiment starts at location *X* and has food at two places, *A* and *B*. The three locations form a triangle, with *X* being closer to *A* than to *B*. A dog on a leash is lead from *X* to *A*, but not allowed to eat, and lead back to *X*. The same process is repeated, taking the dog from *X* to *B* to *X*. Then the dog is released into the field. Although not trained to do so, most dogs travel from *X* to *A*, eat the food, and then take the shortcut from *A* to *B* instead of following the training path *XAXBX*.

For the first phase of the simulation, initialize each cell of a 20 × 20 grid (cognitive map) to have some very low, uniform expectation value and use a diffusion rate parameter of 0.05. Simulate the training session. As discussed in the section "Simulations," diffusion proceeds; each failure results in zeroing out the cell's expectation value; and success sets the value to 1. The second phase allows diffusion to continue in this grid without the dog, say, for 50 time steps. Third, simulate the dog's search for both sources of food. Discuss the results (Reid and Staddon 1998).

b. Run the simulation nine times for diffusion rate parameters of 0.01, 0.05, and 0.1, with Phase 2 times of 50, 60, and 60 time steps each. Discuss the search patterns, why they occur according to the algorithm, and how they parallel experimental results.

5. a. Develop a cellular automaton simulation and visualization of a detour problem in which a blind rat adjusts its path to food in a changed landscape. In one psychological experiment, a blind rat is released from a starting gate into an enclosed area. The starting gate is on the left wall, near the northwest corner, while the goal, food, is on the right wall, near the opposite corner. The area has a barrier wall parallel to the left and right walls, and the barrier has an opening not far from the south end. For five times, the rat is released and allowed to find the food, which is replenished for each training session. Its path becomes increasingly more efficient as it learns its way. Then, the barrier is removed, and for several more times, the rat searches for food. With some variation, the trial rats quickly adjusted their trails to head straight for the food.

For the first phase of the simulation, initialize each cell of a 20 × 20 grid (cognitive map) to have some very low, uniform expectation value, and use a diffusion rate parameter between 0.001 and 0.1. Simulate a training session without food, in which rat completely explores the barrier. Second, simulate the rat's search for food for five times. Each new trial begins immediately after success in the previous trial. Finally, have six trials with food but no barrier.

The computational scientists that developed and analyzed simulations of this experiment wrote, "The difference between smart and less smart subjects lies in their ability to change their maps, not in the level of cognitive processing after they've learned the new map." Discuss the results of your simulation as it applies to this statement (Reid and Staddon 1998).

b. Produce other simulated experiments with different configurations of barriers. Discuss your results.

6. Repeat any of Projects 1–5 with a variation of Equation 1, computation of the diffused expectation value in a cell, that has the diagonal neighbors contributing half as much as the north, east, south, and west neighbors.

7. Repeat any of Projects 1–6 using a stochastic instead of a deterministic approach. Thus, with some probability, an animal moves to the neighbor with the largest expectation value. Otherwise, the creature goes to a random neighbor.

References

Doupe, Alliston J. 1994. "Seeds of Instruction: Hippocampus and Memory in Food-Storing Birds." *Proceedings of the National Academy of Sciences. USA*, 91: 7381–7384.

Gwinner, E. "Circadian and Circannual Programmes in Avian Migration." 1996. *Journal of Experimental Biology*, 199: 39–48.

Maguire, Eleanor A., David G. Gadian, Ingrid S. Johnsrude, Catriona D. Good, John Ashburner, Richard S. J. Frackowiak, and Christopher D. Frith. 2000. "Navigation-Related Structural Change in the Hippocampi of Taxi Drivers." *Proceedings of the National Academy of Sciences. USA*, 97 (8): 4398–4403.

Minsky, Marvin. 1987. *Society of Mind*. 1988. New York: Simon & Schuster.

Reid, Alliston K., and J.E.R. Staddon. 1997. "A Reader for the Cognitive Map." *Information Sciences*, 100. Amsterdam: Elsevier Science, pp. 217–228.

———. 1998. "A Dynamic Route Finder for the Cognitive Map." *Psychological Review*, 105(3): 585–601.

Sherry, David F. 1996. "Middle-Scale Navigation: the Vertebrate Case." *Journal of Experimental Biology*, 199: 163–164.

Sherry, David F., and Sarah J. Duff. 1996. "Behavioral and Neural Bases of Orientation in Food-Storing Birds." *Journal of Experimental Biology,* 199: 165–172.

Staddon, J.E.R., and Ioan M. Chelaru. "A Diffusion-based Guidance System for Autonomous Agents." Duke University. http://people.ee.duke.edu/~lcarin/DeminingMURI/DeminingRouteFinder00.pdf (accessed January 1, 2013)

Wehner, Rudiger. 1996. "Middle-Scale Navigation: The Insect." *Journal of Experimental Biology* , 199: 125–127.

Wehnr, Ruciger, Barbara Michel, and Per Antonsen. 1996. "Visual Navigation in Insects: Coupling of Egocentric and Geocentric Information." *Journal of Experimental Biology*, 199: 129–140.

14.4

Pit Vipers—Hot Bodies, Dead Meat

Prerequisite: One of Module 10.3, "Spreading of Fire"; Module 10.4, "Movement of Ants–Taking the Right Steps"; or Module 10.5, "Biofilms—United They Stand, Divided They Colonize."

Introduction

Why are engineers like Dr. John Pearce at the University of Texas studying pit viper pits? An important reason is that the U.S. Air Force believes that a better understanding of these pits will lead to better missile defense. A snake's pits are on each side of its face between the eyes, and the nostrils are extremely sensitive infrared (heat) detectors. The snake is able to detect heat differences between the background and potential prey that are as small as a thousandth of a degree Celsius. Exactly how they work is not completely understood, but even blind rattlesnakes can strike warm-blooded prey with deadly accuracy. Because visual and heat images are received in the same area of the brain (optic tectum), some suggest that the snake can form a thermal "image" of the size and shape of the prey. The preferred food in the diet of a pit viper is often a small rodent, which is a warm-blooded, or homeothermic, creature. Because such prey usually maintain a body temperature greater than their surroundings, pit vipers have a most effective hunting tool in their pits (Gracheva et al. 2010; Bakken and Krochmal 2007).

Dr. Pearce and his colleagues are attempting to develop mathematical models of this predator-prey relationship. These models attempt to predict heat emitted from the prey at given distances from the snake and the influence these variations in heat emissions have on the pit receptors. The military is hoping that from these models, they can design more precise missile detection systems (University of Texas 2001).

Projects

1. a. Develop a simulation and visualization of a pit viper hunting for food in a cool environment. Have one rodent moving and resting at random. Most mammals have an average body temperature of around 37 ºC. For example, the normal body temperature of a rabbit is 38.3 ºC, and that of a laboratory mouse is 36.9 ºC. Have one pit viper that tends to move toward warmth. In the case of the same temperature in all neighboring cells, the snake selects a direction at random.

b. Running the simulation repeatedly, determine if the temperature of the environment impacts the viper's hunting ability.

2. Repeat Project 1, assuming that the rodent can, but does not consistently, move twice as fast as the snake.

3. Repeat Project 1 with one pit viper and several rodents. When the snake is hungry, it seeks food; and when full, it rests. After eating, the snake remains full for a while.

4. Repeat Project 1 with one pit viper and several rodents. When the snake is hungry, it searches for food; and when full, it seeks a warm place to rest for a while.

5. Repeat one of Projects 1, 3, or 4, in which a rodent avoids a pit viper if the animal detects the snake. However, the prey does not always see or hear its predator. When avoiding the snake, an animal moves twice as fast as when not eluding its predator.

6. a. For thermal regulation in a hot dessert, a snake seeks cooler places, such as under rocks or in rodent boroughs. Develop a simulation and visualization of a snake seeking such places when its body temperature becomes too hot (say, 42 ºC) and tending to stay in the shelter until sufficiently cool. Assume that the rate of heat loss or gain for a snake is proportional to its surface area (Krochmal and Bakken 2003; Krochmal et al. 2004).

b. Plot the snake's temperature versus time. Discuss the results

References

Bakken, George S., and Aaron R. Krochmal. 2007. "The Imaging Properties and Sensitivity of the Facial Pits of Pitvipers as Determined by Optical and Heat-Transfer Analysis." *Journal of Experimental Biology* 210(16): 2801–2810.

Gracheva, Elena O., Nicholas T. Ingolia, Yvonne M. Kelly, Julio F. Cordero-Morales, Gunther Hollopeter, Alexander T. Chesler, Elda E. Sánchez, John C. Perez, Jonathan S. Weissman, and David Julius. 2010. "Molecular Basis of Infrared Detection by Snakes." *Nature* 464 (7291): 1006–1011.

Krochmal, Aaron R., and George S. Bakken. 2003. "Thermoregulation Is the Pits: Use of Thermal Radiation for Retreat Site Selection by Rattlesnakes." *Journal of Experimental Biology*, 206: 2539–2545.

Krochmal, Aaron R., George S. Bakken, and Travis J. LaDuc. 2004. "Heat in Evolution's Kitchen: Evolutionary Perspectives on the Functions and Origin of the Facial Pit of Pitvipers (Viperidae: Crotalinae)." *Journal of Experimental Biology* 207(24): 4231–4238.

University of Texas. 2001. "Heat-Seeking Vipers May Help with U.S. Defense, UT Austin Researcher Finds." Office of Public Affairs, the University of Texas at Austin, May 31. http://www.utexas.edu/news/2001/05/31/nr_vipers/ (accessed January 1, 2013)

MODULE 14.5

Mushroom Fairy Rings—Growing in Circles

Prerequisite: One of Module 10.3, "Spreading of Fire"; Module 10.4, "Movement of Ants—Taking the Right Steps"; or Module 10.5, "Biofilms—United They Stand, Divided They Colonize."

Introduction

If you see a fairy ring
In a field of grass,
Very lightly step around,
Tip-Toe as you pass,
Last night Fairies frolicked there
And they're sleeping somewhere near.
—Author Unkown

Sometimes in a forest or yard, mushrooms seem magically to grow in circles, which we call "fairy rings" (Figure 14.5.1). In this module, projects develop simulations along with animations for the expansion and interactions of such mushroom fairy rings.

You might remember exploring your refrigerator and finding an unmarked container that was pushed against the back wall. Without thinking, you opened the container to find a disgusting mass of green or black slime. So the word *fungus* probably makes you think of something rather unpleasant, like spoiled food, or reminds you of kicking over "toadstools" during your childhood. In either case, if you are like most people, you probably do not know too much about them.

Fungi are in the business of decay. They depend on nutrients from degradation of the organic matter deposited by or from other organisms. Through their ability to break down rather complex organic molecules, fungi are also responsible for returning to the soil a large quantity of nutrients for plant growth that would be unavailable otherwise. Ecosystems are dependent on this recycling of nutrients from the decay of organic matter. We also derive our own direct benefits from fungi—antibiotics,

Figure 14.5.1 Fairy Ring

cheese, beer/wine, truffles, and other edible mushrooms. On the other hand, fungi are responsible for 70% of all major crop diseases and a number of diseases that affect human beings and other animals (Deacon).

Whether we appreciate them or not, we cannot avoid them. In fact, we come into contact with them (or their spores) with every breath we take. Only about 99,000 of the estimated five million species of fungi have even been described (Blackwell 2011). Their ecological, economic, and medical significance makes them worthy objects of human curiosity, and scientists worldwide are busily investigating various aspects of their lives.

What Are Fungi?

To consider a model that involves fungi, we need to understand some of the fundamentals of their biology. Hence, we include a few questions with basic answers.

References are provided with more detailed information. We begin by answering, What are fungi? **Fungi** are multicellular (except for yeasts), spore-producing organisms that depend on absorbing nutrients from their surroundings. The **spores** are reproductive cells that are typically capable of germinating into new fungi. Nutrients are made available by the action of extracellular enzymes secreted by the fungus itself.

What Do Fungi Look Like?

When most people think of a fungus, they think of a mushroom. Mushrooms are clearly one of the more recognizable forms of fungi, but most fungi do not form mushroomlike structures. Amazingly, most of the "body" of a fungus is usually not visible or obvious. Some are spread out over wide areas underground, reaching miles in diameter (Kruszelnicki). The **mushroom** is just the fruiting body of certain types of fungi.

The basic morphological component of all the fungi, except yeasts, is the **hypha**. A hypha is a thin (5- to 10-μm diameter; Deacon), branching tubule. Masses of branching hyphae form the body of the fungus, which we usually call the **mycelium**.

Many of the hyphae are interconnected within this mass. Much of the fungal mass is not visible. The **mushroom** that you do see is an organized mass of hyphae, which become spore-bearing. Hence, mushrooms in mulch or the lawn or the shelf-shaped structures on rotting logs are really only the "tip of the iceberg." They produce the spores, but a much larger mycelium underlies the mushrooms. If we examine the undersurface of the mushroom "cap," we will often find the platelike **gills**, which bear the spores.

How Do Fungi Feed Themselves?

Fungi are not chemo- or photosynthetic, and they have no internal digestive system. So, they must absorb what they need from the environment—water and inorganic and organic nutrients. Many of the organics are too large to be absorbed unaltered. Consequently, fungi produce hosts of extracellular, digestive enzymes. The enzymes degrade the large, and sometimes very complex, organics into molecules they can absorb through their cell wall and plasma membrane.

How Do Fungi Reproduce?

The outgrowth from a spore is by **asexual** cell division. However, most fungi can also reproduce **sexually** through the fusing (mating) of hyphae of opposite mating types. Such sexually produced hyphae can organize into spore-producing structures, such as mushrooms.

How Do Fungi Grow?

Fungi display what is called **apical growth**—that is, they extend or elongate only at the tip of the hyphae, without subsequent increase in diameter. Branches form behind the advancing tip, forming more advancing tips (Lepp and Fagg 2012). A new hypha grows out from a spore and, with its continuous branching, yields a circular growth pattern. Growth can be quite rapid, reaching 1 km of hyphae per day. This rapid growth is made possible because nutrients are rapidly and constantly being delivered toward the tips.

The Problem

Dr. Alan Rayner, an English scientist who has studied forest fungi for years, wrote:

> "I have increasingly come to regard the mycelium as a heterogenous army of hyphal troops, variously equipped for different roles and in varying degrees of communication with one another. Without a commander, other than the dictates of their environmental circumstances, these troops organize themselves into a beautifully open-ended or indeterminate dynamic structure that can continually respond to changing demands. Recall that during its poten-

tially indefinite life, a mycelial army may migrate between energy depots; absorb easily assimilable resources such as sugars; digest refractory resources such as lignocellulose; mate, compete and do battle with neighbours; adjust to changing microclimatic conditions; and reproduce" (Rayner 1991).

Circular or radial growth of the mycelium allows the fungus to move into unexploited areas in its search for nutrients. When the innermost part of the mass has exhausted the resources of that area, that portion becomes expendable. The outer mycelium removes and reabsorbs useful nutrients from the central region and transfers them outward. The innermost hyphae die and decay. This ring-shaped growth pattern can be easily seen on fields or on golf course turf as what is termed a **fairy ring** (Figure 14.5.1). The rate of growth of these rings varies according to the environment (soil texture, moisture, etc.) and species. *Marasmius oreades*, a fairy-ring fungus found extensively in Europe, may increase its radius between 10 and 35 cm per year (Lepp and Fagg 2012). In the Midwest, fairy-ring species, including *M. oreades*, may expand their radii up to 60 cm per year (Illinois Extension Service 1998). It should come as no surprise that property owners and extension agents regard them as diseases.

Fairy rings appear in Shakespeare's *The Tempest* (Act V, Scene I), when Prospero shouts, "you demi-puppets that by moonshine do the green sour ringlets make, whereof the ewe not bites; and you whose pastime is to make midnight mushrumps, that rejoice to hear the solemn curfew. . .." It was not until Dr. William Withering, who also gave the world the heart medication digitalis from the foxglove plant (Kruszelnicki), dug up the buried mycelia of fairy rings in 1792 that we knew that a fungus was the cause of these phenomena. Before this discovery, one of the most common beliefs was that the rings represented a circular path of fairies dancing—hence, the name fairy ring.

Although there are more than 50 species of fungi that can produce fairy rings, the most common causes are one of three species: *Marasmius oreades*, *Agaricus campestris*, or *Chlorophyllum molybdites*. There is some variation in the appearance, depending on which fungus is present. One common type is characterized by three distinct rings—an inner ring of stimulated grass growth, a middle ring of dead or dying grass, and an outer ring of stimulated grass growth. Both of the stimulated zones are probably the result of release of nitrogen from the decaying organic matter. The central ring contains the dense mass of "feeding" hyphae that prevents penetration of water and depletes nutrients (Illinois Extension Service 1998). The fruiting bodies form within this ring or on the margin next to the outer stimulated zone. Some fungi do not produce this "dead zone."

Because of dead zones, two intersecting fairy rings often merge into a figure-8 pattern (Figure 14.5.2). Also, as Figure 14.5.3 depicts, a barrier can induce the mushrooms to grow into an arc.

How Do Fairy Rings Get Started?

A fairy ring grows out from spores or transported soil that contains fragments of mycelium. The circular mass extends from this "nucleus." Soil quality and weather conditions determine the rate of growth.

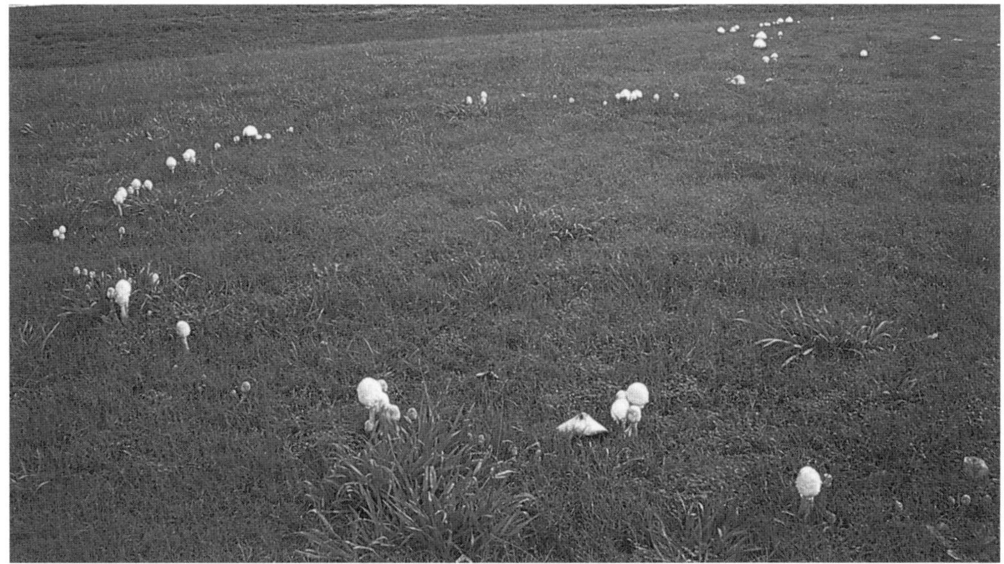

Figure 14.5.2 Intersecting fairy rings

Figure 14.5.3 Arc of mushrooms near a barrier

Initializing the System

In many simulations, we model a grid area under consideration with an $n \times n$ lattice, or a 2D square array of numbers. Each cell in the lattice contains a value representing a characteristic of a corresponding location and a state in the cell's life cycle. For example, in a simulation for the spread of fairy rings, a cell can contain an integer value, 0–9, representing such states as being empty, containing a spore, or having maturing hyphae. Table 14.5.1 gives a list of possible values with associated constants and their meanings.

Table 14.5.1
Possible Cell Values with Associated Constants and Their Meanings

Value	*Constant*	*Meaning: The cell contains*
0	*EMPTY*	**Empty** ground containing no spore or hyphae
1	*SPORE*	At least one **spore**
2	*YOUNG*	**Young hyphae** that cannot form mushrooms yet
3	*MATURING*	**Maturing hyphae** that cannot form mushrooms yet
4	*MUSHROOMS*	Older hyphae with **mushrooms**
5	*OLDER*	Older hyphae with **no mushrooms**
6	*DECAYING*	**Decaying hyphae** with exhausted nutrients
7	*DEAD1*	**Newly dead hyphae** with exhausted nutrients
8	*DEAD2*	Hyphae that have been **dead for a while**
9	*INERT*	**Inert area** where plants cannot grow

To initialize this discrete stochastic system, we can employ the following probability:

probSpore: The probability that a site initially has a spore, or that the grid site is *SPORE* = 1. Thus, *probSpore* is the initial spore density.

Updating Rules

Updating rules apply to different situations. An inert site never can change states; its cell value remains *INERT*. At the next time step, an empty cell (site value *EMPTY*) may or may not have young hyphae (*YOUNG* or *EMPTY*, respectively) growing into it from a neighboring site with young hyphae (*YOUNG*). If a cell contains a spore (*SPORE*), at the next time step the spore may or may not germinate to become young hyphae (*YOUNG* or *SPORE*, respectively). Young hyphae (*YOUNG*) always age to become maturing hyphae (*MATURING*). Maturing hyphae (*MATURING*) may or may not produce mushrooms (*MUSHROOMS* or *OLDER*, respectively); but regardless, these hyphae (*MUSHROOMS* or *OLDER*) eventually exhaust the resources and begin decaying (*DECAYING*). The decaying hyphae (*DECAYING*) die (*DEAD1*). A site with newly dead hyphae (*DEAD1*) cannot support new growth for a while (*DEAD2*) but eventually can (*EMPTY*).

To develop this dynamic, discrete stochastic system, we employ the following probabilities:

probSporeToHyphae: The probability of a spore (site value of *SPORE*) germinating to form young hyphae (*YOUNG*) at the next time step

probMushroom: The probability that maturing hyphae (cell value *MATURING*) produce mushrooms (*MUSHROOMS*) at the next time step

probSpread: The probability that an empty site (cell value *EMPTY*) gets young hyphae (*YOUNG*) at the next time step from a neighbor that has young hyphae

Figure 14.5.4 summarizes the states and transitions of the model in a state diagram. The 10 states, such as *SPORE* and *YOUNG*, give the possible values for a cell.

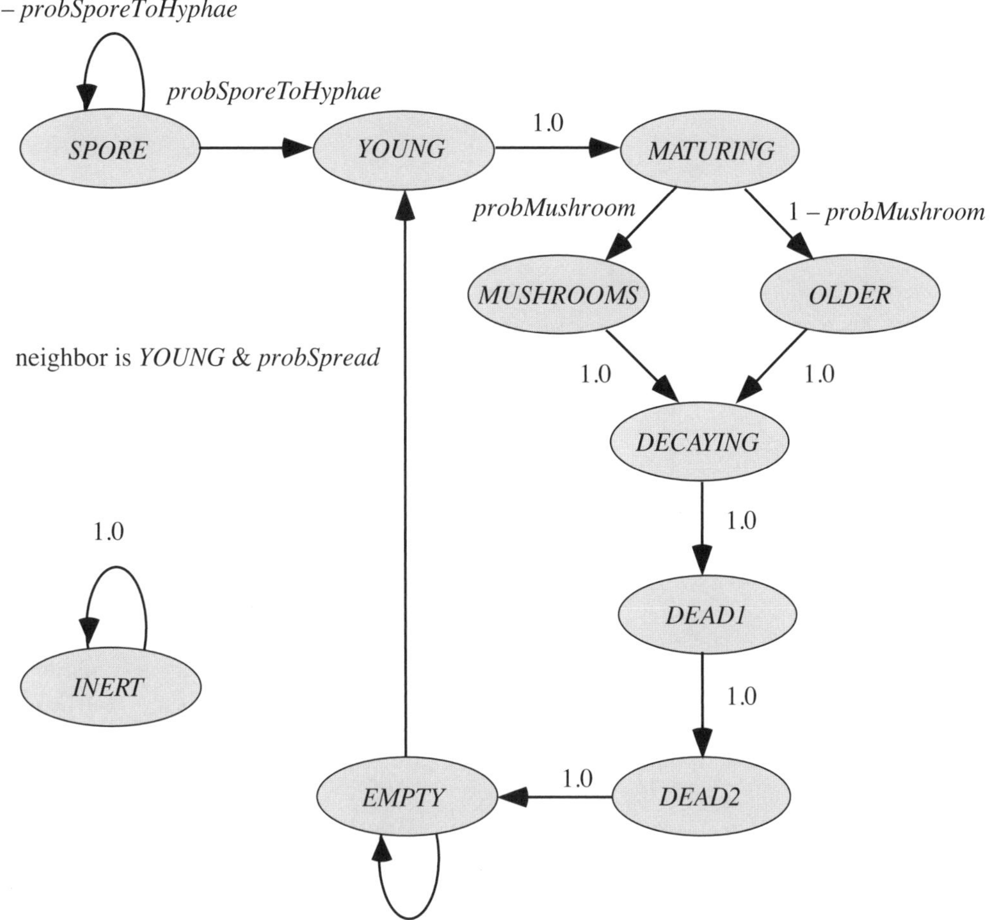

Figure 14.5.4 State diagram for model

The arrows with labels, such as *probSporeToHyphae*, show the probability with which a cell changes from one state to another in subsequent time steps. The arrow from *EMPTY* to *YOUNG* indicates that the situation is more complicated; if a neighbor has the value *YOUNG*, the empty cell becomes *YOUNG* with a probability of *probSpread*.

Display the Simulation

For each lattice in the list that a cellular automaton simulation returns, we generate graphics of a rectangular grid with colors representing the states of the cells, as in Table 14.5.2. Animation of the resulting frames helps us to verify the model and interpret the results.

Table 14.5.2
Possible Cell Values with Associated Constants and Colors

Value	*Constant*	*Colors*
0	*EMPTY*	Light green
1	*SPORE*	Black
2	*YOUNG*	Dark grey
3	*MATURING*	Light grey
4	*MUSHROOMS*	White
5	*OLDER*	Light grey
6	*DECAYING*	Tan
7	*DEAD1*	Brown
8	*DEAD2*	Dark green
9	*INERT*	Yellow

Projects

1. Develop the simulation of this module using absorbing boundary conditions. Include a function to show the situation aboveground. Run the simulation employing various initial grids, as follows:

   a. As described in the module with various values of *probSpore*. Describe the results.

   b. With exactly one spore in the middle. Verify that the figure seems to agree with Figure 14.5.1.

   c. With exactly two spores that are several cells apart toward the middle. Verify that the rings merge into the figure-eight pattern observed in nature, as in Figure 14.5.2.

   d. With exactly one spore and a barrier. Verify that the results appear to agree with the growth pattern in Figure 14.5.3.

2. Do Project 1 where the probability of young hyphae spreading into a site is proportional to the number of neighbors that contain young hyphae.

3. Adjust the simulation of this module so that new spores can form when mushrooms are present. Consider the following two possibilities:

   a. The probability that a cell can obtain a spore at the next time period is equal to the percentage of mushrooms in the grid.

   b. A cell can obtain a spore at the next time period with a specified probability provided one of its neighbors contains a mushroom.

4. Do Project 1 so that the length of time the hyphae are dead is probabilistic; and on the average, they are dead for two time steps.

5. Do Project 1 using periodic boundary conditions.

6. Do Project 1 using reflecting boundary conditions.

7. Do Project 1, where the neighbors of a cell include those cells to the northeast, southeast, southwest, and northwest.

References

Blackwell, Meredith. 2011. "The Fungi: 1, 2, 3. . . 5.1 Million Species?" *American Journal of Botany* 98(3): 426–438.

Deacon, Jim. "The Microbial World—The Fungal Web." Institute of Cell and Molecular Biology and Biology Teaching Organization, University of Edinburgh. Archived. http://www.biology.ed.ac.uk/archive/jdeacon/microbes/fungalwe.htm (accessed January 1, 2013)

Gaylor, Richard J., and Kazume Nishidate. 1996. "Contagion in Excitable Media." *Modeling Nature: Cellular Automata Simulations with Mathematica*. New York: TELOS/Springer-Verlag, pp. 155–171.

Illinois Extension Service. 1998. Department of Crop Sciences, University of Illinois– Urbana-Champaign. "Fairy Rings, Mushrooms and Puffballs." *Report on Plant Disease No. 403*.

Kimball, John W. 2012. "Fungi." http://users.rcn.com/jkimball.ma.ultranet/Biology Pages/F/Fungi.html (accessed January 1, 2013)

Kruszelnicki, Karl S. "Great Moments in Science—Fairy Rings." Karl S. Kruszelnicki Pty Ltd. http://www.abc.net.au/science/k2/moments/s297489.htm (accessed January 1, 2013)

Lepp, Heino, and Murray Fagg. 2012. "The Mycelium." Australian National Botanic Gardens. http://www.anbg.gov.au/fungi/mycelium.html (accessed January 1, 2013)

Rayner, Alan D. M. 1991. "Conflicting Flows: The Dynamics of Mycelial Territoriality." *McIlvainea*, 10: 24-3557-62.

MODULE 14.6

Spread of Disease—Sharing Bad News

Prerequisite: One of Module 10.3, "Spreading of Fire"; Module 10.4, "Movement of Ants–Taking the Right Steps"; or Module 10.5, "Biofilms—United They Stand, Divided They Colonize."

Introduction

The "SIR Model" section of Module 4.3, "Modeling the Spread of SARS—Containing Emerging Disease," considers a model for the spread of disease. The **SIR Model** considers the following population groups: **susceptibles** (**S**) that have no immunity from the disease, **infecteds** (**I**) that have the disease and can spread it to others, and **recovereds** (**R**) that have recovered from the disease and are immune to further infection.

In that module, we considered the spread of disease from a systems dynamics point of view. In this module, projects deal with the spread of disease using the approach of cellular automata.

Exercise

1. Suppose an individual is at each grid point. An individual can be well and susceptible (value *SUSCEPTIBLE* = 0) to a disease, sick with the disease that has two phases (values *PHASE1* = 1 and *PHASE2* = 2), or immune (value *IMMUNE* = 3). Let *probSick* be the probability that the individual initially is sick with a disease. Let *probPhase1* be the probability that initially a sick individual is in Phase 1 of the disease. Suppose, initially, no individual is immune. Write code in a computational tool to initialize the grid.

Projects

1. Develop a simulation with animation for the contagious spreading of the disease described next. Run the simulation for several probabilities and discuss the results.

Suppose in a population an individual can be susceptible, infectious, or immune to a stomach virus. The infection lasts 2 days, and immunity lasts only 5 days before the individual becomes susceptible again. Assume an individual is at each grid point. In the simulation, the value at a grid point can be one of the following:

- 0: susceptible individual
- 1, 2: infectious individual, where the value indicates the day of infection
- 3, 4, 5, 6, 7: immune individual, where the day of immunity is the cell value minus 2. For example, on day 1 of immunity, the cell value is 3.

In the simulation, initialize the grid using the following probabilities:

probSusceptible: the probability the individual is initially susceptible
probInfectious: the probability that an individual that is not susceptible is infectious initially

Uniformly distribute the infected individuals between day 1 and day 2 of the infection. In the initialization, uniformly distribute immune individuals with values 3 through 7. Use constants for the cell values 0 through 7, such as *SUSCEPTIBLE* = 0.

The following rules apply, where the term *neighbor* applies to the cell to the north, east, south, or west:

- If an individual is susceptible and a neighbor is infected, the individual becomes infected.
- The infection lasts for 2 days.
- Immunity lasts for 5 days, after which time the individual again becomes susceptible.

Color the graphic as follows:

- Susceptible: full green
- Infectious: blue; full blue on the first day, pale blue on the second.
- Immune: red; full red on the first day with successively paler shades of red on subsequent days,

After grid initialization, this model is deterministic, because the next state is always determined by the situation. If a susceptible individual is exposed to the virus, that person will definitely get sick for exactly 2 days and be immune for exactly 5 days.

Systematically, run the model for various values of *probSusceptible* and *probInfectious* and discuss the results.

2. Develop a nondeterministic (stochastic) simulation for a situation similar to that in Project 1. In this case, *probCatch* is the probability that a susceptible individual who has a sick neighbor will get sick, and *probBeSusceptible* is the probability that an individual who has been immune for 5 days will become susceptible. Thus, someone who is exposed to the virus might not become sick, and a person might have longer immunity than 5 days. Run the simulation for several probabilities and discuss the results. Try to discover a situation in which an epidemic does not occur, that is, in which the disease does not spread to many people over a short period of time.

3. Develop a nondeterministic (stochastic) simulation for a situation similar to that in Project 1. In this case, the probability that a susceptible individual will get sick is the percentage of sick neighbors.
4. Develop a nondeterministic (stochastic) simulation for a situation similar to that in Project 1. In this case, the probability that a susceptible individual will get sick is the average level of infection of the neighbors. For example, suppose the neighbor to the north is susceptible (level of infection = 0); west is immune (level = 0); south is in the first day of infection and very contagious (level = 1); and west is in the second day of infection and less contagious (level = 0.5). Thus, the probability of the individual becoming sick is $(0 + 0 + 1 + 0.5)/4 = 0.375$. The maximum possible total level is 4 and occurs when all neighbors are in the first day of infection. In this case, the probability of the individual catching the virus is $(1 + 1 + 1 + 1)/4 = 1 = 100\%$.
5. Develop a simulation where initially no individuals are sick; but one individual, "Typhoid Mary," is a carrier who never gets sick. Mary walks at random through the grid, and at each step she changes places with the individual in whose cell she steps. Use a contagion situation as in Project 1, 2, 3, or 4. Color Mary as yellow.

MODULE 14.7

HIV—The Enemy Within

Prerequisite: One of Module 10.3, "Spreading of Fire"; Module 10.4, "Movement of Ants—Taking the Right Step"; or Module 10.5, "Biofilms—United They Stand, Divided They Colonize."

The Developing Epidemic

> As every month went by, I became more convinced that we were dealing with something that was going to be a disaster for society.
>
> —Anthony S. Fauci, M. D., Director, National Institute of Allergy and Infectious Diseases, 1982

> When we see what has happened in Africa, one might think: 'We would have done anything to prevent this—if only we had known.' But we did, and we didn't.
>
> —Peter Piot, UNAIDS Executive Director, 2004

If you had been an attending physician at New York's Bellevue Hospital during the late 1970s, you might have admitted several patients suffering from a fairly rare medical problem—*Pneumocystis* pneumonia. At about the same time, doctors in California were seeing similar patients. Often these patients also were infected opportunistically with cytomegalovirus and/or *Candida albicans* (yeast). Furthermore, there were relatively high occurrences of a fairly rare cancer, Kaposi's sarcoma. The pneumonia and this cancer were almost never seen except in patients with suppressed immune systems, which apparently characterized each of these patients. The first papers reporting on these cases appeared in 1981. So it began—the AIDS epidemic in the United States (CDC 1981; Thebody.com 2007).

The 1981 reports did not gain the public's attention. After all, there were lots of other things going on in the United States and the world. Ronald Reagan replaced Jimmy Carter as president; the Iranian hostages were released; the first shuttle was successfully launched; MTV first appeared on cable; and IBM introduced its first PC. During 1981, the Pope and President Reagan survived assassination attempts, but President Anwar Sadat of Egypt did not (Wikipedia 2012).

Fortunately, these medical reports did alert some in the public health community. In June, the first patient with the new, unnamed disease was seen at the National Institutes of Health (NIH); and in July, The Centers for Disease Control and Protection (CDC) formed a task force on "Kaposi's Sarcoma and Opportunistic Infections." In only a year, there were more than 400 cases and 155 deaths in the United States from this disease, which now had a name—**acquired immune deficiency syndrome** (**AIDS**). The disease was characterized by a defective cell-mediated immune response. The clustering of the rare opportunistic infections and the Kaposi's sarcoma was a result of this weakened immune system, but the cause of the defect was unknown. Substantial evidence, however, pointed to an infectious agent. The unknown agent was apparently acquired sexually, through intravenous drug use, or by transfusion. By the beginning of 1984, 3000 cases of AIDS were reported in the United States, and almost 1300 had died (CDC 1983).

In the spring of 1983, scientists at the Pasteur Institute had isolated a new human retrovirus, LAV (lymphadenopathy-associated virus), but they did not claim it to be the cause of AIDS. In 1984, U.S. scientists at the National Cancer Institute showed that a retrovirus (HTLV III, human T cell leukemia virus III) was the apparent cause of AIDS and later concluded that HTLV III and LAV were the same (CDC 1981; Thebody.com 2007).

Despite earlier predictions of a fast cure, this virus, which we now call HIV, has proved to be a very tricky customer and has not yet succumbed to human genius. By March of 1988, more than 84,000 AIDS cases were reported from 136 countries (CDC 1988).

In 2010, the CDC estimated that 1,148,200 adults ($\geq$ 13 years of age) were living with HIV in the United States, with about 50,000 new cases being diagnosed each year. In 2009, 17,774 people died, making a total of more than 619,000 people with an AIDS diagnosis having died (CDC 2012). The statistics for the world are more dire, especially in sub-Saharan Africa. In 2011, 1.7 million people died from HIV and AIDS-associated diseases, with 2.5 million new cases. In 2011 the world saw an estimated 34 million individuals living with HIV—more than 30.7 million adults and 3.3 million children. In excess of two-thirds of these live in the sub-Sahara (WHO)

Attack on the Immune System

We have learned a lot about HIV and its interactions with the immune system since its discovery. The **human immunodeficiency virus** (**HIV**) is a type of RNA virus called a **retrovirus**. Retroviruses have some important and unusual characteristics that are significant in certain types of cancer and diseases like AIDS. Retrovirus particles are made up of a **core** (RNA + enzymes) surrounded by a **capsid** (protein). These particles are covered with a lipid **envelope**, which fuses with host cell membranes as the virus enters that cell. This envelope contains glycoproteins that recognize and bind to specific receptors on the host cells. These viruses are able to synthesize DNA from their RNA template, and they carry with them an enzyme that helps to insert the newly synthesized **viral DNA (vDNA)** into a chromosome of the host cell. HIV is actually a special type of retrovirus, called a **lentivirus** ("slow viruses"), associated with slow, degenerative disorders. In such infections, typically a considerable amount of time passes between infection and the appearance of major symptoms (NIAID 2011, 2012; Varmus 1988).

HIV infects some incredibly significant cells that play crucial roles in **cell-mediated immunity (CMI)**—CD4+ T-lymphocytes, macrophages, dendritic cells. CMI is one of two major arms of the immune system. It gets its name from its dependence on the effector functions of specific types of immune cells. For example, **CD4+ T-lymphocytes**, also known as **T-helper cells**, help to coordinate immune response through direct interactions with other cells and through the secretion of control chemicals. Some of these chemicals can activate certain cells, which in turn secrete toxins that kill tumor or virally infected cells. Others attract and activate particular white blood cells that engulf invading pathogens. As these cells and lymph tissue become damaged and disabled by HIV, the body becomes progressively more susceptible to various pathogenic agents and cancer. Consequently, death from HIV often occurs from AIDS-associated diseases, rather than from AIDS itself.

Plan of Attack

To enter a cell, the virus must first be able to bind to the cell. Some cells of the body, including certain T-lymphocytes, possess what are termed **CD4** transmembrane receptors. These cells are classified as **CD4+ cells**. Projecting from the HIV envelope is a glycoprotein, **gp120**, which binds to a CD4 receptor like a hand in a glove. Binding induces a conformational change in gp120, which promotes its binding to one of several coreceptor molecules located in the host cell membrane. Once the virus has attached securely to its target, the viral and cell membranes fuse, allowing the virus particle to enter the cell.

Following entry into the cytoplasm of the cell, the capsid is removed from the core of the virus. The core contains RNA and several enzymes, including **reverse transcriptase**. Now activated, reverse transcriptase, using host cell raw materials, synthesizes a double strand of DNA—vDNA; vDNA is transported into the nucleus, where, using an enzyme (integrase), it is inserted into a host cell chromosome. This piece of viral DNA, called a **provirus**, may remain in the chromosome passively **(latent)** or may activate and begin the production of new HIV particles.

The production of new viruses commences with the synthesis of new viral RNA molecules from the vDNA using the host's polymerase. This viral messenger RNA is transported into the cytoplasm, where, using host's ribosomes, enzymes, tRNA, and raw materials, it is translated into HIV **structural proteins** (core, capsid, and envelope) and enzymes. The viral messenger RNA, equivalent to viral genomic RNA, is packaged with core proteins and enzymes into new virus particles near the plasma membrane. Envelope proteins are incorporated into the host's membrane, which encases the new virus as it buds from the cell. The last step involves the cleavage of core proteins and enzymes into shorter pieces by a third viral enzyme (protease). Then, the viruses are infectious, capable of invading new host cells.

The primary targets for HIV are these CD4+ T-lymphocytes, but the virus also may attack other cells important to an immune response. For instance, cells like **macrophages** and **dendritic cells** normally consume pathogens, presenting essential elements from the microbe on their surfaces. These elements can activate various types of T-cells and also stimulate the production of antibodies. Often, the virus does not destroy macrophages and dendritic cells. In this way, significant quantities of virus are concealed safe from destruction in the very cells that are supposed to help protect the body from infection. Dendritic cells associated with the mucosa (lining) of major virus portals (e.g., vagina, vulva, penis, and rectum) pick up viruses and transport them to lymph nodes, which are sites for many types of immune cells (Bugl 2001; Thebody.com 2005; NIAID 2011, 2012)

Simulation of the Attack

Computational scientists are employing cellular automaton (CA) simulations to model the immune system and diseases that attack this system, such as AIDS. With CA's stochastic nature, these scientists can use these simulations to estimate the distribution of the system's behaviors as well as the averages; can easily adjust the complex interactions to study the course of an infection and to consider new scenarios, such as new drug therapies; and can express the components and processes in biological language (Kleinstein and Seiden 2000; Sloot et al. 2002). Moreover, in the case of AIDS, CA can model the infection's two time scales over three phases—weeks for the primary response and years for the clinical latency with deterioration of the immune system and for AIDS—much more easily than an approach with differential equations (Sloot et al.).

Projects

1. a. Develop a cellular automata and visualization of an HIV infection. Each site represents one of the following:

- *healthy*: healthy cell
- *infected-A1*: infected cell that can spread the infection
- *infected-A2*: infected cell in its final state before dying due to immune system intervention
- *dead*: infected cell killed by immune system intervention

The system uses the following probabilities:

- *probHIV*: initial probability (fraction) of *infected-A1* cells
- *probReplace*: probability that a *dead* cell will be replaced by a *healthy* cell at the next time step
- *probInfect*: probability that a new *healthy* cell may be replaced by an *infected-A1* cell

The rules for the system are as follows:

- A *healthy* cell with at least one *infected-A1* neighbor becomes *infected-A1* because of infection due to contact before the immune system can respond.
- A *healthy* cell with *numberOfA2* number of *infected-A2* neighbors, where $3 \le numberOfA2 \le 8$, becomes *infected-A1* because *infected-A2* cells with concentration above some threshold can contaminate a healthy cell.
- All other *healthy* cells remain *healthy*.
- An *infected-A1* cell becomes an *infected-A2* cell after *responseTime*, the number of time steps for the immune system to generate a response to kill the *infected-A1* cell.
- An *infected-A2* cell becomes a *dead* cell.
- A *dead* cell becomes a *healthy* cell at the next time step with a probability of *probReplace* because the immune system has great ability to recover from an infection's immunosuppressant.
- A new, *healthy* cell may be replaced by an *infected-A1* cell with a probability of *probInfect* because new infected cells can come into the system.

Initialize the grid using *probHIV* = 0.05, indicating that during the primary infection, 1 in 100 to 1 in 1000 T-cells harbor viral DNA. Because only 1 in 10^4 to 1 in 10^5 cells in an infected person's peripheral blood express viral proteins, use *probInfect* = 10^{-5}. Because the immune system has great ability to replenish dead cells, use *probReplace* = 0.99. Have *responseTime* be 4 weeks, because the time for the immune system to generate a response to kill the *infected-A1* cell is generally between 2 and 6 weeks. Use 8 neighbors for each site and periodic boundary conditions (dos Santos and Coutinho 2001).

b. Plot the numbers of healthy, infected, and dead cells versus time from 0 through 12 weeks and then from 0 through 12 years. To obtain the data, run the simulation a number of times and compute the appropriate average values.

c. Discuss your results. For the visualization in Part a and the graphs in Part b, identify the stages of the infection and explain your results.

2. a. Revise Project 1 to model an HIV infection in the presence of a drug therapy regime, which attempts to block viral replication within the cells. Assume therapy begins at week 300. A therapy has an associated integer rank level, *rankLevel* ($0 \le rankLevel \le 8$), indicating the effectiveness of the drug, with 0 being the most effective. This rank level models the drug's ability to suppress viral replication and presents a limit to the number of *infected-A1* neighbors that can become infected. At the time of therapy, the first rule in Project 1 changes to be the following (Sloot et al. 2002):

- During drug therapy, a *healthy* cell with *rankLevel* number or more of *infected-A1* neighbors becomes *infected-A1* with a probability of (1 – *probRespond*) * *rankLevel*/8, where *probRespond* is a response-to-therapy-related probability.

b. Consider a *probRespond* function that is a constant probability for a set number of time steps and then becomes a significantly smaller constant. Run the simulation for various values of *rankLevel*, and discuss the results. Discuss the impact of the therapy on the simulation visualization and on the graphs of the numbers of healthy, infected, and dead cells versus time.

c. Repeat Part b employing a decreasing linear function for *probRespond*. Because *probRespond*(*t*) is a probability at time *t* of treatment, its range is between 0.0 and 1.0.

References

Bugl, Paul. 2001. "Cell-Mediated Immunity." From "Immune System," course notes from Epidemics and AIDS. http://uhaweb.hartford.edu/BUGL/immune.htm#cellmed (accessed January 1, 2013)

CDC (Centers for Disease Control and Prevention). 1981. "Kaposi's Sarcoma and *Pneumocystis* Pneumonia Among Homosexual Men—New York City and California." *Mortality and Morbidity Weekly Report*. 30 (July 4): 306–308. http://www.cdc.gov/hiv/resources/reports/mmwr/pdf/mmwr04jul81.pdf (accessed January 1, 2013)

———. 1983. "Acquired Immunodeficiency Syndrome (AIDS)." *Weekly Surveillance Report—United States*. December 22. http://www.cdc.gov/hiv/topics/surveillance/resources/reports/pdf/surveillance83.pdf (accessed January 1, 2013)

———. 1988. "Update: Acquired Immunodeficiency Syndrome (AIDS)", *Worldwide. Mortality and Morbidity Weekly Report*. 37(18): 286–288, 293–295. http://www.cdc.gov/mmwr/preview/mmwrhtml/00000023.htm (accessed January 1, 2013)

———. 2012. "Diagnoses of HIV Infection and AIDS in the United States and Dependent Areas, 2010." *HIV Surveillance Report*, 17(4). http://www.cdc.gov/hiv/surveillance/resources/reports/2010supp_vol17no4/index.htm (accessed January 1, 2013)

dos Santos, R.M.Z., and S. Coutinho. 2001. "Dynamics of HIV Infection: A Cellular Automata Approach." *Physical Review Letters*, 87(16).

Fauci, Anthony S. 1982. "The Syndrome of Kaposi's Sarcoma and Opportunistic Infections: An Epidemiologically Restricted Disorder of Immunoregulation." *Annals of Internal Medicine* (editorial), 96(6, Pt 1) 777–779

Kleinstein, Steven H., and Philip E. Seiden. 2000. "Simulating The Immune System," *Computer Simulations*. July/August: 69–77.

NIAID (National Institute of Allergies and Infectious Diseases). 2011. "HIV/AIDS." http://www.niaid.nih.gov/topics/hivaids/understanding/Pages/Default.aspx (accessed January 1, 2013)

———. 2012. "How HIV Causes AIDS." National Institutes of Health. http://www.niaid.nih.gov/topics/HIVAIDS/Understanding/howHIVCausesAIDS/Pages/cause.aspx (accessed January 1, 2013)

Office of Technology Assessment. 1985. "Review of the Public Health Service's Response to AIDS." U.S. Congress, Washington, DC. http://www.princeton.edu/~ota/disk2/1985/8523/8523.PDF (accessed January 1, 2013)

Piot, Peter. 2004. "AIDS and the Way Forward." A World AIDS Day Address, Woodrow Wilson International Center for Scholars, Washington, DC, November 30.

Sloot, Peter, Fan Chen, and Charles Boucher. "Cellular Automata Model of Drug Therapy for HIV Infection." *Cellular Automata* (2002): 282–293.

Thebody.com. 2005. "The HIV Life Cycle." http://www.thebody.com/content/art40989.html (accessed January 1, 2013)

———. 2007. "A Brief History of HIV." http://www.thebody.com/content/art43596.html (accessed January 1, 2013)

Varmus, Harold. "Retroviruses." 1988. *Science* 240(4858): 1427–1435.

WHO (World Health Organization). "Global Summary of the AIDS Epidemic." http://www.who.int/hiv/data/2012_epi_core_en.png (accessed January 1, 2013)

Wikipedia. 2012. "1981." http://en.wikipedia.org/wiki/1981 (accessed January 1, 2013)

MODULE 14.8

Predator-Prey—"Catch Me If You Can"

Prerequisite: For Projects 1 and 7, one of Module 10.3, "Spreading of Fire"; Module 10.4, "Movement of Ants—Taking the Right Steps"; or Module 10.5, "Biofilms— United They Stand, Divided They Colonize." For Projects 2–6, Module 10.4, "Movement of Ants—Taking the Right Steps."

Introduction

We have already dealt with predator-prey models in Module 4.2 but from a system dynamics point of view. Projects in this module consider the same subject from the perspective of cellular automata.

About 1970, John Conway developed the **Game of Life**, a 2D cellular automaton with rules for the births and deaths of an imaginary species. This program was the first or one of the first to execute on a parallel computer, a system consisting of multiple processors working together to solve a problem. The rules for the mythical life form in the Game of Life are as follows, where the nearest neighbors are the eight surrounding cells:

- A site that is not alive but has exactly three living neighbors has a birth.
- A site that is alive and has exactly two or three living neighbors stays alive.
- All other sites die or stay dead.

Conway carefully developed these rules to enable situations in which patterns grow and evolve over many time steps without becoming stagnant or chaotic. Some patterns, called **life forms**, persist throughout the simulation. **Still lifes** do not change unless other cells interfere, while some life forms exhibit periodic behavior and others move across the grid. Project 1 develops the Game of Life and explores several life forms. Other projects explore more realistic predator-prey environments (Wikipedia 2012).

Projects

Develop a cellular automaton simulation for each of the following.

1. a. Develop a simulation and animation for Conway's Game of Life with periodic boundaries. Run the simulation several times with random initial grids. Then, incorporate the following life forms into grids, and describe their behavior (Wikipedia 2012).

b. Block: A square of four live cells

c. Traffic light: Three consecutive live cells in a row or a column

d. Glider: Three consecutive live cells in a row with another live cell to the north of the leftmost cell and another live cell to the northeast of the latter cell

e. Other life forms, such as those by Eric Weisstein (2002)

2. a. Develop a simulation with visualization in which a cell can be a predator, a prey, or empty. Use eight nearest-neighbor cells. Initialize the grid with a given population density (probability) for a cell being of each type. Each time step of the simulation has two phases: change of state and movement. The rules for change of state are as follows:

- If a prey "meets" (i.e., has as a neighbor) a predator, the predator eats the prey (i.e., the prey's site becomes empty). If more than one predator is encountered, a random predator neighbor is selected to dine.
- A predator dies (i.e., the predator's site becomes empty) when it has gone too long (i.e., a given number of time steps) without food.
- During movement, avoid collision as in the text of Module 10.4, "Movement of Ants—Taking the Right Steps."

b. Run the simulation a number of times for various population densities, and discuss the results.

3. a. Develop a simulation with visualization involving predators, which can be in a hungry or full state, and prey. Use four nearest-neighbor cells. Each cell can contain up to four hungry predators, four full predators, and four prey individuals, so that a cell can hold from none to 12 individuals. Initialize the grid with given population densities (probabilities) for components of each cell. At each time step of the simulation, the animals undergo predation/reproduction, then direction selection, and then movement. The rules for predation/reproduction are applied in the following order (Alfonseca and Ortega 2000):

- If no prey individuals are in the same cell, a hungry predator dies.
- If at least two prey individuals are in a cell, a hungry predator is in the cell, and less than four full predators are in the cell, then a hungry predator eats one prey individual and becomes full.
- If no prey individuals are in the same cell, a full predator becomes hungry.
- If at least two prey individuals are in a cell, a full predator is in the cell, and fewer than three hungry predators are in the cell, then a full preda-

tor eats one prey individual, reproduces a hungry predator, and becomes a hungry predator.

- If two or three prey individuals are in a cell, a prey reproduces.

The rule for direction selection is as follows (Alfonseca and Ortega 2000):

- Each individual turns in a random direction (north, east, south, or west) to which no other individual from that category (hungry predator, full predator, or prey) has turned.

Note that the algorithm avoids collisions by having at most four animals in each category moving in different directions.

b. Graph the population densities of predators and prey versus time.

c. Graph the number of predators versus the number of prey.

4. a. Develop a simulation with visualization involving wolves, sheep, and grass on a grid. A cell is empty or contains one of the following items: a male wolf, a female wolf, a female wolf with pup, a male sheep, a female sheep, a female sheep with lamb, or grass. Associated with each animal is an integer **food ration**, or amount of stored energy from food, up to some maximum value. Assume a population density for each item. The rules are as follows (He et al. 2003):

- A sheep moves into a neighboring empty site, preferring one with grass.
- A lamb leaves its mother and moves into a neighboring empty site. At random this new sheep is a male or female, and its food ration is the same as that of the mother.
- A wolf moves into a neighboring empty site.
- A pup leaves its mother and moves into a neighboring empty site. At random, this new wolf is a male or female, and its food ration is the same as that of the mother.
- If its ration of food is less than the maximum, a sheep eats neighboring grass and increases its ration to the maximum amount.
- If a female sheep has at least a designated amount of food ration (such as 2), is of reproduction age (such as 8), and has a male sheep of reproduction age as a neighbor, she becomes a female sheep with lamb.
- If its ration of food is less than the maximum (such as 3), a wolf eats a neighboring sheep and increases its ration to the maximum amount.
- If a female wolf has at least a certain amount of food ration (such as 2), is of reproduction age (such as 8), and has a male wolf of reproduction age as a neighbor, she becomes a female wolf with pup.
- An independent baby matures in a certain number of time steps, such as 8.
- An animal's food ration decreases by 1 at each time step.
- An animal dies when its food ration becomes 0.
- Grass grows in a certain number of time steps, such as 4.

Avoid collisions as in the text of Module 10.4, "Movement of Ants—Taking the Right Steps." Initialize the grid at random with certain densities of each item and with random food rations and ages for each animal. Run the simulation a number of times, obtaining situations in which the sheep, wolves, and grass coexist with oscillating densities; in which the sheep become extinct; and in which all animals die.

b. Graph the population densities of sheep, wolves, and grass versus time.

c. Adjust the program to run the simulation a number of times, computing and storing the average number of sheep, wolves, and grass at each time step. Plot these averages versus time. Discuss the results.

5. a. Develop a simulation with visualization involving mobile predators and stationary prey. For example, algae that grow on rocks in intertidal areas are a favorite food of some snails. Assume initial population densities for predators and prey. Each predator has a direction to which it turns, a length of time until giving birth (reproduction time), and a length of time until starving (starvation time). The rules are as follows (Gaylord and Nishidate 1996):

- If a predator's reproduction and starvation times are both 0, the predator gives birth and dies. For simplicity, we assume only one child.
- If a predator's starvation time is 0 and reproduction time is positive, the predator dies.
- Prey grows in an empty site with a certain probability.
- If a predator's reproduction time is 0, its starvation time is positive, and a neighboring site with a prey is available, then the predator moves to that site and leaves a child in the old site. Both parent and child get maximum reproduction and starvation times.
- If a predator's reproduction time is 0, its starvation time is positive, and an empty neighboring site is available, then the predator moves to that site with maximum reproduction time and with starvation time decremented by 1 and leaves a child in the old site with maximum reproduction and starvation times.
- If a predator's reproduction and starvation times are positive and a neighboring site with prey is available, then the predator moves to that site with reproduction time decremented by 1 and starvation time set to the maximum.
- If a predator's reproduction and starvation times are positive and an empty neighboring site is available, then the predator moves to that site with its times decremented by 1.
- If a predator does not move, its times decrement by 1.

During movement, avoid collision as in the text of Module 10.4, "Movement of Ants—Taking the Right Steps."

b. Run the simulation and visualization a number of times for various population densities and discuss the results.

c. Adjust the program to run the simulation a number of times, computing and storing the average number of individuals in each species at each time step. Plot these averages versus time, and plot the number of predators versus the number of prey. Adjust the number of predators and prey to obtain graphs that resemble those in Figures 4.2.3 and 4.2.4 of Module 4.2, "Predator-Prey Model."

6. Adjust Project 5 to allow for a prey population in which each individual is mobile and can give birth.

7. a. Simulations can help illuminate ecosystem problems when a species becomes extinct or varies greatly in size. Consider a hierarchy of species, numbered 1, 2, . . ., m, where species i is higher on the food chain than species i –1 and $m \geq 2$. Thus, species i – 1 is food or prey for species i, and species i is predator for species i – 1. Develop a predator-prey cellular automaton simulation and visualization for an ecosystem. Each cell is empty or contains exactly one animal. Initially, with probability $probSpecies_i$, a cell contains an individual from species i, and the sum of these probabilities is less than 1. Use eight surrounding cells as neighbors and the following rules (Yang 2003):

- If a cell has no predator and no prey neighbors, then the cell obeys the rules of Conway's Game of Life (see Project 1).
- If a cell contains a live animal and the predator neighbors outnumber the prey neighbors, then the animal in this cell dies.
- If the prey neighbors outnumber the predator neighbors, then the cell stays or becomes alive.
- If the number of prey neighbors equals the number of predator neighbors, a positive number, then the state of the cell does not change.
- With probability $probDie_i$, a cell containing species i dies.

b. For $m = 2$, adjust the program to run the simulation a number of times, computing and storing the average number of individuals in each species at each time step. Plot these averages versus time, and plot the number of predators versus the number of prey. Adjust the number of predators and prey to obtain graphs that resemble those in Figures 4.2.3 and 4.2.4 of Module 4.2, "Predator-Prey Model."

c. For $m > 2$, run the simulation and visualization several times with $probDie_i = 0$ for all species. Then, after making $probDie_i$ a small, positive number for one species, repeat the experiment. Discuss the results.

d. For $m > 2$, adjust the program to run the simulation a number of times, computing and storing the average number of individuals in each species at each time step. Run the simulation a number of times with $probDie_i = 0$ for all species, and plot the average number of individuals in each species versus time. Then, after making $probDie_i$ a small, positive number for one species, repeat the experiment. Discuss the results.

e. For $m > 2$, adjust the program to run the simulation a number of times, computing and storing the average number of individuals in each species at each time step. Plot the average number of individuals in each species versus time. Then, after increasing $probSpecies_i$ for some species i and adjusting the corresponding probabilities for the other species, repeat the experiment. Discuss the results.

References

Alfonseca, Manuel, and Alfonso Ortega. 2000. "Representation of Some Cellular Automata by Means of Equivalent L Systems." *Complexity International* 7: 1–16.

Gaylord, Richard J., and Kazume Nishidate. 1996. "Predator-Prey Ecosystems." *Modeling Nature: Cellular Automata Simulations with Mathematica*. New York: TELOS/Springer-Verlag, pp. 143–154.

He, Mingfeng, Hongbo Ruan, and Changliang Yu. 2003 "A Predator-Prey Model Based on the Fully Parallel Cellular Automata." International Journal of Modern Physic C, 14(9): 1237-1249. http://arxiv.org/pdf/cond-mat/0305262.pdf (accessed January 1, 2013)

Weisstein, Eric W. 1995–2002. "Eric Weisstein's Treasure Trove of the Life Cellular Automaton." Wolfram Research, Inc. http://www.ericweisstein.com/encyclopedias/life/ (accessed January 1, 2013)

Wikipedia. 2012. "Conway's Game of Life." http://en.wikipedia.org/wiki/Conway%27s_Game_of_Life (accessed January 1, 2013)

Yang, X. 2003. "Characterization of Multispecies Living Ecosystems with Cellular Automata." In *ICAL 2003: Proceedings of the Eighth International Conference on Artificial Life*: 138–141. http://arxiv.org/pdf/1003.5288.pdf (accessed January 1, 2013)

MODULE 14.9

Clouds—Bringing It All Together

Prerequisite: For Projects 1–7, one of Module 10.3, "Spreading of Fire"; Module 10.4, "Movement of Ants–Taking the Right Steps"; or Module 10.5, "Biofilms—United They Stand, Divided They Colonize." For Project 8, the section on "Sequential Algorithm for the N-Body Problem" from Module 12.2, "Parallel Algorithms."

Introduction

The Clouds consign their treasures to the fields;
And, softly shaking on the dimpled pool
Prelusive drops, let all their moisture flow
In large effusion, o'er the freshen'd world.
—James Thomson, *The Seasons,*
"Spring" (lines 27–30)

Clouds. Their endless variety and beauty inspire human beings to dream, to imagine, and to write poetry. They help to cool the earth by reflecting sunlight, but they also help to keep it warm by trapping heat radiated from the earth's surface. But, what are clouds? Scientists tell us that they are collections of water droplets and ice, as well as nonaqueous solids and liquids (Alcorn 2007). During the summer, air near the earth is warmed and rises. As the air rises, it expands and cools, generating relative humidity of 100%, or saturated air. The moisture required for saturation decreases as the air temperature decreases. After saturation is achieved, further cooling triggers water vapor to condense into small droplets. These droplets, and sometimes ice crystals, may form by condensing on suspended aerosols (salt, dust, pollution, etc.). The products of all this condensation are light, fluffy **cumulus clouds**. Known as fair-weather clouds, they are characterized by flat bases and do not give rise to any rain (Alcorn 2007; UIUC 1997; Odman 2004; SSEC 2004; Geerts 2009).

However, if cumulus clouds are overlain by large quantities of cold, unstable air, the warm air may continue to ascend as mighty updrafts. As a consequence, clouds develop further vertically and are transformed into taller, **cumulonimbus** forms. Cumulonimbus clouds, their anvil-shaped tops reaching altitudes exceeding 12,000 m, appear as towers and sometimes in lines called **squall lines**. In these "thunderheads," water droplets continue to condense, and updrafts in the cloud carry the smaller ones upward. The countless droplets collide with each other as they move. Some of these collisions result in droplets merging to form larger droplets. This process is referred to as **coalescence**. The droplets that are large enough begin to fall through the cloud, colliding and coalescing as they tumble downward. Some of these are large enough to make it to the earth's surface. If liquid, we term it rain (UIUC 1997; Brown 2003).

Projects

Develop a cellular automaton simulation for each of the following projects. Be sure during a time step that a droplet is not coalesced into more than one other droplet.

1. **a.** Develop a 2D cellular automaton simulation and visualization of **coalescence** of droplets in a cross section of a cloud. Each grid point contains a number representing the droplet's size or indicating that the cell is empty. At a step of the simulation, each droplet moves in a random direction. If more than one droplet moves into a cell, the droplets coalesce. The size of the new droplet is the sum of their sizes. Initialize the grid with droplets according to a probability that corresponds to the **relative humidity**, which is (partial vapor pressure)/(saturation vapor pressure) for a particular temperature. At initialization, give each droplet a normally distributed random size (Gaylor and Nishidate 1996).
   b. Produce a histogram of the size distribution of droplets at the end of the simulation.
   c. Produce a graph of the average size of droplets versus time. Discuss your findings. What will happen eventually if you allow your simulation to run long enough?
2. The simulation in Project 1 eventually produces one large droplet, which is an unrealistic result (Gaylor and Nishidate 1996).
   a. Repeat Project 1 and achieve a steady-state distribution of droplet sizes by adding small droplets with a designated probability throughout the simulation.
   b. Achieve a steady-state distribution of droplet sizes by removing larger drops with a certain probability.
   c. Achieve a steady-state distribution of droplet sizes using the techniques of Parts a and b.
3. Repeat Project 1 and achieve a steady-state distribution of droplet sizes by breaking each larger droplet into two droplets with a certain probability. The new droplet forms in a random vacant neighbor and is of random size. The sum of the sizes of the two droplets equals the size of the original droplet (Gaylor and Nishidate 1996).
4. Repeat Project 1, including the formation of rain and eight neighbors per cell. Have the ground toward the south. Do not allow medium-sized droplets to move to the north, northeast, or northwest. Medium-large droplets can fall

only to the south, southeast, or southwest. Large droplets can head only toward the south. Remove droplets that travel off the south boundary, indicating that they have fallen to the ground. Continually add small droplets to the grid at random with a certain probability.

5. a. Develop a 2D cellular automaton simulation and visualization of cloud evolution, such as one might view from the ground. Thus, the cross section is a horizontal slice of the cloud. Each cell has three Boolean values (Dobashi et al. 2000):

- *humidity*: true if the cell contains enough water for cloud droplets
- *cloud*: true if the cell has cloud droplets
- *act*: true if the vapor in the cell is ready to transition to cloud droplets

The following are probabilities in the simulation:

- *probHumidity*: probability that a noncloud cell has enough humidity to transition to cloud
- *probExtiction*: probability that a cloud cell becomes a noncloud cell
- *probAct*: probability that a cell that is not ready to act (i.e., to transition from vapor to cloud) becomes ready

The transition rules are as follows:

- If a cell's value of *act* is false, the cell's *humidity* value remains the same at the next time step.
- If *cloud* or *act* is true in a cell, *cloud* is true at the next time step.
- If *act* is not true in a cell but *humidity* is true and at least one neighbor's value of *act* is true, then the cell's value of *act* becomes true.
- With probability *probHumidity*, a cell's *humidity* value becomes true.
- If *humidity* is true, it remains true.
- If *cloud* is true, then with probability *probExtiction*, *cloud* becomes false.
- With probability *probAct*, a cell's *act* value becomes true.
- If *act* is true, it remains true.

Initialize the grid at random with only the middle cell having *cloud* as true.

b. Add wind to your simulation by letting *velocity* be an integer indicating wind velocity from left to right across the grid. Thus, the state of a cell in column j at one time step is the state of cell in column (j – *velocity*) at the previous time step.

6. Develop a 3D version of Project 5.

7. Develop a 2D cellular automaton simulation and visualization of the first step in the formation of precipitation, **condensation,** or **deposition,** of vapor on particles, called **condensation nuclei**, to form droplets. The nuclei are typically tiny (from 0.05 to 0.5 μm in radius) solid particles, such as dust, smoke, or salt. The process, which occurs in high humidity, is fast at first but then slows. Specifically, experiments have shown that the rate of change of the radius (r) of a cloud particle is proportional to $1/r$. For a normal nucleus, which neither attracts nor repels water, if the relative humidity (RH; see Project 1 for definition) is less than 100%, evaporation exceeds condensation so that the droplet shrinks. However, some of the nuclei attract water molecules and thus promote condensation at RH less than 100%. Also, if the RH is equals 100%, with a normal nucleus, the droplet grows. Condensation is important until the cloud particle becomes a cloud droplet with a radius of about 100 μm. At that time, the collision-coalescence process (see Project 1) becomes more significant (Geerts 2009; McCormack 1999).

For the simulation, initially have a cluster of several cells representing the condensation nucleus and other cells picked at random indicating vapor with the number of such vapor particles based on RH. Have vapor particles move at random. If a vapor particle has a neighbor that is part of the growing droplet, the vapor particle condenses on the droplet with a designated probability based on conditions. Repeat the simulation a number of times with various condensation nuclei. Allowing the simulation to run for a while, discuss the ultimate shapes of the droplets.

8. Develop a simulation of the precipitation process in a relatively warm cloud with temperatures above freezing. We call droplets with radii larger than 0.25 mm **raindrops**, and those with radii larger than 2 mm frequently split in two. Suppose an updraft of air pushes smaller droplets up into the cloud. Larger droplets, which have higher terminal velocities, fall. For instance, the terminal speed of a cloud droplet with radius 0.05 mm is 27 cm/s; with a radius of 0.1 mm, the terminal speed of a droplet is 70 cm/s; while a raindrop of radius 1 mm has a terminal speed of 550 cm/s. Some particles that appear on a collision path, such as a large falling **collector drop** and a much smaller rising droplet, do not collide; but the smaller droplet streamlines past the larger one. **Collision efficiency** is $E = d^2/(r_1 + r_2)^2$, where d is the **critical distance**, or distance between the center lines of the drop and droplet, and r_1 and r_2 are the corresponding radii. For a coordinate system in which the xz-plane is horizontal, suppose the collector drop and smaller droplet are at locations (x_1, y_1, z_1) and (x_2, y_2, z_2), respectively. Assume that the collector drop is going straight down and the droplet is going straight up. Then, the critical distance is the distance between the two points projected onto the xz-plane, namely, $\sqrt{(x_2 - x_1)^2 + (z_2 - z_1)^2}$.

Even if collision occurs, coalescence might not occur because sometimes the droplets bounce off each other. Studies indicate that droplets tend to become charged during thunderstorms, and droplets with opposite charges are more likely to coalesce. Collisions can also cause a drop to break apart. Drops involved in the collision/coalescence process usually have radii no larger than 2.5 mm. **Coalescence efficiency** is the portion of collisions that result in the droplets sticking together. Laboratory experiments indicate that if the radius of a collector drop is less than 0.4 mm or the radius of a droplet is less than 0.2 mm, then coalescence efficiency is about 1.0. If the collector drop's radius is between 1 mm and 2.5 mm, then coalescence efficiency is less than 0.2. When the radius of the droplet is about 60% that of the drop, coalescence efficiency is small, but the collision efficiency is close to 1.0. **Collection efficiency** is the product of the collision and coalescence efficiencies (Brown 2003; Geerts 2009).

References

Alcorn, Marion. 2007. "Exercise 10—Clouds." Atmo 202—Atmospheric Science Lab, Department of Atmospheric Sciences, Texas A&M University. http://www.met.tamu.edu/class/atmo202/Exer10dir/clouds.html (accessed January 2, 2013)

Brown, Derek W. 2003. "The Collision and Coalescence Process." From The Plymouth State University Meteorology Program's tutorial "Precipitation: Formation to Measurement." http://vortex.plymouth.edu/precip/precip2aaa.html (accessed January 2, 2013)

Dobashi, Yoshinori, Kazufumi Kaneda, Hideo Yamashita, Tsuyoshi Okita, and Tomoyuki Nishita. 2000. "A Simple, Efficient, Method for Realistic Animation of Clouds." Proceedings of the 27th International Conference on Computer Graphics and Interactive Techniques, pp. 19–28.

Gaylor, Richard J., and Kazume Nishidate. 1996. "Coalescence." *Modeling Nature: Cellular Automata Simulations with Mathematica*. New York: TELOS/Springer-Verlag, pp. 107–112.

Geerts, Bart. 2009. Homepage. ATSC 2000, Introduction to Meteorology. Department of Atmospheric Science. http://www-das.uwyo.edu/~geerts/atsc2000/ (accessed January 2, 2013)

McCormack, John. 1999. Precipitation. Course notes from Atmospheric Sciences 171: Introduction to Meteorology and Climate, University of Arizona. http://www.atmo.arizona.edu/students/courselinks/fall99/atmo171-mcc/atmo171_f99_10.html (accessed January 2, 2013)

Odman, Amy. 2004. "Supercooling, Clouds and Precipitation." General Science 109: Meteorology, Portland Community College. http://spot.pcc.edu/~aodman/supercooling-clouds-precipitation.doc (accessed June 12 2004; site now discontinued)

SSEC (Space Science and Engineering Center). 2004. "Cloud Identification." Satellite Meteorology Course Modules, Cooperative Institute for Meteorological Satellite Studies, University of Wisconsin–Madison. http://cimss.ssec.wisc.edu/satmet/modules/4_clouds/clouds-1.html (accessed January 2, 2013)

UIUC (University of Illinois–Urbana-Champaign). 1997. Weather World 2010. "Clouds and Precipitation." Department of Atmospheric Sciences. http://ww2010.atmos.uiuc.edu/(Gh)/guides/mtr/cld/home.rxml (accessed January 2, 2013)

USGS (United States Geological Survey). 2013. "The Water Cycle: Precipitation." http://ga.water.usgs.gov/edu/watercycleprecipitation.html (accessed January 12, 2013)

MODULE 14.10

Fish Schooling—Hanging Together, Not Separately

Prerequisite: One of Module 10.4, "Movement of Ants—Taking the Right Steps"; Module 11.2, "Agents of Interaction—Steering a Dangerous Course"; or Module 13.5 on "The Next Flu Pandemic—Old Enemy, New Identity."

Introduction

Imagine yourself suspended in the splendid blue of the Caribbean, gazing out over a beautiful, underwater garden. You are diving on one of nature's treasures—a coral reef. These gardens are the most diverse places in the ocean, home to one in four known marine species of plants and animals. You look out over the massive coral heads, decorated with sea fans and whips, various worms, and sea urchins and teeming with fish. A yellowtail damselfish darts in and out of a crevice to protect its territory from rival species. Small herds of parrotfish are grazing loudly on the coral, converting the algae into energy for themselves and the hard skeletal material into sand. Butterfly fish browse the reef for small invertebrates, and a pair of French angelfish munches on some of the many sponges tucked about the reef.

As you glide along the coral walls, you see a small school of blue chromises, moving as if they were articulated parts of one organism. Suddenly, in unison, they scurry away, and you wonder why. Soon, you know why. To your left you see the reason—a beautiful school of jacks are heading for that area of the reef. The chromises have left to avoid this oncoming mass of predators. You wonder at the precision with which this group of 50 swims, turning left, then right, up and down. They seem almost choreographed.

Fish schooling has fascinated human observers for years. Fish schools are social troupes of fish, frequently of comparable age and size, traveling as units, moving in synchrony in the same direction. We wonder why schooling is so common in various fish species (80%) and also how fish are able to coordinate such behavior (Brooks and Yasukawa; Stout).

There are several hypothesized advantages for schooling behavior. Two of the most common explanations are foraging efficiency and protection from predators. Many eyes increase the chances of finding food. Everyone follows those of the group who locate food. They are also able to overwhelm some prey in a group, whereas they would have less chance for success individually. While foraging, there are also many eyes to watch for potential predators. Moreover, while groups of fish are more easily detected by predators, the large group may resemble a single, larger organism and discourage attack (Brooks and Yasukawa; Greenfield-Boyce 2012; Stöcker 1998, 1999; Stout).

Recently, scientists have discovered that fish in their swimming motions create eddies. Schooling fish exploit the energy of eddies created by their neighbors to push them forward (Liao et al. 2003). So, it seems that schooling also decreases energy expenditure for foraging.

How fish are able to coordinate their movements so that they can respond instantaneously to changes in direction and speed of their schoolmates is complex. Most fish, especially those that school, have eyes on the sides of their heads. This location is advantageous to detecting changes in lateral events. Additionally, fish possess a lateral line system on their flanks that is sensitive to pressure changes. Swimming movements of the school generate water displacement that is detected by this system as changes in pressure (Stout).

Simulations

Scientists are attempting to understand biological aggregations, such as fish schools and bird flocks, and to determine pertinent biological and mechanical features and evolutionary behaviors. However, observing individual and group behavior in the laboratory and nature is quite challenging because of the inherent difficulties in 3D tracking of animals in air and water. What they do know has enabled scientists to devise "traffic rules" of an individual animal's response to its neighbors. Using these, computational scientists have developed mathematical models and computer simulations of a group's dynamics. Typically, such a simulation assigns forces that act upon the direction and speed of an individual, while the environment and actions of close neighbors moderate these forces. With these studies, scientists hope to determine the mechanics of relationships between individual behaviors and group spatial patterns. Also, with more detailed observations of aggregations, such as fish schools, and better simulations of observed behaviors, computational scientists hope to determine the behavioral algorithms that some animals, such as fish, employ (Parrish et al. 2002).

Projects

Develop a cellular automaton, agent-based, or matrix-based social-network simulation for each of the following:

1. Develop a 2D simulation and visualization of fish schooling (or bird flocking) behavior. Suppose each fish follows three rules:
   - Cohesion rule: A fish moves toward the mean position of its "closest neighbors."

- Separation rule: A fish does not get closer than some minimum distance to any neighbor.
- Alignment rule: A fish heads in the mean direction to which its "closest neighbors" head.

The separation rule has priority over the other two rules. If possible, a fish swims in the direction to which it is headed. Employ periodic boundaries, and initialize the fish with random positions and orientations.

Because of limited position choices on a grid, for a cellular automaton implementation, use Moore neighborhoods for the closest neighbors; omit the cohesion rule; implement the separation rule by not allowing collision; and consider each fish to have one of eight directions, represented as the integers 1–8, in which it could be headed.

2. Repeat Project 1, considering the boundaries as walls that the fish should avoid. Thus, when a fish moves "close" to a wall, say within two cells of the wall, it turns in a random direction that does not take it closer to the wall.
3. Repeat Project 1 or 2, taking into account the influence of a shark. When not close to fish, the shark moves at random. When close to a fish, the shark moves toward the prey, and the fish moves away from the shark. Have the fish and shark move faster when in close proximity to each other, and have the shark move faster than the fish. If a shark catches a fish, the predator eats the prey.
4. Repeat Project 1, having the fish move from one wall toward part of the opposite wall, which is an entrance to a cave. Fish can go through the entrance to safety. Once in the cave, a fish is no longer a participant in the simulation. Consider the other boundaries as walls that the fish should avoid. Your simulation should also take into account the influence of a shark as in Project 3.
5. Repeat Project 2. Initially, have all fish head in the same direction. After several time steps, turn a certain percentage of the fish in a different direction. Run the simulation a number of times with various percentages. Discuss how part of fish turning in a different direction impacts the behavior of the school (Huse et al. 2000).
6. Repeat any of Projects 1–5, considering only neighbors to the sides of a fish and not taking into account the influence of neighbors to the front and rear, which are out of a fish's field of view.
7. Develop a 2D simulation and visualization of fish schooling behavior in a coral bed. Each fish remains close to its nearest neighbor. Any fish that is beyond some threshold distance from its nearest neighbor is subject to shark attack. In the visualization, use different colors to indicate schooling fish, shark, and loner fish.

References

Brooks, Rebecca L., and Ken Yasukawa. "Schooling Behavior in Fish." Laboratory Exercises in Animal Behavior from Animal Behavior Society. http://animalbehaviorsociety.org:8786/Committees/ABSEducation/laboratory-exercises-in

-animal-behavior/laboratory-exercises-in-animal-behavior-schooling-behavior-in-fishes (accessed January 2, 2013)

Greenfield-Boyce, Neil. 2012. "Swarming Up A Storm: Why Animals School And Flock." Interview of Professor Ian Couzin, Princeton University. *Morning Edition* (National Public Radio). http://m.npr.org/story/158931963?url=/2012/08/17/158931963/swarming-up-a-storm-why-animals-school-and-flock (accessed January 3, 2013)

Huse, Geir, Steve Railsback, and Anders Fernø. 2000. "*Clupeoids*, A Fish Schooling Simulator Based on *Boids*." Humboldt State University. http://www.humboldt.edu/ecomodel/clupeoids.htm (accessed January 2, 2013)

Liao, James C., David N. Beal, George V. Lauder, and Michael S. Triantafyllou. 2003. "Fish Exploiting Vortices Decrease Muscle Activity." *Science*, 302(5650): 1566–1569.

Parrish, Julia K., Steven V. Viscido, and Daniel Grunbaum. 2002. "Organized Fish Schools: An Examination of Emergent Properties." *Biological Bulletin* 202: 296–305.

NetLogo. 1998. "Flocking." NetLogo Models Library. http://ccl.northwestern.edu/netlogo/models/Flocking (accessed January 31, 2013).

Stöcker, Sabine. 1998, "Models for Tuna School Formation." Mathematical Biosciences 156:167–190. http://www.soest.hawaii.edu/PFRP/soest_jimar_rpts/stocker98.pdf (accessed January 3, 2013)

———. 1999, "Models for Tuna School Formation." *Mathematical Biosciences* 156:167–190. http://www.soest.hawaii.edu/pfrp/reprints/stocker.pdf (accessed January 3, 2013)

Stout, Prentice K. "Fish Schooling." Rhode Island Sea Grant Fact Sheet. http://seagrant.gso.uri.edu/factsheets/schooling.html (accessed January 3, 2013)

Tovey, Craig. "Self-Organizing Social Structure in Fish Groups." http://www-2.cs.cmu.edu/~ACO/dimacs/tovey.html (accessed January 3, 2013)

MODULE 14.11

Spaced Out—Native Plants Lose to Exotic Invasives

Prerequisite: One of Module 10.3, "Spreading of Fire"; Module 10.4. "Movement of Ants–Taking the Right Steps"; or Module 10.5, "Biofilms—United They Stand, Divided They Colonize." Project 14 requires one of the preceding modules and Module 13.4, "Probable Cause—Modeling with Markov Chains."

There are no other Everglades in the world. They are unique...in the simplicity, the diversity, the related harmony of the forms of life they enclose. The miracle of the light pours over the green and brown expanse of saw grass and of water, shining and slow-moving below, the grass and water that is the meaning and the central fact of the Everglades of Florida.
—Marjory Stoneman Douglas, *The Everglades: River of Grass*

Introduction

Many tourists to South Florida go to Everglades National Park. The park, composed of about half of the "historical" Everglades, was established in 1947. The historical Everglades consisted of about three million acres, part of an enormous watershed that was more than five million acres. During the early twentieth century, settlers drained much of the wetlands for housing and agriculture. Furthermore, they channeled the water to ensure a constant, domestic supply of water and to control flooding. The upper one-third of the original three million acres is still used primarily to grow sugar cane. Another half million acres was also managed with canals, dams, and dykes for flood control (ETE 2005).

One of the largest wetlands in the world and so unique and ecologically important, the Everglades has been designated a World Heritage Site, an International Biosphere Reserve, and a Wetland of International Importance (NPS Subtropical 2012). Biologically, the highly diverse area is home to many species of plants and animals, including 350 species of birds, and is a convenient resting spot for countless migratory birds. The Everglades is the most important breeding area for wading birds in North America and the largest mangrove ecosystem in the Western Hemisphere (NPS Ecosystem 2012). Moreover, water from the Everglades helps to supply the considerable agricultural activities and drinking water for south Florida. As water passes through this ecosystem, various pollutants and excess nutrients from farms, lawns, and golf courses are removed. Then, this water replenishes the aquifers (NWF 2012). The Everglades has been described as a mosaic of nine interdependent ecosystems, characterized prior to settlement by a slow-moving sheet of water passing through it from Lake Okeechobee to the Florida Bay. This "sheet-flow" is hampered by the many dams and diversions relentlessly encouraged by business and agricultural interests and carried out by the U.S. Army Corps of Engineers. To restore the normal biological processes (e.g., nutrient cycling) and vegetation patterns, the return of the normal sheet-flow is critical (NPS 2012).

The Everglades is threatened by encroaching development, with its unquenchable thirst for water and space and its associated waste, habitat destruction, and pollution. A little progress has been made in recent years to restore some of the natural water flow, which gives us hope that this unique area might survive and help us survive. However, there is a serious threat to the Everglades that is perhaps less obvious—biological pollution by exotic, invasive plant species. Many of these plants were released intentionally, and others have escaped accidentally.

Native to Australia, Melaleuca trees (*Melaleuca quinquenervia*) were introduced to Florida in 1906 to help reduce wetland area and to provide timber. Not really suitable for timber, these trees were then sold as ornamentals. Growing in dry or wet habitats, *Melaleuca* is a prolific seed producer and has spread quickly through the Everglades, forming rather dense forests and forcing out native seedlings and trees. Stressors, like fire from lightening strikes or management burns, often prompt release of millions of seeds, from which lots of thick plots of trees develop. *Melaleuca* now covers about one-half million acres of the Everglades and continues to expand that territory.

Transplanted species of plants are generally kept in check in their native habitat by competition, consumption, or disease. Normal biological and physical checks are not present in novel habitats, and the introduced plants tend to spread without restraint. Generally, the invasive plants are able to out-compete native plants and may prevent the natives from growing and/or reproducing. Not only can this incursion cause reduction in biodiversity, but the animals of the local community also may not be able to find needed food, shelter, and so on. Exotic plants may also alter native networks and processes (e.g., food webs and interactions, biogeochemical cycles, and physical factors/conditions) that are required to maintain the integrity and functional relationships of the community (Stein and Flack 1996).

Invasives, both plant and animal, also affect agricultural and recreational areas. In Florida, losses to agriculture and recreation are estimated to be about $200 million (NPS Nonnative 2012). Nationwide, a study produced by Pimentel et al. (2005) estimates the total economic impact of invasive species to be about $120 billion per year, which does not include the loss of services provided by healthy ecosystems (e.g., purification of air and water). Moreover, not restricted to the United States, the problem of invasives is a worldwide challenge that continues to grow with increasing international exchange and trade.

Competition for Space

To examine the role of spatial configuration in plant community competition, Silvertown et al. (1992) performed cellular automaton simulations of the competition among five grasses, *Lolium perenne*, *Agrostis stolonifera*, *Holcus lanatus*, *Poa trivialis*, and *Cynosurus cristatus*, based on data from Thórhallsdóttir (1990). The latter had performed an invasive species experiment, planting the five grasses in adjacent hexagonal plots. Table 14.11.1 contains the relative biomasses of neighboring species after 18 months. As the sums in the last column indicate, *Agrostis* and *Poa* are the dominant invasive species. However, the sums in the last row indicate that *Lolium*, *Cynosurus*, and *Poa*, in that order, are most likely to be invaded, or replaced. More specifically, the first column values indicate that *Lolium* is most likely to be invaded by *Poa* and least likely to be supplanted by *Cynosurus*. Also, the second row provides evidence that *Agrostis* is more likely to invade *Cynosurus* than to replace *Holcus*.

Table 14.11.1
Rates of replacement (p_{ij}) of native species by invader, which are proportions by biomasses of invaders in native plots 18 months after initiation of an experiment by Thórhallsdóttir (1990; Silvertown et al. 1992)

		Native Species					
		Lolium	*Agrostis*	*Holcus*	*Poa*	*Cynosurus*	**Sum**
Invader	*Lolium*	—	0.02	0.06	0.05	0.03	0.16
	Agrostis	0.23	—	0.09	0.32	0.37	0.81
	Holcus	0.06	0.08	—	0.16	0.09	0.39
	Poa	0.44	0.06	0.06	—	0.11	0.67
	Cynosurus	0.03	0.02	0.03	0.05	—	0.13
	Sum	0.76	0.18	0.24	0.58	0.60	

With this experimental data, the computational scientists used the rate of replacement of a native species j by an invader i, p_{ij} in row i and column j of Table 14.11.1, to estimate the probability that the invader i will replace the current species j in a cell at the next time step. The calculation of the probability and the rule for replacement are as follows.

For a site with species j, a neighbor in its von Neumann neighborhood is picked at random. Suppose that the neighboring cell contains species i and that m is the number of the site's neighbors (1–4) that contain species i. The probability that species i will replace species j in the site at the next time step is the weighed probability $p_{ij}m/4$, where p_{ij} is the proportion in row i and column j of Table 14.11.1.

For example, suppose a site contains *Lolium*, and its neighbors to the north, east, south, and west are *Lolium*, *Poa*, *Agrostis*, and *Poa*, respectively. If the cell to the north with *Lolium* is selected at random, at the next time step the site continues to contain *Lolium*. If the cell with *Agrostis* is picked, the probability that the site will be invaded by *Agrostis* at the next time step is 0.23(1/4) = 0.0575, because only one of the four neighboring cells contains *Agrostis* and the rate of replacement in the *Agrostis* row and *Lolium* column is 0.23. However, if one of the two neighbors growing *Poa* is chosen at random, the probability that the site will be invaded by *Poa* at the next time step is 0.44(2/4) = 0.22. Using this rule, a 40 × 40 grid, von Neumann neighborhoods, 600 simulation steps, and various configurations, the projects will reproduce some of the simulations described in Silvertown et al. (1992) and consider the authors' conclusions.

Projects

1. **a.** Develop a simulation and animation for the five grasses described in the section "Competition for Space." In the initial configuration, have a cell's grass be selected at random with equal probability for each species.
   b. Plot the frequency of each species for each time step.
   c. Run the simulation a number of times, say 100, and determine the average frequency of each species for time steps 0, 100, 200, 300, 400, 500, and 600.
   d. Discuss the results.

In Project 1, grasses initially occurred at random, not in communities. Silvertown et al. (1992) investigated starting community configurations with each species appearing in eight contiguous rows. For Projects 2–4, repeat Project 1 with the indicated configuration.

2. In this initial configuration, the most invasive species, *Agrostis*, appears in the top eight rows of the grid. The next eight rows contain *Holcus*, the species least likely to be invaded by *Agrostis*; that is, in the species with the smallest *p*-value in *Agrostis*' row of Table 14.11.1. We proceed in a similar fashion to obtain *Lolium* for the next eight rows, followed by *Cynosurus* and *Poa*. Each species community is unlikely to invade the species below it.
3. In this initial configuration, the most invasive species, *Agrostis*, appears in the top eight rows of the grid. The next eight rows contain *Lolium*, the species least likely to replace its neighbor to the north; that is, in *Agrostis*' column, *Lolium* is one of the two species with the least *p*-value. Then, from *Lolium*'s column data, we find that *Cynosurus* is least likely to invade *Lolium*. Proceeding in a similar fashion, the bottom two communities are *Holcus* and *Poa*.
4. In this initial configuration, we rank the species from highest to lowest by the overall ability to invade (row total) minus the overall susceptibility to invasion (column total). Thus, the order is as follows: *Agrostis*, *Holcus*, *Poa*, *Cynosurus*, *Lolium*.
5. Develop simulations for each of the configurations in Projects 1–4, and for each generate plots of species frequencies versus the time step. Compare and contrast the results. Discuss the effect that aggregation has on the rate at which stronger competitors are able to push out weaker ones. Discuss the relationships between two dominants and the impact of the presence of a third species on that relationship. Discuss the impact of having low-ranking competitors in close contact with each other.
6. Repeat Project 1 for another pattern of species communities and discuss the results.

7. a. Project 2 from Module 13.4, "Probable Cause: Modeling with Markov Chains," modeled succession in a forest with Markov chains. Using a time step of a generation and Table 13.4.4 (Horn 1975), develop a cellular automaton rule similar to that in section "Competition for Space." A cell can be in an adult-tree or sapling state for any of the species.

b. Develop a cellular automaton simulation and animation using the rule from Part a for the trees. For simplification, you may limit the number of species to be fewer than 11. In the initial configuration, have a cell's species be selected at random proportional to the species count in the last row of Table 13.4.6 (Horn 1975). For example, the sum of the counts on that row is 3286; the species count for BTA (bigtooth aspen) is 104; and 104/3286 = 0.0316. Thus, there is a 3.16% chance that a cell will contain a BTA tree. Once the species is determined, have a 50-50 chance of the tree being an adult or a sapling.

c. Plot the frequency of adult trees for each species for each of 350 time steps.

d. Run the simulation a number of times, say, 100, and determine the average frequency of each species for time steps 0, 50, 100, 150, 200, 250, 300, and 350.

e. Discuss the results. Compare your results to those of Horn's data for several sub-forests of varying ages in Institute Woods in Table 13.4.6 (Horn 1975).

8. Repeat Project 7 using the weighted probabilities discussed in Part c of Project 2 in Module 13.4, "Probable Cause—Modeling with Markov Chains."

9. Repeat Project 7 or 8 using an initial configuration similar to that of Project 2.

10. Repeat Project 7 or 8 using an initial configuration similar to that of Project 3.

11. Repeat Project 7 or 8 using an initial configuration similar to that of Project 4.

12. Using the probabilities of Project 7 or 8, develop simulations for each of the configurations in Projects 7b, 9, 10, and 11; for each, generate plots of species frequencies versus the time step. Compare and contrast the results. Discuss the effect that aggregation has on the rate at which stronger competitors are able to push out weaker ones. Discuss the relationships between two dominants and the impact of the presence of a third species on that relationship. Discuss the impact of having low-ranking competitors in close contact with each other.

13. Repeat Project 7 or 8 for another pattern of species communities and discuss the results.

14. a. Using the Table 14.11.1 data, develop a Markov chain model of this forest's succession and determine the stable equilibrium percentages. Discuss the results.

b. Using some initial distribution of species and the transition matrix from Part a, plot the estimated number of trees of each species for 20 generations. Repeat this work for two other initial distributions. Discuss the results.

References

Douglas, Marjory Stoneman. 1947. *The Everglades: River of Grass*. Sarasota FL: Pineapple Press.

ETE, Exploring the Environment. 2005. "Florida Everglades." http://www.cotf.edu/ete/modules/everglades/FEeverglades1.html (accessed August 16, 2012)

Horn, H. S. 1975. Markovian Properties of Forest Succession. In *Ecology and Evolution of Communities*, M. L. Cody and J. M. Diamond, eds. Cambridge, MA: Harvard University Press, pp. 196–211.

NPS Ecosystem, National Park Service. 2012. "Everglades Ecosystem." http://www.nps.gov/museum/exhibits/ever/ecosystem.html (accessed August16, 2012)

NPS Nonnative, National Park Service. 2012. "Everglades National Park: Nonnative Species." http://www.nps.gov/ever/naturescience/nonnativespecies.htm (accessed August 16, 2012)

NPS Subtropical, National Park Service. 2012. "Everglades National Park: The Largest Subtropical Wilderness." http://www.nps.gov/ever/index.htm (accessed August 16, 2012)

NWF, National Wildlife Federation. 2012. "Everglades." http://www.nwf.org/wildlife/wild-places/everglades.aspx (accessed August 16, 2012)

Pimentel, D., R. Zuniga, and D. Morrison. 2005. "Update on the Environmental and Economic Costs Associated with Alien-Invasive Species in the United States." *Ecological Economics* 52: 273–288.

Silvertown, Jonathan, Senino Holtier, Jeff Johnson, and Pam Dale. 1992. "Cellular Automaton Models of Interspecific Competition For Space—The Effect of Pattern on Process." *J. Ecol.* 80: 527–534.

Stein, Bruce A., and Stephanie R. Flack, eds. 1996. *America's Least Wanted: Alien Species Invasions of U.S. Ecosystems.* The Nature Conservancy, Arlington, VA.

Thórhallsdóttir, T. E. 1990. "The Dynamics of Five Grasses and White Clover in a Simulated Mosaic Sward." *J. Ecol.*, 78: 809–923.

MODULE 14.12

Re-Solving the Problems with Cellular Automaton Simulations

Prerequisite: One of Module 10.3, "Spreading of Fire"; Module 10.4, "Movement of Ants—Taking the Right Steps"; or Module 10.5, "Biofilms—United They Stand, Divided They Colonize."

Introduction

In Modules 14.6 and 14.8, "Spread of Disease" and "Predator-Prey—'Catch Me If You Can,' " respectively, we considered those categories of applications originally solved using system dynamics modeling by then employing cellular automaton (**CA**) simulations. Typically, we can solve problems in a variety of ways, and, often, different approaches illuminate different facets. Moreover, our confidence grows if two solutions arrive at similar conclusions.

In this module, we list other projects, originally solved using agent-based modeling, to approach using CA simulations. The section on "Agent-Based Modeling" from Module 11.2, "Agents of Interaction—Steering a Dangerous Course," discusses the differences between agent-based modeling and cellular automaton simulations. Significantly, at each step, the former sweeps through each agent, while the latter processes each cell of a CA grid. However, many problems previously considered using agent-based modeling, can be solved using CA simulations.

Projects

For each project, do the following parts along with any other indicated parts:

a. Develop a cellular automaton simulation and animate the results.
b. Perform the simulation a number of times, such as 100 times, and average the results.
c. Discuss your results.

For Projects 2–7, also do the following parts:

d–f. Run the simulation repeatedly and produce figures similar to Figures (d) 11.2.3, (e) 11.2.4, and (f) 11.2.5. Discuss the results, including comparisons of your results with those of Module 11.2.

One approach is to have each cell store the type of environment, such as farm or desert; the presence or absence of an animal, such as a beef cow or toad; the state of the animal, including such variables as infection status, weight, and energy; and perhaps other cell characteristics, such as amount of moisture.

Projects 1–7 refer to problems from Module 11.2, "Agents of Interaction—Steering a Dangerous Course."

1. The example discussed in Module 11.2
2. Project 2
3. Project 3
4. Project 4
5. Project 5
6. Project 8
7. Project 9

Projects 8–23 refer to problems from Module 11.4, "Introducing the Cane Toad—Able Invader."

8. The example discussed in Module 11.4
9. Project 1
10. Project 2
11. Project 3
12. Project 4
13. Project 5
14. Project 6
15. Project 7
16. Project 8
17. Project 9
18. Project 10
19. Project 11
20. Project 12
21. Project 13
22. Project 14
23. Project 15
24. Project 16
25. Project 17
26. Project 18

MODULE 14.13

Re-Solving the Problems with Agent-Based Simulations

Prerequisite for Projects 1–44, 53-62, 92–108, 122–151: Module 11.2, "Agents of Interaction—Steering a Dangerous Course"; Prerequisite for Projects 45–52, 63–91, 109–121: Module 11.4, "Introducing the Cane Toad—Able Invader."

Introduction

In the tutorials for agent-based (**AB**) modeling, we considered several problems, such as unconstrained growth, that were originally solved using system dynamics modeling. This different approach of AB simulations follows agents instead of considering populations as a whole. Moreover, AB modeling, which generally includes a greater quantity of detail, is better able to capture complex interactions, to model spatial interactions, and to provide specificity that can aid in decision-making.

As the section "Agent-Based Modeling" from Module 11.2, "Agents of Interaction—Steering a Dangerous Course," indicates, another technique, cellular automaton (CA) simulation, is very similar to agent-based modeling. While an iteration of a CA simulation sweeps through each grid cell, an iteration of an AB simulation updates the state of each agent. Interestingly, virtually every problem that can be solved with CA methods can be solved using AB techniques.

Viewing a problem from another perspective often enhances our understanding of the problem and increases our confidence in the conclusions. In this module, we list projects, originally solved using system dynamics modeling, cellular automaton simulations, or matrix-based modeling, to approach using AB modeling.

Projects

For each project, unless otherwise indicated, have the appropriate agents move at random. Do the following along with any other indicated parts:

a. Develop an agent-based simulation.
b. Plot the number of agents versus time.

c. Perform the simulation a number of times, such as 100 times, and average the results.

d. Discuss your results.

1. With a growth rate of 0.1, model the growth of an organism constrained by space so that only one organism can reside in a cell. Is the graph of the number of organisms versus time similar to the logistic curve discussed in Module 2.3? What is the carrying capacity?

Projects 2–8 refer to problems from Module 2.5.

2. Example in the section "One-Compartment Model of Repeated Doses"
3. Project 2 **4.** Project 3 **5.** Project 4
6. Project 5 **7.** Project 6 **8.** Project 12
9. Example in Module 4.1. Besides BTS and WTS agents, include a *Food* agent that grows appropriately. The BTS should have a greater likelihood than WTS of eating neighboring food. Each type of shark requires a certain amount of energy, which the food provides. Each time step without food diminishes a shark's energy.

Projects 10–14 refer to problems from Module 4.2.

10. Project 2 **11.** Project 3 **12.** Project 6
13. Project 4. Have both predators and prey constrained by space so that only one organism can reside in a cell.
14. Project 12

Projects 15–25 refer to problems from Module 4.3.

15. SIR example of Module 4.3
16. Project 1 **17.** Project 2 **18.** Project 3
19. Project 4 **20.** Project 5 **21.** Project 6
22. Project 8 **23.** Project 10 **24.** Project 11
25. Project 12

Projects 26–30 refer to problems from Module 4.4.

26. Module example **27.** Project 1 **28.** Project 7
29. Project 8 **30.** Project 12

Projects 31–33 refer to problems from Module 4.5.

31. Module example **32.** Project 7 with $n = 2$
33. Project 8
34. Module 7.1, Project 1 **35.** Module 7.1, Project 2
36. Module 7.6, Project 1 **37.** Module 7.10, Project 1
38. Module 7.10, Project 2 **39.** Module 9.2, Project 2
40. Module 9.2, Project 4

Projects 41–44 refer to problems from Module 9.5.

41. Project 4 **42.** Project 6 **43.** Project 7
44. Project 8

Projects 45–52 refer to problems from Module 10.2.

45. Example in module **46.** Project 1 **47.** Project 3
48. Project 4 **49.** Project 5 **50.** Project 6
51. Project 7 **52.** Project 8

Projects 53–62 refer to problems from Module 10.3.

53. Example in module **54.** Project 1 **55.** Project 2
56. Project 3 **57.** Project 4 **58.** Project 5
59. Project 9 **60.** Project 10 **61.** Project 11
62. Project 12

Projects 63–73 refer to problems from Module 10.4.

63. Project 1 **64.** Project 2 **65.** Project 3 **66.** Project 4
67. Project 5 **68.** Project 6 **69.** Project 7 **70.** Project 8
71. Project 12 **72.** Project 13 **73.** Project 14

Projects 74–89 refer to problems from Module 10.5.

74. Module example **75.** Project 1 **76.** Project 2 **77.** Project 3
78. Project 4 **79.** Project 5 **80.** Project 6 **81.** Project 7
82. Project 8 **83.** Project 9 **84.** Project 10 **85.** Project 11
86. Project 12 **87.** Project 13 **88.** Project 14 **89.** Project 16

Projects 92–97 refer to problems from Module 13.3.

92. Age-structured module example **93.** Stage-structured module example
94. Project 1 **95.** Project 2 **96.** Project 3 **97.** Project 6
98. Module 13.4, Project 3 **99.** Module 13.4, Project 4
100. Module 14.1, Project 1 **101.** Module 14.1, Project 3

Projects 102–108 refer to problems from Module 14.2.

102. Project 1 **103.** Project 2 **104.** Project 3 **105.** Project 4
106. Project 7 **107.** Project 8 **108.** Project 12

Projects 109–115 refer to problems from Module 14.3.

109. Project 1 **110.** Project 2 **111.** Project 3 **112.** Project 4
113. Project 5 **114.** Project 6 **115.** Project 7

Projects 116–121 refer to problems from Module 14.4.

116. Project 1 **117.** Project 2 **118.** Project 3 **119.** Project 4
120. Project 5 **121.** Project 6

Projects 122–126 refer to problems from Module 14.5.

122. Project 1 **123.** Project 2 **124.** Project 3 **125.** Project 4
126. Project 7

Projects 127–131 refer to problems from Module 14.6.

127. Project 1 **128.** Project 2 **129.** Project 3 **130.** Project 4
131. Project 5 **132.** Module 14.7, Project 1
133. Module 14.7, Project 2

Projects 134–139 refer to problems from Module 14.8.

134. Project 2 **135.** Project 3 **136.** Project 4 **137.** Project 5
138. Project 6 **139.** Project 7

Projects 140–151 refer to problems from Module 14.11.

140. Project 1
141. Project 2
142. Project 3
143. Project 4
144. Project 5
145. Project 6
146. Project 7
147. Project 8
148. Project 9
149. Project 10
150. Project 11
151. Project 12

MODULE 14.14

Computational Code-Breaking—Deciphering Our Own Mysteries

Prerequisites: Sections on "Proteins," "Nucleic Acids," and "From Genes to Proteins" from Module 7.14, "Control Issues—The Operon Model"; and Module 13.4, "Probable Cause—Modeling with Markov Chains." Project 1 requires only the indicated material from Module 7.14.

Bioinformatics

A newly developing area of computational science, called **bioinformatics**, deals with the organization of biological data, such as in databases, and the analysis of such data, which often makes extensive use of probabilities. Biological systems provide us with complexity that challenges our ability to interpret data. To help unravel these complexities, the Human Genome Project set out to map all the human genome, no simple goal if we consider that our genetic code consists of 20,000 to 25,000 genes, composed of about three billion nucleotides. It is remarkable that the program completed the mapping of the human genome in only 13 years, the last chromosome completed and published in 2006. Now, this tremendous accomplishment seems like the easy part in our attempts to unravel the complexities of ourselves. The data generated by this project, which is now combined with data from the genomes of other organisms, is accumulating with ever-increasing volume and complexity. To analyze these data and derive any understanding will require the development of genomic-scale technologies. Even with such technologies, biological research in this area is likely to take decades.

A few of the research areas of genetics that will be pursued and expanded include the following:

- Gene number, exact locations, and functions
- Gene regulation
- DNA sequence organization
- Chromosomal structure and organization

- Noncoding DNA types, amount, distribution, information content, and functions
- Coordination of gene expression, protein synthesis, and posttranslational events
- Interaction of proteins in complex molecular machines
- Predicted versus experimentally determined gene function
- Evolutionary conservation among organisms
- Protein conservation (structure and function)
- Proteomes (total protein content and function) in organisms
- Correlation of SNPs (single-base DNA variations among individuals) with health and disease
- Disease-susceptibility prediction based on gene sequence variation
- Genes involved in complex traits and multigene diseases
- Complex systems biology, including microbial consortia useful for environmental restoration
- Developmental genetics, genomics

As indicated in "Prerequisites" for this module, Module 7.14, "Control Issues—The Operon Model," contains useful biological background for problems in bioinformatics. In the current module, we approach a number of bioinformatics problems using Markov chains, such as the BLAST algorithm for searching genomic databases and the GeneMark algorithm for locating genes.

Mutations

A mutation in a DNA sequence can occur with the **insertion** or **deletion** of a base or the **substitution** of one base for another. One type of substitution, called a **transition**, occurs between purines, from A to G or from G to A, or between pyrimidines, from T to C, or vice versa. A **transversion** substitution occurs between a purine and a pyrimidine, or vice versa. In a substitution, a transition is much more likely to occur than a transversion.

Locating Genes with Markov Models

The most dependable method of discovering a gene in a new genome is observing a close **homolog**, or a gene from the same ancestral origin, in another organism. However, when homologs to known genes do not exist, we must employ computational methods to help identify genes (Salzberg et al. 1998).

In mammals, the sequence of bases **CG** frequently transforms to (methyl-C)G and then mutates to TG. Thus, the pair CG appears less that we would expect from random occurrences of C and G independently. However, this process of transformation from CG to TG is suppressed in small regions, called **CpG islands**, **upstream** of, or before, many genes; so CpG islands can be employed to locate genes. The "p" in "CpG" indicates a phosphate that links the two bases C and G in DNA. The classical definition of a CpG island is a DNA segment of length 200 that has CG occurring 50% of the time and a ratio of observed-to-expected number of CpG's above 0.6 (Gardiner-Garden and Frommer 1987).

We can use Markov chains to determine whether a short segment of genomic data is from a CpG island or not. First, we use **training sequences** that we know contain CpG islands, called **positive (+) samples**, to derive for each base four probabilities—the probabilities that A, C, G, and T follow the base. For example, consider the sequence ACGTCTATTC, which is exceptionally small for the sake of illustration. To calculate the probability that T is followed by A, written as $P(x_i = \text{A} \mid x_{i\text{-}1} = \text{T})$ or $P(\text{A} \mid \text{T})$, we divide the number of occurrences of TA in the sequence, here 1, by the number of pairs that begin with T, here 4 (TC, TA, TT, and TC). Thus, $P(\text{A} \mid \text{T}) = ¼ = 0.25$ for this sequence. That is, 25% of the time the next base after T is A. Moreover, the sum of the probabilities $P(\text{A} \mid \text{T}) + P(\text{C} \mid \text{T}) + P(\text{G} \mid \text{T}) + P(\text{T} \mid \text{T}) = 0.25 + 0.50 + 0.00 + 0.25 = 1.00$.

Figure 14.14.1a presents a transition matrix for such positive samples determined from 60,000 nucleotides from a database of human DNA sequences with 48 CpG islands. As in the example in the last paragraph, the sum of the elements on each row is 1.00, while the column sum is not necessarily 1.00. In that matrix, the probability of the pair CG (or the probability that G occurs, given that C has just appeared) is 0.274, written as $P_+(x_i = \text{G} \mid x_{i\text{-}1} = \text{C}) = P_+(\text{G} \mid \text{C}) = 0.274$. We also employ training sequences for known **negative (–) samples** to derive another transition matrix, such as in Figure 14.14.1b. Thus, for these training sequences, the probability that the sequence CG occurs in the positive samples with CpG islands is 0.274, while we find that such a sequence is much less likely (probability of $P_-(\text{G} \mid \text{C}) = 0.078$) to occur in the negative samples that do not contain CpG islands.

Figure 14.14.1
Possible transition matrix for (a) positive and (b) negative samples (Durbin et al. 1998)

a

$x_{i\text{-}1}$ \ x_i	+	A	C	G	T
	A	0.180	0.274	0.426	0.120
	C	0.171	0.368	0.274	0.188
	G	0.161	0.339	0.375	0.125
	T	0.079	0.355	0.384	0.182

b

$x_{i\text{-}1}$ \ x_i	–	A	C	G	T
	A	0.300	0.205	0.285	0.210
	C	0.322	0.298	0.078	0.302
	G	0.248	0.246	0.298	0.208
	T	0.177	0.239	0.292	0.292

Quick Review Question 1

Compute the transition matrix using the training sequence ACGTCTATTC.

We can now use Markov chains to determine if a short sequence, $x = (x_1x_2x_3. . .x_n)$ is more likely to come from a positive or a negative sample by considering the ratio of the probability that the sequence is from a positive sample over the probability that the sequence is from a negative sample:

$$\frac{P(x \mid \text{positive model})}{P(x \mid \text{negative model})}$$

If this ratio is greater than 1, the sequence is more likely to be from a CpG island.

To derive the formulas for the numerator and denominator, let us consider a very short sequence of four bases $x = (x_1x_2x_3x_4)$. Regardless of the positive or negative model, the probability that x occurs, $P(x_1x_2x_3x_4)$, is $P(x_4$ occurs after $x_1x_2x_3$ and $x_1x_2x_3$ occurs). As we saw earlier, $P(x_4$ occurs after $x_1x_2x_3$ and $x_1x_2x_3$ occurs) is $P(x_4 | x_1x_2x_3) P(x_1x_2x_3)$, or the probability that x_4 occurs, given that the sequence $x_1x_2x_3$ just appeared, times the probability that $x_1x_2x_3$ occurs. Thus, we have the following:

$$P(x_1x_2x_3x_4) = P(x_4 \mid x_1x_2x_3)P(x_1x_2x_3) \tag{1}$$

Now, with Markov chains, x_4 depends only on the value of its immediate predecessor, x_3, so that $P(x_4 | x_1x_2x_3) = P(x_4 | x_3)$, and we can simplify Equation 1 as follows:

$$P(x_1x_2x_3x_4) = P(x_4 \mid x_3)P(x_1x_2x_3) \tag{2}$$

We then repeat the process to compute $P(x_1x_2x_3)$:

$$\begin{aligned} P(x_1x_2x_3) &= P(x_3 \text{ occurs after } x_1x_2 \text{ and } x_1x_2 \text{ occurs}) \\ &= P(x_3 \mid x_1x_2)P(x_1x_2) = P(x_3 \mid x_2)P(x_1x_2) \end{aligned} \tag{3}$$

Substituting (3) into (1), we have the following:

$$P(x_1x_2x_3x_4) = P(x_4 \mid x_3)P(x_3 \mid x_2)P(x_1x_2) \tag{4}$$

Using the same reasoning, we have

$$P(x_1x_2) = P(x_2 \mid x_1)P(x_1) \tag{5}$$

and, finally,

$$P(x_1x_2x_3x_4) = P(x_4 \mid x_3)P(x_3 \mid x_2)P(x_2 \mid x_1)P(x_1) \tag{6}$$

The probability of the sequence $x_1x_2x_3x_4$ is "unzipped" from right to left as the product of the probability of obtaining x_4 given that x_3 is immediately preceding, the probability of x_3 given x_2 is immediately preceding, the probability of x_2 given x_1 immediately preceding, and the probability of x_1. Generalizing, we have the following formula:

$$P(x_1x_2x_3. . .x_n) = P(x_n \mid x_{n-1})P(x_{n-1} \mid x_{n-2}) \vdots P(x_3 \mid x_2)P(x_2 \mid x_1)P(x_1) \tag{7}$$

The probability of x_1, $P(x_1)$, is the proportion of the time x_1 occurs in a sequence or the total number of occurrences of x_1 over the total number of bases in the sequence. For example, in Quick Review Question 6a of Module 13.4, we determined that base C appears 7 times in the sequence s_1 of 20 bases, so that $P(\text{C}) = 7/20$. We use the training sequences to determine such probabilities. Moreover, the Markov matrices, as in Figure 14.14.1, contain the other probabilities. Again, for the sake of example, suppose we have the probabilities of bases in training sequences that contain CpG islands as in Figure 14.14.2a. Then, we can calculate the probability that the sequence ACGTC is from a CpG island as follows:

$$P_+(\text{ACGTC}) = P_+(\text{C} \mid \text{T})P_+(\text{T} \mid \text{G})P_+(\text{G} \mid \text{C})P_+(\text{C} \mid \text{A})P_+(\text{A})$$

We calculate the first four probabilities using the transition matrix for the positive model in Figure 14.14.1a and the probability of A using Figure 14.14.2a, as follows:

$$\begin{aligned} P_+(\text{ACGTC}) &= P_+(\text{C} \mid \text{T})P_+(\text{T} \mid \text{G})P_+(\text{G} \mid \text{C})P_+(\text{C} \mid \text{A})P_+(\text{A}) \\ &= 0.355 \cdot 0.125 \cdot 0.274 \cdot 0.274 \cdot 0.258 \\ &= 0.00085953 \end{aligned}$$

Figure 14.14.2
Probability of bases for (a) positive (frequencies from gene-rich human chromosome 19) and (b) negative samples (frequencies from reference human genome sequence; *Guide to the Human Genome* 2010)

a	*b*
$P_+(\text{A}) = 0.258$	$P_-(\text{A}) = 0.295$
$P_+(\text{C}) = 0.242$	$P_-(\text{C}) = 0.205$
$P_+(\text{G}) = 0.242$	$P_-(\text{G}) = 0.205$
$P_+(\text{T}) = 0.259$	$P_-(\text{T}) = 0.296$

Similarly, we calculate the probability that ACGTC does not come from a CpG island using probabilities Figures 14.14.1b and 14.14.2b, as follows:

$$\begin{aligned} P_-(\text{ACGTC}) &= P_-(\text{C} \mid \text{T})P_-(\text{T} \mid \text{G})P_-(\text{G} \mid \text{C})P_-(\text{C} \mid \text{A})P_-(\text{A}) \\ &= 0.239 \cdot 0.208 \cdot 0.078 \cdot 0.205 \cdot 0.295 \\ &= 0.00023449 \end{aligned}$$

The calculations indicate a greater probability that ACGTC contains a CpG island than not. Moreover, the quotient of the probabilities being greater than 1 also indicates a CpG island:

$$\frac{P(\text{ACGTC} \mid \text{positive model})}{P(\text{ACGTC} \mid \text{negative model})} = \frac{P_+(\text{ACGTC})}{P_-(\text{ACGTC})} = \frac{0.00085953}{0.00023449} = 3.6655$$

Quick Review Question 2

Using the transition matrices from Figure 14.14.1 and probabilities from Figure 14.14.2, calculate the following:

a. P_+(CCGTCGA)
b. P_-(CCGTCGA)
c. The quotient of Parts a and b
d. Is CCGTCGA more likely to be from a CpG island or not?

However, the sequence ACGTC is much shorter than the usual sequence of 200 to 250 bases. If we were to multiply together 200 probabilities, each less than 1, the result would be on the order of 10^{-200}. To avoid such a small magnitude number, the use of division, and a large number of multiplications, we employ logarithms. With the logarithm of a quotient being the difference of the logarithms, we can replace a division with a subtraction:

$$\ln\left(\frac{P_+(\text{ACGTC})}{P_-(\text{ACGTC})}\right) = \ln\left(P_+(\text{ACGTC})\right) - \ln\left(P_-(\text{ACGTC})\right)$$

Moreover, the log of a product is the sum of the logs:

$$\begin{aligned}\ln(P_+(\text{ACGTC})) &= \ln(0.355 \cdot 0.125 \cdot 0.274 \cdot 0.274 \cdot 0.258)\\ &= \ln(0.355) + \ln(0.125) + \ln(0.274) + \ln(0.274) + \ln(0.258)\\ &= -7.0591\\ \ln(P_-(\text{ACGTC})) &= \ln(0.239 \cdot 0.208 \cdot 0.078 \cdot 0.205 \cdot 0.295)\\ &= \ln(0.239) + \ln(0.208) + \ln(0.078) + \ln(0.205) + \ln(0.295)\\ &= -8.3581\end{aligned}$$

Thus, we have

$$\ln(P_+(\text{ACGTC})) - \ln(P_-(\text{ACGTC})) = -7.0591 - (-8.3581) = 1.2990$$

We then normalize this score by dividing by the length of the sequence to obtain 1.2990/5 = 0.2598. The larger this **length-normalized log-odds score** is the more likely that the sequence is from a CpG island (Tang 2007; Gropl and Huson 2005).

Definition The **length-normalized log-odds score** for a sequence x is

$$\ln\left(\frac{P_+(x)}{P_-(x)}\right)\Big/|x|$$

Quick Review Question 3

Calculate the length-normalized log-odds score for the sequence CCGTCGA of Quick Review Question 2.

GeneMark

The technique of locating genes from the previous section, "Locating Genes with Markov Models," is a **1st-order Markov model** because the method predicts each base using one preceding base in the DNA sequence. For this method, as in Figure 14.14.1a, with positive training sequences that contain CpG islands, $4^2 = 16$ probabilities of base y occurring given base x immediately preceding were calculated. Similarly, as in Figure 14.14.1b, 16 probabilities were obtained using negative training sequences that do not contain such islands. Moreover, as in Figure 14.14.2, the probabilities of each base occurring in a positive sequence and in a negative sequence were required, resulting in an additional 4 + 4 = 8 probabilities.

The gene-finding program **GeneMark**, which is a **5th-order Markov model**, employs five previous bases to predict a base (GeneMark 2012). Compared to the 32 probabilities in Figure 14.14.1, GeneMark must use $4^6 = 4096$ probabilities for positive and 4096 for negative training sequences. Moreover, comparable to Figure 14.14.2, the program must also compute the probability of each sequence of 5 bases occurring in positive and negative training sequence, or $2(4^5) = 2048$ probabilities.

Thus, GeneMark calculates 4096 + 4096 + 2048 = 10,240 probabilities from the training sequences alone. Project 3 discusses the GeneMark algorithm in greater detail.

Projects

1. (Read the section "Mutations" and the sections "Proteins," "Nucleic Acids," and "From Genes to Proteins" from Module 7.14, "Control Issues—The Operon Model." From the text's website, download the PAM1 matrix, *PAM1.dat*, in Table 14.14.1 and the frequency data, *freq.dat*, in Table 14.14.2.) Finding similar sequences in genomic databases can help us determine the biochemistry, physiology, and function of a gene or the protein it produces. In searching such databases, algorithms produce scores that allow us to differentiate sequences that are related to a query sequence from those that are not. One of the main algorithms for database searching is **BLAST** (Basic Local Alignment Search Tool; BLAST 2012), which uses a **PAM** (**Point Accepted Mutations**) scoring matrix. In the 1970s, a research team lead by Margaret Dayhoff carefully studied the evolution of sequences of amino acids. **PAM** or **PAM 1** is the length of time for 1% of the amino acids to mutate. One estimate is that a PAM is about a million years. The **PAM1 matrix** is a Markov chain transition matrix with column and row headings of the amino acids, where entries represent the amount of evolution over one PAM period of time, or for one mutation per hundred amino acids. Thus, the *ij*-element is the probability that the amino acid in the *i*th row will replace the amino acid in the *j*th column after the evolutionary time PAM. A **PAM120 matrix**, which BLAST uses, contains information on the amount of evolution over 120 PAM periods of time. We can obtain this matrix by raising the PAM1 matrix to the 120th power. Use a computational tool as needed to complete the following parts.

a. The values are multiplied by 10,000 for clarity. For example, the element in the first row for Ala (A) and third column for Asn (N) is 3. Thus, the probability that the amino acid Asn mutates to the amino acid Ala in about a million years (one PAM epoch) is 3/10,000 = 0.0003 = 0.03%. Draw a partial state diagram using the four amino acids in the top left corner of the matrix.

b. Calculate PAM120, *M*. The matrix is usually written with each element multiplied by 100 and rounded to the nearest integer.

c. Each element of the **PAM120 scoring matrix**, *S*, which BLAST uses, is obtained using the following formula:

$$S_{ij} = \text{round}(10\ \text{log10}(M_{ij}/f_i))$$

where f_i is the frequency of the amino acid in row *i* and *M* is the PAM120 matrix from Part b. We will compute *S*, called a **log odds-scoring matrix**, using the frequencies in Table 14.14.2. Because we do not know what came first, make this matrix symmetric, using the values on and below the diagonal. For example, the score for a mutation over 120 PAM periods from R to N should be the same as the mutation over that period from N to R (Momand 2006).

Table 14.14.1

PAM1 matrix with values multiplied by 10,000. The element in row *i*, column *j* is the probability that row *i*'s amino acid will replace column *j*'s amino acid in 1 PAM. Adapted from Figure 82 in Dayhoff (1978)

	A	R	N	D	C	Q	E	G	H	I	L	K	M	F	P	S	T	W	Y	V
A	9867	2	9	10	3	8	17	21	2	6	4	2	6	2	22	35	32	0	2	18
R	1	9913	1	0	1	10	0	0	10	3	1	19	4	1	4	6	1	8	0	1
N	4	1	9822	36	0	4	6	6	21	3	1	13	0	1	2	20	9	1	4	1
D	6	0	42	9859	0	6	53	6	4	1	0	3	0	0	1	5	3	0	0	1
C	1	1	0	0	9973	0	0	0	1	1	0	0	0	0	1	5	1	0	3	2
Q	3	9	4	5	0	9876	27	1	23	1	3	6	4	0	6	2	2	0	0	1
E	10	0	7	56	0	35	9865	4	2	3	1	4	1	0	3	4	2	0	1	2
G	21	1	12	11	1	3	7	9935	1	0	1	2	1	1	3	21	3	0	0	5
H	1	8	18	3	1	20	1	0	9912	0	1	1	0	2	3	1	1	1	4	1
I	2	2	3	1	2	1	2	0	0	9872	9	2	12	7	0	1	7	0	1	33
L	3	1	3	0	0	6	1	1	4	22	9947	2	45	13	3	1	3	4	2	15
K	2	37	25	6	0	12	7	2	2	4	1	9926	20	0	3	8	11	0	1	1
M	1	1	0	0	0	2	0	0	0	5	8	4	9874	1	0	1	2	0	0	4
F	1	1	1	0	0	0	0	1	2	8	6	0	4	9946	0	2	1	3	28	0
P	13	5	2	1	1	8	3	2	5	1	2	2	1	1	9926	12	4	0	0	2
S	28	11	34	7	11	4	6	16	2	2	1	7	4	3	17	9840	38	5	2	2
T	22	2	13	4	1	3	2	2	1	11	2	8	6	1	5	32	9871	0	2	9
W	0	2	0	0	0	0	0	0	0	0	0	0	0	1	0	1	0	9976	1	0
Y	1	0	3	0	3	0	1	0	4	1	1	0	0	21	0	1	1	2	9945	1
V	13	2	1	1	3	2	2	3	3	57	11	1	17	1	3	2	10	0	2	9901

Table 14.14.2

Normalized Frequencies of Amino Acids (Nakhleh 2010)

Ala	Arg	Asn	Asp	Cys	Gln	Glu	Gly	His	Ile
8.7%	4.1%	4.0%	4.7%	3.3%	3.8%	5.0%	8.9%	3.4%	3.7%
Leu	**Lys**	**Met**	**Phe**	**Pro**	**Ser**	**Thr**	**Trp**	**Tyr**	**Val**
8.5%	8.1%	1.5%	4.0%	5.1%	7.0%	5.8%	1.0%	3.0%	6.5%

d. Write a function to return the relative position of an amino acid parameter. For example, N is the third amino acid listed in Table 14.14.1, so the function returns 3.

e. Write a function to accept two amino acids, such as N and A, as arguments and to return the corresponding PAM120 score using the PAM120 scoring matrix, *S*, from Part c.

f. The BLAST algorithm searches a database for sequences that have a "good," nongapping local alignment with a segment of the query sequence. The program starts by breaking the query sequence into all possible sequential triplets, or **3-mers**, or **words** of length 3. For example, if the query sequence is s = RHQMN, we have three 3-mers, RHQ, HQM, and QMN. Write a function that has a query sequence parameter and returns a list of all its 3-mers.

g. The BLAST program obtains the evolutionary scores for all possible (20)(20)(20) = 8000 amino acid triplets in relation to each of the 3-mers in the query sequence and compiles a list of all words that have a score greater than or equal to a certain threshold parameter. For example, using the PAM250 scoring matrix in Table 14.14.3, the scoring of QMN relative to pairs QMN, DLL, QSW, and BME is 12, 3, –2, and 8, respectively, as the following computations indicate:

Q		M		N		
Q		M		N		
4	+	6	+	2	=	12
Q		M		N		
D		L		L		
2	+	4	+	(–3)	=	3
Q		M		N		
Q		S		W		
4	+	(–2)	+	(–4)	=	–2
Q		M		N		
B		M		E		
1	+	6	+	1	=	8

Table 14.14.3

The PAM250 scoring matrix for amino acids. B is used when one cannot distinguish between D and N because of amino acid analytical processing. Similarly, Z is used when it is ambiguous whether the amino acid is E or Q. X represents an unknown or nonstandard amino acid. Thus, the matrix has 23 rows and 23 columns.

	A	R	N	D	C	Q	E	G	H	I	L	K	M	F	P	S	T	W	Y	V	B	Z	X
A	2																						
R	–2	6																					
N	0	0	2																				
D	0	–1	2	4																			
C	–2	–4	–4	–5	12																		
Q	0	1	1	2	–5	4																	
E	0	–1	1	3	–5	2	4																
G	1	–3	0	1	–3	–1	0	5															
H	–1	2	2	1	–3	3	1	–2	6														
I	–1	–2	–2	–2	–2	–2	–2	–3	–2	5													
L	–2	–3	–3	–4	–6	–2	–3	–4	–2	2	6												
K	–1	3	1	0	–5	1	0	–2	0	–2	–3	5											
M	–1	0	–2	–3	–5	–1	–2	–3	–2	2	4	0	6										
F	–4	–4	–4	–6	–4	–5	–5	–5	–2	1	2	–5	0	9									
P	1	0	–1	–1	–3	0	–1	–1	0	–2	–3	–1	–2	–5	6								
S	1	0	1	0	0	–1	0	1	–1	–1	–3	0	–2	–3	1	2							
T	1	–1	0	0	–2	–1	0	0	–1	0	–2	0	–1	–3	0	1	3						
W	–6	2	–4	–7	–8	–5	–7	–7	–3	–5	–2	–3	–4	0	–6	–2	–5	17					
Y	–3	–4	–2	–4	0	–4	–4	–5	0	–1	–1	–4	–2	7	–5	–3	–3	0	10				
V	0	–2	–2	–2	–2	–2	–2	–1	–2	4	2	–2	2	–1	–1	–1	0	–6	–2	4			
B	0	–1	2	3	–4	1	2	0	1	–2	–3	1	–2	–5	–1	0	0	–5	–3	–2	2		
Z	0	0	1	3	–5	3	3	–1	2	–2	–3	0	–2	–5	0	0	–1	–6	–4	–2	2	3	
X	0	0	0	0	0	0	0	0	0	0	0	0	0	0	0	0	0	0	0	0	0	0	0
	A	R	N	D	C	Q	E	G	H	I	L	K	M	F	P	S	T	W	Y	V	B	Z	X

If the user picks a threshold value of 5, the program would select QMN and BME and other higher scoring 3-mers but not DLL and QSW as evolutionary matches. Develop a function to accept two 3-mers and to return the evolutionary PAM120 score. These scores will differ from those in Table 14.14.3.

h. Develop a function to have three parameters, a 3-mer (*mer*), a list of 3-mers (*merLst*), and a threshold value (*threshold*), and to return a list of all 3-mers from *merLst* whose evolutionary score relative to *mer* is greater than or equal to *threhold*. For example, as Part g illustrates, if *mer* is QMN, *merLst* is {QMN, DLL, QSW, BME}, and *threshold* is 5, then the function returns {QMN, BME}. Use the PAM120 scoring matrix from Part c.

i. Write a function to return a list of the 8000 possible amino acid triplets.

j. The second step in the BLAST algorithm is to scan the database for locations of high scoring words from the first step (see Part g). For example, the high scoring word BME occurs at location 6 in the sequence NRSQH**BME**LDLDMFPMST. Develop a function that has as parameters a list of 3-mers (*merLst*) and a sequence (*sequence*) and that returns a list of integer starting locations for all occurrences 3-mers from *merLst* in *sequence*.

k. The third step of the BLAST algorithm is to extend each of the seeds in both directions until the subsequence score reaches a maximum value according to the matrix scoring. Using a heuristic, the program stops an extension if the score falls below a certain amount less than the highest score so far. For example, suppose the query sequence is in part. . .SRMCD**RHQ**MNCFPS. . ., and the program located the high-scoring word RHQ in the database sequence . . .NRSQH**RHQ**LDLDMF. . .. Table 14.14.4 shows how we extend from the seed RHQ to find a segment pair (D**RHQ**MN and H**RHQ**LD) with a maximum PAM250 score (1 + 6 + 6 + 4 + 4 + 2 = 23). DRHQMN and HRHQLD are a **locally maximal segment pair**, or a segment from the query sequence and a segment from a database sequence with a score that cannot become larger through shrinking or expanding the segments. We repeat this extension process for all seeds looking for all segment pairs with scores above some threshold. The algorithm is fast in part because it does not consider gaps and uses heuristics involving threshold values.

Develop a function that has parameters of two sequences and an integer starting location and returns a list containing the starting location, length, and PAM 120 score (see Part c) of a locally maximal segment pair.

Table 14.14.4
Finding the locally maximal segment pair from sequences SRMCD**RHQ**MNCFPS and NRSQH**RHQ**LDLDMF, starting at location 6 and using the PAM250 scoring matrix

In query:	S	D	M	C	**D**	**R**	**H**	**Q**	**M**	**N**	C	F	P	S
In database:	N	R	S	Q	**H**	**R**	**H**	**Q**	**L**	**D**	L	D	M	F
PAM250 Score:	1	–1	–2	–5	**1**	**6**	**6**	**4**	**4**	**2**	–6	–6	–2	–3

1. Write a program to implement the BLAST algorithm as presented in this project. Input should include a query sequence, a list of database sequences, and a threshold value. Obtain sequences from a BLAST database at NCBI (BLAST 2012).
2. (From the text's website, download *ProbabilitiesHuman.txt*, which contains the probabilities from Figures 14.14.1 and 14.14.2, respectively. Also, download all or part of the DNA sequence on chromosome 19 of the human genome at http://www.ncbi.nlm.nih.gov/mapview/maps.cgi?ORG=hum&MAPS=ideogr,est,loc&LINKS=ON&VERBOSE=ON&CHR=19.) Employ the techniques of the section on "Locating Genes" to score each subsequence of length 200. Have your program determine the most likely candidates for subsequences being in CpG islands. Determine if your candidates occur in CpG islands as indicated at http://genome.ucsc.edu/cgi-bin/hgTracks?position=chr19:571325-583493&hgsid=264592883&knownGene=pack&hgFind.matches=uc002loy.3. As of this writing, such areas appear in green on the diagram (*Homo Sapiens* 2001, UCSC 2009).
3. (From the text's website, download *AE005174v2.txt* (from *AE005174v2.fas* at http://www.genome.wisc.edu/sequencing/o157.htm), which contains the DNA sequence for *Escherichia coli* (*E. coli*), and *Escherichia_coli_O157H7_plasmid_pO157.txt*, which contains training data generated by generated by GeneMarkSPlusRBS on *E. coli* O157H7, as described in the section "GeneMark" (Borodovsky Laboratory 2005). GeneBank at NCBI contains sequence information on "Escherichia coli O157:H7 EDL933, complete genome" (http://www.ncbi.nlm.nih.gov/nuccore/AE005174). By clicking on any of the *gene* links, create a data file of a sequence of 200 bases immediately before one of the genes and create another data file of a sequence of 200 bases inside a gene (Escherichia coli 2001; Enterohaemorrhagic 2001).) Using the algorithm described in the section "GeneMark" to score each sequence as containing or not containing a CpG island. Employ the first column of data in *Escherichia_coli_O157H7_plasmid_pO157.txt.*
4. (From the text's website, download *Escherichia_coli_O157H7_plasmid_pO157.txt*, described in the previous project.) For **homogeneous Markov models** involving genomic sequences, probabilities are not dependent upon sequence location, while for **inhomogeneous Markov models,** they are. A **reading frame** breaks a sequence of nucleotides into codons. Because we can start the alignment in three possible places on an mRNA strand, three reading frames exist for such a strand. For example, suppose mRNA contains the sequence of bases, AACTGTTAG. . .. We could have the reading frame begin with AAC, as in AAC-TGT-TAG. . .; or one base further with ACT-GTT-AG. . .; or two bases beyond with CTG-TTA-G. . .. Because DNA has two strands, one complementary to the other, we have six possible reading frames from which transcription can occur. As described in the section "GeneMark," develop a program that generates transition matrices and probabilities for the six possible reading frames for training sequences and select the model with the highest score. The GeneMark program considers seven possibilities, these six and a model of noncoding DNA.

Answers to Quick Review Questions

1.

		A	C	G	T
			x_i		
x_{i-1}	A	0.00	0.50	0.00	0.50
	C	0.00	0.00	0.50	0.50
	G	0.00	0.00	0.00	1.00
	T	0.25	0.50	0.00	0.25

2. **a.** P_+(CCGTCGA) = $4.7767 \cdot 10^{-5} = 0.368 \cdot 0.274 \cdot 0.125 \cdot 0.355 \cdot 0.274 \cdot 0.161 \cdot 0.242$
   b. P_-(CCGTCGA) = $4.5822 \cdot 10^{-6} = 0.298 \cdot 0.078 \cdot 0.208 \cdot 0.239 \cdot 0.078 \cdot 0.248 \cdot 0.205$
   c. 10.4245
   d. CCGTCGA is more likely to be from a CpG island because the quotient is greater than 1.
3. $0.3349 = \ln((0.368 \cdot 0.274 \cdot 0.125 \cdot 0.355 \cdot 0.274 \cdot 0.161 \cdot 0.242)/(0.298 \cdot 0.078 \cdot 0.208 \cdot 0.239 \cdot 0.078 \cdot 0.248 \cdot 0.205))/7 = (\ln(0.368) + \ln(0.274) + \ln(0.125) + \ln(0.355) + \ln(0.274) + \ln(0.161) + \ln(0.242) - \ln(0.298) - \ln(0.078) - \ln(0.208) - \ln(0.239) - \ln(0.078) - \ln(0.248) - \ln(0.205))/7$

References

BLAST. 2012. Blast Home at National Center for Biotechnology Information. http://blast.ncbi.nlm.nih.gov/Blast.cgi (accessed January 19, 2012)

Borodovsky Laboratory. 2005. Escherichia_coli_O157H7_plasmid_pO157.mat. School of Biology, Georgia Tech.

Dayhoff M. O., ed. 1978. *Atlas of Protein Sequence and Structure, Suppl 3*, 1978, National Biomedical Research Foundation.

Durbin, R., S. Eddy, A. Krogh, and G. Mitchison. 1998. *Biological Sequence Analysis*. Cambridge, MA: Cambridge University Press.

Enterohaemorrhagic *Escherichia coli* (EHEC) O157:H7. *E. coli* Genome Project. 2001. http://www.genome.wisc.edu/sequencing/o157.htm (accessed May 23, 2012).

Escherichia coli O157:H7 EDL933, complete genome. 2001. NCBI. http://www.ncbi.nlm.nih.gov/nuccore/AE005174 (accessed May 23, 2012)

Gardiner-Garden, M. and M. Frommer. 1987. "CpG islands in vertebrate genomes." *J. Mol. Biol.* 196: 261–282.

GeneMark. 2012. Borodovsky Group, Georgia Institute of Technology. http://exon.gatech.edu/ (accessed August 21, 2012)

Gropl, Clemens and Daniel Huson, 2005. "Hidden Markov Models." 19:15. http://www.inf.fu-berlin.de/inst/ag- bio/FILES/ROOT/Teaching/Lectures/WS0405/aldabi/script-hmm.pdf (accessed November 18, 2011)

Guide to the Human Genome, "Chromosomes and DNA," 2010. Cold Spring Harbor Lab Press. http://www.cshlp.org/ghg5_all/section/dna.shtml (accessed May 5, 2012)

Homo Sapiens (human), Chromosome 19. NCBI Map Viewer. 2001. http://www.ncbi.nlm.nih.gov/mapview/maps.cgi?ORG=hum&MAPS=ideogr,est,loc&LINKS=ON&VERBOSE=ON&CHR=19 (accessed May 23, 2012)

International Human Genome Sequencing Consortium. 2004. "Finishing the euchromatic sequence of the human genome". *Nature*, 431(7011): 931–45.

Momand, Jamil. 2006. "Scoring Matrices," Southern California Bioinformatics Summer Institute. http://instructional1.calstatela.edu/jmomand2/2006/curriculum/ppt/scoring_matrices.ppt (accessed January 17, 2012)

Nakhleh, Luay K. 2010. "Pairwise Sequence Alignment (II)," COMP 571 Bioinformatics: Sequence Analysis. http:// www.cs.rice.edu/~nakhleh/COMP571/Presentation3.ppt (accessed January 17, 2012)

Rupp, Bernhard. 2000. "Protein Structure Basics," UCRL-MI-125269, Lawrence Livermore National Laboratory. Originally accessed in 2000 from http://www.structure.llnl.gov/Xray/tutorial/protein_structure.htm; currently available at http://www.ruppweb.org/Xray/tutorial/protein_structure.htm (accessed October 30, 2011)

Salzberg, S. L., A. L. Delcher, S. Kasif, and O. White. 1998. "Microbial Gene Identification Using Interpolated Markov Models." *Nucleic Acids Research*, 26(2): 544–548.

Tang, Haixu. 2007. "Probabilistic Sequence Modeling II: Markov Chains," *I529: Bioinformatics in Molecular Biology and Genetics: Practical Applications*. darwin.informatics.indiana.edu/col/courses/I529/Lecture/lec-4.ppt (accessed January 9, 2013)

UCSC Genome Browser on Human Feb. 2009 (GRCh37/hg19) Assembly. 2009. http://genome.ucsc.edu/cgi-bin/hgTracks?position=chr19:571325-583493&hgsid=264592883&knownGene=pack&hgFind.matches=uc002loy.3 (accessed May 23, 2012)

MODULE 14.15

Social Networks—Value in Being Well Connected

Prerequisite: Module 13.5, "The Next Flu Pandemic—Old Enemy, New Identity." Part of Project 4 requires Module 8.3, "Empirical Models."

Introduction

Birds belonging to the family Paridae are common on all the continents except for Australia and South America. There are at least 55 species, and if you live in the eastern United States, you will likely recognize some of them—chickadees and titmice. In England, there are some rather famous members of this bird family—the tits, especially the blue tit (*Cyanistes caeruleus*). During the early 1920s people living in Swaythling, a small town in southern England, observed these birds opening the foil lids on bottles of delivered milk. Blue tits are urbanized, curious birds, often attracted to sources of food provided by human inhabitants. They eat insects, seeds, or just about anything nutritious. For instance, they can peel back bark to look for insects, pierce flower bases for nectar, and prepare folded paper for food storage. Relative to other birds, this family contains some rather "cerebral" members. The tits have large brains for their body size, so this lid-opening behavior might not surprise those familiar with them (Lefebvre and Boogert 2010).

For the next 25 years, more sightings of this behavior were reported in various locations in Great Britain, Ireland, and on the continent. Scientists wondered if the proliferation of this behavior was transmitted culturally—did they learn from each other (Fisher and Hinde 1949; Hinde and Fisher 1951)? Actually, many species of birds took advantage of the rich cream in the milk bottles prior to World War I. The bottles then had no lids, so access to this nutritious treat did not require a big brain. However, not all the birds were able to pierce the lids, once they were installed (de Geus 1999).

The very rapid spread of this behavior suggested cultural transmission. To confirm cultural transmission, we need, besides rapid spread, a large spatial range and a spread pattern that displays a logistic (s-shaped) curve—the spread lagging at first (1921–1936), followed by a steep increase (1937–1947), and ending with the curve leveling out. The spread conformed to the first two pattern aspects but did not display a slowdown. Perhaps there were so many places for the behavior to spread that it just kept proliferating from discrete English localities, across Great Britain, and into other countries. In fact, the pattern, which conforms to such a model, should reveal a "ripple effect"—displays of the behavior extending from Swaythling, as the ripples generated by dropping a pebble in water. This ripple configuration was not observed. In fact, no obvious pattern for the spread could be discerned. Scientists have concluded that the behavior pattern reflects the occurrence of the same novel behavior originating independently at different sites. The spreading of the behaviors would take place from each origin by the imitation of the behavior by observers. This more complex explanation of the pattern includes autonomous invention and learning, including social learning (Lefebvre and Boogert 2010).

Interestingly, some of the other species, like English robins, which partook freely of the cream in open bottles, never developed this behavior successfully. It has been observed that robins are more territorial than tits. Tits, after their young have fledged, tend to forage in small flocks, providing more opportunities for social learning. Birds that are very territorial for long periods of the year do not expose themselves to as many opportunities to learn from others (de Geus 1999).

Various tit species are still subjects of study to understand their social interactions. At least two groups at the University of Oxford are conducting field studies designed to describe and understand better the complexities of the birds' social networks. It would seem that, at least with tits, there are such networks that would parallel those of Facebook, Twitter, and so on. One might even say that the tits are all "atwitter."

In a collaborative project between the Departments of Zoology (Edward Grey Institute of Field Ornithology) and Engineering Science (Machine Learning Research Group) at Oxford, scientists collected massive amounts of data for a wild population of great tits (*Parus major*) in Wytham Woods, near Oxford city. During the nesting season, the scientists placed RFID (radio-frequency identification) tags on all nestlings and on captured adult birds. For two subsequent winters, they deployed sunflower seed feeders, spaced at even distances along a grid over the study site. The feeders had data loggers that recorded the visits and the times for specific birds. The data generated over a million records for more than 1200 great tits. From analysis of this spatio-temporal data, the researchers were able to extract the structure of the birds' social network. They hypothesized that birds would visit a feeder *preferentially* in small flocks, and increases in observed density would be equivalent to a gathering *event* of birds that were socially connected. What their data demonstrated, which paralleled field observations, was that the birds did indeed forage with their *friends*. The birds seemingly selected the birds with which they wanted to socialize and even to mate (Psorakis et al. 2012). It almost seems as if the great tits have their own field version of Facebook.

Lucy Aplin, an Australian D.Phil. student, also worked at Oxford on wild populations of several tit species. She led a group of researchers investigating populations of the most common tit species in copses near Wytham Woods. Blue tits (*Cyanistes caeruleus*), great tits (*Parus major*), and marsh tits (*Poecile palustris*), all members of the Paridae, commonly forage in mixed flocks. Birds in the study area were tagged with transponders, and sunflower feeders with data loggers were installed at two locations in each study area. The researchers could gather times for each bird's visit and detect associations. From these data, a matrix of associations, or social network, was constructed. The birds became familiar with the locations of the feeders over the 2-month period, but then, the scientists removed the feeders for 2 weeks. Subsequently, the feeders were repositioned (at night) within the study site in new, unfamiliar locations. After 3 days the feeders were removed, and after 7 days they were returned to new, unfamiliar locations. Data were collected in four trials on the first birds to locate the new food source. The association matrix studies determined non-directional associations of foraging birds, whereas the discovery experiments determined the influx order at the unfamiliar sites. These experiments should be related only if social connections of the birds were important to discovery. In other words, friends tell friends. What the group found was that social connections were definitely involved in discovery. Information passed according to the network of social associations (Aplin et al. 2012). So, at least for these tit species, social connections are a form of social security. It gives an additional perspective for the old idiom, birds of a feather flock together.

Projects

Several of the projects reference Stanford Large Network Dataset Collection (SNAP 2012), a research repository of datasets for large social and information networks. Each referenced SNAP download page contains information about the corresponding dataset.

1. (Download dataset *ca-GrQc.txt.gz* from http://snap.stanford.edu/data/ca-GrQc.txt.gz.) The Stanford Large Network Dataset Collection stores the General Relativity and Quantum Cosmology collaboration network dataset with "scientific collaborations between authors papers submitted to General Relativity and Quantum Cosmology category" (SNAP 2012; ca-GrQc 2012). In this undirected graph, if authors *i* and *j* coauthored a paper, an edge exists between nodes *i* and *j*.

For each of the following parts, calculate the metric, answer the indicated questions, and discuss the results and meaning of the metric in the context of the problem.

a. For this dataset, graph the degree distribution and determine a function that fits this distribution. Is the network scale free? If so, determine the largest hubs.
b. Find a minimal dominating set.
c. Calculate the mean clustering coefficient.
d. Determine the numbers of nodes and edges in the largest connected component.
e. Find the mean shortest-path length and the diameter (see Module 13.5, "The Next Flu Pandemic—Old Enemy, New Identity, Project 2). Does this network exhibit the small-world property?

2. (Download *dophins.gml* from (CASOS 2012) at http://www.casos.cs.cmu.edu/computational_tools/datasets/external/dolphins/index11.php.) "The file *dolphins.gml* contains an undirected social network of frequent associations between 62 dolphins in a community living off Doubtful Sound, New Zealand, as compiled by" Lusseau et al. (2003). Repeat Project 1 for this dataset.

3. (Download *facebook.tar.gz* and *readme-Ego.txt* from http://snap.stanford.edu/data/egonets-Facebook.html.) The Stanford Large Network Dataset Collection stores a "Social circles: Facebook" dataset of **circles**, or **friends lists**, from Facebook (SNAP 2012; ego-Facebook 2012). A social circle is a category of friends, such as family members, schoolmates, or club members. Each file name in the dataset begins with an integer, *nodeId*, indicating the ID of a focal individual, an **ego**, for the associated **ego network**, or personal network, which is undirected. The collection of files available for download at ego-Facebook (2012) contains data for 10 ego networks with ego IDs 0, 107, 348, 414, 686, 698, 1684, 1912, 3437, and 3980 and a total of 4039 Facebook users. In an ego network, the ego is a friend with every other node individual, called an **alter**.

A file ending with *.edges* lists the edges in the ego network for the *nodeId* individual. For example, the first line of *0.edges* contains 236 186. Thus, the individual with ID 0, in this case the ego, is friend with individuals 236 and 186, two of 0's alters. Moreover, the undirected edge (236, 186) indicates that 236 and 186 are friends with each other. Filenames ending with *.circles* name and list the ego's circles. For example, the file *0.circles* indicates that ego 0 has 24 circles of friends (*circle0* through *circle23*), and *circle3* represents a circle with 0 and alters 51, 83, and 237. A *.featname* file lists yes/no feature names in the ego network. To preserve anonymity, features have been **anonymized**, or made anonymous. For example, *0.featname* lists 224 features; features 77 and 78 refer to gender ("anonymized feature 77" and "anonymized feature 78"); and features 24 through 52 relate to the possible schools that network members have attended. The file *0.egofeat* presents in a binary fashion the features of ego 0. With 0 in position 77 and 1 in position 78, the ego is of gender "anonymized feature 78"; and 1 in position 39 indicates a school affiliation of "anonymized feature 39." A file ending in *.feat* has a line representing the features for each alter. Each line begins with the alter's ID and continues with a binary representation of the alter's features. Thus, *0.feat* has 347 lines, one for each of its alters, 1 through 347; and on line 2, 1 for position 78 indicates alter 2 has gender "anonymized feature 78." Consequently, ego 0 and alter 2 are the same gender.

a. For one of the egos, calculate the mean clustering coefficients of each of its circles and of its ego network. Do variations in the values indicate anything about friendships? Discuss the results.

b. One metric for the strength of the tie between two individuals is the number of features they have in common. For one of the egos, calculate the mean tie strength between the ego and its alters and the mean tie strength between the ego and the entire network of 4039 users. Does the ego appear to have more in common with its friends than with the general network population?

c. Calculate the tie strength (see Part b) between one of the egos and each of its alters. Calculate the pairwise mean tie strength among the 10 alters with the highest ego-alter tie strengths, the 10 alters with the lowest ego-alter tie strengths, and the 10 alters in the midrange of ego-alter tie strengths. Discuss the implications of the results.

d. For one of the egos, calculate the mean tie strength (see Part b) between pairs of ego network members. For each of the ego's circles, calculate the mean tie strength between pairs of circle members. Discuss the implications of the results.

e. For one of the egos and each of its circles, calculate the average of each feature. For example, for ego 0 and *circle3* with members 51, 83, 237, the average in position 50 is 0.5; two of the four answered yes to attending school "anonymized feature 50." Repeat the calculation among all the ego network members. Discuss the results.

f. Repeat any of Parts a–e for all the egos. Discuss the overall results.

4. (Download dataset *Wiki-Vote.txt.gz* from http://snap.stanford.edu/data/wiki-Vote.html.) The Stanford Large Network Dataset Collection stores a "Wikipedia vote network" dataset of voter data for Wikipedia administrator elections (SNAP 2012; wiki-Vote 2012). Volunteers around the world collaboratively write articles for Wikipedia, a free online encyclopedia, and users can vote whether a candidate is to become an administrator or not.

After header documentation, each line of the file *wiki-Vote.txt* contains pairs of numbers, such as 110 929, indicating that user 110 cast a vote in the election for candidate user 929.

The file *wikiElec.ElecBs3*, which contains data on 2794 elections and 103,663 votes from 7066 users, uses the following data format:

```
# E: is election successful (1) or not (0)
# T: time election was closed
# U: user id (and username) of editor that is being considered for
     promotion
# N: user id (and username) of the nominator
# V: <vote(1:support, 0:neutral, –1:oppose)> <user_id> <time>
     <username>
```

For example, the following election data indicate that the user with id 929 lost the election with a neutral vote from user id 23, a support vote from 865, and an oppose vote from 110:

```
E    0
T    2006-06-15 04:33:59
U    929    poiuytman
N    –1     UNKNOWN
V    0      23     2005-05-05  00:59:00    cryptoderk
V    1      865    2005-05-05  01:05:00    grace
V    –1     110    2005-05-05  05:17:00    kingturtle
⋮
```

Consider a weighted directional graph with user ids as the nodes, arrows from voters to candidates, and weights indicating the type of vote (1: support, 0: neutral, –1: oppose).

a. Compute the undirected degree distribution, and with empirical modeling and graphing the results, obtain a function that captures the trend of the distribution. Is the undirected network scale free? Identify the hubs.

b. Compute the out-degree distribution, and with empirical modeling and graphing the results, obtain a function that captures the trend of the distribution. Is the directed network scale free? Discuss the results.

c. Compute the out-degree distributions for each of the weights, –1, 0, and 1, and obtain graphs that capture the trends of the distributions. Discuss the results.

d. On the average, do more users cast votes in elections where the candidate is successful or unsuccessful?

References

Aplin, L. M., D. R. Farine, J. Morand-Ferron, and B. C. Sheldon. 2012. "Social Networks Predict Patch Discovery in a Wild Population of Songbirds." *Proceedings of the Royal Society B: Biological Sciences,* 279(1745): 4199–4205.

ca-GrQc, General Relativity and Quantum Cosmology Collaboration Network. 2012. http://snap.stanford.edu/data/ca-GrQc.html (accessed December 18, 2012)

CASOS (Computational Analysis of Social and Organizational Systems). 2012. http://www.casos.cs.cmu.edu/index.php (accessed December 18, 2012)

de Geus, Arie P. 1999. In the Living Company: Growth, Learning, and Longevity in Business, rev. ed. London: Nicholas Brealey.

ego-Facebook, Social circles: Facebook. 2012. http://snap.stanford.edu/data/egonets-Facebook.html (accessed December 18, 2012)

Fisher, James, and Robert A. Hinde. 1949. "The Opening of Milk Bottles by Birds." *British Birds,* 42(347): 57.

Hinde, Robert A., and James Fisher. 1951. "Further Observations on the Opening of Milk Bottles by Birds." *British Birds,* 44: 393–396.

Lefebvre L., and N. J. Boogert . 2010. "Avian Social Learning." In M. D. Breed and J. Moore (eds.), *Encyclopedia of Animal Behavior*, volume 1, pp. 124–130. Oxford: Academic Press.

Lusseau, D., K. Schneider, O. J. Boisseau, P. Haase, E. Slooten, and S. M. Dawson. 2003. "The Bottlenose Dolphin Community of Doubtful Sound Features a Large Proportion of Long-lasting Associations," *Behavioral Ecology and Sociobiology,* 54: 396–405.

Psorakis, Ioannis, Stephen J. Roberts, Iead Rezek, and Ben C. Sheldon. 2012. "Inferring Social Network Structure in Ecological Systems from Spatio-temporal Data Streams." *Journal of The Royal Society Interface,* 9(76): 3055–3066.

SNAP (Stanford Large Network Dataset Collection). 2012. http://snap.stanford.edu (accessed December 18, 2012)

wiki-Vote, 2012. Wikipedia Vote Network. http://snap.stanford.edu/data/wiki-Vote.html (accessed December 19, 2012)

GLOSSARY OF TERMS

AB—Agent-based.
absolute error—Absolute value of the difference betwewen the exact answer and the computer answer.
absolute refractory period—Time during an AP when all the sodium gates are open and an ensuing stimulus, no matter how strong, cannot initiate another AP.
absorbing boundary conditions—Conditions so that all ghost cells of a cellular automaton or agent-based grid have the same constant value.
acceleration due to gravity—Approximately -9.81 m/s^2, where up is considered the positive direction.
acceleration—Rate of change of velocity with respect to time.
action potential (AP)—A very rapid change in the membrane potential along the plasma membrane of a nerve, endocrine, or muscle cell (excitable cells).
adjacency matrix—For a graph with n nodes, an $n \times n$ matrix, where the element in row i and column j indicates the number of edges between node i and node j.
adjacent—In a graph, property of two vertices that have an edge connecting them.
adrenalin—Epinephrine.
aerobic respiration—Cellular respiration, where oxygen serves as the final electron acceptor during the oxidation of organic molecule to obtain energy. This type of respiration uses the Krebs cycle and the electron transport chain, both found in the mitochondria.
agent-based simulation—Simulation technique in which each animal/entity is modeled as an autonomous, decision-making agent that has a state, which is represented by a set of state variables, or attribute values, and behaviors, which control its actions. A method or procedure, which is associated with a class, or breed or group, of agents, is a function that captures some or all of an agent's behavior. Agents often operate in an environment that arranges cells in a rectangular grid.
agent—An autonomous, decision-making entity in an agent-based simulation.
amino acid—Molecule with a central carbon (α carbon), which bonds with 4 chemical groups—an amino group ($-NH_3^+$), a carboxyl group ($-COO^-$), a hydrogen (H), and a variable side-chain (R-group).
amino group— $-NH_3^+$.
ampere (A) —Unit of current for a charge of one coulomb (1 C) to pass through a region in one second (1 s).
amplitude—Of an oscillating function, the maximum value of the function from the horizontal line going through the middle of the function.
angular acceleration—Rate of change of angular velocity.
angular velocity—Rate of change of an angle with respect to time.
anthropogenic—Of human origin.
antiderivative—Function F if $F'(t) = f(t)$, or the derivative of F is f.
antigen-presenting cell—Specialized immune cell that can present antigens on its surface for interaction with T-cells.

antigen-specific T-cell—T-cell, or T-lymphocyte, that bears receptors that bind specifically to an antigen.
antigen—Substance which the body recognizes as foreign and which can bind to specific receptors on immune cells. Many antigens are also immunogens, which can initiate a specific immune response.
AP—Action potential.
apical growth—In fungi, extension or elongation only at the tip of hyphae.
aposematic coloration—Bright coloration often displayed by prey as a warning of their toxic composition.
artificial water point—Water repository created for livestock, such as a trough or dam.
assignment statement—Statement that causes the computer to store the value of an expression in a memory location associated with a variable.
associative property—Property where grouping in addition or multiplication does not matter: $(a + b) + c = a + (b + c)$ and $(ab)c = a(bc)$.
atm—Abbreviation for atmosphere.
atmosphere—Layer of gases surrounding the earth. As a measurement, one atmosphere (1 atm) is the atmospheric pressure at sea level.
ATP synthase—Enzyme, using the energy from an electrochemical gradient, which synthesizes ATP from ADP and phosphate.
atria—In the human heart, the two upper chambers that receive blood returning from the body and lungs.
average velocity—Ratio of the change in position to the change in time.
Avogadro's Number—6.02214×10^{23}, the number of carbon (C) atoms in exactly 12 g of carbon-12 (^{12}C).
AWP—Artificial water point.
axon—Cytoplasmic extension of a neuron that can transmit signals away from the cell body toward other neurons or effector cells.
Barnes-Hut algorithm—Divide-and-conquer parallel algorithm employing clustering that is a simulation of the *N*-Body Problem.
base—Nitrogen base.
base pair—Two bases on different strands of DNA that bond with each other.
basic reproductive number (R_0)—Initial reproductive number with one infectious individual and all others being susceptible.
behavior—In agent-based simulations, actions of an agent.
binary number system—Number system with base 2; used by most computers.
biofilms—Communities of very small organisms that adhere to a surface in an aqueous environment.
bioinformatics—A newly developing area of computational science that deals with the organization of biological data, such as in databases, and the analysis of such data, which often makes extensive use of probabilities.
biomagnification—Cumulative increase in the concentrations of a substance in successively higher levels of the food chain.
biosphere—All living things.
bipartite graph—A graph with vertices partitioned into two sets, where arcs are only between vertices in different sets.
bit—0 or 1 in the binary number system.
BLAST—Basic local alignment search tool.

blood pressure—Hydrostatic (fluid) pressure that moves the blood through the circulation.

BMI—Body mass index.

body mass index (BMI)—Anthropometric measure, defined as 703 multiplied times the weight in pounds and divided by the square of height in inches. BMI is used to indicate if someone is at a healthy weight.

Boyle's law—For gas at a particular temperature, $PV = K$, where P is pressure, V is volume, and K is a constant.

Brownian Motion—The behavior of a molecule suspended in a liquid.

CA—Cellular automaton.

calculus—Mathematics of change.

canopy—The uppermost branches and leaves of a tree that block sunlight. Also, in plant communities, it is the upper portion formed collectively by the plants' crowns.

capacitance—Ability to store charge.

capacitor—Electronic circuit element for storing charge.

capsid—Protein coating that covers the core of a virus.

carbohydrates—Organic molecules composed of the elements carbon (C), hydrogen (H), and oxygen (O) in the ratio of one C to two H to one O.

carbon cycle—Movement of carbon from one earth subsystem to another.

carboxyl group— $-COO^-$.

cardiac output—Product of the stroke volume and the heart rate.

carrying capacity—Maximum population size that an environment can support indefinitely.

CD4+ T-lymphocytes—T-lymphocytes that bear CD4+ receptors, which have important functions in both humoral and cell-mediated immunity. CD4+ cells activate cytotoxic T-cells, B-cells (produce antibodies), and antigen-presenting cells.

cell body—Part of a neuron that contains the nucleus and cytoplasm, not including axons and dendrites.

cell-mediated immunity (CMI)—Part of the immune response, mediated by T-cells, that is directed against virally-infected cells, intracellular bacteria, and cancer cells.

cell—In cellular automaton and agent-based simulations, individual element of a grid.

cellular automaton (plural, automata)—Type of computer simulation that is a dynamic computational model and is discrete in space, state, and time, where space is represented as a regular, finite grid and a discrete state is associated with each grid element (cell).

central place forager—Animal that, after traveling in search for food, always returns to a central place.

central processing unit (CPU)—Part of a computer that performs the arithmetic and logic.

Charles's Law—$PV = nRT$, where P is pressure, V is volume, T is temperature in kelvin (K), n is the number of moles, and R is the constant 0.0832 atm/(mol K).

child—In radioactivity, substance formed by radioactive decay from another substance.

chromosome—A threadlike strand of nucleic acid, with associated proteins, that carries genetic information (genes).

circulatory system—System of interconnected spaces and tubes that transport fluids in multicellular animals.
clustering coefficient—For vertex v, where A is the set of nodes adjacent to node v in graph G, $C(v) = \dfrac{\text{number of edges of } G \text{ in subgraph with set of nodes } A}{\text{number of edges in complete graph with } n(A) \text{ nodes}}$.
clustering—Partitioning data into subsets, or clusters, so that the elements of a cluster have some common trait, such as proximity.
CMI—Cell-mediated immunity.
coalescence efficiency—Product of the collision and coalescence efficiencies; the portion of collisions that results in the droplets of rain sticking together.
coalescence—Merging of rain droplets to form larger droplets.
coarse granularity—Machine with few processors, each executing many instructions simultaneously, so that the ratio of computation time to communication time is large.
codons—Groups of three nucleotides in messenger RNA specifying an amino acid.
coefficient of drag—Constant of proportionality in Newtonian friction.
coenzymes—Organic cofactors that associate with enzymes and help them catalyze.
cognitive map—Series of markers forming geocentric references that are stored in an animal's memory as a neural representation of the animal's environment.
collision efficiency—In clouds, $E = d^2/(r_1 + r_2)^2$, where d is the distance between the center lines of a drop and a droplet, where r_1 and r_2 are the radii of the droplets.
common logarithm—Logarithm to the base 10, usually written $\log n$; $\log n = m$ if and only if $10^m = n$.
community—All the species living in an area.
competition—Struggle between individuals of a population or between species for the same limited resource.
complete—A graph in which exactly one edge exists between each pair of nodes.
component-wise—Vector operation performed component by component, or coordinate by coordinate.
computational science—Emerging interdisciplinary field that is at the intersection of the sciences, computer science, and mathematics.
computer cluster—Small systems with a few processors to supercomputers with thousands of processors having the distributed-memory MIMD architecture.
computer simulation—Having a computer program imitate reality in order to study situations and make decisions.
concurrent processing—Having associated, multiple CPUs working concurrently, or simultaneously.
condensation nuclei—Particles upon which water vapor is deposited to form rain droplets.
condensation—Deposition of vapor on particles (condensation nuclei) to form droplets.
conditional probability—For $P(E_2 \mid E_1)$, the probability of event E_2 given event E_1.
connected—A graph in which there exists a path from any vertex to any other vertex.
connection matrix—For a graph with n nodes, an $n \times n$ matrix, where the element

in row i and column j is 1 if an edge exists between node i and node j and is 0 otherwise.

continuous distribution—Probability distribution with continuous values.

continuous model—Model in which time changes continuously as opposed to discretely.

contractility—Ability to shorten, as in heart muscles.

controlled drug delivery—Combining a polymer with a medicine for release of drug into the body in predetermined manners.

convective current—Current within a medium caused by a difference in temperature.

core—In viruses, the central complex that contains its genetic material (DNA or RNA) and various proteins.

cosine—For a point (x, y) on a circle with radius r and angle t off the positive x-axis, $\cos t = x/r$.

cost function—Function of quantity that returns the total cost in producing the items.

coulomb—Unit of electric charge.

CpG island—Small region where transformation from CG to TG is suppressed, upstream of many genes on a DNA sequence.

CPU—Central processing unit.

C-terminal—End of a protein that has a free carboxyl group.

cumulonimbus—Tall, anvil-shaped clouds.

cumulus clouds—Light, fluffy clouds with flat bases that do not give rise to any rain.

current—Rate of change of charge with respect to time, measured in amperes (A).

Dalton's law—The partial pressure of a gas is the product of the fraction of the gas in the mixture and the total pressure of all gases, excluding water vapor.

data partitioning—In parallel processing, technique where the root process splits data into nonoverlapping subsets and sends each subset to a different process.

decay chain—Sequence of several radioactive decay events, producing in the end a stable product.

decay constant—Constant of proportionality in a unconstrained model for decay.

decimal number system—Number system using the base 10.

decompression sickness—Painful and life-threatening condition experienced by divers who return to the surface of the water too rapidly after deep-water dives; also known as the bends. This condition results from nitrogen bubbles expanding as pressure decreases during ascent.

defibrillator—Medical device that causes a predetermined amount of current to flow across the heart so that normal electrical patterns are restored.

definite integral— $\int_a^b f(t)dt = \lim_{n\to\infty}(\text{left-hand sum}) = \lim_{n\to\infty}(f(t_0)\Delta t + f(t_1)\Delta t + \cdots + f(t_{n-1})\Delta t)$ and $\int_a^b f(t)dt = \lim_{n\to\infty}(\text{right-hand sum}) = \lim_{n\to\infty}(f(t_1)\Delta t + f(t_1)\Delta t + \cdots + f(t_n)\Delta t)$ where $\Delta t = (b - a)/n$ and $a = t_0 < t_1 < \cdots < t_n = b$.

degree distribution—In a graph, $P(k) = n_k/n$, where n is the number of nodes in a graph and n_k is the number of nodes of degree k.

degree of a vertex—In a graph, the number of times the vertex is an endpoint of an edge.

dehydrogenase—Class of enzymes which catalyze oxidation of substrates by removing hydrogen or electrons, often in energy-producing pathways.

dendrite—Cytoplasmic extension of a neuron that can transmit signals toward the cell body along the plasma membrane.

dendrites—Small, treelike structures that form during the process of solidification of crystalline structures; thin, branched extensions from the cell body of a neuron, responsible for reception of neural signals.

dendritic cells— Special types of antigen-presenting cell (APC), located in skin and mucosal membranes that activate T-lymphocytes.

dependent variable—Variable that relies on other variables.

depolarized—Neuron or other excitable cell in which gated channels for Na^+ are opened and Na^+ floods into the cell, temporarily making the inside more positive than the outside.

derivative—For a function *f*(*t*), the derivative at *t* is $f'(t) = \lim_{h\to 0} \frac{f(t+h) - f(t)}{h}$, provided the limit exists; instantaneous rate of change of a function with respect to an independent variable; geometrically, the slope of the tangent line to the curve *f* at *t*.

deterministic behavior—Behavior of systems that is predictable.

deterministic model—Model that is predictable.

diagonal—In an $n \times n$ square matrix *M*, the set of elements $\{m_{11}, m_{22}, \ldots, m_{nn}\}$.

diastolic pressure—Pressure in the arteries as the left ventricle relaxes.

differential calculus—One of the two branches of calculus dealing with problems involving the derivative.

differential equation—Equation containing one or more derivatives.

diffusion rate parameter—Constant, *r*, for rate of diffusion, such as in the model $\Delta site = r\sum_{i-1}^{8}(neighbor_i - site)$, where $0 < r < 1/8 = 0.125$

diffusion-limited aggregation (DLA)—Simulation technique to build a dendritic structure by adding one random walking particle at a time.

diffusion—Dispersal; the spreading of something more widely. In biology, the passive movement of molecules or particles from regions of higher concentration to regions of lower concentration.

directed graph—$G = (V, E)$ consisting of a set *V* of vertices and a set *E* of directed edges connecting pairs of vertices.

discrete distribution—Distribution with discrete values.

discrete model—Model in which time changes in incremental steps.

disintegration constant—Decay constant.

disintegrations per minute (dpm)—A standard expression of radioactive decay.

distributed processing—Several processors, perhaps at great distances from each other, communicating via a network and working concurrently.

distribution of numbers—Description of the portion of times each possible outcome or each possible range of outcomes occurs on the average.

distributive property—$a(b + c) = ab + ac$.

divide-and-conquer algorithm—Algorithm that divides a problem into subproblems of the same form, and then divides theses into subproblems of the same form, etc. The small problems are solved, and the final solution is assembled.

DLA—Diffusion-limited aggregation.

dominant eigenvalue—The largest eigenvalue for a matrix.

dot product—For vectors $\mathbf{x} = (x_1, x_2, \ldots, x_n)$ and $\mathbf{y} = (y_1, y_2, \ldots, y_n)$, $\mathbf{x} \cdot \mathbf{y} = x_1 \cdot y_1 + x_2 \cdot y_2 + \cdots + x_n \cdot y_n$.

double-precision number—Floating-point number using twice as many bits in a computer representation than a single-precision number; typically contains 14 or 15 significant digits and has magnitude between 10^{-308} and 10^{308}.

downwelling—Process where currents move ocean surface water to lower depths.

dpm—Disintegrations per minute; a measure of radioactive decay.

drag coefficient—Coefficient of drag.

drag—Forces that resist the movement of an object through a fluid.

dynamic model—Model that changes with time.

dynamical disease—Disease in which blood cell counts may oscillate, perhaps in an involved or chaotic manner.

e—Part of exponential notation, where *aen* represents $a \times 10^n$, a is a decimal fraction, and n is the exponent; symbol for 2.718281…; base for the natural logarithm.

ecological niche—Complete role that a species plays in an ecosystem.

effector cells—Cells that respond to stimuli, for example, muscle and gland cells. In the immune system, they are cells that are activated to provide protection, e.g. plasma cells, T-helper cells, and cytotoxic T-cells.

eigenvalue—For square matrix M, constant λ, where $M\mathbf{v} = \lambda\mathbf{v}$ for eigenvector $\mathbf{v}$.

eigenvector—For square matrix M, vector $\mathbf{v}$, where $M\mathbf{v} = \lambda\mathbf{v}$ for eigenvalue λ.

elastomer—A polymer that can return to its original shape, following deformation.

electrical potential—Voltage difference across a membrane.

electron transport system—System that removes and passes along electrons and protons from reduced coenzymes. Often the movement of electrons will provide sufficient energy to create a proton gradient, which can be used to synthesize ATP.

electronic potential—Potential energy per unit charge at a point, or the work per unit charge to bring a positive charge from infinity to the point.

embarrassingly parallel algorithm—Algorithm in which computation can be divided into many completely independent parts with virtually no communication.

empirical model—Model employing only data to predict, not explain, a system and consisting of a function that captures the trend of the data.

envelope—Outer coating made up of lipids and proteins that surrounds some virus particles.

environmental subsystem—An interdependent part of the earth's system.

enzyme kinetics—Quantitative study of enzyme activity.

enzyme—Organic catalyst in a chemical reaction for a biological system.

EP—Equilibrium potential.

EPC—Euler's predictor-corrector method, or Runge-Kutta 2 method.

epidemic ratio—Rate of infection over rate of recovery.

epinephrine (adrenalin)—Chemical messenger of the sympathetic nervous system, including the adrenal medulla, released during times of stress. A major effect is to increase heart rate.

equilibrium potential—For any ion, the membrane potential where there is no net diffusion of the ion across the membrane.

equilibrium solution—Solution of a differential equation where the derivative is always zero; solution of a difference equation where the change of the dependent variable is always zero.

equilibrium vector—A vector, **v**, of a Markov chain associated with a transition matrix T, where $T\mathbf{v} = \mathbf{v}$.

Euler's method—Method of numeric integration that estimates $P(t)$ as $P(t - \Delta t) + P'(t - \Delta t)\,\Delta t$, where Δt is the change in t.

Euler's predictor-corrector method (EPC)—Runge-Kutta 2 method.

exploitative competition—Competition where one individual (species) reduces the availability of the resource to the other.

exponential function—Function of the general form $P(t) = P_0 a^{rt}$, where P_0, a, and r are real numbers.

exponential notation—Floating-point number represented as a decimal fraction times a power of 10.

F—Farad.

FAD—Flavin adenine dinucleotide.

fairy ring—Naturally occurring arc of mushrooms arising at the periphery of a radially spreading underground mycelium.

farad (F)—Measure of capacitance, equivalent to having a capacitor hold a charge of 1 C for a potential difference of 1 V across its conductors, or 1 F = 1 C/V.

feedback loop—In a model, a cyclical flow from system to system.

fine granularity—Machine with many processors, each executing relatively few instructions, so that the ratio of computation time to communication time is small.

finite difference equation—Equation of the form (new value) = (old value) + (change in value); a discrete approximation to a differential equation.

finite geometric series—$a^{n-1} + \cdots + a^2 + a^1 + a^0$ for $a \neq 1$ and positive integer n where a is the base.

fishing maximum economic yield (FMEY)—Maximum profit of fishing as it relates to economic yield in the Gordon-Schaefer fishery production model.

fishing maximum sustainable yield (FMSY)—The cost of fishing effort to produce a maximum yield in the Gordon-Schaefer fishery production model.

fitting data—Obtaining a function that roughly goes through a plot of data points and captures the trend of the data.

flavin adenine dinucleotide (FAD)—A major coenzyme involved in oxidation-reduction reactions of metabolism. By taking on two electrons and two protons, it is converted to a reduced form, $FADH_2$.

floating-point number—Real number expressed with a decimal expansion and stored in a computer in a fixed number of bits.

flux—Transfer of carbon or other element from one reservoir to another.

FMEY—Fishing maximum economic yield.

FMSY—Fishing maximum sustainable yield.

free energy—Energy available for cellular work.

functional response— Predator's behavioral reaction to changes in prey density.

fungus (plural, fungi)—Multicellular (except for yeast), spore-producing organism that depends on absorbing nutrients from its surroundings.

gametocyte—Sexual stage of the malarian parasite that circulates freely in a host's blood and that the female mosquito obtains in her blood meal. Male and female gametocytes fuse in the mosquito's gut to form an oocyst, which divides to produce sporozoites.

gated channels—Ion channels that require specific stimuli to open the gates.

Gaussian distribution—Normal distribution.

gene—In cells, a contiguous section of a chromosomal DNA that encodes information to build a protein or an RNA molecule. In RNA viruses, genes are made up of contiguous sections of RNA.

generalist—Predator that may use alternative prey as densities of their primary prey decline. More generally, a species that can live in a wide variety of environmental conditions and exploit varied resources for survival.

genetic code—Represents a correspondence between nucleotide triplets and the amino acids they specify.

genome—A complete set of chromosomes in a cell that contains the organism's hereditary information.

ghost cell—In cellular automaton and agent based simulations, a cell that is part of grid's extension to accommodate boundary conditions.

global simulation variable—A variable that all agents can access in an agent-based simulation.

global warming—Gradual increase in average temperature of the earth's atmosphere and oceans.

glycogen—Branched polymer of glucose used as a storage carbohydrate by animals.

glycolysis—Initial sequence of chemical reactions of glucose oxidation that results in the production of two molecules of pyruvate, ATP, and NADH.

gp 120—Glycoprotein that protrudes from the surface of HIV and binds to CD4+ T cells.

granularity—In a parallel computer system, the ratio of computation time to communication time, which is related to the number of processors. See *coarse granularity* and *fine granularity*.

graph—A directed or undirected graph.

greenhouse effect—Warming of the atmosphere caused by the trapping of infrared radiation by *greenhouse gases*.

greenhouse gas—Atmospheric gas that absorbs infrared radiation, preventing the radiation's loss to space.

grid—A one-, two-, or three-dimensional array of sites, or cells, used in cellular automaton or agent-based simulations.

ground—A circuit reference point; often the negative terminal of a battery.

half-life—In radioactivity, the period of time that it takes for a radioactive substance to decay to half of its original amount; in drug dosage, amount of time for a body to eliminate half of the drug.

henry (H)—Unit of measure of inductance, where 1 H = 1 V s/A (volt second/amp).

Henry's law—For the amount of any gas in a liquid at a particular temperature, $V_g/V_L = sP_g$, where V_g is gas volume, V_L is liquid volume, s is the solubility coefficient for the gas in that liquid, and P_g is the pressure of the gas.

hill-climbing process—Process in cellular automaton or agent-based simulations by which at each time step, an animal/entity moves to a neighboring cell that has the highest value.

HIV—Human immunodeficiency virus.

homeothermic—Term used to describe an animal that can maintain its core body temperature at a nearly constant level regardless of the environmental temperature.

homolog—A gene from the same ancestral origin as that from another species; each member of a pair of chromosomes.

Hooke's law—Within the elastic limit of a spring, $F = -ks$, where F is the applied force, k is the spring constant, and s is the displacement (distance) from the spring's equilibrium position.

host—Organism whose body supplies nourishment and shelter for another.

hubs—Nodes with high degrees in scale-free networks.

human immunodeficiency virus (HIV)—Type of RNA virus belonging to the retroviruses that is the causative agent of AIDS (acquired immune deficiency syndrome).

hydrosphere—All water of the earth, including bodies of water, ice, and water vapor in the atmosphere.

hyperpolarization—Making the membrane potential even more negative.

hypha (plural hyphae)—Fungal filament that is the basic morphological component of all fungi, except yeasts.

hypnozoite—Dormant life stage of some species of malarian parasites that develop from merozoites. Once out of dormancy, they may reinvade other liver cells, where they produce more merozoites.

ideal gas laws—Laws that describe the behaviors of an ideal gas.

ideal gas—Gas in which the volume of its atoms is insignificant in comparison to the total volume of the gas and in which atom interactions are negligible except for the energy and momentum exchanged during collisions.

impulse—Product of the thrust and the length of time of force application.

incident—In a graph, an edge e is incident to vertex v if v is an endpoint of e.

indefinite integral—$\int f(t)dt = F(t) + C$, where $F'(t) = f(t)$ and C is an arbitrary constant.

independent events—Events where the occurrence of one event has no impact on the occurrence of the other.

independent variable—Variable on which other variables depend.

individual fishing quota (IFQ)—Part of the total allowable catch that must be owned by people who fish in order to participate in fishing operations. These quotas are properties, which may be bought and sold.

individual-based simulation—Agent-based simulation.

inductance—Ability, measured in henrys, of a circuit element to store energy and oppose changes in current flowing through it.

inductor—Circuit element, such as a coil of wire, that dampens sudden changes in current.

infected—Individual within a population that has a disease and can spread it to others.

initial condition—Value of the dependent variable when the independent variable is zero.

instantaneous velocity—For position $s(t)$ at time $t = b$, the number the average velocity, $\dfrac{s(b) - s(b - \Delta t)}{\Delta t}$, approaches as Δt comes closer and closer to 0 (provided the ratio approaches a number).

integral calculus—Branch of calculus that deals with problems involving the integral.

interference competition—Direct interaction between individuals (species), where one interferes with or denies access to a resource.

interpolation— Computing intermediate data values between existing data values.

ion channels—Special channels in cell membranes only permitting certain ions through when they are open.

isolated point—In a graph, a point whose degree is 0.

isolation—Separation of a person who has a communicable disease (e.g., a SARS patient) from those who are healthy and susceptible.

keystone predator—Dominant predator; predator that has a major influence on the community structure.

kinetic friction—Force that tends to slow a body in motion.

Kirchhoff's current law—Sum of the currents into a junction equals the sum of the currents out of that junction.

Kirchhoff's voltage law—In a closed loop, the sum of the changes in voltage is zero.

Krebs's cycle—Pathway in energy metabolism in which electrons are removed from pyruvate and placed onto oxidized coenzymes.

lactate fermentation—Following glycolysis, the conversion of pyruvate into lactate, which reoxidizes coenzyme NADH to NAD^+, which is needed for glycolysis to continue.

large-scale navigation—Guided movement of an animal over long distances.

lattice—Grid.

launching rectangle— Surrounding a developing simulated dendritic structure in diffusion-limited aggregation, a rectangle from which new particles are released.

law of mass action—A model for a solution in equilibrium in which the rate of change of a reaction is proportional to the product of the concentrations of interacting molecules.

leak channels—Ion channels that are essentially open all the time.

length of a path—In a graph, the number of edges.

Leslie matrix—A matrix of the following form, where all entries F_i and P_i are non-negative:

$$\begin{bmatrix} F_1 & F_2 & F_3 & \cdots & F_{n-1} & F_n \\ P_1 & 0 & 0 & \cdots & 0 & 0 \\ 0 & P_2 & 0 & \cdots & 0 & 0 \\ \vdots & \vdots & \vdots & & \vdots & \vdots \\ 0 & 0 & 0 & \cdots & P_{n-1} & 0 \end{bmatrix}.$$

ligand-gated channels—Ion channels open in response to binding specifically to a chemical signal.

limit—Generally, a number approached by $f(x)$ as x approaches some number c.

linear combination—Linear combination of x_1, x_2, ..., and x_n is the sum $a_1x_1 + a_2x_2 + \cdots + a_nx_n$, where $a_1, a_2, \ldots, a_n$ are constants.

linear congruential method—Method to generate pseudorandom integers from 0 up to, but not including, the *modulus* using $r_0 = seed$ and $r_n = (multiplier \times r_{n-1} + increment) \bmod modulus$, for $n > 0$, where *seed*, *modulus*, and *multiplier* are positive integers, and *increment* is a nonnegative integer.

linear damping—For the motion of a pendulum, the assumption that damping force is proportional to the angular velocity.

linear equation—$a_1x_1 + a_2x_2 + \cdots + a_nx_n = c$, where a_i and c are numbers for $i = 1, 2, \ldots, n$.

linear function—Function whose graph is a nonvertical straight line, which has the form $y = mx + b$, where m is the slope and b is the y-intercept.

linear least-squares regression—Line that "best captures" the trend of the data, $(x_1, y_1), (x_2, y_2), \ldots, (x_n, y_n)$; line $y = mx + b$, where $b = \dfrac{\sum x_i^2 \sum y_i - \sum x_i y_i \sum x_i}{n\sum x_i^2 - \left(\sum x_i\right)^2}$ and $m = \dfrac{n\sum x_i y_i - \sum x_i \sum y_i}{n\sum x_i^2 - \left(\sum x_i\right)^2}$.

linear regression—Linear least-squares regression.

linear speedup—In high performance computing, situation where the speedup factor for n processes is n.

lipophilic—Term used to describe a substance that can combine with or dissolve in fat (lipid).

lithosphere—Outer solid part of the earth, including the crust and topmost mantle.

loading dose—Initial dosage of a drug that is much higher than the maintenance dosage.

logarithm—Logarithm to the base b of n, written $\log_b n$, is m if and only if $b^m = n$.

logarithmic function—Function, $\log_b(n) = m$, that is the inverse of the exponential function $g(x) = b^x$, so that $b^m = n$.

logical operator—Symbol (such as *AND, OR*, and *NOT*) used to combine or negate expressions that are true or false.

logistic function—Function to model population, $P(t) = \dfrac{MP_0}{\left(M - P_0\right)e^{-rt} + P_0}$, where M is the carrying capacity, P_0 is the initial population, r is the continuous growth rate, and t is time.

loop—Segment of an algorithm that is executed repeatedly.

lymphocyte—White blood cell, either a B-cell or a type of T-cell. Functions include regulation of the immune response, production of antibodies, cytotoxicity.

macrophage—Specialized white blood cell that ingests cellular debris and pathogens. It degrades the ingested material, processes it, and displays foreign substances on its surface. T-cells can interact with the presented material and activate an immune response.

magnitude— For a floating-point number expressed in normalized exponential notation as $a \times 10^n$, 10^n.

Malthusian model—Unconstrained model.

mantissa—Significand.

marginal cost—Instantaneous rate of change of the cost with respect to quantity.

marginal revenue—Instantaneous rate of change of revenue with respect to quantity.

Markov chain—A sequence of variables $X_1, X_2, X_3, \ldots$ in which the value of any variable, X_{n+1}, depends only on the value of its immediate predecessor, X_n.

Markov matrix—Transition matrix.

matrix (plural, matrices)—A rectangular array of numbers.

maximum therapeutic concentration (MTC)—Largest amount of a drug that is helpful without having dangerous or intolerable side effects.

mean arterial pressure—Average pressure during an aortic pulse cycle.

MEC—Minimum effective concentration.

membrane potential—Electrical potential.

merozoite—Life stage of malarian parasite produced asexually from sporozoites or other merozoites, released into the blood, where it may infect other host cells.

message-passing multiprocessor—System in which the processors communicate through message passing.
messenger RNA—A molecule of RNA that carries genetic information from DNA in the nucleus to a ribosome for protein synthesis.
method—In agent-based modeling, a function that captures some or all of an agent's behavior.
methylation—Chemical addition of a methyl group (CH_3) to another atom or molecule.
methylmercury—Organic form of mercury, synthesized from metallic or elemental mercury by sulfate-reducing bacteria in sediments. Aquatic organisms absorb this form of mercury, which tend to accumulate in the top components of a food chain (fish).
Michaelis-Menten constant— Measure of the affinity of an enzyme for a particular substrate; the concentration of substrate, where the velocity of the reaction is equal to half of the maximum velocity (V_{max}).
Michaelis-Menten equation—Equation that describes the relationship in an enzymatic reaction between substrate concentration [S] and reaction velocity (v), as follows: $v = \frac{V_{max}[S]}{K_m + [S]}$, V_{max} is the maximum velocity.
microstructure—During solidification of crystalline structures, complex formed by interconnections of dendrites.
MIMD—Multiple instruction streams, multiple data streams computer architecture.
minimum dominating set—For a people-locations graph, a smallest set of locations that a given proportion of the population visits.
minimum effective concentration (MEC)—Least amount of drug that is helpful.
mitochondria—Membrane-bound cellular compartments that are the sites of most energy production in aerobic (using oxygen) cells.
mod—Function to return the remainder in integer division.
modeling—Application of methods to analyze complex, real-world problems in order to make predictions about what might happen with various actions.
modulus— Divisor when calculating the remainder in integer division for the mod function.
molar—One mole/liter (mol/L).
molarity—The concentration of a solute in a solvent; the number of moles of solute per liter of solution.
mole (mol)—The quantity of a substance containing 6.02214×10^{23} units (atoms, molecules, or some other unit); molecular weight of a substance in grams.
monomer—Chemical building block of a polymer.
monosaccharide—Carbohydrate composed of a single sugar unit, which is often made up of a five- or six-carbon skeleton.
Monte Carlo simulation—Simulation model involving an element of chance.
Moore neighborhood—In a two-dimensional grid of a cellular automaton or agent-based simulation, a set of eight surrounding cells and the site itself.
most significant digit—Leftmost of the significant digits of a number.
motor neuron—Neuron that carries signals away from the central nervous system (brain and spinal cord) to effector cells (e.g., muscle cells, glands) and "effects" a response (e.g., muscle contraction, secretion).

mRNA—Messenger RNA.
MTC—Maximum therapeutic concentration.
multicompartment model—In modeling drug dosage, representation of the body with more than one compartment.
multiprocessor—Computer system with more than one processor.
mushroom—Fruiting body of certain types of fungi.
mutually exclusive events—Events that cannot occur at the same time.
mycelium—Body of a fungus.
myocardial infarction—Medical term for a heart attack.
***N*-body problem**—Problem concerning the interactions and movements of a number of objects, or bodies, in space.
N-terminal—End of a protein that has a free amino group.
NAD^+ (nicotinamide adenine dinucleotide)—Major coenzyme involved in oxidation-reduction reactions of metabolism. NAD+ accepts 2 electrons and 1 proton to become NADH.
natural logarithm—Logarithm to the base e, usually written $\ln n$; $\ln n = m$ if and only if $e^m = n$.
neighbor—In a cellular automaton or agent-based simulation, one of the cells that surrounds a lattice site.
neurons—Functional nerve cells (those that conduct signals).
Newton (N)—Measure of force, where $1\ \text{N} = 1\ \text{kg m/s}^2$.
Newton's gravitational constant—$G = 6.67 \times 10^{-11}\text{m}^3\text{kg}^{-1}\text{s}^{-2}$ in $F = \frac{Gm_1m_2}{r^2}$, the magnitude of the gravitational force between two bodies with masses m_1 and m_2 at a distance r apart.
Newton's third law of motion—For every action, there is an equal and opposite reaction.
Newton's law of heating and cooling—Rate of change of the temperature of an object with respect to time is proportional to the difference between the temperatures of the object and of its surroundings.
Newton's second law of motion—Force F acting a body of mass m gives the body acceleration a according to the formula $F = ma$.
Newtonian friction—Model for drag on a larger object moving through a fluid which is expressed as $F = 0.5CDAv^2$, where C is a constant of proportionality (the coefficient of drag, or drag coefficient) related to the shape of the object, D is the density of the fluid, and A is the object's projected area in direction of movement.
nitrogen bases—Adenine (A), guanine (G), cytosine (C), and thymine (T) in DNA or uracil (U) in RNA.
nitrogen narcosis— Sudden feeing of judgment-impairing euphoria experienced by divers resulting from increased residual nitrogen in the blood.
node—Point in a graph.
normal distribution—A distribution of numbers with a probability density function $\frac{1}{\sqrt{2\pi\sigma}}e^{-(x-\mu)^2/(2\sigma)}$, where μ is the mean and s is the standard deviation.
normalized number—Number in exponential notation with the decimal point immediately preceding the first nonzero digit.
nucleic acids—DNA and RNA; single or double chain of nucleotides.
nucleotide—A compound molecule made up of a sugar (either deoxyribose for DNA or ribose for RNA), a phosphate, and a nitrogen base.

octtree—Tree with each node having branches descending to at most eight nodes; in the Barnes-Hut Algorithm, each node corresponds to a subcube in the 3D partitioning process.

ohms (Ω)—Measure of the resistance of a resistor; 1 Ω = 1 V/A (volt/amp); the resistance of a circuit in which a potential difference of 1 V produces a current of 1 A.

one-compartment model—In modeling drug dosage, simplified representation of the body as one homogenous compartment, where distribution is instantaneous.

one-term model—In empirical modeling, a function with one independent variable that can capture the trend of data whose plot is always concave up or always concave down.

optic tectum—Roof of the midbrain that receives visual images. It also receives "heat images" in pit vipers.

overflow—Error condition that occurs when there are not enough bits to express a value in a computer.

oxidation—Removal of electrons or hydrogens from a molecule.

oxidative phosphorylation—Production of ATP using the proton gradient established by the electron transport system.

pacemaker—Group of cells (sinoatrial node) located in the right atrium of the heart that conducts impulses through the right and left atria, signaling these chambers to contract and pump blood into the ventricles.

PAM—The length of time for 1% of the amino acids to mutate.

PAM1 matrix—A Markov chain transition matrix with column and row headings of the amino acids where entries represent the amount of evolution over one PAM period of time, or for one mutation per hundred amino acids.

parallel processing—Collection of connected processors in close physical proximity working concurrently.

parameter sweeping—The execution of a model for each element in a set of parameters or of collections of parameters to observe the resulting change in model behavior.

parasympathetic nervous system—Subdivision of the autonomic nervous system that generally opposes the effects of the sympathetic nervous system. Major effects of parasympathetic activity include increasing digestive activities and decreasing heart rate.

parent—In radioactivity, substance from which a second substance forms by radioactive decay.

path integration—When an animal takes all the angles and distances it experiences on a foraging trip and integrates them into a mean home vector.

path—In a graph, the sequence of vertices and edges from one vertex to another.

peptide bond—Bond formed through the interaction of an amino group of one amino acid with the carboxyl group of another; bond that links amino acids.

period—Length of time to complete a full cycle of an oscillating function, such as a function describing the motion of a spring or pendulum.

periodic boundary conditions—Conditions so that a border cell of a cellular automaton or agent-based grid is considered to have the corresponding cell on the opposite border as a neighbor. Grid considered to wrap as a torus, so that on a grid row, the leftmost site has the rightmost cell as its west neighbor and vice versa;

similarly, on a column, the topmost site has the bottommost cell as its north neighbor, and vice versa.

pheromone—Chemical produced by animals that send specific signals to other members of the same species.

pinkeye—A highly infectious disease that afflicts cattle and other animals.

pits—Pair of surface concavities, located between the eyes and nostrils of certain snakes (pit vipers), that house extremely sensitive innervated "heat detectors" for hunting.

pixel—Picture element; dot on a computer monitor's screen.

planktonic—Solitary microbe that is not part of a biofilm; groups of microorganisms that float in fresh and saltwater environments.

plasma—Fluid portion of the blood, containing suspended blood cells and clotting factors.

plastic—Long-chain carbon polymer that has a high degree of cross linking.

platelet—Blood cell that helps the blood to clot.

Poiseuille's Equation—Model for blood flow in an arteriole: $Q = \dfrac{\pi r^4 \Delta P}{8\eta L}$, where Q is the blood flow through a vessel over time, r is the cross-sectional area of the vessel, ΔP is the pressure gradient, h is the viscosity of the blood, and L is the vessel length.

polarized—Resting neuron or other excitable cell in which the outside is more positive than the inside.

polymer—Class of chemical compounds composed of repeating chemical building blocks.

polymerization—Process of polymer synthesis.

polynomial function of degree *n*—Function of the form $f(x) = a_n x^n + \cdots + a_1 x + a_0$, where $a_n, \ldots, a_1, a_0$ are real numbers and n is a nonnegative integer.

postconditions—Conditions that describe the state of the system when the function finishes executing, any error conditions, and the information the function returns or otherwise communicates.

potential difference—Difference in electronic potential between two points.

power law—A function f, where if $f(x)$ is proportional to x^b for some constant b.

precision—In mathematics, number of significant digits in an number.

preconditions—Conditions that must be true for the function to behave properly.

predation—When one species (predator) kills and consumes another species (prey).

predator—Organism that consumes another organism (prey) for food.

predictor variable—In linear regression, the independent variable.

pressure—Weight of matter per unit area.

prey—Organism that is consumed by another organism (predator) for food.

probabilistic behavior—Behavior of systems with an element of chance.

probabilistic model—Model that exhibits random effects.

probability function—For a discrete distribution, returns the probability of occurrence of a particular argument; for a continuous distribution, indicates the probability that a given outcome falls inside a specific range of values.

probability matrix—Transition matrix.

probability of an event—The chance of the event's occurrence, a number between 0 and 1, inclusive.

probability vector—A vector whose components are nonnegative and sum to 1.

procedure—Method.
process—Task or a piece of a program that executes separately.
processor—Part of a computer that performs the arithmetic and logic.
profit—Total gain from producing and selling a given quantity of items.
protein synthesis—The production of proteins.
proteins—Basic molecules of life, made up of chains of amino acids and performing many critical functions in the cell and organism.
prototyping—Implementation of a preliminary version of a solution or part of a solution.
provirus—Viral DNA that may remain in the chromosome passively (latent) or may activate and begin the production of new viral particles, as in HIV.
pseudocode—A structured outline, in English, of a computer program's design.
pseudorandom numbers—Random numbers.
pulmonary circulation—Circulatory loop that transports blood to and from the lungs.
purine—Base A or G.
pyrimidine—Base C, T, or U.
pyruvate—Three-carbon product of glycolysis, resulting from the splitting and oxidation of glucose.
quadratic function—Function of the form $f(x) = a_2x^2 + a_1x + a_0$, where a_2, a_1, and a_0 are real numbers.
quadtree—Tree with each node having branches descending to at most four nodes; in the Barnes-Hut algorithm, each node corresponds to a subsquare in the 2D partitioning process.
quarantine— Limitation on freedom of movement of an individual for a period of time to prevent spread of a contagious disease to other susceptible members of a population.
R-group—A variable side-chain in an amino acid.
radiative forcing—For global warming, increased infrared absorption and warming as a result of increased concentration of greenhouse gases.
random numbers—Part of a sequence of numbers that appear random but that an algorithm actually produces.
random walk—The apparently random movement of an entity, taking single steps in apparently random directions.
rate of absorption—For gas absorption by tissues, $dP_{tissue}/dt = k(P_{lungs} - P_{tissue})$, where P_{lungs} is the partial pressure of the gas in the lungs, P_{tissue} is the partial pressure of the gas in the tissue, and $k = \ln(2)/t_{half}$, where t_{half} is the time for the tissue to absorb or release half of the partial difference of the gas.
real number—Number that can be expressed with a decimal expansion and used to measure continuous quantities.
recovered—An individual no longer having a disease and immune to further infection in an SIR model.
recursion—Process of a function or task calling itself.
red blood cell—Blood cell used for oxygen transport between the lungs and tissues.
reference dose—Amount of a substance that may be ingested on a daily basis for a lifetime with no adverse effects on health.

refinement—Adding more complexity, such as eliminating simplifying assumptions, to a model.
reflecting boundary conditions—Conditions where the boundary tends to propagate the current local situation. Values on the original first row occur again on the new first row of ghost cells, and similar situations occur on the last row and the first and last columns.
reflective boundary conditions—Reflecting boundary conditions.
relative error—Absolute error divided by the absolute value of the exact answer, provided the exact answer is not zero.
relative humidity—Ratio of the actual water vapor pressure to the vapor pressure that would occur if the air were saturated at the same ambient temperature.
relative refractory period—Time during an AP when the sodium channels become inactivated, the potassium channels begin to close, and an AP can be initiated, if a sufficiently strong stimulus is applied.
relative sensitivity—Sensitivity.
repolarization—The process in which voltage-gated potassium channels open, and potassium ions diffuse out of the neuron, helping to make the internal potential to become more negative.
reproductive number—Expected number of secondary infectious cases resulting from an average infectious case once an epidemic is in progress.
reservoir—Portion of the atmosphere, hydrosphere, lithosphere, or biosphere where various forms of an element of a biogeochemical cycle are stored; any living or non-living agent that serves as a continuing source of infection.
resistance—Measurement of the ability of a resistor to reduce the flow of charges.
resistor—Device used to control current in electrical circuits by imparting resistance.
respiratory distress syndrome—Inflammatory disease of the lung, characterized by a sudden onset of edema and respiratory failure.
response variable—In linear regression, the dependent variable.
resting potential—Electrical potential of a resting nerve cell.
retrovirus—Type of enveloped, RNA virus that uses reverse transcriptase to convert its RNA into DNA in the host cell.
revenue—Total amount of income from selling a given quantity of items.
reverse transcriptase—Enzyme used by retroviruses to form a complementary DNA sequence (cDNA) from viral RNA.
ribosome—Molecular machine that contains the apparatus to translate groups of codons into specific amino acid sequences.
***RLC* circuit**—Electrical circuit with a resistor, an inductor, and a capacitor.
RNA polymerase—An enzyme that synthesizes RNA from a DNA template.
root process—In high-performance computing, a process that commands all other processes.
root—In mathematics, unique top node in a (rooted) tree; in high-performance computing, root process.
round down—For a normalized number, to truncate the significand to the desired number of significant digits.
round up—For a normalized number, to truncate the significand to the desired number of significant digits, adding one to the last of the remaining significant digits.
round-off error—Problem of not having enough bits to store an entire floating-point

number and approximating the result to the nearest number that can be represented.

round—For a normalized number, to round to precision k is usually to round down if the $(k + 1)$st significant digit is less than 5 and to round up otherwise.

RP—Resting potential.

rules—In a cellular automaton simulation, specify local relationships and indicate how cells are to change state; regulate the behavior of the system.

Runge-Kutta 2 method—Method of numeric integration that employs a correction to each Euler's method estimate.

Runge-Kutta 4 method—Method of numeric integration, where each approximation is a weighted average of four estimates.

scalability—Capability of a computer system with expanded hardware resources to exhibit better performance.

scale-free networks—Networks that follow the power law $P(k) \propto k^{-r}$ with $r > 1$.

schooling—Behavior of certain species of aquatic animals to swim in large groups for protection against predators or for foraging.

scientific notation—Exponential notation in which the decimal point is placed immediately after the first nonzero digit.

seed—In a method for generating pseudorandom numbers, an initial value; in diffusion-limited aggregation, an initial simulated dendritic structure.

self-avoiding walk (SAW)—A random walk that does not cross itself, that is, a walk that does not travel through the same cell twice.

sensitivity—The sensitivity of λ to parameter P in a transition matrix is the instantaneous rate of change of λ with respect to P, which is approximately $\dfrac{\lambda_{new} - \lambda}{P_{new} - P}$, where P_{new} is the new value of P close to P and λ_{new} is the resulting new value of λ.

sequential processing—Single processor working on one program.

serum—Fluid portion of the blood that remains after the blood clots.

shared memory multiprocessor—System in which two or more processors communicate through shared memory.

significand—For a floating-point number expressed in exponential notation as $a \times 10^n$, the integer formed by dropping the decimal point from a.

significant digits—For a floating-point number, all digits except leading zeros; for an integer, all digits except leading and trailing zeros.

simple harmonic oscillator—System that satisfies the following properties: The system oscillates around an equilibrium position. The equilibrium position is the point at which no net force exists. The restoring force is proportional to the displacement. The restoring force is in the opposite direction of the displacement. The motion is periodic. All damping effects are neglected.

simple pendulum—Pendulum where the mass for the bob is concentrated at a point, the stiff string has no mass, and friction does not exist.

sine—For point (x, y) on a circle with radius r and angle t off the positive x-axis, $\sin t = y/r$.

single-precision number—Floating-point number using half as many bits in a computer representation as a double-precision number; typically contains 6 or 7 significant digits and has magnitude between 10^{-38} and 10^{38}.

sink—In a biogeochemical cycle, the destination of an element coming from a source.

SIR model—Spread of disease model that considers the susceptible, infected, and recovered population groups.
site—Cell.
size of a graph—The number of nodes in the graph.
size of a vector—The number of elements.
slope—For a nonvertical line, a change in y over the corresponding change in x.
small-world property—A graph property in which the average shortest path length is on the order of magnitude of log n or smaller, where n is the number of nodes.
solidification—Becoming solid.
soma—Cell body.
source—In a biogeochemical cycle, the origin of an element as it flows to a sink.
specific impulse—Impulse per pound of burned fuel, or the quotient of impulse and the change in the fuel's weight.
speed—Magnitude of velocity; the absolute value of the change in position with respect to time.
speedup factor—Execution time of an algorithm on a sequential computer over execution time of a comparable algorithm on a system with multiple processors.
spore—In fungi, a reproductive cell that typically is capable of germinating into a new individual.
sporozoite—Life stage of malarian parasite that accumulates in the salivary glands of the female mosquito. A female mosquito inoculates her human host with this stage, which invades liver cells.
squall line—Line of storms with a well-developed gust front at the leading edge.
square matrix—An $n \times n$ matrix.
stable solution— Solution q to a differential equation dP/dt or a difference equation ΔP, where there is an interval (a, b) containing q, such that if the initial population is in that interval, then $P(t)$ is finite for all $t > 0$ and $\lim_{t \to \infty} P(t) = q$.
standard deviation—In statistics, a measure of the amount of variation of data from the mean; for a normal distribution of data, about 68.3% of the numbers are within one standard deviation of the mean.
standard error bars—Symmetric error bars that are two standard deviations in length.
state—In a cellular automaton simulation, characteristics of a cell.
static model—Model that does not consider time.
steady-state vector—Equilibrium vector.
stochastic behavior—Behavior of systems with an element of chance.
stochastic model—Model that exhibits random effects.
Stokes' friction—Model for friction on the particle that is approximately proportional to its velocity.
stroke volume—Volume of blood that the left ventricle ejects upon contraction.
subdiagonal—In an $n \times n$ square matrix M, the set of elements $\{m_{21}, m_{32}, \ldots, m_{n(n-1)}\}$.
subgraph—S is a subgraph of graph G if S is a graph and every node and edge of S is in G.
substrate—Molecule that is acted upon by an enzyme; surface for growth.
substrate-level phosphorylation—Synthesis of ATP from ADP and P, where the

P used to make ATP has come from an organic compound that has a higher energy level than does ATP.

susceptible—Individual within a population that has no immunity to a disease.

sympathetic nervous system—Subdivision of the autonomic nervous system responsible for mobilizing the body's energy when stressed or aroused. Major effects of sympathetic activity include raising of blood pressure and increasing heart rate.

synapse—Junction between terminal buttons and other neurons or effector cells.

systemic circulation—Circulatory loop carrying blood to and from the body (except for the lungs).

systemic vascular resistance—Resistance or impediment of the blood vessels in the systemic circulation to the flow of blood.

systolic pressure—Highest pressure exerted as the left ventricle contracts.

T-cell (T-lymphocyte)—White blood cell that matures in the thymus and bears antigen-specific receptors on its surface. T-cells participate in both the cell-mediated and humoral immune responses.

T-helper cell—White blood cell (lymphocyte) that mediates both cell-mediated immunity and antibody production. It is a cell that displays one of two protein structures on the surface of a human cell, allowing HIV to attach, enter, and thus infect the cell.

T-lymphocyte—T-cell.

tangent—For point (x, y) on the unit circle and angle t off the positive x-axis, $\tan t = y/x$.

terminal buttons—At the branched ends of an axon; interface with the other neurons or effector cells by way of the synapse.

terminal speed—Constant speed a falling object reaches due to friction.

therapeutic range—Drug concentration between the minimum effective concentration and the maximum therapeutic concentration.

thermal conduction—Heat transfer in an object due to a temperature gradient.

thrust—Mechanical force caused by the acceleration of a mass of gas and in the opposite direction to gas flow.

transcription—The synthesis of RNA.

transient equilibrium—In radioactive decay from *substanceA* to *substanceB*, when the ratio of the mass of *substanceB* to the mass of *substanceA* is almost constant.

transition matrix—A matrix in which all the entries are nonnegative and the sum of the elements in each column (or each row) is 1.

transition rules—In a cellular automaton simulation, specify local relationships and indicate how cells are to change state; regulate the behavior of the system.

transition—In DNA mutation, a type of substitution that occurs between purines or between pyrimidines.

translation—Conversion on the ribosome of a mRNA sequence into a sequence of amino acids for the protein.

transmission constant—Infection rate indicating the infectiousness of a disease.

transversion—In DNA mutation, a type of substitution that occurs between a purine and a pyrimidine, or vice versa.

tree—In mathematics, a rooted tree, often called a tree, is a connected, hierarchical structure of nodes (points), which contain information, and edges connecting them that has no cycles and has a unique node, the root, at the top.

triplet—A sequence of three nucleotides.

truncate—For a normalized number, to chop off all digits of the mantissa beyond the desired number of significant digits.

truncation error—Error that occurs when a truncated, or finite, sum is used as an approximation for the sum of an infinite series.

two-compartment model—In modeling drug dosage, representation of the body as two chambers (e.g., gastrointestinal tract and blood).

type-1 predator functional response—Response where the predator consumes a constant proportion of prey, regardless of prey density.

type-2 predator functional response—Response where the predator consumes less as it nears satiation, which determines the upper limit on consumption.

type-3 predator functional response—Response where predation increases slowly at low prey density, increases rapidly at higher densities, but levels off at satiation, even if prey density continues to increase.

unconstrained model—Model where no limiting factor exists; unconstrained or exponential growth/decay model, where the rate of change of an amount is directly proportional to the amount.

underflow—Error condition that occurs when the result of a computation is too small for a computer to represent.

undirected graph—$G = (V, E)$ consisting of a set V of vertices and a set E of edges connecting pairs of vertices.

uniform distribution—Discrete distribution in which all possible outcomes have an equal chance of occurring; continuous distribution in which equal-length intervals of outcomes have an equal chance of occurring.

unstable solution—Solution, q, to a differential equation dP/dt or a difference equation ΔP that is not stable.

upwelling—When deep currents bring cool, nutrient-rich bottom ocean water to the surface.

V—Volt.

validation—Process that establishes if a system satisfies the problem's requirements.

vasoconstriction—Decrease in blood vessel diameter.

vasodilation—Increase in blood vessel diameter.

vector—In biology, an animal that transmits a pathogen, or something that causes a disease, to another animal; in mathematics, ordered n-tuple of numbers, $(v_1, v_2, \ldots, v_n)$.

velocity—Instantaneous velocity.

venous return—Flow of blood to the heart.

ventricle—In the human heart, one of two lower chambers that pump blood to the body and the lungs.

ventricular fibrillation—Chaotic electrical disturbance in the heart ventricle.

verification—Process which determines if a solution works correctly.

vertex (plural vertices)—Point in a graph.

V_{max}—Limit of the rate of an enzymatic reaction.

volt (V)—Unit of measure of potential difference.

voltage difference—Difference in electronic potential between two points.
voltage-gated—Ion channels whose gates open due to changes in voltage.
voltage—For a point in an electrical circuit, the voltage difference between a point and a circuit reference point, the ground.
von Neumann neighborhood—In a two-dimensional grid of a cellular automaton or agent-based simulation, a set of cells directly to the north, east, south, and west of a grid site and the site itself.
weight—The force on an object due to gravity.
white blood cell—Blood cell that is part of the body's defense mechanism against infections.
y-intercept—Value of y when $x = 0$ in the equation of a line, $y = mx + b$; where the graph of a function crosses the y-axis.

ANSWERS TO SELECTED EXERCISES

Chapter 2

Module 2.2

2. a. $15000e^{0.02(20)} = 22377.4$
b. 24.9, 15024.9; 24.94, 15049.8; 24.98, 15074.8
7. a. $Q = Q_0\, e^{-0.0239016t}$

Module 2.3

1. b. $P(t) = \dfrac{e^{rt} MP_0}{M - P_0 + e^{rt} P_0}$ **2. a.** $\pi/2 + \pi n$, where n is an integer

Module 2.5

4. b. 0.099021
5. b. Assuming *elimination_constant* = 0.0315, 50.90 mg

Chapter 3

Module 3.1

1. a. $v = ds/dt$, $a = d^2s/dt^2 = dv/dt = -9.81$ m/s^2, $s_0 = 11$ m. $v_0 = 15$ m/s
b. $v = -9.8t + 15$, $s = -4.9t^2 + 15t + 11$

Module 3.2

2. a. $m\dfrac{d^2s}{dt^2} = -km$

Module 3.3

1. a. 1   **b.** θ **c.** $F = -mg\theta$ **e.** $d^2(\theta)/dt^2 = -g\theta/l$

Module 3.4

2. $dv/dt = -I_{sp}\, g\, (dm/dt)/m + g - 0.5CDAv^2/m$, where C is the drag coefficient, A is the rocket's cross-sectional area, and the density of the atmosphere is $D = 1.225e^{-0.1385y}$, where y is altitude < 100 km.

Chapter 4

Module 4.1

1. a. $dW/dt = aW - bWB$, $dB/dt = cB - dWB$, $W_0 = 20$, $B_0 = 15$
b. $a = bB$ and $c = dW$, where b and d are any nonnegative real numbers

Module 4.2

2. $\Delta s = (k_s * s(t - \Delta t) - k_{hs} * h(t - \Delta t) * s(t - \Delta t)) * \Delta t$
$\Delta h = (k_{sh} * s(t - \Delta t) * h(t - \Delta t) - k_{sh} * s(t - \Delta t) * h(t - \Delta t)^2/M) * \Delta t$

Module 4.3

2. $dS_Q/dt = qk(1 - b)I_U S/N_0 - uS_Q$ **11. a.** $3^{10} = 59{,}049$

Module 4.4

2. $d(human\_hosts)/dt = (prob\_bit)(prob\_vector)(uninfected\_humans)$
$- (recovery\_rate)(human\_hosts)$
$- (malaria\_induced\_death\_rate)(human\_hosts)$
$- (immunity\_rate)(human\_hosts)$

Module 4.5

1. $d[\mathrm{E}]/dt = -d[\mathrm{ES}]/dt$ **6. a.** $d[\mathrm{S}]/dt = -k_1[\mathrm{E}][\mathrm{S}]^n + nk_2[\mathrm{ES}]$

Chapter 5

Module 5.2

1. 0.6385×10^5 **13.** Magnitude = 10^{25}, precision = 6 **16.** 4, 6
22. 0.1×10^{-5} to 0.999×10^5 **24. a.** 0.001 **b.** 0.0160% **c.** 0.009
d. 0.144% **27. a.** $0.36000000094 \times 10^5$ **b.** 0.36000×10^5
c. 0.94×10^{-4} **d.** 0.26×10^{-8} **29.** 6.23; 12.4; 0.625%
34. a. $-1/3 = -0.33\overline{3}$ **b.** –0.4161

Chapter 6

Module 6.2

1. $P_2 = 324$ at $t = 16$ h

Module 6.3

1. Starting with $P_1 = 212$ at $t = 8$ h, $P_2 = 449.44$ at $t = 16$ h

Module 6.4

1. Starting with $P_1 = 222.24$ at $t = 8$ h, $P_2 = 493.906$ at $t = 16$ h

Chapter 9

Module 9.2

1. 6 **4 a.** 219, 244, 36 **b.** 0.627507, 0.69914, 0.103152
10. 20.0 * *rand* + 6.0 **14.** INT(21.0 * *rand* + 6.0)

Chapter 10

Module 10.3

1. Have two constants for burning, such as *BURNING1* and *BURNING2*. Replace the rule for *BURNING* with two rules: If the site is *BURNING1*, then return *BURNING2*. If the site is *BURNING2*, then return *EMPTY*.

4. *spread* has 11 parameters: *site*, *N*, *NE*, *E*, *SE*, *S*, *SW*, *W*, *NW*, *probLightning* and *probImmune*. The rule for determining if a tree will burn at the next time step begins as follows: if *site* is *TREE* and (*N*, *NE*, *E*, *SE*, *S*, *SW*, *W*, or *NW* is *BURNING*)

Module 10.4

2. The ant goes back and forth between its first cell and the cell it initially selected at random from its neighbors, none of which had any chemical at the start of the simulation. As the ant leaves a cell, the amount of chemical increases by *DEPOSIT*. Thus, after the first time step, the cell from which it just came has the maximum amount of chemical in the neighborhood.

Chapter 12

Module 12.1

3. 0.25 log *n*

Module 12.2

1. Algorithm for Process i in Calculation of Scalar Product av: return av_i
7. The divide phase is identical to that for the divide-and-conquer algorithm for adding numbers. The conquer phase is the same except that each process returns the number of occurrences of a particular element in its subarray, which the receiving process adds to its number of occurrences.
9. **a.** $r_n = (81r_{n-1}) \bmod 349$, for $n > 0$

Chapter 13

Module 13.2

1. (21, –28, 56, 0) **4.** (–6, 0, 29, 2) **12.** 125
27. $a = 8, b = 4, c = ½$ **28. a.** 2 × 3 **b.** 0, 3, no such element, –2

29. $\begin{bmatrix} 18 & 9 & -6 \\ 0 & -24 & 12 \end{bmatrix}$ **32.** $\begin{bmatrix} 5 & 5 & 1 \\ -7 & -6 & 5 \end{bmatrix}$ **44. a.** 100

49. b. $\begin{bmatrix} 2 & 10 \\ 14 & 6 \end{bmatrix}$ **e.** $D_4 = \begin{bmatrix} 0 & 8 & 2 & 10 \\ 12 & 4 & 14 & 6 \\ 3 & 11 & 1 & 9 \\ 15 & 7 & 13 & 5 \end{bmatrix}$ **50.** $\begin{bmatrix} 6 & 0 & 8 \\ 18 & 2 & 16 \end{bmatrix}$

64. $\begin{bmatrix} 1.33 & 2.99 \\ 3.93 & 7.23 \end{bmatrix}$ **68.** $\begin{bmatrix} 5 & 0 \\ 0 & 7 \end{bmatrix}$ **72.** $\begin{bmatrix} x + 2y \\ 3x + 4y \end{bmatrix}$

Module 13.3

1. **a.** Shrinking **b.** By 1.6% per year
   c. 47.47%, 21.01%, 17.58%, and 13.94% because sum = –2.35 + (–1.04) + (–0.87) + (–0.69) = –4.95 and (–2.35, –1.04, –0.87, –0.69)/(–4.95) = (0.4747, 0.2101, 0.1758, 0.1394)
2. **b.** (2030, 652, 287)
3. **b.** 0.4583 dominant eigenvalue and the annual growth rate; decline
   d. 1.3417 because the dominant eigenvalue of the Leslie matrix with 0.1 – 0.1(0.1) = 0.9(0.1) = 0.09 replacing 0.1 is 0.4449 and (0.4449 – 0.4583) / (0.09 – 0.1) = 1.34
4. **c.** 32.53% = 0.3253 = 1 – 0.6747
   d. 21.44% = 0.2144 = 1 – (0.7370 + 0.0486) because stage 2 turtles survive and remain at stage 2, survive and advance to stage 3, or die in any one year.
5. **a.** 0.1 is the probability that in one year a stage 1 animal survives and advances to stage 2. 0.2 is the probability that in one year a stage 2 animal survives and remains at stage 2. (Other descriptions are left to the reader.)

Module 13.4

1. a. $\begin{bmatrix} 0.84 & 0.55 & 0.40 & 0.15 \\ 0.04 & 0.26 & 0.06 & 0.07 \\ 0.03 & 0.03 & 0.35 & 0.02 \\ 0.09 & 0.16 & 0.19 & 0.76 \end{bmatrix}$

b. (0.25, 0.25, 0.25, 0.25)

2. a. $1 - 3\alpha$

b. $\begin{bmatrix} 1-3\alpha & \alpha & \alpha & \alpha \\ \alpha & 1-3\alpha & \alpha & \alpha \\ \alpha & \alpha & 1-3\alpha & \alpha \\ \alpha & \alpha & \alpha & 1-3\alpha \end{bmatrix}$

d. $\begin{bmatrix} .01 & .03 & .03 & .03 \\ .03 & .01 & .03 & .03 \\ .03 & .03 & .01 & .03 \\ .03 & .03 & .03 & .01 \end{bmatrix}$

with (0.25, 0.25, 0.25, 0.25) ultimate distribution of bases

3. a. $1 - \alpha - 2\beta$

INDEX